Advances in Intelligent Systems and Computing

Volume 1117

The series "Advances in Intelligent Systems and Computing" contains publications on theory, applications, and design methods of Intelligent Systems and Intelligent Computing. Virtually all disciplines such as engineering, natural sciences, computer and information science, ICT, economics, business, e-commerce, environment, healthcare, life science are covered. The list of topics spans all the areas of modern intelligent systems and computing such as: computational intelligence, soft computing including neural networks, fuzzy systems, evolutionary computing and the fusion of these paradigms, social intelligence, ambient intelligence, computational neuroscience, artificial life, virtual worlds and society, cognitive science and systems, Perception and Vision, DNA and immune based systems, self-organizing and adaptive systems, e-Learning and teaching, human-centered and human-centric computing, recommender systems, intelligent control, robotics and mechatronics including human-machine teaming, knowledge-based paradigms, learning paradigms, machine ethics, intelligent data analysis, knowledge management, intelligent agents, intelligent decision making and support, intelligent network security, trust management, interactive entertainment, Web intelligence and multimedia.

The publications within "Advances in Intelligent Systems and Computing" are primarily proceedings of important conferences, symposia and congresses. They cover significant recent developments in the field, both of a foundational and applicable character. An important characteristic feature of the series is the short publication time and world-wide distribution. This permits a rapid and broad dissemination of research results.

** Indexing: The books of this series are submitted to ISI Proceedings, EI-Compendex, DBLP, SCOPUS, Google Scholar and Springerlink **

More information about this series at http://www.springer.com/series/11156

Mohammed Atiquzzaman · Neil Yen ·
Zheng Xu
Editors

Big Data Analytics for Cyber-Physical System in Smart City

BDCPS 2019, 28–29 December 2019,
Shenyang, China

Set 2

 Springer

Editors
Mohammed Atiquzzaman
School of Computer Science
University of Oklahoma
Norman, OK, USA

Neil Yen
University of Aizu
Fukushima, Japan

Zheng Xu
Shanghai University
Shanghai, China

ISSN 2194-5357   ISSN 2194-5365   (electronic)
Advances in Intelligent Systems and Computing
ISBN 978-981-15-2567-4   ISBN 978-981-15-2568-1   (eBook)
https://doi.org/10.1007/978-981-15-2568-1

This Springer imprint is published by the registered company Springer Nature Singapore Pte Ltd.
The registered company address is: 152 Beach Road, #21-01/04 Gateway East, Singapore 189721, Singapore

Foreword

With the rapid development of big data and current popular information technology, the problems include how to efficiently use systems to generate all the different kinds of new network intelligence and how to dynamically collect urban information. In this context, Internet of things and powerful computers can simulate urban operations while operating with reasonable safety regulations. However, achieving sustainable development for a new urban generation currently requires major breakthroughs to solve a series of practical problems facing cities.

A smart city involves a wide use of information technology for multidimensional aggregation. The development of smart cities is a new concept. Using Internet of things technology on the Internet, networking, and other advanced technology, all types of cities will use intelligent sensor placement to create object-linked information integration. Then, using intelligent analysis to integrate the collected information along with the Internet and other networking, the system can provide analyses that meet the demand for intelligent communications and decision support. This concept represents the way smart cities will think.

Cyber-physical system (CPS) as a multidimensional and complex system is a comprehensive calculation, network, and physical environment. Through the combination of computing technology, communication technology, and control technology, the close integration of the information world and the physical world is realized. IOT not only is closely related to people's life and social development, but also has a wide application in military affairs, including aerospace, military reconnaissance, intelligence grid system, intelligent transportation, intelligent medical, environmental monitoring, industrial control, etc. Intelligent medical system as a typical application of IOT will be used as a node of medical equipment to provide real-time, safe, and reliable medical services for people in wired or wireless way. In the intelligent transportation system, road, bridge, intersection, traffic signal, and other key information will be monitored in real time. The vast amount of information is analyzed, released, and calculated by the system, so that the road vehicles can share road information in real time. Personnel of road management can observe and monitor the real-time situation of the key sections in the system and even release the information to guide the vehicle so as to improve the existing urban traffic conditions. The Internet of things, which has been widely used in the industry, is a simple application of IOT. It can realize the function of object identification, positioning, and monitoring through the access to the network.

BDSPS 2019 which is held on December 28–29, 2019, Shenyang, China, is dedicated to address the challenges in the areas of CPS, thereby presenting a consolidated view to the interested researchers in the related fields. The conference looks for significant contributions to CPS in theoretical and practical aspects.

Each paper was reviewed by at least two independent experts. The conference would not have been a reality without the contributions of the authors. We sincerely thank all the authors for their valuable contributions. We would like to express our appreciation to all members of the Program Committee for their valuable efforts in the review process that helped us to guarantee the highest quality of the selected papers for the conference.

We would like to express our thanks to our distinguished keynote speakers, Professor Bo Fei, Shanghai University of Medicine & Health Sciences, China, and Professor Tiejun Cui, Shenyang Ligong University, China. We would also like to acknowledge the strong support of Shenyang Ligong University, as well as the general chairs, publication chairs, organizing chairs, program committee members, and all volunteers.

Our special thanks are due also to the editors of Springer book series "Advances in Intelligent Systems and Computing," Dr. Thomas Ditzinger, Dr. Ramesh Nath Premnath, and Arumugam Deivasigamani for their assistance throughout the publication process.

Organization

General Chairs

Tharam Dillon   La Trobe University, Australia
Bo Fei   Shanghai University of Medicine & Health Sciences, China

Program Committee Chairs

Mohammed Atiquzzaman   University of Oklahoma, USA
Zheng Xu   Shanghai University, China
Neil Yen   University of Aizu, Japan

Publication Chairs

Juan Du   Shanghai University, China
Ranran Liu   The University of Manchester, UK
Xinzhi Wang   Tsinghua University, China

Publicity Chairs

Junyu Xuan   University of Technology Sydney, Australia
Vijayan Sugumaran   Oakland University, USA
Yu-Wei Chan   Providence University, Taiwan, China

Local Organizing Chairs

Qingjun Wang   Shenyang Ligong University, China
Chang Liu   Shenyang Ligong University, China

Program Committee Members

William Bradley Glisson	University of South Alabama, USA
George Grispos	University of Limerick, Ireland
Abdullah Azfar	KPMG, Sydney, Australia
Aniello Castiglione	Università di Salerno, Italy
Wei Wang	The University of Texas at San Antonio, USA
Neil Yen	University of Aizu, Japan
Meng Yu	The University of Texas at San Antonio, USA
Shunxiang Zhang	Anhui University of Science and Technology, China
Guangli Zhu	Anhui University of Science and Technology, China
Tao Liao	Anhui University of Science and Technology, China
Xiaobo Yin	Anhui University of Science and Technology, China
Xiangfeng Luo	Shanghai University, China
Xiao Wei	Shanghai University, China
Huan Du	Shanghai University, China
Zhiguo Yan	Fudan University, China
Rick Church	UC Santa Barbara, USA
Tom Cova	The University of Utah, USA
Susan Cutter	University of South Carolina, USA
Zhiming Ding	Beijing University of Technology, China
Yong Ge	University of North Carolina at Charlotte, USA
T. V. Geetha	Anna University, India
Danhuai Guo	Computer Network Information Center, Chinese Academy of Sciences, China
Jianping Fang	University of North Carolina at Charlotte, USA
Jianhui Li	Computer Network Information Center, Chinese Academy of Sciences, China
Yi Liu	Tsinghua University, China
Foluso Ladeinde	SUNU Korea
Kuien Liu	Pivotal Inc., USA
Feng Lu	Institute of Geographic Sciences and Natural Resources Research, Chinese Academy of Sciences, China
Ricardo J. Soares Magalhaes	The University of Queensland, Australia
D. Manjula	Anna University, India
Alan Murray	Drexel University, USA
S. Murugan	Sathyabama Institute of Science and Technology, India
Yasuhide Okuyama	University of Kitakyushu, Japan
S. Padmavathi	Amrita University, India

Latha Parameswaran	Amrita University, India
S. Suresh	SRM University, India
Wei Xu	Renmin University of China, China
Chaowei Phil Yang	George Mason University, USA
Enwu Yin	China CDC, USA
Hengshu Zhu	Baidu Inc., China
Morshed Chowdhury	Deakin University, Australia
Min Hu	Shanghai University, China
Gang Luo	Shanghai University, China
Juan Chen	Shanghai University, China
Qigang Liu	Shanghai University, China

Contents

Algorithm of the Risk Cost of Not Supplied Energy in Unitization Planning of Power Grid

Qiang Li[1], Lu Gao[2], Wei Li[1], Zhihang Qin[1], and Yujing Zhang[2(✉)]

[1] State Grid Wuhan Electric Power Supply Company, Wuhan 430012, China
[2] Beijing SGITG-Accenture Information Technology Co., Ltd.,
Beijing 100032, China
yujingl19830@163.com

Abstract. With the development of grid distribution network planning technology, it has become a reality to carry out reliability investment analysis through big data of distribution network operation. The risk cost quantification of power shortage based on grid distribution network planning makes the development of grid change from the traditional goal-oriented to problem-oriented. Through the reasonable quantification of cost caused by shortage of electric supply, the user portrait characteristics of grid planning are brought into play, and accurate positioning of distribution network investment planning is made. The research and application of the cost quantification of short power supply can integrate the quality of power supply into the evaluation of investment requirement and provide a reliable theoretical basis for the establishment of a reasonable electric transmission and distribution price. At the same time, the research uses the lack of power supply cost quantification to reflect the power consumption service expectations of different industries or users of different scales in the same industry, which can realize the quantitative portrait of the user's power consumption behavior and demand, so as to guide the power companies to carry out quality and high value-added integrated energy services.

Keywords: Big data · Risk cost · Energy Not Supplied · Unitization planning

1 Introduction

With the construction of ubiquitous power Internet of things, ubiquitous power data continues to enrich, power grid operation perception ability continues to strengthen, particle size is more refined, so that the relationship between power grid investment and equipment assets can be refined and analyzed through the cost-benefit curve. The grid-based distribution network planning method further relates the investment development direction of distribution network to the grid electricity consumption behavior characteristics. The introduction of the index of short electric supply describes the risk cost that may be caused by power grid outage. Taking the short supply as the cost quantitative research and combining with grid-based planning will further expand the research depth of planning investment from the perspective of risk control and differential investment and quantify new service business model of power grid.

© Springer Nature Singapore Pte Ltd. 2020
M. Atiquzzaman et al. (Eds.): BDCPS 2019, AISC 1117, pp. 1013–1020, 2020.
https://doi.org/10.1007/978-981-15-2568-1_139

The concept research of power shortage has been carried out for a long time, and fruitful research results have been obtained. The research of Hashemi-Dezaki and Askarian-Abyaneh proves that the grid topology structure of power shortage can be optimized through grid ring network structure [1]; Hashemi Dezaki and Askarian Abyaneh and others have studied how to reduce the power outage risk of distribution grid lines through investment [2]; Banerjee and Islam have jointly studied the power outage risk. The technical method of optimizing the reliability of power supply by over distributed generation equipment [3]; in the aspect of subsection optimization of distribution grid network, in 1999, Celli and Pilo jointly published the research results on subsection positioning and rapid optimization of distribution network structure [4]. In these studies, the concept of short power supply as a technical evaluation index is combined with the characteristics of power grid structure and electrical operation. From the perspective of technology and topology, the influence of distribution network equipment and structure on the power capacity of users is analyzed. However, the above research does not combine the characteristics of user behavior profile, regional development characteristics and power supply shortage in grid planning.

In the aspect of operation cost optimization of grid distribution network, the research on the selection of optimal conductor section by Celli and Pilo laid the foundation for the research on the minimum operation cost of distribution network with radiation structure by Chandramohan and Atturulu [5]. Da Silva, Pereira, and Mantovani (2004b) In the aspect of distribution automation protection, elaborated the research on equipment data to improve reliability [6, 7]. Combined with the analysis method of distribution automation data in their early research results, the data link between operation and maintenance data and reliability operation characteristic analysis of distribution network is basically formed [8]. The research results on improving distribution network reliability by using capacitor equipment in Etemadi [9] provide help for the research on service income of energy storage equipment. In 2005, Popović and Greatbanks et al. co-authored a paper on the improvement of the operation cost of distribution network by small capacity power generation equipment from the perspective of the linkage between distributed generation equipment and distribution automation system [10]. Trebolle and Gómez published two articles successively in 2010, showing the influence of distributed generation equipment on distribution network planning [11] and distribution network reliability planning [12], among which the cost factors are discussed. The analysis is of great significance for this study.

The research on the lack of power supply in China has not been developed for a long time, and there has been a research gap. Especially in the aspect of the cost quantification of the short supply power, the unified purchase and sale price under the planned economy mode has been used for a long time, which does not distinguish the characteristics of users, and does not reflect that the short supply power is the basic core concept of the loss of users' power failure. In this study, the grid planning characteristics are taken as the research's starting point. Through the analysis of the distribution grid network users' industrial characteristics and the calculation of the unique outage loss risk cost distribution, the lack of power supply is taken as the research object of cost quantification.

2 Data and Algorithm

2.1 The Data

Two types of data are applied in the process of studying the algorithm of power shortage cost quantification: power cut data of distribution automation system (DMS) and power user risk loss survey data, in which the data of sub distribution automation system includes the current and voltage data of all line switches, and the data of power distribution automation system can be obtained by power Rate calculation $P = UI$ to carry out power outage load calculation; in terms of user survey data, 1000 study sample users are used, and Monte Carlo method is used to extract from 15 consecutive distribution grids. According to the development characteristics of different regions, the sample number is adjusted according to industry, industry and commerce, residents and administrative offices (Table 1):

Table 1. Proportion of user research in different types of regions

Catalog	Urban core area	Urban suburb	Industrial development zone
Industry	0	20%	60%
Industry and commerce	40%	20%	20%
Residents	10%	40%	10%
Administrative offices	50%	20%	10%

Users in various categories are classified according to different industries to ensure that each industry is divided into three sub-categories: large, medium and small according to the scale of the industry, and each sub-category guarantees at least 30 sample users are included.

The calculation example of this research institute includes 9 planning grids which belong to the suburb of Wuhan City, 1000 user survey samples, of which 796 sample data are available. User survey data include user's reported installed capacity, user's annual power consumption, production loss estimation of 600 ms, 1 s, 1 min, 3 min, 10 min, 30 min, 60 min, 2-4 h, more than 8 h blackout, and also include the risk level of emergency power generation equipment shutdown installed by users themselves and the annual operation and maintenance cost.

2.2 The Algorithm

(1) Risk Algorithm of Equipment Outage
    Based on the collected data of distribution automation, the outage event data of each line switch is divided into three data: outage load, outage duration and outage frequency. The annual outage risk frequency and duration of per hundred kilometers (hundred sets) of equipment is calculated by extracting the information of conductor section of the down-stream feeder, switch type and other equipment.
(2) Calculation of Risk Electricity Quantity in Planning Grid Outage

Taking every 5 years as a calculation cycle, taking the switch from the disjunctor in substation to the next line switch as a line segment, selecting the most typical cross-sectional area of the wire in the grid as the benchmark, and calculating the length of all line segments to the benchmark length according to the unit of the benchmark cross-sectional area length. Based on the data of smart meter, the power curve $\sum_1^n Pn$ is multiplied by the benchmark length L, and the reference load moment of grid, PLN, is calculated. The outage events in the planning unitization are classified according to the distribution lines and the total PLN of each unitization, so as to calculate the annual outage time risk, frequency risk and risk power per megawatt kilometer (MW km) of each unitization.

(3) Normal Distribution of The Risk Cost of Electricity Shortage in Different Industries Which Comply with Markov Process

The risk cost of power shortage of users in different industries or different scales in the same industry follows the Markov process distribution which is positively related to the change of power consumption scale of users, and the risk cost of power shortage of users of the same scale in the same industry follows the normal distribution which is positively related to the change of power outage duration. According to the user survey data, with the outage time as the X axis, the risk cost of short supply as the Y axis and the industry as the Z axis, all the user survey sample data are filled in three dimensions to form a v = f(t) curve, and the centre of gravity of each normal distribution is calculated. Using the proportion of GDP of each industry in the planning grid as the adjustment coefficient β, the V = f(t) curve is treated linearly. The cost of short supply corresponding to the midpoint of the curve is the quantified value of the cost of the average short supply of the planning grid.

3   Empirical Results

3.1   The Risk of Power Cut off According to Equipment

Table 2 shows the results calculated according to the equipment outage risk.

Table 2. Calculation results of outage risk by equipment classification

Equipment	Power outage probability (times/100 km per year)	Power outage period (min/100 km per year)
Over bare wires	5.80	56
Over insulated wire	4.01	70
Cable	0.98	180
Switch	2.10	104

3.2   The Result of Planning Unitization Not Supplied Energy Per PLN

To calculate the outage risk of reference load based on grid unitization can simplify the analysis and calculation of constructive topology of grid. Use the calculated equipment outage risk results in Table 2 to calculate the reference load moment outage risk level of each grid, results are shown in Table 3.

Table 3. Planning unitization power cut off risk level result

Unitization	Power cut off duration per MW km (minutes)	Cut off puissance per MW km (MW)	Not supplied energy per MW km (MWh)
No. 1	43	6.0	4.30
No. 2	55	3.01	2.76
No. 3	26	2.3	0.99
No. 4	13	1.13	0.24
No. 5	107	1.4	2.50
No. 6	94	1.7	2.66
No. 7	43	0.98	0.70
No. 8	37	0.89	0.55
No. 9	35	5.6	3.27

According to the results calculated in Table 3, the annual reference power shortage of each grid can be further calculated, which is the cost quantification calculation condition in Table 6.

3.3   Risk Cost of Not Supplied Energy Result

User survey shows that the possible risk loss cost of various users due to power outage is affected by different factors, and the conclusion by industry is shown in Table 4.

Table 4. Sensitivity coefficient of outage risk cost by industry

Industry	1–3 min	3–10 min	10–30 min	30–120 min	120–180 min	180–240 min
Data center	0.3	0.2	0.125	0.125	0.125	0.125
Commercial center	0.125	0.125	0.125	0.125	0.3	0.2
Industrial and commercial outlets	0.125	0.125	0.125	0.125	0.3	0.2
Restaurants	0.125	0.125	0.125	0.125	0.3	0.2
Machining	0.3	0.2	0.125	0.125	0.125	0.125
Residents	0.3	0.2	0.125	0.125	0.125	0.125

Calculate the risk loss of power shortage of users of different scales in the same industry according to the Markov process, and the calculated quantitative results of power shortage cost of each industry are shown in Table 5.

Table 5. Quantitative calculation results of risk cost of electric shortage by industry

Industry	Quantification of power loss cost (RMB/kW)	Quantification of electric shortage cost (RMB/kWh)
Data center	20.9	14.26
Commercial center	13.7	56.7
Industrial and commercial outlets	19.1	58.3
Restaurants	0.05	49.9
Machining	2.4	34.96
Residents	0	7.8

Based on the annual benchmark power shortage of each planning grid unitization calculated in Table 3 and the proportion of GDP of each grid by industry, adjust the proportion of power shortage of different planning unitization, then calculate the quantitative cost of power shortage loss of each grid unitization. The calculation results are shown in Table 6.

Table 6. Risk cost of not supplied energy result

Unitization	Risk cost of not supplied energy per MW km (Yuan/MW km)
No. 1	160390
No. 2	102948
No. 3	36927
No. 4	8952
No. 5	93250
No. 6	99218
No. 7	26110
No. 8	20515
No. 9	121971

The result shows in Table 6 that due to the difference of industry distribution and grid structure, the potential loss of each grid under power outage is positively related to the benchmark load distance of PL_N, which indicates that the risk cost quantification of power shortage can be used to guide the grid planning and investment decision-making, and reduce the risk cost of power grid outage by lowering the benchmark load distance PL_N through optimization of grid structure and block connection switch.

4 Results

By analyzing the characteristics of grid planning unitization, this study makes a cost quantitative study on the social loss caused by power outage and the index of Energy Not Supplied in grid investment. The obtained results can be applied to the planning of unitization of grid distribution network and optimize the configuration of line switches. First, it reduces the PL_N to lower the loss from power outage. Second, it reduces users' loss suffered from power outage and improve the reliability of power supply by configuring the optimal number of line segments and interconnection switches. Quantification of energy shortage cost based on grid planning unitization reflects the power demand characteristics of unitization and transform the power grid operational reliability index into cost index. It then can be directly applied to distribution network investment decision-making and transmission and distribution price verification. It solves the problem of cost comparison in previous technical schemes, and fundamentally consolidate the precise investment planning of distribution network.

With the transformation and innovation of the operation mode of power grid company, the application of new energy storage technology is becoming increasing important. The planning of energy storage commercial projects on the user side has various benefits. First, it can delay the investment of distribution network, reduce the debt-asset ratio, and improve the operation vitality of the company; second, it can optimize the load distribution of the line from the load side, so as to reduce the benchmark load distance PLN; third, the energy storage equipment has the corresponding advantages of fast speed, which can meet the demand of high reliability users (such as users from precision machining and data center sector) for power safety, and reduce their losses caused by power outage. By using the quantitative algorithm for the cost of Energy Not Supplied based on grid planning unitization to make the cost-benefit analysis of energy storage, V2G and other new power service equipment and modes can enable the power companies to carry out precise investment planning for energy storage business applications, so as to guide the investment planning of multi-operation projects such as energy storage equipment operation and maintenance services, financial leasing services, power supply reliability insurance services.

References

1. Hashemi-Dezaki, H., Askarian-Abyaneh, H., Haeri-Khiavi, H.: Reliability optimization of electrical distribution systems using internal loops to minimize energy not-supplied (ENS). J. Appl. Res. Technol. **13**(3), 416–424 (2015)
2. Hashemi-Dezaki, H., Askarian-Abyaneh, H., Agheli, A., Hosseinian, S.H., Mazlumi, K., Nafisi, H.: Optimized investment to decrease the failure rate of distribution lines in order to improve SAIFI. In: IEEE Conference 4th International Power Engineering and Optimization Conference (PEOCO 2010), Shah Alam, Selangor, Malaysia (2010)
3. Banerjee, B., Islam, S.M.: Reliability based optimum location of distributed generation. Int. J. Electr. Power Energy Syst. **33**, 1470–1478 (2011)
4. Celli, G., Pilo, F.: Optimal sectionalizing switches allocation in distribution networks. IEEE Trans. Power Deliv. **14**, 1167–1172 (1999)

5. Chandramohan, S., Atturulu, N., Devi, R.K., Venkatesh, B.: Operating cost minimization of a radial distribution system in a deregulated electricity market through reconfiguration using NSGA method. Int. J. Electr. Power Energy Syst. **32**, 126–132 (2010)
6. Da Silva, L.G.W., Pereira, R.A.F., Mantovani, J.R.S.: Optimized allocation of sectionalizing switches and control and protection devices for reliability indices improvement in distribution systems. In: Transmission and Distribution Conference and Exposition: Latin America, IEEE/PES, pp. 51–56. IEEE (2004b)
7. Da Silva, L.G.W., Pereira, R.A.F., Abbad, J.R., Mantovani, J.R.S.: Optimised placement of control and protective devices in electric distribution systems through reactive tabu search algorithm. Electr. Power Syst. Res. **78**, 372–381 (2008)
8. Da Silva, L.G., Pereira, R.A., Mantovani, J.R.: Allocation of protective devices in distribution circuits using nonlinear programming models and genetic algorithms. Electr. Power Syst. Res. **69**, 77–84 (2004)
9. Etemadi, A.H., Fotuhi-Firuzabad, M.: Distribution system reliability enhancement using optimal capacitor placement. IET Gener. Transm. Distrib. **2**, 51–59 (2008)
10. Popović, D.H., Greatbanks, J.A., Begović, M., Pregelj, A.: Placement of distributed generators and reclosers for distribution network security and reliability. Int. J. Electr. Power Energy Syst. **27**, 398–408 (2005)
11. Trebolle, D., Gómez, T.: Reliability options in distribution planning using distributed generation. Latin Am. Trans. IEEE (Revista IEEE America Latina) **8**, 557–564 (2010)
12. Trebolle, D., Gómez, T., Cossent, R., Frías, P.: Distribution planning with reliability options for distributed generation. Electr. Power Syst. Res. **80**, 222–229 (2010)

BP Neural Network-Based Product Quality Risk Prediction

Yingcheng Xu[1], Fei Pei[1(✉)], Qian Wu[1], and Bisong Liu[2]

[1] Quality Research Branch, National Institute of Standardization,
Beijing 100191, China
15810265898@126.com, springblue410@126.com,
sqshan8@163.com
[2] National Institute of Standardization, Beijing 100191, China
yxyangcnis@163.com

Abstract. This paper aims to characterize the quality risk of products through the number of injured people. Based on time series data about the number of product-injured people, a stable BP neural network model is established to make a scientific prediction about the number of people injured by products and provide methodological support for product quality risk management and control.

Keywords: Consumer products · Web · Influence evaluation

1 Introduction

Product quality and safety is directly concerned with physical health and life safety of the general public and even more with government stability and social harmony and stability. Once products are caught in quality issues, substantial social impact may be produced. To strengthen product quality and safety regulation, prevent and control product quality and safety risk, establish corresponding analyzing and warning mechanisms, and perform scientific analysis on status quo of product quality and safety and potential issues are both effective and necessary for promoting the harmonious development of society. In present stage, the foundation for domestic studies on product quality and safety risk prediction and warning remains weak. Related studies are mostly qualitative. There is a paucity of research achievements that are based on quantitative analysis of the product quality and safety risk prediction [1, 4, 8–10], such as Lachapelle et al., Hooshmand et al., Jalili et al., Tsai et al. and Chen et al. Featuring high predicting accuracy and outstanding learning capacity, Siman-Tov et al. and Rotondo et al. present BP neural network can be used to erect a prediction model bridging data with risk in order to predict the number of injured on future time series according to given original indicator sample data within a future point of time [2, 5]. Xu et al., Yang et al. and Richard et al. analyze the information integration and transmission model of multi-source data. The product quality and safety regulators are thus able to draft risk plans in accordance with the number of damaged and implement effective regulation on products [3, 6, 7].

2 Structure and Working Principle of BP Neural Network

BP neural network, shaped like a network of transmission, is composed of input layer, middle layer, and output layer. Network layers are connected, while elements on the same layer remain unconnected at all. The so-called BP neural network is realized through two processes, namely forward information transmission from input to output and backward information correction from output to input. In the middle layer, there hides an action function so that input is processed here to generate output result which is compared with target output. Then, hidden weight is returned for correction. Process is input again after being corrected until the difference between trained output and target output becomes satisfactory. In this way, the network is constructed. The structure of BP neural network is demonstrated in Fig. 1.

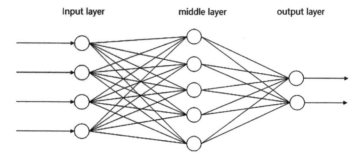

Fig. 1. BP neural network model

3 Implementation of BP Neural Network Risk Prediction Model

The prediction principle of BP neural network is as follows. BP neural network is initialized first, and program is invoked to fit the training effect. Parameters are constantly adjusted until they meet the precision requirements. After parameters are set, historical data are used as input to train the created neural network. The network can regulate the weights of different layers through derivation after automatically learning the implicit rules of samples. If derivative at weight node is above 0, it means the weight is a bit high and should be properly reduced; otherwise, it should be properly raised. The training effect can be optimal after undergoing the process above for multiple times. Afterwards, the input data that need prediction should be substituted into the neural network of function learning, so neural network can make prediction on the basis of learned nonlinear mapping relationship. Aerated entertainment products are used below as an example to introduce how BP neural network risk warning model is implemented.

3.1 Sample Design

The sample data used in present BP neural network risk warning model are from National Electronic Injury Surveillance System (NEISS) of U.S. During sample training, the input data are number of emergency cases injured by aerated entertainment products. For sample data, t, t + 1 and t + 2 are used as input data, whereas t + 3 as output data. The samples to be predicted are number of emergency cases damaged by aerated entertainment facilities in 2014. The goal of this prediction model is to determine the predicting relationship between original indicator sample data and future data so as to predict the number of emergency cases in future years through the neural network.

3.2 Program Implementation

Data are imported into MATLAB to be initialized. BP neural network is then created and parameters are set for training the network. Prediction and analysis are developed. The trained network is saved. Specific program design process is as follows:

(a) **Import and initialize data**

Data are imported into MATLAB and renamed. As the data in present case are small in size, it is feasible to directly set P as input data and T as output data, and the data can be directly inputted in form of matrix into MATLAB program of BP neural network. Data initialization preprocessing is a job to be done before the establishment of the neural network, and data should be initialized to be within the range of [−1, 1]. The initial data in present study undergo dimensionless processing by max and min method. In MATLAB, the simulation tool minmax (data) can be used to initialize the training data and change them into an Rx2-dimensional matrix of max and min values from elements of each input group (R groups in total) required in command. Suppose q is the Rx2-dimensional vector matrix composed of max and min values of trained sample data.

(b) **Create BP network**

newff is defined in MATLAV as a command for initializing one BP neural network. This process involves setting of multiple parameters, such as input matrix, number of hidden layers, number of nodes, transmission function or training function used among different layers.

For example, net = newff (P, Q, [1, 5], {'logsig', 'purelin'}). In this way, a 5-step BP neural network containing both input and output layers is successfully created.

(c) **Train BP neural network**

A common function train in the neural network toolkit of MATLAB is used to train the neural network. The format of function invoking is as follows: Net = train (net, P, T) where net indicates function-returned value and trained neural network, P indicates network's input signal, and T means the goal of network. Net means BP neural network. There are two performance parameters, namely Net.trainparam.epochs (max training epochs of network) and Net.trainparam.goal (goal of network training). The present network training should be performed for 1,000 times, and training error should be lower than 1e−007. The network should be trained through train command.

(d) **Make prediction with trained network**

sim function in the neural network toolkit of MATLAB is employed to achieve simulated prediction on the trained network. The format of function calling is y = sim (net, p_test) where net means trained network and p_test means to-be-simulated data.

The trained network is simulated with sim function: Y = sim (net, data1) where data1 means the predicting sample data that undergo uniform processing, net means BP neural network, Y means network output.

(e) **Save the network**

Apply command save ('filename', 'net') to save the trained network for future prediction purpose.

4 Experimental Cases and Results Analysis

4.1 Experimental Data

Using NEISS of U.S. Consumer Product Safety Association as the data source, the present study reviews the reports about aerated entertainment-injured cases during 2003–2014 published in the system. The data from 2003 to 2011 are used for modeling, while that from 2012 to 2014 for model verification. As for details about the injuries, see Table 1.

Table 1. Predicted number of people injured by aerated entertainment facilities during 2003–2014

Year	Measured value	Year	Measured value
2003	172	2009	297
2004	163	2010	367
2005	202	2011	477
2006	209	2012	521
2007	246	2013	608
2008	319	2014	582

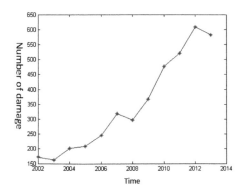

Fig. 2. Variation trend in number of the injured

The variation tendency of people injured by aerated entertainment facilities during 2002–2013 is shown in Fig. 2.

4.2   Result Analysis

After the procedures described above are completed, the system will return the graphs of network training effect. A comparison of trained network with target output is made in Fig. 3.

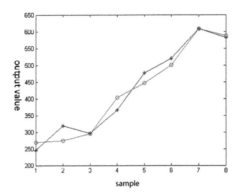

Fig. 3. Comparison of trained BP neural network with target output

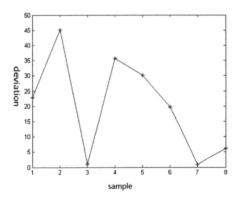

Fig. 4. Errors between predicted and target values of BP neural network

From the figure, we can find that network training is relatively accurate. As for precision, the network training parameters are well configured and the training effect is satisfactory.

In view of the feedback from trained network, the maximum error is 45 while the minimum one is 1. When the trained predicting data are compared with target values as shown in Fig. 4, BP network training turns to be satisfactory in effect and high in predicting accuracy.

Undesirable network training could bear imperfect fitting image and significant predicting error. In such a case, it may be considered to adjust the learning step length in initially created network so as to correct the predicting outcome.

5 Conclusion

In this paper, BP neural network model is applied to the quality risk prediction for aerated entertainment products. The model is fully utilized to simulate the biological mechanisms and find out the potential rules and features of matters so as to predict the number of injured. Judged from predicted outcome, the model is close to actual data and has high predicting accuracy. This testifies its feasibility and reasonableness. It is hopeful in providing technical support for relevant regulatory divisions to strengthen product risk prediction and warning.

Acknowledgments. We would like to acknowledge that this research is supported and funded by the National Science Foundation of China under Grant No. 91646122 and 91746202, the Basic Scientific Research Business Projects 552018Y-5927-2018.

References

1. Lachapelle, U., Noland, R.B., Von Hagen, L.A.: Teaching children about bicycle safety- an evaluation of the New Jersey. Accid. Anal. Prev. **52**, 237–249 (2013)
2. Siman-Tov, M., Jaffe, D.H.: Bicycle injuries- a matter of mechanism and age. Accid. Anal. Prev. **44**, 135–139 (2012)
3. Richard, J.-B., Thélot, B., Beck, F.: Evolution of bicycle helmet use and its determinants in France. Accid. Anal. Prev. **60**, 113–120 (2013)
4. Hooshmand, J., Hotz, G., et al.: BikeSafe: evaluating a bicycle safety program for middle school aged children. Accid. Anal. Prev. **66**, 182–186 (2014)
5. Rotondo, A., Young, P., Geraghty, J.: Quality risk prediction at a non-sampling station machine in a multi-product, multi-stage, parallel processing manufacturing system subjected to sequence disorder and multiple stream effects. Ann. Oper. Res. **209**, 255–277 (2013)
6. Xu, Y., Wang, L., Xu, B.: An information integration and transmission model of multi-source data for product quality and safety. Inf. Syst. Front. **18**(4), 1–22 (2016)
7. Yang, F., Zhang, R., Yao, Y., et al.: Locating the propagation source on complex networks with propagation centrality algorithm. Knowl.-Based Syst. **100**(C), 112–123 (2016)
8. Jalili, M.: Social power and opinion formation in complex networks. Phys. A **392**(4), 959–966 (2013)
9. Tsai, F.S., Chan, K.L.: Redundancy and novelty mining in the business blogosphere. Learn. Organ. **17**(6), 490–499 (2010)
10. Chen, Y., Tsai, F.S., Chan, K.L.: Machine learning techniques for business Blog search and mining. Expert Syst. Appl. **35**(3), 581–590 (2008)

Comprehensive Information Management Analysis of Construction Project Based on BIM

Huixiang Zhang[1], Jian Wei[2], Jianheng Jiao[2], and Yunpeng Gao[1(✉)]

[1] China Electric Power Enterprise Association Electric Power Development Research Institute, Beijing 100000, China
1220600327@qq.com
[2] Inner Mongolia Electric Power Information and Communication Center, Hohhot 010000, Inner Mongolia, China

Abstract. Informatization management of construction projects refers to the application of information technology in various stages of the project construction process based on Computer technology, Internet technology, Internet of Things technology and Communication technology. Collect, store, and process information from different parties, and control various aspects such as project cost, schedule, quality, and safety to improve management efficiency and level. Apply building information models to construction projects and combine construction projects with modern information technology to realize the information management of all participants, the whole goal, the whole factor and the whole life cycle of the construction project. Exploring the path and method of realizing comprehensive informationization of construction projects from three aspects: combining modern information technology, Digital Twin and BIM-based collaboration. The realization of comprehensive information management of BIM-based construction projects is of great significance for better promoting the application of BIM and promoting the development of intelligent buildings and information management of construction projects in China.

Keywords: Building information modeling · Construction project management · Comprehensive information · Intelligent buildings

1 Introduction

The management informationization of construction projects refers to the process of continuously applying information technology, collecting project information, and developing information resources to continuously improve the management level of construction projects during the various stages of construction projects. At present, the application scope of China's construction project management informationization is narrow, and it cannot be fully applied to every link. The system process is not clear and there is no unified standard. In addition, China has not yet formed an informatization management system that can be used for reference [1, 2].

As a life-cycle project information management model, BIM works collaboratively with data as the center, providing a data sharing platform for all participants, and each participant can exchange information in this virtual environment to realize information

M. Atiquzzaman et al. (Eds.): BDCPS 2019, AISC 1117, pp. 1027–1033, 2020.
https://doi.org/10.1007/978-981-15-2568-1_141

integration shared. The information management of engineering projects based on BIM technology can help save construction time, reduce construction and operation and maintenance expenses, and improve project management quality and building quality. The realization of comprehensive information management of engineering projects based on BIM is of great significance for improving the efficiency of project management and improving the information management level of engineering projects [3, 4].

This paper analyzes the comprehensive information management of BIM-based engineering projects, and conducts research on comprehensive information management from three aspects: combining modern information technology, Digital Twin and BIM-based collaboration, and provides management ideas about BIM-based engineering projects.

2   The Content of Comprehensive Information

Traditional project construction information management practices often lead to information splitting, sharing inefficiency, information storage distortion and transmission delay, serious loss of effective information, and inability to effectively integrate and utilize information because of the backwardness of communication methods. The popularity of information technology and the application of BIM technology have brought new ideas to the construction of project information management.

Drawing on Hall's "three-dimensional structure system" theory, constructing a three-dimensional framework for information management of construction projects. The framework consists of three dimensions: the full participant dimension, the full feature dimension, and the full goal dimension. They are unified in engineering projects and are the decomposition of engineering information in different dimensions. As shown in Fig. 1, the left frame is at the beginning of the project and the right frame is at the end of the project. The "time channel" between the two represents the entire life cycle of the entire construction project. The comprehensive information management of construction projects is to manage the information represented by the framework at all time points of the construction project, that is, to effectively manage the information of the total factor dimension, the full participant dimension and the full target dimension throughout the life cycle.

Comprehensive information management of construction projects based on BIM, organically integrate information between different stages of construction projects, different professions and different elements to realize information sharing and integration, and accurately grasp the implementation of engineering projects based on these information, and achieve Effective management of construction projects, improve project management quality and efficiency, and promote the realization of construction project objectives.

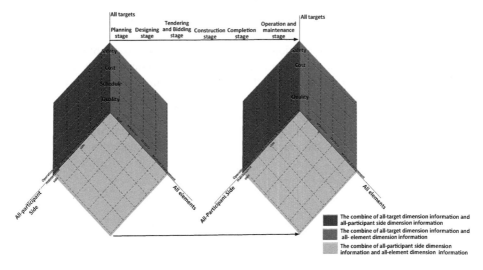

Fig. 1. 3D schematic diagram of comprehensive information management of construction projects

3   The Realization Path of BIM-Based Project Management Comprehensive Information

3.1   Combining Modern Information Technology

The combination of construction projects and information technology will bring new opportunities to the development of the construction industry and create a new era of smart construction in the construction industry. Take the new project as an example, and analyze the combination of the two simply.

In the early stage of the new construction project, through the artificial intelligence analysis and simulation of the big data related to the construction project, the AI system proposed multiple sets of solutions for the builders in a short time. In the early planning of the construction project, BIM technology and GIS technology are combined to establish a digital terrain model of the construction site through on-site measurement, photogrammetry or aerial measurement, which truly reflects the situation on the construction site [5–7].

In the construction phase, the BIM-based smart site management system can effectively realize the management of the construction site. Around the construction process, through the use of information technology to establish an interconnected, intelligent, scientific information management system for construction projects, improve the level of engineering management information. For example, at the construction site, BIM technology and IoT technology are combined, and chips or stickers are placed in materials and equipment, one object and one code, and the code is linked, and the material is scanned in and out of the material. You can get information about the quantity and location of materials in the BIM platform, and perform data query and analysis [8].

With the advancement of the project progress, the data information generated by the entire construction project is also increasing, and the types are more complete. As shown in Fig. 2, the data information is collected and merged to establish a back-end database, which can be divided according to the information of each stage or the information of each participant. Upload the database data information to the BIM project collaborative cloud platform for storage, analysis, processing, integration, and through the combination with mobile communication, you can browse, query, and manage the construction project progress and project construction through the APP on the mobile device. Improve the efficiency of project management.

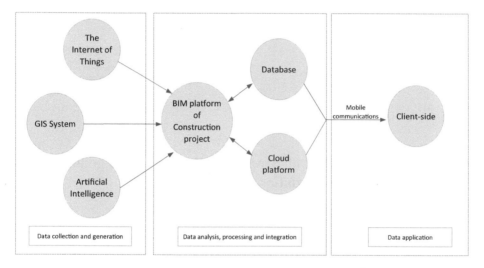

Fig. 2. Combination of construction projects and information technology

With big data being the background of the times, the Internet of Things, cloud computing, "3S" technology, VR/AR technology, and BIM technology have developed rapidly and matured, building a smart environment for project construction and operation management. The cross-border integration of BIM with artificial intelligence, cloud computing, Internet of Things, GIS and other information technology enables resources to be redeployed and effectively utilized, which will greatly promote the development of information management in the construction industry.

3.2   Digital Twin

Digital twin refers to the use of physical models, sensor updates, operational history and other data, integration of multi-disciplinary, multi-physical, multi-scale, multi-probability simulation process, and complete mapping in the virtual space, reflecting the corresponding physical equipment life cycle process [9, 10]. Apply the digital twin concept to the construction project lifecycle management application to realize the construction project information management of the whole life cycle.

At each stage of the construction project, each participating unit will generate relevant information of the party into the project data through digital hygiene and give the data to the BIM information model, and continuously enrich the data in the BIM database. For example, in the previous planning stage, the construction unit analyzes the social environment, policy environment, and market environment in which the project is located through environmental surveys, and conducts an argumentation analysis on the scale, composition, function, and development of the project, and also builds highly realistic 3D visualization virtual model in the virtual space, analyzing the safety factor of the design scheme, whether the structure layout is reasonable and energy consumption (Fig. 3).

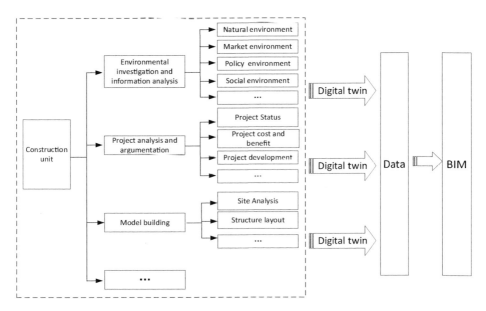

Fig. 3. Construction unit carries out digital hygiene at the project planning stage

Through the digital twinning technology, the two-way real mapping and real-time interaction between the physical engineering and the virtual twinning project realizes the full integration and integration of the whole factor, the whole process and the whole business data between the physical engineering and the virtual twinning project, forming a project-based digital twinning project, Data assets such as digital archives lay a solid foundation for realizing multi-project, enterprise-level and group-level data assets.

3.3 BIM Collaboration Platform

Based on BIM technology, each participant introduces data information to the BIM platform at different stages, establishes and maintains the BIM information model, and uses the engineering information platform to summarize the project information of each project participant, and combines the information model to organize and store. Information integration and sharing, eliminating information silos. This makes the information transmission of each participant faster, greatly reduces the cost of information transmission, and fundamentally solves the information gap between the application systems at various stages of the project life cycle (Fig. 4).

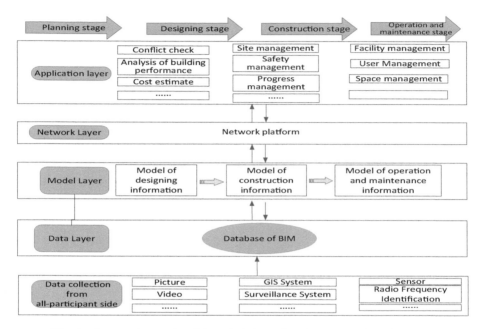

Fig. 4. Construction project life cycle information management based on BIM

Based on BIM technology, all participants participate in the construction of information management platform, build information platform and share information base, thus breaking the information barrier between the participants and realizing the comprehensive integration and sharing of the life cycle information of the construction project. Each participant can accurately grasp the project information and achieve effective management of project information.

4   Conclusion

The rapid development of information technology has made the information between the horizontal and vertical directions of the construction industry more symmetrical. The cross-border integration of BIM with cloud computing, big data and other information technology has brought new vitality and vitality to the construction industry. The BIM parameterized model and the standard unified and interrelated data enable the participants to maintain a high degree of operability between the different phases of the project life cycle, and easy to achieve information exchange and joint management.

Based on the comprehensive information management of BIM-based construction projects, on the basis of the combination of BIM technology and modern information technology, the information of the whole process of the project is collected and transmitted by various technical means, and the whole life cycle of the construction project, all participants, Edit, store, transfer, analyze, and integrate multi-dimensional attribute information such as the whole environment and all elements, comprehensively control and control various aspects such as project schedule, quality, cost, and safety, improve project integration and informationization level, and realize control At the same time as project cost, shorten construction period and improve construction quality objectives, it will bring more advanced and convenient management tools to the project. This is of great significance for continuously improving the level of engineering management information, promoting green construction and intelligent construction, and promoting the development of the construction industry.

References

1. Li, X.: Informatization construction of project management. Constr. Des. Eng. (02), 209–210 (2019)
2. Yuan, F.: Analysis of the development of building management informatization. Constr. Des. Eng. (16), 249–250 (2019)
3. Cao, X., Yan, Y., Bao, Y., You, B.: Research on application of BIM technology in information collaboration of prefabricated building. Constr. Econ. **40**(09), 85–89 (2019)
4. Li, Q., Li, Y.: Research on comprehensive cost management of highway engineering life cycle based on BIM. Highw. Eng. **44**(03), 264–269 (2019)
5. Wang, J., Lei, H., Chong, H.Y., et al.: A cooperative system of GIS and BIM for traffic planning: a high-rise building case study. In: International Conference on Cooperative Design (2014)
6. Wu, B., Zhang, S.: Integration of GIS and BIM for indoor geovisual analytics. Int. Arch. Photogramm. Remote Sens. Spat. Inf. Sci. **XLI-B2**, 455–458 (2016)
7. Wang, Q.K., Li, P., Xiao, Y.P., et al.: Integration of GIS and BIM in metro construction. Appl. Mech. Mater. **608–609**, 698–702 (2014)
8. Lu, L.: Research on RFID intelligent entrance guard system in the Internet of Things. Inf. Technol. **37**(07), 87–90 (2013)
9. Glaessgen, E., Stargel, D.: The digital twin paradigm for future NASA and US air force vehicles. In: Reston: the 53rd Structures, Structural Dynamics and Materials Conference, pp. 1–14 (2012)
10. Liu, D., Guo, K., Wang, B., Peng, Y.: Summary and perspective survey on digital twin technology. Chin. J. Sci. Instrum. **39**(11), 1–10 (2018)

Risk Analysis of P2P Lending Platforms Based on Fuzzy AHP

Rui Li, Jianzhong Lei$^{(\boxtimes)}$, and Li Tan

School of Mathematics and Finance, Hunan University of Humanities,
Science and Technology, Loudi 417000, Hunan, China
3082316537@qq.com

Abstract. P2P lending platforms is an important operating mode in China's
Internet financial system. It also causes huge credit risks following the explosive
growth. Based on fuzzy AHP, this paper makes weight calculation and empirical
analysis on P2P lending platforms risk. The results show that the risk rating of
P2P lending platforms is very high, among which credit risk and technical risk
are the leading factors affecting P2P lending platforms Eventually.

Keywords: P2P lending platform · Fuzzy AHP · Risk analysis

1 Introduction

Peer-to-peer (P2P) lending platform is a virtual financial service intermediary, through
which investors and borrowers can accomplish the P2P credit loan. In 2007, PPDai
which is the first lending platform company was established in China. P2P lending
platform began to flourish, due to its low threshold, small amount and efficiency, the
amount of online turnover was also rising rapidly, it promoted the development of
China's Internet finance, and the financing difficulty of small enterprises and personal
loans was eased by providing credit available to the long tail crowd and organization.
At the same time, a large number of investors got considerable benefits. However, in
recent years, the high bad debt rate of P2P lending platform and a barrage of closed tide
has a huge impact on P2P lending industry, also exposed the lack of P2P lending
platform's credit evaluation system, the information asymmetry brought by the space
crossing and high concealment, and the hidden danger of the capital safety under the
operational errors, etc., therefore, there is some important practical significance in
building P2P lending risk evaluation system and quantifying the risk.

2 Literature Review

Due to the lack of supervision from the similar traditional Banks, P2P lending platforms serve as shadow Banks [1] and help reduce the information asymmetry in the
lending process [2]. The development of such platforms and their financing models will
profoundly change the traditional lending market [3]. The healthy development of P2P
lending requires the establishment of trust mechanism [4], while the consideration

M. Atiquzzaman et al. (Eds.): BDCPS 2019, AISC 1117, pp. 1034–1041, 2020.
https://doi.org/10.1007/978-981-15-2568-1_142

about the risk of lending platform is mainly distinguished from the borrowers' behavior characteristics, such as credit rating [5], friends on the platform [6], and appearance [7].

The research of domestic scholars major in P2P lending risk characteristics, determinants, risk identification and so on. such as Liao Li used all credit data of RenRen Dai, the study found that when the marketization of interest rate formation mechanism is not completely marketed, to some extent investors can recognize borrower default risk that is not included in the rates [8]; Sun et al. studied the risk characteristics of P2P lending platforms. By constructing platform operation indicators, binary regression model was used to distinguish the platforms in difficulty of withdrawing cash from those in normal operation [9]. Ba et al. analyzed the risk of bank run on P2P lending platforms from the perspective of creditor's right transfer [10]. Shi et al. differentiated the risks of P2P online lending platforms with machine learning methods and tested the advantages and disadvantages of different machine learning methods [11]; Zheng et al. empirically analyzed the determinants of interest rates of P2P lending platforms [12]; Liu et al. analysised the core elements of the P2P lending and its corresponding risk [13]; Zhou conducted an empirical analysis on the credit rating system of borrowers, the effectiveness of adverse selection and the possible moral hazard caused by the lending interest rate. The results showed that the default risk of P2P platform borrowers was significantly negatively correlated with the credit rating of borrowers, while significantly positively correlated with the lending interest rate [14].

3 Model Setting

3.1 Establishing a P2P Lending Platform Risk Assessment System

(See Table 1).

Table 1. P2P lending platform risk assessment indicators

Evaluation object	Primary indicator	Secondary indicators
P2P lending platform risk-B	Liquidity risk-B_1	Whether to support bond transfer-B_{11}
		Top ten investors to be included in the ratio-B_{12}
		Top ten borrowers to pay off-B_{13}
		Loan concentration-B_{14}
		Term index-B_{15}
	Operational risk-B_2	Customer size-B_{21}
		Capital leverage-B_{22}
		Interest rate level-B_{23}
	Credit risk-B_3	Credit information abuse risk-B_{31}
		Internal fraud risk-B_{32}
		External fraud risk-B_{33}
		Credit risk-B_{34}

(continued)

Table 1. (*continued*)

Evaluation object	Primary indicator	Secondary indicators
	Information transparency risk-B_4	Completion of public information-B_{41}
		Media index-B_{42}
		Search index-B_{43}
	Legal risks-B_5	Regulatory vacancy risk-B_{51}
		Subject qualification risk-B_{52}
		Network money laundering risk-B_{53}
		Legal and legal absence risk-B_{54}
	Technical risk-B_6	Technical strength risk-B_{61}
		Network system security risk-B_{62}
		Data transmission security risk-B63
		Virus infection risk-B_{64}

3.2   Calculating the Weight of Index

Based on Satty's score of 1–9, this paper comprehensively considers 30 expert opinions and obtains the judgment matrix of the first-level indicators, as shown below in Table 2.

Table 2. P2P lending platform risk evaluation judgment matrix

B	B_1	B_2	B_3	B_4	B_5	B_6
B_1	1	1/2	1	1	1/3	1/3
B_2	2	1	1/2	1	1/2	1/2
B_3	1	2	1	2	3	3
B_4	1	1	1/2	1	2	1/2
B_5	3	2	1/3	2	1	3
B_6	1/4	1/2	1/3	1/2	1/3	1

So the judgment matrix B of the risk of the P2P lending platform is

$$B = \begin{pmatrix} 1 & 1/2 & 1 & 1 & 1/3 & 1/3 \\ 2 & 1 & 1/2 & 1 & 1/2 & 1/2 \\ 1 & 2 & 1 & 2 & 3 & 3 \\ 1 & 1 & 1/2 & 1 & 2 & 1/2 \\ 3 & 2 & 1/3 & 2 & 1 & 3 \\ 1/4 & 1/2 & 1/3 & 1/2 & 1/3 & 1 \end{pmatrix}$$

According to AHP, we figure out that the weight distribution of the criterion layer is $W = (0.1018, 0.1308, 0.2994, 0.1468, 0.2493, 0.072)$.

3.2.1 Secondary Indicator Weight

(a) Liquidity risk. The judgment matrix and evaluation index of liquidity risk are

$$B_1 = \begin{pmatrix} 1 & 3 & 3 & 1/4 & 1/2 \\ 1/3 & 1 & 1 & 1/3 & 1/4 \\ 1/3 & 1 & 1 & 1/3 & 1/4 \\ 4 & 3 & 3 & 1 & 2 \\ 2 & 4 & 4 & 1/2 & 1 \end{pmatrix}$$

According to AHP, the weight distribution of the secondary indicators under the liquidity risk indicator is $W_1 = (0.168, 0.0801, 0.0801, 0.386, 0.2857)$.

(b) Operational risk. The operational risk judgment matrix and evaluation indicators are

$$B_2 = \begin{pmatrix} 1 & 1/2 & 2 \\ 2 & 1 & 3 \\ 1/2 & 1/3 & 1 \end{pmatrix}$$

According to AHP, the weight distribution of the secondary indicators under the operational risk indicator is $W_2 = (0.297, 0.5396, 0.1634)$.

(c) Credit risk. The credit risk judgment matrix and evaluation indicators are

$$B_3 = \begin{pmatrix} 1 & 1/4 & 1/4 & 1 \\ 4 & 1 & 2 & 4 \\ 4 & 1/2 & 1 & 4 \\ 1 & 1/4 & 1/4 & 1 \end{pmatrix}$$

the weight distribution of the secondary indicators under the credit risk indicator is $W_3 = (0.0988, 0.47, 0.3324, 0.0988)$.

(d) Information transparency risk. The information transparency risk judgment matrix and evaluation indicators are

$$B_4 = \begin{pmatrix} 1 & 1/3 & 4 \\ 3 & 1 & 5 \\ 1/4 & 1/5 & 1 \end{pmatrix}$$

According to AHP, Through the consistency test, the weight distribution of the secondary indicators under the information transparency risk indicator is $W_4 = (0.2797, 0.6267, 0.0936)$.

(e) Legal risk. The legal risk judgment matrix and evaluation indicators are

$$B_5 = \begin{pmatrix} 1 & 5 & 4 & 2 \\ 1/5 & 1 & 1/4 & 1/3 \\ 1/4 & 4 & 1 & 1/2 \\ 1/2 & 3 & 2 & 1 \end{pmatrix}$$

The weight distribution of the secondary indicators under the legal risk indicator is $W_5 = (0.4999, 0.0714, 0.1671, 0.2616)$.

(f) Technical risks. The technical risk judgment matrix and evaluation indicators are

$$B_6 = \begin{pmatrix} 1 & 1/4 & 1/3 & 1/2 \\ 4 & 1 & 3 & 2 \\ 3 & 1/3 & 1 & 3 \\ 2 & 1/2 & 1/3 & 1 \end{pmatrix}$$

the weight distribution of the secondary indicators under the technical risk indicator is $W_6 = (0.0953, 0.4668, 0.2776, 0.1603)$.

4 Fuzzy Comprehensive Evaluation

Invite 30 professionals who are familiar with P2P lending to rate the level of each single factor risk. The evaluation level and membership degree of each risk indicator are shown in Table 3 below.

Then the membership matrix R_i is:

Table 3. Fuzzy level evaluation of financial risk assessment indicators for P2P lending platform

Primary indicator (weight)	Secondary indicator (weight)	Membership				
		Very high	High	Medium	Low	Very low
Liquidity risk-B_1 (0.1018)	Whether to support bond transfer-B_{11}	0.3	0.1	0.25	0.25	0.1
	Top ten investors to be included in the ratio-B_{12}	0.35	0.05	0.25	0.25	0.1
	Top ten borrowers to pay off-B_{13}	0.05	0.35	0.25	0.25	0.1
	Loan concentration-B_{14}	0.25	0.3	0.15	0.25	0.05
	Term index-B_{15}	0.25	0.1	0.3	0.2	0.15
Operational risk-B_2 (0.1308)	Customer size-B_{21}	0.3	0.25	0.25	0.15	0.05
	Capital leverage-B_{22}	0.5	0.2	0.2	0.1	0
	Interest rate level-B_{23}	0.2	0.25	0.15	0.25	0.15
Credit risk-B_3 (0. 2994)	Credit information abuse risk -B_{31}	0.5	0.3	0.2	0	0
	Internal fraud risk-B_{32}	0.7	0.2	0.1	0	0
	External fraud risk-B_{33}	0.6	0.2	0.1	0.1	0
	Credit risk-B_{34}	0.5	0.3	0.1	0.1	0
Information transparency risk-B_4 (0.1468)	Completion of public information-B_{41}	0.4	0.3	0.2	0.05	0.05
	Media index-B_{42}	0.4	0.35	0.15	0.05	0.05
	Search index-B_{43}	0.5	0.15	0.2	0.15	0

(continued)

Table 3. (*continued*)

Primary indicator (weight)	Secondary indicator (weight)	Membership				
		Very high	High	Medium	Low	Very low
Legal risk-B_5 (0.2493)	Regulatory vacancy risk-B_{51}	0.6	0.2	0.15	0.5	0
	Subject qualification risk-B_{52}	0.4	0.3	0.2	0	0.1
	Network money laundering risk-B_{53}	0.5	0.2	0.2	0.1	0
	Legal and legal absence risk-B_{54}	0.5	0.4	0.05	0.05	0
Technical risk-B_6 (0.072)	Technical strength risk-B_{61}	0	0.1	0.25	0.35	0.3
	Network system security risk-B_{62}	0	0.4	0.25	0.15	0.1
	Data transmission security risk-B_{63}	0	0.25	0.35	0.25	0.05
	Virus infection risk-B_{64}	0	0.25	0.25	0.25	0.25

$$R_1 = \begin{pmatrix} 0.3 & 0.1 & 0.25 & 0.25 & 0.1 \\ 0.35 & 0.05 & 0.25 & 0.25 & 0.1 \\ 0.05 & 0.35 & 0.25 & 0.25 & 0.1 \\ 0.25 & 0.3 & 0.15 & 0.25 & 0.05 \\ 0.25 & 0.1 & 0.3 & 0.2 & 0.15 \end{pmatrix}$$

$$R_2 = \begin{pmatrix} 0.3 & 0.25 & 0.25 & 0.15 & 0.05 \\ 0.5 & 0.2 & 0.2 & 0.1 & 0 \\ 0.2 & 0.25 & 0.15 & 0.25 & 0.15 \end{pmatrix}$$

$$R_3 = \begin{pmatrix} 0.5 & 0.3 & 0.2 & 0 & 0 \\ 0.7 & 0.2 & 0.1 & 0 & 0 \\ 0.6 & 0.2 & 0.1 & 0.1 & 0 \\ 0.5 & 0.3 & 0.1 & 0.1 & 0 \end{pmatrix}$$

$$R_4 = \begin{pmatrix} 0.4 & 0.3 & 0.25 & 0.05 & 0.05 \\ 0.4 & 0.35 & 0.15 & 0.05 & 0.05 \\ 0.5 & 0.15 & 0.2 & 0.15 & 0 \end{pmatrix}$$

$$R_5 = \begin{pmatrix} 0.6 & 0.2 & 0.15 & 0.5 & 0 \\ 0.4 & 0.3 & 0.2 & 0 & 0.1 \\ 0.5 & 0.2 & 0.2 & 0.1 & 0 \\ 0.5 & 0.4 & 0.05 & 0.05 & 0 \end{pmatrix}$$

$$R_6 = \begin{pmatrix} 0 & 0.1 & 0.25 & 0.35 & 0.3 \\ 0 & 0.4 & 0.25 & 0.15 & 0.1 \\ 0 & 0.25 & 0.35 & 0.25 & 0.05 \\ 0 & 0.25 & 0.25 & 0.25 & 0.25 \end{pmatrix}$$

From the fuzzy comprehensive evaluation formula, $T_i = W_i \cdot R_i$, the evaluation results of each secondary index are:

$$T_1 = W_1 \cdot R_1 = (0.2504, 0.1932, 0.2257, 0.2357, 0.0950)$$
$$T_2 = W_2 \cdot R_2 = (0.2787, 0.2418, 0.1879, 0.1958, 0.0958)$$
$$T_3 = W_3 \cdot R_3 = (0.6272, 0.2198, 0.1099, 0.0431, 0)$$
$$T_4 = W_4 \cdot R_4 = (0.4094, 0.3173, 0.1687, 0.0594, 0.0453)$$
$$T_5 = W_5 \cdot R_5 = (0.5429, 0.2595, 0.1358, 0.2797, 0.0071)$$
$$T_6 = W_6 \cdot R_6 = (0, 0.3057, 0.2778, 0.2129, 0.1292)$$

The evaluation matrix for the secondary indicators is:

$$T = \begin{pmatrix} 0.2504 & 0.1932 & 0.2257 & 0.2357 & 0.0950 \\ 0.2787 & 0.2418 & 0.1879 & 0.1958 & 0.0958 \\ 0.6272 & 0.2198 & 0.1099 & 0.0431 & 0 \\ 0.4094 & 0.3173 & 0.1687 & 0.0594 & 0.0453 \\ 0.5429 & 0.2595 & 0.1358 & 0.2797 & 0.0071 \\ 0 & 0.3057 & 0.2778 & 0.2129 & 0.1292 \end{pmatrix}$$

Since the weight distribution of the criteria layer indicators is $W = (0.1018, 0.1308, 0.2994, 0.1468, 0.2493, 0.072)$.

According to the

$$D = W \cdot T \tag{1}$$

We can figure out that $D = (0.4452, 0.2504, 0.1591, 0.1563, 0.0399)$. According to the principle of maximum subordination, the Internet financial risk $d = \max(d_1, d_2, d_3, \ldots, d_n) = 0.4452$, corresponding to the first risk level "very high", so the overall evaluation level of the P2P lending platform risk is very high.

Acknowledgements. This work is supported by Youth foundation of Hunan University of Humanities and Science and Technology, the Research Foundation of Education Committee of Hunan University of Humanities, Science and Technology (RKJGY1714).

References

1. Emekter, R., Yanbin, T., Jirasakuldech, B., et al.: Evaluating credit risk and loan performance in online Peer-to-Peer (P2P) lending. Appl. Econ. **47**(1), 54–70 (2015)
2. Berger, S.C., Gleisner, F.: Emergence of financial intermediaries in electronic markets: the case of online P2P lending. BuR Bus. Res. J. **2**(1), 39–65 (2009)

3. Wang, H., Greiner, M., Aronson, J.E.: People-to-People lending: the emerging e-commerce transformation of a financial market. In: Value Creation in E-Business Management, pp. 182–195. Springer, Berlin (2009)
4. Greiner, M.E., Wang, H.: Building Consumer-to-consumer trust in e-finance marketplaces: an empirical analysis. Int. J. Electron. Commer. **15**(2), 105–136 (2010)
5. Puro, L., Teich, J.E., Wallenius, H., Wallenius, J.: Borrower decision aid for People-to-People lending. Decis. Support Syst. **49**(1), 52–60 (2010)
6. Lin, M., Prabhala, N.R., Viswanathan, S.: Judging borrowers by the company they keep: friendship networks and information asymmetry in online Peer-to-Peer lending. Manag. Sci. **59**(1), 17–35 (2013)
7. Ravina, E.: Love and loans: the Effect of beauty and personal characteristics in credit markets. Working Paper. Columbia University (2012)
8. Liao, L., Li, M., Wang, Z.: The intelligent investor: not-fully-marketized interest rate and risk identify-evidence from P2P lending. Econ. Res. J. **7**, 125–137 (2014)
9. Sun, B., Niu, C., Zhao, X., Jing, W.: Identifying financial distress: the risk characteristics of Peer to Peer lending platforms. J. Cent. Univ. Finance Econ. **07**, 4 (2016)
10. Ba, S., Bai, H., Li, Y.: P2P online lending platform run-off risk: based on the perspective of credit transfer. Res. Financ. Econ. Issues (2019)
11. Shi, Y., Zhang, B., Yan, J.: P2P online lending platform risk screening research. Stat. Decis. (2018)
12. Zheng, Y., Chen, X., Xin, Y.: The determination of interest Rate of China's P2P online lending: an empirical study based on cross-section data of cross-platform. Contemp. Finance Econ. (2018)
13. Liu, H., Shen, Q.: Research on risk and supervision of P2P network lending in China. Res. Financ. Econ. Issues (2015)
14. Zhou, J., Zhao, Z., Xu, Z.: Research on default risk of P2P platform in China. Price: Theory Pract. (2016)

Animation Feature Changes Based on Big Data Analysis

Dong Shao$^{(\boxtimes)}$

Dalian Neusoft University of Information, Dalian, China
Shaodong@neusoft.edu.cn

Abstract. With the development of The Times and the progress of science and technology, the rapid application of the Internet, human beings have stepped into the era of big data. Massive data and unlimited information have become the typical characteristics of the era of big data. It is against this background that the animation industry has undergone a profound transformation. In animation just the moment, it as a kind of magic, in the form of an ill-deserved satisfy people's aesthetic needs, with the development of media technology and animation, animation gradually endowed with long-term cultural mission, and are being developed into a cultural transmission carrier, development since modern times, the stride development of science and technology make the animation is changing forms and the longing for aesthetic, are efforts to the standard of art in the age of big data show the charm of animation. Taking the development of The Times as the main line, this paper discusses the transformation of animation communication and the essential characteristics of art in a divergent way, and conducts in-depth research on the animated films that have appeared in front of the people in recent years, so as to have a profound understanding of the development and evolution of animation in the era of big data.

Keywords: Big data era · Characteristics of the times · Animation industry · Characteristics of the change

1 Introduction

In the 21st century, as a period of great development of knowledge and information, data can be said to be everywhere, and everyone has become a part of the information source [1]. In such an era, data has become an important social resource and means of production. By analyzing and mining data, new knowledge can be acquired and value can be created. Study, work, production, investment, financial management, management are inseparable from data.

The definition of big data is quite different from that of traditional data. As far as we know, the concept of big data is very fuzzy and uncertain [2]. The relatively uniform resonance is that big data has four significant characteristics: large Volume of data, high Velocity of data processing, Variety of data, and low Value of data, which are the four significant characteristics that we emphasize [3, 4]. Victor mayer-schonberg, who has the reputation of "the first business application of big data", said in his book the era of big data that the information storm brought by big data is revolutionising our work,

M. Atiquzzaman et al. (Eds.): BDCPS 2019, AISC 1117, pp. 1042–1048, 2020.
https://doi.org/10.1007/978-981-15-2568-1_143

life and thoughts, and the era of big data has opened the door to a significant transformation of the era [5, 6]. He used three chapters in the book to discuss the business transformation, thinking transformation and management transformation in the era of big data. Therefore, big data is an important feature of this era, which has a profound impact on people's production and life [7]. Naturally, the field of animation is also heavily influenced by big data. First of all, traditional animation has changed the way of visual expression. Hand-drawn animation began to develop into full three-dimensional computer animation; Animation production is no longer a one-way transmission, but a two-way choice between the producer and the audience. Secondly, in the era of big data, the aesthetic standard variation, a lot of people also significantly increase demand for culture, therefore, animation works want to win the love of the masses, need to focus on the theme of the animation design and creative original planning, highlight the cultural mission in the process and time feeling, create a personalized animation audio-visual experience [8]. In fact, the earliest research on animation can be traced back to the beginning of the last century. However, the research on animation in that period only rested on simple introduction of animation text documents or inspiration of animation producers, while the technical research on animation production was very rare [9]. Really is from the perspective of animation begin to pay close attention to since the 1950s, Mr George dole's (1950–2015) of the cinema in the world, was an early step is quite influential writings of the cinema, the book will be animated film as a special chapter elucidated, given the animated film legal identity, after that, both at home and abroad study of animation just like spring up of [10].

In order to analyze animation features more accurately and objectively, this paper adopts inductive analysis method to further refine animation features through inductive reasoning based on the history, aesthetics and theory of animation, and provide corresponding Suggestions for the development of animation industry on this basis. As an important part of the history of science and technology, this paper creatively analyzes and summarizes the characteristics, motivations and future trends of animation industry and its technology development, which can bring reference significance to the development of animation industry in the new stage.

2   Method

2.1   Basic Ideas and Principles of Inductive Analysis

The research method of inductive reasoning is a reasoning process from individual cases to general cases. It expands from a certain degree of view on individual things to a larger range of views, and enumerates the interpretation of general principles and principles by specific examples. All things in nature and society exist in the individual, in the particular, and through the individual. The general is in the concrete objects and phenomena, and therefore the general can only be understood by knowing the individual. When people explain a larger thing, they summarize and generalize all kinds of general principles or principles from the individual and special things, and then they can start from these principles and draw a conclusion about the individual things. First, in inductive reasoning, the explainer not only applies inductive reasoning, but also deductive reasoning. In people's thinking of explanation, induction and deduction are complementary, interrelated and inseparable. Second, inductive reasoning is probabilistic except that the relation between premises and conclusion is inevitable in complete inductive reasoning. The principle of inductive analysis is: first prove that the proposition is true at a certain starting point, and then prove that the process from an eigenvalue to the next concrete value is valid. Once both of these points have been verified, any value can be proved in reverse by repeated use of this method.

2.2   Detailed Research Methods

The primary method of this paper is logical analysis, but in the specific research process also borrowed other excellent research methods to help this paper. Firstly, this paper borrows the literature method. In the process of research, the author, with the help of the paper database and the literature resources of the library, retrieves the relevant contents of animation features in the era of big data, and sorts out a large number of paper books, papers and electronic journals that have guiding significance and value for the development of this subject. On this basis, the research data obtained are classified and sorted out, and the contents of the literature with guiding significance and value are screened out to assist the demonstration of this topic. Second use case analysis, the influential in recent years in China, some achievements of animation works were analyzed, and through its approach, themes, and aesthetic connotation deeply on the aspects such as discussion, to summarize the progress made in our country's animation industry in recent years, as well as the existing problems and future development trend and direction. Finally, the research method used is the network survey method. With the help of the network resources on the network, the characteristics of animation and its changes can be collected and sorted out, so as to apply these rich materials to my own research practice and enrich the theoretical research of the subject. In general, these three research methods have laid a good foundation for the development of this study and provided a guarantee for the smooth completion of this study.

3   Experiment

This article attempts through to comb the development history of Chinese animation features, analysis of animation features in different historical stages of in China, summarizes the achievements, and summarizes the performance art of animation methods, analyzes the different historical stages of art thoughts and ACTS on the influence of the animation, and animation in different historical periods in the value orientation of social culture and art. The final goal of this paper is to return to what kind of transition process animation characteristics have undergone in the era of big data, and to make a prediction on the future development trend.

This paper takes the characteristics of animation as the research object and systematically sorts out the development history of animation in China. Through searching and sorting out the works of Chinese animation, it summarizes the artistic charm and characteristics of animation, and compares and analyzes the expression methods combining animation features with big data. By referring to the research results of media communication, animation production principle, animation development history, literature historical connotation and other related aspects, the characteristics of animation were comprehensively studied under the background of the era of big data, and the in-depth artistic charm of animation was further explored from the perspective of development.

4 Discuss

4.1 Changes in Artistic Styles

The formation of style is a sign of the maturity of an art. The formation of style is influenced by the objective and specific conditions of the time and the historical environment. Artists' creations show different styles under different social backgrounds. At the same time, the creation styles of these artists also reflect the profound impression of that era.

(1) Animation style characteristics of the 20th century

The creation of animation was closely connected with the spirit of The Times in the early days. Even the earliest animation creators showed their high spirit and impressive sense of national mission. Before the founding of the People's Republic of China, the creation of animation themes are a lot of anti-japanese patriotic propaganda. For example, wan brothers created "united effort" and "compatriots wake up quickly" and so on, and qian jiateng painting creation "farmhouse music" and so on. In the early 1950s, influenced by the Soviet model, Chinese animation also imitated more forms of Soviet animation. The best example of this imitation is "why are crows black", in which the animated characters are so heavily designed that many foreigners at the Venice film festival thought it was a Soviet cartoon. However, this cartoon occupies an important position in the history of Chinese animation. It is the first Chinese cartoon to win an international prize and the first color cartoon in China.

(2) Animation style characteristics in the era of big data

With the advent of the era of big data, the development of animation industry has entered a brand-new period. According to relevant data, by the end of 2018, the total output value of China's animation industry had reached 145.2 billion yuan, with a growth rate of over 23%, higher than the growth rate of added value of China's cultural industry. As the big data age increase gradually to a driving force for the development of animation industry, China's animation industry is striding towards the Internet as the core, across different shape, a new era of media, cross-industry integration development, in addition, in the development process of our country's animation industry show there are a lot of niche business, mainly published by the film and television animation, animation stage, animation industry, animation, theme parks, animation, animation derivatives authorization and emerge in the era of big data and mobile Internet animation, etc. In the process of research, this paper sorted out the changes of animation style characteristics, and the results are shown in Table 1 below. It can be seen from the table that since the era of big data, animation performance style is developing towards the direction of technicalization and integration.

Table 1. Sorting out the changes of animation art style

Twentieth century			Period of big data		
	Emerging number	Contemporaneous ratio		Emerging number	Contemporaneous ratio
Modelling sex	132	27.91%	Modelling sex	216	19.69%
Technical	56	11.84%	Technical	117	10.67%
Assumption	38	8.03%	Assumption	67	6.11%
Comprehensive	172	36.36%	Comprehensive	313	28.53%
Fashion	75	15.86%	Fashion	384	35.00%

Data in the table are obtained by the author after consulting the data

4.2   Change of Formal Features

(1) Animation forms in the 20th century

In 1953, the world's first ink-inspired cartoon, tadpoles tadpoles look for their mother, was released. The idea of carving copper plates made the world's first etching-style cartoon "night in the mountains". Perhaps it is this desire to make beautiful pictures move that produces the styles of each art form from a different kind of painting. These only forms constitute the 20th century animation expression form all.

(2) Animation expression in the era of big data

Since entering the era of big data, animation industry has realized leapfrog development driven by the Internet. The scale of the animation industry has maintained a steady growth momentum. For example, although the growth rate in 2018 is slower than that of last year, the total output value of the animation industry has reached 169.2 billion yuan. At the same time, in addition to a number of powerful animation enterprises gradually emerging and forming their own development characteristics, animation derivative market, Internet animation, film and television animation and other diverse animation products have also been rapidly developed, the expression form of animation industry is constantly developing, the content presents diversified development. The market share of animation in the era of big data is shown in Fig. 1 below. The data in the figure comes from the website of the state administration of culture. Animation theme parks, animation derivatives and animation authorization account for a large proportion, about 65–70% of the total.

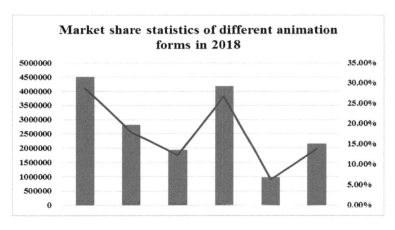

Fig. 1. Market share statistics of different animation forms in 2018

5   Conclusion

The study of this paper shows that the boundary area of the fixed industry shaped by the traditional characteristics of animation is being blurred with the advent of the era of big data, and the animation industry is moving from separation to the integration of industries based on digital integration. It is found that the characteristics of cross-media, cross-form and cross-industry integration are increasingly strong. Therefore, animation industry workers need according to the requirements of The Times development characteristics, purposefully absorbs the traditional culture to modern animation creation element, and according to the aesthetic characteristics of contemporary people, for the traditional animation creative inheritance and innovation, on the basis of the technical force, with the new animation makes modern animation creation to adapt to the development of era, show the conform to the era of big data features of animation, so that the animation industry can go higher and further.

References

1. Ping, D., Tu, C.: Quadrangulations of Animation Sequence. Int. J. Pattern Recogn. Artif. Intell. **31**(11), 17–21 (2017)
2. Munawwarah, M., Anwar, S., Sunarya, Y.: How to develop electrochemistry SETS-based interactive E-book? J. Phys: Conf. Ser. **895**(1), 12–14 (2017)
3. Li, C., Sun, R., Dai, Y.: Intelligent exhibition platform of Chinese ancient farming virtual scene based on Unity3D. Trans. Chin. Soc. Agric. Eng. **33**(45), 308–314 (2017)
4. Yun, J., Son, M., Choi, B.: Physically inspired, interactive lightning generation. Comput. Anim. Virtual Worlds **28**(12), 1760 (2017)
5. Cong, C.: Behavioral decision in development of henan animation industry under "Internet+" environment. Neuroquantology **16**(5), 61–73 (2018)
6. Sammond, N.: Birth of an Industry: Blackface Minstrelsy and the Rise of American Animation. J. South. Hist. **83**(2), 174–176 (2017)

7. Timmons, N.: Birth of An Industry: Blackface Minstrelsy and the Rise of American Animation. Nicholas Sammond. Duke University Press, Durham (2017). J. Popular Cult. **50** (1), 200–202

8. Lin-Hi, N., Blumberg, I.: The power(lessness) of industry self-regulation to promote responsible labor standards: insights from the chinese toy industry. J. Bus. Ethics **143**(4), 1–17 (2017)

9. Hui, L., Deng, S., Jian, C.: Semantic framework for interactive animation generation and its application in virtual shadow play performance. Virtual Reality **22**(1), 1–17 (2018)

10. Mammar, A.: Modeling a landing gear system in Event-B. Int. J. Softw. Tools Technol. Transf. **19**(2), 167–186 (2017)

Discussion of Innovated Training Management Based on Big Data

Hui Wang$^{(\boxtimes)}$ and He Ma

Department of Information and Business Administration, Dalian Neusoft
University of Information, Dalian, Liaoning, China
wanghui_xg@neusoft.edu.cn

Abstract. The challenges of being a training manager or a trainer are always including information of training needs and statistics reflecting training effectiveness. And these problems are always difficult to solve because they need collect and analyze valuable information from amount of data. The technology of big data seems to be a perfect breakthrough. Based on big data technology, we are able to provide the staff with customized courses and more convenient ways of learning. At the same time, we collect information of learning activities and transform it into evaluation indicator of training effectiveness and improve training service. Therefore it is an inevitable trending of innovative training management. At the same time, with the help of e-HR platform, the sharing of knowledge and information within the company become more convenient. Staffs will be more comfortable to share, communicate and obtain what they truly need in the learning community. It is a better way to construct a learning organization. This article brings out the method of training management based on big data and designs a model of training management system.

Keywords: Big data · Training management · e-HR platform

1 Introduction

Currently, human capital management (HCM) has become the main topic that replace the word human recourse management (HRM). To increase the value of human capital, talent development become more important than ever [10]. As the main and most necessary method of talent development, training management is facing more responsibilities and challenges.

Compared with traditional training, now training begins to focus on the needs of employees, so the current commonly used training management mode often starts from the analysis of training demands. Each business department and even each employee has personalized training demands. Normally, training demands are investigated from the organizational level, the work level and the individual level. The training demands are extracted by synthesizing the demands of the three levels. In theory, this can take into account the demands of all aspects, but in the process of implementation, the workload is large and requires the close cooperation of the senior management and various business departments to complete [9]. Even if the time-consuming and labor-consuming research work is completed, it is necessary to unwrap the layers of

© Springer Nature Singapore Pte Ltd. 2020
M. Atiquzzaman et al. (Eds.): BDCPS 2019, AISC 1117, pp. 1049–1053, 2020.
https://doi.org/10.1007/978-981-15-2568-1_144

requirements and find the "pain points" that satisfy all three levels. Although many training managers have realized that the training mode should be diversified in order to stimulate learners' motivation and improve the learning effect, except for some skills training, most of the training still adopts the training mode with trainers-centered. The enthusiasm of training is mainly shaped by the personal influence of trainers but not the learning contents themselves. At the same time, both on-the-job training and off-job training will affect employees' daily work schedule to some extent.

To organize a training successfully, the training managers need the support of top managers in the organization. However, no matter how active the atmosphere is, it cannot represent the effect of the training. Especially for top managers, what they need to see is the input-output ratio of training. The commonly used evaluation model also reminds the training managers that the post-training satisfaction survey and the evaluation results of the students are only parts of the evaluation. The change of employee behavior in the work, as well as the results of the contribution to the company need specific data. And it also need to eliminate the influence of some factors influencing the effect of training in order to improve the reliability and validity of training effect evaluation. So, it is not easy to give top manager a good-looking data represents the output of training and because of this, to earn the support of top managers become quite challenging.

2 Innovative Training Management Based on Big Data

Traditional training demand analysis is time-consuming and labor consuming, while training demand analysis based on big data will be able to collect all kinds of behavioral data of employees through digitized office space in an invisible way, form a powerful database, conduct data mining from it, and extract training requirements that meet the needs of the organization and work [1]. The acquisition of these scattered data is very difficult in traditional human resource management. However, in the Internet era, the implementation of information-based human resource management is indispensable, which enables employees to digitize various information and collect data "buried" in the process of information-based process construction. The cooperation with cloud platform and shared human resource management is able to update dynamic data automatically, and thus further promoted the possibility of training demand analysis based on big data [3, 6].

2.1 Establishing e-HR Platform

Enterprises should establish a large platform of e-HR to support the needs of training management [4]. In the platform, the logical structure based on big data is established. The bottom layer is the system layer, as in data collection layer, which collects data through various systems. This requires system interconnection and data interchange within the enterprise. The second layer is the data layer. Enterprises can use Spark, Hadoop and other technical support services to store these complex data, including pictures, comments and other unstructured data. In addition, through ETL time-sharing data extraction service and OGG real-time data synchronization service, the data is processed through the data mechanism. The next layer above is the analysis layer. The data are sorted and concluded to the analysis layer and classified according to the requirements of human resource management modules. The top is the display layer. In addition to being able to show results, it can also provide some self-service [8].

2.2 Training Data Collection

In aspect of data collection, due to the characteristics of 4 V, big data has a large volume of data, variety of data types, high velocity, and a high value [2]. Therefore, in this process, enterprises need to establish a comprehensive system of talent database and collect all the data of employees. It includes the basic information of employees, their work process, work results, employees' performance appraisal information, rewards and punishments, etc., as well as the data of employees in the previous training process, including the training participated by employees, performance in the training process, satisfaction after training, assessment and change records after training, etc. [7]. At the same time, it is also possible to ask employees to record their personal work conditions and actively convert their information into data. In the context of big data, the training should also deeply explore the needs of employees. Employees recorded by the server should use search engines to retrieve the information generated by the retrieval item information, and take the initiative to push accurate information and potential resources needed in the future for employees by using technologies such as data mining, clustering analysis and relevant analysis.

In addition to the basic data of employees that need to be collected and extracted as training needs of employees, data at the organizational and work level can also be collected directly. This work needs to cooperate with performance management to transform the strategic goals at the organizational level into quantitative goals and decompose accordingly. The data at the work level will also quantify the requirements of work tasks based on the decomposition of organizational goals. The data is then dynamically matched with the employee and job level data. Finally, training needs are formed.

Without doubt, it is difficult to find correlations among these huge amounts of data to reflect training needs. This requires the human resource management department to cooperate closely with the information technology department according to the management experience, make use of efficient data mining technology, and complete the extraction of requirements according to certain rules. In the process of extracting requirements, clustering analysis technology can be used to divide employees into several categories according to certain rules. For example, according to the position sequence, results of performance appraisal, rank and so on. Through the similarity of people to extract similar needs, and then take targeted training measures, in order to maximize their enthusiasm.

In addition, decision tree technology can be fully utilized. For example, after data selection, data cleaning, data induction and data transformation, the ID3 algorithm in the decision tree is used to establish an employee classification model, design classification rules, and find the main characteristics of employees whose performance declines, which can be transformed into training needs. At the same time, using this model, we can also predict the potential employees with unsatisfactory performance, and then identify the training objects. This method is also suitable for analyzing the characteristics of high-performing employees, digging out potential high-performing employees, and carrying out targeted training to accelerate the development of core employees [5] (Fig. 1).

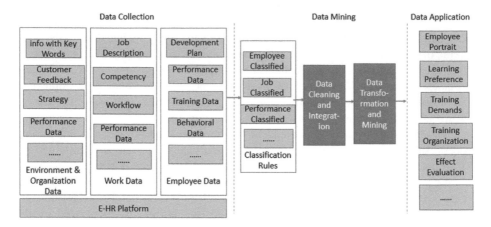

Fig. 1. Data management

2.3 Effect Evaluation Based on Big Data

Training effect evaluation based on big data should be practiced more online, that is, the training process should be completed through e-learning, Moocs, APP or other platforms. Therefore, relevant platforms should collect all kinds of quantitative and qualitative data and information according to specific training effect evaluation objectives. Information to be collected should include: the students' course selection, learning time, place, frequency and duration, learning behavior, such as pause, play-back, exit, etc., and also should include test result, discussions, questions, etc. The last is the evaluation of training results and students' satisfaction feedback. Finally, the analysis of these data and information should be combined with the characteristics of the students, such as the original foundation, training participated in and the feedback of the training received. They will be cross-compared and analyzed, and the final conclusion will be reached after the factors affecting the effectiveness and validity of the training evaluation were excluded.

The evaluation of training effect also needs to collect the data of behavior and result levels through their future work performance to evaluate the effect of behavior and result. In other words, training managers and data analysis are far from finished after the training. The data platform will continue to collect the relevant data of employees' work behaviors and results, analyze the differences before and after the training, and eliminate the interference items, so as to obtain a more comprehensive training effect.

3   Conclusions

How to collect and analyze massive data and find training "pain points", give employees a just-in-time training, and be able to extract effective information through massive data to reflect the training effect. This work will solve the current difficulties in enterprise training management. At the same time, managers of information and human resources are facing great challenges. In the future, human resource management will be combined with more information methods to meet the needs of big data transformation.

Acknowledgements. This work was supported by 2018dlskyb223, LNKX2018-2019C37, L18ARK001.

References

1. Wang, H., Chan, P., Petrikat, D.: The big data challenge in human resource management. J. Acad. Bus. Econ. **17**(4), 23–26 (2017)
2. McAfee, A., Brynjolfsson, E.: Big data: the management revolution. Harv. Bus. Rev. **90**(10), 35–39 (2012)
3. Li, J., Ren, Q.: Big data and human resources management: the rise of talent analytics. Soc. Sci. **8**(10), 67–71 (2019)
4. Calvard, T.S., Jeske, D.: Cross-functional reorganization: human resources, information technology, and big data analytics. In: Academy of Management Proceedings, vol. 2017, no. 1, pp. 45–50 (2017)
5. Picciano, A.G.: The evolution of big data and learning analytics in american higher education. J. Asynchronous Learn. Netw. **16**(4), 9–13 (2012)
6. Yun, H., Xing, A., Jing, X.: Reflection on enterprise human resources management reform in the age of big data. Econ. Res. Ref. **63**, 26–32 (2014)
7. Liu, M., Kang, J., Zou, W., et al.: Using data to understand how to better design adaptive learning. Technol. Knowl. Learn. **22**, 271–298 (2017)
8. Birjali, M., Beni-Hssane, A., Erritali, M.: A novel adaptive e-learning model based on Big Data by using competence-based knowledge and social learner activities. Appl. Soft Comput. **69**, 14–32 (2018)
9. Madhavan, K., Richey, M.C.: Problems in big data analytics in learning. J. Eng. Educ. **105**(1), 34–38 (2015)
10. Brock, T.R.: Performance analytics: the missing big data link between learning analytics and business analytics. Perform. Improv. **56**(7), 6–16 (2017)

Analysis of Computer Network Information Security in the Era of Big Data

Zhenxing Bian[(✉)]

Shandong Polytechnic College, Jining, Shandong, China
bianzx01@163.com

Abstract. In the era of big data, computer network has brought great convenience to people's work and life, but also brought security risks in information. Therefore, the analysis of network information security plays an important role in the development and use of computer network. Under the background of big data, this paper puts forward three research hypotheses from three different factors of technology, personnel and environment, and constructs the evaluation model of computer network information security. By using the structural equation with good adaptability to test the research hypothesis, it is found that the correlation coefficients of the research hypothesis have no significant difference, and the model hypothesis is all valid. In this paper, entropy method is also introduced, and an index weight model is proposed. Experiments show that the index weight model based on entropy method is reasonable.

Keywords: Big data · Computer network · Information security · Entropy method

1 Introduction

With the rapid development of Internet technology and computer technology, computer network has become one of the essential tools for people's daily life, work, learning and so on [1]. In the era of big data, there are countless data generated by computer network, among which the leakage and loss of important information may bring huge economic losses [2]. Therefore, the analysis of computer network information security problems can better predict and judge the hidden dangers brought by network information security, and provide a reference for making a reasonable network defense. This paper analyzes the security of computer network information in the era of big data, and provides a new way to solve the problem of network information security.

At present, there are many scholars and researchers on the computer network information security research. In [3], the author improves the method of evaluating network security based on Hidden Markov model, and the result shows that the method can reflect the trend of network security reasonably. In [4], after analyzing and quantifying the network information security elements, the author also describes the network security secret vector, network security integrity vector and network security availability vector. In [5], the author introduces network security technology, including authentication, data encryption technology, firewall technology, intrusion detection system (IDS), anti-virus technology and virtual private network (VPN). In [6], the

© Springer Nature Singapore Pte Ltd. 2020
M. Atiquzzaman et al. (Eds.): BDCPS 2019, AISC 1117, pp. 1054–1059, 2020.
https://doi.org/10.1007/978-981-15-2568-1_145

author puts forward a multi-attribute decision-making problem of computer network information security evaluation based on triangular fuzzy information, and verifies the rationality of the method through experiments. In [7], the author empirically studies the relationship between online victimization and user activity, as well as personal information security perception on social network services (SNS). The results show that network users need to properly improve their information security awareness. In [8], the author proposes a computer security event processing algorithm to prevent and eliminate the consequences of network attacks on information resources. In [9], the author puts forward the classification method of safety risk assessment, which will provide reliable value for future research in the field of safety risk assessment. In [10], the author puts forward a new safety learning model based on many experiences and lessons.

Under the background of big data era, this paper analyzes the information security of computer network. This paper first introduces the hidden dangers of computer network information security in the era of big data, then puts forward three different research hypotheses and constructs the evaluation model of computer network information security, finally introduces entropy method and proposes the index weight model based on entropy method.

2 Method

2.1 Network Security in the Era of Big Data

In the era of big data, with the emergence of massive data, the update speed of information is faster and faster, and information is developing from one to multiple. In recent years, with the rapid development of computer network, the data generated in it can not be counted. However, the information security problems caused by some reasons will cause huge losses to individuals, enterprises and governments. At present, the hidden dangers of computer network information security in the era of big data mainly include the following aspects:

(1) impact of natural disasters: the data on the network needs to be stored in the hardware. After the natural disasters, the damage or failure of the hardware equipment will result in the loss of data information.
(2) the impact of virus invasion: hackers will spread virus on the network and invade the computer to disclose important information. If the information system design is not perfect, it will easily become the target of hackers, and be changed and controlled at will.

2.2 Evaluation Model of Computer Network Information Security

Under the background of big data, the evaluation indexes of computer network information security become more and more complex, which are not only affected by hardware and software, but also by various other factors. Therefore, the establishment of computer network information security evaluation model can help to analyze network security problems, so as to provide better management and defense strategies.

Based on the three elements of environment, technology and personnel, this paper puts forward three research hypotheses: (H1) environment has a positive impact on network information security; (H2) technology has a positive impact on network information security; (H3) personnel has a positive impact on network information security. In order to verify the research hypothesis, the computer network information security evaluation model is established as shown in Fig. 1.

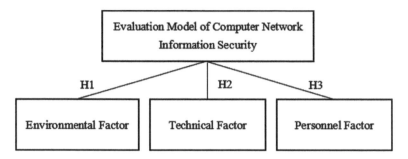

Fig. 1. Evaluation model of computer network information security

2.3   Index Weight Model Based on Entropy Method

In information theory, the greater the entropy, the greater the degree of dispersion, the greater the uncertainty; otherwise, the smaller the degree of dispersion, the less the uncertainty. Entropy method is to use the property of entropy to evaluate the dispersion degree of the index. By calculating the entropy value to get the weight of the index, we can judge the influence of the index on the comprehensive evaluation. If the probability of the event is Pi(i = 1, 2, …, m), there are n evaluation items, m evaluation indexes, and the original index data matrix is $R = (r_{ij})_{n * m}$, then the entropy value formula is as follows:

$$E = -\sum_{j=1}^{n} \frac{r_{ij}}{\sum_{i=1}^{n} r_{ij}} \ln \frac{r_{ij}}{\sum_{i=1}^{n} r_{ij}} \quad (1)$$

When the probability of all events is equal, Pi = 1/N, I = 1, 2, …, N, the maximum entropy value is:

$$E_{\max} = \ln n \quad (2)$$

Based on the entropy method, the process of the proposed evaluation index weight model is as follows:

(1) index standardization: because the dimensions, orders of magnitude and positive and negative orientations of individual indexes in the evaluation system are inconsistent, it is necessary to standardize the original sample data before calculation. Let the ideal value of the positive index be max x_i, and the negative

index be min x_i. The approach degree X^* of each index is expressed as the ratio of each index to the ideal value of the corresponding index.

(2) according to the standardized matrix P, the index specific gravity P_{ij} is obtained.
(3) determine the entropy value e_j of the index.
(4) determine the effective value d_j of the index.
(5) determine the weight W_j of the index.

3 Data

Due to the wide range of computer network information security, in order to ensure the effectiveness of the experiment and the authenticity of the experimental data, this paper standardized the data collected by the reptile technology. In addition, 10 experts from universities and enterprises were invited during the experiment. According to the expert experience and the judgment matrix obtained by the method in this paper, it is from the qualitative and quantitative point of view, respectively. The experimental data collected, screened and processed have certain rationality and scientificity, which can better verify the method in this paper.

4 Result

Result 1: Hypothesis Test
In order to verify the research hypothesis, this paper uses the structural equation with good adaptability to test. The relevant parameters obtained from the structural equation are shown in Table 1.

Table 1. Relevant parameters of research hypothesis

Research hypothesis	Standard error	Non-standardized coefficient	Standardization coefficient
H1	0.110	0.371	0.257
H2	0.082	0.223	0.235
H3	0.151	0.491	0.331

It can be seen from Table 1 that the correlation coefficients of research hypotheses do not have significant differences, so the model hypotheses are all tenable, that is, (H1) environmental factors have a positive impact on network information security, with an impact coefficient of 0.257; (H2) technical factors have a positive impact on network information security, with an impact coefficient of 0.235; (H3) personnel factors have a positive impact on network information security The coefficient is 0.331. The results show that it is feasible and reasonable to build a computer network information security evaluation model from the perspective of environment, technology and personnel, and the size of the influence coefficient can provide a reference for the subsequent index weight model.

Result 2: Model evaluation results
According to the process of index weight model based on entropy method, the result of evaluation model is calculated. Sort out the entropy weight of each evaluation index, and the data obtained is shown in Fig. 2.

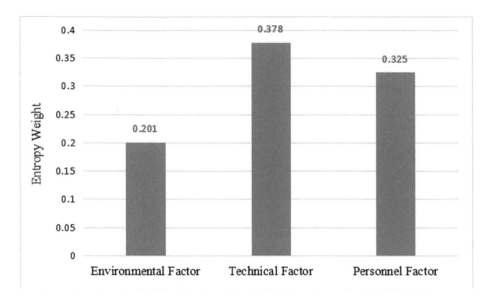

Fig. 2. Entropy weight of model evaluation

From Fig. 2, it can be seen that the weight of technological factors of entropy, human factors and environmental factors are increased from small to large respectively. That is to say, in the evaluation model, the order of environmental factors, human factors and technical factors is increasing, which also has an increasing impact on the information security of the computer network. This result is consistent with the previous research hypothesis, so the entropy based index weight model proposed in tlus paper is reasonable to some extent.

In the era of big data, network information security is mainly affected by technology, personnel, environment. Therefore, in view of information security problems, the main countermeasures should also be from these three perspectives: (1) pay attention to independent innovation of information security technology, make full use of current cloud computing, Internet of things, big data and other technologies, in information transmission, system design, security defense and other aspects Apply. (2) cultivate high-tech professional and technical personnel to inject new forces into the field of information security. (3) improve the information security awareness of network users, create a good network environment, and use database technology to regularly maintain and back up data.

5   Conclusion

With the advent of the era of big data, computer network plays an increasingly important role in people's life. At the same time, there are also hidden dangers of network information security, such as information leakage, virus intrusion and so on. This paper analyzes the problem of computer network information security in the era of big data, constructs the evaluation model of computer network information security from three angles of environment, personnel and technology, and proposes an evaluation index weight model by entropy method. The results show that the index weight model based on entropy method is reasonable.

References

1. Purohit, L., Kumar, S.: Web services in the Internet of Things and smart cities: a case study on classification techniques. IEEE Consum. Electron. Mag. **8**(2), 39–43 (2019)
2. Camacho, D.: Bio-inspired clustering: basic features and future trends in the era of Big Data. In: IEEE International Conference on Cybernetics (2015)
3. Xi, R.R., Yun, X.C., Zhang, Y.Z., Hao, Z.Y.: An improved quantitative evaluation method for network security. Chin. J. Comput. **38**(4), 749–758 (2015)
4. Pawar, M.V., Anuradha, J.: Network security and types of attacks in network. Procedia Comput. Sci. **48**, 503–506 (2015)
5. Yan, F., Jian-Wen, Y., Lin, C.: Computer network security and technology research. In: 2015 Seventh International Conference on Measuring Technology and Mechatronics Automation, pp. 293–296. IEEE (2015)
6. Guo, J.C., Fan, D., Che, H.Y., Duan, Y.N., Wang, H.S., Zhang, D.W.: An approach to network security evaluation of computer network information system with triangular fuzzy information. J. Intell. Fuzzy Syst. **28**(5), 2029–2035 (2015)
7. Saridakis, G., Benson, V., Ezingeard, J.N., Tennakoon, H.: Individual information security, user behaviour and cyber victimisation: an empirical study of social networking users. Technol. Forecast. Soc. Change **102**, 320–330 (2016)
8. Vorobiev, E.G., Petrenko, S.A., Kovaleva, I.V., Abrosimov, I.K.: Analysis of computer security incidents using fuzzy logic. In: 2017 XX IEEE International Conference on Soft Computing and Measurements (SCM), pp. 369–371. IEEE (2017)
9. Shameli-Sendi, A., Aghababaei-Barzegar, R., Cheriet, M.: Taxonomy of information security risk assessment (ISRA). Comput. Secur. **57**, 14–30 (2016)
10. Ahmad, A., Maynard, S.B., Shanks, G.: A case analysis of information systems and security incident responses. Int. J. Inf. Manag. **35**(6), 717–723 (2015)

Characteristics of International Trade Benefits Based on the Cluster Analysis of Compatibility Relation

Ting Dai[✉]

Jiangxi College of Foreign Studies, Nanchang, China
29505285@qq.com

Abstract. The sustainable development of international trade is related to the success or failure of reform and opening up. This paper constructs a set of evaluation system of sustainable development of international trade based on c-means clustering algorithm. Clustering analysis is an important research content in data mining, the traditional clustering algorithm can be divided into two kinds of hard clustering and fuzzy clustering, puts forward a method based on object set on the compatible relationship of clustering algorithm, this algorithm through great compatible clusters to classifying the data object sets, making the same object can belong to different clusters, and each cluster has its own unique member object, and is different from both hard clustering is different from fuzzy clustering effect of clustering. First through a summary of existing research results, and gives out the evaluation index system of sustainable development of international trade, then the argument is based on c-means clustering analysis of the scientific nature and necessity of the evaluation of the sustainable development of foreign trade, and gives the corresponding evaluation method, through the evaluation of the sustainable development of foreign trade in the 15 region in empirical analysis, the feasibility and practicability of this method was verified. Finally, according to the characteristics of the sustainable development of foreign trade, the paper puts forward some countermeasures and Suggestions to strengthen the sustainable development of foreign trade.

Keywords: C-means clustering algorithm · Data mining · Characteristics of income from international trade · Sustainable development

1 Introduction

The balance of benefits of international trade should be the rational use of resources in the region while maintaining and improving the coordinated development of the society, ecology and environment in the region, so as to improve the trade efficiency, enhance the competitiveness of the regional economy and drive the sustainable and healthy development of the regional economy [1]. The income balance of international trade mainly refers to the steady growth of the scale of foreign trade in the regional economy. The industrial structure of foreign trade can realize the optimal allocation of resources, and while ensuring the sustained growth of the income of foreign trade in the region, the society, ecology, environment and resources in the region can achieve

M. Atiquzzaman et al. (Eds.): BDCPS 2019, AISC 1117, pp. 1060–1067, 2020.
https://doi.org/10.1007/978-981-15-2568-1_146

coordinated development [2]. The benefits of international trade balance is a dynamic process, establish the sustainable development of foreign trade, must constantly improve and upgrade, which requires the region's foreign trade for dynamic evaluation of sustainable development, through the evaluation to master foreign trade sustainable development present situation and the insufficiency, to better to improve the regional trade development and promotion [3].

Research on evaluation of the development of international trade benefits, also is one of research focuses in academic circles in recent years, domestic some scholars from the point of view of strategic analysis of the necessity and important significance for the sustainable development of foreign trade, foreign trade sustainable development evaluation index system is expounded the design of the orientation and basic principles of preliminary constructed by the total scale of international trade, service/proportion of trade in goods, trade income rate, economic, trade, technology, trade, seven aspects, such as ecological resources, trade benefit index of evaluation system of foreign trade, and each index is calculated separately and instructions. On this basis, the vector autoregression model was used to discuss the specific factors affecting the sustainable development of foreign trade, and a set of evaluation index system consisting of economic benefits, social benefits and ecological benefits was designed from the two aspects of export and import [4–6]. The results show that there are still some problems in the international trade balance system, such as unreasonable trade system structure, poor coordination among economy, society and environment [7].

Based on the analysis of the existing evaluation results of sustainable development of international trade, it can be seen that at the present stage, the main research direction is still focused on the construction of evaluation index system and selection of evaluation methods, and there are still some deficiencies in the feedback and analysis of evaluation results [8]. And in the choice of foreign trade sustainable development evaluation method, most scholars are according to the degree of foreign trade sustainable development for sorting and optimal method, without considering the foreign trade sustainable development in the area for sorting and preferential at the same time, should also be according to the different stages of development of foreign trade development status classification and grading, and aims at different levels and categories [9, 10] for different development strategies.

This paper mainly analyzes the sustainable development of international trade earnings in different regions and USES c-means clustering method to evaluate and cluster the sustainable development of foreign trade in different regions. Through the classification and ranking of different regions within the region, the sustainable development of international trade in various regions within the region can be graded and ranked, which is of more targeted and guiding significance in the formulation of sustainable development countermeasures of international trade earnings in the later stage.

2   Research Methods

2.1   Clustering Analysis Algorithm

Clustering analysis algorithm is a group statistical analysis technique which divides the research objects into relatively homogeneous ones. There are many clustering methods. In this paper, Euclidean distance square is used as the distance measure to describe the degree of closeness between samples. Then, in the calculation, this paper USES the sum of deviation squares method and average connection method in c-means clustering analysis to cluster the samples respectively. By comparing the classification results of the two methods, it is found that there is little difference in the results. Therefore, this paper only discusses the clustering method of the sum of squares of deviations in this problem. First of all, data is transformed before clustering, usually using standardized transformation. The main reason for this is that the three indicators selected in this paper have different units of measurement. According to the object of study, make specific analysis to choose the appropriate distance. In this paper, Euclidean distance square is selected, and the formula is as follows:

$$d_{ij} = \sum_{k=1}^{p} (x_{ik} - x_{jk})^2 \quad (i, j = 1, 2, \ldots n) \tag{1}$$

Secondly, the method is obtained according to the principle of anova. If the classification is reasonable, the sum of deviation squares between the same kinds of samples is small, and the sum of deviation squares between classes is large. Assuming that class Gp and class Gq are merged into a new class Gr, the recursive formula of the distance between Gr and any class Gi is:

$$D_{ir} = \frac{n_i + n_p}{n_r + n_i} . D_{ip}^2 + \frac{n_i + n_q}{n_r + n_i} . D_{ip}^2 - \frac{n_i}{n_r + n_i} D_{pq}^2 \tag{2}$$

It is proved by practice that the classification effect is better by using the sum of deviation squares.

2.2   Evaluation Indicator System and Indicator Weight Determination Method

International trade gains development evaluation system, the construction of evaluation index is the first step to make scientific decision, based on the evaluation of the sustainable development of the existing international trade analysis and summarization of the research results, combined with the basic characteristics of sustainable development and international trade needs, on the premise of following the principles of sustainable development, must also be in the process of selecting the evaluation index, evaluation index must ensure that systematic and targeted unity between, between various indicators is to remain independent and to have a certain relevance, index selection should not only has certain theoretical value but also should satisfy certain feasibility principle of data acquisition.

There are many research results on the method of determining index weight. In general, it can be divided into three kinds: subjective weighting method, objective weighting method and combination weighting method. The objective weighting law is to use a certain data model to calculate the weight of the evaluation index through the actual evaluation data of the sustainable development of international trade. Combined weighting law is a method to obtain the weight of indexes by combining the experience of evaluation experts in the subjective weighting method and the development law of things existing in the objective weighting method. It is not difficult to find that the combination weighting method is relatively scientific and reasonable. It takes into account both subjective factors and objective factors, and is the main method to determine the weight of the evaluation index. Based on this, this paper chooses the combination weighting method as the method to determine the weight of the evaluation index of the development of international trade gains.

2.3   Evaluation Method Based on C-means Cluster Analysis

In the process of multi-attribute decision making, it is not only necessary to sort and select the best of various schemes, but also in many cases, it is necessary to carry out clustering analysis on schemes, and analyze the similarity and correlation between schemes through the classification of schemes, so as to make targeted evaluation and improvement on them. Similarly, in the process of the evaluation of the sustainable development of international trade, especially for more area to evaluate the sustainable development of international trade, not only to the sustainable development of different regions in the comparison and sorting, get high degree of regional sustainable development, should also be classified analysis was carried out on the similar degree of regional sustainable development. As a result, not only can effectively obtain the area of foreign trade sustainable development degree of the sorting result, also can classify the regions for the region in the international trade earnings in the process of sustainable development countermeasure research, according to the area of different grade or class different, targeted policy measures and suggestions.

3   Experiment

Based on c-means clustering analysis, analyzing the characteristic of the international trade gains its main idea is through the c-means clustering method, first the n region clustering, the situation of the trade development will be divided into K classes, and through the class center weighted comprehensive value to achieve between classes, according to the class in regional trade development comprehensive weighted value to achieve the sort inside the class, the ranking results finally got all the region's trade development. In this way, it not only classifies regions according to the relevance of trade development in all regions, but also realizes the ranking and selection of the sustainable development of regional foreign trade.

The following steps are given for the development evaluation algorithm of trade gains based on c-means clustering:

Step 1: determine the number of classes and set the initial cluster center. First of all, the final number of classification shall be determined according to the number of regions within the region and the level of trade development, generally no less than two, and no more than seven. The sustainable development of foreign trade in n regions is finally determined to be divided into K categories, and the standardized evaluation vectors of K regions are randomly selected to form the initial clustering center.

Step 2: calculate the square sum of the distance between the region and the class center. The distance square sum between Yi, the evaluation vector of sustainable development of foreign trade in each region, and each center vector is calculated respectively, where ZJK represents the value of the k index of the j clustering center after the mth iteration.

Step 3: classification. According to the Step 2 the distance to calculate the sum of squares of eta classifying scheme, classification principle is the evaluation of the sustainable development of foreign trade and class centre vector distance minimum sum of squares, the region is divided into the corresponding class accordingly. If have eta IP eta or less IQ, indicates the ith a region's foreign trade sustainable development evaluation vector and the distance from the centre of the first p value is smaller than the first q class center distance, is the ith a region should be classified as the first p class. By comparison one by one, the regional classification domain after the mth classification iteration can be obtained.

Step 4: recalculate the cluster center. The evaluation vector of sustainable development of foreign trade after m times clustering is recalculated to obtain a new clustering center. Generally, the mean value of the evaluation vector of the sustainable development of foreign trade in each region obtained after m times of clustering is used as the new class center for the next iteration of clustering analysis. Let there are Ni regions contained in the ith class after m times of clustering, and the set of contained regions is Si (m).

Step 5: judgment. Given discriminant threshold epsilon > 0, the end index $J = \sum$ indicates in clustering analysis of n area, the first m + 1 cluster of cluster center with the first m times clustering center of the cluster is the same, has reached the optimal clustering effect, can terminate the classification, the output of the final classification result, if $J = \sum 2$ continued iteration step.

Step 6: sort all regions by the above steps. After the realization of in-class and inter-class ranking, the ranking of foreign trade sustainable development evaluation results of all regions in the whole region can be realized.

4 Discuss

4.1 Empirical Analysis

The development status of foreign trade earnings of 7 randomly selected international regions was evaluated. The purpose was not only to rank the development status of trade earnings of these countries and regions, but also to classify these countries and regions according to their development relevance. Through data collection and standardized processing, the statistical description of the finally obtained standardized data is shown in Table 1.

Table 1. Statistical description of trade gains of 7 countries or regions

	N	Range	Minimum	Maximum	Mean Std.	Deviation
	Statistic	Statistic	Statistic	Statistic	Std. error	Statistic
V1	15	0.4629	0.5269	0.9539	0.03348	0.14235
V2	15	0.4742	0.5269	0.9894	0.04532	0.12385
V3	15	0.4602	0.5392	0.9829	0.03492	0.14584
V4	15	0.4494	0.5327	0.9827	0.04217	0.15392
V5	15	0.3997	0.5382	0.9709	0.03528	0.14589
V6	15	0.4484	0.5225	0.9686	0.03539	0.14376
V7	15	0.4377	0.5309	0.9673	0.04385	0.12527

Considering that the seven regions should be classified and sorted, the final decision was made to divide the seven regions into four categories. The iterative process was obtained through the analysis of standardized data, as shown in Table 2 below.

Table 2. Number of C mean clustering iterations

Iteration historya				
Iteration	Change in cluster centers			
	1	2	3	4
Dimension (1)	0.234	0.263	0.291	0.167
Dimension (2)	0.321	0.398	0.412	0.237

After further analysis, the clustering center position vector finally divided into 4 categories was obtained. The clustering result is shown in Fig. 1 below.

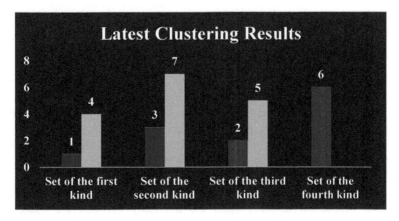

Fig. 1. Clustering results

It can be found from the above figure that the set of the first category is {1, 4}, the set of the second category is {3, 7}, the set of the third category is {2, 5}, and the set of the fourth category is {6}. Then, the combination of analytic hierarchy process and entropy weight method was used for calculation, and the weight vectors of 7 evaluation indicators were obtained as follows: W = (0.1504, 0.1635, 0.1729, 0.1259, 0.1090, 0.1485, 0.1297).

Similarly, for each classified area in the collection, calculation of each country or region weighted comprehensive evaluation value of trade development, get the first region in the set {1, 4} the weighted value is: zeta 1 = 0.7836; Set for 3, 7} {the second region, the regional trade development for the integrated evaluation: zeta 2 = 0.8272; Zeta 2; In third class collection for {2, 5}, regional trade development for the integrated evaluation: zeta 3 = 0.8571; Set for {6} fourth class region, the regional trade development for the integrated evaluation: zeta 4 = 0.7087, the comprehensive evaluation value pictorial diagram are shown in Fig. 2 below.

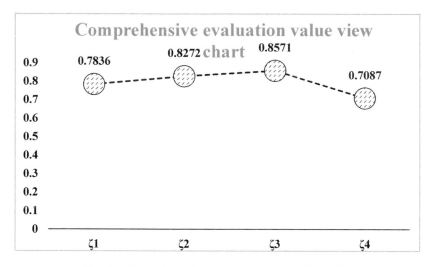

Fig. 2. Direct view of comprehensive evaluation value

4.2 Analysis of Earnings Characteristics

Through the above experiments and discussions, the following important characteristics of the benefits of international trade can be obtained:

First, considering the external and internal clustering, on the foreign trade advantage we can get the following conclusion: under the condition of production globalization, dramatic changes have taken place in the international division of production, as countries in the global production network of production, the object of international trade is no longer a complete product but refined into "process" trade.

Second, through the internal cluster analysis, it can be found that the division of international production division blurs the boundary between the manufacturing industry and the service industry. More importantly, the development of the manufacturing industry needs the input of the service industry.

Third, given the elaboration of international division of labor and the development of global value chain, the measurement of value-added trade can help countries better understand their specific advantages in the global production chain, better participate in international division of labor, and better benefit from global production.

5   Conclusion

Proposed in this paper using c-means clustering, classification, according to the development of the regional trade benefits association in regional classification and sorting, it is helpful to analyze the intra-regional trade development and guidance, in the analysis of characteristics of trade development to overcome the past focus on sorting, division does not pay attention to the analysis of the defects. In addition, this paper also emphasizes that it is necessary to continuously strengthen and promote the sustainable development of foreign trade, and scientific analysis of the sustainable development of trade is the key link to ensure its sustainable development. Through the analysis of income characteristics, we can grasp and monitor the new situation of international trade income in time, and provide decision basis for countries and regions to make corresponding macro policies.

References

1. Ziesemer, T.: Renewable electricity supply, infrastructure, and gains from international trade in electric current. Appl. Econ. Lett. **2**(15), 1–5 (2019)
2. Kılınç, U.: Assessing productivity gains from international trade in a small open economy. Open Econ. Rev. **29**(5), 953–980 (2018)
3. Buongiorno, J., Johnston, C., Zhu, S.: An assessment of gains and losses from international trade in the forest sector. For. Policy Econ. **80**(93), 209–217 (2017)
4. Soo, K.T.: International trade and the division of labor. Rev. Int. Econ. **145**(4), 320–330 (2017)
5. Ahsan, R.N., Ghosh, A., Mitra, D.: International trade and unionization: evidence from India. Can. J. Econ./Rev. Can. Déconomique **50**(2), 398–425 (2017)
6. Wolford, S., Kim, M.: Alliances and the high politics of international trade. Polit. Sci. Res. Methods **5**(4), 1–25 (2017)
7. Duan, Y., Jiang, X.: Temporal change of China's pollution terms of trade and its determinants. Ecol. Econ. **132**(22), 31–44 (2017)
8. Egger, P.H., Nigai, S.: Sources of heterogeneous gains from trade: income differences and non-homothetic preferences. Rev. Int. Econ. **26**(5), 1021–1039 (2018)
9. Shin, W., Ahn, D.: Trade gains from legal rulings in the WTO dispute settlement system. Soc. Sci. Electron. Publ. **18**(1), 1–31 (2018)
10. Irlacher, M., Unger, F.: Capital market imperfections and trade liberalization in general equilibrium. J. Econ. Behav. Organ. **145**, 402–423 (2018)

Educational Reform Under Computer Network

Bingjie Liu[✉]

Jiangxi Vocational and Technical College of Communication, Nanchang, China
66863378@qq.com

Abstract. At present, the mode of social and economic development in China is changing from planned economy to market economy. Competition among industries is becoming more and more fierce. Enterprises are demanding more and more talents. The application level of computer technology has become one of the important indicators for enterprises to assess talents, which puts forward stricter requirements for basic computer network education. However, there are many problems in the basic education of computer network in our country, which can not meet the needs of actual development, so that the level of computer application can not be effectively improved. The purpose of this paper is to discuss the problems existing in the basic education of computer network in Colleges and universities, and put forward corresponding reform strategies, in order to improve the basic education level of computer network in an all-round way and promote the overall development of students' comprehensive quality. This paper will adopt the research method of concrete analysis of specific problems, do sampling survey and carry out data analysis to draw conclusions. The results of this study show that we should actively implement the reform of computer network basic education, adjust and optimize the curriculum, teaching content and teaching methods to ensure that computer network basic education can provide better services.

Keywords: Computer network · Basic education · Reform and innovation · Problem strategy

1 Introduction

Modern information technology, with computer technology as its core, has developed rapidly since the 1960s. Its wide application fields and profound impact on all aspects of human society are unprecedented [1, 2]. Especially in the 1990s, the combination and development of computer technology, network technology and multimedia technology has made it the core of the third information revolution. As far as the current development situation is concerned, due to the continuous improvement of computer performance and price decline, as well as the low-cost, fast and flexible use of computer networks represented by the Internet as information and communication projects, computer networks have become ubiquitous, ranging from research institutes, educational institutions, industrial units, commercial departments to more and more ordinary families [3]. Its application scope extends to online information query, online medical

© Springer Nature Singapore Pte Ltd. 2020
M. Atiquzzaman et al. (Eds.): BDCPS 2019, AISC 1117, pp. 1068–1074, 2020.
https://doi.org/10.1007/978-981-15-2568-1_147

treatment, online education, online shopping, online games and so on. The "virtual university" and "virtual factory" based on computer network have also begun to emerge [4].

With the continuous renewal of information processing technology, computer networks have begun to popularize and develop throughout the country, such as education, finance, satellite, communications, transportation and other important departments have widely used the network system, especially the opening of the "China Education and Research Network", playing the main theme of China's education from traditional mode to modernization [5]. Computer network refers to a computer system that connects geographically dispersed and autonomous computer systems through a certain communication medium to achieve data communication and resource sharing. The "International Computer Interconnection Network" which interconnects these networks in a certain organizational way realizes the function of receiving and sending information among users worldwide. It realizes people's dream of "getting any information anywhere, anytime" [6, 7]. In a sense, the Internet is reducing the world to a "global village". At present, China Education and Research Network (CERNET) is under the supervision of the State Education Commission. The State Education Commission currently has jurisdiction over two international Internet exports. One is through the Internet of Beijing University of Chemical Technology, which is joined by Tokyo University of Science in Japan, and the other is the China Education and Research Network Demonstration Project (CERNEL). In December 1995, CERNET completed 10 64KDDN lines one year ahead of schedule. At present, there are about 108 colleges and universities connected to CERNET. It is estimated that the number of people using the Internet in China is about 50,000 [8]. Faced with the fact that the application of computer and its network has increasingly become a basic tool in all walks of life and even in daily life, educational institutions, which are responsible for training talents for the future, must carry out corresponding reforms in teaching methods, teaching organization, management mode and the requirements for the knowledge structure of talents so as to adapt to the tremendous progress brought about by technological progress Change.

The rapid growth of computer network not only greatly influences and changes the teaching methods in the above aspects, but also poses a severe challenge to the development of China's education. How to seize this opportunity and make China's education develop rapidly and healthily is an unavoidable problem for every educator [9]. With the development of computer network, the information superhighway extends to all corners of the society, effectively driving human society from "industrialized society" to "information society", and the value of education inevitably undergoes substantial changes. This paper uses the research method of concrete analysis of specific problems, makes sample survey and carries out data analysis, and draws a conclusion that people need education to roam freely in the tide of information. Only by constantly supplementing and mastering new knowledge and technology, can they base themselves on society and seek development. Thus, the value of education is highlighted [10]. In order to meet the needs of education in the information society and realize and increase the value of education, it is necessary to reform education from mechanism to goal.

2 Methodology

2.1 Research Train of Thought

According to the 42nd Statistical Report on the Development of Internet in China issued by the China Internet Information Center, the scale of Chinese netizens reached 802 million by June 2018, with a penetration rate of 57.7%. In the first half of 2018, 29.68 million netizens were added, an increase of 3.8% over the end of 2017. Therefore, it is necessary to strengthen the training of computer network application talents. Computer network has now become the core course of computer specialty in Colleges and universities. There are many theoretical knowledge points, fast technology updating, and students have difficulty in understanding. We must consolidate and understand theoretical knowledge through matching teaching reforms. How to improve the teaching concept, standardize the content of experimental teaching, establish an experimental teaching system matching theoretical teaching, and better serve the cultivation of applied innovative talents is the most important issue to be considered in the current computer network teaching. Computer network is a professional course introducing computer network equipment and technology. It can lay a foundation for students to learn computer network related knowledge in the future. It plays an important role in the computer education system. The research idea of this paper is to discuss the reform of teaching mode in terms of contents, methods and evaluation methods in education reform, so that computer network technology, as the core technology of communication engineering, can be widely used in all aspects of life, improve students' practical operation ability in an all-round way, and improve students' comprehensive application ability of network knowledge and their adaptability to future posts.

2.2 Research Methods

Using grey statistical method to screen important factors and using whitening function to process data can avoid compromise caused by different opinions or the influence of abnormal values (extreme values) on survey results. This study screens important influencing factors through the evaluation index system of computer network literacy. From the evaluation index system of computer network literacy ability, the principal component score function is obtained, as shown in formula (1).

$$nk(j) = \sum n(ij) + FK(ij) \tag{1}$$

According to the weight of the contribution rate of each principal component factor, the evaluation model of College Student's computer network literacy ability under the mobile Internet environment is obtained, as shown in formula (2).

$$Factor = 0.243 Factor1 + 0.120 Factor = 0.243 Factor2 + 0.166 Factor3 \tag{2}$$

The factor whose eigenvalue is greater than 1 is chosen as the principal component to ensure that the principal component factor has sufficient explanatory power. Because the original variable load value is not explained enough, the original principal component load matrix is transformed by orthogonal transformation of variance maximum method. The first principal component is mainly composed of data organization and management, the second principal component is mainly composed of data knowledge and skills, the third principal component is mainly composed of data awareness, and the fourth principal component is mainly composed of data expression and interpretation.

3   Experiment

Taking a college student in China as the sample of the questionnaire, the first round of small sample questionnaire survey was conducted. The subjects covered undergraduates and postgraduates from universities and scientific research institutes. A total of 80 questionnaires were distributed and 74 valid questionnaires were obtained. The validity rate of the questionnaires was 92.5%. The reliability and validity of the preliminary data items of the questionnaire were tested by SPS Road 22.0 software. Cronbach's α coefficient was used to test the internal consistency and cohesion degree of variables of the questionnaire. The items with single item coefficient less than 0.6 were excluded. The KMO was used to sample the adaptability test to determine whether the sample data passed the validity test results and was suitable for factor analysis. The larger the KMO value, the more common factors among variables, the more suitable for factor analysis. The analysis results show that the KMO value of sample data reaches 0.737, the Bartlett spherical test value is 4335.025, and the degree of freedom is significant, indicating that there are common factors among the correlation matrices, which is suitable for exploratory factor analysis. Factor analysis uses a small number of factors to describe multi-dimensional indicators, reveals the structural characteristics of the internal relationship between factors, optimizes the load factor after rotation, and determines the content of the final test scale based on the load number > 0.4. Delete and modify some items of the evaluation index system, and determine to extract 4 first-level indicators and 11 second-level indicators.

4   Discuss

4.1   Visual Display of Data

According to the data of the evaluation index system of College Students' computer network literacy under the mobile Internet environment, there is little difference in data literacy awareness among different grades of college students in the mobile Internet environment, which is at a relatively low level. The first-level indicators include computer network awareness, which is measured by ethics and norms awareness, sensitivity, demand and exploration; computer network knowledge and skills, which emphasize basic knowledge and analysis methods, analysis and application software, and security management capabilities; computer network organization and management capabilities, which are measured by search and acquisition, evaluation and application capabilities, sharing and production capabilities. Evaluation; Computer network expression and interpretation application metadata display, visualization ability to evaluate (Table 1).

Table 1. Evaluation index system in mobile internet environment

Level indicators	The secondary indicators	Assignment
Computer network awareness (10%)	Ethics and normative consciousness (X_1)	4.275
	Sensitivity (X_2)	3.09
	Demand and inquiry (X_3)	2.132
Computer network knowledge and skills (40%)	Knowledge and analytical methods (X_4)	4.13
	Analysis and application software (X_5)	3.17
	The safety management (X_6)	2.76
Organization and management of computer network (30%)	Find and retrieve (X_7)	3.008
	Evaluate and apply capabilities (X_8)	2.43
	Sharing and production (X_9)	4.33
Computer network expression and interpretation (20%)	Metadata presentation (X_{10})	2.211
	Visualization (X_{11})	3.21

4.2 Analysis and Discussion

Through data analysis, it can be concluded that there are three main reform strategies: first, the use of modular curriculum system. In view of the drawbacks of the current basic computer network education, the model curriculum system is designed. According to the students' professional characteristics, interests and development needs, several modules are formulated. Each module contains several courses with strong relevance. The hierarchical teaching method is adopted to ensure that all students can gain in the basic computer network education. Some module courses belong to compulsory courses, all students must participate in, elective courses need students to choose according to their own needs. This modular curriculum system organically integrates students' major with computer network technology, fully affirms students' teaching subject status, can greatly mobilize students' learning enthusiasm and enhance their learning confidence, students' enthusiasm for basic computer network education will also rise unprecedentedly, in this case, teaching efficiency and teaching quality will be greatly improved. Increase in magnitude. Second, reform the teaching method and examination mode. Teachers should attach importance to improving their teaching level, refer to and draw lessons from advanced teaching ideas and methods at home and abroad, carry out a comprehensive reform of teaching methods and examination modes, strengthen communication and interaction with students, improve students' classroom participation, and test the actual teaching effect through assessment, so as to facilitate the continuous adjustment and optimization of teaching contents and methods. Third, optimize the curriculum. First of all, non-computer majors focus on the application of computers, so that they can quickly grasp practical skills is the best way, instead of sticking to the strict study of professional terms and concepts, they should have their own language belonging to basic computer network education. Secondly, combined with modular curriculum system, optimize the content of basic computer courses, and guide students to improve in the right direction. Some important courses, such as operating system, discrete mathematics and programming, are included in the modular curriculum system, and the contents with strong applicability and emphasis on professional theory are chosen so that these courses can be easily accepted by students (Fig. 1).

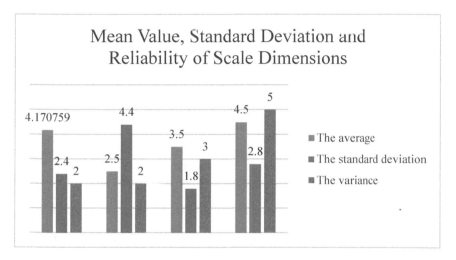

Fig. 1. Average, standard deviation and reliability of scale dimensions

5 Conclusion

This research shows that computer network technology is closely related to people's life and work, and has been widely used in the production and management activities of enterprises. No matter which industry is in urgent need of high-level computer application talents. Based on this, we must carry out the reform of computer network basic education, adopt modular curriculum system, optimize curriculum settings, improve teaching methods and evaluation system, so as to stimulate students' subjective initiative, mobilize students' inherent potential and promote the comprehensive development of students' comprehensive ability, so as to achieve the goal of training high-level talents. "The development of education must be adapted to the development of society", which is a basic law of the development of education. The revolution of human information exchange technology caused by the progress of science and technology will have a tremendous impact on all levels of society, thus bringing about tremendous changes, which has become the common understanding of most people.

Therefore, it is the basic guarantee for the effectiveness and vitality of the current educational reform to analyze and predict these changes as far as possible and to examine the inadaptability of our existing educational framework comprehensively and carefully.

References

1. Tinker, R.: Netcourses reform education using the power of the internet. Book Rep. **17**(1), 44–45 (2018)
2. Shcherbakov, A.Y.: New cryptography and computer security teaching methods using network technologies. Sci. Tech. Inf. Process. **44**(4), 305–307 (2017)
3. Nelson, E.A., Mcguire, A.L.: The need for medical education reform: genomics and the changing nature of health information. Genome Med. **2**(3), 18–21 (2018)
4. Ferren, A.S.: General education reform and the computer revolution. J. Gen. Educ. **42**(3), 164–177 (2017)
5. Natarajan, R.: An Indian perspective on engineering education reform. J. Eng. Educ. **97**(4), 395–396 (2018)
6. Andersen, V.N.: Transparency and openness: a reform or education policy? Scand. Polit. Stud. **30**(1), 38–60 (2017)
7. Pittman, J.: Empowering learners through digital education in urban environments: using internet-based teacher education portals. Teachers **2002**(1), 48–53 (2018)
8. Andersen, V.N.: Transparency and openness: a reform or education policy? Chin. Sci. Bull. **62**(30), 514–527 (2017)
9. Kober, N., Rentner, D.S.: State education agency funding and staffing in the education reform era. Mach. Des. **33**(5), 16–31 (2017)
10. Wojcicka, M.: Reform in polish vocational education. Eur. J. Vocat. Train. **77**(3), 70–76 (2018)

Precision Marketing Method and Strategy Based on Big Data Analysis in E-Commerce Environment

Lei Fu[(✉)]

Shenyang Polytechnic College, Shenyang, China
1292755032@qq.com

Abstract. With the rapid development of e-commerce, a large amount of consumption data has been generated, and consumers have also put forward personalized requirements for the purchased goods, so the traditional marketing model has not been able to better meet the needs of customers. Therefore, it is very important to study the precision marketing method in the e-commerce environment. This paper first expounds the significance of precision marketing, then summarizes the process of data mining in e-commerce, and proposes a precision marketing model based on big data analysis. Finally, from the aspects of timeliness of data, multi-dimensional data collection, and improvement of user cohesion, the paper gives the precise marketing strategy in the e-commerce environment.

Keywords: E-Commerce · Big data analysis · Precision marketing · Methods and strategies

1 Introduction

With the development of information technology, shopping is not only offline, but also online, which forms a good e-commerce environment [1]. In such a case, the online sales mode generates a large amount of transaction and comment scoring data [2]. The traditional marketing methods have been unable to keep up with the pace of consumers. If these consumption data can be used to achieve accurate marketing of online customers, it will undoubtedly promote consumption and improve the customer shopping experience, which will be conducive to the long-term healthy development of e-commerce platform [3, 4].

There has been a lot of research on precision marketing. Precise marketing theory was first put forward in 2005 by Philip Kotler, the father of modern marketing. His view is that precise marketing is a more accurate and measurable enterprise marketing communication that can produce high investment report [5]. There are also many combinations of big data and e-commerce. Traditional supply chain management can't keep up with the development of e-commerce at all. In [6], the author uses big data to analyze the customer demand of e-commerce supply chain. In [7], the author constructs the e-commerce marketing mode of agricultural products based on the investigation of the attributes of agricultural products and customer preferences. In [8], the author

© Springer Nature Singapore Pte Ltd. 2020
M. Atiquzzaman et al. (Eds.): BDCPS 2019, AISC 1117, pp. 1075–1080, 2020.
https://doi.org/10.1007/978-981-15-2568-1_148

combines the precision marketing data source system based on big data, studies the connotation and thought of precision marketing, conducts enterprise application around big data, introduces data standardization and model quality, finds the basic methods to promote data standardization, and provides reference for building data source system based on big data. In [9], based on the fuzzy method and neural network model, combined with big data, the author constructs the model of marine logistics warehousing center and optimizes the precise marketing strategy. In [10], the author uses big data technology to solve the problem of precision medicine and provide appropriate dose for each patient. Big data technology has achieved good results in all aspects, so big data analysis is considered to be used in e-commerce precision marketing.

Based on the idea of big data analysis, this paper studies the precision marketing methods and strategies under the e-commerce environment. To provide more scientific decision-making for e-commerce enterprises, in order to effectively improve the sales volume and profits, so as to maximize the profits of sellers, and also help to improve the user experience of customers.

2   Method

2.1   Precision Marketing Overview

Big data is a precious resource of precision marketing. Through big data analysis, precision marketing of e-commerce can be realized. In the past, in terms of marketing, it was difficult for sellers to understand all kinds of needs of consumers, which resulted in the phenomenon that goods were difficult to sell. For companies, there are many types of customers, such as loyal customers, potential customers and customers with low loyalty. In addition, different age, gender, occupation and other aspects of customers, the consumer demand is often different. In the past, marketing used to find the target audience by the way of spreading the net all over the place, which had little effect. Precision marketing on the basis of big data is to transfer targeted information according to different needs and characteristics of customers, search for more accurate target customers, launch marketing for them, and transmit data through the Internet, so that the effect of precision marketing can be fed back in time.

2.2   Data Mining Process in E-Commerce

In e-commerce, the process of data mining is generally divided into three parts as follows:

(1) Data preparation

The purpose of data preparation is to test the quality of data. It starts from data collection, removes invalid information to ensure the effectiveness of data mining, and conducts preliminary analysis after screening to form meaningful processing assumptions. This step involves data integration, data selection and data preprocessing. Data integration is to select the required data from multiple databases for consolidation; data selection is to select, organize, transform and classify the extracted data set; data preprocessing is to filter, clean, integrate, transform and filter the data. Data preparation not only improves the quality, but also makes up for the shortcomings of data mining itself.

(2) Data mining

The operation of data mining is to obtain meaningful data through appropriate mining methods according to mining objectives. There are many methods of data mining, such as memory based reasoning, shopping basket analysis, clustering analysis, discriminant analysis and so on.

(3) Expression and Interpretation of Results

After the first two steps, we can get the conclusion of the goal. Next, we need to sort out the data and extract valuable information. Through the way of decision support tools, it is provided to decision makers. If the decision-maker is not satisfied with the information submitted, the above process shall be repeated until the information available to support the decision is obtained.

2.3   Customer Selection Forecast

Here assume n customers and J items, customers can choose not to buy when they choose to buy. That is to say, customer n (n = 1, 2, ..., N) when selecting from J commodities, there are j + 1 selection methods, which are j = 1, 2, ... J. For the i-th customer facing j + 1 choice, the utility of choice j is assumed to be:

$$U_{nj} = X_n'\beta_j + \varepsilon_{nj}, \quad n = 1, 2, \cdots, N; j = 1, 2, \cdots, J; \tag{1}$$

Among them, X_n' is a vector, representing the internal characteristics of customers, and β_j is also a vector, representing the parameter corresponding to the internal characteristics of customers when the j-th product is selected, and ε_{nj} is an independent random term, which obeys $\varepsilon_{nj} \sim N(0, 1)$.

P_{nj} is used to express the probability that customer n chooses commodity j. then according to the principle of utility maximization, the probability that customer n chooses commodity j is:

$$P_{nj} = P(U_{nj} > U_{nk}, \ k \neq j) \tag{2}$$

3   Experiment

In order to obtain the data of the seller of e-commerce platform, the crawler technology is used to crawl the sales data. In the process of crawling web page data, the first choice is to search "cosmetics" in the search interface of Taobao web page, and then copy the address to the crawler page. Enter the product details page, and then collect the required data. The data to be collected include the seller's shop name, product name, price, quantity sold, etc. After the completion of data collection, the redundant data is eliminated. Finally, the data is sorted out again, and the information after sorting includes: store name, store rating, rated price, price at the time of sale, sales quantity, number of favorable comments, number of poor comments, payment method, etc. Finally, 25136 pieces of effective data information are obtained.

4   Results and Strategy

According to the previous description, the e-commerce platform precision marketing model based on big data analysis constructed in this paper is shown in Fig. 1.

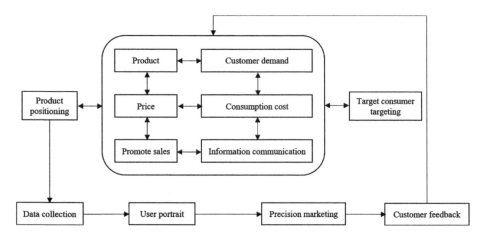

Fig. 1. Precision marketing model of e-commerce platform based on big data analysis

In Fig. 1, in the marketing model, it can be seen that the price and promotion of products are remote corresponding to the customer demand, consumption cost and information communication centered on consumers. Through this kind of big data analysis, we can find the shortcomings and advantages of our own business platform, so that we can clear the positioning of our products and understand the target consumer groups, which is the purpose of precision marketing.

In the "cosmetics" stores, the consumption amount and consumption frequency of corresponding customers are analyzed to determine members of different levels. Tap the customer's demand and promote sales. According to different segments of members (super members, VIP members, general members), develop accurate marketing strategies to improve customer value and sales. According to the internal data of a cosmetics store, from the transaction limit analysis, as shown in Fig. 2.

Based on the transaction volume, 57% of customers account for 0–100 dollars, 37% for 100–500 dollars, and 6% for 500 dollars. We can know the actual consumption situation of consumers by distinguishing the transaction amount. Therefore, when carrying out the corresponding marketing activities, the threshold value of transaction amount can be effectively set. At the same time, we can provide senior services for senior members.

Aiming at the precision marketing based on big data analysis in the current e-commerce environment, the following strategies are given here:

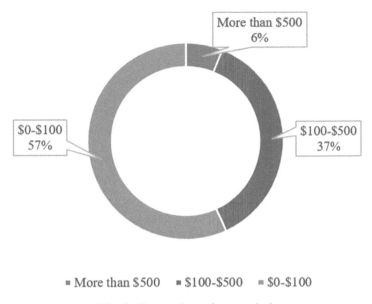

Fig. 2. Transaction volume analysis

(1) Enhance the timeliness of data. The timeliness of data reflects the constantly updated and rapidly generated data sets. Then the information implied in these data sets keeps pace with the times. The market and customer information obtained through big data is more accurate. When it is advertising, accurate and real-time market information can help enterprises to more accurately carry out advertising, and more accurately spread advertising to target customers. The exact match between the advertisement and the customers is to improve the efficiency of the advertisement.

(2) Multi-dimensional data collection. Precision marketing is based on big data. The size and quality of data are directly related to the difficulty of data value mining, and even affect the effectiveness of the whole precision marketing model.

(3) Build a precision marketing service system. Through big data analysis, users have accurate portraits, which can improve user cohesion through membership and other ways, and communicate with customers in a timely manner. The core point is to continuously improve the experience satisfaction of the target consumers.

5   Conclusion

The past marketing model has been unable to keep up with the needs of customers. The continuous development of information technology has created an important environmental foundation for the operation of big data, and the original extensive marketing mode has gradually changed to the precise marketing mode. Precision marketing based on big data analysis in e-commerce environment can promote consumption upgrading, improve customer experience satisfaction, and ease the difficulty of marketing personnel in finding target customers. It is hoped that the precision marketing method and strategy based on big data analysis proposed in this paper can provide reference for related work.

References

1. Yang, Y., Gao, G.: Overview of online shopping experience research. E-commerce **12**, 41–43 (2017)
2. Ren, J.: Strategy analysis of online reviews in book sales under the background of big data. Editing J. **4**, 42–46 (2016)
3. Fen, S.: Empirical analysis of factors affecting consumer acceptance of corporate big data precision marketing. J. Xihua Univ. (Philos. Soc. Sci.) **37**(6), 51–63 (2018)
4. Zhu, B.: Research on precision marketing of traditional retail industry based on big data. Manag. Obs. **689**(18), 24–25 (2018)
5. Fu, X., Chen, Y.: Theory and application: research review and prospect of accurate marketing. Mod. Mark. **7**(03), 71 (2017)
6. Li, L., Chi, T., Hao, T., Yu, T.: Customer demand analysis of the electronic commerce supply chain using Big Data. Ann. Oper. Res. **268**(1–2), 113–128 (2018)
7. Chen, X., Li, Y., Chen, M., Li, L.: Research on the precise marketing model of Shiyan agricultural products e-commerce under the background of big data. J. Hubei Ind. Vocat. Tech. Coll. **31**(2), 31–34 (2018)
8. Zhao, S., Ma, J.: Research on precision marketing data source system based on big data. Int. J. Adv. Media Commun. **7**(2), 93–100 (2017)
9. Xiao, K., Hu, X.: Study on maritime logistics warehousing center model and precision marketing strategy optimization based on fuzzy method and neural network model. Pol. Marit. Res. **24**(s2), 30–38 (2017)
10. Peck, R.W.: The right dose for every patient: a key step for precision medicine. Nat. Rev. Drug Discov. **15**(3), 145–146 (2016)

Analysis of Financial Education Demand Based on College Students' Entrepreneurship

Qiong Zhang and Hui-yong Guo$^{(\boxtimes)}$

School of Economics, Wuhan Donghu University, Wuhan, China
515275934@qq.com, 345430543@qq.com

Abstract. "Public entrepreneurship and innovation" can alleviate the employ-ment pressure of college students, effectively promote the transformation of China's economic development mode, maintain the sustained and steady growth of China's economy, and promote the vertical flow of society and fairness and justice, and increase the income of residents. Nowadays, the primary problem faced by college entrepreneurs and the focus of entrepreneurial issues actively explored by relevant departments in the country 8 are still the bottleneck of the shortage of venture capital. Therefore, it is particularly urgent for the govern-ment to accelerate the support of financial institutions for college students' entrepreneurship. This paper discusses the current situation of college students' entrepreneurial financial support system. Through the analysis of the problems existing in the needs of college students' entrepreneurial financial education, this paper puts forward some suggestions for the needs of college students' entre-preneurial financial education.

Keywords: College student entrepreneurship · Financial education · Demand analysis

1 Introduction

With the increasing employment pressure, self-employment has become an important career choice for more and more college students [1]. China mainly focuses on the relevant policy system, entrepreneurship education, or the study of entrepreneurship education and entrepreneurial environment that encourages university students to start their own businesses, and the impact of entrepreneurial environment on entrepreneurs. From the point of view of individual policy of college students, the incentive policy for self-employment of college students The research is relatively rare, and there are many researches on the difficulty of college students' entrepreneurship [2–4]. At present, China is in the transition period of the economic system. Globalization leads the information age, competition is everywhere, the market is changing rapidly, and those who are prepared can obtain valuable opportunities [5, 6]. Under this circumstance, schools and relevant departments should strengthen the cultivation of the financial education needs of students' entrepreneurship and establish the ability of students to start their own businesses [7, 8]. The establishment of the entrepreneurial financial system is a hot topic in recent years [9, 10]. It truly implements the establishment of various related systems, perfects various credit mechanisms and management

M. Atiquzzaman et al. (Eds.): BDCPS 2019, AISC 1117, pp. 1081–1088, 2020.
https://doi.org/10.1007/978-981-15-2568-1_149

mechanisms, and facilitates the entrepreneurial environment of the whole society, so that start-ups are no longer in terms of loans and other issues.

2 Discussion on the Support System for College Students' Entrepreneurial Finance

2.1 The Content of College Students' Entrepreneurship

The entrepreneurial group of college students is composed of two groups of college students and university graduates. There is a certain difference between college students' entrepreneurship and what we usually call entrepreneurship. It has advantages and there are also many objective defects. Because college students have high-quality college education resources, young, passionate, thinking and broader vision. There are obvious advantages in entrepreneurial subjects and entrepreneurial motives. However, in terms of entrepreneurial content and coping with entrepreneurial risks, due to lack of social experience and practical experience, and the professional and knowledge that is not enough to support the entire entrepreneurial process, it requires multi-faceted support from the family, the education sector, social organizations, and government policies. According to the status quo, in the case of college students' entrepreneurship, the government department needs to take the lead and lead role, mainly from the perspective of entrepreneurial environment optimization, to support college students in entrepreneurship guidance, social atmosphere construction and policy support.

2.2 Current Status of College Students' Entrepreneurial Financial Support Policies

Funding is an indispensable part of entrepreneurial activities, especially in the early stages of entrepreneurship. The college student entrepreneurship microfinance project was jointly launched by the Central Committee of the Communist Youth League and the State Development. The government established a special guarantee fund, and the relevant institutions have learned about the entrepreneurial loan applicants. They believe that they have the entrepreneurial quality and have certain self-owned funds, fixed residences and entrepreneurial establishments, etc., to provide loan guarantees, starting funds and liquidity from commercial banks to qualified unemployed persons or partnerships to start businesses. However, because the competent business of many departments is not entrepreneurial work, especially for college students' entrepreneurial work, there is a lack of special funds to support college students' entrepreneurship, and they cannot effectively carry out corresponding work. At the same time, because the policy of college students' entrepreneurial support is implemented by multiple departments, it is easy for various institutions to communicate effectively. They are independent, and the cohesiveness and complementarity between policies are insufficient. It is difficult to form an effective policy support system. However, the support policy defines the object of implementation as individual industrial and commercial households, which invisibly raises the "threshold" in the object, and rejects the entrepreneurs who apply for the company's college students, making it difficult for college students' financial support policies to fall.

2.3 Financial Ecosystem Theory

Financial support is the most important part of financial support, but the financial support system should not only be understood as financial support, but also includes financial markets and financial regulatory systems. Some scholars believe that financial ecology refers to the sum of various factors that are related to financial survival and development, including social and political, economic, legal and credit environments closely related to the development of the financial industry, as well as financial markets and finance. Products and other factors within the financial system, and these factors form a system of mutual influence through the capital medium and the credit chain. Referring to the definition of financial ecosystem, it is considered that the university student entrepreneurship financial support system is a collaborative development dynamic balance system consisting of college students' entrepreneurial enterprises and their entrepreneurial financial support environment. From the perspective of financial support, for college students in the start-up period, all kinds of financial support channels should be unblocked, and a rich and complete entrepreneurial financial support system should be established to provide support and guarantee for college students' entrepreneurship.

3 Analysis of the Problems Existing in the Needs of College Students' Entrepreneurial Financial Education

3.1 The Demand for College Students' Entrepreneurial Financial Education Is Backward

Although college students' entrepreneurs have a strong demand for financial education, their participation in financial education is relatively low, which is related to the low popularity of financial education and the insufficient attention of entrepreneurs to financial education. At present, the main platforms for financial education in China include colleges, network training, and lectures organized by relevant departments. The platform of colleges and universities is obviously the most used financial education channel for college entrepreneurs. The proportion of college entrepreneurs who plan to participate in financial education is close to 40%. They believe that financial education is beneficial to guiding entrepreneurial practice on the one hand, and has the opportunity to meet entrepreneurs or students with financial and financial expertise on the other hand. Undergraduate entrepreneurs without this plan said that many financial education is often out of touch with practice, "actuality" is not enough, and entrepreneurial activities themselves have taken up a lot of time, and they are involved in financial education.

3.2   Lack of College Students' Entrepreneurial Funds and Self-fear

According to the survey, as shown in Fig. 1. The number of entrepreneurs in developed countries is relatively high compared with the total number of graduates, while the proportion in China is relatively low. This shows that for both college entrepreneurs and other entrepreneurs, that is, all Entrepreneurs, especially for college students, are still the biggest reason for entrepreneurship, and the shortage of venture capital is highlighted. Affected by the current imperfect policies, capital investment still unable to meet the needs of market entrepreneurship and the general lack of entrepreneurial experience of college students, although the central government and local governments have consciously promoted the implementation of college students' entrepreneurial plans, as the main pressure to ease the employment pressure of college students. One of the ways, but now college students are still at a lower level of development. The lack of entrepreneurial funds for college students has become the main shortcoming to limit the entrepreneurial development of college students.

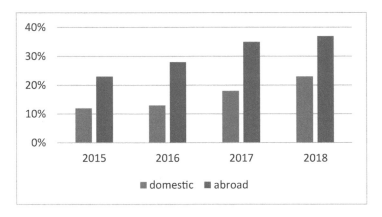

Fig. 1. Proportion of graduates from home and abroad

As shown in Fig. 2, it can be seen that most of the university student respondents feel that their lack of relevant experience is an important reason for not choosing to invest in the entrepreneurial industry after graduation. It can be seen that lack of experience makes them fearful and thus does not stand still. Before, fearing entrepreneurship.

Therefore, the requirements of entrepreneurs for individuals themselves are extremely high. Entrepreneurs need to have a keen eye for business opportunities, strong communication and communication skills, and can still make a calm and rapid benefit in various complicated situations. Judgment and decision-making of enterprise development, and can independently face and deal with various difficulties and obstacles in the process of entrepreneurship. Although the college student group already has certain professional knowledge, but lacks practical experience, they can not blindly choose to follow the trend and join the heat wave of entrepreneurship, but realize that "experience

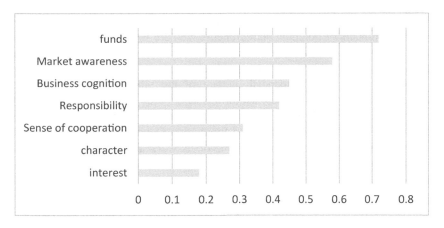

Fig. 2. Factors affecting college students' entrepreneurship

is not enough", which can also be regarded as based on the reality. A more rational choice.

3.3 College Students' Entrepreneurship Loans Are Cumbersome and Small in Amount

According to the network data survey, most of the respondents did not apply for small loans because the approval procedures for microfinance loans were relatively numerous and the process was cumbersome. The final loan amount was only a small part. During the investigation, one of Li said that he had specific products and entrepreneurial intentions. After understanding the microfinance process, he thought it was too much trouble. He needed to go through the school to apply for relevant certificates and "college students' employment and entrepreneurship certificate". The relationship is to find a guarantee for the Ministry of Foreign Affairs, and then submit the application, and finally can only borrow 2–3 million yuan. In comparison, he is more willing to choose to borrow money from relatives and friends as the starting capital. It is not enough for the funding needs of graduated college students. Therefore, only about 6.7% of the applications for microfinance loans are submitted in the student entrepreneurship microfinance service window.

4 Suggestions on the Needs of College Students' Entrepreneurial Financial Education

4.1 Construction of Financial Education System

The reform of the entrepreneurship education system should pay attention to the accumulation of entrepreneurship, skills and practice in the non-graduate school year of college students. At the same time, pay more attention to the entrepreneurship education of college students who have a certain foundation in practical experience, capital and contacts. Entrepreneurship courses are offered to former university freshmen, and then to entrepreneurs or potential entrepreneurs who are in need of college graduation. The government should incorporate financial education into the national education system as soon as possible, so that the public can receive financial education from an early age and improve the financial literacy of the whole people. The education department should unite financial institutions and other substantive departments to promote financial education in a multi-faceted manner. In addition, the newly established Volkswagen Entrepreneurship and Innovation Department of the State Council can incorporate financial education into one of its work and actively play the role of financial education in mass entrepreneurship. To achieve good financial education results, we also need concerted efforts from all walks of life. At present, China's financial education relies more on colleges and universities, and should try to carry out various forms of entrepreneurial financial education. For example, encourage the establishment of training institutions for entrepreneurial financial education, teach and exchange financial experience; set up financial education columns in newspapers, television and other media, hire experts to explain financial knowledge about entrepreneurship; conduct financial education lectures through financial institutions such as banks and securities. Wait.

4.2   Cultivate College Students' Education on Financial Management

In the face of the proposition of university physical finance, we can't help but have contradictions. College students generally have no income, and only rely on the fixed living expenses provided by the family. However, as a fashionable "new human", the desire for consumption is extremely inflated, which is also the consumer credit can be on campus. One of the reasons for the rise, but unexpected to our research, college students who are keen on loan consumption are also very interested in investment and financial management. According to the survey data, 70% of the three college students have plans to manage their finances, and 40% of the respondents said that they will buy as long as they are reasonable. As we all know, financial management is comprehensive and systematic. Accounting, savings, and investment are all One aspect of financial management, the demand for college students should also be multi-faceted. In the era of financial management, students have become a pivotal group among financial users. Advance consumption and financial planning are not contradictory. Rational planning and balancing of their own consumption, financial management, leverage and credit are becoming a compulsory subject for contemporary college students. University physique will surely become an important part of the campus financial ecosystem.

4.3   Construction of College Students' Entrepreneurial Financial Support Policy System

Entrepreneurship of college students is an indispensable measure for employment guidance in higher education institutions in China. As a young and energetic group with little social experience, college students' resources, social perspective, limited wealth resources, and strong mobility, support for college students' entrepreneurship should be different from other groups. Therefore, in the special work for college students to support entrepreneurship, the government should do a good job in guiding and propagating the overall work, implement it at all levels of management units, and at the same time set up special venture capital, and set up special departments to manage the relevant work. In the current implementation of the entrepreneurial favorable policy, the proportion of total loans supported by college students' entrepreneurial loans is quite low. The reason for the analysis is that the current conditions of college students' entrepreneurial loans are often difficult to meet the bank's loan requirements. In order to enrich the channels and methods for college students' venture capital support, the government should not be limited to setting up special funds, but also need to encourage other channels to enter the field of college students' entrepreneurial financing. In colleges and universities, many innovative fund committees have sprung up, donated by well-known corporate alumni, and have set up considerable entrepreneurial encouragement funds. After carrying out the innovation, feasibility and social benefit evaluation of the undergraduate research projects or pre-research projects of college students, the credit base conditions of the comprehensive entrepreneurial team will be given corresponding incentive funds for the project, and the loan funds will establish a support system for entrepreneurial funds. The government-funded special earmarked investment funds; secondly, the higher education department and the colleges and universities across the country to implement the implementation and support; finally, the participation and support of the financial departments at all levels headed by the People's Bank of China.

5 Conclusion

In the face of the new economic situation, as long as financial education institutions continue to improve their education, training methods and teaching directions, and carry out teaching activities based on the financial needs of college students, and strengthen exchanges and cooperation between domestic and foreign enterprises and schools, It can effectively improve the personal quality of students and play an important role in the financial education construction of college students' entrepreneurship. At the same time, local governments should increase their support for college students' entrepreneurial needs, and realize their own dreams and social values through self-selected entrepreneurial projects.

Acknowledgments. This research is funded by the school-level teaching research project of Wuhan Donghu University(A talent training research of real estate valuation based on the integration of industry and education) (Item Number: 34).

References

1. Gutter, M.S., Copur, Z., Garrison, S.: Do the financial behaviors of college students vary by their state's financial education policies? In: International Handbook of Financial Literacy. Springer Singapore (2016)

2. Chen, X., Yur-Austin, J.: College challenge to ensure "timely graduation": understanding college students' mindsets during the financial crisis. J. Educ. Bus. **91**(1), 32–37 (2016)
3. Fan, L., Chatterjee, S.: Application of situational stimuli for examining the effectiveness of financial education: a behavioral finance perspective. J. Behav. Exp. Finan. **17**, 68–75 (2018). S2214635017301119
4. Moon, U.J., Bouchey, H.A., Kim, J.E.: Factors associated with rural high school students' financial plans for meeting their college costs. Soc. Sci. Electron. Publishing **10**(2), 23–52 (2017)
5. Williams, T.: Closing the gender financial literacy gap for college students. Women High. Educ. **25**(10), 6–13 (2016)
6. Chi, X., Hawk, S.T., Winter, S., et al.: The effect of comprehensive sexual education program on sexual health knowledge and sexual attitude among college students in Southwest China. Asia-Pac. J. Public Health/Asia-Pac. Acad. Consort. Public Health **27**(2), 2049–2066 (2015)
7. Means, D.R., Bryant, I., Crutchfield, S., et al.: Building bridges: college to career for underrepresented college students. J. Coll. Student Dev. **57**(1), 95–98 (2016)
8. Vatanartiran, S., Karadeniz, S.: A needs analysis for technology integration plan: challenges and needs of teachers. Contemp. Educ. Psychol. **6**(3), 206–220 (2015)
9. Chu, Z., Wang, Z., Xiao, J.J., et al.: Financial literacy, portfolio choice and financial well-being. Soc. Indic. Res. **132**(2), 799–820 (2017)
10. Cheng, Y., Liu, W., Lu, J.: Financing innovation in the Yangtze river economic belt: rationale and impact on firm growth and foreign trade. Can. Public Policy **43**(S2), S122–S135 (2017)

Safety Management of Assembled Construction Site Based on Internet of Things Technology

Shanshan Li[(✉)]

Chongqing Real Estate College, Chongqing, China
Lubianinfo@foxmail.com

Abstract. As a new production method, assembled construction has many advantages such as high construction efficiency, short construction period, low labor cost and low energy consumption. It has attracted extensive attention from scholars at home and abroad. In the safety management of fabricated building construction, the links of major safety hazards mainly include seven parts of assembled component shipment, on-site storage, hoisting operation, temporary support system, frontier high-level operation protection, temporary power consumption, safety education and training. There have been many problems, and it has been difficult to meet the needs of large dataset and real time. Because radio frequency identification (RFID) technology has data non-contact reading, no manual intervention, not limited to line of sight, simultaneous reading and writing of multiple radio frequency tags, long-distance recognition, fast and accurate data acquisition and input, etc. This provides a new solution to solve existing tracking problems. Through actual research, the experimental results of the dynamic reading, multi-label simultaneous reading, and pasting of different positions in the actual application of electronic tags were tested and analyzed.

Keywords: IoT · Assembled construction · Construction site · RFID

1 Introduction

Assembled construction, also known as industrial buildings and modular buildings, has a wide range of connotations [1, 2]. On the one hand, the assembled construction is an innovation of the construction method, which moves part of the work of the on-site construction to the factory, produces the component products by means of large-scale production, and assembles the buildings through on-site assembly [3, 4]. On the other hand, the assembled construction method represents the integration and development of the entire industry chain of the construction industry. Through comparative research with traditional buildings, it is generally believed that assembled constructions can improve production efficiency, improve project quality, shorten construction time, reduce labor, and reduce energy consumption [5]. RFID technology, one of the core technologies of the Internet of things (IoT) has been applied to the management process of the construction industry. Relevant researches have been carried out on quality management, schedule management and logistics management [6, 7]. Among them, there are many innovative applications and development prospects of RFID technology

© Springer Nature Singapore Pte Ltd. 2020
M. Atiquzzaman et al. (Eds.): BDCPS 2019, AISC 1117, pp. 1089–1096, 2020.
https://doi.org/10.1007/978-981-15-2568-1_150

in the life cycle management of assembled constructions. Due to the immature development of emerging technologies and the fact that there are few examples, the application of IoT in fabricated buildings has many limitations, which makes the integration of the two fail to maximize the value of utilization [8, 9]. For example, RFID technology is currently focused on tracking and recording information in the construction industry, and there is not much deep mining and analysis of data that can truly guide management decisions. Vigorously developing assembled constructions play an important role in promoting the rapid advancement of urbanization strategy and promoting the transformation and upgrading of the economy and society. Under this condition, the internet connection of assembled constructions is the general trend [10, 11]. Applying IoT technology to the construction site of an assembled construction is a powerful guarantee for real-time interaction of information. Communication equipment and construction site resources can realize orderly construction and improves the safety of the construction site.

The IoT is a dynamic network infrastructure with self-organizing capabilities. It can realize the synergy and interaction between things and people, people and things, things and people without human intervention. It is called the third time in the information industry [12]. The wave has been vigorously developed and promoted by countries all over the world in recent decades. The research and development level of Internet of Things technology is at the forefront of the world, and all aspects of research have achieved success. In recent years, IoT has been widely used in intelligent buildings, construction safety management, transportation and logistics, etc. The development and achievements of the Internet of Things in all aspects are undoubtedly. It provides new solutions and means to solve the problems. The construction site applied radio frequency technology for material management, but did not integrate the information well, and did not truly realize the industrialization and intelligent management of the building. Here, the integrated concrete structure is taken as the research object. IoT technology is used to construct the model, and an effective and feasible IoT architecture is developed to realize an Internet of Things based on four levels: sensing layer, transport layer, application layer and decision layer. The health monitoring management structure provides a theoretical basis and method for the intelligentization of the integrated industrial production and installation process in the future.

RFID can achieve non-contact automatic identification of multi-targets and moving targets, which is particularly concerned by the logistics industry. Especially in the process of automatic sorting and loading and unloading of the logistics distribution center, the goods move with the conveying equipment, and the cargo information needs to be dynamically identified. The RFID technology can meet the needs of the logistics industry for information dynamic gathering, and the reading rate is more directly affected. The rest of this paper is organized as follows. Section 2 discusses layer strategies, followed by the model design. The results and discussions are present in Sect. 3. Section 4 concludes the paper with summary and future research directions.

2 IoT Layer Strategies

2.1 Layer of Data Acquisition

The components need to be tracked in real time during transportation. One IoT application model with intelligent service and real-time regulation of fabricated concrete logistics management should be constructed. To ensure the construction progress, a unique electronic code is installed on the component. The RFID reader reads the information. The various components of the vehicle are reflected in real time to the monitoring base of the construction site through the wireless communication network, so that the on-site management personnel can be based on the specific situation to determine the construction plan. The basic working principle of the RFID system is:

(1) Transmitting a certain of the radio frequency signal through the antenna and generates an induced current when the electronic tag enters the effective working area,

(2) Activating the tag and the sending out self-encoded information through the antenna,

(3) Receiving antenna of the receiver receives the carrier signal of the tag, and then transmits it to the reader signal processing module via the antenna adjuster, demodulates and decodes the received signal,

(4) Sending the valid information to the background host system for related processing,

(5) Updating the host system according to the logic.

The operation recognizes the identity of the tag, performs corresponding processing and control for different settings, and issues a command signal to control the reader to complete the reading and writing operation. In the RFID speed test, the test antenna is suspended above the gantry. The height between the antenna and the label is named h, and the distance between the vertical projection point and the label is named d. The characteristics of the RFID system constrain the maximum reading distance of the tag is d_{max}. If the tag can be successfully read, the three sides must satisfy the equation:

$$\sqrt{h^2 + d^2} \leq d_{max} \tag{1}$$

The tag must communicate with the reader through the duration of the reader antenna. In the case where the maximum reading distance of the system is satisfied, the tag can be successfully read within the effective reading length d of the field. If the label moves at speed v, the time through the distance passes is:

$$\Delta t = d/v \tag{2}$$

2.2 Layer of Information Transmission

In the component manufacturing process, the quality sensing monitoring network is mainly responsible for the data transmission of the raw materials, the manufacturing process and the quality of the finished components. According to the transmission object and coverage, it is divided into the self-organizing network and the production workshop of the sensor sensing nodes in the workshop. The communication network with the construction site, the construction site and the government communication network, the network topology is shown in Fig. 1(a). According to the needs of the assembled concrete in the four stages of production, transportation, assembly and operation, the transmission network is divided into two layers, including the manufacturer's monitoring network and the construction site monitoring network. Both include a quality sensing monitoring network, a video image monitoring network, and a transportation logistics monitoring network. The difference is that the construction site monitoring network also includes a component crane installation monitoring network. In the component hoisting assembly process, the quality sensing monitoring network is mainly responsible for the overall quality data transmission during component hoisting. The main part includes the component installation node and the site management network, the component installation node and the government department's information network. Communication network and network topology are shown in Fig. 1(b).

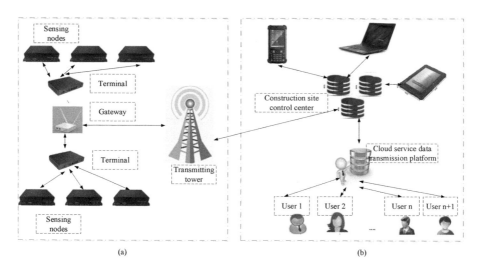

(a)   (b)

Fig. 1. Quality sensing detection network topology

2.3 Layer of Application

Assembled integrated concrete control management requires the construction of a four-party (construction, construction, supervision and government) monitoring center. The first three parties are mainly responsible for real-time monitoring on the site, and multiple detection systems to detect and report various types of information in the production and construction process are used. The base control center will then store, summarize and process the information. Use the internet to upload to the government's big data service to provide information and evidence for government departments' inspections. The government departments' various data are reflected on the big screen to realize real-time monitoring of various projects, and each project base and government departments are composed. The big data service platform includes cloud computing, file storage system Haystack, NoSQL database system, retrieval and query system as the basic platform and supporting technology. The main function is to organize, summarize and classify the information of the integrated concrete manufacturer and each project.

3   Results and Discussion

3.1   Label Quantity Test

There are two different basic communication methods in a multi-tag RFID system: data transmission from the reader to the tag and data transmission from the tag to the reader. The first type of communication is also known as "radio broadcasting". The second communication method is that data with multiple tags in the range of the reader is simultaneously transmitted to the reader, and the communication form is called multiple access. Each path has a defined path capacity, and the path capacity allocated to each user must be such that when multiple tags simultaneously transmit data to a single reader, mutual interference, such as collision, does not occur. In a communication system, each communication path has a prescribed path capacity, that is, a total transmission bandwidth. This path capacity is determined by the maximum data rate of this communication path. The path capacity assigned to each tag must satisfy the condition that mutual interference cannot occur when multiple tags simultaneously transmit data to a single reader. For an inductively coupled RFID system, only the receiving portion of the reader acts as a common path for all tags within the reader's range to transmit data to the reader. The maximum data rate is based on the effective bandwidth of the tag antenna. In addition, the processing speed of the reader is determined. When multiple tags appear in the same area, the system will reduce the efficiency in the process of handling conflicts, and it requires a longer dwell time than a single tag. When the reading time is much longer, the speed is higher with more tags. Experiments show that when multi-labels are dynamically read, the speed of label movement and the number of labels can affect the label reading effect. When the speed of label movement is higher, the number of multi-label readings is less. When the size of reading areas is larger, the speed of label movement is lower, and the effective reading of the label is also lower (Fig. 2 showed the Relationship between tag read rate and total tag number).

3.2   Label Position Test

The reader operates at 915 MHz and the reader emits in a power of 0.1 W. Using different packaging material labels, the tape conveyor speed is 1 m/s. The test label is placed on the top and front of the tote (the same direction of motion). The reading rate when one side is on the side (perpendicular to the side in the direction of motion). Each group was tested by 200 times, and the average reading rate of each label was obtained. The relationship between the label reading rate and the label position is shown in

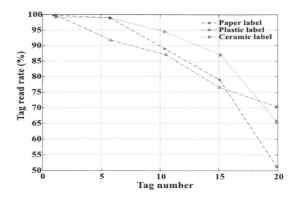

Fig. 2. Relationship between tag read rate and total tag number

Fig. 3. It can be seen that the label attached to the front of the tote has a higher reading rate than the top and side, and the ceramic label reading rate is higher than that of the paper label and the plastic label. Within the finite length d of the range of the field, the time for the high-speed movement of the label to pass through the area is shortened and the number of readings is reduced. At higher speeds, the dwell time of the tag in this area becomes shorter. Due to the side label, the effective area of the antenna cuts less magnetic field lines of the radiation field. Therefore, the number of times the tag can be activated is reduced, and the tag reading rate rapidly decreases. The reading rate is lower than the front and top labels. The results show that the dynamic reading rate of the tag is related to the position of the tag. The location of the label attachment affects the label reading rate, and the reading rate for the reader is the highest. The relative position of the label and the reader is different, and the relative angle between the label and the reader is different. The change of the angle has a great influence on the label reading rate. In order to clarify the influence of the relative angle of the label and the reader on the label reading rate, the angle test is designed.

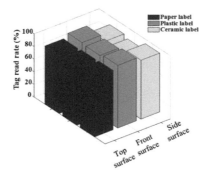

Fig. 3. Relationship between tag read rate and tag location

3.3   Power Test

The reader operates at 915 MHz and adjusts the reader's transmit power by using different packaging material labels. The dynamic reading rate of a single label attached to the top surface of the tote at a tape conveyor speed of 1 m/s is tested. The relationship between the label reading rate and the transmission power is shown in Fig. 4. Each group was tested by 200 times to obtain an average reading rate. It is known from Fig. 4 that the reader's transmit power increases and the tag reading rate increases. The labels of the three materials increased with the transmission power, and the reading rate increased. Compared with the three materials, the paper label reading rate is low. The reader operates at 915 MHz and the reader emits power at 0.1 w. The tag-to-reader distance is changed in steps of 0.1 m to test the reading rate of the tag at different distances.

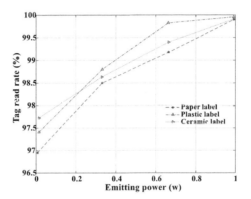

Fig. 4. Relationship between tag read rate and emitting power

4   Conclusion

The emergence of assembled buildings is a good solution to the many drawbacks of traditional construction methods, and the application of the IoT in assembled buildings has further developed the quality management system of assembled buildings. Starting from the core technology of the IoT, this study focuses on the specific application of IoT in the assembled building construction site from the aspects of component unloading, storage and hoisting for the application of the Internet of Things in the assembly building. It realizes the monitoring and management of multiple angles in the completely assembled project construction process. Thus promoting the popularization of the application of the assembled building in the construction is necessary. Finally, completing the transformation and upgrading of the construction industry is realized. The information is transmitted to the information center through the receiving device. It can track information from RFID signals, such as remote access to complex situations on and off the construction site and inside and outside buildings. The status of the components is automatically collected by RFID information transmission, construction progress, concealed works. The management personnel can control the situation on the spot in real time. Finally, the experimental results of the dynamic reading, the simultaneous reading of multiple tags, and the reading of different positions in the actual application of electronic tags were tested. The test results were analysed to conclude that the increasing number of tags can decrease the reading rates.

References

1. Zhong, R.Y., Peng, Y., Xue, F., et al.: Prefabricated construction enabled by the Internet-of-Things. Autom. Constr. **76**, 59–70 (2017)
2. Skibniewski, M.J.: Research trends in information technology applications in construction safety engineering and management. Front. Eng. Manag. **1**(3), 246–259 (2015)
3. Li, C.Z., Xue, F., Li, X., et al.: An Internet of Things-enabled BIM platform for on-site assembly services in prefabricated construction. Autom. Constr. **89**, 146–161 (2018)
4. Lu, W., Huang, G.Q., Li, H.: Scenarios for applying RFID technology in construction project management. Autom. Constr. **20**(2), 101–106 (2011)
5. Fang, S., Da Xu, L., Zhu, Y., et al.: An integrated system for regional environmental monitoring and management based on internet of things. IEEE Trans. Ind. Inf. **10**(2), 1596–1605 (2014)
6. Zhong, D., Lv, H., Han, J., et al.: A practical application combining wireless sensor networks and internet of things: safety management system for tower crane groups. Sensors **14**(8), 13794–13814 (2014)
7. Niu, Y., Lu, W., Chen, K., et al.: Smart construction objects. J. Comput. Civil Eng. **30**(4), 04015070 (2015)
8. Zhou, C., Luo, H., Fang, W., et al.: Cyber-physical-system-based safety monitoring for blind hoisting with the internet of things: a case study. Autom. Constr. **97**, 138–150 (2019)
9. Dave, B., Kubler, S., Främling, K., et al.: Opportunities for enhanced lean construction management using Internet of Things standards. Autom. Constr. **61**, 86–97 (2016)
10. Tang, S., Shelden, D.R., Eastman, C.M., et al.: A review of building information modeling (BIM) and the internet of things (IoT) devices integration: present status and future trends. Autom. Constr. **101**, 127–139 (2019)
11. Duan, Y.E.: Research on integrated information platform of agricultural supply chain management based on Internet of Things. JSW **6**(5), 944–950 (2011)
12. Oesterreich, T.D., Teuteberg, F.: Understanding the implications of digitisation and automation in the context of Industry 4.0: a triangulation approach and elements of a research agenda for the construction industry. Comput. Ind. **83**, 121–139 (2016)

Teaching Introduction to Computer Science with the Mixed Method

Jianguo Yang[✉], Hong Zheng, and Weibin Guo

School of Information Science and Engineering, East China University
of Science and Technology, Shanghai, China
yangjg@ecust.edu.cn

Abstract. Introduction to Computer Science is an introductory course for undergraduates majoring in computer science and technology, which plays an important role in the disciplinary teaching system. Using a mixed teaching method, the course aims to improve students' practical abilities with the combination of theory teaching and practice.

Keywords: Practice teaching · Introduction to Computer Science · Teaching methods

1 Introduction

Introduction to Computer Science is an introductory course for undergraduates majoring in computer science and also an introduction to the complete knowledge system of computer majors. The overarching goal of the course is to help freshmen to understand what problems can be solved by computer specialized knowledge, what should be learned, how to learn, to stimulate students' interest in computer specialized knowledge, and ultimately to help them with easier transition into the follow-up courses [1].

The scattered knowledge in Introduction to Computer Science can be very difficult for first-year University students who do not have the basis of professional knowledge, and especially for students from less developed remote areas of China who have not come into contact with the information technology course. Based on the review of the teaching experience of computer science in recent years, it is found that students may lack a comprehensive understanding of computer science and technology, and their critical thinking skills, innovative ability and analytical problem-solving ability is poor. Worst of all, many, upon graduation, cannot find the appropriate professional learning method. Students often complain about boring professional courses [2]. The problem, however, cannot be entirely attributed to the students. On the contrary, teachers should make full use of the resources of the information society, change the traditional education mode and adapt to the development of the society.

In the past, the emphasis of teaching was on the acquisition of theoretical knowledge instead of the strategic knowledge for solving practical problems [3]. At present, there has been a great deal of discussion on the contents and teaching methods of computer introduction course. In recent years, the teaching content reform of

© Springer Nature Singapore Pte Ltd. 2020
M. Atiquzzaman et al. (Eds.): BDCPS 2019, AISC 1117, pp. 1097–1102, 2020.
https://doi.org/10.1007/978-981-15-2568-1_151

computer introduction course based on computational thinking has been put forward, which promotes the development of introduction to computer courses.

At present, problems exist in teaching with regard to both teaching methods and assessment. First, with regard to teaching methods, viewing textbooks and teaching references as the center of teaching, teachers are in charge of everything, while students are only passive containers to be indoctrinated. In this context, students' autonomous learning ability cannot be developed. Second, in terms of assessment, the assessment criteria is single and absolute that emphasizes test results while neglecting the learning process.

Therefore, the present teaching reform focuses on teaching models, including the contents, teaching methods and curriculum assessment. The driving force of the reform is to carry out the student-oriented theoretical teaching process [1]. The project-driven practice teaching modules was carried out and the examination system of the course was developed according to the characteristics of the teaching process.

2   The Teaching Mode

In fact, computer technology is more about its practical operations than theories. Computer professionals needed by corporations must be equipped with strong hands-on abilities. For the education of computer major students, we should pay more attention to practice, more hands-on operations, and abilities to combine theory with practice. In order to comply with the requirements of the society and strengthen the competitive ability of the society, we should try our best to improve the quality of the students, cultivate their strong computer operation ability, and guide them to learn to think alone and cooperate in a team [4].

College students have a strong curiosity about the objective world. At the same time, they also have self-actualization needs, i.e. the hope to grasps the solid specialized knowledge and to become the valuable person. In view of these psychological needs of college students, undergraduate teaching must integrate theory with practice. One of the effective ways is to increase the proportion of practical teaching so that practical teaching and theoretical teaching complement each other, so to develop students' professional interest and knowledge.

Introduction to Computer Science, as a basic course for computer majors, plays an important role in the curriculum system of this major. Therefore, it is necessary to set up the teaching contents of the course reasonably so that the skills can meet the needs of teaching objectives, and easy for the students to grasp [5]. Due to the difference in the level of students' computer knowledge, the whole group of students should be taken into account when designing and implementing the teaching concept, teaching content and course assessment [6].

2.1   Teaching Design

To fully reflect the strong practical orientation of the course, so that students can apply the theory to the practice process, it is necessary to consider two points. First, in the teaching process of basic computer knowledge, students will be transformed into active learners with practical goals [7]. Second, considering the characteristics of Introduction to Computer Science, we need to familiarize our students with the basic computer skills and to cultivate students' interest in computer science at the same time.

Overall, we need to arouse students' interest in computer science and cover the contents from the historical developments of computer to the present-day applications. We need to ensure that students master the knowledge and skills, as well as professional ethics. At the same time, we may introduce cutting-edge knowledge in the field and some outstanding figures.

The course aims to allow the students to master the operations of the computer system and the use of the basic soft wares. Examples include the use of Linux, Office and Dreamweaver operations. This part is the practical teaching. After teaching the basic knowledge, it is necessary to carry out the teaching of basic operating skills in order to achieve the goal of combining theory with practice.

2.2   Course Contents

2.2.1   Diverse Course Contents

The main teaching methods of Introduction to computer Science are courseware and network communication. When designing courseware, we should pay attention to interest. Courseware with more images, multimedia resources and interactive links, can effectively improve the quality of classroom teaching. For example, if you're talking about computer virus, you can use the recent Bitcoin ransomware as an example, not only to introduce the concept of the virus, characteristics, harms and prevention methods, but also to introduce the hot bitcoin, blockchain technology [8]. When it comes to knowledge related to follow-up courses, teachers need to be careful to let students understand the connections between courses, while avoiding specific in-depth explanations. Overall, teachers need to aware that teaching methods should depend on the contents.

Introduction to Computer Science, being the first specialized course for computer science majors, serves the function of guidance and demonstration. In order to stimulate students' curiosity and initiatives, we need to provide students with extra-curricular reference materials and questions in addition to the basic assignments in each teaching unit. These materials also help to encourage students to take the initiative to understand the professional aspects of computer problems and progress. During each class, students are required to complete online assignments to help them grasp the content and focus of the lecture. The teacher can also go through online homework at any time to understand the progress of students. Students are also encouraged to participate actively in the discussion of innovation projects or competition projects of senior students and learn about the learning experiences of their seniors majoring in computer science.

By combining online assignments with offline discussion, the course allows students to improve their interest in learning, enhance their sense of participation, and gain access to computer knowledge more effectively. It also helps students to have a better understanding of their major and future career plans. In addition, interesting and contemporary extra-curricular reading assignments, as well as group discussions can better stimulate students' enthusiasm for learning.

For example, in the chapter Algorithms and Data Structure, introducing the classic case of "100RMB to buy 100 chicken" allows students to experience the beauty of the Algorithm. In learning about big data and other relevant knowledge, teachers can ask students to think about the real situation of using the campus dining card, so that they can understand the process behind the campus card. In the Multimedia Technology section, image processing experiments allow students to learn the use of Photoshop. Operation of specific images can leave a deeper impression on students. In the section of Web Design, the teacher can ask students to work in groups. Through teacher-guided group discussion, and group competition, students can complete the learning of web design software. In sum, the practical teaching of this course is of great significance to the establishment of students' professional thoughts, learning goals and the construction of learning methods [9].

In addition to in-class theory learning and practice, extracurricular learning mainly focuses on practice. Students are asked to work on a small project using HTML5. By working on a project, they can not only learn about computers, but also develop some teamwork skills, which can be very helpful for students to get a job after graduation.

2.2.2   Teaching Classroom Miniaturization

Introduction to Computer Science sets up a platform for network-based micro-classroom and through a variety of terminal accesses, mobile learning can be achieved. Learning resources are carefully designed before each class for each specific knowledge point (e.g. key points, difficulties, etc.) or a teaching section (e.g. learning activities, processes, tasks, etc.). The mini-lectures include the following features:

(1) Prepare focused contents. For the knowledge points of each lecture, the teacher should raise several questions for students to answer. Students need to take each lecture seriously and by previewing they can be in a better state for classroom learning.

(2) Keep questions short and simple. Teachers should limit the number of questions to 5–7, so that students can complete the task within 10 min. In order to arouse learners' interest, questions should be simple and straightforward. Long and complex questions may cause students to focus on mobile devices for extended period of time so they lose patience. Thus, it will be difficult to continue mobile learning due to the boring contents.

(3) Provide timely feedback on the questions. Students will get feedback immediately after submission so that they can evaluate their own learning of the current course and the problems to be noticed. For teachers, feedback allows them to evaluate the overall knowledge of their students in the class in order to make changes accordingly.

(4) Serve students' needs of learning. Based on the 'learner-centered' instruction, the course is designed for students' efficient autonomous learning. After students' logging onto the micro class platform, the teacher can keep track of students' learning progress.

3   Teaching Resources Platform

Based on the contents of Introduction to Computer Science, the auxiliary teaching materials are developed. The main contents of the network course include: course syllabus, course contents, electronic teaching plan, exercises, simulated test questions, multimedia courseware, experiments, course evaluation and feedback.

The purpose of establishing online course platform is to facilitate students' online autonomous learning with quiz, group discussion and homework submission. The platform gives full play to the main role of students in the learning process [10].

Multimedia is adopted in class teaching. To arouse students' interest and enhance their hands-on abilities, step-by-step teaching and learning is implemented. In this way, the teaching process is a two-way interactive process, which requires both teachers' "teaching" and students' "learning". Through the network and other platforms, teachers and students have more communication, which contributes to the learning atmosphere of mutual learning and discussion.

4   The Course Assessment System

The traditional examination methods cannot reveal students' learning outcome in the teaching process. Student's initiative and interest cannot be aroused in this context. Therefore, course examination systems should be developed according to the characteristics of the course [11]. We have formulated a two-time assessment system with phased and double mutual assessment. The specific measures are as follows:

(1) The course examination consists of two parts: final examination (70%) and course participation (homework, homework practice and attendance) (30%).
(2) The assessment of practical achievement homework consists of group achievement and individual achievement, with each accounting for 50% of the total. The scores of first quiz will be counted as group score, while the second quiz results will be taken as individual scores.
(3) The team score consists of the average of the team achievements at each stage and the final score of the completed project (including whether the system meets the functional and non-functional requirements according to the requirements specification, whether the documentation is comprehensive, reasonable, standard, etc.). The periodic performance of the group is assessed jointly by the teachers and other project teams outside the group.
(4) Individual performance is assessed by the teacher on the basis of the performance of each member of the team in response to the tasks completed by the team member in the project development. It shall also examine the organizational and managerial capabilities of its project development and give appropriate added scores.

Through answering questions in class, studying micro-class, doing exercises in each chapter and practicing in the course, teachers can track students' learning state, provide supervision, so that every student can meet the teaching requirements.

5   Conclusion

Introduction to Computer Science is a compulsory course for students majoring in computer science. It plays an important role in the whole disciplinary teaching system. Emphasizing practice can help students combine theoretical knowledge with practice, and improve their professional skills and innovative spirit in the process of practice. Students practice their skills in designing, implementation, document organization, writing and other aspects of the ability. The practice-oriented course also helps to improve students' team spirit, interpersonal communication, organization and management skills.

Acknowledgements. FOUND projects: Reform and Construction of Undergraduate Experimental Practice Teaching in East China University of Science and Technology; and MOOC "Introduction of Computer Science" course construction. "Action Plan for Innovation on Science and Technology" Projects of Shanghai (project No: 16511101000).

References

1. Donham, P.: Introduction to Computer Science. Cognella Academic Publishing, San Diego (2018)
2. Cuéllar, M.P., Pegalajar, M.C.: Design and implementation of intelligent systems with LEGO Mindstorms for undergraduate computer engineers. Comput. Appl. Eng. Educ. **22**(1), 153–166 (2014)
3. McIlroy, M.L.: Introduction to Computer Science. CreateSpace Independent P (2016)
4. Bau, D., Gray, J., Kelleher, C., Sheldon, J., Turbak, F.: Learnable programming: blocks and beyond. Commun. ACM **60**, 72–80 (2017)
5. Blikstein, P., Worsley, M., Piech, C., Sahami, M., Cooper, S., Koller, D.: Programming pluralism: using learning analytics to detect patterns in the learning of computer programming. J. Learn. Sci. **23**, 561–599 (2014)
6. Hwang, J.E., Kim, N.J., Song, M., et al.: Individual class evaluation and effective teaching characteristics in integrated curricula. BMC Med. Educ. **17**(1), 252 (2017)
7. Hartford, W., Nimmon, L., Stenfors, T.: Frontline learning of medical teaching: "you pick up as you go through work and practice". BMC Med. Educ. **17**(1), 171 (2017)
8. Ji, X., Hu, B.: Research on the teaching system of the university computer foundation. MATEC Web Conf. **61,** 4 (2016)
9. Lepareur, C., Grangeat, M.: Teacher collaboration's influence on inquiry-based science teaching methods. Educ. Inquiry **9**(4), 363–379 (2018)
10. Guzdial, M.: Balancing teaching CS efficiently with motivating students. Commun. ACM **60** (6), 10–11 (2017)
11. Felszeghy, S., Pasonen-Seppänen, S., Koskela, A., et al.: Using online game-based platforms to improve student performance and engagement in histology teaching. BMC Med. Educ. **19** (1), 273 (2019)

Food Quality Testing Application Based on Non-destructive Testing Technology

Xiaomin Shang[✉] and Qiong Liu

Jilin Engineering Normal University, Changchun 130052, Jilin, China
Tougao_005@126.com

Abstract. People's livelihood is the question of our country's follow-up. Improving people's livelihood is to protect people's living standards and provide people with a food quality and safe living environment. However, at present, China's research on food testing is still very scarce. The main purpose of this paper is to use non-destructive testing technology to detect food quality and quality, and to ensure a food safety issue. This paper mainly uses potato as the research object, mainly researches on the internal quality information detection and rapid identification of variety information of potato, and analyzes the five different non-destructive testing techniques, and summarizes the principle and application of non-destructive testing technology.

Keywords: Nondestructive testing technology · Food quality · Safety testing

1 Introduction

The safety of food is related to people's health, strict and effective detection of food safety, and the protection of food quality is also a guarantee for people's health. The main components of the potato are water, starch, protein and reducing sugar. When the food is processed with the potato, the quality of the potato product will be affected, which may cause damage to the hollow, potato, and black heart, while indirectly affecting the potato. Edible, product sales. Therefore, non-destructive testing of potato quality, ensuring the good quality of potatoes, and promoting the market competitiveness and economic benefits of potatoes are extremely important.

With the importance of non-destructive testing technology being widely recognized worldwide, many researchers have explored non-destructive testing techniques in recent years. In 2016, Wu et al. [1] solved the problem of excessive penetration depth in traditional harmonic excitation eddy current testing, and analyzed the sensors, signal processing methods, detection theory models and practical applications to overcome the signal of magnetically permeable materials. The problem of large difference in characteristics has found that the pulse eddy current detection method can better measure the attenuation of eddy current in the structure. In 2016, Qi et al. [2] in order to study the technical research on the freshness detection of chilled pork, synthesize the existing development status and application of related technologies such as infrared spectrum analysis, machine vision, multi-sensor fusion, etc. Molecular spectroscopy combined with chemometrics studies have shown that it is feasible to perform non-destructive testing of pork freshness based on THz spectroscopy. In 2016, Sun et al. [3]

© Springer Nature Singapore Pte Ltd. 2020
M. Atiquzzaman et al. (Eds.): BDCPS 2019, AISC 1117, pp. 1103–1109, 2020.
https://doi.org/10.1007/978-981-15-2568-1_152

proposed to construct a non-destructive testing technique in order to detect the fat content of eggs. In order to calculate the fat content more clearly in the experiment, the egg sample was determined by acid hydrolysis method, and the infrared spectrum nondestructive testing technology was constructed by using the characteristics of diffuse reflection. It was found that the method can effectively calculate the fat of egg. The content. In 2018, Zheng et al. [4] conducted research on infrared wave detection technology, selected some commonly used items that can cause infrared heat, and studied the development of infrared thermal wave non-destructive testing technology from the aspects of noise reduction, enhancement, and defect processing. The study found that the technology has an effective detection effect.

In 2016, Liu et al. [5] conducted research on digital polymerase chain reaction, based on its application in the field of food safety, and explored the field of digital polymerase chain from the fields of genetically modified components and food source microbial detection. The application of the reaction technology in the field of food safety testing has great development prospects. In 2016, Wang et al. [6] studied the multi-residue immunoassay technology for food safety, and used hybrid-hybridoma technology to analyze the small molecule pollutants in food and explored the mechanism of Bs MAb-based generation. Research on food safety immunization. In 2018, Tang et al. [7] used a fluorescent-labeled immune layer with high sensitivity and convenience in order to study new food detection technology. The analysis of organic dye fluorescence immunoassay showed that the technology is effective for food detection. In 2018, Wang et al. [8] combined the electrochemical immunosensor with food safety detection in order to comply with the hot spot of electrochemical development, using its characteristics of convenience, low cost and simplicity, and found that the sensor can be well combined with immunity. The high selectivity of the assay and the high sensitivity of the electrochemical analysis give it a good development prospect.

This thesis takes potato as the research object, mainly focusing on the detection of potato products, including NIR spectroscopy detection technology, ultrasonic non-destructive testing technology, computer vision detection and analysis technology, nuclear magnetic resonance spectroscopy technology and hyperspectral image technology. The detection method is used to rapidly identify the quality of potato [9, 10]. Then from the factors that cause product hazards and the principle of non-destructive testing technology, the actual detection and application of the method is studied [11].

2   Methods

2.1   Near-Infrared Spectroscopy Analysis Technology

This detection method is a technique for rapidly testing the composition of meat using infrared light. Different chemical substances will leave different spectral curves in the spectrum under the detection of infrared light, and the product composition can be quickly determined through this. The main indicators of potato, such as protein and starch, can be obtained by near-infrared spectroscopy, which is simple and convenient. However, in actual conditions, factors such as temperature and transmittance are extremely easy to improve the accuracy of the detection index. At the same time, the high cost also makes many merchants would rather lose some of the integrity of the meat than the non-destructive testing.

2.2   Ultrasonic Non-destructive Testing Analysis Technology

This method is to judge the pros and cons of the potato by the reflection wavelength of the sound wave. After the low sound wave enters the potato, the phenomenon of refraction and reflection is continuously generated. Through the change of the acoustic wave parameter data, it is easy to determine the chemical composition and protein of the potato. Because of its high efficiency, it is usually used for high-volume pipeline inspection, which can quickly achieve product quality grading. The energy of the sound waves is very low, and the internal damage to the potato is small and does not change the traits of the potato. At the same time, the ultrasonic detection technology is easy to manually detect, and can be widely applied to the rapid detection platform of the production line. It has the advantages of strong adaptability and sensitive detection, and is easy to promote in food quality detection.

2.3   Nuclear Magnetic Resonance Spectroscopy

The nuclear magnetic resonance spectroscopy technique is a chemical analysis technique. By emitting a beam of radio frequency magnetic field, the parameters of the magnetic nucleus absorption in the potato are recorded, and the quality of the potato is determined according to the parameters, and the tissue of the sample can be specifically understood from different angles. Internally, this method of testing is time consuming, difficult, and inefficient. It is generally only suitable for use in biomedical applications, and is not suitable for commercial testing.

2.4   Computer Vision Detection and Analysis Technology

Computer vision detection and analysis technology refers to the use of image sensors to acquire images, and to change these images into numbers, and then analyze the images and numbers to achieve a result of detecting the samples. It has the advantage of high efficiency and is currently used in the detection process of food selection. At present, some people have used computer vision technology to measure and analyze the quality fluctuation and grading of potatoes, and it is possible to detect PSE potatoes more clearly.

2.5   Hyperspectral Image Technology

Hyperspectral image technology refers to a new spectrum detection technology developed on the basis of hyperspectral remote sensing imaging technology. This spectrum detection technology has a wider range of applications. It is used in the sample detection process to produce specific images for different samples, so that the potato can be better judged. Some scholars have used hyperspectral imaging technology to make a simple prediction of the total number of bacteria in fresh pork. This technique was successfully used to determine the potato, indicating that this method can be widely used in the quality detection process.

3 Results and Discuss

3.1 Main Factors Affecting Food Safety

There are two reasons for food safety problems caused by the current situation of food safety in China. One is the self-factor of food, which is very prone to food-borne diseases caused by harmful bacteria. According to the Ministry of Health's notification of major food poisoning in the 1–3 quarters of 2004, there were 5,849 microbial poisonings in the 1–3 quarters of the country. Most of the collective food poisoning in China was caused by microorganisms. One is a human factor. At present, the food safety situation in China is worrying. From the whitening of rice whitening agents to the chemical ripening of fruits, food safety issues have become particularly prominent.

Harmful chemicals have a major impact on food safety. First, the safety status of food sources is worrying. The large amount and unscientific use of pesticides and fertilizers have caused excessive residues of harmful substances such as pesticides and fertilizers in agricultural products. Second, the abuse of compound feeds has caused antibiotics, hormones and other Hazardous substances remain in poultry, livestock and dairy products and aquatic products; third, heavy metals exceed the standard in agricultural and poultry products. According to the survey, chemical poisoning has accounted for more than 40% of major toxic events in China.

In addition, food storage problems also cause food safety hazards. Because there are no measures and regulations to effectively control pollution during the storage and transportation of food, it is extremely easy to cause the growth of food bacteria. People who eat degraded food will cause harm.

At the same time, the safety status of genetically modified foods is worrying. In China, genetically modified foods have already been placed on people's tables. However, from the research on genetically modified foods at home and abroad, genetically modified foods may damage the human immune system, produce allergic syndromes or produce toxicity, and have unknown harm to humans.

3.2 Principles of Non-destructive Testing Technology

The basic principle of non-destructive testing technology is to inject energy into the target detection object, so that the energy passes through the detected object to generate energy changes, and analyze the quality of the detected object according to the degree of energy change. This is the basic principle of non-destructive testing technology.

In fact, until now, non-destructive testing technology has not yet a clear concept, and its definition is different in different industries. In short, the same concept explanation is a non-damaging detection method for the detected object. Regardless of the industry, such detection means can be used as non-destructive testing as long as there is no change in the chemical properties of the object before detection. Non-destructive testing is mainly composed of three parts, namely information acquisition, processing and signal control. The information collection process plays a key role in the detection results. The information processing mainly relies on the computer to analyze the collected data. The signal control mainly displays the detected results to facilitate the next process. Non-destructive testing technology is the most widely used in fruit grading at this stage. Non-destructive testing technology has great advantages compared with other testing technologies. It has low cost, high detection efficiency, low technical requirements and good detection effect. The advantages make it the focus of research and development. Non-destructive testing technology mainly involves information technology science, biotechnology, materials science and sensing technology. With the rapid development of science and technology, non-destructive testing technology is also constantly developing, and the content is also constantly enriched.

3.3   Application of Non-destructive Testing Technology

According to Fig. 1, it can be clearly found that the number of times the non-destructive testing technique is applied increases as the year increases. At present, non-destructive testing technologies mainly include optical methods, magnetic methods, mechanical methods and other methods. The optical method is mainly applied to the analysis of some specific components, impurity analysis, product defects, etc., such as a box of citrus, if there is a citrus rot, it will quickly affect other citrus, other citrus will quickly rot Off, the non-destructive testing technology can be used to classify the citrus, and the damage is detected well, which not only reduces the detection cost, but also improves the detection efficiency and improves the economic benefit. Non-destructive testing technology is also widely used in detecting the maturity of persimmons. Under the exposure of persimmons to strong light, the persimmons can be observed according to the difference in transmittance. Infrared spectroscopy was used to detect the sweetness and hardness of apples. Based on the correlation analysis of spectral data, the mathematical model of apple quality was established, and the sugar content of apple was analyzed and predicted by photometer. At present, the main application results of the electromagnetic method are related to electronic resonance. The mechanics method is mainly to apply some mechanical characteristics of agricultural products (such as the density, color, external morphology, nuclear magnetic resonance, odor, color, etc.) of the agricultural products for quality testing. For example, in determining the maturity of some fruits, it is sufficient to judge the fruit. There are a wide range of applications for internal defects and whether canned foods are fresh. The detection principle in detecting maturity is mainly based on the principle that the acoustic characteristics generated by different fruit strikes vary with the flesh. The difference in the waveform of the sonic attenuation can predict the maturity of different parts of the fruit. The waveform can be analyzed using the corresponding computer equipment to determine the maturity of the fruit.

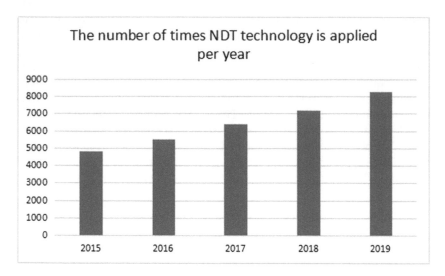

Fig. 1. Number of times the annual non-destructive testing technology is applied

4 Conclusion

With the continuous development of science and technology and economic society, the content of non-destructive testing technology is also enriched, which plays an irreplaceable role in the development of industry and agriculture. It has played a role in improving food quality, improving work efficiency and improving economic efficiency. A very important role, especially the application of non-destructive testing technology in food quality and safety testing, will provide important theoretical and practical significance to accelerate the development of China's food processing industry.

From an economic point of view, strengthening the detection of foreign matter in food is of great significance to the development of China's food industry. With the comprehensive development of China's economy and society, some excellent food companies not only have a huge domestic market share, but also emerged in the international food field. The biggest obstacle for some excellent food companies in China to expand the international market is not tariffs and intellectual property rights, but food safety. In particular, the food safety problems that have emerged in China in recent years have caused the international community to have a very serious negative impact on the food safety produced in China. In order to make the food industry in our country develop healthily and vigorously, food production enterprises must strictly test the foreign matter contained in food before leaving the factory. This is the last key step for food enterprises to ensure food safety.

Acknowledgements. Doctoral Research Initiation Jilin Engineering Normal University (BSKJ201861).

References

1. Wu, X., Zhang, Q., Shen, G.: Review of pulsed eddy current nondestructive testing technology. J. Instr. Path **37**(8), 1698–1712 (2016)
2. Qi, L., Zhao, W., Zhao, M.: Current status of nondestructive testing technology for cold fresh pork and prospect of thz detection technology. Food Mech. **32**(9), 219–224 (2016)
3. Sun, Y., Yin, C., Li, Z., et al.: Nondestructive testing technology for egg fat content in near infrared spectroscopy. Food Ind. (9), 177–180 (2016)
4. Zheng, K., Jiang, H., Chen, L.: Research status and development of infrared thermal wave nondestructive testing technology. Infrared Technol. **40**(305(5)), 5–15 (2018)
5. Liu, J., Liu, E., Xie, L., et al.: Research and application progress of digital polymerase chain reaction technology in food safety testing. Food Sci. **37**(17), 275–280 (2016)
6. Wang, F., Yu, W., Wang, H., et al.: Research progress in preparation of BsMAb by hybrid-hybridoma technique and its application in food safety testing. Food Sci. **37**(13), 251–256 (2016)
7. Tang, Y., Zhang, H., Cui, Y., et al.: Progress in the application of fluorescent labeling immunochromatography in food safety testing. Sci. Technol. Food Ind. **39**(2), 314–319 (2018)
8. Wang, Y., Xie, S., Chen, D., et al.: Research progress of electrochemical immunosensors in food safety testing. J. Anim. Husbandry Vet. Med. (7) (2018)
9. Hamidi, S., Shahbazpanahi, S.: Sparse signal recovery based imaging in the presence of mode conversion with application to non-destructive testing. IEEE Trans. Signal Process. **64**(5), 1352–1364 (2016)
10. Wang, F., Xi, Z., Ju, Y., et al.: Research progress of grain quality nondestructive testing methods. IFIP Adv. Inf. Commun. Technol. **393**, 255–262 (2016)
11. Huixiang, M.A., Zhou, J., Zhao, R., et al.: Non-destructive testing of steel bar stress based on metal magnetic memory technology. J. Jiangsu Univ. (2018)

Innovation of Financial Shared Service Center Based on Artificial Intelligence

Yanchang Zhang[1,2(✉)]

[1] College of International Vocational Education, Shanghai 201209, China
yczhang@sspu.edu.cn
[2] Shanghai Polytechnic University, Shanghai 201209, China

Abstract. With the development of the global economy and the further improvement of artificial intelligence research, the relationship between artificial intelligence and finance is increasingly close. The booming financial sharing services have greatly improved the efficiency of relevant financial work. At present, the global financial sharing service platform has excellent performance in reducing labor cost and improving operation efficiency, but it has not fully exerted the function of creating value of financial work. The establishment of financial sharing requires a large amount of data research, which requires the help of artificial intelligence to carry out relevant data statistics. Using big data and artificial intelligence technology to improve the financial sharing service platform can further liberate human resources and promote capital integration, which is an important way for the financial sharing service center to play its role in creating value. Finally, the improvement work should properly deal with the balance between centralization and decentralization, actively extend the business and expand the professional skills of financial personnel. However, we must see its two sides in the development process and avoid the problems in the research.

Keywords: Artificial intelligence · Financial sharing · Big data · Operational efficiency

1 Introduction

According to the dictionary of artificial intelligence in China, artificial intelligence is defined as "making the computer system simulate the intelligent activities of human beings and completing the tasks that can only be completed by human intelligence". Artificial intelligence is a new technical science that researches and develops theories, methods, technologies and application systems used to simulate, extend and extend human intelligence [1]. Artificial intelligence is widely applied. In recent years, as the global economy has entered a more complex stage of development, the basic framework and ecosystem of the financial industry are undergoing comprehensive and profound changes. With the rapid development of economic globalization and Internet technology, the running speed of modern society is accelerating, and the life cycle of products, enterprises and even industries is getting shorter [2, 3]. This has been accompanied by an increase in the size, complexity and speed of finance.

© Springer Nature Singapore Pte Ltd. 2020
M. Atiquzzaman et al. (Eds.): BDCPS 2019, AISC 1117, pp. 1110–1116, 2020.
https://doi.org/10.1007/978-981-15-2568-1_153

In this context, domestic and even global financial work is facing a severe test. Finance has become a global whole, resulting in a complex and complex financial work. Rely on spelling, spell flow channel, spell gains the extensive development of age has in the past, building, new technology, intelligent operation mode is becoming the main aims of the financial industry, financial service mode for open sharing, decentralized system of financial trust, scene and financial integration development, financial innovation is more and more relying on the powerful computation ability is becoming a new characteristics of financial innovation and development, and artificial intelligence technology is the important tool of power transform the operation mode of financial institutions [4, 5]. Therefore, various financial institutions at home and abroad have turned their attention to the field of artificial intelligence technology. Innovative financial services, such as face brushing payment, intelligent investment and customer service, have emerged at the historic moment and become an important force to promote the innovative development and inclusive development of China's financial industry. The financial sharing service center has solved the problems of scattered work, redundant institutions, high costs, difficult management and control, slow response speed and high financial risks caused by the traditional financial work methods, greatly improved the operating efficiency of global finance, and made capital operation become light and agile [6].

At home and abroad in this aspect of the research is gradually deepening, but there are still many deficiencies. For example, lack of experimental data, too wide coverage leads to difficulties in implementation, insufficient experimental experience and so on.

As a result, the establishment of financial sharing center lacks scientific nature and rigor to a certain extent [7, 8]. Research on financial sharing based on artificial intelligence is the requirement of the rapid development of Internet information. The financial sharing center established on such a special basis will certainly be able to withstand the test of The Times, so as to cope with various difficulties from the new era [9, 10]. At the same time, it also has certain theoretical guiding significance for the formulation of financial programs.

2 Method

2.1 Evolutionary Algorithm of Artificial Intelligence

Evolutionary algorithms simulate the evolution of organisms in nature. Darwin's theory of evolution pointed out that "natural selection leads to the survival of the fittest". Evolution explains almost everything. "why is this creature like this? That kind of thing. Organisms that are more adapted to their environment are more likely to leave their chromosomes behind. So, in the computer, we simulate the biological selection. Although the researchers who do evolutionary calculation know that evolutionary algorithm is suitable for solving optimization problems without analytic objective function, most evolutionary algorithms tend to assume that the objective function is known when they are designed. The general form of evolutionary algorithm is as follows:

$$D(t) = D(t-1) + H \quad (l,n) \tag{1}$$

Where $D(t)$ represents a target function in the data cluster.

Relatively speaking, our research and application of data-driven evolutionary optimization are relatively early. Why do artificial intelligence drive optimization? This is because there are many optimization problems in the real world that cannot be described by analytic mathematical formulas, and their performance can only be verified by simulation or experiment. The general evolutionary optimization algorithm needs to solve the challenge mainly lies in the problem contains a lot of local optimization, large-scale, multi-objective, strong constraints and uncertainty, and artificial intelligence-driven optimization is bound to face the challenge from the data.

2.2   Combined Sort Algorithm

To explore the details of artificial intelligence algorithms in the financial sharing center, it is not enough to only rely on understanding the evolutionary algorithm of artificial intelligence, or even to grasp the basic technology roughly. To further understand the mystery of artificial intelligence, we must also have a certain understanding of the combined sort algorithm. In the statistical calculation of this algorithm, all the appearing elements search for the most probable parameter location value in the probabilistic model, where the model relies on undiscovered potential variables. The first step is to calculate the expectation and calculate the maximum possible estimate of the hidden variable by using the existing estimate. The second step is to maximize the maximum possible value obtained in the first step to calculate the value of the parameter. The algorithm attempts to find the maximum flow from a traffic network. Its advantage is defined as finding the value of such a flow. The maximum flow problem can be regarded as a specific case of a more complex network flow problem. The sorting of element positions in this algorithm is updated according to the following formula:

$$v_i^d = \omega^* v_i^d + c_1 * r_1 * \left(\rho_i^d - \chi_i^d\right) + c_2 * r_2 * \left(\rho_g^d - \chi_i^d\right) \tag{2}$$

When we solve the financial sharing problem with merge sort algorithm, we can get a series of models, some of which may be the best interpretable, some of which may be the most accurate, and some of which may be overfitting. When you look at all the different models you can pick a few of them. This is one of the advantages of using evolutionary algorithms to address financial sharing. In general, using evolutionary algorithms to help build Shared centers offers more possibilities. At the same time, it can be used for parameter optimization, structure optimization, as well as model interpretability and security. This provides sufficient data support for the establishment of financial sharing center.

3   Experiment

Based on the big data in the era of artificial intelligence, the smooth and scientific conduct of experiments needs to integrate a large number of relevant data. In this experiment, we consulted various data websites at home and abroad, investigated the financial industry, and obtained first-hand information. The use of artificial intelligence to collate and analyze all kinds of data collected has strong scientificity and rigor, which accelerates the scientific establishment of financial sharing center. The specific experimental statistical data are shown in Table 1, and the specific experimental steps are as follows:

Firstly, based on relevant financial knowledge and knowledge, the required financial and economic indicators were screened through the website of the National Bureau of Statistics, the website of the People's Bank of China, global database and other authoritative official websites, and a preliminary judgment was made according to the current domestic and international financial situation.

Secondly, further collect the required industry information through the financial industry website, mainly including the industry development status, industry competition analysis, industry development prospects analysis. And through journals, data network and other investigation of the development of artificial intelligence.

Then, through financial situation judgment and industry analysis, and combined with the relevant data of current artificial intelligence development, the above formula method was used for calculation, and the success of the establishment of financial sharing center was taken as the judgment standard.

Finally, based on the relevant conclusions obtained from the experiment and the practical development, Suggestions are provided for the establishment of the financial sharing center.

Table 1. Statistics of the application of artificial intelligence in the financial industry

Application of Artificial Intelligence in Financial Industry									
	2011	2012	2013	2014	2015	2016	2017	2018	First half of 2019
Data rate	23%	28%	34%	35%	37%	41%	44%	45%	46%
Utilization rate	26.50%	26.70%	28.90%	31.00%	32.40%	33.60%	35.00%	35.87%	36.20%
Appropriate rate	17.80%	17.90%	18.30%	20.40%	21.00%	30.60%	34.00%	37.20%	41.60%

*Data came from the National Bureau of Statistics website

4   Discuss

4.1   Analysis of Financial Risks Brought by Artificial Intelligence

Some people believe that a series of risks brought by artificial intelligence to the financial industry can be divided into several aspects: first, systematic risks. The homogenization judgment of artificial intelligence algorithm based on the market situation also tends to aggravate the spread and spread of market risks. The second is the risk of information and privacy disclosure, which requires financial institutions to collect, collate and make use of a large amount of various financial information, which is highly sensitive and unique. Once disclosed, it will definitely infringe on the privacy of users and may threaten other financial and non-financial applications of users. Third, technical risk, the complexity of the artificial intelligence technology is high, the parameter variables involved, complicated operation process, operation result is closely related to the training data, algorithm of judgment, most results consistent with actual situation or anastomosis, but there are also part of the judgment, the predicted results are not accurate, this will bring uncertainty to the corresponding financial business. In addition, artificial intelligence usually builds research and judgment models based on historical data. When major rule changes or policy turns occur, the prediction effect of the model will be greatly reduced due to the low correlation between historical data and future situation. The artificial intelligence system also has the decision-making transparency low, to the user interpretability poor and so on insufficiency. In order to make AI better serve the financial sharing center, we must pay attention to these problems. People's views on artificial intelligence are shown in Fig. 1. The statistical data in the figure is obtained by the author after consulting materials.

What Influence Will AI Effect Financial Industry.

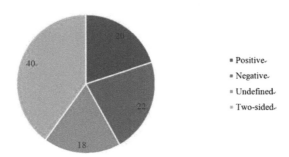

Fig. 1. Data on the impact of artificial intelligence on financial sharing service centers

4.2 Innovation Mechanism

The impact of AI on the establishment of financial sharing centers is two-sided. Big data and artificial intelligence have broadened the scope of financial sharing services and further strengthened the centralized control ability of financial sharing service centers. However, it also brings a lot of problems. At present, as for the application of artificial intelligence technology in the financial field, there is still a lack of relatively perfect regulation rules. Once there are business or service disputes, there will be a series of regulatory problems. In reality, artificial intelligence terminal products involve a variety of technologies and hardware equipment, when the failure and loss, how to deal with? Among them, although higher centralized control can improve the overall management efficiency, the loss of flexibility may increase the difficulty of financial development. On the other hand, although higher flexibility helps financial level growth, the corresponding operating costs will also rise significantly. Therefore, a good organizational design should strike a balance between centralized management and decentralized management, especially pay attention to the difference between the scope of Shared services and the scope of authority, and reasonably define the limits of authority between financial Shared services center and relevant departments. Second, attention should be paid to further extend to relevant financial services to promote the integration of industry and capital. The application of big data and artificial intelligence tools will further liberate manpower and cut costs. However, it should be noted that the introduction of new tools is intended to form new capabilities and functions. If the new tools are used blindly to reduce labor costs and cut down the relevant financial personnel and departments, then there is a risk of deepening financial separation. Third, attention should be paid to the expansion of professional skills of financial personnel. The function upgrade of financial sharing service center will promote big data analysis ability to become an important professional skill for financial workers, especially for financial accounting workers to be close to data scientists.

5 Conclusion

Based on the relationship between artificial intelligence and the establishment of financial sharing center, this paper describes its development status, contact methods and the disadvantages and advantages of financial sharing. Through analysis, it is not difficult to draw the conclusion that the establishment of financial sharing center is the general trend of modern economic development and plays a certain role in promoting the global economy. However, its establishment must rely on artificial intelligence to complete. At the same time, we must have a clear understanding of the possible impacts during and after the establishment of the financial sector, and formulate countermeasures to promote the stable development of the global financial industry.

References

1. Heaton, J.B., Polson, N.G., Witte, J.H.: Deep learning for finance: deep portfolios. Appl. Stoch. Models Bus. Ind. **21**(1), 3–12 (2017)
2. Staff, C.: Artificial intelligence. Commun. ACM **60**(2), 10–11 (2017)
3. Hassabis, D., Kumaran, D., Summerfield, C.: Neuroscience-inspired artificial intelligence. Neuron **95**(2), 245–258 (2017)
4. Moravčík, M., Schmid, M., Burch, N.: DeepStack: expert-level artificial intelligence in heads-up no-limit poker. Science **356**(6337), 508–533 (2017)
5. Du, X.: What's in a surname the effect of auditor-CEO surname sharing on financial misstatement. J. Bus. Ethics **43**(3), 1–26 (2017)
6. Bao, H., Ponds, E.H.M., Schumacher, J.M.: Multi-period risk sharing under financial fairness. Insur. Math. Econ. **72**, 49–66 (2017)
7. Tchamyou, V.S., Asongu, S.A.: Information sharing and financial sector development in Africa. Soc. Sci. Electron. Publishing **18**(1), 24–49 (2017)

8. Hu, K.H., Wei, J., Tzeng, G.H.: Improving China's regional financial center modernization development using a new hybrid MADM model. Technol. Econ. Dev. Econ. **24**(6), 1–38 (2018)
9. Day, M.Y., Lin, J.T., Chen, Y.C.: Artificial intelligence for conversational robo-advisor. **435** (213), 1057–1064 (2018)
10. Tokic, D.: BlackRock robo-advisor 4.0: when artificial intelligence replaces human discretion. Strateg. Change **27**(4), 285–290 (2018)

Support Resource Optimization Based on Utilization Efficiency

Ou Qi[1]([⊠]), Yanli Wang[2], Wenhua Shi[3], and Lei Zhang[4]

[1] Army Academy of Amored Forces, Changchun, China
haikuotiankongru@163.com
[2] PLA 32256 Troops, Kunming, Yunnan, China
[3] The Army of 95795, Guilin, China
[4] Changchun Military Representative Office of Shenyang Military
Representative Bureau of Ground Force, Changchun, China

Abstract. The rationality of resource planning is the key to guarantee. In order to achieve the final effect of resource planning, reduce the wastage in the process of support and maximize the effectiveness of support, the complex resource planning problem is decomposed by analysis method, and the non-linear programming technology is used to improve the scientificity of planning and increase the resistance to risk in the process of support. The results show that the planning in this paper can improve the effect of resource security.

Keywords: Resource planning · Nonlinearity · Risk

1 Introduction

Support resource planning is an important branch in the field of equipment support and one of the key links affecting the effectiveness of equipment support. Mainly used in industrial production, natural resources protection, military combat effectiveness enhancement and non-war military operations [1–3]. Good resource planning will save necessary safeguard resources, shorten the system construction time, improve the overall construction effect, and attract the favor of experts at home and abroad [4–8].

The army's participation in non-war military operations such as earthquake relief and flood relief is an important means to ensure the safety of people's lives and property. At the same time, the army also bears the important mission of protecting the country and the family. In order not to affect the military's war equipment support capability, the non-war military operation equipment support capability of our army is mainly based on the existing one [9–11]. On the basis of the equipment support system of war military operations, some equipment support equipment and equipment for non-war military operations are added to form. With the increasingly heavy tasks of non-war military operations in China, this simple way of generating equipment support capability can not meet the needs of non-war military operations in China [14, 15]. It is necessary to plan the equipment support system of non-war military operations as a whole in order to form a suit suitable for the type and severity of natural disasters in China. Standby support system.

© Springer Nature Singapore Pte Ltd. 2020
M. Atiquzzaman et al. (Eds.): BDCPS 2019, AISC 1117, pp. 1117–1121, 2020.
https://doi.org/10.1007/978-981-15-2568-1_154

However, the above model does not consider the problem that equipment support of non-war military operations relies on equipment support basis of war military operations. It is directly applied to equipment support resource planning of Non-war Military operations, which will lead to the problem of duplicate construction of equipment support system of non-war military operations and directly increase the rules of military support system. Mould, not only inconvenient for command, but also brings great trouble to peacetime maintenance.

In order to solve this problem, this paper puts forward a method of equipment support resource planning for non-war military operations based on IDEF0-QFD, establishes a functional model of equipment support system for non-war military operations using IDEF0, realizes the division of military capability for non-war military operations as a whole, and realizes military operations in war using QFD. On the basis of operational equipment support, the resource planning of equipment support capability in Non-war Military Operations saves support resources and improves the effect of equipment support construction in Non-war Military operations.

2   Model of Engineering Equipment Support Plan

IDEF0 is a method used to analyze the relationship between function modules and sub-modules of the system. IDEF0 is used to model the system. It can achieve independent and non-repetitive partition of system functions.

QFD is a tool to implement quality function deployment. It guarantees that the requirements of engineering equipment support run through the planning process. The transformation of house of quality guarantees the continuity and consistency of support resource planning. QFD maps the requirement to the support resource planning through the transformation of demand-planning, so that the engineering equipment support resource planning can meet the equipment support requirements of Non-war Military operations.

In order to realize the effective planning of engineering equipment support resources, the IDEF0 model is used to analyze the support requirements of engineering equipment for non-war military operations, and the analysis results are fed back to the QFD quality house, so that the model can realize the analysis of uncertain factors, ensure the accuracy and stability of the analysis results, and improve the pertinence of the application of the model.

IDEF0-QFD makes use of IDEF0 model module-sub-module relationship analysis function to realize independent and non-repetitive division of engineering equipment support capability. It uses QFD to precipitate tasks to fulfill army support requirements, compares traditional engineering equipment support requirements in Non-war Military Operations tasks, and combines existing engineering equipment support resources and support capabilities. We should strive to achieve resource conservation under existing conditions.

3 Support Capability Model

Through IDEF0 capability planning, the support capability of engineering equipment support system is divided into independent and unrelated indicators, but these indicators can not guide the planning of equipment support resources. In order to realize the saving of resources under the condition of maximizing the capacity indicators, the quality function of the support resource capability indicators is developed and transformed into the quality function of the support resource capability indicators. Specific operational resource control.

3.1 Support Capability QFD Planning Purpose

The planning of equipment support capability construction is based on the index of system capability, and it is a process of combining and allocating system elements. It is also a process of planning and determining the optimal planning around the elements that meet the goal of engineering equipment support to the greatest extent.

3.2 Establishment of Resource Guarantee System

The system is an independent unit with certain functions and requirements. According to the characteristics of engineering equipment support in Non-war Military operations, engineering equipment support is divided into support command system, maintenance support system, equipment storage and supply system and support force system from the perspective of system.

The system is an independent unit with certain functions and requirements. According to the characteristics of engineering equipment support in Non-war Military operations, engineering equipment support is divided into support command system, maintenance support system, equipment storage and supply system and support force system from the perspective of system.

The determination of parameter values of elements is a key link in the process of system optimization, which is to find a set of parameter values of elements of a system $x_1, x_2, \cdots .x_m$, To achieve the highest level of system capability $\max V(y_1, y_2, \cdots, y_n)$. $V(y_1, y_2, \cdots, y_n)$ It is a multivariable function composed of the level value of a system's capability, which synthesizes the level of each system's capability and consists of each level. $V_i(y_i)(i = 1, 2, \cdots n)$ Composition, i.e.

$$V(y_1, y_2, \cdots, y_n) = \sum_{i=1}^{n} w_i V_i(y_i) \tag{1}$$

If the level of system capability index is expressed by scale 1–5, when the level of system capability index is high, the following can be constructed.

$$V_i(y_i) = 0.25y_i - 0.25 \tag{2}$$

In order to improve the accuracy of the model, considering the linear regression of the model, the above model is fuzzified.

$$\tilde{Y}_i = f(X, \tilde{A}_j) = \tilde{A}_{i0} + \tilde{A}_{i1}x_1 + \cdots + \tilde{A}_{im}x_m \tag{3}$$

In the formula, it is the output vector of the first engineering equipment support capability. It is the input vector of system elements and deterministic data. It is a set of fuzzy coefficients. The whole problem can be described as Under a certain fitness criterion, a set of fuzzy parameters can be found. For the first system capability index of the benchmark comparative system, Eq. (4) fits best.

$$\tilde{A}_{ij} = (a_{ijC}, a_{ijS}) \tag{4}$$

Formula (4) can be recorded as:

$$\tilde{Y}_i = f_i(X, \tilde{A}_i) = (f_{iC}(X), f_{iS}(X)) \tag{5}$$

among

$$f_{iC}(X) = a_{i0C} + a_{i1C}x_1 + \cdots + a_{imC}x_m \tag{6}$$

From the house of quality, it can be seen that there is a strong correlation between the efficiency of command agencies and the staffing of commanders, professional support ability and tenacious survivability. It is necessary to strengthen its construction directly and make it slightly higher than the requirements of previous non-war military operations in order to meet the requirements of Non-war Military operations. The correlation between degree and commander's educational level and equipment support capability is weak, which can slightly reduce the construction intensity and make it slightly lower than the requirement value of previous tasks; the requirement for command scale of support personnel is higher, but the correlation with capability is very low, which is due to the low level of other four factors and the lack of experience in dealing with non-war military operations. It is slightly stronger than the current level.

4 Conclusion

The application shows that by embedding QFD into IDEF0 model, the effect of resource planning can be improved, the support capability of existing support resources can be re-planned, the waste of resources caused by re-planning can be avoided, and the combination optimization of equipment support capability of war military operations and equipment support capability of non-war military operations can be realized.

References

1. Lee, Y., Kim, J., Ahmed, U., Kim, C., Lee, Y.-W.: Multi-objective optimization of Organic Rankine Cycle (ORC) design considering exergy efficiency and inherent safety for LNG cold energy utilization. J. Loss Prev. Process Ind. **58**, 90–101 (2019)
2. Qutub, M., Vattappillil, A., Govindan, P.: Optimizing efficiency of testing, reporting and utilization of antimicrobials in diagnostic microbiology. J. Infect. Public Health **12**(2), 299 (2019)
3. Reddy, V., Badamjav, O., Meyer, D., Behr, B.: Pilot utilization of convolutional neural networks to improve the efficiency of fertilization checks. Fertil. Steril. **111**(4), e5 (2019)
4. Chen, J., Yang, S., Qian, Y.: A novel path for carbon-rich resource utilization with lower emission and higher efficiency: an integrated process of coal gasification and coking to methanol production. Energy **177**, 304–318 (2019)
5. Jayawickrama, S.M., Han, Z., Kido, S., Nakashima, N., Fujigaya, T.: Enhanced platinum utilization efficiency of polymer-coated carbon black as an electrocatalyst in polymer electrolyte membrane fuel cells. Electrochim. Acta **312**, 349–357 (2019)
6. Sanaiha, Y., Kavianpour, B., Mardock, A., Khoury, H., Downey, P., Rudasill, S., Benharash, P.: Rehospitalization and resource use after inpatient admission for extracorporeal life support in the United States. Surgery **166**, 829–834 (2019)
7. Bodas-Freitas, I.-M., Corrocher, N.: The use of external support and the benefits of the adoption of resource efficiency practices: an empirical analysis of european SMEs. Energy Policy **132**, 75–82 (2019)
8. Benuto, L.T., Singer, J., Gonzalez, F., Newlands, R., Hooft, S.: Supporting those who provide support: work-related resources and secondary traumatic stress among victim advocates. Saf. Health Work. **10**(3) (2019)
9. Kelly, C.M., Strauss, K., Arnold, J., Stride, C.: The relationship between leisure activities and psychological resources that support a sustainable career: the role of leisure seriousness and work-leisure similarity. J. Vocat. Behav. 103340 (2019)
10. Karras, J., Dubé, E., Danchin, M., Kaufman, J., Seale, H.: A scoping review examining the availability of dialogue-based resources to support healthcare providers engagement with vaccine hesitant individuals. Vaccine (2019)
11. Goldsmith, J.V., Wittenberg, E., Terui, S., Kim, H., Umi, S.: Providing support for caregiver communication burden: assessing the plain language planner resource as a nursing intervention. Semin. Oncol. Nurs. **35**(4) (2019)
12. Lopez-Zafra, E., Ramos-Álvarez, M.M., El Ghoudani, K., Luque-Reca, O., Augusto-Landa, J.M., Zarhbouch, B., Alaoui, S., Cortés-Denia, D., Pulido-Martos, M.: Social support and emotional intelligence as protective resources for well-being in moroccan adolescents. Front. Psychol. **10** (2019)
13. Whitlatch, C.J., Heid, A.R., Femia, E.E., Orsulic-Jeras, S., Szabo, S., Zarit, S.H.: The support, health, activities, resources, and education program for early stage dementia: results from a randomized controlled trial. Dementia (London, England) **18**(6), 2122–2139 (2019)
14. Robinson, J.M., Renfro, C.P., Shockley, S.J., Blalock, S.J., Watkins, A.K., Ferreri, S.P.: Training and toolkit resources to support implementation of a community pharmacy fall prevention service. Pharmacy (Basel, Switzerland) **7**(3), 113 (2019)
15. Fuller, M.G., Vaucher, Y.E., Bann, C.M., Das, A., Vohr, B.R: Lack of social support as measured by the Family Resource Scale screening tool is associated with early adverse cognitive outcome in extremely low birth weight children. J. Perinatol. **39**, 1546–1554 (2019). Official Journal of the California Perinatal Association

Ideological-Political Education of Colleges and Schools in the Large-Data Times

Huizi Fang[(⊠)]

Hubei Business College, Wuhan 430079, Hubei, China
2893207180@qq.com

Abstract. In recent years, education is becoming more and more important in the society. Meanwhile, ideological-political education in colleges and schools has attracted the attention of schools and parents. Because ideological politics learning is the basis of learning in various disciplines, but ideological and political learning is very tedious and boring. Traditional ideological politics education is always carried out in the form of meetings, talks and so on, and the results achieved are more and more important. Therefore, this paper studies the ideological-political education in the large-data times, analyzes the traits of large data times and the new connotation and traits of the ideological politics education in universities in the large-data times, and puts forward the countermeasures to strengthen the ideological politics education in colleges and schools in the large-data times. First, universities should actively improve the application of big data technology in the education process, and second, establish a high school. University ideological politics education big data sharing platform, and finally come to the conclusion that ideological politics education in the large-data times can improve the ideological and political level of students, so that students have more enthusiasm for ideological and political learning.

Keywords: Big data · Ideological politics education · Traditional ideological politics education · Sharing platform

1 Introduction

Large data means the data collection that cannot be grabbed, managed and processed by conventional software tools at a specific time interval. Large data has five traits: large amount, high speed, diversity, low value density and authenticity. It has no statistical sampling method, but only observes and tracks what happened. The use of large data tends to be analysis and prediction, user behavior analysis or some other advanced data analysis methods.

With the development of society and economy, education becomes very important in people's mind. As the most important part of education, ideological politics education must be different from the traditional ideological politics education, but following the steps of the large data times, constantly changing and improving. In [1], the author analyzes the introduction and necessity of ideological politics education in colleges and schools from the perspective of ecological science, and constructs the introduction of ecological principles in ideological politics education in colleges and

© Springer Nature Singapore Pte Ltd. 2020
M. Atiquzzaman et al. (Eds.): BDCPS 2019, AISC 1117, pp. 1122–1127, 2020.
https://doi.org/10.1007/978-981-15-2568-1_155

schools. The results show that the application of this method to ideological politics education in colleges and schools can achieve better results. In [2], the author explores the cognitive mechanism of ideological and political work in colleges and schools through the experimental research on the cognitive process of college students, so as to realize the micro research of ideological-political education at the individual cognitive level. This has become a new breakthrough point in the field of ideological and political work research methods. In [3], the author used the detailed rules and regulations of more than ten schools for reference and inquired about the discipline mechanism and dynamics of these schools. It emphasizes how the misplaced school education seeks to re socialize minority students in the values, ethics and norms of the mainstream Han society. Its goal is to create a national comprador elite, act as the spokesperson for the power of the Han/CPC, and provide an example for the Uighurs and Tibetans who have not yet fully accepted the norms of the majority. In [4], the author believes that the curriculum reform starts with the participation of the United Nations initiative, because Turkey's human rights and democracy issues are not recognized when the curriculum reform is launched. In [5], the author discusses the causes of the special problem of qualitative research ethics in the times of large data, and summarizes many urgent challenges and chances brought by the era of big data, especially the challenges and opportunities for qualitative research. In [6], the author first discusses the next generation sequencing data, and suggests that the traditional parallel mechanism progressive assembly is an interesting method to solve the large-scale question. In [7], the purpose of the author is to propose a new perspective to understand the phenomenon of network behavior, namely, privacy paradox, that is, to worry about the preservation of personal data and content, but pay little attention to the disclosure of these information and content, so as to introduce a new definition of E-man. In [8], the author points out that smart city construction is not only a specific application of big data in infrastructure construction, but also a development highlight of a new round of digital economy and new and old energy conversion. In [9], this paper attempts to provide a new perspective on the transformation and challenges of Internet public opinion in the times of large data, and proposes the following countermeasures: (1) build a government platform for Internet public opinion, and improve information gathering. (2) use the open data policy to improve the credibility of the government. (3) cultivate more professional talents in R & D and technological innovation to maintain sustainable development. In [10], with the development of new technology, data collection and analysis are becoming more and more important in various application fields. The purpose of the author is to introduce the health online medical parents system: a special big data platform supporting multiple applications.

This paper discusses the ideological politics education in colleges and schools in the times of large data, discusses the new connotation and traits of ideological politics education in colleges and schools in the times of large data, and concludes that ideological politics education in colleges and schools can improve the ideological and political level of students in the times of large data.

2 Large Data Times and Ideological Politics Education

2.1 Large Data Times

The times of large data is now widely used in a field. The so-called times of large data refers to a new data era developed on the basis of some skills such as computer, internet of things, cloud computing, etc. at present, large data is widely used in business, medical, education and other fields. Every object has its essence, large data is no exception, and the essence of big data is complete data information. Many scholars use the information of various dimensions and angles to record the behavior track of some things and judge their original traits. Therefore, it can be understood that although the definition of big data has not yet formed a unified conclusion in the current academic circles, the research on big data by scholars offers a significant theoretical basis for the further application of big data technology. Large data has four main traits: large amount of data, special types of data, fast data processing speed and low value density. Many researchers apply big data to various fields of research, and this paper is really to explore some changes and countermeasures of ideological politics education in the times of large data.

2.2 New Connotation and Traits of Ideological Politics Education in Colleges and Schools in the Times of Large Data

In large data times, the ideological politics education in colleges and schools must be different from the traditional ideological politics education. Based on the times of large data, ideological politics education in colleges and schools is equivalent to an innovation in this context. This innovation is not only reflected in the process of applying ideological-political education technology, but also reflected in the higher requirements for teachers, because in the large-data times, old teachers need to constantly update their old knowledge, learn more useful information and technology, to face no information age of breaking change. In addition, the emergence of the era of big data has also changed the way of thinking of students. Students can use large data to collect some useful information and strengthen the study of Ideological and political. From the perspective of the times of large data, ideological politics education in colleges and schools pursues seeking truth from facts, and can obtain valuable education information through large data. Therefore, the content of ideological politics education in colleges and schools has been enriched and developed through big data.

In the times of large data, ideological politics education in colleges and schools has some new traits, the main traits are as follows: first, the data-based information source. Traditional ideological politics education has no rich data, so it cannot provide valuable information resources for ideological politics education. In the times of large data, ideological politics education is different. In people's daily life, there are a lot of rich data, which can be applied to ideological politics education. Second, there are scientific decision-making methods. In the times of large data, ideological politics education records the whole and random process, which is totally different from the traditional ideological-political education, because it can further reveal the facts, so that we can make more scientific ideological-political education decisions based on these data.

3   Countermeasures for Strengthening Ideological-Political Education in Colleges and Schools in the Times of Large Data

3.1   Colleges and Schools Actively Improve the Use of Large Data Technology in the Process of Education

At present, large data is used in all aspects of college education, and large data is the key to data use. How to improve the application of large data technology in the process of education is a problem that college s should pay attention to. First of all, we can hire some special large data network technology talents to promote the application of large data technology in universities. Secondly, in the context of large data times, schools should take improving the level of ideological-political education as their own teaching goal. To achieve this goal, colleges and schools should carry out active technical exploration, and make some innovation and improvement on the methods and approaches of education. There are some improved methods and approaches as follows: (1) build the best technology of searching data. (2) explore data prediction technology. (3) using data to carry out intelligent learning.

3.2   Establish a Big Data Sharing Platform for Ideological Politics Education in Colleges and Schools

In today's society, there are many kinds of big data resources, and they are increasingly rich. However, big data still faces some problems, such as data segmentation, lack of communication and so on. These problems hinder the ideological politics education in colleges and schools to obtain ideological information. In view of these problems, universities should expand the opening of data, let the government, enterprises, schools and other industries enjoy data sharing, and provide a big data sharing platform for ideological politics education in universities. The specific countermeasures are as follows: (1) each university should establish an information sharing platform to promote the data sharing and the integration of information resources. (2) build a new media platform for ideological politics education in colleges and schools, and apply QQ, WEIBO, WECHAT and other new media to ideological-political education in colleges and schools. The use of these platforms can improve students' vision to a large extent and stimulate students' enthusiasm for ideological and political learning. (3) if colleges want to further implement the ideological-political education, they should establish a partnership with data companies, because this can attract some professionals to join in the ideological-political education, so that the development of Ideological-political education in colleges and schools is better.

4   Achievements of Ideological Politics Education in Colleges and Schools in the Times of Large Data

Compared with the traditional ideological-political education in colleges and schools, the ideological-political education in colleges and schools in the large-data times can further improve the ideological and political level of students, because the traditional ideological-political education is usually carried out in the form of lectures, talks, meetings, etc., while the ideological politics education in the times of large data achieves the form of data sharing through some new media platforms. So that students can receive ideological-political education anytime and anywhere. In addition, in the large-data times, students are more active and active in ideological politics learning, because in many data, they can choose what they are interested in to learn and understand. We compared the average ideological politics scores of students in the same class. One is traditional ideological politics education, the other is ideological-political education in the times of large data, further explaining the role of ideological-political education in the large-data times. See Fig. 1 for the specific results.

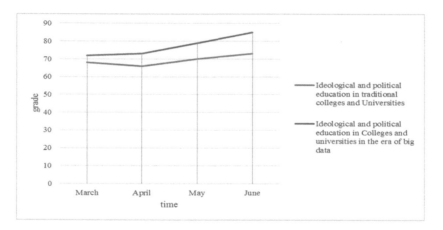

Fig. 1. The effect of ideological-political education in the large-data times

5   Conclusion

Based on the times of large data, this paper studies the ideological politics education in colleges and schools. Firstly, this paper discusses the research of some scholars at home and abroad on the ideological-political education in universities and the research of large data. Then, it expounds the concept and traits of large data, and analyzes the new connotation and traits of ideological politics education in colleges and schools in the times of large data. These new connotations and traits are related to the traditional ideological politics education is not the same. Finally, this paper offers how to strengthen the ideological politics education in colleges and schools in the times of large data. The above research shows that the ideological politics education in the times of large data can further improve the ideological politics level of students.

Acknowledgement. This article is supported by <Multi-dimensional class management in colleges and universities promotes comprehensive education—From the perspective of "parent class"> the project, the project number is 2018XGJPF3010.

References

1. Yan, Z.: The role of introduction to ecological science in ideological-political education in universities. EKOLOJI **28**(108), 517–521 (2019)
2. Xia, X., Chen, R.: Conventional motivation and practice of ideological and political work inuniversities based on cognitive ecological science. EKOLOJI **28**(108), 427–431 (2019)
3. Leibold, J., Grose, T.A.: Cultural and political disciplining inside China's dislocated minority schooling system. Asian Stud. Rev. **43**(1), 16–35 (2019)
4. Sen, A.: Militarisation of citizenship education curriculum in Turkey. J. Peace Educ. **16**(1), 78–103 (2019)
5. Hesse, A., Glenna, L., Hinrichs, C., Chiles, R., Sachs, C.: Qualitative research ethics in the large data times. Am. Behav. Sci. **63**(5), 560–583 (2019)
6. Guo, M., Zou, Q.: Perspectives of bioinformatics in large data times. Curr. Genomics **20**(2), 79 (2019)
7. Vassallo, M.: The privacy paradox in the large data times? No thanks, we are the e-people: the e-people in the large data times. Int. J. Cyber Behav. Psychol. Learn. (IJCBPL) **9**(3), 32–47 (2019)
8. Wei, L., Hu, Y.: Research on new and old kinetic energy transformation supported by smart city construction in large data times. J. Adv. Comput. Intell. Intell. Inform. **23**(1), 102–106 (2019)
9. Chen, J.: Transformations, opportunities and challenges: how government copes with online public opinion in large data times. Open J. Soc. Sci. **7**(4), 424–437 (2019)
10. Amato, F., Marrone, S., Moscato, V., Piantadosi, G., Picariello, A., Sansone, C.: HOLMeS: eHealth in the big data and deep learning era. Information **10**(2), 34 (2019)

A Remote Chinese Medicine Diagnosis and Treatment System

Jiayue Wang, Dong Xie[✉], and Shilin Li

Information School, Hunan University of Humanities, Science and Technology,
Loudi 417000, Hunan, China
287566288@qq.com

Abstract. This paper presents a remote Chinese medicine system based on TM32F103, and patients can communicate with doctors by video. In addition, the design can use a single-chip controlled heart rate sensor to measure the patient's basic information, such as heart rate, blood oxygen, etc., and upload it to the server by the WIFI module and display it on the doctor's side. Doctors can judge the patient's condition by looking at the patient, asking him about the condition, listening to the patient's breath and speed of speech, and his own feelings about the condition, and watching the patient's basic information from the server.

Keywords: Chinese medicine system · TM32F103 · Wifi

1 Introduction

In recent years, Chinese rapid economic development makes people's life quality greatly improved. People often pay the most attention to the healthy problem.

The 51 microcontroller is a more commonly used 8-bit microcontroller. It is based on typical structure and perfect bus special register centralized management, a large number of logical bit operation functions and control-oriented rich instruction system, simple development, cheap, universal good. However, its speed is not fast. It only executes a single-cycle instruction during 12 shock cycles, and it is more suitable to control simple devices with little information. In addition, it also consumes relatively large power.

The STM32F103 series microcontroller integrates ADC, rich timer and I/O port and other functions, and it is also low power consumption, stable performance, reliable. It can use 64 K bytes of built-in SRAM to achieve multi-functional control of the system, high-speed data acquisition and processing, as well as control of the motor [7].

This paper will propose a remote Chinese medicine system, and this system can make patients communicate with doctors by video [1]. As a result, we will select the STM32F103 series microcontroller to design this system according to these above analyses. Moreover, we will select WIFI module to realize multi-data transmission for video, audio, complex commands, select TFTLCD to display information, select linear device 2940 to form the regulator circuit, and select the field effect tube 9926B chip to form the motor drive module.

© Springer Nature Singapore Pte Ltd. 2020
M. Atiquzzaman et al. (Eds.): BDCPS 2019, AISC 1117, pp. 1128–1133, 2020.
https://doi.org/10.1007/978-981-15-2568-1_156

2   System Design

Patient-side clients use STM32F103 to control max30102 heart rate sensor to collect information such as heart rate and blood oxygen. The patient-side clients configure WIFI module to connect server ports for uploading data to the system server after the connection is completed. On the server side, we use the socket server's listening port to create a socket connection to upload data to doctor's side clients. Doctors judge a patient's condition by comparing information uploaded in real time [2–4].

The system uses Java and socket, to realize these functions, and these mainly enable patients to choose their own doctors, as well as doctors and patients chat. Moreover, doctors can understand the patient's condition and give the appropriate treatment plan and recommendations through the "look and see", according to the information uploaded by the patient [8].

This design makes full use of the Internet of Things, combining the Internet of Things with healthcare, and enabling medical intelligence [5]. The design consists of the following parts: the user-side part, the server part, the information collection part, the main control part, the information upload part.

Each part is relatively independent and closely linked to each other, and works together to form an intelligent telemedicine diagnosis and treatment system.

Intelligent: the realization of patients and doctors from a distance. The sensor on the patient side feeds the measured data back to the microcontroller, which uploads the data to the server via the WiFi module. At the same time, the server connects the patient side to the doctor's side for video calls. Doctors can judge the patient's illness by data and the patient's face, the tone of his voice, the state of mind, etc., in order to achieve the effect of remote Chinese medicine diagnosis and treatment [6].

Popularity: Just download our app and see the doctor online for remote treatment. When a doctor receives a request for a remote consultation, he can start the consultation by clicking "Start the consultation" to help patients with special reasons. The difficulty and focus of this design is to realize remote video calls and data upload and processing through server technology.

The user-side section of this design provides the following features: doctor and patient registration account, view of online doctors and users, and have current patients and doctors treat remotely.

The server-side part is divided into three small parts, the first for patients and doctors to provide chat services, the second for data upload and data reception to provide an interface, the third for video calls to provide an interface.

The information collection part consists mainly of temperature sensors and heart rate sensors, which measure the basic information of the patient [10].

The main control section is responsible for the transmission of the whole system data and the maintenance of the stable operation of the system.

The information upload part is mainly composed of WiFi module, Which connects to the system server via wireless hotspot and uploads the data collected by the master part [9] (Figs. 1, 2, 3 and 4).

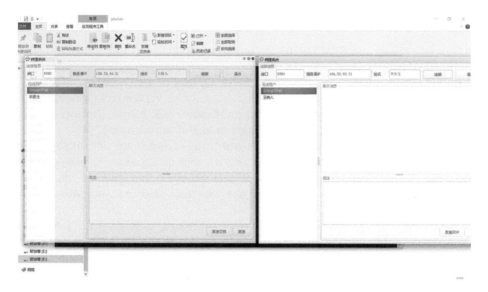

Fig. 1. A online view

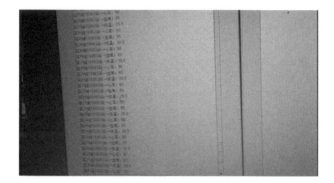

Fig. 2. A doctor-side client

Fig. 3. A video between a doctor and a patient

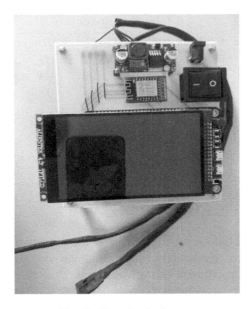

Fig. 4. The circuit diagram

3 Data Analysis

We execute data analysis respect to three aspects such as heart rate, blood oxygen and temperature shown as following two tables (Tables 1 and 2).

Table 1. Test data

Sequence	Heart rate bpm	Blood oxygen %	Temperature/°C
1	96	95	35.5
2	96	97	35.6
3	90	95	35.5
4	93	91	35.1
5	91	90	35.7
6	70	94	35.4
7	88	95	36.1
8	95	98	36.5
Reference	60–100	95–100	36–37

Table 2. Mistakes

Sequence	Heart rate bpm	Blood oxygen %	Temperature/°C
1	0	0	0.5
2	0	0	0.4
3	0	0	0.5
4	0	4	0.9
5	0	5	0.3
6	0	1	0.6
7	0	0	0
8	0	0	0
Mistake	0	1.25	0.4

The data above is only a small part of test data. According to the many measurements we have taken, repeated comparison with the normal range, as well as to the regular hospital and survey, these basically meet our ideal goal.

4 Conclusion

This paper will propose a remote Chinese medicine system, which can achieve remote diagnosis and treatment. It greatly close the distance between doctors and patients, medical intelligence implies that our living standard sings continuously. We are constantly to intelligent development, in the development of science and technology pay attention to their health.

Acknowledgements. This work is supported by the Research Foundation of Education Committee of Hunan Province, China (18C0896), the Foundation of Hunan University of Humanities, Science and Technology (2018).

References

1. Meng, Q.: Remote Chinese medicine diagnosis and treatment system patent (2010)
2. Xiao, G.: ARM embedded Development Examples – System Design Based on STM32. Electronic Industry Press (2013)
3. Fan, M.: Discussion on the design of intelligent garbage bins based on microcontroller control system. Electronic test (2017)
4. Perez-Gaspar, L.A., Caballero-Morales, S.O.: Multimodal emotion recognition with evolutionary computation for human-robot interaction. Expert Syst. Appl. **66**, 42–61 (2016)
5. Guan, S., Shen, L., Cao, R.: Research on design strategy of sorting garbage bins based on the CREATE action funnel. In: Proceedings of the 2nd International Conference on Intelligent Human Systems Integration (IHSI 2019): Integrating People and Intelligent Systems, February 7–10, 2019, San Diego, California, USA. Intelligent Human Systems Integration 2019. Springer, Cham (2019)

6. Zhang, Z.: Design of LED light control system based on mobile phone WiFi. Comput. Program. Ski. Maint. **401**(11), 70–72 (2018)
7. Chen, C., He, L., Li, C.: Wireless charging car control system based on STM32. Telecom Express **09**, 35–38 (2019)
8. Jiang, Z.: Design and implementation of a remote system for power monitoring system testing. Microcomput. Appl. **35**(20), 5–8+12 (2016)
9. Hang, T., Zheng, Y., Qian, K., Wu, C., Yang, Z., Zhou, X., Liu, Y., Chen, G.: WiSH: WiFi-based real-time human detection. Tsinghua Sci. Technol. **24**(05), 615–629 (2019)
10. Buvik, A., Bugge, E., Knutsen, G., Småbrekke, A., Wilsgaard, T.: Patient reported outcomes with remote orthopaedic consultations by telemedicine: a randomised controlled trial. J. Telemed. Telecare **25**(8), 483 (2019)

Building the Influence Evaluation Model of Web Quality Information for Consumer Products

Yingcheng Xu[1], Haiyan Wang[2(✉)], Ya Li[1], and Ruyi Ye[1]

[1] Quality Research Branch, National Institute of Standardization,
Beijing 100191, China
15810265898@126.com, 15201290448@163.com,
yery@cnis.ac.cn
[2] School of Management and E-Business, Zhejiang Gongshang University,
Hangzhou 310018, China
23730759@qq.com

Abstract. This paper establishes the influence evaluation model of Web quality information for consumer products based on four dissemination channels, such as web portals, forum, blogs and microblog etc. What's more, it combines entropy weight method with gray theory to put forward the multilevel gray entropy Comprehensive Influence Evaluation Model of Web Quality Information for Consumer Products. Finally, the paper takes "Anxin Floor Event" as an example to carry out the empirical research so as to verify the validity and feasibility of evaluation model.

Keywords: Quality · Information · Evaluation model

1 Introduction

With the development of Web 2.0 technology, quality information of consumer products can be obtained not only from the traditional web portals but also from the newly-developing social media, such as blogs, forum, microblog and social network etc. Zhang et al. [5] presents being different with the model of former traditional media disseminating information, the Web-based information dissemination of consumer products is characterized by its strong crypticity, as well as its large possibility of giving rise to significant public opinion incidents. If it fails to conduct influence evaluation to the quality information of consumer products disseminated on the Web in time, determine its severity degree and propose effective intervention and control measures; it may cause the devastating consequences. The main contributions of this paper lie in two aspects: firstly, it puts forward the Influence Evaluation Index System of Web Quality Information for Consumer Products; secondly, it establishes multilevel Influence Evaluation Model for Consumer Products based on Gray Entropy. At present, there are so many research achievements on influence evaluation of Web information. Cha et al., Cao et al. and Yang et al. analyze the scope and speed of information dissemination in the social network, presents the influence model of users in Twitter

© Springer Nature Singapore Pte Ltd. 2020
M. Atiquzzaman et al. (Eds.): BDCPS 2019, AISC 1117, pp. 1134–1142, 2020.
https://doi.org/10.1007/978-981-15-2568-1_157

and points out that the more fans do not mean the greater influence [2, 8, 9]. Chen et al. takes advantage of Linear Threshold Model to propose Node Activation Threshold-Based Heuristic Algorithm [4]. Kempe et al., Bornmann et al. and Goyal et al. aim at the issue of influence maximization for social network and raises the diffusion model [1, 3, 6]. Tian et al., Eyal et al. and Zhang et al. present the quantification method for the influence of public opinion leader in the community [5, 7, 10]. Although many researchers have been conducted on influence evaluation of Web information, the results are still far from satisfying. This paper focuses on influence evaluation of Web information about consumer product quality and safety based on gray entropy.

2   Building the Influence Evaluation Index System

Table 1. Influence Evaluation Index System of Web Information

Influence evaluation system of web information for consumer products	First level index	Second level index
	Web portals	Quantity of occurrence of dependent event information
		Number of participation
		Number of comments
		Participation resonance
		Index for importance of information source
	Forum	Quantity of occurrence of dependent event information
		Forum post pageviews
		Number of comments
		Index for importance of information source
	Blogs	Quantity of occurrence of dependent event information
		Click rate of blog articles
		Number of recommendation of blog articles
		Number of comments
		Index for importance of information source
	Microblog	Quantity of occurrence of relevant contents
		Number of forwarding
		Number of comments
		Geographic distribution degree
		Number of certified user
		Number of original user

Major dissemination channels for Web-based quality information of consumer products mainly include web portals, forum, blogs and microblog which belong to first level index; the subordinate sub-item that could represent its characteristics are second level index which basically can be divided into index of event information, index of netizen's attention, index of information source and index of information receivers. Event information index refers to quantity of occurrence of dependent event information in

the four dissemination channels; netizen's attention index covers number of participation, forum post pageviews, click rate of blog articles, number of comments, participation resonance, number of recommendation of blog articles and number of
forwarding; information source index means the importance of information source; and
information receivers index includes geographic distribution degree, number of certified user and number of original user. The index system is shown as Table 1.

3 Building the Influence Evaluation Model Based on Grey Entropy

Process for influence evaluation of Web quality information for consumer products
based on gray entropy includes the following six steps:

Step 1: Judgment matrix constituted by evaluation indexes.
Use the matrix to represent the characteristic value of m_j evaluation indices for all n
evaluation targets of information influence. The matrix is shown as follows:

$$x_{ij}^h = \begin{bmatrix} x_{m_1 1}^h & x_{m_1 2}^h & \cdots & x_{m_1 n}^h \\ x_{m_2 1}^h & x_{m_2 1}^h & \cdots & x_{m_2 n}^h \\ & & \cdots & \\ x_{m_i 1}^h & x_{m_i 2}^h & \cdots & x_{m_i n}^h \end{bmatrix}$$

In which, $i = m_1, m_2. \ldots .m_i$ $h = 1, 2, \cdots, n$; $j = 1, 2, \cdots, n$
If the index becomes better and better as the value grows bigger and bigger, it shall
be calculated in accordance with formula (1) as below:

$$r_{ij}^h = \frac{x_{ij}^h - \wedge x_{ij}^h}{\vee x_{ij}^h - \wedge x_{ij}^h} \tag{1}$$

If the index becomes better and better as the value grows smaller and smaller, it
shall be calculated in accordance with formula (2) as below:

$$r_{ij}^h = \frac{\vee x_{ij}^h - x_{ij}^h}{\vee x_{ij}^h - \wedge x_{ij}^h} \tag{2}$$

In the formula, $\vee, \wedge$ are symbols respectively indicating taking bigger value or
smaller value.

After normative processing, the matrix is shown as below:

$$
r_{ij}^h = \begin{bmatrix}
r_{m_1 1}^h & r_{m_1 2}^h & \cdots & r_{m_1 n}^h \\
r_{m_2 1}^h & r_{m_2 2}^h & \cdots & r_{m_2 n}^h \\
& & \cdots & \\
r_{m_i 1}^h & r_{m_i 2}^h & \cdots & r_{m_i n}^h
\end{bmatrix}
$$

Step 2: Determine correlation coefficient for various evaluation targets systems and reference sequence.

Because the influences of Web information on n information dissemination channels (correspond to n sub-systems) is featured by comparatively relativity. The largest extent influence in the number h sub-system is in terms of m_i information dissemination indices. Therefore, it shall select the idealized and optimal program firstly which could ensure every dissemination index maintain its biggest value. It can be denoted as:

$$
F_i^h = [f_{m_1}^h, f_{m_2}^h, \ldots, f_{m_i}^h]
$$

In the formula, $f_i^h = \max(r_{i1}^h, r_{i2}^h, \ldots, r_{in}^h)$, $i = m_1, m_2, \ldots, m_i$, that is to say, f_i in F_i^h is the maximum value of a certain dissemination index for all influence evaluation targets on the dissemination channel and it can be regarded as the standard extreme reference program. Taking it as the reference sequence, n programs shall be taken as comparative sequence respectively. Similarity degree of data geometrical relationship between reference sequence and comparative sequence is usually measured by the correlation degree. The correlation coefficient between j comparative sequence and i index in F_i^h reference sequence can be donated by $\varepsilon_{Fj}^h(i)$ and it can be calculated in accordance with formula (3) as below:

$$
\varepsilon_{Fj}^h(i) = \frac{\min\limits_{j} \min\limits_{i} \left| f_i^h - r_{ij}^h \right| + \rho \max\limits_{j} \max\limits_{i} \left| f_i^h - r_{ij}^h \right|}{\left| f_i^h - r_{ij}^h \right| + \rho \max\limits_{j} \max\limits_{i} \left| f_i^h - r_{ij}^h \right|}
\tag{3}
$$

In which, $j = 1, 2, \ldots, n$; $i = m_1, m_2, \ldots, m_i$, in the formula (6–3), $\rho \in [0, 1]$, in general, $\rho = 0.5$.

Step 3: Determine the weight of each evaluation index.

According to Entropy theory, if there has a big difference of a certain index value between every evaluation target and the entropy is less, it indicates that the amount of effective information provided by this index is large and its weight is supposed to large as well. On the contrary, if there has a small difference of a certain index value between every evaluation target and the entropy is large, it indicates that the amount of effective information provided by this index is small and its weight is supposed to small as well. When the certain index values between every evaluation target are the same, the entropy reaches its maximum value which means that there is not any available information of the index and it shall be removed from the evaluation index.

Steps for determining the weight of each evaluation index with the entropy are as follows:

(1) The entropy for the number V_h sub-system can be defined as:

$$H_{vh} = -\frac{1}{\ln n}\sum_{j=1}^{n}\frac{\varepsilon_{Fj}^{h}(i)}{\sum\limits_{j=1}^{n}\varepsilon_{Fj}^{h}(i)}\ln\frac{\varepsilon_{Fj}^{h}(i)}{\sum\limits_{j=1}^{n}\varepsilon_{Fj}^{h}(i)} \tag{4}$$

In the formula, $j = 1, 2, \ldots, n$, $i = m_1, m_2, \ldots, m_i$.

(2) Weight is calculated in accordance with formula (6–5) as below:

$$\omega_{vh} = \frac{1}{\sum\limits_{j=1}^{m}(1 - H_{vh})}(1 - H_{vh}) \tag{5}$$

In the formula, $0 \leq \omega_{vh} \leq 1$, $\sum\limits_{i=m_1}^{m_i}\omega_i = 1$.

Weight vector constituted by m_i index weight in the number V_h sub-system is represented as below:

$$\omega_{m_i}^{vh} = [\omega_{m_1}^{vh}, \omega_{m_2}^{vh}, \ldots, \omega_{m_i}^{vh}]$$

Step 4:   Make up correlation degree matrix.

Calculate the correlation degree between each reference sequence and comparative sequence in accordance with the recombination of every index weight and correlation coefficient obtained in Step 2:

$$R_j^{vh} = \omega_{m_i}^{vh}\varepsilon_{Fj}^{h}(i), j = 1, 2, \ldots, n \tag{6}$$

Calculate correlation degree for V_h sub-systems separately and establish correlation degree matrix constituted by n influence information evaluation targets of V_h sub-systems:

$$R_{Vh\times n} = \begin{bmatrix} R_1^1 & R_2^1 & \cdots & R_2^1 \\ R_1^2 & R_2^2 & \cdots & R_n^2 \\ & & \cdots & \\ R_1^{Vh} & R_2^{Vh} & \cdots & R_2^{Vh} \end{bmatrix}$$

Step 5: Determine the weight of sub-system.
Select the reference sequence from correlation degree matrix determined by Step 4 and the rest are comparative sequences. Work out the correlation coefficient $\varepsilon_{Fj}(v)$ between reference sequence and comparative sequence. Then calculate the weight of V_h sub-systems by entropy weight method:

$$\omega_v = [\omega_{v_1}, \omega_{v_2}, \ldots \omega_{v_h}] \text{ and } \sum_{v_1}^{v_h} \omega_v = 1$$

Step 6: Determine the correlation degree.
Recombine the weight of sub-system and correlation coefficient obtained in Step 5 to work out the correlation degree for comprehensive evaluation of information influence. The size of correlation degree determines the degree of information influence.
    Correlation degree of system's comprehensive evaluation is as below:

$$R_j = \omega_v \cdot \varepsilon_{Fj}(v), v = v_1, v_2, \ldots, v_h \tag{7}$$

4   Case Study

Aiming at "Anxin Floor Event" disclosed on the Internet, this paper conducts researches on information influence of the event on the network. Through four kinds of network information dissemination channels, such as web portals, forum, blogs and microblog etc and on the basis of index system established in advance, data for almost one month are collected successively to build up multilevel comprehensive evaluation model based on gray entropy and finally acquire the evaluation results of information influence.

4.1   Model Calculation

If the whole Web is one complex system, the four network dissemination channels can be regarded as four sub-systems. Tracking to weight change of four sub-systems could keep pace with the change of four channels' function reflected in the process of information influence dissemination. After evaluating four sub-systems, add them up to obtain correlation coefficient which could represent the quantification value of event information influence and its variations.

(1) Normalization Processing
Carry out normalization processing to every index in line with formula (1) and formula (2). Basically, indices for research objects in this paper belong to the type that the index becomes better and better as the value grows bigger and bigger, such as number of comments and number of forwarding etc. When these indices become bigger and bigger, it indicates that the information influence grows stronger and stronger. As a result, it shall calculate in accordance with formula (1) during normalization computing.

(2) Calculate correlation coefficient between each evaluation target system and reference sequence.

Calculate correlation coefficient according to formula (3). Take the data ten days ago as example and the calculation results are shown in Table 2:

Through the correlation coefficient matrix, use formula (4) to work out the entropy of sub-system; then, use formula (5) to calculate weight vector of every evaluation element for sub-system of each dissemination channels, $\omega = \{0.06, 0.07, 0.07, 0.06, 0.06, 0.09, 0.06, 0.09, 0.08, 0.06, 0.07, 0.06, 0.06, 0.05, 0.05, 0.08\}$; at last, recombine weight vector with correlation coefficient matrix by formula (6) to acquire the first layer correlation degree vector quantity for each reference sequence and comparative sequence, that is to say $\{0.06, 0.18, 0.11, 0.09, 0.07, 0.07, 0.07, 0.09, 0.07, 0.06\}$. With the same methods, evaluation index weight for the rest three sub-systems could be determined as well, respectively concluding three weight vectors, as well as three first layer correlation coefficient vectors.

With the same methods, select the reference sequence from the matrix constituted by four groups of first layer correlation degree vectors and take the matrix itself as comparative sequence; re-calculate the second layer correlation coefficient for reference sequence and comparative sequence; utilize entropy weight method to work out the weight of four sub-systems and the calculation results are $\{0.14, 0.18, 0.28, 0.4\}$; recombine the vector with second layer correlation coefficient and finally conclude the array of correlation degree for comprehensive evaluation of information influence with the value of $\{0.46, 0.94, 0.68, 0.47, 0.44, 0.42, 0.44, 0.47, 0.43, 0.42, 0.42, 0.41\}$.

4.2   Result Analysis

In order to represent the influence evaluation results of "Anxin Floor Event" information on the Web more intuitively, conduct normalization processing to 32 comprehensive evaluation correlation degree and the result is shown in Fig. 1.

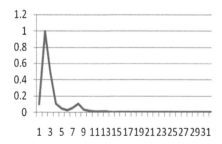

Fig. 1.   Information influence of "anxin floor event"

As we can see from Fig. 1, in the first three days, the influence of event information increases intensively and declines immediately soon afterwards; one week later, there is a rebound within a narrow range but then declines again and reduces to the minimum level and tends to level off in the end. The calculation result conforms to the actual situation. It's exactly the several days after the explosion of event that netizens discuss it with the strongest enthusiasm. But then the influence of event reduces at once. After one week, it's the time for relevant enterprise to deliver the clarification to the public which causes a rebound within a narrow range. However, it's obvious that the influence is inferior to the influence at the time of event explosion. Later on, the influence tends to the minimum level and has relatively stable condition which indicates that the influence of the event on the Web lasts short duration.

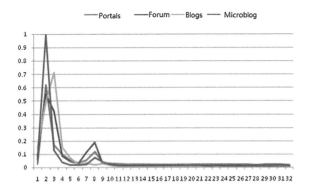

Fig. 2. Influence of dissemination channel of "anxin floor event"

The tracking to weight of influence of four major network information dissemination channels could reflect the influence change of four major channels to information dissemination. Tracking results are shown as Fig. 2.

As it is shown in Fig. 2, in the first three days after event explosion, as a newly-born network information dissemination channel, the function of microblog for information dissemination has already exceeded traditional web portals and it be followed by forum, the blogs come last. Besides, the influence of blogs obviously lags behind microblog, web portals and forum but it can catch up from behind. In the following days, the influence of microblog sharply reduces and ranks to the last place among four channels which means that its influence has a strong time sensitive. After that, the microblog never ranks to the preceding places. From the slope aspect, the influence of forum reduces at the slowest speed. At the stage of rebounding within a narrow range, it can be seen that forum has the strongest influence, following by web portals, and it reflects that the forum works as an excellent carrier for netizens' continuous discussion.

5 Conclusion

Aiming at typical consumer quality and safety events, the paper studies on the dissemination of relevant information on the Web; analyzes the four dissemination channels that include web portals, forum, blogs and microblog; proposes the comprehensive evaluation method to information influence; establishes the comprehensive evaluation model based on gray entropy; evaluates the level of Web information influence on the basis of Web information influence evaluation system; conducts empirical analysis to consumer quality and safety events; and at the same time carries out comparative analysis to mean weighted evaluation model and TOPSIS evaluation model to verify the effectiveness of the model. The paper takes most of quantitative indices into consideration and has no regard for qualitative indices. Web information influence can be effected not only by quantitative indices, such as information content of dependent events and number of netizens who take in the interaction; but also by social factors like regional informatization network environment, political environment and policy environment in local places etc.

Acknowledgments. We would like to acknowledge that this research is supported and funded by the National Science Foundation of China under Grant No. 91646122 and 91746202, the Basic Scientific Research Business Projects 552018Y-5927-2018.

References

1. Bornmann, L., Haunschild, R.: t factor: metric for measuring impact on Twitter. Malays. J. Libr. Inf. Sci. **13**, 13–20 (2016)
2. Cha, M., Haddadi, H., Benevenuto, F., Gummadi, K.P.: Measuring user influence in Twitter: the million follower fallacy. In: ICWSM 2010 (2010)
3. Kempe, D., Kleinberg, J., Tardos, É.: Influential nodes in a diffusion model for social networks. In: Proceedings of the 32nd International Conference on Automata, Languages and Programming, pp. 1127–1138 (2005)
4. Chen, H., Wang, Y.: Threshold-based heuristic algorithm for influence maximization. J. Comput. Res. Dev. **49**(10), 2181–2188 (2012)
5. Zhang, W., Wang, B., He, H.: Public opinion leader community mining based on the heterogeneous network. Acta Electron. Sin. **40**(10), 1927–1932 (2012)
6. Goyal, A., Bonchi, F., Lakshmanan, L.V.: Learning inuence probabilities in social networks. In: Proceedings of the Third ACM International Conference on Web Search and Data Mining, WSDM 2010, pp. 241–250 (2010)
7. Even-Dar, E., Shapira, A.: A note on maximizing the spread of influence in social networks. Inf. Process. Lett. **111**(4), 184–187 (2011)
8. Cao, T., Wu, X., Wang, S., et al.: Maximizing influence spread in modular social networks by optimal resource allocation. Expert Syst. Appl. **38**(10), 13128–13135 (2011)
9. Yang, W., Dai, R., Cui, X.: Model for Internet news force evaluation based on information retrieval technologies. J. Softw. **20**(9), 2397–2406 (2009)
10. Tian, J., Wang, Y., Feng, X.: A new hybrid algorithm for influence maximization in social networks. Chin. J. Comput. **34**(10), 1956–1965 (2011)

Informatization Construction of Minority Languages and Characters Under the Background of Big Data

Na Li[✉]

College of Arts and Sciences, Yunnan Normal University,
Kunming, Yunnan, China
jiansuoinfo@aliyun.com

Abstract. China is a multi-ethnic socialist country, and the relations between various ethnic groups are united. To treat minority languages, we need to respect their characteristics, promote minority culture, and develop minority education. The informatization construction of minority languages is conducive to the protection and development of rich national language resources. This paper briefly introduces the basic connotation of big data and the related characteristics of minority languages. Based on the analysis of the status quo of minority languages in the context of big data, this paper discusses how to promote minority languages in the context of big data. Borrowing big data with existing information, resources and technology, in-depth excavation, rational construction of minority language and language resources, and speeding up the construction of minority text information platform; combining strict cultural propaganda and minority language features to establish strict norms To build a website for minority languages and texts, and accelerate the development of software for minority languages, with a view to realizing the informationization of minority languages in the true sense.

Keywords: Big data · Minority national languages · Information construction · Countermeasures

1 Introduction

Big Data refers to the massive data generated, which has a wide variety of information values and "becomes a source of new inventions and services". At present, big data applications have covered many fields such as politics, economy, society, and humanities [1]. In recent years, people have begun to incorporate the protection of traditional culture into the big data framework through digital and networked methods, including digital conversion of texts and images, data storage, database construction of cultural resources, statistical analysis of data associations and government decision support [2]. As the most important information carrier, language and writing is a communication tool repeatedly used in all social activities and one of the most basic objects of information exchange [3]. In the development of minority language information technology, the normative standard of language and characters has become the premise of information technology development and application, and the guarantee of

M. Atiquzzaman et al. (Eds.): BDCPS 2019, AISC 1117, pp. 1143–1151, 2020.
https://doi.org/10.1007/978-981-15-2568-1_158

the effective operation of information systems. The informatization construction of minority languages has started late and is slow to develop [4]. It is limited by factors such as technology, capital and policy in the construction of basic resources and later development and utilization [5]. The research and formulation of minority language information processing standards can promote the standardization and modernization of the national language itself, so that it can reduce or even eliminate chaos regularly, which is conducive to the application of computer big data technology [6]. The modernization of national language information processing research methods provides norms for the development and application of information technology [7]. Therefore, the research on the construction of information technology of minority languages in the context of big data is of great significance for promoting the application of national language information processing technology and promoting the development of various ethnic minority undertakings in China.

2   An Overview of Big Data and Minority Languages

2.1   The Basic Connotation and Characteristics of Big Data

Big data is generally used to define and describe the vast amounts of data generated by the era of information explosions, which are so large that they cannot be obtained and organized into information that can be used by people in a reasonable amount of time. In essence, the so-called big data refers to the large or complex data size from which data sets that conform to the regularity of the development of things can be mined [8]. With technologies such as data storage, knowledge discovery, visual presentation, association rules, classification clustering, and decision support, new technologies are needed to capture, store, manage, analyze, and apply these data. In terms of the types of big data technologies, it obviously has the characteristics of fast data processing speed and high data storage quality. As long as the corresponding software conforms to the nature of the work, users can process huge amounts of data information anytime and anywhere through electronic mobile devices [9]. From the perspective of popularization, media information such as pictures, videos, documents, and data tables can be summarized in the range that big data technology can handle.

2.2   Minority Language Features

Text is the symbol of written language. After it is produced, it overcomes the time and space restrictions of language, expands information dissemination, increases cultural accumulation, and has a tremendous impetus to the development of society [10]. The existing minority languages have been developed in the long-term use process after experiencing the test of language function competition in history. These language developments have become the mother tongue of the family, which is determined by objective and practical needs, and has its objective necessity. All ethnic groups love the language of their own nation, regard language as an important component of national identity, regard it as a valuable asset of the nation, and have a unified desire and consciousness for the national language. Nowadays, except for a few ethnic groups that have differently transferred to Chinese, most ethnic groups use their own language as the main communication tool of daily life; some ethnic groups use traditional Chinese characters, while others use traditional language as they develop. New text based on text. The language of ethnic minorities plays an important role in the social life of ethnic minorities, and the language of other ethnic groups cannot be completely replaced.

2.3   Minority Language Classification

At present, there are roughly three types of use of minority languages:

(1) The first type, such as Tibetan, Mongolian, Uyghur, Kazakh, North Korean, etc., has large settlements with a population of more than one million and a long history of writing. Their language is not only in the family, but also in the neighborhood. In addition, it is used in various fields of politics, economy, culture and education of the nation, even in some neighboring or other ethnic groups living together.

(2) The second type, such as the Dai and the Dai, although they also have a large area of settlement, with traditional characters, but the text has no uniform norms, the dialect is also different, and the language of the nation is not as good as the society. Mongolia, Tibet, Uyghur, Kazakh, North Korea and other ethnic groups are so extensive. Zhuang language and proverbs, Lahu language, Jingpo language, and Shiwa language have different internal language, the common language of the whole nation has not yet formed, the scope of use of the text is also small, and the language usage is close to proverbs and proverbs.

(3) The third type, the language of the nation is only used in the daily life of the nation. In political life and school education, the language of other ethnic groups is used (mainly Chinese, and some places also use other minority languages); Words with the same language generally use Chinese characters. There are quite a few minority languages belonging to this type, accounting for more than three-quarters of the total number of languages, and the population accounts for more than half of the total minority population (see in Table 1).

2.4   The Practical Significance of Big Data Technology to the Informationization of Minority Languages and Characters

With the accelerating process of industrialization, urbanization and marketization throughout the country, and the advent of a new era of informationization, the localities are self-sufficient, self-sufficient, and self-sufficient, and there are fewer and fewer situations, and the pattern of mutual interaction and interdependence is increasing. The market for information integration across the country calls for a more unified, more versatile and more standardized language in the whole society. The national common language and characters are used more and more people, the frequency of use is getting higher and higher, and the scope of use is wider and wider; some minority languages are relatively shrinking. The endangerment or extinction of some minority languages is often caused by the interaction of many factors. For example, the number of native

Table 1. Classification of minority languages

Classification	First type	Second type	Third type
Race	Tibetan, Mongolian, Uighur, Kazakh, Korean, etc.	Yi and Yi	Minority people using Chinese characters
Language text used	Have traditional text, language	There are traditional words, and the dialects are also different	Mainly Chinese
Use the range of language text	In addition to being used within the family, between friends and relatives in the neighborhood, it can also be used in various fields of official politics, economy, culture and education	The scope of use of the text is also small, and the common language common to the whole nation has not yet formed	Used in daily life within the nation, the use of the population accounts for more than half of the total minority population

speakers is reduced, the passage area is reduced, the social function is declining, and there is no written text; the language attitude of the parents of the nation is changed, and the children are not willing to let their children learn the native language and choose to enter the Chinese language school. In the context of big data technology, through the establishment of websites and databases of minority languages, using digital recording methods such as audio and video, the carrier of language and text resources can be transferred from people to new media such as databases, which can prevent "death of people". The phenomenon of "interest" is also conducive to the preservation and inheritance of the language of ethnic minorities, and is also a reflection of the protection of traditional Chinese culture.

3   The Status Quo of Minority Languages

3.1   The Use of Ethnic Languages as a Native Language Is Reduced or Even Endangered

Because of the social and economic development, like many countries in the world, the population of some minority languages in China is also decreasing year by year. Some weak minority languages, such as Manchu, Hezhe, Tujia, Yi, Yi, Tatar, Oroqian, Ewenki, Yugu, etc. are in an endangered state. The Manchu population is over 10.72 million. About 100 people who can understand Manchu are now able to speak about 50 old people. The usage rate of Manchu is 0.0006%. There are more than 5,200 people in the Hezhe family. Only 10 or more elderly people who are 60 years old or older who can communicate in Hezhe language, if they can talk a little, cannot count more than 50 people. The usage rate of Hezhe is 0.07%. In the 1980s, there were 800,000 Tujia people, and the use rate of Tujia language was 25%. In 2000, there were more than

8 million Tujia people and the population was 60,000. The use rate of Tujia language is 0.086%. There are more than 750,000 Yi people and about 1,500 people who can speak the proverb. The proverb usage rate is 0.13%. There are more than 600,000 Yi people, and only 6,000 people can use the proverb. The proverb usage rate is 1.36%. The Tatar population is nearly 5,000, and less than 1,000 people use Tatar. The Tatar language usage rate is 20%. There are more than 8,200 people in the Oroqen family. Only a few hundred people will speak the Oroqen language. The young people under the age of 20 will not. Most of these ethnic minority languages that are on the verge of disappearing do not have corresponding texts. They are only kept in spoken language forms such as folk songs and legends, and they are passed on by word of mouth. Some have no communicative function and can only become the language of a few elderly people. This indicates that "in the not too distant future (within decades), the language of these peoples will inevitably disappear".

3.2   Reduced Number of Students Choosing to Study Ethnic Languages

With the development of the market economy and the acceleration of the urbanization process, the exchanges between various ethnic groups have become more and more extensive and frequent. Many minority primary and middle school students and their parents believe that Chinese literature is coming in the market competition, career choice and future market competition. In the past ten years, the number and number of publications of minority nationality books in China have shown a growing trend, but the proportion of national books in the total number of books in the country has shown a downward trend. According to statistics, in 1980, the national books accounted for 8.9% of the total national books, but only accounted for 4% in 1991, and fell to less than 2% after 2005. It can be seen that the population of ethnic minorities in China is not large, the readership is not large, and the inheritance of minority languages will face a narrower and narrower situation.

4   The Strategy of Informatization Construction of Minority Languages and Characters Under the Background of Big Data

4.1   Accelerate the Construction of Minority Characters Information Platform

The construction of information platform for minority languages and language, with the help of local minority language resources and the support of big data technology, can attract more technical talents in language and information processing. Absorb some of the information technology, understand the international code declaration, coding character set construction, font standard formulation, keyboard standard formulation, and pay more attention to minority languages. Combining the characteristics of minority languages, we will set up a multi-disciplinary research team to help ethnic minorities complete the construction of language and text information platform faster and better (see in Fig. 1).

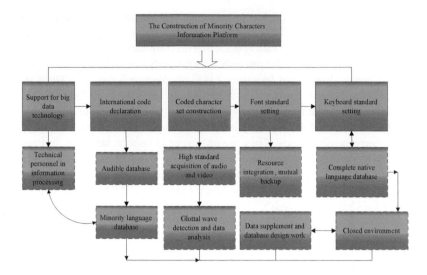

Fig. 1. The construction of minority characters information platform

4.2 Develop Strict Normative Standards and Rationally Construct a Minority Language and Language Resource Library

Borrowing the cutting-edge information technology of big data development, using modern technology to quickly obtain resource information and collecting high-quality data that meets the requirements, it can facilitate the construction of minority language and language resources. There are some existing resources in minority languages. For example, ethnic minorities with traditional Chinese characters have a large number of national texts, and using image recognition and other technologies can convert paper documents into electronic resources for storage and utilization. For the construction of minority language and language resources, it is necessary to integrate existing language resources and achieve efficient data conversion. The libraries of minority languages and languages that have been built and are being built are mostly collected by experts and scholars and collected by hand. The collected corpus, although highly accurate, is time-consuming and laborious, and the existing online real-time crawling technology can assist Experts and scholars quickly obtain corpus and update the corpus in real time to ensure the effectiveness, dynamics and balance of the national language resource library. Hire a team of experienced companies in database construction and maintenance to assist the builders of minority language and language resources to establish strict normative standards, and establish a unified special agency to supervise and implement, so that problems that may be encountered in the use of minority language resources in the future Try to avoid the least (see in Fig. 2). The construction of the minority language and language resources database is still at the initial exploration stage. In order to facilitate the development and utilization of minority language and language resources in the future, and the interaction and sharing between resource libraries, strict normative standards should be formulated in the initial stage of construction. The future operating environment is very necessary. The data storage of ethnic minority language resource database construction needs to be secure and reliable. The establishment of big data technology development gathering area provides space and equipment resources for data storage, and also maintains the later data to ensure the minority language and language resource library. Long-term operation offers the possibility. In addition, pay attention to the development of the search function of the minority language resource library, not only to do the basic field search, but also to develop a more practical advanced function search, so that the established minority text resource library can be applied to the maximum extent.

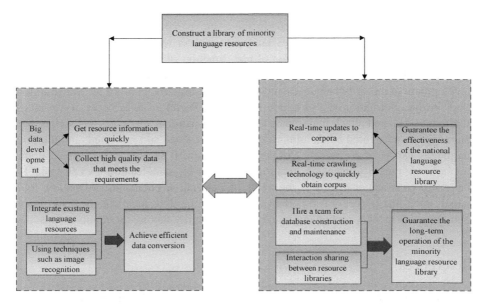

Fig. 2. Reasonable construction of a minority language resource library

4.3 Building a Minority Language Website

Integrate the resources for the construction of minority languages in the early stage, and use the talents and technology developed by the big data industry to build a unique minority language based on the information construction of minority languages and the content of minority cultural propaganda. The website is also an important and intuitive embodiment of the informatization of minority languages. The monolingual website of ethnic languages consisting only of minority languages is not common, and the website journalists and editors are trained in information technology to make them become compound talents. The website is easy to do. Judging from the current situation, there are mainly several types of websites involving minority languages: one is a multi-provincial language-based website that is jointly built by many provinces; the other is a national language that is published by the folk society and the masses. The website of the website, these ethnic language texts mainly exist in the Chinese language website or the bilingual website of the Chinese and the Han, mainly the translation of ethnic literary works, the introduction of national cultural knowledge and the content of author reviews; The official media publishes a website containing national language text pages. These websites use Chinese as the main publishing language and have minority language text pages. Most of these minority language web pages are nationalities that have published the Chinese language information content. Chinese translation, the topic is mostly political, and the information is lagging behind.

5   Conclusions

In summary, language and writing are also an important part of national work, and ethnic equality can be implemented. In the context of big data, the challenges and opportunities of minority languages are coexisting in terms of cultural inheritance, population use, scope of use, degree of use, function of use, status of language and protection of rights and interests. In the context of big data technology, the use and development of minority languages and scripts is an important part of the integration of work in the information age of ethnic minorities. Therefore, it is necessary to attach importance to it and attach importance to the use of national languages and scripts. The language of endangered ethnic minorities is better rescued and protected, and minority languages can be more widely developed and used. It is precisely because of this emphasis on minority languages that the support of big data technology can make He has made great achievements in the fields of translation, publishing, education, journalism, etc., and played an important role in safeguarding national unity and maintaining social stability, achieving harmony and stability among all ethnic groups.

References

1. Tan, Q., Guo, W., Sun, P.: The research on the construction of monitoring and evaluation system for the operation of marine economy in Liaoning Province. Asian Agric. Res. **7**(1), 39–43 (2015)
2. Jian, L.I., Wang, M., Yao, R., et al.: Optimization of collision detection algorithm based on hybrid hierarchical bounding box under background of big data. J. Jilin Univ. **55**(3), 673–678 (2017)
3. Mulder, H.A.: Is GxE a burden or a blessing? Opportunities for genomic selection and big data. J. Anim. Breed. Genet. **134**(6), 435–436 (2017)
4. Zhang, M.: Agricultural and rural informatization construction in China. Asian Agric. Res. **07**(10), 22–24 (2015)
5. Liu, Y.X., Wang, H.: Green development of western cities from the perspective of informatization and greenization. Ecol. Econ. (3), 4 (2016)
6. Li, H., Zhang, Q., Zheng, Z.: Research on enterprise radical innovation based on machine learning in big data background. J. Supercomput. **14**, 1–15 (2018)
7. Feng, X., Wei, W., Bekele, T.M., et al.: Big scholarly data: a survey. IEEE Trans. Big Data **3**(1), 18–35 (2017)
8. Marquez, J.L.J., Carrasco, I.G., Cuadrado, J.L.L.: Challenges and opportunities in analytic-predictive environments of big data and natural language processing for social network rating systems. IEEE Lat. Am. Trans. **16**(2), 592–597 (2018)

9. Timmer, B., Hickson, L., Launer, S.: Hearing aid use and mild hearing impairment: learnings from big data. J. Am. Acad. Audiol. **28**(8), 731–741 (2017)
10. Kamali, P., Curiel, D., van Veldhuisen, C.L., et al.: Trends in immediate breast reconstruction and early complication rates among older women: a big data analysis. J. Surg. Oncol. **115**(7), 870–877 (2017)

Quantity Calculation of Robots Based on Queuing Theory in Intelligent Warehouse

Quan Jiao$^{(\boxtimes)}$ and Yuebo Wang

Automotive Sergeant School, Military Traffic Academy, Bengbu 233011, China
JQuan0107@163.com

Abstract. In the intelligent warehouse, the quantity of intelligent handling robot configuration is not scientific, causing adverse shelf handling and increasing waiting time, which reduces the picking efficiency and increases the cost. Thus, in this paper, the problem of quantity demand analysis of robots in intelligent warehouse is described and simplified to queue theory. Then, considering the factors of the numbers of order, the efficiency of picking and the cost of equipment, the model of quantity demand analysis of robots has been established. Based on the queue theory, the main index and the objective function of the model is calculated with different number of robots in intelligent warehouse, by the simulation program in MATLAB. So that the optimal number configuration of robots can be obtained, solving the problem of low picking efficiency and high cost during the homework peak period.

Keywords: Queuing theory · Intelligent warehouse · Conveying robot · Quantity analysis

1 Introduction

In the intelligent warehouse, the picking operation mainly adopts the method of "cargo-to-person", in which the robot carries the moving shelf (pallet) to the picking platform, and then the picking operation is completed manually [1]. This picking method has high efficiency, saves manpower effectively and reduces operation cost. However, in the process of operation in the logistics center of electric business, the phenomenon of low picking efficiency and high cost has risen. In addition to the robot path planning, the main problem relies on the unscientific quantity configuration of intelligent handling robot, which leads to adverse shelf handling and increasing waiting time, and makes the order picking low efficiency.

Thus, it is necessary to analyze the demand for mobile robots in the intelligent warehouse, and establish a calculation model considering the orders number, picking efficiency, cost, etc. Through solving the calculation, we can obtain reasonable robots number, and solve the problems such as the decrease in picking efficiency and the increase in cost in the peak period of operation.

Presently, the research on the handling robots in the intelligent warehouse, mainly focus on the design and implementation of the handling robot [2, 3], the cooperative scheduling of multi-task robots [4, 5], the optimization of warehouse layout [6, 7], and

© Springer Nature Singapore Pte Ltd. 2020
M. Atiquzzaman et al. (Eds.): BDCPS 2019, AISC 1117, pp. 1152–1157, 2020.
https://doi.org/10.1007/978-981-15-2568-1_159

the algorithm of robot path planning [8–10], while the research on quantity calculation of robots in intelligent warehouse is rare.

Firstly, this paper describes the problem of quantitative demand analysis of robots in intelligent warehouse, which can be simplified to queue for multiple services desks by abstracting orders and handling robots to customers and service desk. Secondly, this paper establishes the analysis model of robot quantity demand, from the perspective of picking cost, considering time cost, distance cost, and equipment cost. Thirdly, the main index and its calculation formula in the process of quantitative analysis of intelligent warehouse robots are given, based on the basic theory of queuing theory. At last, the simulation program was established by MATLAB, to calculate the system indexes and objective functions of some intelligent warehouse with different number of robots, and obtain the optimal configuration number of robots.

2 Problem Description

The intelligent warehouse generally adopts the 'goods on delivery' picking method. Therefore, the main task of the handling robot is to move the goods to the picking personnel's operation area according to the orders assigned by the system. If the orders arrived in the warehouse are considered as customers, then a handling robot can be seen as a service desk, the calculation of handling robots can be abstracted to the problem of multi-channel waiting queue, where multi-channel can be seen as multiple service desks, that is, multiple robots in the intelligent warehouse. Multichannel queuing problem has the same characteristics as the single channel standard model.

In the intelligent warehouse, the quantity analysis of robots in intelligent warehouse majors on the service situation of orders with different number of handling robots. In the premise of order picking requirements, the goals of problem can be expressed as minimizing the waiting time in the system or the order length of waiting for chosen, meanwhile trying to improve the utilization rate of the robot, avoiding the idle resources, to achieve high investment benefit. Therefore, in the process of research, it is assumed that the artificial picking platform can always meet the picking requirements, and only the robot service orders are considered. At this point, the quantity demand problem of warehouse robots can be abstracted into a multi-desk queuing problem, that is, M/M/C model.

3 Quantitative Demand Analysis Model of Intelligent Warehouse Robots

From the perspective of picking cost, the objective function of calculating the demand quantity of intelligent carrying robot is established. Usually, the cost of picking includes time cost, distance cost and equipment cost.

Let t_N represents the average service time of the order when there are N handling robots in the system. The longer the service time, the higher the time cost. t_{min} is the minimum service time for the system. Let $t_s = \frac{t_N}{t_{min}}$, and t_s represents the ratio between the average service time of the current system and the minimum service time of the system. The larger the ratio is, the longer the average service time of each order is, which means, the higher the time cost is.

The equipment cost is embodied in the picking efficiency of the system, which can be measured by the average idle rate p_0 of each robot in the system, with $p_0 = 1 - r$, where r represents the working probability of a single robot and P_0 reflects the idle degree of each robot in the system. The larger the value is, the lower the utilization degree of intelligent handling robot is, the higher the picking cost is.

In summary, the objective function of the quantity analysis model of demand of intelligent transport robot can be established as follows:

$$minF = t_s + P_0 \tag{1}$$

Assuming that, the arrival of the order obeys Poisson Distribution, based on the theory of multi-desk queuing system, and the picking operation cost of the system can be calculated when the system takes different numbers N of intelligent handling robots. When F is the minimum, the N is the optimal number.

4   Calculation of the Number of Robots Based on Queuing Theory

Assuming that, the orders arriving in the warehouse obey Poisson Distribution, the arrival rate is λ; There is C robots in the warehouse, and each robot works independently of each other, with the service rate μ of each robot is the same, then the average service rate of the whole warehouse of the order is $C\mu$. Let $\rho = \frac{\lambda}{C\mu}$ as the service strength of the system or the average utilization rate of the service organization, Obviously, only when $\rho < 1$, then the queue will not be infinite.

The related symbol definitions are listed below:

λ: order arrival rate; μ: service rate of a single robot; C: number of robots; ρ: system service intensity; L_s: average order; L_q: average number of waiting orders; W_s: average order stay time; W_q: average queuing time for orders.

Defined P_n as the probability that there is N orders in the system. According to the queuing theory, the following theory can be obtained:

$$P_0 = \left[\sum_{k=0}^{C-1} \frac{1}{k!} \left(\frac{\lambda}{\mu}\right) + \frac{1}{C!}\frac{1}{1-\rho}\left(\frac{\lambda}{\mu}\right)^C \right]^{-1} \tag{2}$$

$$P_n = \begin{cases} \frac{1}{n!}\left(\frac{\lambda}{\mu}\right)^n P_0 & n \le C \\ \frac{1}{C!C^{n-C}}\left(\frac{\lambda}{\mu}\right) P_0 & n > C \end{cases} \tag{3}$$

The calculation formula of the operating index of the system is as follows:

$$L_s = L_q + \frac{\lambda}{\mu} \tag{4}$$

$$L_q = \sum_{n=C+1}^{\infty} (n - C)P_n = \frac{(C\rho)^C \rho}{C!(1-\rho)^2} P_0 \tag{5}$$

$$W_q = \frac{L_q}{\lambda} \tag{6}$$

$$W_s = \frac{L_s}{\lambda} \tag{7}$$

In the process of quantity analysis of intelligent warehouse robots, the above indexes are mainly analyzed :

(1) average captain L_s, which refers to the average order quantity in the system, including the sum of the orders waiting and being selected.
(2) average queue length L_q, which refers to the number of orders waiting for picking in the system.
(3) average stay time W_s, which refers to the average time from the arrival of the order to the completion of picking.
(4) average waiting time W_q, refers to the average time from the arrival of the order to the picking.

5 Case Analysis

In the simulation experiment, it is assumed that the average number of orders arriving per hour is 625, and the number of orders that a single handling robot can pick and complete per hour is 20, that is $\lambda = 0.0016$, $\mu = 0.05$. At the same time, the following assumptions are made in the simulation experiment:

(1) Only one type of goods per order;
(2) The selection of each commodity on one shelf can be satisfied at one time. That is, the picking task of each order can be completed by a carrying robot by moving the shelf once.

The simulation program was established by MATLAB, and the queuing theory was used for calculation. When different Numbers of robots were taken from the intelligent warehouse, various indexes of the system were shown in Table 1 below.

Table 1. Various indexes of the system with different Numbers of robots

C	L_s	L_q	W_s	W_q	P_0
32	66.62	35.37	6.40	3.40	0.00
33	43.30	12.05	4.16	1.16	0.00
34	37.27	6.02	3.58	0.58	0.00
35	34.67	3.42	3.33	0.33	0.01
36	33.31	2.06	3.20	0.20	0.07
37	32.53	1.28	3.12	0.12	0.12
38	32.06	0.81	3.08	0.08	0.16
39	31.76	0.51	3.05	0.05	0.19
C	L_s	L_q	W_s	W_q	P_0
40	31.58	0.33	3.03	0.03	0.21
41	31.46	0.21	3.02	0.02	0.23
42	31.38	0.13	3.01	0.01	0.25
43	31.33	0.08	3.01	0.01	0.27
44	31.30	0.05	3.00	0.00	0.29
45	31.28	0.03	3.00	0.00	0.30
46	31.27	0.02	3.00	0.00	0.32
47	31.26	0.01	3.00	0.00	0.33
48	31.26	0.01	3.00	0.00	0.35
49	31.25	0.00	3.00	0.00	0.36

Based on the demand analysis model for the number of intelligent warehouse robots established above, the objective function and various indexes in the system with different number of robots can be obtained as shown in Table 2 below.

Table 2. Objective function and system indexes under different robot Numbers

C	L_s	L_q	W_s	W_q	P_0	$t_s(t_{\min}=3)$	F
32	35.37	66.62	6.40	3.40	0.00	2.13	2.13
33	12.05	43.30	4.16	1.16	0.00	1.39	1.39
34	6.02	37.27	3.58	0.58	0.00	1.19	1.19
35	3.42	34.67	3.33	0.33	0.01	1.11	1.12
36	2.06	33.31	3.20	0.20	0.07	1.07	1.14
37	1.28	32.53	3.12	0.12	0.12	1.04	1.16
38	0.81	32.06	3.08	0.08	0.16	1.03	1.19
39	0.51	31.76	3.05	0.05	0.19	1.02	1.21
40	0.33	31.58	3.03	0.03	0.21	1.01	1.22
41	0.21	31.46	3.02	0.02	0.23	1.01	1.24
42	0.13	31.38	3.01	0.01	0.25	1.00	1.25
43	0.08	31.33	3.01	0.01	0.27	1.00	1.27
44	0.05	31.30	3.00	0.00	0.29	1.00	1.29
45	0.03	31.28	3.00	0.00	0.30	1.00	1.30
46	0.02	31.27	3.00	0.00	0.32	1.00	1.32
47	0.01	31.26	3.00	0.00	0.33	1.00	1.33

As can be seen from Table 2, according to the proposed warehouse robot quantity demand analysis model, when the order arrival rate $\lambda = 0.0016$, and the service rate $\mu = 0.05$ of the robot, the objective function value of the robot number C = 35 is the minimum value F = 1.12. It can be seen comprehensively from the selection efficiency and cost, the optimal robot configuration number is 35.

6   Conclusion

Based on the comprehensive consideration of order quantity, picking efficiency and robot purchase cost, an intelligent warehouse robot quantity demand analysis model is established. Based on the queuing theory, the average length of the system L_s, the average duration of the order W_s and other indicators are analyzed, and the objective function of the model is calculated, so that the optimal robot number configuration can be considered under the factors of picking efficiency and picking cost. It can not only meet the demand of picking orders in the intelligent warehouse, but also avoid the increasing cost caused by long waiting time. Thus, it is certain that each robot can be fully utilized, excessive investment and waste of resources caused by can idle robots be avoided.

References

1. Pan, C.: Simulation study on picking path planning of warehousing logistics robot. North China University (2017)
2. Wang, X.: Design and implementation of intelligent and agile storage system autonomous handling robot. School of Control Science and Engineering. Shandong University (2017)
3. Jin, Q., Ren, J., Yuan, M.: Design of automatic sorting and handling robot system in complex competition tasks. Light. Ind. Mach. **37**(2), 17–23 (2019)
4. Gu, D., Zheng, W.: Overview of multi-mobile robot cooperative handling technology. J. Intell. Syst. **14**(1), 20–27 (2019)
5. Chen, Q.Q., Qian, T.H., Zhang, S.Z.: Multi-robot warehouse scheduling based on reinforcement learning. Mod. Electron. Technol. **42**(14), 165–168 (2019)
6. Guo, Y.: Study on standby strategy and warehouse layout optimization of intelligent storage system. School of Management, Huazhong University of Science and Technology (2016)
7. Lingxing, Z., Liangwei, Z.: Optimization simulation of intelligent storage system based on Flexsim. Softw. Guide **17**(4), 161–163 (2018)
8. Zhang, D., Sun, X., Fu, S., et al.: Multi-robot collaborative path planning method in intelligent warehouse. Comput. Integr. Manuf. Syst. **24**(2), 410–418 (2018)
9. Tang, S., Liu, G., Wang, Z.: Path planning and realization of picking robot in e-commerce warehouse. Logist. Technol. **37**, 85–87 (2018)
10. Lu, W., Lei, J., Shao, Y.: Path planning of mobile robot based on improved A* algorithm. J. Orarmament Eng. **40**(4), 197–201 (2019)

Some Thoughts on 'Intelligent Campus' of Military Academy

Jun Liu(✉), Xuexin Zhang, and Meng Jin

Basic Department, Changchun Institute of Engineering and Technology,
Changchun, China
332552083@qq.com, 876728255@qq.com, 525268710@qq.com

Abstract. In recent years, smart campus is gradually prevailing in our country, many colleges and universities and even primary and secondary schools have carried out the construction of 'Smart campus'. In recent years, military academies have begun to carry out exploratory construction of 'Smart campus', but the pace of construction is slow due to various reasons. This paper puts forward some Suggestions and reflections on the construction of 'Smart campus' in military academies.

Keywords: Intelligent campus · Military education · Intelligent terminal

1 Introduction

Smarter Planet came into being in 2008 when IBM came up with its Smarter Planet strategy. In 2010, zhejiang university proposed to build an exciting intelligent campus in the 12th five-year plan of informatization [1–4]. In this school, such a blueprint is drawn: ubiquitous online learning, innovative online research, transparent and efficient school management, colorful campus culture, convenient and thoughtful campus life [5–8]. Nanjing university of posts and telecommunications, tongji university and other universities have also started to build smart campuses [9–11]. In recent years, a considerable number of colleges and universities across the country have carried out the construction of 'Smart campus' and achieved some results. However, a large number of colleges and universities blindly follow the trend in the process of construction, which is not practical and wastes a lot of resources. So, what is a "Smart campus"?

2 What Is 'Smart Campus'?

Currently, there is no unified standard definition of smart campus. For its connotation and characteristics, experts and scholars in different research fields give different definitions with different emphases:

From the perspective of the Internet of things, in order to highlight its intelligent perception function, the smart campus is positioned as a campus that integrates

M. Atiquzzaman et al. (Eds.): BDCPS 2019, AISC 1117, pp. 1158–1163, 2020.
https://doi.org/10.1007/978-981-15-2568-1_160

information technology highly, deeply integrates information application, and constructs a campus with the characteristics of network, information and intelligence with extensive perception of information terminals.

From the perspective of smart education, smart campus is regarded as a learning space divided according to different scales, just like smart classrooms and smart terminals. It is an integral part of smart education and serves smart education.

From the Angle of management wisdom campus is defined as a campus model, mainly through the use of related technologies such as cloud computing, Internet of things to change all kinds of resources of teachers, students and campus interactive way, the school's teaching and scientific research, management services, all kinds of resources and application system of process reengineering and integration, improve the accuracy of all kinds of application interaction, flexibility, and response speed, so as to realize the intelligence services and management. Comprehensive most scholars, the core of 'Intelligent campus' is based on people-oriented, it with learners as the core goal of innovative development, pay attention to the top-level design and the combination of the actual demand, will be a new generation of information technology as a tool and means of omni-directional integration of campus information resources, is the fusion of the new mode of teaching, management and service, is the advanced form of education informatization. The connotation relationship of smart campus in universities is shown in Fig. 1, and the functional system of smart campus is shown in Fig. 2.

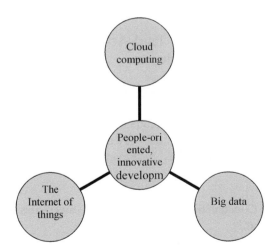

Fig. 1. Connotation relation of smart campus

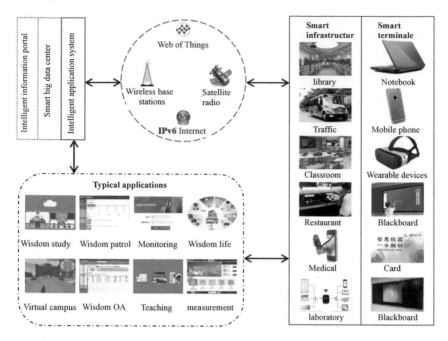

Fig. 2. Function system of smart campus

3   Problems That Should Be Paid Attention to in the 'Smart Campus' of Military Academies

Can be seen from the above meaning 'Intelligent campus', 'Intelligent campus' is based on the Internet, mobile Internet, big data and cloud computing platform built by a variety of mature, the underlying foundation of the construction of the relatively strong, but the above resources at present can only say that just getting started in the army, to the construction of 'Intelligent campus' to strengthen infrastructure construction in the first place.

3.1   The Network Construction of Colleges and Universities Should Be Promoted Steadily

With the gradual advancement of digitized informationized military, the existing military network resources are far from satisfied with the informationized process of colleges and universities, and have become an important factor restricting the modernization of military academies. Therefore, it is extremely urgent to accelerate the networking process of colleges and universities. However, the network is a double-edged sword. While promoting the construction of the network, the security of the network must be ensured. When constructing the mobile campus network, we cannot blindly seek for new things.

3.2 The Construction and Management Organizations Shall Have Clear Rights and Responsibilities

The existing digital construction of military academies is often the independent construction of each department, you build your, I build mine, mutual data is not Shared, agreement is not general, resulting in the so-called 'digital' is only the department of the digital, even after the completion of not only did not improve the learning and work efficiency of teachers and students, but also increased the new burden. Therefore, it is necessary to clarify the rights and responsibilities of the construction of 'Smart campus' in colleges and universities. Who will build it? Who is going to tube? Make it clear that construction and maintenance should be carried out by the same department, so as to avoid disputes in the future.

3.3 Late Upgrade and Maintenance Should Follow

The construction of 'Smart campus' itself is a long-term project, affected by various factors, the construction of the initial stage will inevitably be ill-considered phenomenon. Therefore, the late maintenance and improvement is a far-reaching and long-term process. Just as the system needs to be constantly patched and improved, the construction of 'smart campus' also needs to be constantly updated and improved. Unlike many current military information systems, they are no longer maintained and upgraded after the completion of construction, and they do not check the gaps and make up for the new problems. As a result, the user experience is poor, which gradually becomes marginal and finally leads to the failure of the project.

3.4 Conduct Comprehensive Research Before Construction

At present, many unsuccessful 'smart campus' cases in China are constructed without comprehensive investigation and understanding of the learning and work demands of teachers and students directly by referring to other cases, which is good, but not required by teachers and students, resulting in low utilization rate and mere formality. Before the construction of 'Smart campus', a comprehensive investigation should be conducted to understand the demands of teachers and students in study, work and life. Only in this way can we make great efforts to solve the problems urgently so as to achieve targeted goals and truly obtain the user experience of teachers and students.

3.5 Digital Resources Shall Be Constructed Simultaneously with Hardware Facilities

Many unsuccessful cases only focus on how to building hardware equipment, built a huge framework, but ignores the construction of digital resources, users are often not available when using resources effectively, in the long term, cannot bring users convenient system, there will be no more market makes 'wisdom'.

3.6 'People First' or 'System First'

The essence of 'Smart campus' is people-oriented, but to grasp the essence, there are still many system constraints. If we ignore the essence of human-oriented construction in order to consider the system, we will deviate from the direction in the process of construction and eventually become the product of the next form. Therefore, we should coordinate actively, strengthen the demonstration, and abolish the system that is unreasonable or has not adapted to the development of The Times. Only in this way can we avoid constraints in the process of construction.

4   Conclusion

The construction of 'Smart campus' in military academy is not only a new stage of campus information construction, but also a new reform of teaching, management and service mode of military academy under the background of the whole army colleges striving to create 'Double first-class' colleges. We should take this opportunity to clarify the connotation and denotation of the concept of wisdom campus, as well as the relationship between the related concepts, based on the academy's own reality and advantageous for the top-level design, at the same time, the innovation system mechanism, in order to promote the wisdom campus construction, and take this opportunity to accelerate the high level industry characteristic point to create the world first-class university and the first-class discipline, promote our troops fighting capacity generation. In the foreseeable future, smart campus will continue to promote the development of military education information, and further promote the realization of education modernization.

References

1. Wang, S.: Positioning accuracy analysis of ball screw feed system. Dalian University of Technology, Master thesis (2006)
2. Ding, H.: Research on positioning error dynamic measurement and compensation system of x-y table. Shandong University, Master thesis (2008)
3. Zhang, F.: Error analysis and compensation technology research on planar near-field measurement of ultra-low sidelobe antenna. Xi'an University of Electronic Science and Technology, Doctoral dissertation (1999)
4. Jiao, H., Chen, W., Wang, C.: Analysis and discussion on the error analysis of scanning frame for large scale compression field test. Journal of Beijing Technology and Business University (2006)
5. Shi, M., Kong, X.: Discussion on the design of foundation drawing of machine tool to improve the accuracy of machine tool. Standardization and Quality of Machinery Industry (2011)
6. Sun, X., Yan, K., Ding, G.: Analysis on positioning accuracy of machine tool by comprehensive error of ball screw. Mach. Manuf. Technol. (2008)
7. Lei, T.: Automatically detects the thickness of the non-contact development system. University of Electronic Science and Technology, Master's thesis (2013)

8. Zhang, F.: Error analysis and compensation ultralow sidelobe antenna planar near field measurement. Xi'an Electronic Science and Technology, Doctoral thesis (1999)
9. Li, K., Hu, Y., Yuan, F.: Simulator flatness measurement and compensation of Spatial Target Based on weighted least squares. Opt. Electron. Eng. (2015)
10. Wng, X.: Using a laser tracker measuring straightness long way. Appl. Opt. **34**(4) (2013)
11. Liu, C., Li, H.: Straightness space is based on the principle of coordinate conversion error evaluation. Mod. Manuf. Eng. (2013)

Thoughts on Human Resource Management of Enterprises in the Era of Big Data

Xin Rao and Bixia Fan[✉]

Hubei Business College, Wuhan 430079, China
fanbixia@hbc.edu.cn

Abstract. In recent years, the three words of artificial intelligence, big data and driverless have become more and more popular. Therefore, in the aspect of enterprise human resource management, we also need to follow up the changes of the times and innovate constantly. With the advent of the era of big data, the human resource management model and the optimization and upgrading of management organizations are also promoted. Using the latest technology big data to optimize human resource management has undoubtedly become an important direction and choice of enterprise reform in the era of big data. The improvement of big data technology in human resource management can be in many aspects, such as in personnel recruitment, employee performance appraisal, employee skill training, and the reverse of employee leave and shift arrangement, which have very important application value. Therefore, based on big data technology, this paper studies the human resource management of enterprises.

Keywords: Big data · Human resources · Management change · Model innovation

1 Introduction

With the development of science and technology, especially artificial intelligence, big data, cloud computing and driverless technology, etc., the country has achieved unprecedented development. In the era of big data, in the face of billions of megabytes of data generated every day, how to deal with and make good use of these huge data will become a magic weapon for enterprises and even the country. At the enterprise level, through the analysis of data, we can understand the preferences of the public, and use enterprises to formulate relevant research directions; at the national level, through data mining and data analysis, we can understand the basic situation of the people, and through online speech, we can obtain the ideological dynamics of the people, which provides a good reference value for further development of relevant policies.

In recent years, the research of big data is very hot, and the related application papers emerge in endlessly. In document [1], the author applies big data analysis technology to the platform of smart farm, which makes the management of farm more efficient and the production capacity greatly improved. In document [2], the author analyzes the role of big data in decision-making analysis in academia and industry. In document [3], the author applies big data analysis to policy problems, and helps

M. Atiquzzaman et al. (Eds.): BDCPS 2019, AISC 1117, pp. 1164–1169, 2020.
https://doi.org/10.1007/978-981-15-2568-1_161

decision makers to predict and evaluate the effect of policy through data analysis, so as to achieve the best effect. In reference [4], the author investigates big data, which not only provides a global view of major big data technologies, but also provides a comparison based on different system layers (such as data storage layer, data processing layer, data query layer, data access layer and management layer), and classifies the main technical features, advantages, limitations and uses of big data. Discussion. In literature [5], the author studies the security and privacy of big data, and finds that in the process of data collection, storage and use, it is easy to cause personal information leakage, and it is difficult to identify data. How to ensure the security and privacy protection of big data has become one of the hot issues in the current research. Starting from big data, this paper analyzes the security of data and puts forward the security and privacy protection strategies of big data. In literature [6], the author provides a series of reasonable expectations for the role of these two technologies in health care from the research of big data and machine learning. Of course, in recent years, the research on human resource management is also endless. In literature [7], the author proposes another approach to human resource management, which prioritizes practices aimed at improving well-being and establishing positive employment relationships, and proposes that both of these elements are essential. It provides evidence to support the choice of practices and believes that these practices also have the potential to improve individual and organizational performance. In reference [8], the author reviewed the concept of human resource management (HRM) and the related concepts of strategic HRM and international HRM, and determined that the integration of senior experts in the senior management team is a key issue, and provided data on the degree of integration in Europe. In document [9], the author first briefly reviews the conceptual logic that links human resource management (HRM) practices with corporate outcomes, with the aim of highlighting different approaches to RBV in SHRM and strategic human capital documents. Then, we propose a conceptual model to show that human resource management practice is not a simple lever for enterprises to create sustainable competitive advantage, as most strategic human capital research institutes assume. In reference [10], the author puts forward a conceptual model, in which it is clear that the HRM system based on knowledge-based human resource management practice will affect the company's intellectual capital, thus producing higher innovation performance.

This paper analyzes the current enterprise human resource management problems through big data, using the mainstream big data technology to improve the level and efficiency of enterprise human resource management.

2   Method

2.1   Big Data in the Era of Big Data

In recent years, the theoretical research and application of big data has set off a huge wave in academic circles, political circles and capital circles at home and abroad, which indicates the real arrival of the era of big data. As early as 1980, Alvin, a famous American scholar, wrote in a book called "the third wave" that big data will succeed in the third wave, but at that time, the level of science and technology was limited, big data lacked the support of necessary computing power and data sources, and the world at that time did not fully have the conditions to embrace the era of big data. In the early 1990s, bill, the father of data warehouse, put forward the theory of big data for the first time. In his opinion, the difference between big data and data warehouse is that data warehouse may be more like a process of expression, which is the integration, storage, processing and analysis of business data distributed in various departments of the enterprise. In 2005, John and Kristen wrote in their book "all inclusive data" that the arrival of big data will definitely affect the development of enterprises and people's lives. After that, with the advent of the Internet era, the popularity of social media and smart phones provides strong support for data sources, and the form of data also presents a trend of diversification, scale and diversification. In September 2008, the concept of "big data" was put forward in nature and published as an album.

2.2 Understanding of Big Data

At present, there are different views and standards for the definition of big data in industry and academia. Only from the definition of big data itself, the more authoritative explanation can refer to the description on Wikipedia: big data, or huge data, refers to the data collection that can not be grabbed, managed and processed by conventional software tools in the allowed time. Scholar McKinsey defines big data as: big data refers to those huge data sets whose size exceeds the ability of acquisition, storage, management and analysis of conventional database tools. It should be noted that big data does not only refer to the large amount of data. For example, if the size exceeds a specific TB, it can be called big data. This understanding is obviously unreasonable.

There are also many explanations for the description of the characteristics of big data. In 2001, scholar Doug pointed out that there will be three challenges and opportunities in the growing process of data, the first is massive, the second is high-speed, and the third is diversity. These three characteristics are the characteristics of the early definition of big data. After that, the researchers extended the three features of big data to four features. For example, scholars Brian and Boris jointly published a report called "CIO, please use big data to expand the digital vision", in which they added a variability to the characteristics of big data. In the big data era written by Victor, he said that big data should have the characteristics of low value density and high business value in addition to the three early characteristics. In the process of attribute analysis of big data, IBM proposes that big data also needs to be authentic.

In terms of the attribute level of big data, big data should include two attributes: technical attribute and social attribute. Big data is also a technical term in the IT industry at the beginning, which embodies more of its technical attributes at the beginning. However, more scholars believe that big data shows some columns that can represent the most advanced modern information technology system, mainly in the current cloud technology, distributed processing technology, perception technology, data storage technology and other technical fields. At the same time, big data also has social attributes. For example, in Albert's book "breaking out: new thinking of foreseeing the future in the era of big data", it points out that in the era of big data, data mobility is very large, and everyone can become self-Media, and data presents personalized development. Big data is widely connected with the laws of people and society, and has social attributes.

3 Experiment

3.1 Big Data Thinking

Through big data analysis of human resource data, appropriate transformation of the original fixed management thinking mode, formation of advanced big data thinking mode, and effective improvement of the existing management mode and management methods, has become an important challenge for enterprises in human resource management to meet the era of big data. Big data thinking mainly includes the following aspects: human resource managers must have keen observation and foresight for the change of talent demand at the strategic level; regard big data technology as the core analysis technology of enterprises; human resource management needs to effectively apply the seven characteristics of big data to talent absorption and talent cultivation.

3.2 Enrich Human Resource Management Means

The application of big data to human resource management will promote the innovation of human resource management and the upgrading of decision-making. Victor said: big data is the root of people's new cognition and value creation; big data is also an important method to change the market environment, organizational structure and the relationship between government and citizens, and stressed that the essence of big data is actually prediction. Through the multi-dimensional warehouse function of big data technology, the talent data is modeled to improve the efficiency of enterprise management.

4 Result

Result 1: Lower turnover

By using big data to analyze human resource data, we upgraded the original management model. The improved management mode is mainly reflected in optimizing the performance appraisal, attendance methods and skills improvement training of employees. Through a one-year data analysis of the implementation of the improved management method, we find that the turnover rate of the company is greatly reduced. As shown in Fig. 1 below:

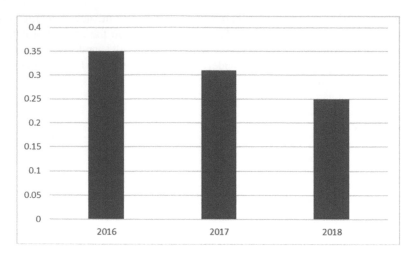

Fig. 1. Statistics of turnover rate

Result 2: Employee satisfaction improvement.

By using big data to analyze human resource data, we upgraded the original management model. The improved management mode is mainly reflected in optimizing the performance appraisal, attendance methods and skills improvement training of employees. Through a one-year data analysis of the implementation of the improved management method, we found that the company's employee satisfaction has been greatly improved. Figure 2 below:

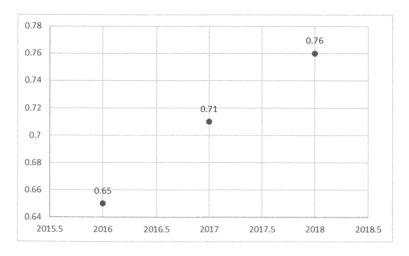

Fig. 2. Satisfaction statistics

5 Conclusion

The era of big data has come, and enterprise management should also follow the trend, using big data technology to update management mode and optimize organizational structure. This paper analyzes the application of big data in human resource management of enterprises, and through the implementation of optimized management mode, confirms the effectiveness of management under big data analysis. Of course, while big data benefits human beings, there are also many hidden dangers, such as personal privacy. Through big data analysis, personal behavior habits and other privacy are mined out by data, which is very worthy of attention.

References

1. Wolfert, S., Ge, L., Verdouw, C., et al.: Big data in smart farming–a review. Agric. Syst. **153**, 69–80 (2017)
2. Sivarajah, U., Kamal, M.M., Irani, Z., et al.: Critical analysis of Big Data challenges and analytical methods. J. Bus. Res. **70**, 263–286 (2017)
3. Athey, S.: Beyond prediction: using Big Data for policy problems. Science **355**(6324), 483–485 (2017)
4. Oussous, A., Benjelloun, F.Z., Lahcen, A.A., et al.: Big Data technologies: a survey. J. King Saud Univ.-Comput. Inf. Sci. **30**(4), 431–448 (2018)
5. Zhang, D.: Big Data security and privacy protection. In: 8th International Conference on Management and Computer Science (ICMCS 2018). Atlantis Press (2018)
6. Beam, A.L., Kohane, I.S.: Big Data and machine learning in health care. JAMA **319**(13), 1317–1318 (2018)
7. Guest, D.E.: Human resource management and employee well-being: towards a new analytic framework. Hum. Resour. Manag. J. **27**(1), 22–38 (2017)
8. Brewster, C.: The integration of human resource management and corporate strategy. In: Policy and Practice in European Human Resource Management, pp. 22–35. Routledge (2017)
9. Delery, J.E., Roumpi, D.: Strategic human resource management, human capital and competitive advantage: is the field going in circles? Hum. Resour. Manag. J. **27**(1), 1–21 (2017)
10. Kianto, A., Sáenz, J., Aramburu, N.: Knowledge-based human resource management practices, intellectual capital and innovation. J. Bus. Res. **81**, 11–20 (2017)

Intelligent Outdoor Trash Can

Jiayue Wang, Dong Xie[✉], and Shilin Li

Information School, Hunan University of Humanities, Science and Technology,
Loudi 417000, Hunan, China
287566288@qq.com

Abstract. With the continuous development of intelligence, the trash cans in life can no longer meet the needs of people. This paper introduces the design of intelligent trash can based on STM32 single-chip microcomputer. The design is mainly through infrared detection module, GPRS module, human body sensing module, GPS module and so on. The trash can realizes the intelligent flip cover, senses the human body to avoid obstacles, and the mobile APP checks whether the trash can is full. The analysis of the trash can shows that the design can improve the efficiency of garbage disposal, facilitate the treatment of people's daily garbage, and has good use value.

Keywords: Intelligent · App · Single chip microcomputer · Trash can

1 Introduction

With the continuous improvement of the material level and the accelerating pace of life, we will inevitably have a lot of domestic garbage in our daily lives. garbage also need to be disposed of. At this time, the trash cans play a big role. In our daily life, the dustbin may become an indispensable part of our life. However, there is a lot of domestic trash can in our country. If the garbage is not cleaned up in time in tourist areas, the overflow trash will give off a foul smell. To manually check the status of the trash can requires a lot of manpower and material resources. To this end it is necessary to know the status of the trash can in real time [1].

To solve the above problem this paper design a practical very strong at the same time around the GPRS wireless communication, to the trash can has a capacity of real-time state reminder and the positioning of bin location design, makes every effort to not only can satisfy the above function, image and beautiful, can and the surrounding environment be in harmony are an organic whole, become a part of the beautify the environment.

From the current situation of smart trash, smart trash still needs to be improved. Therefore, this paper designs a multi-purpose and multi-functional outdoor intelligent trash can, which allows users to monitor the status of the garbage in the trash can in real time, determine whether the garbage needs to be removed according to its status, and achieve the garbage full alarm and remind, GPRS wireless communication and use two-dimensional code to locate the location information of each trash can.

M. Atiquzzaman et al. (Eds.): BDCPS 2019, AISC 1117, pp. 1170–1176, 2020.
https://doi.org/10.1007/978-981-15-2568-1_162

2   The System Design of Intelligent Outdoor Trash Can

The system consists of single-chip microcomputer STM32F103 as the minimum system, the whole system includes advertising part, infrared detection module, alarm part, two-dimensional code, mobile APP control part, cloud server, user, GPRS module, GPS module, human body sensing module, night illumination part composition. The system block diagram is shown in Fig. 1.

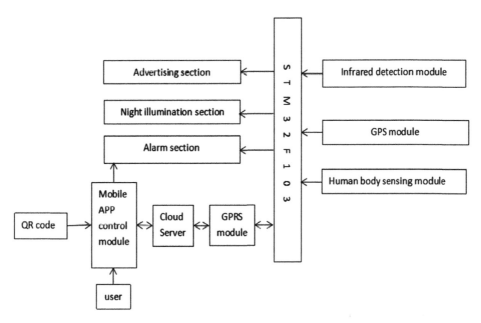

Fig. 1. Overall system block diagram

The communication module between mobile APP and microcontroller includes mobile APP control terminal, cloud server, GPRS module and microcontroller STM32F103. The control terminal of mobile APP obtains GPRS and microcontroller for communication to obtain the state of garbage in the trash can, its geographical position and the amount of power, so as to monitor the whole state of the trash can real time [2]. The communication structure between APP and MCU is shown in Fig. 2 [10].

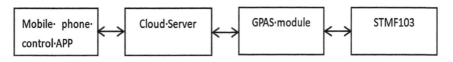

Fig. 2. Communication structure between APP and MCU

3 Hardware Design

3.1 Circuit Design of the Main Control Part

The STM32F103 MCU circuit used in this system is shown in Fig. 3 [9].

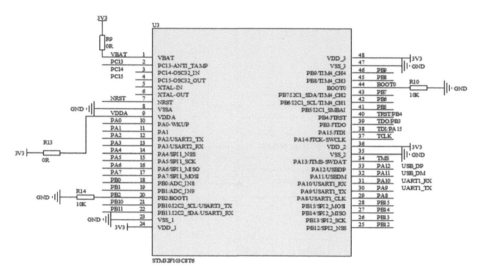

Fig. 3. Microcontroller STM32F103 circuit diagram

3.2 Microcontroller Minimum System

The smallest system of the single chip microcomputer is to let the program inside the single chip microcomputer run, the minimum configuration required [3]. The minimum system consists of power supply, crystal oscillator circuit and reset circuit.

3.3 Liquid Crystal Display Circuit

Nowadays, we can see all kinds of display screens, digital tubes, LED, LCD, etc. This design adopting LCD12864, LCD12864 is a kind of 4-bit/8-bit parallel, 2-wire or 3-wire serial interface mode. This paper USES the module flexible interface mode [4], and also USES its simple, convenient operation instructions, can form the whole Chinese human-computer interaction graphical interface. The LCD circuit adopted in this paper is shown in Fig. 4.

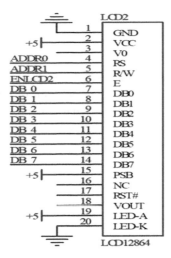

Fig. 4. The LCD circuit diagram

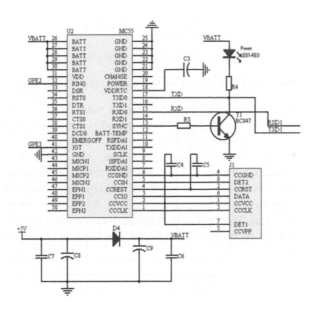

Fig. 5. GPRS circuit diagram

3.4 GPRS Circuit

Wireless communication GPRS module circuit, the resistance in the circuit R5 is the current limiting resistance, when there is a clock signal generated, the transistor on, LED light, there is synchronous signal generated, GPRS circuit using 5 V power supply for power supply, when the power is off, C3 energy storage components can give real-time clock circuit power supply [5]. Its circuit is shown in Fig. 5.

4   Software Design

4.1   Main Program Design

Software design part of this paper can be divided into main program and each module subroutine. The main function of the main program of this paper is to complete the initialization of each module. In order to make the subroutine of each module realize its functions smoothly, LCD12864 Display subroutine, button scan subroutine. The main program flow chart is shown in Fig. 6 [7].

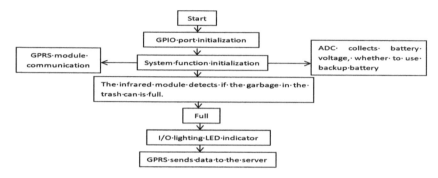

Fig. 6. Main program flow charts

4.2   Find the Program Flow Chart of the Untreated Trash Can

When the trash placed in crowded public places or some famous scenic spots, every holiday, as a result of the sanitation workers workload is bigger, personnel deployment, causing the trash can waste state not to clean up in time. [8] this will affect our living environment and the visitors of the mood, in this paper, the research design of this kind of new intelligent outdoor trash can he a real-time software can find untreated trash cans garbage to the dustbin of state judgment, it steps is to find the untreated bin program is initialized first, then look at the map, to determine whether a trash can is full, if the trash can is full, The light next to the trash can is red, otherwise the light next to the trash can will not be on [6]. The flowchart for finding an untreated trash can is shown in Fig. 7.

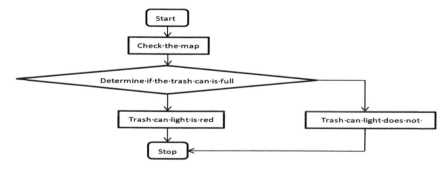

Fig. 7. Find the flow chart of the untreated trash can program

5 Test Results

The design through the design of the hardware circuit part of the outdoor smart trash can, the software part design, using the STM32F103C8T6 microcontroller main control chip to realize the intelligent function of the trash can, through hardware circuit debugging and software debugging. This design tested three sets of data. Through testing, there are several infrared detection trash cans full, whether the mobile APP is alarming, whether the GPS positioning is accurate, and the real state in the trash.

According to the survey results in the above table, we should conduct many measurements. The above table is a small part of our knowledge, which basically meets our ideal goals. It can realize the intelligent management of the trash can, which greatly makes our trash can become Humanization implies that our living standards are constantly improving. We are constantly developing towards intelligence and comforting ourselves while beautifying the environment. The data obtained through testing are shown in Table 1.

Table 1. Data test

Test data sets	1	2	3
There are several infrared detection trash cans full	4	2	5
Whether the mobile APP alarms the police	Yes	No	Yes
Whether GPS positioning is accurate	Accurate	Deviation is not big	Accurate
Real situation	The garbage is full	Garbage under	The garbage is full

6 Conclusion

The design of this paper is based on the design of smart home, and further proposes the design method of researching new intelligent trash can. This paper realizes the intelligent control of outdoor trash can. Its implementation mainly utilizes sensor technology and automatic control technology. The innovation is to let the trash can be personalized to the user, that is, to solve some of the drawbacks of the ordinary trash can according to the user's needs, for example, the problem of fixed docking of the traditional trash can. The user can operate through the APP interface, and can realize various functions such as automatic obstacle avoidance. In addition, the trash can is equipped with alarm device, sensor device, mobile phone terminal control device and voice control device. The whole system realizes automatic opening and closing of the lid, voice control and other functions, so that people can interact with objects and further improve the shortcomings of traditional trash cans. The outdoor smart trash can really change our life with technology. Its development has taken a big step forward for the Intelligent of home furnishing, contributed to the further development of intelligent city, and opened up a new idea for the comprehensive realization of smart city in the future.

Acknowledgements. This work is supported by the Research Foundation of Education Committee of Hunan Province, China (18C0896), the Foundation of Hunan University of Humanities, Science and Technology (2018).

References

1. Qiu, S., Zhu, Y., Liang, D.: Design of mobile environment monitoring and analysis system based on GPRS. In: IOP Conference Series: Materials Science and Engineering, vol. 490, no. 6 (2019)
2. Chen, Qiang: Design of smart home system based on STM32 single chip microcomputer. Electron. Prod. **23**, 87–88 (2018)
3. Gu, Y.: Study on automatic recognition method of two-dimensional code of elevator control system. Autom. Instrum. (09), 97–100 (2019)
4. Ge, X., Chen, X., Sun, H., Chen, Y., Liu, C.: Design of intelligent waste bin. Mod. Ind. Econ. Informatiz. **8**(16), 38–39 (2018)
5. Zhou, H., Xu, J.: Design of a new type of intelligent trash can. J. Guangdong Univ. Technol. (03), 85–88, 94 (2006)
6. Mao, X., Guo, H., Zhu, W., Song, X.: Intelligent infrared induction control system. Electron. Meas. Technol. (02), 45–49 (2005)
7. Li, S., Zeng, R.: Design of data acquisition system based on STM32 infrared sensor. Ind. Control Comput. **31**(08), 30–31 + 33 (2018)
8. Sun, J., Zhu, B.: Automobile entry and start system based on single-chip Bluetooth module. Shanxi Arch. **45**(10), 253–254 (2019)
9. Xu, Z.: A human body safety warning device. Rural Electr. **27**(03), 27–28 (2019)
10. Wang, Hanqing: Application of GPRS wireless communication technology in expressway electromechanical system. Comput. Knowl. Technol. **13**(20), 208–209 (2017)

Analysis of Backward Payment Model in Traffic Service

Zhe Chen[1] and Yanrong Yan[2(✉)]

[1] School of Economics, Guangdong Ocean University, Zhanjiang, China
[2] Beibu Gulf Economic Research Center, Guangdong Ocean University Cunjin College, Zhanjiang, China
jaoshow@qq.com

Abstract. Backward payment is an emerging traffic operation business model for operators. At present, there is a lack of research on this model in China. This paper discusses the backward payment model base on cost-benefit relation and customer value, analyzes the internal operation principle and interest balance relationship of the model, and briefly describes the development trend of the telecom industry.

Keywords: Back payment · Telecom operator · Traffic · Customer value

1 Introduction

Backward payment, similar to the telecom operator 800 phone mode, operators use their own communication channels to provide traffic resources to mobile terminal users, as a communication bridge between enterprises and users, this business model takes enterprises as payers. [1] The biggest feature is that the traffic cost generated by the terminal users is passed on to the enterprise, and the mobile terminal users have access to information provided by the enterprise at zero cost. Compared with the traditional traffic product, the backward payment mode can remain same service quality and reduce the user cost. Therefore, it is more attraction to mobile terminal users.

2 Characteristics of Backward Payment Mode

The backward payment model is not an innovation in the telecom industry. There are many large supermarkets in metropolis provide free shuttle service to customers. When passengers arrive at the destination city, they usually get a map that highlights the information of the operator or sponsor. These "free" services or products have the common characteristics: Firstly, the cost of the product or service is much lower than the profit generated by the customer's consumption; secondly, the customer behavior patterns are obvious and easy to distinguish; finally, the probability of "free rider" is not high due to the patterns [2].

© Springer Nature Singapore Pte Ltd. 2020
M. Atiquzzaman et al. (Eds.): BDCPS 2019, AISC 1117, pp. 1177–1180, 2020.
https://doi.org/10.1007/978-981-15-2568-1_163

3   Relationship Analysis Among Operators, Users and Enterprises

3.1   Motivation Analysis of Operators

In the process of traditional operator traffic operation, mobile terminal users purchase online traffic from operators for browsing websites, downloading files, watching videos and other online behavior. [3, 4] Among them, information acquisition (watching news, blogging, using WeChat, etc.) as a basic online behavior of mobile terminal users, merging occupies the largest share of the cost of user traffic. In addition to operators, mobile terminal users are not the only beneficiaries in the process of using network information. Internet companies, as information providers, also gain benefits including advertising revenue and offline product sales due to information publicity.

Modern market management use various means to analyze customers' personality and behavior patterns, for expanding profit space. [5] Backward payment model is a successful case of the application of the theory above. Usually, operators are the sellers of information pipeline service and mobile terminal users are the buyers of traffic transactions. The cost of traffic is equal to the customer cognitive value. When the customer value is zero, the operator's revenue can be maximized. In the backward payment model, Internet information providers are traffic buyers. When customer cognitive value of Internet enterprises equals the traffic cost, the operator's revenue can also reach the apex [6].

3.2   Customer Cognitive Value of Internet Enterprises

If the customer's cognitive benefit is much higher than the customer's cognitive price, then the customer will be willing to pay for the product or service, and the customer value will be higher. [7] On the contrary, if the customer is not interested in the product or service, the customer value will be low. Customer cognitive value of Internet enterprises comes from the precise placement of advertisements. The purpose is to change customers' behavior intentions and make their business partners or their own other products beneficial. For example, mobile terminal users click on news channels to get national policy news. As a common operation of the netizens, Internet enterprises cannot determine the customer behavior patterns, and information cannot be precisely delivered at this time either. [8] Mobile terminal users who use keyword "mobile phone cost performance ratio" in Internet search is likely to become consumers of a mobile phone brand. At the same time, Internet enterprises pushing a brand information on the mobile terminal users will greatly improve the accuracy of the delivery. A certain proportion of the mobile terminal user group will purchase a specific brand mobile phones, and the revenue of mobile phone sales enterprises will partly be used to fill the operating costs of Internet enterprises which provide information services.

3.3   Operation Analysis of Backward Payment in Internet Enterprises

Whether it is a traditional model or a backward payment model, the Internet companies that provide information are designed to maximize the benefits of operating revenue.

The most important thing for most companies is the revenue of advertising. [9] It depends on two aspects, one is the user scale of browsing; another is the number of views converted into the volume ratio of the goods. [10] As far as possible, the company strives for more customers to improve the number of view through marketing. In addition, customers who browse will be accurately marketed and become a factual customer under mature operation mode. Focusing on the objectives above, the company's products can be divided into two categories. The first one is a kind of information product that enhances the perceived value in order to meet the general needs of customers and win the market competition. Another one is a kind of information products, for instance soft-text advertisement that are aimed at profit, customer are divided into different groups so the transaction volume will be higher. The former exists to meet customer needs, and the latter exists to change customer intention. While the former one fulfilling customer needs in order to reduce operating costs, the latter one needs to accurately analyze customer group delivery in order to reduce the information acquisition cost of non-target customers. (There is a flood of information in the current Internet market, and customers need to spend a lot on filtering before archiving useful information).

3.4   Comparison of Customer Cognitive Value Between Mobile Terminal Users and Internet Enterprises

Affected by market failure, China's Internet information provision market is far from complete competition. The benefits obtained by mobile terminal users and Internet companies in the process of acquiring information are not consistent. Therefore, in the process of traffic transactions with operators, the customer cognitive value is not equivalent between Internet companies and mobile terminal users.

In the traditional carrier traffic products, the mobile terminal user pays traffic fee to obtain the information on the Internet. But individual access to the information on the Internet does not guarantee that the customer cognitive value of the mobile terminal user can cover the payment cost. In practice, it is the infinite loop game that provide the guarantee. As described above, Internet enterprise can use content and cost control under backward payment mode to cover the payment cost with higher customer cognitive value.

However, due to the difficulty in accurately analyzing and locating the general mobile terminal user group, the operator cannot apply the price discrimination strategy to the mobile terminal user, so as to realize the customer cognitive value infinitely close to the traffic payment cost. Instead, operators can conduct effective price discrimination against Internet companies. There are two main reasons: Firstly, the failure of the telecom market. Three major domestic operators have almost controlled all market shares and have strong pricing advantages. Secondly, operators master the mobile terminal user resources. The acquisition of traffic information is generally under authorization from the operators. The operator can effectively guide the traffic of the mobile terminal user (for example, the pre-installed software, customized mobile phones in operator's Sales, customer service staff sales system), so the operator's partner Internet companies can take the benefit. In the backward payment mode, operators can re-pricing so that the traffic payment cost tends to equal the customer cognitive value of the enterprise.

4 Conclusion

In the model of backward payment, operators, mobile terminal users and Internet enterprises are not in zero-sum games. Due to the reduction of traffic cost, the utility of mobile terminal users' online behavior is improved, so as the consumption of Internet products. After adding traffic cost in operation process of Internet enterprises, the consumption of two kinds of products mentioned above has raised correspondingly. It can be concluded that the profit of Internet enterprises are always positive. Under the backward payment mode, operators have more flexible pricing strategies, so that the traffic fee keeps close to the buyer's customer cognitive value, so as to realize the increase of revenue. Overall, the backward payment traffic management model result in better resource allocation and lower transaction costs than traditional one.

References

1. Ikeda, K., Bernstein, M.S.: Pay it backward: per-task payments on crowdsourcing platforms reduce productivity. In: Proceedings of the 2016 CHI Conference on Human Factors in Computing Systems, pp. 4111–4121. ACM (2016)
2. Ickler, H., Schülke, S., Wilfling, S., et al.: New challenges in e-commerce: how social commerce influences the customer process. In: Proceedings of the 5th National Conference on Computing and Information Technology, NCCIT 2009, pp. 51–57 (2009)
3. Chau, P.Y.K., Cole, M., Massey, A.P., et al.: Cultural differences in the online behavior of consumers. Assoc. Comput. Machinery. Commun. ACM 45(10), 138 (2002)
4. Thubert, P., Lorrain, J.: Method and system for improving traffic operation in an internet environment. U.S. Patent 6,603,769, 5 August 2003
5. Cheung, K.W., Kwok, J.T., Law, M.H., et al.: Mining customer product ratings for personalized marketing. Decis. Support Syst. 35(2), 231–243 (2003)
6. Raivisto, T., Gustafsson, P.: Method and system for sharing transmission revenue between mobile operators and content providers. U.S. Patent 6,968,175, 22 November 2005
7. Yang, Z., Peterson, R.T.: Customer perceived value, satisfaction, and loyalty: the role of switching costs. Psychol. Mark. 21(10), 799–822 (2004)
8. Kwan, I.S.Y., Fong, J., Wong, H.K.: An e-customer behavior model with online analytical mining for internet marketing planning. Decis. Support Syst. 41(1), 189–204 (2005)
9. Drèze, X., Hussherr, F.X.: Internet advertising: Is anybody watching? J. Interact. Mark. 17(4), 8–23 (2003)
10. Robinson, H., Wysocka, A., Hand, C.: Internet advertising effectiveness: the effect of design on click-through rates for banner ads. Int. J. Advert. 26(4), 527–541 (2007)

Computer Simulation System Optimization of the Queuing System Model Based on the Integration of the Priority Service

Xiaohui Tian[(✉)]

School of Network Security and Informatization, Weinan Normal University,
Weinan 714099, Shaanxi, China
txh@wnu.edu.cn

Abstract. In this paper, under the environment of computer simulation system, the M/G/1 queuing system models corrected under the influence of class I false alarm event and class III missed detection event. The exact solutions for the first and second moments of the primary user and secondary user data transmission time are presented, and the (M/G/1-Revised) queuing system model is obtained. Secondly, in the multi-priority secondary user context, a kind of queuing system model based on the integration of the priority service is put forward. And the proof procedure for the system model to minimize the priority service is provided. The simulation and numerical calculation results have verified the effectiveness of M/G/1-R. And BM/G/1-R can obtain the minimum number of interruptions in the priority service and the secondary user data transmission in the computer simulation system.

Keywords: Computer simulation system · Queuing system model · Priority service · Minimization

1 Introduction

How to establish a queuing simulation system that is capable of providing low priority services and number of data interruptions to the secondary user (SU) in the computer simulation system (CSS) environment has become one of the most urgent problems to be solved in the computer systems [1, 2]. At present, the CSS queuing simulation models can be divided into four categories: Two-dimensional Markov model [3], Bernoulli model [4], M/M/1 model [5] and M/G/1 model [6]. Among the four models, only M/G/1 model can be applied to different service time distribution scenarios [7]. In addition, all the four models are integrated with the Preemptive Resume (PR), that is, If PU occurs when SU accepts service, it hangs up immediately and wait for PU service to finish and then continue the data transmission [8, 9]. In the PR mechanism, it is assumed that SU can completely avoid the same frequency interference with PU, and the false alarm probability and missed detection probability are 0 [10].

In this paper, the class III missed detection event is defined on the basis of the existing class I false alarm event and the class II missed detection event. According to the model of PRPM/G/1 queuing system (referred to as M/G/1-C, M/G/1-Classical for

© Springer Nature Singapore Pte Ltd. 2020
M. Atiquzzaman et al. (Eds.): BDCPS 2019, AISC 1117, pp. 1181–1190, 2020.
https://doi.org/10.1007/978-981-15-2568-1_164

short), the influence of the four classes of events on the PU and SU data transmission time is analyzed. And the exact solutions to the first moment and second moment of PU and SU data transmission time are deduced. The modified M/G/1 queuing system model is obtained to solve the problem of the connection between the actual CSS environment and the M/G/1 model. Secondly, in accordance with the multi-priority M/G/1 model, the unit time loss priority service is integrated, and the multi-priority queuing system model is designed using the method of bubble sort. It can determine the new queuing order according to the unit time loss priority service and the system busy rate of each priority user, so as to provide unified solutions to the problem of the number of interruptions in the SU data transmission and the minimization of the priority service.

2   Queuing System Model

The BM/G/1-R put forward in this paper does not allow the SU with high priority level to interrupt the data transmission of the SU with low priority level. And the FCFS queuing mode is implemented for the SU with the same priority level. The arrival of the PU and the SU with each priority level complies with the Poisson distribution of the mean value $\lambda_1, \lambda_2, \cdots, \lambda_N$, and the unit of measurement is arrival/frame. The mean values of the data transmission time of the PU and SU are $E[X_1], E[X_2], \cdots, E[X_N]$, and respectively, the unit of measurement is frame/arrival. The system structure is shown in Fig. 1.

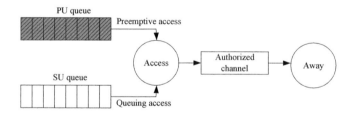

Fig. 1. Queuing system model

In accordance with the 802.22 standard, the SU adopts the PR transmission mode and the periodic spectrum sensing mode in the data transmission process, in which T_S stands for the spectrum sensing time, T_D stands for the data transmission time, and $T = T_S + T_D$ stands for the frame length. SU may be influenced by the PU preemptive simulation in the data transmission process, and the data transmission will be divided into several segments.

3   Construction of the Queuing System Model Based on the Integration of Priority Service

3.1   Forecast Accuracy

The influence of the CSS environment on the seamless connection between the users of high and low priority levels in the traditional queuing system model can be divided into two categories. One category is the SU virtual delay due to the false alarm event, and the other category is the same frequency interference of PU and SU due to the missed detection event, which also lead to the data retransmission, as shown in Fig. 2. For the sake of simplicity, firstly, let all SU priority levels be the same, and the effects of the four classes of events on the PU and SU data transmission are discussed. Finally, the differences of the multi-priority classification and the two priority classification are discussed. In the next section, the definition and the probability of occurrence of these four classes of events are introduced, respectively.

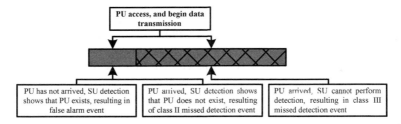

Fig. 2.   Class I false alarm events and class III missed detection events

Class I false alarm event: The length of the PU queue is zero, and the false alarm event occurs during the spectrum detection in the SU data transmission process. And the probability of the occurrence of false alarm event is as the following

$$P_{fa} = P_f \cdot P_r(Q_p = 0) \tag{1}$$

In which Q_p stands for the length of the PU queue, $P_r(Q_p = 0) = 1 - \lambda_p$ $E[X_P] = 1 - \rho_P$, where ρ_P stands for the PU system busy rate. Therefore, the following can be obtained

$$P_{fa} = P_f(1 - \rho_P) \tag{2}$$

The missed detection events sensed by the CSS spectrum can be divided into two categories.

The first class of missed detection events can be divided into two sub-categories: (1) PU is using the channel, the SU queue is empty, the SU arrives during the PU data transmission, spectrum detection is conducted and missed detection event occurs. (2) SU is using the channel, the PU queue is empty, the PU arrives within the spectrum detection time T_S, and missed detection event occurs.

The probability of class I missed detection event is as the following

$$
\begin{aligned}
P_{m1} &= Pr(Q_S = 0) \cdot (1 - exp(-\lambda_S T))P_m \\
&+ Pr(Q_P = 0) \cdot (1 - exp(-\lambda_P T_S))P_m \\
&= [(1 - \rho_S) \cdot (1 - exp(-\lambda_S T)) \\
&+ (1 - \rho_S) \cdot (1 - exp(-\lambda_P T_S))] \cdot P_m
\end{aligned}
\tag{3}
$$

In which Q_S stands for the SU queue length, $\rho_S = \lambda_P E[X_S]$ stands for the system busy rate of the SU, λ_S and $E[X_S]$ stand for the mean value of the arrival strength of the SU and the time of data transmission, respectively.

Class II missed detection event: PU and SU queues are not empty. PU is using the channel. The SU at the head of the queue shows missed detection event. The probability of the occurrence of class II missed detection event is as the following

$$
P_{m2} = Pr(Q_S \neq 0) \cdot Pr(Q_P \neq 0) \cdot P_m = \rho_P \cdot \rho_S \cdot P_m
\tag{4}
$$

Class III missed detection events: The PU queue is empty, the SU is in the process of data transmission, the PU does not arrive within T_S. However, it arrives within T_D, resulting in the same frequency interference between the SU and the PU. The probability for the occurrence of the class III missed detection event is as the following

$$
\begin{aligned}
P_{m3} &= exp(-\lambda_P T_S) \cdot (1 - exp(-\lambda_P T_D)) \cdot Pr(Q_P = 0) \\
&= exp(-\lambda_P T_S) \cdot (1 - exp(-\lambda_P T_D)) \cdot (1 - \rho_P)
\end{aligned}
\tag{5}
$$

For the SU, all four classes of events will result in prolonged transmission of SU data. For the PU, as false alarm events will not cause data retransmission, the class III missed detection event will prolong the PU data transmission time.

3.2   Data Transmission of Users at All Levels in the Multi Priority Context

The two priority preemptive queuing system models are a special case of the multi-priority queuing system model, which requires the correction of the first moment and second moments of the PU and SU data transmission time obtained. For the SU, the probability P_{fa} for the occurrence of false alarm event is only related to ρ_P. Therefore, P_{fa} is still expressed by the Eq. (2). And the probability for the occurrence of the class III missed detection event is only related to T_S, T_D and λ_P, ρ_P. Therefore, P_{m3} is still expressed by the Eq. (5). The probability of occurrence of the class I and the class II missed detection events are related to the data transmission characteristics of the secondary users of different priority levels. In the Eq. (3), $Pr(Q_P = 0) \cdot (1 - exp(-\lambda_P T_S)) P_m$ is not related to the SU. Hence the occurrence of the first sub class event of the class I missed detection event of the secondary user at the j-th priority level is equivalent to the case that the number of SU in the SU queue with the priority level greater than or equal to j is zero. In addition, the user with the priority level of one frame of the PU within the data transmission time greater than j does not show up. And the SU that is greater than or equal to the j-th priority level shows up with the probability of the occurrence of the missed detection event. Hence, the following can be obtained

$$
\begin{aligned}
P_{m1,j} &= Pr(Q_P = 0) \cdot (1 - exp(-\lambda_P T_S)) \cdot P_m \\
&+ Pr(Q_{S,i \leq j} = 0) \cdot (1 - exp(-\lambda_j T)) \cdot P_m \cdot \prod_{i=1}^{j-1} exp(-\lambda_i T) \\
&= (1 - exp(-\lambda_j T)) \cdot \left[\prod_{i=1}^{j-1} exp(-\lambda_i T) \right] \cdot P_m \cdot \prod_{i=2}^{j} (1 - \rho_i), i \geq 2, j \geq 2
\end{aligned} \tag{6}
$$

The occurrence of the class II missed detection event of the secondary user at the j-th priority level is equivalent to that the number of SUs in the SU queue at the priority level greater than j is zero, the number of SUs in the SU queue at the j-th priority level is non-zero, the length of the PU queue is non-zero, and the probability for the occurrence of the missed detection effect is as the following.

$$
\begin{aligned}
P_{m2,j} &= Pr(Q_{S,i \leq j} = 0) \cdot Pr(Q_{S,i=j} \neq 0) \cdot Pr(Q_P \neq 0) \cdot P_m| \\
&= \rho_1 \cdot P_m \cdot \rho_j \prod_{i=2}^{j} (1 - \rho_i), i \geq 2, j \geq 2
\end{aligned} \tag{7}
$$

4   A Kind of Priority Service Queuing Method B-M/G/1-R

Since PU has the highest priority level, and the queuing mode of SU will not cause effect on the system residence time, the priority service function σ can be set to the cost paid for all the SUs joining the queue until the service is completed

$$
\sigma = \sum_{k=2}^{N} \xi_k T_k \tag{8}
$$

$\xi_k T_k$ stands for the overhead of the user with the k priority level, and σ stands for the priority service function.

4.1   User Resodemce Time and Priority Service Function

For the users at the k-th priority level, the residence time in the system consists of two parts: (1) The time for the completion of all the user data prior to waiting in the queue, which is denoted as $t_{1,k}$; (2) The time of data transmission itself and the delay caused by the preemptive simulation of the PU, which is denoted as $t_{2,k}$, $T_k = t_{1,k} + t_{2,k}$. And $t_{1,k}$ can be expressed as the following

$$
t_{1,k} = \frac{E\left[(\tilde{X})^2 \right]}{2 \left(1 - \sum_{i<k} \rho_i \right) \left(1 - \sum_{i \leq k} \rho_i \right)} \tag{9}
$$

In which $E\left[(\tilde{X})^2 \right] = \sum_{i=1}^{N} \lambda_i E\left[(\tilde{X}_i)^2 \right]$. Then $t_{2,k}$ can be expressed as the following

$$t_{2,k} = E[(\tilde{X}_k)] + \lambda_1 E[(\tilde{X}_1)]_{t_{2,k}} = \frac{E[(\tilde{X}_k)]}{1 - \lambda_1 E[(\tilde{X}_1)]} \tag{10}$$

Then such priority service function σ can be expressed as the following

$$\begin{aligned}
\sigma &= \sum_{k=2}^{N} \xi_k T_k \\
&= \sum_{k=2}^{N} \frac{\xi_k E[(\tilde{X})^2]}{2(1 - \sum_{i<k} \rho_i)(1 - \sum_{i \leq k} \rho_i)} \\
&+ \sum_{k=2}^{N} \frac{\xi_k E[(\tilde{X}_k)]}{1 - \lambda_1 E[(\tilde{X}_1)]}
\end{aligned} \tag{11}$$

The second part of the Eq. (11) is not related to the order of queuing, hence only $t_{1,k}$ in the σ is related to the queuing order. Therefore, the minimization of σ is equivalent to the following

$$min\{\sigma\} = min\left\{ \sum_{k=2}^{N} \frac{\xi_k E[(\tilde{X})^2]}{2(1 - \sum_{i<k} \rho_i)(1 - \sum_{i \leq k} \rho_i)} \right\} \tag{12}$$

4.2   SU Bubbling Queuing Principle

It is assumed that the two SUs with the adjacent priority levels in the $U_2, \cdots, U_N$ are U_i and U_{i+1}. To facilitate expression, assuming that $a = U_i$, $b = U_{i+1}$. Herein the method of adjacent comparison is applied to determine whether $a > b$ or $a < b$ can minimize σ. $\sigma_{a>b}$ is used to stand for the priority service letter obtained in the priority level $a > b$ after the conversion, and $\sigma_{a<b}$ stands for the priority service function obtained in the priority level $a < b$ after the conversion. Since the comparison of a, b does not have an effect on the priority ranking order of $U_2, \cdots, U_{i-1}$ and $U_{i+2}, \cdots, U_N$, $\sigma_{a>b} - \sigma_{a<b}$ can be represented as the following

$$\begin{aligned}
\sigma_{a<b} - \sigma_{a>b} &= \xi_a \frac{\lambda E[(\tilde{X})^2]}{\eta(\eta - \rho_a)} + \xi_b \frac{\lambda E[(\tilde{X})^2]}{(\eta - \rho_a)(\eta - \rho_a - \rho_b)} - \xi_a \frac{\lambda E[(\tilde{X})^2]}{(\eta - \rho_b)(\eta - \rho_a - \rho_b)} - \xi_b \frac{\lambda E[(\tilde{X})^2]}{\eta(\eta - \rho_b)} \\
&= \frac{\lambda E[(\tilde{X})^2]\{\xi_a(\eta - \rho_b)(\eta - \rho_a - \rho_b) + \xi_b\eta(\eta - \rho_b) - \xi_a\eta(\eta - \rho_a) - \xi_b(\eta - \rho_a)(\eta - \rho_a - \rho_b)\}}{\eta(\eta - \rho_a)(\eta - \rho_b)(\eta - \rho_a - \rho_b)} \\
&= \frac{\lambda E[(\tilde{X})^2](\rho_a\xi_b - \xi_a\rho_b)(2\eta - \rho_a - \rho_b)}{\eta(\eta - \rho_a)(\eta - \rho_b)(\eta - \rho_a - \rho_b)}
\end{aligned}$$

$$\tag{13}$$

In   which, $\eta = 1 - \sum_{j=1}^{i=1} \rho_j$.   As $\lambda E[(\tilde{X})^2] > 0$, $(2\eta - \rho_a - \rho_b) > 0$, $\eta(\eta - \rho_a)(\eta - \rho_b)(\eta - \rho_a - \rho_b)$, the positive or negative of $\sigma_{a>b} - \sigma_{a<b}$ depends on

the positive or negative of $(\rho_a \xi_b - \xi_a \rho_b)$. Therefore, the following conclusion can be obtained:

$$\begin{cases} \sigma_{a<b} < \sigma_{a>b}, & \xi_a/\rho_a > \xi_b/\rho_b \\ \sigma_{a<b} > \sigma_{a>b}, & \xi_a/\rho_a < \xi_b/\rho_b \end{cases} \tag{14}$$

It is determined that the users of the adjacent priority levels are converted in accordance with ξ/ρ. And the method of bubble sort is adopted to obtain the priority ordering of all the SUs by comparison in multiple rounds in turn. Therefore, this kind of multi-priority queuing method is referred to as the priority service queuing method.

5 Experiment and Analysis

5.1 Simulation Parameters

In order to effectively evaluate the analysis model put forward in this paper, the data transmission time of the established CSS system is an integer number of frames. In accordance with the 802.22 standard, the frame length $T = 10$ ms, the single same spectrum detection time consumption $T_S = 0.15$ frame, and the single frame data transmission time $T_D = 0.85$ frame .Users of four different kinds of priority levels are considered. Among them, the first priority level is PU, and the other three are SU. The arrival strength of the four types of users is in line with the Poisson distribution with the mean value of $\lambda_1, \lambda_2, \lambda_3, \lambda_4$, respectively. For $U_2, U_3, U_4, P_m \in [0, 1], P_f \in [0, 1]$.

In order to verify that the model is suitable for the multiple transmission time distributions, two distributions, that is, the truncated Pareto distribution and the negative exponential distribution, are considered here. These two distributions can properly perform fitting for the actual data and voice transmission. And the truncated Pareto distribution can be expressed as the following

$$f_x(x) = \begin{cases} \alpha K^\alpha / x^{\alpha+1}, & K \leq x \leq m \\ K^\alpha / m^\alpha, & x = m \end{cases} \tag{15}$$

The parameter settings are as the following: the flow parameter $\alpha = 1.1$, the scale parameter $K = 81.5$, and truncation upper limit parameter $m = 66666$ bytes. In accordance with the above settings, the mean value of single data transmission of PU is 480 bytes. For fairness, it is assumed that the mean value of the data transmission volume of PU corresponding to the negative exponential distribution is 480bytes as well. When the data rate of PU is 19.2Kbps, the mean value of PU transmission time under the condition of negative exponential distribution and truncated Pareto distribution is 24frames/arrival. Take $\rho_1 \in [0, 0.8]$, hence $\zeta = [0, 0.0333]$ arrival/frame. The single-frame loss priority services of the three levels of secondary users are $\zeta = [3/6, 2/6, 1/6]$, respectively. And the mean value of the data transmission time of the secondary users are all $E[X_2] = E[X_3] = E[X_4] = 20$ (arrival/frame). The arrival strength λi of SUs with different priority levels is randomly selected in line with the condition $\sum_{i=1}^{4} \rho_i \langle 1$.

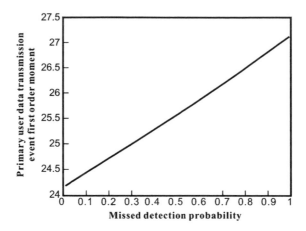

Fig. 3. $E[\tilde{X}_P]$ under the different effects of P_m

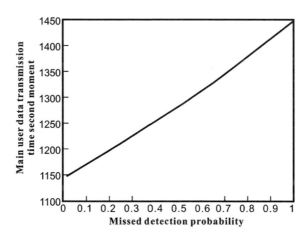

Fig. 4. $E\left[(\tilde{X}_P)^2\right]$ under the difference effects of P_m

5.2 Analysis of PU and SU Data Transmission Time in the Queuing System Model

In this section, the system is set to have only two priority levels, that is, PU and SU, and the arrival strength of PU and SU are $\lambda_P = 0.01$ and $\lambda_S \in = 0.02$, respectively. By comparing the first and second moments of the data transmission between PU and SU under different conditions of P_m and P_f, the effect of the actual CSS environment is analyzed. The first and second moments of the PU and SU data transmissions are expressed as $E[\tilde{X}_P]$, $E\left[(\tilde{X}_P)^2\right]$, $E[\tilde{X}_S]$ 和 $E\left[(\tilde{X}_S)^2\right]$. In accordance with the setting in the Sect. 5.1, $E[X_P] = 24$, $E[X_S] = 20$. From the Eqs. (10) and (12), it can be obtained that $E\left[(X_P)^2\right] = 1128$, $E\left[(X_S)^2\right] = 780$.

The $E[\tilde{X}_P], E[(\tilde{X}_P)^2]$ under the influence of the different P_m are shown in Figs. 3 and 4. The following conclusions can be drawn from Figs. 3 and 4: (1) When $P_m=0$, $E[\tilde{X}_P] - E[X_P] = 0.2$, $E[(\tilde{X}_P)^2] - E[(X_P)^2] = 17.1$; (2) $E[\tilde{X}_P]$ and $E[(\tilde{X}_P)^2]$ increase with the increase of P_m; (3) When $P_m = 0$, $E[\tilde{X}_P] - E[X_P] = 3.13$, $E[(\tilde{X}_P)^2] - E[(X_P)^2] = 317.2$. Cause analysis: (1) It can be known from the Eq. (5) that, P_{m3} is not related to P_m, but only related to T_S, T_D, λ_P and μ_P; while these four variables are fixed. The value of P_{m3} is fixed, and the length of the PU data transmission delay time length is also fixed. Hence even in the case of perfect sensing, it will still lead to the extension of the PU data transmission time; (2) From the Eqs. (3) and (4), it can be known that P_{m1} is proportional to P_{m1} and P_m. As shown in the Eqs. (12)–(14), $E[\tilde{X}_P]$ and $E[(\tilde{X}_P)^2]$ are the increasing function of P_{m1} and P_{m1}. Hence $E[\dot{X}_P]$ and $E[(\tilde{X}_P)^2]$ increase with the increase of P_m. At the same time, when P_m reaches the maximum value 1, $E[\tilde{X}_P], E[(\tilde{X}_P)^2]$ also reach the maximum value.

6   Conclusions

It is necessary to consider more than just the effects of the frame structure, the period detection method, the false alarm event and the missed detection event defined by 802.22 standard in the queuing system model of the computer system. In this paper, the class III missed detection event is first defined, and then the exact solutions for the first and second moments of the primary user and secondary user data transmission time are derived under the influence of the class I false alarm event and the class III missed detection event. And the M/G/1-R queuing system model is obtained, which has solved the cohesion problem of the M/G/1 model and the computer system. Secondly, a kind of B-M/G/1-R queuing method is put forward using the method similar to the bubble sort in the multi-priority secondary user context. The numerical analysis and the simulation results have verified that the BM/G/1-R is able to minimize the number of interruptions in the multi-level user data transmission in the real computer system environment.

Acknowledgement. This research was supported by the following projects.

1. Research on the training path of college students' computational thinking ability in general education course JG201624.

2. Master's degree in electronic information (computer technology) construction project 18TSXK06.

3. Exploration on the integration practice of computer industry, teaching and research in the construction of "new engineering" 201802314001.

References

1. Sippola, E., David, F., Sandra, P.: Temperature program optimization by computer simulation for the capillary GC analysis of fatty acid methyl esters on biscyanopropyl siloxane phases. J. Sep. Sci. **16**(2), 95–100 (2015)
2. Salzmann, D., Diederich, A.: Setting priorities in preventative services. J. Public Health **21**(6), 515–522 (2013)
3. Chalekar, A.A., Daphal, S.A., Somatkar, A.A., Chinchanikar, S.S.: Minimization of investment casting defects by using computer simulation - a case study. J. CSSustacean Biol. **27**(4), 529–533 (2015)
4. Drgan, V., Kotnik, D., Novič, M.: Optimization of gradient profiles in ion-exchange chromatography using computer simulation programs. Anal. Chim. Acta **705**(1), 315–321 (2011)
5. Ma, C., Xu, X., Wang, Z.: The quantitative evaluation on the value-oriented priority of service elements. IEEE, pp. 257–261 (2011)
6. Lee, Z., Wang, Y., Zhou, W.: A dynamic priority scheduling algorithm on service request scheduling in cloud computing. IEEE **9**, 4665–4669 (2011)
7. Attar, A., Raissi, S., Khalili-Damghani, K.: A simulation-based optimization approach for free distributed repairable multi-state availability-redundancy allocation problems. Reliab. Eng. Syst. Saf. **157**, 177–191 (2017)
8. Krätzig, W.B.: Physics, computer simulation and optimization of thermo-fluidmechanical processes of solar updraft power plants. Sol. Energy **98**(4), 2–11 (2013)
9. Alfieri, A., Matta, A., Pedrielli, G.: Mathematical programming models for joint simulation–optimization applied to closed queueing networks. Ann. Oper. Res. **231**(1), 105–127 (2015)
10. Schwartz, H.J.: The computer simulation of automobile use patterns for defining battery requirements for electric cars. IEEE Trans. Veh. Technol. **26**(2), 118–122 (2013)

Application of Artificial Intelligence Technology in Future Equipment Control System

Chao Tu[1], Hangyu Wu[2], and Lei Qiu[3(✉)]

[1] Vehicle Engineering Department, Changchun Institute of Engineering and Technology, Changchun, China
fdn_198591@163.com
[2] Audit Centre of Changchun, Changchun, China
8845510@qq.com
[3] Vehicle Engineering Department, Wuhan Armory School, Wuhan, China
332552083@qq.com

Abstract. With the continuous progress of intelligent and unmanned weapons and equipment, the battlefield is becoming increasingly complex and the application of armored equipment is becoming more diversified, which requires the equipment systems to update and recombine with the changes of the battlefield. Based on the existing technology, this paper prospects the future development trend of intelligent fire control system.

Keywords: Control system · Artificial intelligence · Image recognition

1 Introduction

With the gradual development of science and technology and weapons and equipment, the battlefield of today has been more and more inclined to "sea, land, air, space and electricity" and other multi-dimensional integrated operations [1–3]. For armored equipment, the threat from all levels is increasing, especially the development of individual anti-armor weapons and smart ammunition, which further squeeze the living space of armored vehicles [4, 5]. Therefore, the application of armored vehicles needs to be more flexible and diversified, and the battlefield adaptability needs to be stronger to play a greater role in the future battlefield. This puts forward higher requirements for the vehicle fire control system of armored equipment [6, 7]. The fire control system needs to be further intelligent in order to be able to perform well in the multi-dimensional battlefield in the future.

2 Intelligent Fire Control System

2.1 Better Battlefield Perception

The future battlefield threat from multiple levels, access to information of today's fire control system depends on the charge system and communication system, and soldier,

© Springer Nature Singapore Pte Ltd. 2020
M. Atiquzzaman et al. (Eds.): BDCPS 2019, AISC 1117, pp. 1191–1196, 2020.
https://doi.org/10.1007/978-981-15-2568-1_165

battlefield information nodes, error information, therefore, the fire control system itself needs to have good battlefield awareness, can get the target information in the first time, grasp the battlefield [8–11].

1. Upper anti-stabilizing image type vehicle gun length mirror. In order to have better battlefield perception ability, the fire control system needs to have complete observation equipment. The long field of view and the vehicle gunner's mirror with thermal image function can observe the battlefield in day and night and in all kinds of weather (Fig. 1).

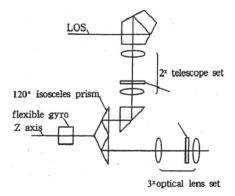

Fig. 1. Diagram of upper mirror

2. Intelligent target identification system. It can scan the battlefield with hologram, screen suspected targets intelligently, identify enemy and enemy, and judge the threat degree of targets intelligently, so as to provide effective information support for fire attack (Fig. 2).

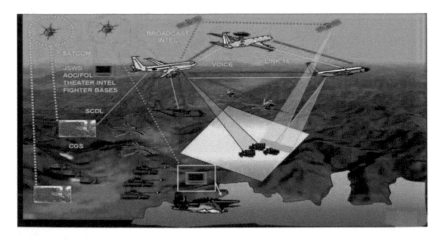

Fig. 2. Intelligent battlefield identification

2.2   Coordination Between Fire Control System and Command and Control System Is More Smooth

Armored equipment fire control system is different from suppression weapon fire control system, fire control system with the command and control system, no effective information sharing and exchange between armored equipment fire control system is only a fast and accurate turned fire shooting device, in order to better play to the advantages of and extension of fire control system, intelligent fire control system should have the following functions.

1. Information sharing with the charging system. The target detected and identified by the fire control system should be automatically sent to the charging system and Shared to the entire command level, thus realizing the information sharing mode of one-car discovery and all-discovery. Further fire strikes are coordinated to bring the units together (Fig. 3).

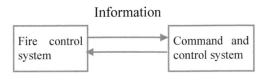

Fig. 3. Information sharing

2. The charging system can control the fire control system of each vehicle. Sur-mounted artillery can only be conducted by the vehicle commander on the current combat vehicle, and the external vehicle cannot directly command and control the fire of the combat vehicle. In the future, the battlefield will have a faster pace, leaving less reaction time for the crew. Therefore, the charging system is required to command the artillery commander to fire with the function of system-wide sur-mounted artillery.

2.3   Multi-functional Armored Equipment

The design of the Israeli merkava not only includes the functions of the main battle tank, but also takes into account the functions of armored personnel carriers, battlefield ambulances, mobile infantry fortress and other functions, helping merkava adapt to a variety of combat modes and make the battlefield more adaptable. In the future battlefield, it is difficult to have large-scale armored group operations. More often, they cooperate with other arms on a small scale to carry out combat missions. According to different missions, each unit will allocate one or several armored vehicles to carry out various types of missions or cope with various possible crises. This requires the armored vehicle of the future to have not only a strong fire support capability, but also the flexibility to move quickly to any position on the battlefield and provide necessary cover for other units when necessary. Therefore, the future fire control system should take into account multiple shooting modes.

1. Integration of indirect sight. At present, as a direct aiming weapon system, the armored vehicle weapon system does not have the calculation ability of indirect aiming, but only carries on the rough aiming and firing according to semi-automatic or manual method. Based on the integration of fire control system and charging system in the future, the fire calculation model of direct fire weapon is embedded, which makes direct fire weapon have the capability of automatic calculation of fire shooting. In this way, the armored tank unit can be converted into a self-propelled artillery unit when necessary to carry out fire attack beyond the range of sight, which makes the use of armored vehicles more diverse and enables the commander to have more flexibility in the use of fire (Fig. 4).

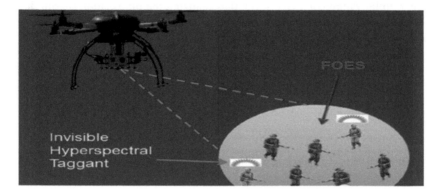

Fig. 4. Intelligent identification

2. The connection between smart ammunition and fire control is closer. With the rapid development from traditional ammunition to new smart ammunition, more and more information is carried on the ammunition, and new requirements will be put forward for the fire control system in the whole process of loading, launching, flying and detonating the smart ammunition. Traditional fire control system for ammunition loading and ignition operation has been far from meeting the requirements of new intelligent ammunition, fire control function should be expanded accordingly, the establishment of the corresponding ammunition data link equipment, and intelligent ammunition for high real-time and large capacity of information interaction.

2.4 Generalization and Modularization

At present, there are various types of fire control system for armored equipment, and almost every vehicle type has its own fire control system, which not only brings barriers for multiple vehicle types to cooperate with each other, but also makes it more inconvenient for troops to use, maintain and repair in battle.

1. Generalization. With the gradual maturity of all-electric gun control, the future armored equipment fire control system will definitely realize all-electric control.

Different models should use the same type of fire control system, most of the components should be universal, and the core of a small part should be consistent except for different adjustments and fixtures for easy installation. This will greatly reduce the workload of users and maintenance personnel, in a shorter time to form combat effectiveness.

2. Modularization. The existing fire control system is low in integration, single in function and large in volume, which occupies a small space of armored equipment and greatly reduces the combat efficiency. With the continuous development of chip technology, the whole core processing components of the future fire control system will be smaller and smaller in size and quantity. Only the signal generation, signal processing and signal execution components are more convenient to replace directly after failure or damage, greatly reducing the number of cables in the turret. It improves the combat efficiency and makes the battlefield repair of fire control system more rapid.

2.5 Unmanned

Unmanned turret fire control can be realized through the driver's multimedia terminal or remote control outside the vehicle.

3 Conclusion

Fire control system has experienced the transformation from mechanization to digitalization. With the continuous development of intelligent technology, intelligent fire control system is not far away from us. In the future, the war form will gradually change from semi-automatic combat platform to automatic or remote control combat platform. The battle for a local war is like a game, which can be decided directly in the command center of both sides. The change of war pattern requires us to aim at the needs of future wars in advance and prepare well for them so as to remain invincible in future operations.

References

1. Ding, H.: Research on positioning error dynamic measurement and compensation system of x-y table. Shandong University, Master thesis (2008)
2. Jiao, H., Chen, W., Wang, C.: Analysis and discussion on the error analysis of scanning frame for large scale compression field test. Journal of Beijing Technology and Business University (2006)
3. Shi, M., Kong, X.: Discussion on the design of foundation drawing of machine tool to improve the accuracy of machine tool. Standardization and Quality of Machinery Industry (2011)
4. Sun, X., Yan, K., Ding, G.: Analysis on positioning accuracy of machine tool by comprehensive error of ball screw. Machinery Manufacturing Technology (2008)
5. Lei, T.: Automatically detects the thickness of the non-contact development system. University of Electronic Science and Technology, Master's thesis (2013)

6. Yue, S., Guo, J.: Overview of foreign army weapon and equipment development in 2016. Equipment/technology (2017)
7. Xue, C., Huang, X., Zhu, X., Jiang, Y.: Status and development trend of unmanned systems in Foreign military. Radar and Countermeasures (2016)
8. Han, L., Wang, B.: Development status and enlightenment of unmanned logistics equipment. Military aspect (2016)
9. Liu, C.: New development of unmanned remote control turret. Chariot Zongheng (2018)
10. Han, Z.: Application of artificial intelligence technology in fault detection of airborne fire control system. J. Qingdao Univ. (2001)
11. Jing, Y., Wang, X., Zhang, Z.: Development trend of suppression weapon fire control system. Fire power and command control (2012)

On the Application of Large Data Technology in B2C E-Commerce Precision Marketing Mode

Gang Xu and Qiaolin Chen[✉]

Hubei Business College, Wuhan 430079, China
chenqiaolin@hbc.edu.cn

Abstract. In recent years, big data analysis method has been widely used in many fields, such as environmental art design, computer technology application enterprise investment and so on. With the popularization of online consumption and payment, big data technology is more and more widely used in e-commerce. E-commerce is mainly divided into three types, but this paper mainly analyzes the application of big data technology in B2C e-commerce precision marketing pattern. Firstly, the notion and features of B2C e-commerce as well as the concept and characteristics of precision marketing mode are described. Secondly, this paper further analyzes some challenges faced by B2C e-commerce in the times of large data. Finally, taking China Amazon B2C electronic commerce as an example, this paper briefly analyzes the ratio of B2C e-commerce in China's online retail B2C market share in the third quarter of 2017, which is 4.1%. This shows that B2C e-commerce is developing rapidly under the background of large data era.

Keywords: B2C electronic commerce · Precision marketing mode · Large data · China Amazon

1 Introduction

Large data means the data collection that cannot be grabbed, managed and processed by conventional software tools at a specific time interval. Large data has five characteristics: large amount, high speed, diversity, low value density and authenticity. It has no statistical sampling method, but only observes and tracks what happened. The use of large data tends to be analysis and prediction, user behavior analysis or some other advanced data analysis methods.

With the burgeon of economics, the progress of science and technology, big data technology is applied in all aspects and fields. Especially in the B2C electronic commerce precision marketing mode. In [1], the author discusses the implicit hypothesis in more and more literatures, that is, the use of social media in B2B and B2C literature is fundamentally different. The results show that B2B social media use is different from B2C, hybrid and B2B2C business model methods. Specifically, B2B members believe that the overall efficiency of social media as a channel is low, and that social media is less important for relationship oriented use than other business models. In [2], the

© Springer Nature Singapore Pte Ltd. 2020
M. Atiquzzaman et al. (Eds.): BDCPS 2019, AISC 1117, pp. 1197–1203, 2020.
https://doi.org/10.1007/978-981-15-2568-1_166

author studies the moderating effect of consumer's social demographic characteristics on online shopping behavior, uses the existing literature to conduct a questionnaire survey, uses the questionnaire survey method to collect 337 effective records, uses the association rule mining method to analyze, and determines several interesting rules representing the preferences of online consumers. The results show that the behavior of online consumers in Turkey is related to its demographic characteristics. In [3], this paper analyzes the questions of enterprise reputation evaluation in the current e-commerce environment, and establishes an online enterprise reputation evaluation index system in the B2C electronic commerce environment. In [4], the author estimates the willingness to act as a collector (Supplier) and purchase a consolidation service (demand) to deliver/ pick up goods in the last mile of B2C e-commerce. The results help to understand and quantify the e-commerce potential of this freight strategy in urban environment, and provide a good knowledge base for local policy makers for their future development. In [5], the author chooses an empirical method and conducts quantitative research in nearly 300 participants to reveal the differences between the capital decisions of start-ups located in B2B or B2C markets. The results show that there are indeed investment differences between B2B and B2C companies, which indicates that other success factors are needed in these cases. In [6], the author developed a mixed integer programming model to determine the location of the incompetent distribution center (UDCL) in B2C electronic commerce. Based on the characteristics of B2C electronic commerce enterprise distribution system, the model considers the influence of multi commodity supply cost. In [7], the intention of this paper is to find out the traits of the two countries' samples when small and micro enterprises expand their channels. The results show that it is difficult for microenterprises to balance technology, business and customer needs with capabilities and resources. In [8], the author uses a survey method (n = 281) and tests the key premise related to self -disclosure proposed by the B2C relationship stage theory of e-commerce. The research model determines the following preconditions for self -disclosure: attractiveness, perceived reward, switching cost, participation and trust. The results show that trust and perceived reward explain the huge difference of self - disclosure intention in online B2C environment. In [9], the author investigated the influence of DM tools on consumers' online impulse buying propensity (obit), that is, the intervention of AD and CD, GDR and EL. In [10], the author aims to contribute by fully understanding the scope of implementing sustainable marketing means for SMEs in the food and beverage industry in Europe. The focus will be on identifying differences between companies operating in business to business (B2B) and business to customer (B2C) environments. The results show that both B2B and B2C groups have implemented sustainable marketing means to some extent.

Based on big data, this paper simply analyzes the use of large data in B2C electronic commerce precision marketing mode.

2 B2C Electronic Commerce and Precision Marketing

2.1 B2C E-Commerce

Electronic commerce can be divided into many parts, such as enterprise to enterprise electronic commerce B2B and consumer to consumer electronic commerce C2C. What this article is about is indeed enterprise to consumer electronic commerce B2C. B2C indicates that enterprises directly face consumers, and enterprises can directly but not indirectly provide their products to customers on the Internet. If customers have needs, they can also provide corresponding services. B2C electronic commerce is more direct and specific than B2B and C2C. B2C e-commerce pattern is the earliest business mode born on the basis of the Internet. Since the B2C business mode, the consumption mode of customers has changed dramatically. It is not the traditional offline consumption mode, but the consumption and shopping can be carried out through the Internet. The usual payment mode can also choose online payment, which can greatly save consumers' time, and make consumer life more convenient. Compared with other electronic commerce, B2C electronic commerce has obvious characteristics: firstly, the efficiency of transaction will be greatly improved, because before the B2C e-commerce mode, people only consume offline, and the transaction efficiency is very low, while B2C e-commerce uses the Internet as a good platform to store a lot of customer information, and carry out classification and archiving, so efficiency will be improved. Secondly, it has the nature of long-distance transaction. Consumers only need to pay online to complete the consumption, instead of face-to-face transaction, which greatly shortens the distance between time and space.

2.2 Precision Marketing

Precision marketing, just like its name, emphasizes "accuracy" and "accuracy". This marketing mode mainly focuses on customers, and provides customers with satisfactory products to meet their needs, which can be unlimited in time and place. There are three meanings in precision marketing: the first is to have a precise marketing idea, know what you want to market, and provide these marketing products to what customers. The second is to implement the system guarantee and measurable means of precision marketing. Third, precision marketing can reduce the cost and achieve the goal of the enterprise. To sum up, the main content of precision marketing is that enterprises first obtain customers' information and their needs through various information technology means, and enterprises provide corresponding products for these needs of customers. The characteristics of precision marketing are divided into the following two points: first of all, we can accurately find our own customers, because precision marketing through technical means to find their own potential customers, and provide corresponding products according to the needs of customers. Secondly, it can reduce the communication cost of enterprises, because precision marketing has grasped the customer information in advance, and effectively classified different customers, which will greatly reduce the cost.

3   Challenges Brought by Big Data to B2C E-Commerce Precision Marketing

Everything has two sides, advantages and disadvantages. Big data is no exception. It brings a lot of convenience and challenges to B2C e-commerce. Because large data has a lot of information and data, B2C e-commerce is bound to face many technical processing problems and privacy protection problems under the background of big data. The specific challenges are as follows: the first one is that it becomes more difficult to store and quickly access big data. We know that B2C e-commerce must rely on data to survive, because B2C e-commerce is analyzed, compared and tracked by enterprises according to the needs of customers. In the context of large data, the amount of data becomes larger and larger, which will inevitably lead to storage. It becomes more difficult to store and quickly access big data. The second challenge is information security and privacy. In the era of big data, there are not only many kinds of data information, but also many kinds of data information. Once the customer's information is lost, it will inevitably disclose their privacy, so this is a great challenge for B2C e-commerce.

4   Practical Case Analysis: The Application of China Amazon B2C E-Commerce in Big Data

Amazon China is a B2C electronic commerce website, which is one of the most effective electronic commerce websites in China at present. There are many kinds of commodities sold on this website, including those in study and life. Since 2006, Amazon China has started to transfer its store from offline to online. After more than ten years of development, Amazon China's "online mall" can provide users with a variety of products, including books, electronics, furniture, clothing, etc., covering all aspects of customer life and learning. Compared with other e-commerce websites, the development of amazon.com in China is relatively late, but its results are very significant. According to a survey, the transaction scale of China's online retail B2C market in the third quarter of 2017 was 964.52 billion RMB, an increase of 34.5% year on year. The specific results are shown in Figs. 1 and 2.

From the perspective of market share, TMALL is the first, JINGDONG is the second, VIPSHOP is the third, GOME is the fourth, and Amazon is the fifth. The specific market share is as follows: TMALL accounts for 60.5% of the total turnover, with a growth of 42.3% over the same period; JINGDONG accounts for 24.6%, with a growth of 24% over the same period; VIPSHOP accounts for 6.5% and GOME accounts for 4.3%. See the Fig. 3 below for details.

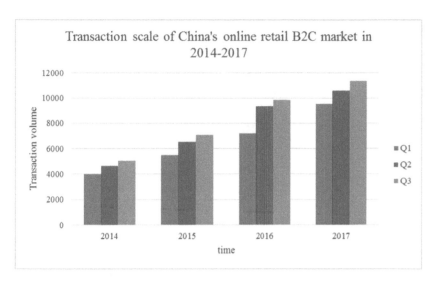

Fig. 1. Transaction scale of China's online retail B2C market in 2014–2017

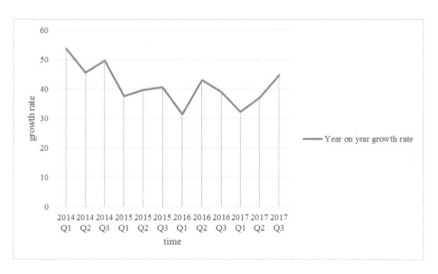

Fig. 2. Year on year growth rate of China's online retail B2C market transaction scale from 2014 to 2017

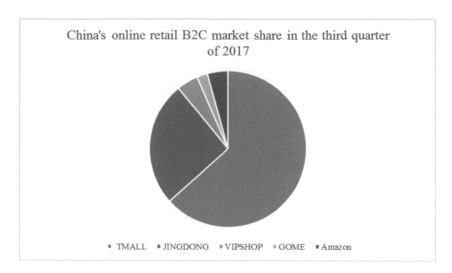

Fig. 3. China's online retail B2C market share in the third quarter of 2017

5   Conclusion

On the basis of big data, this paper analyzes the use of large data in B2C electronic commerce precision marketing mode. Firstly, it introduces the notion, characteristics and traits of B2C electronic commerce and precision marketing mode. Big data will not only bring some opportunities to B2C electronic commerce, but also bring some challenges. This paper specifically analyzes what big data will bring to B2C e-commerce. Challenge, this paper takes China's Amazon B2C electronic commerce as an example to briefly describe the application of this electronic commerce in the times of large data.

References

1. Iankova, S., Davies, I., Archer-Brown, C., Marder, B., Yau, A.: A comparison of social media marketing between B2B, B2C and mixed business models. Ind. Mark. Manag. **81**, 169–179 (2019)
2. Kahraman, N., Tunga, M.A., Ayvaz, S., Salman, Y.B.: Understanding the purchase BEHAXIOUR of Turkish consumers in B2C E-Commerce. Int. J. Intell. Syst. Appl. Eng. **7** (1), 52–59 (2019)
3. Zhang, F., Xu, Z.: Research on the evaluation index system of enterprise's online credit in B2C E-commerce. Int. J. Sci. **8**(06), 117–121 (2019)
4. Gatta, V., Marcucci, E., Nigio, M., Serafini, S.: Sustainable urban freight transport adopting public transport-based CROWDSHIPPING for B2C deliveries. Eur. Transp. Res. Rev. **11**(1), 13 (2019)
5. Jovanovic, T., Brem, A., Voigt, K.I.: Who invests why? An analysis of investment decisions in B2B or B2C equity CROWDFUNDING projects. Int. J. Entrep. Small Bus. **37**(1), 71–86 (2019)

6. Zhang, H., Beltran-ROYO, C., Wang, B., Zhang, Z.: Two-phase semi-LAGRANGIAN relaxation for solving the UNCAPACITATED distribution centers location problem for B2C E-commerce. Comput. Optim. Appl. **72**(3), 827–848 (2019)

7. Sell, A., Walden, P., Jeansson, J., Lundqvist, S., Marcusson, L.: Go digital: B2C microenterprise channel expansions. J. Electron. Commer. Res. **20**(2), 75–90 (2019)

8. Campbell, D.E.: A relational build-up model of consumer intention to self-disclose personal information in e-commerce B2C relationships. AIS Trans. Hum.-Comput. Interact. **11**(1), 33–53 (2019)

9. Alam, M.S.A., Wang, D., Waheed, A.: Impact of digital marketing on consumers' impulsive online buying tendencies with intervening effect of gender and education: B2C emerging promotional tools. Int. J. Enterp. Inf. Syst. (IJEIS) **15**(3), 44–59 (2019)

10. Rudawska, E.: Sustainable marketing strategy in food and drink industry: a comparative analysis of B2B and B2C SMEs operating in Europe. J. Bus. Ind. Mark. **34**(4), 875–890 (2019)

Middle School Mathematics Teaching Based on Multiple Intelligence Theory

Changwen Zhang[✉] and Changhui Liu

School of Mathematical Sciences, University of Jinan, Jinan, China
hzq-1230@163.com

Abstract. In the process of China's social development, more and more diverse talents and creative talents are needed. A high score does not mean a strong ability, and the saying that "achievement is the most important" has been gradually eliminated. Under such social needs, American scholar Gardner's theory of multiple intelligence has gradually entered the field of vision of the scholars of educational research in our country. Gardner pays attention to the human intelligence should not only have one, we should pay attention to the cultivation of multiple intelligence of students, especially the cultivation of students' creativity, to adapt to the students' personality, so that students can develop in an all-round way. In this paper, the general teaching mode of the theory of multiple intelligence is summarized, good teaching suggestions and teaching strategies in view of this mode are put forward.

Keywords: Theory of multiple intelligence · Gardner · Creativity cultivation

1 Introduction

Under the background of economic globalization, the competition between our country and other countries depends on the strength of our country's comprehensive national strength. The essence of such competition is the comprehensive strength competition among countries' high-level talents, especially the innovation and creation ability of high-level talents [1]. Therefore, we should carry out innovation and creativity education extensively as the core plan of mathematics education in middle schools in our country. The study, the author from the perspective of the theory of multiple intelligences to promote the innovation of the middle school mathematics education teaching ability development, it is the embodiment of the connotation of education reform in our country, the requirement of a new round of world science and technology revolution, the rapid economic and social development of all countries to the domestic high level talented person's demand trend, has essential significance for deepening the education management research. The cultivation and development of middle school mathematics education creativity based on the theory of multiple intelligences is the most basic way to cultivate creative high-level talents, and it is the most elite strength and advantage accumulated for promoting our country's innovation strength to drive the rapid development strategy.

© Springer Nature Singapore Pte Ltd. 2020
M. Atiquzzaman et al. (Eds.): BDCPS 2019, AISC 1117, pp. 1204–1208, 2020.
https://doi.org/10.1007/978-981-15-2568-1_167

2   The Main Content of the Theory of Multiple Intelligences

2.1   Concept of Multiple Intelligences

The theory of multiple intelligences was put forward by Gardner, a psychologist at Harvard University. Multiple intelligences can be divided into seven categories: linguistic intelligence, motor intelligence, interpersonal intelligence, logic intelligence, music intelligence, spatial intelligence and introspective intelligence [2] (Table 1).

Table 1. Summary table of the concepts of multiple intelligences

Scholars and ages	Definition of intelligence
Term an (1921)	Intelligence is the ability to think abstractly
Boring (1923)	Ability measured by intelligence tests
Burt (1940)	Intelligence is innate cognitive ability
Stoddard (1943)	Intelligence is a person who solves current problems and prepares for the future
Ferguson (1956)	Intelligence is the ability of a person to learn and accumulate experience by moving from one situation to another
Wechsler (1958)	Intelligence is the ability to perform purposeful actions, the ability to think rationally and deal effectively with one's surroundings
Robinson and Robinson (1965)	Intelligence is the ability to perform purposeful actions, think rationally, and deal effectively with problems
Gardner (1979)	Intelligence is the ability to create work that is valued in each culture
Sternberg (1985)	Intelligence is the ability to learn, to get along with others, to read, to respond and to work
Good and Brophy (1990)	Intelligence is the ability to discover known knowledge or information and to build knowledge
Armstrang (1999)	Intelligence is the ability to cope with new situations and avoid repeating the same mistakes
Clark (2002)	Intelligence is the result of the interactive development of various brain functions, and the genetic form and environment provide opportunities for interaction, either promoted or suppressed
Moberg (2009)	Intelligence is dynamic and can be improved through education

The theory of multiple intelligences is the most advanced theory form of intelligence research and development at the present stage. Its influence and application are very extensive. It has been recognized by people to a large extent [3].

2.2   Statistics and Analysis of Some Intelligent Low-Score Samples

Some students with lower scores in language intelligence, spatial intelligence and logical mathematics intelligence were extracted for comparative analysis, and the results were shown in Table 2.

Table 2. Score table of multiple intelligence with lower scores in linguistic intelligence, spatial intelligence and logical mathematics intelligence

Serial number	Language functions	Logical mathematical intelligence	Spatial intelligence	Natural observation intelligence	Self cognitive intelligence	Self cognitive intelligence
29	11	11	12	15	15	17
36	13	10	14	15	13	16
61	15	17	9	16	17	15
86	14	18	10	17	15	18
92	13	11	11	10	14	17
108	15	11	13	15	14	16
112	10	14	12	10	15	17
136	11	12	11	15	16	18
153	14	11	12	15	13	15
166	12	10	13	14	16	15

By Table 2 it can be seen that although some students in linguistic intelligence, logical-mathematical intelligence and spatial intelligence score is low, but the students in other intelligence score is not too low, some even are quite outstanding, the relatively high intelligence also each are not identical in different student body, fully illustrates the differences between each student's intelligence degree is not the same, each has his strong point, and there will always be more prominent on the one hand, which requires teachers to understand the characteristics of students, students don't one-sided evaluation [4].

3   Teaching Strategy Based on Multiple Intelligences to Cultivate Students' Creativity

3.1   A Hierarchical and Classified Teaching System Should Be Built to Highlight Individuality

In our country, for many years, the middle school classes have been arranged and the teaching method has been adopted parallel classes. We propose that the method of creativity cultivation under the perspective of multiple intelligence should be given the same education to students with different interests, hobbies, personalities and levels. The reform of the new examination system will certainly change the original homogeneous teaching method and gradually push it forward. The future deployment of classified and hierarchical teaching system will surely be successful. Such reforms are not completely overturned the previous teaching mode, is relying on the original class teaching administration, to interest is different, different hobby, different personality, different levels of students, to set up a classification in the teaching goal, and puts forward different teaching, improving teaching methods, classification design teaching content, design different teaching progress, fully develop the students' intelligence diversity [5].

3.2   Build Various Platforms for Middle School Students to Promote Creativity Development

First, the types of student activities held in schools should be diversified. For example, innovation competitions, art festivals, clubs and organizations based on various interests, and different platforms serve students to find suitable activities to develop their potential. Second, we should use developmental education to promote cultural integration at home and abroad. It is important to note that the questions raised should be in line with the students' own cognitive level and can be solved, but also have certain challenges [6].

3.3   Provide Students with Sufficient Opportunities for Independent Thinking

Do not exclude students to expose their various difficulties in the process of learning, even if students' thinking is wrong can accept, experience wrong thinking is a kind of growth, students can according to their own understanding to state the problem of the topic, conclusion, and reasoning out the conditions available; This is to give students time to actively participate in exploring and solving problems and actively building knowledge system.

3.4   Encourage Students to Ask Questions

After we create a math situation for students, students should be encouraged to actively express their own ideas, real conclusions, and actively experience and try to solve the problem. Students will have a more profound understanding of new knowledge and deepen their understanding [7]. As the guide and guide of students' learning, teachers mainly lead students to learn and inspire their learning process. According to the development of students' cognitive level and the relationship between the new and old knowledge, they should be encouraged to explore the problem actively and arouse students' thinking on the problem. Finally, it can optimize students' knowledge system and make progress in solving mathematical problems [8].

3.5   Diversified Problem-Solving Strategies

Different students always adopt different ways of thinking and solving the same problem. This tells us that it is unreasonable to let students form a fixed problem-solving strategy when teaching. We should pay attention to the cultivation of students' multiple thinking, multiple intelligence, bold attempts, careful demonstration, and finally draw their own conclusions [9]. To view and analyze problems from different perspectives; According to different problems, use different methods to solve, such as: segmentation, local substitution, drawing and so on. Flexible application, live learning to use, let the knowledge in the mind movement, learning will become easy. Teachers appropriate encouragement, will let the process of learning less detours [10].

4 Conclusion

The theory of multiple intelligence is designed to understand what students do well, rather than to provide a measure of intelligence, and students can't always be judged on their report cards. It is our task to learn to make rational use of this theory to provide help for our daily teaching. The author hopes to be guided by the theory of multiple intelligence and pay attention to cultivating students' creativity. More learning is needed to achieve this goal. There have been a lot of studies, but there are still a few experiences that can be applied in the actual classroom. We need more teachers to try new methods bravely, so as to make real progress in practice. However, at present, there is a lack of detailed teaching plans and textbooks for various subjects related to the theory of multiple intelligence [11]. We urgently need staff engaged in relevant research to develop a set of scientific theoretical textbooks. In this way, the front-line teachers will have more evidence to work on and achieve the goal of applying multiple intelligence to practical teaching.

Acknowledgments. This research was supported by Teaching Research Project of Qilu Institute of Technology (Grant No. JG201823).

References

1. Xu, H.: High school mathematics teaching research and practice based on multiple intelligence theory. Liaoning Normal University (2018)
2. Li, C.: Application of multiple intelligences theory in junior middle school mathematics teaching. Guangzhou University (2016)
3. Wang, G.: Review and reflection on the theoretical research of multiple intelligences in China in the past 20 years. Comp. Res. Cult. Innov. 2(31), 163–165 (2018)
4. Jiang, H.: Research on the development path and countermeasures of creativity based on multiple intelligence in high school. Zhejiang University (2017)
5. Feng, Y.: Problems and countermeasures of multiple intelligence diagnosis. Mod. Educ. Technol. 05, 27–33 (2007)
6. Na, K., Li, C.: Translation. Liberating Asian Students' Creativity. China Light Industry Press, Beijing (2005)
7. Li, C., Fan, J., Zhang, Q.: Application and significance of group cooperative learning model based on multiple intelligence theory in middle school teaching practice. Educ. Teach. Res. 29(8), 94–98 (2015)
8. Song, Y., Zhou, J.: Modern teaching evaluation view based on multiple intelligences theory. J. Luoyang Inst. Technol. (Soc. Sci. Ed.) 19(4), 109–112 (2004)
9. Li, Y.: Research on Mathematics Teaching and Practice in Middle Schools. Higher Education Press, Beijing (2001)
10. Wa, C.H.: Investigation on the current situation of mathematics teaching methods in middle schools. Huazhong Normal University (2001)
11. Hu, Z.: The cultivation of creative thinking ability in high school mathematics teaching. J. Hunan Educ. Coll. 7, 35–48 (2011)

Ultra HD Video and Audio Broadcast System for Shared Living Room

Jinbao Song[1]([✉]), Junli Deng[2], Xin Xin[3], and Xiaoguang Wu[3]

[1] Department of Scientific Research, Communication University of China,
Beijing, People's Republic of China
songjinbao@cuc.edu.cn
[2] Faculty of Science and Technology, School of Information Engineering,
Communication University of China, Beijing, People's Republic of China
[3] Beijing HeXinChenGuang Information Technology Co., Ltd.,
Beijing, People's Republic of China

Abstract. HD video and high reproduction audio are two important playback indicators. The 8K ultra-high-definition video player developed by our team has reached a practical level. The multi-channel sound system (DMS) is a new multi-channel sound-recovery system based on the sound field optimization principle and Huygens principle developed by CUC. Through experiments and theoretical analysis, this paper studies the operation mechanism of ultra-high definition shared living room audio and video playback system, summarizes the common techniques of audio and video synchronization at home and abroad, and provides a feasible synchronous implementation method, which makes the audio and video synchronization error range. To meet the requirement of relevant national standards, the ultra-high-definition video and audio broadcast system shared living room is realized.

Keywords: DMS · Video and audio synchronization · 8K Ultra-high-definition

1 Introduction

In recent years, terminal companies such as BOE have developed 8K ultra-high-definition TV sets. Products on 8K filming system have also been listed. CUC has successfully developed four major platforms for recording, producing, synthesizing and broadcasting. It is called DMS(Dynamic Matrix Sound) system. Compared with the current US dolby system, DTS system, Japan's NHK22.2 system and Europe's Barco 11.1 system [1], the DMS sound system has the advantage that advanced dynamic matrix technology realizes adaptive sound field reconstruction and various expansions. The YCbCr422 has a color depth of 10/12bit and 8K@60p YCbCr [2, 3].

The development of TV technology has experienced from analog to digital, from standard definition to high definition. The current development stage is entering from high definition to ultra high definition. With the continuous development of digital media and high-definition technology, ultra-high-definition television is becoming

© Springer Nature Singapore Pte Ltd. 2020
M. Atiquzzaman et al. (Eds.): BDCPS 2019, AISC 1117, pp. 1209–1214, 2020.
https://doi.org/10.1007/978-981-15-2568-1_168

more and more popular. All the events of the 2018 World Cup are broadcast in ultra-high-definition format for the first time. According to the data in April 2018, the total number of global TV channels is 11700, HDTV channels account for about 27%, and the total number of Ultra HD TV channels is 125 [4].

In the mono era, there is no concept of sound field, because the listener's perception of the sound source is a fixed point when only one channel is output. The two-channel stereo that was put into use in the 1950s, which firstly introduced the concept of sound field. When two output channels are given a difference in sound level, the listener will feel that the sound source is at a point where it does not exist. Changing the sound level difference will also change the position of the virtual point accordingly. In addition, there are many channels of audio systems [5]. The DMS (Dynamic Matrix Sound) sound system independently developed by CUC is a multi-channel sound-recovering system based on Huygens principle. The system picks up through a microphone array, using multiple inputs. The multiple output multi-solution system combines a limited sound source with an infinite sound field to give the listener a completely new auditory effect [6].

So far, the trend of audio and video is 8K and multi-channel.

We believe that the UHD shared living room is a good 8K practical choice. It can break through the bandwidth bottleneck and give full play to the advantages of 8K products. So that viewers can even get more than the cinema playback level, a higher audiovisual experience.

Through experimental and theoretical analysis, this paper studies the operation mechanism of ultra-high-definition video and audio shared playback system, reviews the common techniques of audio and video synchronization at home and abroad, and provides a feasible synchronous implementation method, which makes the audio and video synchronization meet the relevant national standards. The ultra-HD shared audio and video playback system is realized.

2 Related Work

In 2012, the International Telecommunication Union published the international standard ITU-R BT.2020 on ultra-high-definition televisions, in which the categories of ultra-high-definition televisions were divided into 4K TVs and 8K TVs according to the physical resolution of images [7]. In 2013, BOE demonstrated its 98-inch 8K display, marking the entry of 8K display devices into the mass production era [8]. At the World Cup in Brazil in June 2014, the NHK Broadcasting Association of Japan broadcasted 9 games including the Japanese team to the 8K standard [9], and realized the transcoding of 4K/12bit images into 2K/8bit real-time transmission.

There are solutions for adding playback equipment in commercial sound systems, mainly Dolby [10].

This paper studies the integrated application of 8K ultra HD video and audio broadcast system for shared living room.

3 Overview of Ultra HD Video and Audio Broadcast System for Shared Living Room

3.1 8K Ultra HD Video Broadcast System

The 8K uncompressed video player is based on the new ITU standard. The 8K-HDR player can play 7680 × 4320 resolution uncompressed ultra high definition video signal. It can play 60 frames per second video signal to meet the requirements. It can choose to play 10/12 bit color depth YCbCr422 video signal or RGB444 video signal 8-bit color depth. It can choose the video signal that matches BT.2020 gamut standard or BT. The 709 color gamut standard allows you to play HDR video signals that conform to the PQ standard or the HLG standard (Fig. 1).

Fig. 1. 8K uncompressed video signal player

3.2 8K Ultra HD Video and Audio Broadcast System for Shared Living Room

Based on Qt4, SMPlayer is a graphical front-end interface (GUI) of MPlayer. It can support most multimedia files. It is free and open source. It is GPL-compliant and can be installed across platforms. The DMS audio player is based on Open source SMPlayer for improvement and development.

The DMS audio playback system is divided into five parts: FPGA, CPLD, DSP, backplane (power/AD conversion), DA output board.

TinyXML is an open source XML parsing library that can be directly added to the project written in C and C++ language. It is small and powerful, and can be compiled in Windows or Linux. This parsing library model makes it easy to facilitate the XML tree by parsing the XML file and then generating the DOM model in memory.

The DOM model, the Document Object Model, divides the entire document into multiple elements and uses a tree structure to represent the order relationship between these elements and the nested inclusion relationship.

The audio files of 30 channels are read by the DMS audio player in a specific XML format (Fig. 2):

Fig. 2. Audio file XML organization

4   Ultra HD Video and Audio Broadcast System for Shared Living Room Synchronization Implementation

4.1   Feasibility Demonstration of Synchronization Method

Because the sampling rate of audio is 48 kHz, the time for launching a large packet is 13.33 ms. After collecting the time interval of 10,000 large packets, the average value is 13.33 ms, which proves that the computer clock accuracy is the same as that of the audio processor.

According to the above situation, the following audio and video synchronization methods can be proposed:

In the DMS audio player, a reference time is defined. The time of the first big packet is saved and used as the relative zero time. Then, each time a large packet is sent, the clock value is collected. The time value subtracts the time corresponding to zero time, and the obtained time continuation value is sent to the video end. The time continuation value is used as the relative time stamp, and the video end is synchronized with the audio based on the time continuation value.

Specifically, the frame rate of the 8K video player is 60FPS (Frames Per Second). The time for playing one frame is 1/60 = 16.67 ms. This is the frame number that the video player should broadcast. When the continuation time value is 1000 ms, the number of frames that the 8K video player should play: 1000 ms/16.67 ms = 60 frames. That is to say, the video player should play 60 frames when the duration value is 1000 ms. The ratio of the video playback frames number to the audio playback frames number is 1:800.

According to the national standard, the synchronous evaluation method adopted in this paper is that the difference between the audio timestamp and the video timestamp should be between −160 ms and +65 ms. The audio is slower than the video is negative, and the video is slower than the audio is positive.

In theory, the video player and the DMS audio playback system are in the same environment, the same local area network, and the communication conditions are good. When the video player and the audio player communicate in the Gigabit network environment, the communication delay is negligible.

When calculating the number of frames broadcasted by the video player, the division operation may be performed. The rounding may cause an error, but the error does not exceed 16.67 ms. For example, when it is calculated that it should play to 59.99 frames, the program will clear the decimals and think that it should play to 59 frames.

From the above analysis, theoretically, the error caused by this synchronization method is between −16.67 ms and 0 ms, which is in line with national standards.

4.2   The Synchronization Effect

The school host uses a wireless network card to establish a wireless LAN socket connection with the notebook, and sends a timing code to the video end every second. The video end is synchronized according to the current code. The experimental results are as follows, and the subjective test results are good. The real machine and the real network are used to connect two subsystem programs to play the 8K audio and video files. The detailed test uses the methods of manual observation, packet capture and GUI response time to establish the connection, play, pause connection and other functions. The test results show that the triggering and response between the two systems is fast, the communication is normal, the audio and video synchronization method is reliable, and the ultra-high-definition video and audio broadcast system for shared living room is realized (Fig. 3).

Fig. 3. The demonstration effect

5   Conclusion

In this paper, through the problems encountered in the process of compiling and debugging the synchronization program of DMS audio playback system and 8K video player, the experimental and theoretical analysis of DMS audio playback system working mechanism was carried out. The reason for the emergence of the video synchronization scheme is proposed. The feasible audio and video synchronization scheme is implemented. The ultra-high-definition video and audio broadcast system for shared living room is realized.

Acknowledgments. This research was supported by the Fundamental Research Fund of the Central University (No. CUC2019T008) and Social Science Research Project of the National Radio and Television Administration (No. GDT1913).

References

1. Hamasaki, K., Nishiguchi, T., Okumura, R.: 22.2 multi-channel sound system for ultra high definition television (UHDTV). SMPTE Motion Imaging J. **117**(3), 40–49 (2008)
2. Hamasaki, K., Hiyama, I.K.: Developed 22. 2 multi-channel sound system. Broadcast. Technol. (25), 9–13 (2006)
3. Xie, X.: Principle of Stereo Sound, pp. 2258–2260. Science Press, Beijing (1981)
4. Research on microphone array recording method based on Lv Xiaoshi and Jin Cong's wave formation theory. Electro-Acoust. Technol. **36**(8), 43–45 (2012)
5. Nan, C., Li, B., Wu, Y.: Audio-video synchronization technology in dynamic network environment. Comput. Syst. Appl. **21**(11), 120–122 (2012)
6. Yang, H.: Optimal design of audio and video synchronization in streaming media. Nanjing University of Technology (2016)
7. Qi, C.: Research and implementation of audio-video synchronization. Harbin University of Technology (2018)
8. The world's first 10.5 generation line product for the 8K BOE panel market [EB/OL]. http://www.sohu.com/a/211610441_115565,2017
9. Editorial Department: Research on 8K Ultra HD TV Technology in Japan - Based on 8K Ultra HD (8K shv) Research. Film Telev. Prod. **1**(9), 31 (2015)
10. The strategic analysis report emphasizes price declines, improved upgrade capabilities and higher consumer awareness as key drivers of ultra-high-definition TV ingestion [EB/OL] (2015). http://www.strategyan-alytics.com/default.aspx?Mod=pressreleaseviewer&a0=5513

Relationship Between Science and Technology Finance and Economic Growth—Based on VAR Model

Jiangyan Huang[✉]

School of Economics, Shanghai University, Shanghai 200444, China
1749199408@qq.com

Abstract. The development of high technology industry is an vital position of economic growth, and effective financial support is a necessary condition for the development of high technology industry. This paper establishes a VAR model based on the investment in science and technology finance and the output value of high technology industry in Shanghai from 2004 to 2016, and studies the relationship between the development of science and technology finance and economic growth in Shanghai. And then the paper puts forward the prospect of creating the "Shanghai model" of financial support for science and technology innovation to promote the rapid and good growth of Shanghai economy.

Keywords: Science and technology finance · VAR model · High technology industry · Economic growth

1 Introduction

Science and technology is the first productive force and finance is the primary driving force [1]. Science and technology innovation and the development of financial market as well as their integration are playing an increasingly important role in economic construction and social progress. How to speed up the development of science and technology finance, use science and technology finance to promote the comprehensive improvement of independent innovation ability and enhance the core competitiveness of Shanghai is one of the research hotspots in recent years.

This paper, by using the output value of Shanghai high technology industry from 2004 to 2016 as the presentative indicator of economic growth, the use of government capital, enterprise capital, loan institution capital, capital market capital as the indicators to build financial input system of science and technology and to establish the VAR model to carry on the empirical analysis, to guide the financial development of science and technology and promote economic growth in Shanghai.

The rest of this paper is written as follows. The second part explains the selection of input and output index variables of science and technology finance in Shanghai and establishes an econometric model. The third part conducts an empirical test based on the determined index variables. The fourth part draws the relevant conclusion and puts forward the suggestion to the Shanghai science and technology finance investment to promote the economic growth.

© Springer Nature Singapore Pte Ltd. 2020
M. Atiquzzaman et al. (Eds.): BDCPS 2019, AISC 1117, pp. 1215–1222, 2020.
https://doi.org/10.1007/978-981-15-2568-1_169

2 The Selection of Variables and the Construction of the Model

The system of science and technology finance in this paper is based on the ecological community subsystem of science and technology finance of Zhang Yuxi (2018), which includes six parts: government agencies, science and technology innovation enterprises, science and technology loan institutions, science and technology capital market, venture capital investment and science and technology guarantee institutions. In view of the great uncertainty of venture capital and the lack of development time and data of science and technology guarantee in China, the impact of these two indicators is not considered. The output index of science and technology finance, that is, the representative index of economic growth, the paper selects the output value of high technology industry. Therefore, this paper will study the impact of Shanghai science and technology financial investment on economic growth from four aspects: government capital, enterprise capital, loan institution capital and capital market capital [2].

Government capital (gov). Scientific and technological innovation is an exploratory activity with large early investment, high sunk cost and strong uncertainty. In general, R&D activities have strong positive externalities, and its spillover effect may lead to market failure [3]. Government's investment in scientific and technological innovation can be used as a part of the index of scientific and technological financial investment while making up for the market failure. Therefore, this paper measures the financial support provided by the government for scientific and technological innovation by using the internal expenditure of R/D of the Shanghai municipal government.

Enterprise capital(ent). Scientific and technological innovation enterprises usually improve the competitiveness of themselves by increasing the investment in R&D activities [4], so this paper measures the financial support provided by enterprises for scientific and technological innovation by using the internal expenditure of R&D of Shanghai enterprises.

Loan institution capital (loa). In the system of science and technology finance, science and technology loan institutions mainly provide indirect debt financing for science and technology innovation enterprises [5]. Science and technology loan institutions in China usually refer to commercial banks, policy banks and other banking financial institutions. In consideration of the availability of data, this paper measures the support of science and technology from loan institutions by using the RMB loan balance of Shanghai financial institutions.

Capital market capital (cap). Science and technology capital market provides long-term capital support for science and technology innovation enterprises [6]. In China, the main board of stock market and the sub-markets of science and technology innovation enterprises server constitute with different life cycles and play an important part of science and technology financial investment. In the paper, the stock transaction amount of listed companies in Shanghai is used to measure the support intensity of capital market for science and technology innovation.

The output value of high technology industry(ion). The intelligence, innovation and strategy of high technology industry play an extremely vital role in the evolution of society and economy [7], which is an effective representative index of science and technology output in Shanghai, we can measure the economic growth to a certain extent [8].

The models are established by government capital (gov), enterprise capital (ent), loan institution capital (loa), capital market capital (cap) and high-tech industry's output value (ion). For the sake of eliminating the influence of heteroscedasticity, the logarithmic modeling of variables is obtained:

$$\text{lnion} = \alpha + \beta_1 * \text{lngov} + \beta_2 * \text{lnent} + \beta_3 * \text{lnloa} + \beta_4 * \text{lncap} + \mu \qquad (1)$$

α is intercept term, μ is random variable.

3 Empirical Analysis

The paper selects the relevant data of input indexes and output indexes of science and technology finance in Shanghai during the 13 years from 2004 to 2016. The input indexes include government capital (gov), enterprise capital (ent), loan institution capital (loa), capital market capital (cap). The representative index of Shanghai's economic growth is the output value of high technology industry (ion). The relevant data of government capital and enterprise capital come from *the Statistical Yearbook of Science and Technology of China,* the relevant data of loan institution capital are found in the website of the National Bureau of Statistics, the relevant data of capital market capital are found in *the Statistical Yearbook of China Securities and Futures*, and the data related to the output value of high technology industry come from Shanghai Science and Technology Statistics website. In the paper, the measurement software stata is used to build the model.

3.1 Correlation Analysis Between Variables

Before building the VAR model, the correlation analysis of variable lngov, lnent, lnloa, lncap and lnion was carried out to understand the correlation between the variables.

As can be seen from the result, the correlation between different variables including explanatory and explained variables is significant at the level of 5%. The correlation coefficients between explanatory variables lngov, lnent, lnloa and lncap and explained variables lnion are 0.9165, 0.9191, 0.8693 and 0.7733, respectively, indicating that each explanatory variable can effectively explain the explained variables. The correlation analysis can preliminarily draw an conclusion that the changes of government capital, enterprise capital, loan institution capital and capital market capital will lead to the change of the output value of Shanghai's high technology industry and then affect the economic growth of Shanghai.

3.2 Stability Test of Variables

For avoiding the pseudo-regression phenomenon of the established model, the paper first uses ADF unit root test to see the stability of government capital, enterprise capital, loan institution capital, capital market capital and the output value of high technology industry in Shanghai. The results are presented in Table 1. The test results mean that lnion, lngov, lnent, lnloa and lncap are stationary sequences, while lnloa is non-stationary sequences, but though the first order difference, $\Delta lnion$, $\Delta lngov$, $\Delta lnent$, $\Delta lnloa$, $\Delta lncap$ are stationary sequences, indicating that these variables are all single integral sequences of the first order, and the cointegration test can be done.

Table 1. Unit root test results of variables

	ADF test statistics	1% critical value	5% critical value	10% critical value	Conclusion
lnion	−3.232	−2.896	−1.860	−1.397	stable
Δlnion	−1.864	−2.998	−1.895	−1.415	stable
lngov	−1.681	−2.896	−1.860	−1.397	stable
Δlngov	−2.845	−2.998	−1.895	−1.415	stable
lnent	−2.294	−2.896	−1.860	−1.397	stable
Δlnent	−3.528	−2.998	−1.895	−1.415	stable
lnloa	−1.211	−2.896	−1.860	−1.397	uneven
ΔLnloa	−1.578	−2.998	−1.895	−1.415	stable
lncap	−3.257	−2.896	−1.860	−1.397	stable
Δlncap	−1.814	−2.998	−1.895	−1.415	stable

Note: If the ADF test statistic is not more than any of the three critical values, the variable is stable at the significant level of the corresponding critical value.

3.3   Co-integration Test

The results of trace test show that there are only two linear independent cointegration vectors (identified by "*" in Table 2), and the maximum eigenvalue test also means that the original hypothesis of "cointegration rank 0" ($65.4534 > 33.46$) is wrong at the level of 5%. the original hypothesis of "cointegration rank 1" ($28.6345 > 27.07$) is wrong at the level of 5%, but the original hypothesis of "cointegration rank 2" ($17.3473 < 20.97$) is right. Therefore, the cointegration rank is 2. To sum up, the variables lnion, lngov, lnent, lnloa, lncap have a equilibrium relationship in a long term.

Table 2. Johansen cointegration test results

Maximum rank	Trace statistic	5% critical value	Max statistic	5% critical value
0	122.8571	68.52	65.4534	33.46
1	57.4037	47.21	28.6345	27.07
2	28.7693*	29.68	17.3473	20.97
3	11.422	15.41	8.9694	14.07

3.4   The Stationarity Test of VAR Model

In progress of constructing a VAR model, it is necessary to determine the lag order of variables, more based on information criteria to determine. As far as this paper is concerned, in view of the limitation of the number of samples, the lag period of 1 is selected to build the VAR model.

In the VAR model, there are n*k roots, n refers to the number of variables, k refers to optimal lag period. In this model, there are five variables and the optimal lag period is 1, so there should be five points in the unit circle, and exactly five points in Fig. 1 are in the unit circle, and this means that the VAR model with lag period of 1 is stable.

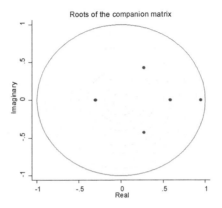

Fig. 1. Check diagram of unit root of VAR model

3.5   Construction of VAR Model

Based on the previous analysis, the VAR model is established. The government capital (gov), the enterprise capital (ent), the loan institution capital (loa), the capital market capital (cap) and the high technology industry's output value (ion) are used to determine the investment index of Shanghai's system of science and technology finance from 2004 to 2016. The regression results are listed in Table 3.

Table 3. lnion, lngov, lnent, lnloa, lncap vector autoregression model result

	lnion	lngov	lnent	lnloa	lncap
R^2	0.9497	0.9536	0.9972	0.9972	0.6324

It can be seen from Table 3 that the goodness of fit of VAR model with variables lnion, lngov, lnent, lnloa and lncap is very high, and its determinant coefficient R^2 is above 0.5. In the VAR model with the explained variable lnion, the determinant coefficient R^2 is 0.9497, which lays a foundation for further analysis of the role of science and technology finance investment in promoting the development of high technology industry.

3.6 Variance Decomposition Test

The variance decomposition test of Shanghai high technology industry's output value is mainly to further analyze the contribution of government capital (lngov), enterprise capital (lnent), loan institution capital (lnloa) and capital market capital (lncap) to the high technology industry's output value (lnion) in the science and technology financial system. The results of variance decomposition are listed in Table 4:

Table 4. Variance decomposition Result of lnion by lngov, lnent, lnloa and lncap

Peroid	lnion	lngov	lnent	lnloa	lncap
1	1	0	0	0	0
2	0.741348	0.162639	0.00019	0.053077	0.042746
3	0.632534	0.164202	0.011496	0.10656	0.085208
4	0.595813	0.159519	0.02292	0.121446	0.100302
5	0.588075	0.157706	0.027422	0.123691	0.103106
6	0.58674	0.157458	0.028414	0.123933	0.103455
7	0.586316	0.157483	0.028625	0.124043	0.103533
8	0.586005	0.157527	0.028724	0.124157	0.103586
9	0.585773	0.157534	0.028812	0.124254	0.103626
10	0.585627	0.157526	0.028887	0.124314	0.103646

As can be seen from Table 4, the fluctuation of output value (lnion) of high technology industry is most affected by its own impact (above 58% in each peroid). The contribution degree of period 1 is 100%, and the contribution value is 74.1348%, 63.2534% and 59.5813% respectively in the next few periods. In addition, the impact of government capital (lngov), loan institution capital (lnloa) and capital market capital (lncap) is also greater, and the impact of lngov is more significant. The contribution of lngov to lnion has reached more than 15% since period 2, while the contribution of lnloa to lnion has reached 10% since period 2 and has stabilized to more than 12% in each subsequent period. Compared with lngov, lnloa and lncap, the impact of lnent on lnion is the least, which is stable at 2.8% for a long time. Through the variance decomposition results of lngov, lnent, lnloa and lncap on lnion, it can be found that government capital, enterprise capital, loan institution capital and capital market capital in Shanghai all promote the output value of high technology industry, and the promotion effect of government capital is more significant. But over time, the role of load institution capital and capital market capital will rise. The contribution of loan institution capital increased from 5.3077% in period 2 to 12.4314% in period 10, and the contribution of capital market capital increased from 4.2746% in period 2 to 10.3646% in period 10. In contrast, the role of enterprise capital in promoting the development of high technology industry is weak, but in the long run, the contribution of enterprise capital is obviously increased, from 0.019% in period 2 to 2.8887% in period 10. The contribution of government capital on the whole is large, but in the long run, it shows a downward trend, from 16.2639% in period 2 to 15.7526% in period 10.

4 Conclusions and Recommendations

First of all, Shanghai's system of science and technology finance is a vital basis for the development of high technology industry. The empirical results show that the investment of science and technology finance is beneficial to improve the output value of high technology industry. Compared to the other three indicators, the contribution of government capital to the development of high-tech industry is the largest, for promoting the development of high technology industry, the Shanghai municipal government should increase the capital investment in science and technology finance to act as a guiding role of government capital. In the process of the development of high technology industry, the input of capital from enterprises plays a much smaller role, but in a longer period of time, the input of enterprise capital will be further brought into play, so enterprises need to properly increase the investment in R/D. With the further development of high technology industry, the demand for capital will be greater and greater. Government capital and enterprise capital are difficult to meet the needs, so the capital of loan institutions and capital market are also necessary. Therefore, Shanghai should actively guide loan institutions and capital market to give financial support to high-tech industry.

Secondly, a stable relationship between science and technology finance and the development of high technology industry in Shanghai [9] in a long term is shown. The empirical results show that the four indicators government capital, enterprise capital, loan institution capital and capital market capital that represent the investment of science and technology all play a certain part in promoting the development of high technology industry. In different stages of high technology industry'development, different input indicators play different roles, and shows the characteristics of complementarity and symbiosis among the four indicators.

Shanghai is actively building a scientific innovation center and an international financial center [10]. Science and technology and finance are two important engines in economic development, how to grasp the relationship scientifically between the tow, and strengthen the combination of each other is of great significance to promote Shanghai's economic growth.

References

1. Cao, H., You, J., Lu, R.: An empirical study on the development index of science and technology finance in China. Natl. Manag. Sci. **19**, 134–140 (2011). (in Chinese)
2. Wang, Q., Shi, X.: Analysis on the definition, connotation and practice of science and technology finance. Shanghai Finance **9**, 112–119 (2013). (in Chinese)
3. Zhai, H., Fang, F.: Research on the development of regional science and technology finance, R&D input and Enterprise growth-based on empirical evidence of listed companies in strategic emerging industries. Sci. Technol. Prog. Countmeasures **31**(5), 34–38 (2014). (in Chinese)
4. Du, J., Liang, L., Lu, H.: Research on regional science and technology financial efficiency in China based on three-stage DEA model analysis. Econ. Financ. Res. **31**(6), 84–93 (2016). (in Chinese)

5. Zhang, M., Zhao, X.: The connotation and function of science and technology financial center and its practice in Shanghai. Sci. Manag. Res. **34**(4), 101–105 (2016). (in Chinese)
6. Chen, X., Cheng, Y.: A study on the influence of science and technology finance on the development of high-tech enterprises in Fujian Province. J. Fujian Inst. Adm. **5**, 79–90 (2017)
7. Zhang, Y., Zhang, Q.: Dynamic comprehensive evaluation of regional science and technology financial ecosystem. Sci. Res. **36**(13), 1963–1974 (2018). (in Chinese)
8. Juan, P., Yuxi, Z.: An empirical study on the synergy between scientific and technological innovation and science and technology financial system. Financ. Theory Math. **5**, 43–48 (2018). (in Chinese)
9. Luo, Q., Zhu, S.: Thoughts on the development of science and technology finance in Shenzhen in the new period a study on the efficiency of science and technology finance input and output. Sci. Technol. Manag. **16**, 74–80 (2018). (in Chinese)
10. Tao, J., Yingjun, S.: An empirical study on the relationship between science and technology finance and economic growth in Shanghai-based on VEC model. Sci. Technol. Manag. **20** (5), 72–76 (2018). (in Chinese)

Countermeasures for the Innovative Development of China's Sports Industry Under the Background of Big Data

Chenliang Deng[✉], Zhaohua Tang, and Zhangzhi Zhao

Sports Department, University of Electronic Science and Technology of China,
Chengdu 610054, China
850620188@qq.com

Abstract. The traditional sports industry structure has undergone radical changes under the background of big data. Big data with the top technology level provides numerous sports consumers with more personalized, customized, precise and other services that the traditional sports industry cannot imagine. Big data under the background of development of sports industry should speed up the industrialization of sports work practice, aim at sports market positioning, execution concept updating and solid talent pool, take the initiative to improve sports personalized content customization platform, software and information service as the core to build sports industry chain, industry focus on the core elements to optimize the perfect sports industry development environment, the value of using big data to grasp the sports industry development opportunities.

Keywords: Big Data · Sports industry · Industrial transformation · Development countermeasures

1 Introduction

Big Data can also be called a huge amounts of Data, huge amounts of Data, information, etc., first put forward by the scientific community, it refers to the amount of Data involved, the scale is huge, made up of many small Data Set (Data Set), the size of a single Data Set on from several trillion (TB) to several trillion (PB) range, can by artificial in a reasonable amount of time to capture, management, processing, interpretation and organized for the human information [1]. The analysis of big data can start from the four dimensions of data volume, speed, diversity and authenticity. With the development to today, computer science has not only made remarkable progress in the volume of computers, but also made great breakthroughs in the four aspects. For example, the Internet of things makes people and things closely connected, eliminating the low efficiency and high cost of material use. Cloud computing collects the terminal computing of users in a big data processing center with faster computing, which saves the hardware cost of each user and improves the computing speed [2]. These huge breakthroughs have been fully utilized in the context of big data, making big data have the highest scientific and technological level of mankind so far, and have incomparable scientific and technological advantages.

© Springer Nature Singapore Pte Ltd. 2020
M. Atiquzzaman et al. (Eds.): BDCPS 2019, AISC 1117, pp. 1223–1229, 2020.
https://doi.org/10.1007/978-981-15-2568-1_170

Big data has been widely used in all walks of life and penetrated into every field of life [3]. Big data is an important factor of production in the future and an important driving force to help human productivity. Data analysis for the importance of sports is self-evident, in the case of NBA, detailed data for each player, such as height, weight, arm exhibition, bounce, high touch, shooting, 3-point percentage, the number of blocks, the action and time, and so on, as long as the search for the name of the player, can show the comprehensive data in the database data, basically is composed of data, from head to foot is players training guidance, is also on the team measured a player the players deal reference index. Under the background of big data, the importance of data has been raised to an unprecedented height. The sports industry needs to seize the opportunity of this era make full use of data analysis and take advantage of the opportunity to develop.

2 Sports Industry Is Facing Great Changes Under the Background of Big Data

2.1 Traditional Sports Industry with High Investment and Low Production Will Be Changed

Is a traditional labor intensive manufacturing sports products manufacturing, in the early stage of development, very little contact with competitive sports, basically is designed for the national athletes only producing sports products and exercise equipment, but due to the small sports management mechanism of binding, all kinds of capital investment, money collecting faster, so have maintained a good momentum of development. But competitive performance, there is no such a loose policy environment, selection and training of athletes, sports organization, from beginning to end by the central government or local physical culture and sports authority dockyard, killed the vitality of the socialist market, no release of the socialist market potential, only rely on the government's efforts to personnel, without wisdom, sports economy still so weak and it is had buried the bane of malformed development [4]. The sports, fitness and leisure industry is struggling to develop in the gap between the big hands of the government. Although the government intends to give priority to official management and promote the industry from top to bottom, the rent-seeking resistance is too great and the market is still not formed.

2.2 Combination of Traditional Sports Industry Structure Is Facing Optimization

China's traditional sports industry structure is not reasonable, the level is not high, leading to the sports industry is greatly bound, it is difficult to let go of development. From the current situation, China's sports industry is not able to meet people's basic sports consumption needs, the main reason is the unreasonable structure of sports industry. This is mainly reflected in the fact that the leading industries are not sports, fitness and leisure industry and sports management industry, while the sports intermediary industry and sports training industry are relatively weak and cannot take on important responsibilities. There is no positive interaction between various sports branches and they cannot develop in a coordinated way [5]. Therefore, in the context of big data, the sports industry must carry out industrial structure optimization, improve policies and regulations, enlarge and strengthen the sports products industry, upgrade the organization and management industry, and make overall planning among industries, industries, regions and governments to promote the sustainable development of the sports industry.

2.3   Traditional Sports Industry Business Model Has Been Overturned

Most of the traditional sports industry business models are dominated by the design and manufacturing links. For example, in the event design, the selling point and hot spot of the event should be determined. Special economic rules can also be used as hot spots, such as world e-sports competitions; In addition, the popularity of the athletes, the place of the competition, the time of the competition can become the hot spot of the event [6]. Manufacturing in the sports industry is also lagging behind, decoupled from the needs of sports consumers. The products produced by enterprises are far from what consumers really want. Not only the production of sports products is backward, but also the output of professional athletes and sports stars are eager for quick success and instant benefit. Liu xiang, for example, has been inundated with commercials since he won the gold medal at the Athens Olympics, destroying a generation of sports stars. In the context of big data, the sports industry is dominated by sales and consumption, focusing on the needs of consumers. Sports industry is a comprehensive industry spanning manufacturing and service industries. That is to say, sports consumer goods not only include the manufacturing of various sports products, such as ball products, sportswear, fitness equipment, etc. It also includes various services, such as fitness coach services, swimming lifesaving services, sports training services and so on. Sports consumption in the context of big data should be more targeted, so it is necessary to analyze big data to find the right market [7]. Instead of blindly throwing sports goods into the immature market, like the traditional sports industry, which relies on game tickets and manufacturing behind closed doors. The emerging sports industry also has event souvenirs. For example, every major event has its own mascots. The Beijing Olympic Games have fuwa, and the guangzhou Asian games have their own five RAMS mascots. In the context of big data, the consumption link of the value chain of sports industry is a whole set of processes, starting from sports consumers watching matches, watching sports programs, and enjoying fitness services, etc.

3   Strategies for Innovative Development of Sports Industry Under the Background of Big Data

3.1   Actively Promote the Application of Big Data to Accelerate the Active Practice of Sports Industrialization

Big data has its own life cycle. To promote the application of big data, it is necessary to develop the key technologies of big data around these life cycles, and take data analysis technology as the core of application practice. We will take strengthening business intelligence, artificial intelligence and machine learning as the center of theoretical research and technology research and development in the field of big data to lay a solid foundation for development. Unstructured data processing technologies and visualization technologies, the Internet of things, cloud computing, etc. should be integrated to form mature and available solutions. Take the optimization of sports industry structure as an opportunity to accelerate the demonstration application of big data. The first is to guide the participation of enterprises in the sports industry in the fields with large amount of data, such as athletes, sports product manufacturing, sports consumer preferences, competition design, sports leisure industry model, and vigorously develop integrated business circle models such as data collection and utilization monitoring, business decision-making, and rapid implementation. Instead, it aims to meet the service needs of domestic and foreign sports fans, consumers and spectators, and promote the optimization of big data in sports industry structure as soon as possible, as well as the application of personal-oriented interactive platform, entertainment link and life and fitness, and continuously upgrade the service development level of data processing software and hardware.

3.2   Targeting Sports Market Positioning and Precision Marketing Based on Big Data Technology

The most advanced big data technology is to collect, process and analyze super large amount of data to obtain effective market research in a reasonable time, instead of months or even years of empirical market research, which saves enterprises' labor cost, capital cost and more importantly, time cost. When these sunk costs are saved, they can promote the rapid promotion and development of enterprises and lay a solid foundation for future competition. The trend of consumer demand is the vein of market positioning, which is the orientation and the business process inevitably designed by scientific and reasonable enterprise organizations. Enterprise decision-making is no longer a head-scratching project, but after detailed analysis of big data technology, data driven enterprise decision-making has gradually become the main means of enterprise organization. The implementation of precision marketing can understand consumers' willingness to pay more quickly and deeply than competitors, and launch the most competitive products in the first time, no matter it is organized events or actual sports products [8]. It must be clear that the huge value of big data will change with the change of space and time. Only by constantly updating the data with the change of time can the value of big data be brought into full play and the accuracy of data analysis be guaranteed.

3.3   Update the Concept of Keeping Promise and Promote the Future of Sports Industry with Solid Talent Reserve

The concept renewal should be thoroughly implemented not only in the sports industry, but also in other industries, because this is a rapidly changing information age, if the change is not rapid enough, it may be like the fate of Nokia's fate, the giant of the mobile phone industry suddenly collapsed. This is especially true in the sports industry under the background of big data. We should constantly update our ideas, absorb the advantages of traditional sports industry, abandon the disadvantages of traditional sports industry, and seize the data analysis to provide better sports consumer goods for sports consumers faster [9]. In addition, under the background of big data, the necessary condition is talent reserve. Only with sufficient talent resources can we make good use of big data analysis technology and obtain the latest technological update under the background of big data.

3.4 Actively Improve the Personalized Sports Content Customization Platform to Show Customer Value

Today's consumers, because of the rapid pace of life, as well as the rapid rise of mobile terminals, consumption has become everywhere at any time. A sports goods deal can be done at a few hours in the morning, on the subway, anywhere, anytime. The sports industry has become the dominant market for housing, at this time, highlighting customer value has become the number one goal of the sports industry. In order to highlight customer value, it is necessary to improve the personalized customization platform, so that every consumer of sports products can feel their uniqueness and importance. Every consumer in the sports industry is pursuing self and individual development, which inevitably requires the sports industry to use big data to analyze the future product content design and brand positioning.

3.5 Build the Sports Industry Chain with Software and Information Services as the Core

Hardware is the foundation, while software and information services are the superstructure. Only by building excellent software and information services on the basis of hardware can the level of business intelligence of sports industry be reflected. The most effective and fastest way is to promote in-depth cooperation between enterprises, universities and colleges, with practical utilization as the core and research as the auxiliary. Software and hardware enterprises and service enterprises should also cooperate closely on the improvement and application of new technologies to ensure that the hardware is hard enough, the software is smooth enough and the service is comprehensive enough. In order to form a benign ecosystem and continue to develop in the right direction, it is necessary to accelerate the integration of local information services and prevent foreign high-tech information enterprises from forming a monopoly, leading to a bottleneck in the development of sports industry under the background of big data [10].

3.6 Optimize and Improve the Development Environment of Sports Industry by Focusing on the Core Elements of the Industry

The development of big data technology needs good policy and regulation environment, fair competition market environment and abundant talent reserve environment. The core of the sports industry is exactly like this. Big data and core elements of the sports industry are gathered together and mutually integrated. More financial support should be given to big data projects, such as cloud computing engineering, software service engineering, high-tech engineering, and education resources. We need to ensure a competitive environment in the market, so that the best can stand out and the inefficient and wasteful ones can exit. It is necessary to timely enrich the talent pool and master the latest big data analysis technology to ensure the continuous optimization of the big data development environment [11].

3.7   Exploit Big Data Value to Grasp the Opportunity of Sports Industry Development

Demand drives development. The biggest driving mechanism for the development of big data to promote the sports industry is the development of the sports industry itself. After all, big data analysis technology is a technology to promote the development, rather than an ontology to be developed. At present, one of the driving forces for the development of big data is that the development of big data has brought us positive influences from various aspects. Taking e-commerce platform as an example, the sales volume of sports consumer goods has increased in the share of e-commerce platform, and the physical store consumption of traditional sports industry has been gradually replaced by e-commerce platform [12]. And this is a very small aspect of big data technology, and there are many other aspects that big data technology plays an irreplaceable role. In order to seize the opportunity of sports development and usher in the spring of sports industry development, the value of using big data should be explored in depth.

Acknowledgments. This research was supported by Basic Scientific Research Operating Expenses Project of the Central University of China (Grant No. ZYGX2019J143).

References

1. Lynch, C.: Big data: how do your data grow. Nature **455**, 28–29 (2008). (in USA)
2. Chen, R.: Challenges, values and countermeasures in the era of big data. Mob. Commun. **17**, 14–15 (2012). (in Chinese)
3. Editorial Department of this Magazine: Big data management: concepts, technologies and challenges. Comput. Res. Dev. **50**, 146–169 (2013). (in Chinese)
4. Zhang, Z., Cai, M.: Discussion on maritime big data. China Maritime **12**, 39–41 (2015). (in Chinese)
5. Zhang, R.: Research on the optimization of China's sports industry structure. Journal of Physical Education **18**, 21–26 (2011). (in Chinese)
6. Guo, Y., Luo, J., Song, Z.: Research on the influence mechanism of large-scale sports events on urban visual image. J. Hangzhou Norm. Univ. **12**, 278–283 (2013). (in Chinese)
7. Zhao, Y.H., Huang, Y.H.: Exploring big data applications for public resource transaction. J. Phys: Conf. Ser. **1087**, 1–8 (2018). (in England)
8. Renato, A., Konstantinos, P.: Sports analytics in the era of big data: moving toward the next frontier. Big Data **7**, 1–2 (2019). (in USA)
9. Spaaij, R., Thiel, A.: Big data: critical questions for sport and society. Eur. J. Sport. Soc. **14**, 1–4 (2017). (in England)

10. Rein, R., Memmert, D.: Big data and tactical analysis in elite soccer: future challenges and opportunities for sports science. SpringerPlus **5**, 1–13 (2016). (in USA)
11. Greenbaum, D.: Wuz you robbed? Concerns with using big data analytics in sports. Am. J. Bioeth. **18**, 32–33 (2018). (in USA)
12. Dong, Y., Zhong, J., Ding, F.: Research on the development path of China's sports industry in the age of big data. Sports Cult. Guide **12**, 76–81 (2018). (in Chinese)

Application of Association Rules Algorithm in Teaching Reform Under the Background of Internet Plus Era – Taking the Teaching Reform of International Trade as an Example

Feiyan Zhong[(⊠)]

Guangzhou City Construction College, Guangzhou 510925, China
45847192@qq.com

Abstract. Under the background of internet plus era, the teaching reform of international trade major is imperative, but the reform is very difficult. Therefore, an algorithm based on association rules was established to calculate and analyze the related issues of teaching reform. In this way, the difficulty of artificial reform can be simplified, and the goal of reform can be achieved. Through the test and experiment of algorithm, it is proved that our algorithm has great advantages. It not only has short time and high accuracy, but also is an algorithm that has been utilized well, which is suitable for our teaching reform.

Keywords: Internet · International trade · Association rules · Data mining

1 Introduction

In today's society, the internet plus era has come, with rapid development of cross-border e-commerce. But at present, the teaching of international trade major in China already can't keep up with the pace of social development, So it's necessary to make appropriate reforms to use social resources better and cultivate more useful talents for society [1]. In addition, computer technology develops rapidly, and it has been a shortcut of our development to use computer algorithms for reform and calculation. Computer algorithms can be used in many fields [2]. In particular, the association rule algorithm is the computer algorithm which is more suitable for the reform. For the use of association rules, there are still some omissions. For computer algorithms, each algorithm has some shortcomings and deficiencies. The shortcoming is not well understood at present, and constant research is still needed to make up for the deficiency in the algorithm [3].

According to the highly developed computer technology, an algorithm model based on association rules is established to analyze the relevant data of the teaching reform in this paper. The work of curriculum reform of international trade major is liberated from human labor by using the powerful computing ability of computer technology and various data processing methods in association rules, and the powerful computer technology is used for calculation and analysis, which simplifies the amount of labor, and also makes the analysis of data more accurate [4]. Another very important aspect is

© Springer Nature Singapore Pte Ltd. 2020
M. Atiquzzaman et al. (Eds.): BDCPS 2019, AISC 1117, pp. 1230–1238, 2020.
https://doi.org/10.1007/978-981-15-2568-1_171

that the deficiency of the algorithm can be made up and analyzed through the research of association rules algorithm, thus contributing to the development and research of association rules algorithm in the future. It can also provide some help for teaching reform [5].

2 State of the Art

The data mining algorithm of association rule started very early in foreign countries, and it has been the focus of foreign scholars. However, domestic research is very late, and it was brought up in the late 90s of the last century and developed slowly. However, foreign scholars and scientists have carried out detailed research and analysis on the association rules with the support of the developed computer technology [6]. So that the association rules develops rapidly. From the 80s of last century to the beginning of this century, the academic seminar on the topic of association rules has been held for fourteen times. What's more noticeable is that the use of association rule data mining in foreign countries has been divorced from the theoretical research stage, has begun a substantive probation period, and has played a huge role in education, medicine and other departments. It has achieved remarkable achievements in teaching reform [7].

At present, the research on association rules is still not very deep in China. Foreign countries have begun to put it into practice, but our country is still at theoretical research stage [8]. This is mainly because many scholars only focus on theoretical research but don't care too much about practical application. As a research topic, it has been studied by colleges and universities continuously. Because of the limitation of computer technology, the association rule algorithm developed too slowly, but the constant development of computer technology in recent years also promoted the research progress of association rules [9]. The research of association rules has been extended to many fields and has achieved some success. It is believed that the domestic research speed will catch up with foreign countries in the near future and reach the world class level [10].

3 Methodology

3.1 Calculation Steps of Association Rule Data Mining Algorithm

For the teaching reform, the association rule is a calculation rule that is very suitable for application. The association rules algorithm is used to study the teaching reform of international trade major under the background of internet plus era in details. But the association rules algorithm is combined with data mining, which can make our calculation more accurate and fast. For the use of association rules algorithm, the first thing is data mining. First of all, the data mining analysis is conducted on the data that we should pay attention to, and then the association rules are calculated, which is a safer way to deal with it. But data mining can't be interconnected without certain order and steps, so that the data excavated can be analyzed and sorted out in calculation, and the data can be summed up and associated when association rules are calculated. The data mining steps of association rules can be divided into two steps. In the first step, based on the minimum support degree, the highest frequency item sets are found in a large amount of data that we associate with. In the second step, the association rules are deduced from the high frequency item sets according to the minimum confidence level. Figure 1 is the mining sequence of our data mining.

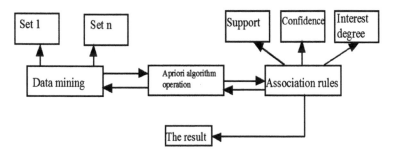

Fig. 1. Data mining calculation steps and methods

Through the information transfer path in above figure, it can be seen that our data is input into the data mining model to do simple data mining firstly, and then the data is calculated and analyzed simply through the data mining calculation rules set. Finally, data calculated by data mining is input to the Apriori algorithm phase in the next node. This part is the key step of using association rules algorithm. After data mining, the data are analyzed step by step here, and all the useless information is ruled out until the data left is highly correlated data. This part efferent data is highly concentrated, which is highly centralized by us. Then the data can be transmitted to the next stage, and the detailed study of association rules is carried out. Through the confidence, interest-ingness and other computing rules, the data we really want is rearranged and assembled, and then the calculation is conducted with the Apriori algorithm until the frequency of the data reaches our requirements. Finally, the data that meets our requirements are exported. This method of calculating back and forth is helpful to improve the accuracy of the data.

3.2   Research on the Algorithm of Association Rule Data Mining

Association rules and data mining are usually inseparable, and so are the applications in this paper, which is determined by the nature of data mining and association rules. Data mining is responsible for mining data carefully and doing deep analysis on a lot of data. The deep connection of many data is very close, and it's possible that the connections can't be found by simple associations and observations. Only through the form of data mining can the deep meaning of data be dug out, which is also a very important reason of data mining. Another point is that data mining can also analyze the connotation of the data simply, and the irrelevant data can be excluded in the data mining phase, which can also reduce the difficulty of association rules calculation in the next phase. It's very easy that the association rules calculate the data after data mining. For the data that finishes the data mining, the association can be calculated easily, and the weight of all kinds of information in association rules can be analyzed, which is very important. Another difficulty is the calculation of the threshold of association rules. The algorithm research of association rules is analyzed and optimized strictly.

Next, the concrete formula and research content of association rule algorithm are introduced. The first thing introduced is the concept of "item". Items are the basis of association rules, and what the association rules calculate is the relationships among these items and the causal relationship between them as well as the importance comparison. Association rule is to associate one item with another and calculate them. Many items form a set, which is the same as the set in mathematics. The item set represents $I = (i_1, i_2, i_3 \cdots i_n)$, and the data item set represents $X = (k_1, k_2, k_3 \cdots k_n)$. The expression method of transaction item set is $D = (t_1, t_2, t_3 \cdots t_n)$. The introduction of the basic information about items is finished, and then the research of association rule algorithm calculation is started below.

The concept of "the support degree of item set" is introduced, and the support degree is the most important measurement standard in association rules, which has a great influence on the calculation results. The support degree of transaction D is $Suppert(X)$, and the probability is $P(X)$. The support degree contains probability, and the formula of support degree is defined as the following formula:

$$Suppert(X) = \frac{Sup(X)}{|D|} \tag{1}$$

The above formula $Sup(X)$ is the support degree of data set X.

The frequent item set is a measure unit, which is usually used to describe a threshold for describing transaction density. The frequent item set can be written in the form of min sup. In terms of the concept of support, this is the minimal support degree. Compared with $Suppert(X)$, the frequent item set is less than or equal to $Suppert(X)$. In the definition and calculation process of association rules, it can be found that in a transaction, any two item sets are associated with each other, but the size of the association is different. How to describe the strength of this association? The two concepts of support degree and credibility are introduced, and the two concepts are used to restrict. The associations needed are left, and the irrelevant or unnecessary associations are excluded. This allows the data to be included in the required range. Below is the introduction of the two concepts of credibility and support degree.

Before introducing the support degree and credibility, the concept of association rules should be introduced first. Association rules are similar to the implication expression $X \Rightarrow Y$. The X and Y are two disjoint item sets. The support degree and credibility are the parameter description used to express the strength of association rules.

The support degree: the support degree of item set $(X \cup Y)$ is called the support degree of $X \Rightarrow Y$, and it can be interpreted as the percentage of the $(X \cup Y)$ in the data, which is the probability size. The formula is expressed as follows:

$$Suppert(X \Rightarrow Y) = \frac{Sup(X \cup Y)}{|D|} \tag{2}$$

The credibility: The credibility in association rules is the ratios of the number of transactions that contain both X and Y to the number of transactions that contain X in the database, namely, the conditional probability $P(Y|X)$, and its manifestation is as follows:

$$Confidence(X \Rightarrow Y) = \frac{Sup(X \cup Y)}{Sup(X)} \tag{3}$$

The above is all the calculation formula about correlation algorithm. Many common problems can be solved by the above formula. However, this calculation method is still not enough for this paper, and it is difficult to meet the requirements of education reform. In addition, this calculation can lead to a very serious problem, that is, sometimes it is found that in the calculation the support degree and credibility are very high, but the relevance this item to the object of the study is not very strong, which is because the restrictions are not enough. Therefore, a new concept is introduced, that is interest degree. The emergence of interestingness can filter out many irrelevant factors automatically, plays an important role in improving the accuracy of calculation, can improve the anti-jamming ability of the association rules algorithm, makes the main research object more prominent, and reduces the computing time of association algorithm. The formula of interest degree is defined as follows:

$$I = Interest(X \Rightarrow Y) = \frac{confidence(X \Rightarrow Y) - Support(y)}{max\{confidence(X \Rightarrow Y), Support(y)\}} \tag{4}$$

According to this formula, it can be determined that the greater I than 0, the higher the interest degree is. The smaller I than 0, the lower the interest degree is. The above is all the rules and methods of calculation in this paper, but this method can only be realized by inputting it into the computer. Figure 2 is the computational steps in the computer designed.

In addition, when the model is built in the computer, the requirement of choosing database is also very high, and the database selection for different computations is summarized in Table 1 below.

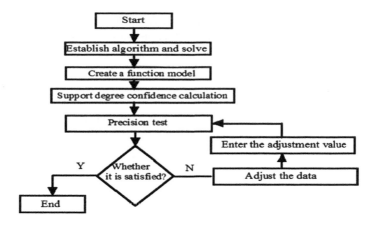

Fig. 2. The calculation steps in the compute

Table 1. Association rules data mining algorithm in the project management application database selection table

Amount of information	Number of concurrent data access records (per/sec)	System stability requirements	Maintenance level	Candidate database
Medium (20–100)	5000–50000	Moderate	General	SQL Server, MySOL
Large (100–1000)	50000–200000	Relatively high	Better	SQLServer, Oracle
Very large (over 1000)	More than 200000	High	Good	Oracle

4   Result Analysis and Discussion

After the calculation rules of the algorithm and the establishment of the computer model are finished, it is necessary to test the use of the algorithm. Through test, the problems or deficiencies in algorithms are found for later modification and maintenance. It can also prove the superiority of the algorithm.

A computational experiment of five projects is established, the items are A/B/C/D/E, and the weight of each part is set as 0.9, 0.6, 0.5, 0.4 and 0.2. It is assumed that the minimum support degree σ is 0.35, the threshold τ is 3. The maximum possible length of state frequent item set is 4. Then, the data mining is tested and calculated. Firstly, the first iteration is made on the parameters, and the number of expenditures of the five sets is obtained. According to the formula: support degree = support number/total number of transactions, C_1 is obtained. Then, according to the minimum transaction support degree set 0.56, the set L_1 is obtained. The calculation set table is shown in Table 2. The calculation process of the following table is as follows, and the weighted temporal support degree of item set A is $S(\{A\}) = W(A) \times \frac{3}{6} = 0.45 \geq 0.35$.

So item set A is a frequent item set, for item set B, $S(\{B\}) = W(B) \times \frac{3}{6} = 0.3 \leq 0.35$. So B is not a frequent item set. With this method, the results of the following table are obtained eventually. The frequent item sets of weighted state are $L_1 = \{\{A\}, \{C\}, \{D\}\}$.

Table 2. Calculation of the data collection table

C1			C2			C3		
Item set	Support count	Life cycle	Item set	Support count	Life cycle	Item set	Support count	Life cycle
A	3	[1, 6]	AB	2	[1, 6]	ABE	2	[1, 6]
B	3	[1, 6]	AC	2	[1, 6]			
C	3	[1, 3]	AE	2	[1, 6]			
D	4	[2, 5]	BE	2	[1, 6]			
E	3	[1, 6]						

The above experiments show that the algorithm not only can calculate the reasonable results, but also has advantages over other methods. In this paper, the calculation steps of association rules algorithm are much shorter than those of other algorithms, so that the computation time is shortened greatly. The advantages of the algorithm used in this paper are prominent, and the algorithm has good practicability compared with other algorithms. Through calculation test, it is found that there are many advantages of the algorithm used in this paper, and the summary is in Table 3.

Table 3. Comparison of the algorithm and the general algorithm

Algorithm class	Advantage	Disadvantages
Algorithm	Calculate a small amount	The calculation rules are cumbersome
	High accuracy	
	Eliminate, do not repeat the calculation	Set up more trouble
General algorithm	Longer use	Accuracy at the end
	Computationally cumbersome	Calculate for a long time

In addition, when support degree is different, the percentages of important item sets are also compared. The objects of the comparison are the association rules algorithm used in this paper and the traditional algorithm. The histogram of the two columns' percentages of important items is made for analyzing, as shown in Fig. 3.

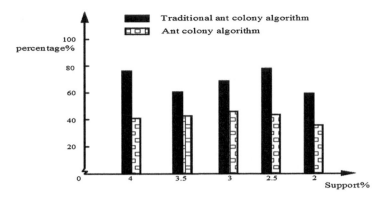

Fig. 3. The two important items set percentage of the histogram

From the figure above, it can be seen that no matter how much the support degree is, the percentage of important items calculated by the association rules algorithm in this paper is higher than the traditional algorithm, which shows that this algorithm has great advantages and good practicability. The results of data mining by the algorithm used in this paper have more practical value, and can provide powerful help for policy makers. Through the analysis of the above experimental results, it can be concluded that the association rules algorithm used in this paper has a strong use value and accuracy, and can calculate more results in shorter time. Not only the time used is short, but also the calculation result is more accurate than the traditional algorithm. It's an algorithm that works well.

5   Conclusions

The continuous development of computer technology has brought unlimited possibilities. With today's advanced computer technology, computers can be used to do a lot of work that people can't do or can't do very well, such as the teaching reform of international trade major under the background of internet plus era. At the present stage of international trade major, there are many places that should be paid attention to for a good education reform. These things are often done by human beings, which can cause a lot of oversight and errors. Therefore, a model using association rules algorithm was established in this paper to promote the progress of the reform. Through the test of association rules algorithm, it is found that the minimum support degree σ is 0.35, and the threshold τ is 3, and the maximum possible length of state frequent item set is 4, the algorithm can calculate the calculation results at a very fast speed through a simplified algorithm. Then, the contrast experiments on the percentage of important item sets with different support degree are also carried out. Through the test, it can be found that the percentage of important items calculated by the algorithm used in this paper is around 70%, while the percentage of important items calculated by the traditional algorithm is about 45%. The algorithm used in this paper has great advantages. Despite this, we still need to continue to work hard, and better algorithms should be worked out for the benefit of society.

Acknowledgments
Foundation Project: Teaching Reform Project of Business Specialty Teaching Steering Committee of Guangdong Higher Vocational Education—Under the Background of Internet Plus Era, Research and Exploration on the Training Mode of Modern Apprenticeship Talents in International Trade Major (Project number: YSYJZW2017YB76).

References

1. Sahoo, J., Das, A.K., Goswami, A.: An efficient approach for mining association rules from high utility itemsets. Expert Syst. Appl. **42**(13), 5754–5778 (2015)
2. Zhang, L., Lu, Z.: Applications of association rule mining in Teaching Evaluation. In: Proceedings of the 2018 3rd International Conference on Humanities Science, Management and Education Technology (HSMET 2018) (2018)
3. Ghafari, S.M., Tjortjis, C.: A survey on association rules mining using heuristics. Wiley Interdisc. Rev.: Data Min. Knowl. Discov. **9**(4) (2019)
4. Indira, K., Kanmani, S.: Mining association rules using hybrid genetic algorithm and particle swarm optimisation algorithm. Int. J. Data Anal. Techn. Strat. **7**(1), 59–76 (2015)
5. Sivanthiya, T., Sumathi, G.: An ontological approach for mining association rules from transactional dataset. Int. J. Eng. Res. Appl. **05**(01), 1313–1316 (2015)
6. Ait-Mlouk, A., Gharnati, F., Agouti, T.: Multi-agent-based modeling for extracting relevant association rules using a multi-criteria analysis approach. Vietnam J. Comput. Sci. **3**(4), 235–245 (2016)
7. Martín, D., Alcalá-Fdez, J., Rosete, A., et al.: NICGAR: a Niching Genetic Algorithm to mine a diverse set of interesting quantitative association rules. Inf. Sci. **355–356**, 208–228 (2016)
8. Fan, F.L., Yao, C.L., Yu, X., et al.: Research on association rules of mining algorithm based on temporal constraint. J. Am. Soc. Hypertens. **2**(6), 403–409 (2015)
9. Liraki, Z., Harounabadi, A., Mirabedini, J.: Predicting the users' navigation patterns in web, using weighted association rules and users' navigation information. Int. J. Comput. Appl. **110**(12), 16–21 (2015)
10. Onan, A., Bal, V., Bayam, B.Y.: The use of data mining for strategic management: a case study on mining association rules in student information system/Upotreba rudarenja podataka u strateškom menadžmentu: analiza slučaja upotrebe pravila pridruživanja rudarenja podataka u informac **18**(1), 41–70 (2016)

Impact of Cloud-Based Mechanism on Supply Chain Performance with Supply Disruption Risk

Jianchang Lu and Siqian Wu[⊠]

North China Electric Power University, Baoding 071003, China
15717199855@163.com

Abstract. Supply chain is virtually an open complex giant system. Given nonlinearity, uncertainty and dynamics in supply chains, the case of supply disruptions caused by natural disasters, terrorist attacks and other kinds of accidents is increasing. A throughput-based analysis reveals that cloud-based mechanism improves the information sharing and quality controlling during the process of supply chain management when supply disruption occurs. In this regard, we investigate how supply disruption affects supply chain performance in traditional supply chain and cloud-based supply chain system. Applying system dynamics simulation, this paper established a two-echelon supply chain simulation mode and analyzes how each of these sub-structures responds to a supply disruption. Result shows that cloud-based system is an advanced supply chain management model, compared to traditional mode, it can effectively decline inventory, reduce cost, improve to be out of stock and raise service levels when experiencing supply disruption.

Keywords: Supply chain · Disruptions · Cloud-based system · System dynamics

1 Introduction

In modern society, the flow of material in global supply chains could be devastated by man-made disasters or unexpected natural, such as earthquakes, economic crisis, terrorist attack, floods, hurricanes and fires [1]. Many scholars have explicitly indicated that supply disruptions have significant impacts on both short-term and long-term financial performances [2]. Extensive research suggests that organizational silos and lacking of collaborations are the main barriers for effective risk management programs [2].

Cloud computing (CC) technology allows for timely, dynamic, transparent and cost-effective information sharing of the whole supply chain [4], to overcome the weaknesses of traditional IT systems, mangers have started adopting cloud computing (CC) technology [3].

CC has achieved the information sharing and quality controlling to reduce inventory costs of the entire supply chain, and to improve customer service levels [3]. However, a few studies address the information sharing and quality controlling

© Springer Nature Singapore Pte Ltd. 2020
M. Atiquzzaman et al. (Eds.): BDCPS 2019, AISC 1117, pp. 1239–1244, 2020.
https://doi.org/10.1007/978-981-15-2568-1_172

mechanism on supply disruptions, this paper focuses on how information sharing and quality controlling mechanism reduces the losses of supply disruption in two-echelon stage supply chain system.

2 Model Descriptions

2.1 Conditional Management Model

In order to research the effect of disruptions on a traditional model and a CC model, we set up a Two-Echelon supply chain, including one supplier and one retailer, and made the simulation on VENSIM PLE platform [3].

First, we analyze the relationship between traditional supply chain. The information flowing between the members is not identical [5]. As the supply chain members enterprise are independent interest entity, each sector only takes into account of maximizing their own profit, the orders of every link of supply chain depend on sales forecasting of every link. The retailer receives customer demand information [6].

2.2 Cloud-Based Management Model

Figure 1 shows the flow of product and information between each member for cloud-based information sharing and quality controlling [4]. Cloud-based supply chain are defined as "full collaborative supply chain" meaning that node enterprises in the SC share the inventory information and critical demand through a scalable Cloud Computing platform [5].

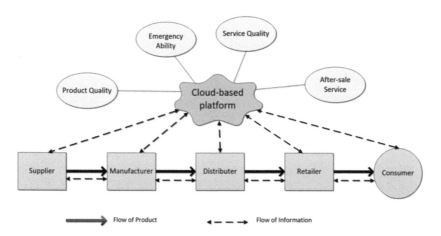

Fig. 1. The stock flow diagram of a Cloud-based supply chain

In cloud-based structure, both the supplier and the retailer could timely receive information of customer demand. Supply chain partners can check each other's planning forecasting and replenishment, and adjust procurement strategies by using electronic data interchange [6]. Thus, demand and information are instantly shared through CC technology.

3   Numerical Illustration and Simulation Analysis

We assume that the order rate and sales rate would be zero when disruption occurred. The equations of disruption are listed in Table 1, which means the value of the order rate and sales rate will be zero between the *mth* period and the nth period, namely, the process of disruption lasts for *mth* periods.

Table 1. Equations of three supply chain disruptions

Type	Equations
Supply disruption	=if then else(time ≤ m, Vendor Shortage Rate/Vendor Inventory Adjustment Time, if then else(time > n, Vendor Shortage Rate/Vendor Inventory Adjustment Time, 0))

In this numerical study, we assume stochastic market demand follows a normal distribution with a standard deviation of 5, the impacts of the disruption uncertainty over the order rate and sales rate are ambiguous. The SD models are then set up to analyze and compare the performance of a traditional and a cloud-based supply chain when supply disruptions occur at Two-Echelon supply chain [7].

This study examines the information sharing and quality controlling system on resilience performances across different performance indicators in a supply chain comprising a vendor and a retailer, the indexes are classified into inventory levels, shortage rate and profitability, which has been valued a lot for carrying out supply chain performance evaluation and reflecting the result feedback [8].

3.1   Inventory Levels

Analyzed excessive the traditional supply chain current inventory control mode in the presence of the vendor and retailer inventory, inventory fluctuations, we notice significant variability in stock levels, especially at the vendor, inventory levels swung up and down in part because of the bullwhip effect [9].

Both of the vendor's and retailer's stock face volatile swings when disruptions occur, reflects on Fig. 2, the shortage and backlog of stocks have occurred frequently, reducing the competitiveness of enterprises in traditional supply chain. As can be seen from Fig. 3, the variability of inventory levels across the supply chain is reduced effectively compared to the inventory levels in the traditional setting, which are verified by simulation study that CC mode performs better than traditional mode in restrain inventory variation and bullwhip effect.

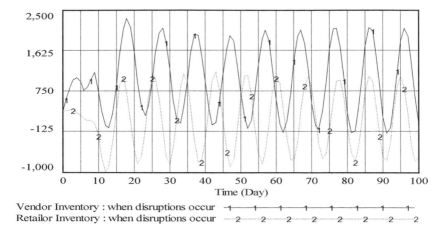

Fig. 2. Inventory levels in traditional supply chain after disruptions occur

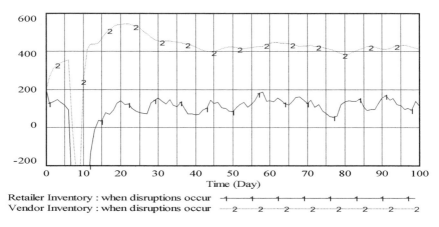

Fig. 3. Inventory levels in Cloud-based supply chain after disruptions occur

3.2   Profitability

In this subsection, the impacts of supply disruption on the profitability of each member and the entire supply chain are stimulated and compared [10]. Figure 4 display Vendor's and Retailer's Profit Comparison in the traditional supply chain, Fig. 4 reveal the Vendor's and Retailer's Profit Comparison in the cloud-based supply chains.

Through the simulation, we can observe that the supply disruption has a direct impact on both supplier's and retailer's profits. Gross profit for traditional supply chain fell 16.39% after experiencing supply chain, while in the cloud-based supply chain the gross profits decreased nearly 15.29%, as a result, the profit of vendor and retailer reached their lowest levels, which is shown by Fig. 4.

In terms of profitability, there is a striking difference between a cloud-based system and a traditional supply chain system. As seen from the figure and table above, the

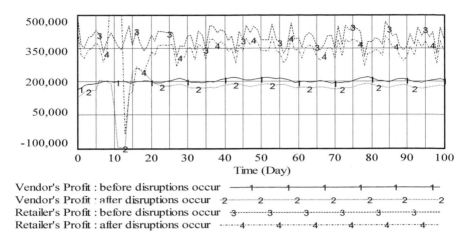

Vendor's Profit : before disruptions occur   ——1———1———1———1———1———1—
Vendor's Profit : after disruptions occur   ·2·······2·······2·······2·······2·······2·······2
Retailer's Profit : before disruptions occur   ·3·········3··········3·········3·········3·········3········
Retailer's Profit : after disruptions occur   ·····4········4·······4········4········4········4···

Fig. 4. Profit Comparison in Cloud-based supply chain

average operating profit of vendor and retailer in cloud-based supply chain increased by 34.60% and 29.67% respectively when compared with traditional one, the gross profit of supply chain grew by 31.33%. Even with a supply disruption lasting for 10 days, net profit for the whole supply chain still rose by 21.39%, vendor's earnings is up 25.72%, and retailer's earnings is up 18.92% contrast to traditional mode (Table 2).

Table 2. Test-Statistics between traditional mode and CC mode

		Before disruption		After disruption	
		Vendor	Retailer	Vendor	Retailer
Traditional	Mean	127670.31	278086.93	132608	278808.11
	Std. deviation	57722.272	77607.45	8967.6	28033.817
	Min	−81688	−53002	111910	214241
CC	Mean	171875.01	342967.94	202752	396373.96
	Std. deviation	5123.726	77117.292	8466.8	37597.66
	Min	−94639	−38417	169446	310121
Mann-Whitney U. Sig		0.038	0.015	0.036	0.024
Moses. Sig		0.020	0.005	0.019	0.018
Wald-Wolfowitz. Sig		0.017	0.011	0.018	0.026
Reject "$\mu 1 = \mu 2$"?		YES	YES	YES	YES

The results indicate that CC model is a cooperative model in which supplier and retailer can maximize the total profit of them. Supply chain members are strengthening their cooperation to achieve a win-win result by implementing information sharing and quality controlling technology. By analysis and comparison, it was concluded that cloud-based system can effectively reduce negative impacts of supply disruption on profit performance.

4 Concluding Remarks and Future Research Aspect

In order to prove the effectiveness and advantage of cloud-based supply chain, by implementing information and quality controlling mechanism, we employed system dynamics and made the simulation experiment to compare the effect of supply disruption on SC performance which includes inventory levels, shortage rate, and profitability. Statistics' results show that traditional mode is dominated by the cloud-based one, as to supply chain performance under supply disruption risks. Through the comparative analysis of the CC model and traditional model, we found traditional strategy takes on worse performance. However, cloud-based mechanism is a kind of advanced supply chain management mode, in which mode vendors and retailers perform well on reducing shortages rates and total inventory, and increasing customer demand fill rate when supply is disrupted, remarkably, the most significant competitive edges in cloud-based system lies in the strengthens of squeezing costs and improving margins.

References

1. Sawik, T.: Disruption mitigation and recovery in supply chains using portfolio approach. Omega **84**, 232–248 (2019)
2. Behzadi, G., O'Sullivan, M.J., Olsen, T.L., Zhang, A.: Agribusiness supply chain risk management: a review of quantitative decision models. Omega **79**, 21–42 (2018)
3. Vilko, J., Ritala, P., Hallikas, J.: Risk management abilities in multimodal maritime supply chains: visibility and control perspectives. Accid. Anal. Prev. **123**, 469–481 (2019)
4. Kochan, C.G., Nowicki, D.R., Sauser, B., Randall, W.S.: Impact of cloud based information sharing on hospital supply chain performance: a system dynamics framework. Int. J. Prod. Econ. **195**, 168–185 (2018)
5. Novais, L., Maqueira, J.M., Ortiz-Bas, Á.: A systematic literature review of cloud computing use in supply chain integration. Comput. Ind. Eng. **129**, 296–314 (2019)
6. Büyüközkan, G., Göçer, F.: Digital Supply Chain: literature review and a proposed framework for future research. Comput. Ind. **97**, 157–177 (2018)
7. David, Z., Gnimpieba, R., Nait-Sidi-Moh, A., Durand, D., Fortin, J.: Using Internet of Things technologies for a collaborative supply chain: application to tracking of pallets and containers. Procedia Comput. Sci. **56**, 550–557 (2016)
8. Zheng, M., Kan, W., Sun, C., Pan, E.: Optimal decisions for a two-echelon supply chain with capacity and demand information. Adv. Eng. Inform. **39**, 248–258 (2019)
9. Feng, Y.: System dynamics modeling for supply chain information sharing. Phys. Procedia **25**, 1463–1469 (2012)
10. Wilson, M.C.: The impact of transportation disruptions on supply chain performance. Transp. Res. Part E: Logist. Transp. Rev. **43**, 295–320 (2007)

Application of Big Data in Smart Marketing of All-for-One Tourism

Hao Wang[1,2](✉)

[1] Tianjin University, Tianjin, China
wh771201@126.com
[2] Hospitality Institute of Sanya, Sanya, China

Abstract. In recent years, with the rise of "Big Data", all kinds of industry have turned around with the help of "Internet+", and the deep integration of tourism and "Internet+" has also conformed to this trend of the times. With the advent of the mass tourism era, big data has played an increasingly important role in the tourism industry. With the development of big data technology, smart tourism has become a hot spot in tourism information construction. In addition to integrating tourism resources and strengthening infrastructure construction, how to make destination tourism marketing has become a new topic in front of the government tourism authorities.

Keywords: Big data · Smart marketing · All-for-one tourism

1 Introduction

Nowadays, China's tourism industry is booming and has become a national strategic industry, big data has played an increasingly important role in the tourism industry. All-for-one tourism development no longer relies solely on perceptual experience, but needs to rely on big data for decision making. In January 2017, at the National Tourism Work Conference, the National Tourism Administration required the provinces to "use big data to judge the development of the industry" and "start the tourism big data project". Because of the emphasis on big data by big data strategies and tourism management departments at the national level, and the strong demand for data in the tourism industry, "big data" has become an extremely hot topic in the tourism industry. [1] Almost all tourism-related institutions are discussing, using and selling big data. Each organization serves different fields and has different data sources and application models. In the era of mobile Internet, we have reason to believe that those who can reach the mobile terminal and has mastered the user behavior, has the ability to lead the entire industry. In response to the two functions of the domestic tourism management department—public management and service, the application of big data has been greatly developed. [2] For the marketing of tourism destinations, the application is still blank in current. There is no doubt that the use of big data to improve the effectiveness of tourism destination marketing has an infinite prospect.

2 The Practical Significance of Big Data in Helping the Development of All-for-One Tourism

Analyzing from different angles, it is of great practical significance to use big data to promote the all-for-one tourism practice and development.

2.1 Analysis from the Perspective of Supply

All-for-one tourism is a supply-side reform, and big data brings strategic positioning, precision marketing and business innovation to tourism destinations. Commodity and logistics information in the Internet of Things, interaction information between people in the Internet, and location information are the three main sources of big data. Big data has broken through the narrow vision of tourist attractions and bring changes at least two aspects, one is to help strategic positioning and precision marketing of tourism destinations, and the other is to contribute to the innovation of tourism and products. In addition, when all-for-one tourism launches a specific area as a complete tourist destination, the position strategy is very significant. [3] In the past, marketing positioning tends to focus on tourism resource endowment and historical accumulation, while big data sorts out the amount, structure, interests, route and scenic spots preferences of tourists, which helps tourism destinations to accurately locate tourists, competitors and resources.

As a supply-side structural reform, all-for-one tourism inevitably involves the changes in tourism status and the innovation and optimization of tourism products. The combination of tourism and the Internet has spawned new areas such as virtual tourism, customized tourism or tourism O2O and other new formats, which not only break through the traditional operating modes of the industry, but also bring greater tourists value and added value.

2.2 Analysis from the Perspective of Supply

All-for-one tourism is a supply-side reform, and big data brings strategic positioning, precision marketing and business innovation to tourism destinations. Commodity and logistics information in the Internet of Things, interaction information between people in the Internet, and location information are the three main sources of big data. Big data has broken through the narrow vision of tourist attractions and bring changes at least two aspects, one is to help strategic positioning and precision marketing of tourism destinations, and the other is to contribute to the innovation of tourism and products. [4] Big data is critical in the segmentation of the tourism market, the identification of target markets, and the development of tourism strategies. In addition, when all-for-one tourism launches a specific area as a complete tourist destination, the position strategy is very significant. In the past, marketing positioning tends to focus on tourism resource endowment and historical accumulation, while big data sorts out the amount, structure, interests, route and scenic spots preferences of tourists, which helps tourism destinations to accurately locate tourists, competitors and resources.

As a supply-side structural reform, all-for-one tourism inevitably involves the changes in tourism status and the innovation and optimization of tourism products. The combination of tourism and the Internet has spawned new areas such as virtual tourism, customized tourism or tourism O2O and other new formats, which not only break through the traditional operating modes of the industry, but also bring greater tourists value and added value.

2.3   Analysis from the Perspective of Demand

The purpose of supply-side reforms is to serve tourists ultimately, and big data should be based on building smart tourism for value-oriented ecosystems. The ultimate goal of all-for-one tourism is to provide visitors with more services and experiences. [5] For example, Forbidden City provides visitors with a better three-dimensional consumption experience through digital museum reconstruction and network channel strategies. Currently, consumer behavior has evolved from the traditional AIDMA model (attention, interest, desire, memory, action) to the AISAS model which with network traits (attention, interest, search, action, share). When the new consumption behavior model is generated, only by capturing the characteristics of this model and adopting the corresponding marketing strategy can it become a real winner.

Under the "Tourism + Internet", it is very important to establish a smart tourism concept for a value-oriented ecosystem. As experts pointed out, the use of massive data heralds a new round of productivity growth and the arrival of consumer surplus. In the information age and the era of interconnection, it is necessary to change the logic of commodity-lead, and to turn the idea into service-oriented logic. [6] This requires the idea of building big data for tourists, using big data systems and platforms to build a full ecological value chain which serves tourists.

2.4   Analysis from the Perspective of Industrial Integration

All-for-one tourism emphasizes resource allocation. The resource allocation in the era of big data is actually using the quantity, dimension and breadth of data, comprehensively analyzing various types of information, and allocating resources with the optimal principle. For example, the user trace recovery analysis method can reconstruct the economic and industrial development, transportation location, tourism resources, tourist market and other tourism conditions of the entire region in the data space.

Under "Tourism + Internet", the use of big data to break the barriers of industrial development requires strengthening the construction of big data analysis and processing platforms and innovative data analysis methods. Big data involves a wide range of systems, and the system is huge and complex. It is a double-edged sword. If it is not well done, it will invade the privacy of citizens. It must rely on the authority of government agencies, establish a big data sharing package with the government as the leading factor, and strengthen the science and technology to protect information security through legislation.

2.5   Analysis from the Perspective of Industrial Supervision

Big data helps the service and regulation of market subjects. In order to achieve tourists' satisfaction, all-for-one tourism must try to avoid and reduce the "negative feelings" of tourists. The "negative feelings" of tourists may come from all aspects of "food, hospitality, travel, visit, shopping, entertaining", which requires improving the overall governance capacity and optimizing the tourism environment. [7] The State Council issued the document "Several Opinions on Strengthening the Service and Supervision of Market Subjects by Using Big Data", requiring local governments to use big data to strengthen service and supervision of market entities.

Big data has great potential for industry regulation. With the help of big data, relevant government departments can fully acquire and use information, more accurately understand the needs of market entities, and improve the pertinence and effectiveness of services and supervision. On the one hand, efficient use of modern information technology, social data resources and socialized information services can strengthen social supervision and play the positive role of the public in regulating behavior of market subjects, and on the other hand can reduce administrative supervision costs.

3   The Status Quo and Problems of Smart Marketing in All-for-One Tourism

Traditional tourism marketing lacks brand positioning for tourism destinations, and marketing entities are basically tourism resources of a single scenic spot. With the promotion of all-for-one tourism, the marketing of tourism destinations has become the core of the work of local tourism bureaus. [8] However, the marketing tools that attract tourists are still promoted through distribution channels, held the tourism promotion conferences or concentrated advertising on the television, newspapers, radio and other offline media. This is fully reflected in the 2016 public tendering marketing project of the Provincial Tourism Commission, which is famous for its leadership in big data construction. Due to the lack of effective monitoring methods, it is impossible to effectively evaluate the promotion effect of these marketing means on the tourism destination brand. On the other hand, though the various digital marketing tools brought by the Internet, such as SEM, social media interactive communication and accurate marketing using big data, have become mainstream, but the advertising service providers still use more resources to FMCG, car, finance or real estate. At the same time, due to the lack of tourism insights, these marketing companies simply copy or transplant the experience of other industries, this cannot adapt to the characteristics of the tourism industry which the industrial chain is long and the resources are scattered, lacking solutions focus on sales transformation. For the tourism management department, it is obviously difficult to fully understand the emerging marketing methods of the Internet, let alone apply these methods to the relatively long-term and systematic work of tourism destination brand marketing.

Applying big data to tourism destination brand marketing is undoubtedly the road one must follow. Through big data, we can analyze the market competition of destinations, identify the brand positioning, integrate marketing resources for interactive communication, and monitor the effects, then establishing a marketing system suitable for sustainable development of destinations. [9] However, in the tourism management department, the technical department such as the information center is usually responsible for the construction of smart tourism, and this has made the application of smart tourism focus on industry management and public services. There are two reasons: First, the management needs of these two aspects are not much different from the conventional IT formalization, big data is just an enrich means. Second, the information center cannot get a clear demand from the business unit because destination marketing is new to any travel authority.

The role of big data in smart travel is important, but its application is still in its infancy. Currently, service providers on the market are mainly companies that provide data sources, and simply develop tourism application products based on their own data. Most of the data scenarios are applications such as passenger flow analysis, visitor portraits, and public opinion to meet management and service needs. These data scenarios can attract the attention of superior leaders and the media attractive for visualization. Builders of smart tourism are more willing to use budgets or funds to purchase a variety of data and set up large screens.

4 The Suggestion on Application of Big Data in Smart Tourism Marketing

From the specific functions of the big data application in smart marketing, the following suggestions can be made.

4.1 Take Marketing Needs as the Basic Starting Point

The function of smart tourism marketing applications should start with marketing needs rather than data. This requires in-depth insight into the travel industry and marketing. It is impossible to develop by copy nor by work behind closed doors. It is necessary to thoroughly investigate the marketing needs of the tourism industry. Starting from management tools and connect big data with actual needs through real-tum monitoring, data analysis, data visualization and other application scenarios.

4.2 Take Data Analysis as the Main Function

The function of smart marketing application products should be based on data analysis. Simple data statistics, listing, and display are not the purpose of data analysis. [10] Data analysis should explore the value of tourism marketing. From the perspective of all-for-one tourism, it is the primary task to analyze the regional tourism consumption structure and realize dynamic monitoring methods for tourists consumption, self-driving tour, etc., to provide decision-making basis for tourism managers.

4.3   Take Application Scenario Requirements Meeting as the Standard

The development of big data has provided more and more choices for smart tourism. However, due to the lack of standards, there are misunderstandings which is to pursuit "big and complete" on the data sources. For example, the same type of data which uses multiple data sources at the same time, the reason behind this is that the application requirements, that is, the goal of data analysis, are not used as the standard for measuring data sources. It is not only a waste of funds, but also a difficulty for the subsequent construction of smart tourism. In the choice of data source, we must consider long-term planning, implementation in stages. [11] In the first, to define the structure of the application layer, and then determine the construction of the entire data layer. Secondly, to fully adopt new technologies, such as cloud computing, big data processing, and the relative platforms, to ensure that smart travel systems are secure, high-performance, and salable.

5   Conclusion

In terms of concept, both the tourism management organization and the builder of smart tourism need to reverse the thinking that the management is significant while the marketing is unimportant. In the relationship in management, service and marketing, we need to clarify that there is no precedence of the three. The purpose of developing all-for-one tourism is to integrate local tourism resources to increase the attractiveness of tourism destination. Smart tourism should also be built in accordance with this goal. This is also in line with the function setting of the Tourism Bureau—responsible for tourism marketing in the area under its administration. The purpose of improving the management and service level of the tourism industry is also to serve tourists. The artificial separation of the three relationships will inevitably lead to an imbalance in the development of destination tourism. Therefore, planning smart tourism should meet the needs of management, service and marketing simultaneously. It not only ensures the effectiveness of smart tourism, but also improves the efficiency of IT resources for big data is reusable in these three areas.

Acknowledgments. This research was supported by 2018 Hainan Province Philosophy and Social Science Planning Project (Grant NO. HNSK (YB)18-100), and 2019 Sanya Philosophy and Social Sciences Funding Project (Grant NO. SYSK2019-15).

References

1. Fan, J.: Data Open and Sharing China Unicom Big Data, p. 11 (2016). (in Chinese)
2. Li, Z.: Special research on big data analysis of overall planning of Anhui tourism area, p. 2 (2017). (in Chinese)
3. Huang, C., Huang, Z.: Big data opens a new chapter in global tourism. Inf. Constr. **9**, 52–53 (2017). (in Chinese)

4. Lu, S., Zhang, M.: Innovation research on hotel status and management model in the background of big data—taking star hotels in Changsha as an example. Technol. Market **2**, 99–101 (2016). (in Chinese)
5. Feng, G.: Analysis of the business strategy of single hotel in the era of big data. Natl. Bus. **23**, 32–33 (2016). (in Chinese)
6. Friedman, T.: Revolution hits the universities. New York Times **26** (2013)
7. Baggaley, J.: MOOC postscript. Distance Educ. **35**, 126–132 (2014)
8. Ajzen, I.: The theory of planned behavior. Res. Nurs. Health **14**, 137–144 (1991)
9. Venkatesh, V., Bala, H.: Technology acceptance model and a research agenda on interventions. Decis. Sci. **39**, 273–315 (2008)
10. Lacaille, L.: Theory of Reasoned Action (2013)
11. Bandura, A., Bandura, S., Bandura, A.: Social Foundation of Thoughts and Actions: A Social Cognitive Theory. Prentice-Hall, Upper Saddle River (1986)

Human Parameter Extraction Model Based on K-nearest Neighbor and Application in Jewelry Modeling Design

Guoyan Yuan[(✉)]

Jilin Engineering Normal University, Changchun 130052, Jilin, China
Tougao_006@163.com

Abstract. Computer technology has been widely used in various fields. In this paper, we try to use new computer technology and new ideas in jewelry modeling design. By constructing a classification model of human parameters based on K-nearest neighbor algorithm, we can apply it to modern jewelry design and production. Experiments were conducted to classify and detect the contours of different types of human figures, such as frontal human body, side human body, male and female. The results show that the classification model has a high detection accuracy in human body parameter recognition, especially in the detection of frontal human body image parameters, the recognition accuracy percentage reaches 96.5%, and good classification results are achieved. The classification results of male and female parameters are 93.14% and 91.3% respectively. Through these good human contour classification results, the human body parameters extracted by classification are applied to jewelry modeling design by using data analysis and graphics generation capabilities, and a new idea of jewelry design is developed.

Keywords: K-nearest neighbor algorithm · Human body parameter extraction · Parametric design

1 Introduction

In recent years, design has been regarded as a new innovation driving force besides technology and market. How to help enterprises acquire users preferences through product meaning innovation has aroused extensive research interest of researchers [1]. How to integrate the ecological, aesthetic and economic elements into the design without forgetting the final product function of the design [2].

Jewelry design and production is a process of precious raw materials and low processing loss. The traditional manual mode cannot meet the actual needs of enterprises, and the computer technology can solve this practical problem [3, 4]. Previous research on jewelry design mainly focused on the technology in the process of jewelry design, such as the use of tools including computer-aided design, rapid prototyping and other design technologies. However, the understanding of jewelry design practice itself, including research methods and theoretical basis, is limited [5]. The development of artificial intelligence technology in algorithmic design, application and hard calculation and parameter optimization in jewelry modeling design and casting [6]. In order to find

© Springer Nature Singapore Pte Ltd. 2020
M. Atiquzzaman et al. (Eds.): BDCPS 2019, AISC 1117, pp. 1252–1259, 2020.
https://doi.org/10.1007/978-981-15-2568-1_174

different methods suitable for image classification and recognition, many researchers use machine learning to classify jewelry images [7]. Combining the computer aided design and rapid prototyping technology in the field of computer technology with the actual production demand, the "concurrent engineering" from "product development" to "product promotion" to "product production" in jewelry industry has been realized [8, 9]. The optimization of process parameters is one of the ways to realize the ideal quality of jewelry modeling. Based on aesthetics and manufacturing principles, some researchers have proposed an automatic line decomposition algorithm, which divides a graph into a small number of lines. Random optimization method is used to produce good results for a variety of different inputs [10]. Mathematical models and algorithms can provide computational support for the design of traditional jewelry, simplify the design method of complex traditional decorative design in this process, and combine with the design of jewelry shape to cope with the new development of jewelry industry [11].

In this paper, K-nearest neighbor algorithm is used to classify and fuse different human body images and extract relevant parameters. The image board of jewelry design is determined by parametric jewelry design. The jewelry works of bionic design are used for reference. The cultural connotation and humanistic emotion behind the objects are embodied through the relevant jewelry design.

2 Method

2.1 Human Body Feature Extraction and Selection

In order to extract meaningful parameter components from human body image region, feature extraction and selection are included or implied in the process of human body parameter recognition. Feature extraction and selection is one of the key steps in human body parameter recognition model, as shown in Fig. 1. This process is to extract and select the simplest and most effective feature vectors from the original collected or preprocessed data for human body recognition and classification, so as to provide guarantee for the subsequent realization of the effect of parameter recognition and classification.

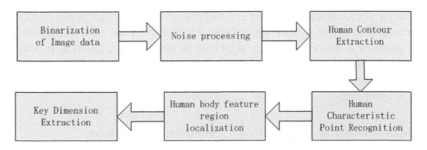

Fig. 1. Body feature extraction process based on image

Once there are many clutters or the edges of the image data are not smooth, in order to reduce the noise in the image data, the image data are pre-processed by morphological algorithm, the unnecessary information and the key information needed for enhancement are eliminated, and then the feature points are searched by feature detection. Methods The feature points of human contour were extracted automatically.

2.2   K-nearest Neighbor Algorithm

K-nearest neighbor classification algorithm is the most classical and simplest classification method in the field of pattern classification. It can handle various problems flexibly without any training, and achieve good performance. It is the most common nearest neighbor classification method, so it has a very wide range of applications. K nearest neighbor classification algorithm uses Euclidean distance to measure "proximity". The Euclidean distance of two points or tuples $x_1 = (x_{11}, x_{12}, \ldots, x_{1n})$ and $x_2 = (x_{21}, x_{22}, \ldots, x_{2n})$ is:

$$\text{dist}(x_1, x_2) = \sqrt{\sum_{i=1}^{n} (x_{1i} - x_{2i})^2} \tag{1}$$

That is to say, for each numerical attribute, the D-value between the attributes values of tuple x_1 and x_2 is taken, and the square root is taken after accumulating the square of the D-value.

In order to prevent the weights of attributes with larger initial range from smaller initial range, Before calculating formula (1), it is necessary to transform the V value of attribute A into the v' in [0, 1] interval by formula (2) to normalize the value of each attribute.

$$v' = \frac{v - min_A}{max_A - min_A} \tag{2}$$

max_A and min_A represent the maximum and minimum values of attribute A.

K-nearest neighbor classification algorithm is an analogy-based learning algorithm. By comparing a given test tuple with a similar training tuple, the K-nearest neighbor classification algorithm matches the current input variables with the historical data values of similar input variables, in which the input variables are called eigenvectors. Each feature vector describes a point in a multivariable space called feature space. If the feature vector contains n attribute values, the feature space is n-dimensional. The output value of K-nearest neighbor algorithm is defined as a function related to the known output history of input eigenvectors with similar eigenvectors. As the number of training elements tends to be infinite and K = 1, the error rate does not exceed twice the Bayesian error rate.

2.3   Measurement of Human Body Circumference Dimension

After the two-dimensional size information of the human body is obtained by processing the front and side images of the subjects, the human body circumference size is calculated according to the method of statistics or curve fitting. The process of transforming the two-dimensional data from image processing into the human body's circumference size by various mathematical methods is called circumference fitting. After circumference fitting, it is expressed by plane curve. The length of curve is the circumference size of human body. Fitting refers to finding an approximate expression of an unknown function from a lot of disorderly data. The expression does not need to pass through all the data. Curve fitting is the process of curve solving.

In this paper, the width and thickness of the known girth are calculated by binary quadratic polynomial regression method to get the girth size. The least square method is used to establish the normal equations. Then the regression coefficients are obtained based on the least mean square error principle. The regression equation can be obtained by introducing the equation. The equation includes the product of width and thickness, and the square of width and thickness as variables.

Assuming that X_1, X_2 and Y represent chest width, chest thickness and chest circumference respectively, the regression formula for calculating chest circumference by using least squares method is as follows:

$$Y(X_1, X_2, X_3, X_4, X_5) = A + BX_1 + CX_2 + DX_3 + EX_4 + FX_5; \qquad (3)$$

Among them, $X_3 = X_1^2, X_4 = X_2^2, X_5 = X_1X_2$;

In this paper, T^2 measurement sample is used to calculate the human body circumference size, and the optimal fitting regression equation is obtained as follows.

Chest circumference regression model:

$$W_b = 127.83 - 2.113x_1 - 4.241x_2 - 0.054x_3 - 0.065x_4 + 0.282x_5 \quad \text{(Male)}$$

$$W_b = -7.678 + 1.853x_1 + 1.748x_2 + 0.004x_3 + 0.005x_4 - 0.011x_5 \quad \text{(Female)}$$

Waist circumference regression model:

$$W_b = 101.99 - 4.796x_1 + 0.292x_2 + 0.014x_3 - 0.152x_4 + 0.277x_5 \quad \text{(Male)}$$

$$W_b = 1.924 + 1.756x_1 + 1.122x_2 + 0.008x_4 - 0.002x_5 \quad \text{(Female)}$$

Hip circumference regression model:

$$W_b = 217.02 - 9.516x_1 - 0.326x_2 + 0.156x_3 + 0.052x_4 - 0.024x_5 \quad \text{(Male)}$$

$$W_b = 4.051 + 2.561x_1 - 0.026x_3 + 0.042x_5 \quad \text{(Female)}$$

2.4   Parametric Design of Jewelry

Parametric design is a new design method. Designers first design the corresponding rules and logic, and then treat the design and guide the relevant modeling design from a systematic and logical point of view. In jewelry modelling design, parametric design refers to the jewelry designer transforming parameter variables and variable data information into images through computer technology and forming parametric or parametric models. When design conditions and design ideas change, new results can be obtained by modifying the parametric models. At the same time, when the value of information changes, only the value of input information variable is changed, and new results can be obtained without changing the parameter model. At the same time, when the value of information changes, only the value of input information variable is changed, and new results can be obtained without changing the parameter model. Parametric design technology can be applied in jewelry sketch design, dimension-driven graphics modification function and other fields. Parametric design can also be an effective means for initial design, product modeling, multi-design scheme comparison, modification of serialization design and dynamic design.

In this paper, parametric design method is applied to jewelry design, and parametric design of non-linear human body lines is used to simulate human behavior in complex world and the interaction between human bodies with reference to computer technology. All the main factors affecting the design are organized as parametric variables and input together for calculation. The prototype of jewelry design is simulated in the machine. In order to establish the basic shape of jewelry design, when the basic graphic logic of the jewelry design prototype is established, the jewelry ontology is drawn and modeled. The most important role of parametric design is to select graphics through controllable data, so as to achieve the satisfaction of designers.

3   Experimental Simulation

In this paper, K-nearest neighbor algorithm is used to extract feature points and measure the size of the human body in the subject image. The feature points are classified and the classified pictures are taken as input parameters of a necklace design. Then, the necklace is designed by establishing a parametric design model based on the extracted human body feature elements, and finally the necklace is made. Relevant Necklace finished products. The experimental process mainly revolves around the extraction of front and side human feature points and size measurement based on single background.

3.1   Sample Establishment

This study classify human images based on K-nearest neighbor classifier, so 302 human images under different background are selected as the sample set of this experiment, and the collected images are marked according to the key feature points such as face, chest, waist, hip. Thirty-two human pictures were randomly selected as training set, and the remaining 20% as test set. Finally, 242 training sets and 60 test sets were obtained.

3.2   Establishment of K-nearest Neighbor Classification Model

K-nearest neighbor classifier is built for training set, and the classification results are verified by test data. This paper evaluates the effect of the classifier by using the correct rate and the correct classification results, and evaluates the descriptive effect of the extracted principal components on human parameters. The whole classification process is as follows:

(1) Preprocessing the image data.
(2) Choose appropriate memory unit training tuple and test tuple.
(3) Setting parameter k, the value of K is usually odd. The more training tuples, the greater the value of K.
(4) Store the nearest neighbor's training tuple and set the priority queue of size k, which is arranged from large to small by distance.
(5) K tuples are randomly selected from the training tuples, and the K ancestors are taken as the initial nearest neighbor tuples. The distance from the test tuples to the K tuples is calculated. Finally, the labels and distances of the training tuples are stored in the priority queue.
(6) The test tuple set is traversed and the maximum distance between L and priority queue is compared. If L is greater than or equal to the maximum distance, the test tuple is discarded and the next training tuple is traversed. If it is less than the maximum distance, the current tuple is saved in the priority queue.
(7) Complete the traversal, calculate the majority of the K training tuples of the priority queue, and regard them as the categories of the test tuples.
(8) Continue to set different K values for re-training, calculate the error rate and take the minimum K value of the error rate.

4   Experimental Results and Analysis

4.1   Determine the Value of K

In K-nearest neighbor algorithm, the selection of K value has a great influence on the experimental results. The selection of K value is odd generally. However, if the K value is too small, the classification results are vulnerable to noise, and if the K value is too large, the classification results will be affected because the nearest subset contains too many error points. So this experiment starts with k = 1 and estimates the error rate of the classifier by using the test set. Repeat the process, increasing K by 2 and adding two neighbors each time. The K value of the minimum error rate is selected as the best value for this experiment. The selection of K is shown in Table 1.

Table 1. Accuracy comparison of classification results with different K values

Index	K = 1	K = 3	K = 5	K = 7	K = 9	K = 11	K = 13
Classification accuracy	73.3	71.7	76.7	75	79.1	76.7	73.3
Classification accuracy number	44	43	46	45	48	46	44

It can be found from Table 1 that the average correct rate is the highest when k = 9. In order to achieve good classification results, the value of K is 9 in this study.

4.2   Performance Evaluation of Human Body Parameter Classification

The accuracy, extraction time, extraction times and other indicators were used to evaluate the effect of k-nearest neighbor classifier on human parameters classification in this study. The results are shown in Table 2.

Table 2. Classification effect of human parameters based on K-nearest neighbor

Human body image type	Accuracy of feature point extraction	Search efficiency	
		Extraction time (s)	Search times (frequency)
Male	93.14%	126	26
Female	91.3%	265	22
Frontal body	96.5%	8.51	19
Side body	95.1%	4.1	7

The comparative analysis of the experimental results shows that the accuracy of extracting feature points is higher by detecting the contour of different types of human figures, such as frontal human body, lateral human body, male and female, respectively, based on K-nearest neighbor classification. However, the number of iterations of male human body is more than that of female human body, and the extraction time of male human body feature points is less than that of female human body. At the same time, because there are more feature points in the detection of men and women, the accuracy rate is lower than that of detecting human contour only from the front and side, the extraction time is higher, and the number of iterations is more, as shown in Fig. 2.

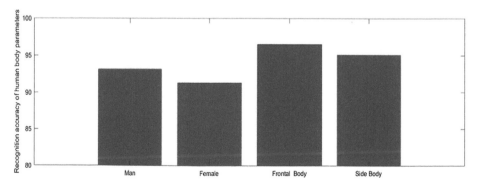

Fig. 2. Comparison of recognition correctness rates of different human image types

5   Conclusion

In this paper, K-nearest neighbor classification algorithm is used to extract human body feature points and measure the size of single background human body images. The extracted human body contour parameters are applied to jewelry design through parametric design method. A new idea of jewelry design is put forward, which breaks through the original design. The experimental results show that the K-nearest neighbor classification algorithm has fast extraction speed, high accuracy and good classification effect for human parameters. This research fully utilizes the data analysis and graphics generation ability of computer, and provides a new method for the design process of jewelry.

Acknowledgements.   Jilin province social science foundation project (Item No. 2019B158).

References

1. Wang, H.: User classified algorithm and model based gold jewelry design definition. Cluster Comput. (1), 1–13 (2018)
2. Boginskaya, E.A., Sotnikov, N.N.: Multifunctional elements for energy efficient jewelry design. In: Иностранный язык в контексте проблем профессиональной коммуникации: материалы II Международной научной конференции, 27–29 апреля 2015 г., Томск.— Томск, p. 27 (2015)
3. Xia, J.: Research on application of computer technologies in jewelry process. Res. Mod. High. Educ. 3(1), 113–119 (2017)
4. Sharma, S., Singh, Y., Virk, G.S.: Study on designing and development of ornamental products with special reference to the role of CAD in it. Int. J. Curr. Eng. Technol. 2(4), 443–447 (2012)
5. Rajili, N.A.M., Olander, E., Warell, A.: Characteristics of jewellery design: an initial review. In: Research into Design Across Boundaries, ICoRD 2015, vol. 2, pp. 613–619. Springer, New Delhi (2015)
6. Wannarumon, S.: Reviews of computer-aided technologies for jewelry design and casting. Naresuan Univ. Eng. J. 6(1), 45–56 (2011)
7. Singh, V., Kaewprapha, P.: A comparative experiment in classifying jewelry images using convolutional neural networks. Sci. Technol. Asia 7–17 (2018)
8. Kaixuan, W., AiNi, G.: Analysis of jewelry design based on computer graphics simulation technology. Int. J. Hybrid Inf. Technol. 9(11), 425–436 (2016)
9. Baturynska, I., Semeniuta, O., Martinsen, K.: Optimization of process parameters for powder bed fusion additive manufacturing by combination of machine learning and finite element method: a conceptual framework. Procedia CIRP 67, 227–232 (2018)
10. Iarussi, E., Li, W., Bousseau, A.: WrapIt: computer-assisted crafting of wire wrapped jewelry. ACM Trans. Graph. (TOG) 34(6), 221 (2015)
11. Goel, V.K., Khanduja, D., Garg, T.K., Tandon, P.: Computational support to design and fabrication of traditional Indian jewelry. Comput.-Aided Des. Appl. 12(4), 457–464 (2015)

Comprehensive Evaluation of Node Importance in Complex Networks

Jundi Wang$^{(\boxtimes)}$, Zhixun Zhang, and Huaizu Kui

Lanzhou Institute of Technology, Lanzhou, Gansu, China
511378759@qq.com

Abstract. Identifying important nodes quickly and effectively in complex networks is one of the effective ways to control the propagation process of networks. Node importance ranking is the main method to identify important nodes. In this paper, the SIR model is used to simulate the network propagation process based on seven node importance sorting algorithms. The performance of the algorithm is evaluated from aspects of resolution and accuracy. In the experiment, seven algorithms are compared and analyzed in three theoretical networks and seven real networks using the above two evaluation criteria. The network characteristics suitable for different algorithms are obtained, which has great reference value for the application of important node sorting algorithm in complex networks.

Keywords: Complex networks · Node importance ranking · SIR spreading model · Resolution · Accuracy

1 Introduction

In real life, most complex systems, such as human social structure, power network system, aviation system and so on, can be abstracted as complex networks [1]. Nodes and edges are the basic components of complex networks. Any node is connected to each other to form a network. The connection mode between nodes determines the efficiency of information transmission and the basic topological structure of the network [2]. The importance of nodes in the network is reflected by their position in the network. Nodes that can decide the structure and performance of the whole network are called important nodes [3]. From viewpoint of theory, finding out the important nodes and controlling them can promote or curb the propagation phenomenon occurring on the network [4]. From viewpoint of reality, many problems in identification of important nodes need to be solved, such as the protection of hubs in power network system, the identification of the source of infectious diseases in large-scale outbreaks to control the spread, the identification of a small number of influential users in microblog network [5], etc.

In this paper, seven ranking algorithms of node importance are studied, including Degree Centrality, Eigenvector Centrality, k-shell, mixed degree decomposition, coreness centrality, Close Centrality and Betweenness Centrality. The SIR model is used to simulate the network propagation process, and the performance of importance sorting algorithm is evaluated from aspects of resolution and accuracy. In terms of

© Springer Nature Singapore Pte Ltd. 2020
M. Atiquzzaman et al. (Eds.): BDCPS 2019, AISC 1117, pp. 1260–1267, 2020.
https://doi.org/10.1007/978-981-15-2568-1_175

resolution measurement criteria, each algorithm is tested in different networks, and its resolution is compared and analyzed. In terms of accuracy measurement criteria, 10 different propagation probabilities are selected around the theoretical propagation threshold for testing.

All the seven main algorithms mentioned in this paper are applicable to connected networks. Because of using relevance performance of testing algorithm, there is some uncertainty for the disconnected network. In this paper, all the networks used in the algorithm are undirected and unweighted network.

2 Theory and Method

2.1 Topological Characteristics of Complex Networks

With more in-depth research on complex networks, many scholars have proposed many characteristic metrics for complex networks in order to reflect the same characteristics of all kinds of complex networks. In order to judge and weigh the accuracy and effectiveness of various complex network evolution models, this paper mainly discusses three characteristics include degree and degree distribution, clustering coefficient, and average path length of network.

2.2 SIR Model

SIR (Susceptible-Infected-Recovered) model is an infectious disease model established by Kermack and McKendrick on the basis of dynamic theory [6]. Nowadays, SIR model is still used in the relevant theoretical research and is constantly improved. In SIR model, all individuals in the network have three states: susceptibility state, infection state and recovery state. Susceptibility state ($S(t)$) means that at time t, a person is in healthy state but can be infected by a disease. Infection status ($I(t)$) means that at time t, a person has been infected and can infected others. If all individuals in the SIR model recover, they will be recovered permanently and not be affected by outside. In the above statement, $S(t)$, $I(t)$ and $R(t)$ respectively represent the number of individuals in the three states at time t. From the model principle can obtain: $S(t) + I(t) + R(t) = n$. n represents the total number of vertices in the network.

The propagation progress of SIR model can be divided into two stages. In the first stage, the susceptible individuals will be infected after contacting with the infected individuals, assuming that the probability of disease transmission is α and the disease can only be transmitted to susceptible individuals if there is contact between the infected individuals and the susceptible individuals. In the second stage, the individuals in the infected state will change to the restored state with probability β. When the infected individuals become recovered, they will not be infected by other infected individuals or infected by other susceptible individuals. At time t, the proportion of susceptible individuals was $S(t)/n$, the number of infected individuals is $I(t)$, the number of susceptible individuals is reduced as the following formula [7].

$$\frac{ds}{dt} = -\alpha \frac{S(t)I(t)}{n} \tag{1}$$

At time t, the number of recovered individuals is expressed as follow

$$\frac{dR}{dt} = \beta \frac{I(t)}{n} \tag{2}$$

The number of infected individuals at t time is given as follow

$$\frac{dI}{dt} = \alpha \frac{S(t)I(t)}{n} - \beta \frac{I(t)}{n} \tag{3}$$

AT time t, the percentages of individuals in susceptible, infected and restored states are given as follow

$$s(t) = S(t)/n \tag{4}$$

$$i(t) = I(t)/n \tag{5}$$

$$r(t) = R(t)/n \tag{6}$$

In which S(t) + I(t) + R(t) = 1

3 Sorting Algorithm Based on the Importance of Neighbor Nodes

In order to meet the comprehensive requirements of index selection, representative algorithms of neighbor node importance were selected based on the research results at home and abroad. Including Degree Centrality, Eigenvector Centrality, k-shell, mixed degree decomposition, coreness centrality.

In this paper, two evaluation methods, resolution ratio and accuracy, are used to evaluate the performance of the ranking method of node importance.

1. **Resolution Ratio**

Resolution ratio refers to the ability of a node importance ranking method to distinguish the propagation influence of different nodes in the network [8]. The method used here [6]

$$f(r_A) = 1 - \frac{\sum_{i=1}^{R} N_i^2}{R \cdot N^2} \tag{7}$$

Among them, r_A represents the ranking result A obtained by the ranking method of node importance; R represent for the number of elements in r_A; N_i represents the number of elements I existing in r_A; N represents all nodes in the network.

Theorem 1: When the number of nodes in the network is N $\rightarrow \infty$ and R = N, then $f(r_A) = 1$.

From Theorem 1, if each node in the ranking result of node importance has different values then the resolution obtained by the ranking method A of node importance is relatively the highest. At the same time, the value of $f(r_A)$ will become larger, the size of the network is getting bigger. When the value of $f(r_A)$ is infinitely close to 1, this value is the upper limit of $f(r_A)$.

Theorem 2: When R = 1, then $f(r_A) = 0$.

From Theorem 2, if each node in the ranking result of node importance has the same value then all nodes are assigned the same level. At the same time, the resolution obtained by node importance sorting method A is relatively lower. When $f(r_A) = 0$, it is the minimum value of $f(r_A)$. The conclusions are as follows: when $f(r_A)$ equals 1, it represents the optimal resolution value.

On the basis of the above evaluation criteria, Δ function is assumed to represent the difference between the resolution obtained by some sort of node importance sorting method and the optimal resolution 1.

$$\Delta(r_A) = 1 - f(r_A) \tag{8}$$

It is obvious from the above formula that the smaller the value of Δ function is, the higher the resolution of the ranking method of node importance is.

2. Accuracy

Accuracy is defined as the propagation influence of a node obtained by an algorithm of ranking the importance of a node is the same as the actual propagation influence of the node [9]. If we want to effectively control the propagation process occurring on the network, we need to accurately evaluate the propagation influence of all nodes in the network. The propagation influence of a node is defined as the total number of other nodes that can be affected by the node when the propagation process stops after a period of propagation [10]. This paper uses SIR model, and uses SIR model to calculate the propagation influence of all nodes in the whole network. In this paper, in order to explore the influence of different propagation probability on the propagation influence of nodes, as well as the change of the propagation influence around the theoretical propagation threshold (β_c) [6], ten different propagation probabilities will be selected. The range of propagation probability is (0, 1). When the selected propagation probability is less than the theoretical propagation threshold, the propagation process will disappear in a small area of the network. When the selected propagation probability is larger than the theoretical propagation threshold, the whole propagation process will spread rapidly in the network. Generally, the propagation thresholds of networks with different topologies are different. Here, the propagation probability of $\beta \in [\beta_c - 0.05, \beta_c + 0.05]$, and the interval is 0.01, a total of 10 different propagation probabilities are selected. It is stipulated that the recovery probability $\lambda = 1$, that is, the node in the infected state will eventually become the recovery state. If there are no infected nodes in the network, the whole propagation process is completed.

4   Experiments and Results Analysis

In order to test the performance of node importance ranking algorithm, experiments were carried out on three synthetic networks and seven real networks. Artificial synthetic network includes ER random network, small-world network and BA scale-free network. The seven real networks are dolphins network, PDZBase network, karate network, netscience network, lesmis network, football network and jazz network.

4.1   Experimental Settings

In terms of accuracy measurement, unlike the traditional method, which chooses a smaller propagation probability, this paper selects 10 different propagation probabilities from the vicinity of the theoretical propagation threshold to test the node sorting algorithm.

The scale and topological properties of synthetic and real networks are shown in Table 1.

Table 1. Scale and topology of 10 networks

Network	N	E	$<k>$	$<k^2>$	A	H	C	β_c
Dolphins	62	159	5.13	34.90	−0.04	1.33	0.31	0.15
Football	115	613	10.66	114.43	0.16	1.01	0.41	0.09
Jazz	198	2742	27.70	1070.24	0.02	1.40	0.52	0.03
Karate	34	78	4.59	35.65	−0.48	1.69	0.26	0.13
Lesmis	77	254	6.60	79.53	−0.17	1.83	0.50	0.08
Netscience	379	914	4.82	38.69	−0.08	1.66	0.43	0.12
PDZBase	161	209	2.60	15.25	−0.47	2.26	0.003	0.17
ER	200	483	4.83	28.06	0.02	−0.01	1.20	0.17
WS	200	400	4	16.66	0.26	−0.03	1.04	0.24
BA	200	397	3.97	21.45	0.03	0.11	1.36	0.19

In Table 1, N represents the size of nodes in the network, E represents the number of edges in the network, $<k>$ represents the mean of node degree in the network, $<k^2>$ represents the mean represents of the second order degree of a network node, A represents Network Coefficient, H represents the heterogeneity of network degree, C represents network cluster coefficient, β_c represents the theoretical propagation threshold of the network.

4.2 Analysis of Resolution

According to the definition of resolution mentioned above, the results are shown in Table 2. The resolution of eigenvector centrality algorithm is higher than that of other seven algorithms. Because the eigenvector centrality algorithm has the highest resolution in BA network, karate network, dolphins network and jazz network. Eigenvector centrality algorithm is second only to coreness centrality algorithm in ER network. Eigenvector centrality algorithm is second only to median centrality algorithm in WS network. In lesmis network eigenvector centrality algorithm is second only to Close Centrality algorithm. The eigenvector centrality algorithm is also at the top of the rest of the networks. According to the above analysis results, the resolution of eigenvector centrality algorithm is the highest. At the same time, the resolution of K-shell decomposition algorithm is the lowest among the seven algorithms. In Table 2 according to the Δ function we can know K-shell decomposition algorithm is always the lowest or next to last in each network. Therefore, the K-shell decomposition algorithm has the lowest resolution.

Table 2. Δ value of seven node importance sorting algorithms in 10 networks

Network	$\Delta(r_{DC})$	$\Delta(r_{BC})$	$\Delta(r_{ks})$	$\Delta(r_{EVC})$	$\Delta(r_{MDD})$	$\Delta(r_{cnc})$	$\Delta(r_{CC})$
ER	1.19e−02	4.39e−05	2.37e−01	2.5e−05	2.74e−03	**1.05e−04**	7.21e−05
WS	5.38e−02	**2.54e−05**	3.17e−01	2.5e−05	2.4e−02	1.85e−03	5.61e−05
BA	1.74e−02	**2.5e−05**	1e+00	**2.5e−05**	7.5e−03	7.71e−04	6.03e−05
Karate	1.67e−02	7e−03	7.92e−03	**1.79e−03**	1.05e−02	2.13e−03	3.98e−03
Dolphins	8.58e−03	6.46−04	9.9e−02	**2.6e−04**	2.81e−03	4.39e−04	6.78e−04
Lesmis	6.06e−03	9.96−03	1.86e−02	3.24e−04	3.83e−03	6.52e−04	**1.03e−03**
Football	6.7e−02	7.56e−05	4.91e−01	7.56e−05	2.26e−02	4.6e−04	7.82e−04
PDZBase	2.46e−02	4.06e−03	1.51e−01	7.51e−05	1.25e−02	**1.04e−03**	2.51e−04
Jazz	3.57e−04	6.08e−05	5.39e−03	**2.79e−05**	8.8e−05	2.83e−05	8.76e−05
Netscience	6.1e−03	1.08e−03	2.51e−02	7.39e−06	1.78e−03	4.34e−05	**1.79−05**

Note: The bold part represents the optimal value.

4.3 Analysis of Accuracy Result

The results are shown in Table 3. In terms of the accuracy of the algorithm, it can be obtained that the close centrality algorithm has better performance and higher accuracy in a class of network with higher co-matching coefficient and higher theoretical propagation threshold, and is suitable for this kind of network. Mixed degree decomposition exhibits excellent performance in a class of networks with low network heterogeneity. The lower the degree heterogeneity, the higher the accuracy and the better the performance. Therefore, it is suitable for such networks (e.g. ER stochastic networks). K-shell algorithm performs better in a class of networks with higher clustering coefficients and lower theoretical propagation thresholds, and its accuracy is also higher. It is suitable for such networks (such as BA scale-free network, jazz network). Betweenness Centrality algorithm is suitable for such networks (such as karate network, lesmis network, netscience network) because of its better performance and higher accuracy in a class of networks with lower matching coefficient, higher degree heterogeneity and lower theoretical propagation threshold. Degree Centrality algorithm is suitable for such networks (such as Dolphins network, football network, PDZBase network) because of its better performance and higher accuracy in a class of networks with high degree of homogeneity and heterogeneity. Eigenvector Centrality algorithm performs well in a class of networks with high clustering coefficient and high theoretical propagation threshold, and is more suitable for such networks (such as BA scale-free networks). Coreness centrality algorithm has better performance and higher accuracy in a class of networks with higher co-allocation coefficients and lower degree of heterogeneity, so it is suitable for such networks (e.g. ER stochastic networks).

Table 3. Accuracy comparison of seven node importance sorting algorithms in 10 networks

Network	Optimal algorithm name	Propagation threshold
ER	Mixed degree decomposition	0.172
WS	Close centrality	0.24
BA	k-shell	0.185
Karate	Betweenness centrality	0.129
Dolphins	Degree centrality	0.152
Lesmis	Betweenness centrality	0.083
Football	Degree centrality	0.093
PDZBase	Degree centrality	0.170
Jazz	k-shell	0.026
Netscience	Betweenness Centrality	0.125

5   Conclusion

Node importance ranking is one of the main ways to achieve network propagation control. In this paper, the current mainstream node importance sorting algorithms are studied in depth, and seven current mainstream node importance sorting algorithms are tested by simulation technology. In terms of resolution, the resolution of the eigenvector centrality algorithm is higher than that of the other six algorithms. In terms of accuracy, the network characteristics suitable for the seven algorithms are given, which provides a reference for the application of important node sorting algorithms in complex networks.

Acknowledgments. The paper is supported by: (1) Scientific research project of colleges and universities in gansu province under grant NO. 2018B-059 (2) National Social Science Fund Project under grant NO. 15XMZ035 (3) the Science and Technology Foundation of Gansu Provice (Grant No. 18JR3RA228) (4) Science and Technology project of Lanzhou (Grant No. 2018-4-56).

References

1. He, D.: Complex System and Complex Network. Higher Education Press, Beijing (2009). (in Chinese)
2. Gu, Y., Zhu, Z.: Node ranking in complex networks based on LeaderRank and modes similaritya. J. Univ. Electron. Sci. Technol. China **46**(2), 441–448 (2017). (in Chinese)
3. Chen, C., Sun, L.: Node importance assessment in complex networks. J. Southwest Jiaotong Univ. **44**(03), 426–429 (2009). (in Chinese)
4. Liu, J.: Advanced PID Control and MATLAB Simulation, 2nd edn. Electronics Industry Press, Beijing (2003). (in Chinese)
5. Li, J., Ren, Q.: Study on supply air temperature forecast and changing machine dew point for variable air volume system. Build. Energy Environ. **27**(4), 29–32 (2008). (in Chinese)
6. Yang, F.: Research on vital nodes identification and propagation source location in complex networks. Ph.D. Lanzhou University, Lanzhou, China (2017)
7. Sun, R.: Summary of node importance assessment methods in network public opinion. Appl. Res. Comput. **29**(10), 3606–3608+3628 (2012). (in Chinese)
8. Zhang, X., Zhu, J., Wang, Q., et al.: Identifying influential nodes in complex networks with community structure. Knowl.-Based Syst. **42**(2), 74–84 (2013). (in Chinese)
9. Zhu, T., Wang, B., Wu, B., et al.: Maximizing the spread of influence ranking in social networks. Informationences **278**, 535–544 (2014). (in Chinese)
10. Dong, W., Zhang, W., Tan, C.W.: Rooting out the rumor culprit from suspects. Comput. Sci. 2671–2675 (2013). (in Chinese)

Advances in Metagenomics Based on High-Throughput Sequencing and Big Data Mining

Na Wang[1], Kun Li[2], Yangyang Song[1], and Jinguo Wang[3(✉)]

[1] Department of Anesthesiology, The First Hospital of Jilin University,
Changchun, China
[2] Department of Urology, Tengzhou Central People's Hospital,
Tengzhou, China
[3] Department of Urology, The First Hospital of Jilin University,
Changchun, China
wangjinguolily@163.com

Abstract. The study of life science and the development of biotechnology have produced new requirements for DNA sequencing technology. Next-generation sequencing which is developed on the basis of traditional Sanger sequencing technology is widely applied in many fields for its advantages of high-throughput and low-cost. Next-generation sequencing is called high-throughput sequencing, because it is characterized by the large amount of sequence data produced in a single operation. With the improvement of sequencing technology and the development of genomics, the determination of the DNA sequence of a single species has been unable to meet the development of science, so a new research, metagenomics has been put forward.

Keywords: Metagenomics · High-throughput · Sequencing · Big data mining

1 Introduction

The base of metagenomics is high-throughput sequencing and big data mining. High-throughput sequencing has provided a large amount of data for metagenomics research to fully explore the principles contained in the data [1]. Big data mining has three major features characteristics (3V), including volume of data, velocity of processing the data and variability of data sources [2]. In terms of metagenomics research, the 3V feature of big data is reflected in the large amount of data of research objects. Normally, a single microbial community metagenomic sequencing will involve hundreds of microbial species, and the relevant high-throughput sequencing data output is huge. Therefore, the metagenomics can objectively, comprehensively and rapidly analyze the structure and function of biological hybrid systems [3]. This research has gradually applied into many research fields.

M. Atiquzzaman et al. (Eds.): BDCPS 2019, AISC 1117, pp. 1268–1272, 2020.
https://doi.org/10.1007/978-981-15-2568-1_176

2   Advantages and Characteristics of Metagenomics

In fact, the driving force behind the metagenomics approach was a scientific failure. Handelsman and other researchers point out that they can't grow the vast majority of microbes in the lab and have no idea what conditions they need to grow. Microbes live in almost every habitat, whether it's soil or water, and if you try to bring them back to the lab, you might be able to grow something, but they're not usually dominant microbes, and you can isolate DNA from microbes, because DNA is the way to understand these communities [4]. Tringe is referring to the microbial community and its environment. In an orchard, for example, researchers can study microbial communities in some soil crops, disrupt cell walls and cell membranes, fuse the genomes of different bacteria and analyze their functions to test the antibiotics used in apple trees to see if the genes make them ineffective. The research will help gardeners know how effective they are with antibiotics and allow researchers to identify genes associated with resistance [5].

2.1   Wide Application of Metagenomics

When metagenomics was first proposed, the research object was microbial community, that is, the genome of microbial community in environmental samples was sequenced to obtain the sequence of required functional genes, microbial diversity and the relationship between them and the environment [6]. The genomics approach mainly includes two relatively independent but closely complementary means: by the evolution of the amplification marks assay and genome-wide analysis of sequence, the former using specific primers to molecular phylogenetic markers (such as 16 s rRNA biomarkers) amplification, and by sequencing to identify biological community species composition and their relative abundance quantitatively. The latter measures have all the DNA sequences in the system, theoretically providing all the genome information, including evolutionary markers [7].

Therefore, the meta-genomics method can objectively, comprehensively and rapidly analyze the structure and function of biological hybrid systems. This research idea has gradually penetrated into many research fields, including the study of biological communities such as soil, sea, human oral cavity and gastrointestinal tract [8, 9] (Fig. 1).

2.2   Alignment Free Alignment Method

The alignment free alignment method based on k-tuple frequency statistics is largely improved compared with the alignment based method. First, Alignment is independent of the reference sequence database, overcoming Alignment errors caused by missing reference sequences or sequencing errors in the reference database. There is no need for highly complicated sequence alignment process, which greatly saves the time of data calculation by computer [9]. Now, k-tuple frequency statistics method has been widely used in biological computing problems, and has achieved very good results in the comparative analysis of microbial communities [10]. K-tuple's collation-free methods are widely used in biological fields, such as genome-wide evolutionary studies,

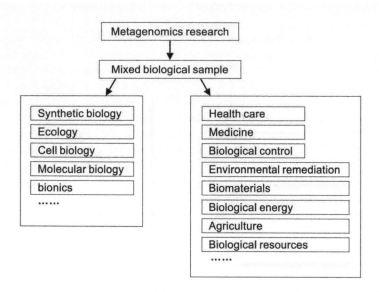

Fig. 1. The relationship of metagenomics and many other research fields

identification of genome sequence control modules, and sequence assembly of macro genome data. In the field of biological information, the sequence of a genome can be expressed by the frequency of reading segments of m length. Previous studies have shown that the statistical data, analyzing the characteristic of the genome sequence of the same sample short has strong stability, the characteristics of frequency distribution in samples of different species, short k-tuple this distribution curve and presents the obvious difference [11]. This is based on k-tuple without comparison method is widely applied to the metagenomic cancer gene types of comparative analysis provides a theoretical basis.

3   Advances in Medical Science

3.1   Search for Differences Between Cancer Genomes

Among many learning algorithms for distinguishing different samples, unsupervised clustering is the most commonly used method for comparing different clusters. In essence, unsupervised clustering is to find clustering clusters of different samples through biological laws within samples. In recent years, the application of the method of k-tuple frequency statistics clustering algorithm to compare the microbial community has received a lot of ideal results, in this aspect research focuses on statistical genome short k-tuple sequence frequency, characteristic vector of background model through a combination of genome and thin place distance method, to the cancer gene of different unsupervised clustering [12]. Two to four nucleotide sequences have been used to identify different macrogenomes and to search for differences between cancer genomes.

3.2   Macro Genomics Research Methods

Metagenomics, also known as environmental genomics, is an emerging discipline in the context of the rapid development of high-throughput sequencing technology in recent years. In human cancer samples were analysed, based on the comparative method of macro genomics is still at the exploration, and in the process of exploring the micro world, many researchers begin to abandon the cancer samples were obtained through the artificial cultivation, start looking for a kind of cultivation can be independent of the method to study the cancer cells, and this is how macro genomics research methods [13]. Initially, the genome referred to in this discipline refers to the nucleotide sequence extracted directly from a specific environment, rather than isolated culture from the laboratory, which is called the macro genome sample.

3.3   Composition of Microbial Communities

These studies explored the effects of different environmental gradients on the composition of microbial communities from the perspective of unsupervised clustering, and compared the performance of different measures of heterogeneity. Although the frequency statistics method based on k-tuple has achieved very satisfactory results in comparing oncogenes with normal tissues, there are still many defects based on the above model [14]. When combining with the phase difference measurement method of background model to calculate the phase difference between samples, a fixed order model is adopted, because it has low model adaptation and high computational complexity.

4   Conclusion

Currently, based on this short method, only the population statistical model of k-tuple distribution can be established to find out the relationship between communities and measure their overall heterogeneity. But specifically which feature sequences, which oncogenes or gene sequences are responsible for this difference and grouping of sample categories are problems that the k-tuple statistical model cannot solve. Unsupervised clustering method in the microbial community in the comparison does not complete, and for unsupervised clustering to obtain samples of the category, through supervised pattern classification to further identify the specificity of the different types of high-throughput sequencing data to depict different cancer gene features, and to find biomarkers to provide important reference information.

Acknowledgments. This research was supported by The First Hospital of Jilin University.

References

1. Coghlan, M.L., Haile, J., Houston, J., et al.: Deep sequencing of plant and animal DNA contained within traditional Chinese. Biodiv. Sci. **11**, 207–212 (2001)
2. Medicines reveals legality issues and health safety concerns. PLoS Genet. **8**, e1002657 (2012)
3. Chen, S.L., Yao, H., Han, J.P., et al.: Validation of the ITS2 region as a novel DNA barcode for identifying medicinal plant species. PLoS One **5**, e8613 (2010)
4. Chase, M.W., Salamin, N., Wilkinson, M., et al.: Land plants and DNA barcodes: short-term and long-term goals. Philos. Trans. R. Soc. Lond. Biol. Sci. **360**, 1889–1895 (2005)
5. Ning, S.P., Yan, H.F., Hao, G., et al.: Current advances of DNA barcoding study in plants. Biodiv. Sci. **16**, 417–425 (2008)
6. Lou, S.K., Wong, K.L., Li, M., et al.: An integrated web medicinal materials DNA database: MMDBD (Medicinal Materials DNA Barcode Database). BMC Genom. **11**, 402 (2010)
7. Pang, X., Song, J., Zhu, Y., et al.: Applying plant DNA barcodes for Rosaceae species identification. Cladistics **27**, 1289–1298 (2010)
8. Teeling, H., Glockner, F.O.: Current opportunities and challenges in microbial metagenome analysis – a bioinformatic perspective. Brief. Bioinform. **13**, 728–742 (2012)
9. Cheng, X., Su, X., Chen, X., et al.: Biological ingredient analysis of traditional Chinese medicine preparation based on high-throughput sequencing: the story for Liuwei Dihuang Wan. Sci. Rep. **4**, 5147 (2014)
10. Sommer, M.O., Church, G.M., Dantas, G.: A functional metagenomic approach for expanding the synthetic biology toolbox for biomass conversion. Mol. Syst. Biol. **6**, 360 (2010)
11. DeSantis, T.Z., Hugenholtz, P., Larsen, N., et al.: Greengenes, a chimera-checked 16S rRNA gene database and workbench compatible with ARB. Appl. Environ. Microbiol. **72**, 5069–5072 (2006)
12. http://www.ncbi.nlm.nih.gov/nuccore/
13. Hebert, P.D., Cywinska, A., Ball, S.L.: Biological identifications through DNA barcodes. Proc. Biol. Sci. **270**, 313–321 (2003)
14. Cheng, X., Chen, X., Su, X., et al.: DNA extraction protocol for biological ingredient analysis of Liuwei Dihuang Wan. Genom. Proteom. Bioinform. **12**, 137–143 (2014)

Unrestricted Handwritten Character Segmentation Method Based on Connected Domain and Dripping Algorithm

Minjing Peng[1,2] and Baoqiang Yang[1,2(✉)]

[1] School of Economics and Management, Wuyi University, Jiangmen, Guangdong, China
18292722059@163.com
[2] Engineering Technology Center of E-commerce Augmented Reality of Guangdong Province, Jiangmen 529020, Guangdong, China

Abstract. Aiming at the problem of low accuracy of unrestricted handwritten character segmentation, an algorithm based on connected domain algorithm and drip algorithm is proposed. The algorithm first uses the connected domain algorithm to segment the whole character, and then uses BP neural network to find the sticky characters. The drip algorithm is used to segment the sticky characters. Combining the two algorithms can realize the segmentation of the mixed handwritten characters of the confined characters and the non-adhesive characters in the general scene. The experimental results show that compared with the single segmentation algorithm, the segmentation accuracy of the unrestricted handwritten characters reaches 89.5%.

Keywords: Handwritten characters · Connected domain algorithm · Dripping algorithm

1 Introduction

The recognition of handwritten characters is a difficult problem in the field of optical character recognition. Unrestricted handwritten character recognition is a hot spot that people are generally concerned about [1]. The main reason is that there are many irregularities such as adhesion, distortion, and tilting of non-restricted handwritten characters relative to printed characters [2]. In actual engineering, optical character recognition usually analyzes the image of the image first, classifies the text and pattern in the image, divides the text area into lines, divides the text line into a single character line by line, and then sends it to the recognizer for identification of a single character. In terms of text line segmentation, Bluche proposes a method based on handwritten text recognition, which can automatically perform page analysis on images, aligning columns, lines, words and characters [3]. Kim et al. proposed training a full convolutional neural network (FCN) to predict the structure of text lines in a document image [4]. Separating a single character from an image is the premise and key of optical character recognition [5]. Many people have done a lot of research on this problem for many years. The commonly used segmentation method is vertical projection method [6, 7],

© Springer Nature Singapore Pte Ltd. 2020
M. Atiquzzaman et al. (Eds.): BDCPS 2019, AISC 1117, pp. 1273–1279, 2020.
https://doi.org/10.1007/978-981-15-2568-1_177

connected domain [8], drip algorithm [9, 10], cluster analysis method [11], template matching method [12], and so on. The vertical projection method is to vertically project the pixels of a character. Choudhary et al. propose a vertical segmentation technique for cutting a single character from a handwritten cursive [6]. A local minimum value appears at the gap of the character, and the minimum value is selected. Segmentation, for non-adhesive, non-tilted characters works well, but can not accurately segment characters that are stuck or skewed seriously; connected domain algorithm divides a connected region as a whole, it can split oblique and irregular characters, but For sticky characters it often fails. The drip algorithm simulates the trajectory of water droplets to achieve the segmentation of sticky characters, but does not effectively segment non-adhesive characters. The cluster analysis method is based on the principle that the same word forms a connected domain, and then combines the a priori relationship of each character to realize the pair of characters. It can achieve good segmentation for Chinese characters. The template matching method is for relatively regular characters. Segmentation, it can better solve the problem of character disconnection, but it does not perform well for the segmentation of the left and right borders; the guarantee of the effect of each segmentation method needs to meet certain conditions, and the actual situation is often a single character appearing in a handwritten string, a series of complex scenes, such as sticky characters, slanted characters, noise, etc., as shown in Fig. 1.

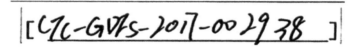

Fig. 1. An example of a common handwritten string

A single character segmentation method is difficult to meet the needs. For unconstrained string segmentation, this paper proposes a combination of connected domain and drip algorithm to learn from each other. Firstly, the characters in the image are segmented by the connected-domain algorithm, and then BP neural network is used to classify the single characters and adhesions in the segmentation results. The drip algorithm divides the concatenated characters twice, and the segmentation results are more than a single segmentation. The accuracy of the algorithm has been improved.

2 Segmentation Algorithm Based on Connected Domain Handwritten Characters

2.1 Connected Domain Algorithm

The connected domain of an image refers to a certain relationship between a pixel and a surrounding pixel. According to the pixel position relationship, it is generally divided into 4 neighborhoods and 8 neighborhoods, where 4 neighborhoods refer to the upper and lower sides of the pixel x, left and right pixels, 8 neighborhoods refer to the top left, top right, bottom left, right bottom and 8 points in addition to these four points. For image denoising and normalization, each image is processed into a 0, 1 matrix (Fig. 2).

0	1	0
1	x	1
0	1	0

1	1	1
1	x	1
1	1	1

Fig. 2. 4 neighborhood and 8 neighborhood

Taking the 4-connected domain as an example, the connected-domain algorithm scans from top to bottom until the first black pixel is marked, and then scans the top, bottom, left, and right pixels of the pixel to see if it is equal to the pixel. As long as at least one of the four pixels is equal, the same mark is made, and then the unmarked adjacent pixels are scanned, and the new unmarked and connected pixels are no longer found in the same area. Point, then a connected domain is found, continuously Repeat this process until all pixel points of the image have been scanned.

In the actual application, it is difficult to remove all the noise when preprocessing the pictures of handwritten characters. There are often some noises that cannot be removed. The connected domain algorithm also divides the noise into a region when segmenting, and the connected domain pairs single characters and oblique characters. Both can be more accurately segmented, but the sticky characters are split together (Fig. 3).

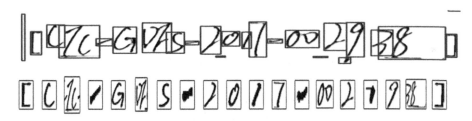

Fig. 3. Connected domain algorithm preliminary segmentation of strings

2.2   BP Neural Network Character Classification

Because the connected domain algorithm can't segment the sticky characters, and the drip algorithm can only split the sticky characters, the sticky characters must be separated, and then the drip algorithm is used to segment the sticky characters. This is a typical two-class problem. BP neural network does not need to be used. The mathematical relationship between the input and the output is determined in advance, and the distribution law of the data can be learned only by self-learning, and the input value can be output by calculating the desired value, there by setting the threshold to realize the classification of the glued characters. Horizontal projection of each character, because the width of the glued characters and the non-adhesive characters are inconsistent, the length of the array obtained by the projection is inconsistent, and the uniform length processing is adopted, which is convenient for inputting the BP neural network. Make 2000 single-character and sticky-character data sets, labeled 0 and 1. After training, the classification of the sticky characters is realized. The stuck characters are found out from the connected domain and stored in a new folder, ready for the segmentation of the drip algorithm, as shown in the following Fig. 4:

Fig. 4. BP neural network classification of sticky characters

2.3   Drip Algorithm

The drip algorithm simulates the trajectory of water droplets falling from a high point to a low position to achieve the cutting of the glue characters. The water droplets start to scroll down from the top of the character along the contour of the character, and can not penetrate when the groove is encountered. Going down, you can complete the cutting of the sticky characters according to the trajectory of the water drop. As shown in the figure below, 1 represents black pixels and 0 represents white pixels (Fig. 5).

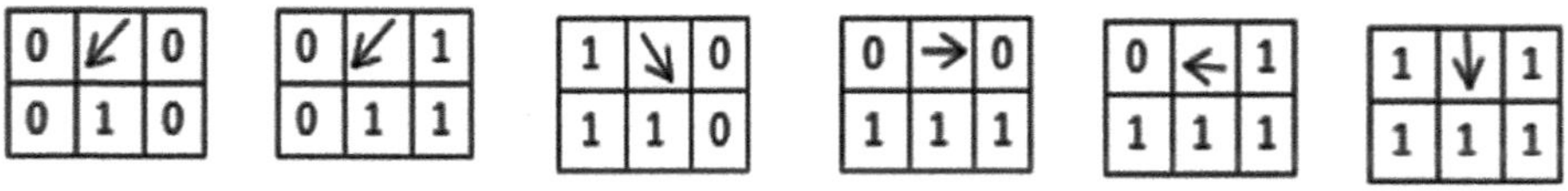

0	↙	0
0	1	0

0	↙	1
0	1	1

1	↘	0
1	1	0

0	→	0
1	1	1

0	←	1
1	1	1

1	↓	1
1	1	1

Fig. 5. Drop rule of the drip algorithm

The drip algorithm can segment two sticky characters, especially those with oblique and sticky characters. It can't be split by algorithms such as projection algorithm or connected domain, but can only be cut into two halves for three or more characters. The design uses BP neural network to detect the segmentation result again. If there are sticky characters, it is taken out again and sent to the drip algorithm for segmentation (Fig. 6).

Fig. 6. Drip algorithm splits the trajectory of sticky characters

Combining the connected domain algorithm with the drip algorithm can realize the segmentation of unconstrained handwritten characters. Another problem is that the order of handwritten characters cannot be changed by segmentation. For this problem, the first segmentation is followed by naming, after the second segmentation. Add a corresponding number or letter to the end of a split, and wait until the final result of the quadratic split is returned to the result of a split (Fig. 7).

Fig. 7. Combines two algorithms to segment the results

3 Analysis of Experimental Results

The experimental data used in this experiment is to collect 500 ticket numbers, which are free to write, and have various common handwritten character types such as single, oblique, overlapping, and adhesive, including English letters, Arabic numerals, horizontal bars, and brackets (Fig. 8).

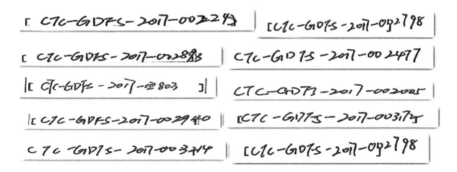

Fig. 8. Test data set sample

Under the same test samples above, the vertical projection method, the connected domain algorithm, the drip algorithm and our proposed connected domain and drip algorithm are used to test the results of the segmentation. It is found that the algorithm is better than the literature [6, 8, 9] algorithm segmentation accuracy has been greatly improved. Due to the increased computational complexity of the two algorithms, the average time taken to complete a single string segmentation is slightly increased, and the accuracy (p) and segmentation time (T) are used to calculate the performance of several character segmentation algorithms. Use n to indicate the number of correctly split characters, and N to indicate the total number of characters. Accuracy (Table 1):

$$p = n/N \tag{1}$$

Table 1. Comparison of several character segmentation methods

Segmentation method	Average time/s	Accuracy/%
Vertical projection method	0.57	65.6
Connected domain method	0.93	78.3
Drip method	0.78	85.7
Ours	0.84	89.5

4 Conclusion

The quality of character segmentation directly affects the accuracy of recognition. In practice, most handwritten characters are unrestricted, and the accuracy of a single segmentation algorithm is low. In this paper, the connected domain algorithm is combined with the drip algorithm. The method firstly divides the connected domain, and then uses BP neural network to classify the segmentation results into two types: sticky characters and non-adhesive characters. The glued characters after sorting are segmented by a drip algorithm. The applicability of this algorithm is greatly enhanced, effectively overcoming the limitations of the single segmentation algorithm, effectively improving the accuracy of character segmentation, and laying a good foundation for subsequent handwritten character recognition. However, there are some shortcomings. If there is a break in strokes in a character writing, the connected-domain algorithm will divide them into two or more characters, which will cause segmentation errors. How to improve the accuracy of segmentation will be the next step of research.

Acknowledgments. This work is supported by National Science Foundation of China (Grant: 71203162), Science and Technology Planning Project of Guangdong Province (Grant: 2014B040404072), Natural Science Foundation of Guangdong Province (Grant: 2015A030313642), Innovation Project of Wuyi University (Grant: 2014KTSCX128 and 2015KTSCX144).

References

1. Gao, L., Zhang, X., Zhi, T., et al.: A sequence labeling based approach for character segmentation of historical documents. In: Iapr International Workshop on Document Analysis Systems (2018)
2. Zhao, B., Su, H., Xia, S.: An unconstrained handwritten digit string segmentation method. Chin. J. Inform. Sci. **03**, 22–29 (1998)
3. Bluche, T., Stutzmann, D., Kermorvant, C.: Automatic handwritten character segmentation for paleographical character shape analysis. In: Document Analysis Systems (2016)
4. Vo, Q.N., Kim, S.H., Yang, H.J., et al.: Text line segmentation using a fully convolutional network in handwritten document images. IET Image Proc. **12**(3), 438–446 (2018)
5. Choudhary, A., Kumar, V.: A robust technique for handwritten words segmentation into individual characters. In: Speech and Language Processing for Human-Machine Communications, pp. 99–106 (2018)

6. Choudhary, A.: A new character segmentation approach for off-line cursive handwritten words. Proc. Comput. Sci. **17**, 88–95 (2013)
7. Ran, L.: A method of vehicle license plate segmentation based on vertical projection. Commun. Technol. **45**(04), 89–91+98 (2012)
8. Zhu, Y., Qiu, J., Yang, C.: A license plate character segmentation method based on improved connected domain algorithm. J. Hangzhou Dianzi Univ. (Nat. Sci. Edn.) **36**(02), 48–51 (2016)
9. Li, X., Gao, W.: Method of blocking character segmentation in verification code based on dripping algorithm. Comput. Eng. Appl. **50**(01), 163–166 (2014)
10. Zhang, C., Zhai, Z., Xiao, B., Guo, J.: A dripping algorithm for handwritten digit string segmentation of bank notes. J. Beijing Univ. Posts Telecommun. (01), 13–16 (2006)
11. Chen, L., Huang, X., Wang, M., Li, W.: A license plate character segmentation method based on cluster analysis. Comput. Eng. Appl. (06), 221–222+256 (2002)
12. Cui, W., Cui, Y., Wang, Z., Gong, L., Liu, M., Tan, C.: A license plate character segmentation algorithm based on template matching and vertical projection. J. Qiqihar Univ. (Nat. Sci. Edn.) **31**(06), 12 (2015)

Reconstruction of Brand Communication on Media Integration in the 5G Digital Era

Li Wang[✉]

Art and Design College, GuiLin University of Electronic Technology,
Liuhe Road, Qixing District, Guilin 541004, Guangxi, China
42265301@qq.com

Abstract. In China, the 5G digital era means that unbuffered video technology has become a reality, and media convergence scope is becoming more in-depth. The development of media technology has produced corresponding changes and influences on brand communication which reflect on content production and communication strategies. Media integration is reshaping new smart brands.

Keywords: Media integration · Brand communication · 5G

1 Introduction

Nowadays, the rapid development of digital technology has become an unstoppable trend and trend, subverting the traditional media pattern. The use of media is embedded in everyone's life, and it brings more opportunities and challenges to brand communication. According to the 43rd Statistical Report on China's Internet Development Status released by China Internet Network Information Center, as of December 2018, the number of Internet users in China reached 829 million, with a penetration rate of 59.6% [1]. The number of mobile Internet users in China reached 187 million, and the proportion of Internet users accessing the Internet through mobile phones was as high as 98.6%. The Internet demographic dividend peaked, and the incremental competition entered the era of stock competition. Since 2016, the social network of "two micrometers and one end", which meas We-chat, blog, mobile client, has refreshed the form of media communication. At present, mobile phones have become the most important Internet terminal tools in our daily life. The impact of the development of digital technology on brand communication is trans-formative. It has greatly enhanced the timeliness of brand information production and release, and the brand content has been greatly enriched. The diversification of brand subjects and the interaction of audiences have been greatly enhanced. The performance is greatly improved. In the eyes of Philip Kotler and Kevin Keller, Branding is the power to give products and service brands. And the same time, media technology will be the source of power.

The 5G digital era has arrived. In June 2019, four operators of China Telecom, China Mobile, China Unicom and Radio and Television took the lead in obtaining the official 5G commercial license issued by the Ministry of Industry and Information Technology, and approved four companies to operate the "fifth generation digital cellular mobile communication service". The 5G digital era means that unbuffered

M. Atiquzzaman et al. (Eds.): BDCPS 2019, AISC 1117, pp. 1280–1286, 2020.
https://doi.org/10.1007/978-981-15-2568-1_178

streaming and Virtual Reality will become a common fact, an era of Internet of Everything is coming soon, and innovative applications of 5G+4K+AI+VR+AR technology [2]. The media will bring new experiences and innovative opportunities for the spread of brand stories. Scholar Yu Guoming proposed in the "Opportunities and Essentials of Media Development in the 5G Era": The arrival of 5G has opened the "second half" of Internet development. If we say that in the "first half" of Internet development, traffic is power, and who gets the attention resources of consumers can get good communication effects. Then, the contest of the "second half" is reflected in how brands use immersive communication to connect brands and consumers by constructing appropriate scenarios.

The development of digital technology has broken through the boundaries of media, forming media convergence. Media convergence refers to various media expressions such as text, sound, video, animation, VR, AR, etc., as well as newspapers, radio, television and Weibo, WeChat. The client and the like muti-multidimensional present the information content. The experience that the media brings to the user will present more comprehensive and deeper information content for the user. In the 5G era, we can understand media convergence from three levels.

2 Media Integration in the 5G Digital Era

2.1 The Integration of Media Content and Functions

Virtual reality technology enables the integration of brand content and media functions. Recently, Tencent's ads "Dashan Children's Autumn and Winter Fashion Show" has used 360-degree interactive virtual reality technology to convey the attention of children in poverty-stricken areas. The promotion of the Forbidden City "cultural creative design" has broken through the model aimed at sales in the Forbidden City. It is sold in the online stores of the Forbidden City Taobao, the Forbidden City Tmall, the Forbidden City Mall, etc., pushed on WeChat, Weibo, and the public number, 2016, H5 See you through the Forbidden City, CCTV documentary "I am in the Forbidden City to repair cultural relics" [3] continued to follow up. The contents of the Forbidden City in different channels form the sharing and integration of online and offline, physical sales and network virtual scenes.

2.2 The Integration of Media Production and Consumption

This level focuses on the transformation of media operations management and the transformation of consumer usage logic brought about by technological changes. Different media cooperate with each other to participate in content production and drive the migration and integration of media audiences. For example, Shenzhen Satellite TV and Ali jointly launched a super-release conference to create a scene-based entertainment marketing that realizes the deep integration of TV and e-commerce, audience and consumers. The content flow of TV media and online media, horizontal cooperation between different industries, and mutual migration of different audiences to consumer audiences have been realized.

2.3   The Structural Transformation of Social Relations

This refers to the ambiguity and even collapse of the social intercourse that was originally divided, and the transformation of some of the original social relations. Haier washing machine users and Xiaomi users can become part of the brand's product development and operation, and the identity of the user can be converted to the product manager. The application of the barrage technology has greatly enhanced the enthusiasm of participatory consumption as a young person, and various evaluations and opinions on brand content have become a social activity with commercial value [4].

3   How Media Integration Reshaping the Pattern of Traditional Brand Communication

3.1   From the Perspective of the Communicator, the Communicator Has Changed from the Information Processor at the Front Desk to the Precise Analysis and Planning of Media Operations and Consumer Behavior in the Background

Everyone may have more or less such experience. The last second is still discussing with the classmates where to go to eat at noon. On Taobao, we will add snacks to the shopping cart. The next day, the homepage will be overwhelming. If you receive information about similar products, the traces of the search will become the target content of the homepage of the web page. Is this coincidence? This is because with the penetration of digital media in everyday life, we generate massive amounts of data in the process of using media, while big data technology translates human behavior into quantifiable data nodes. Thus providing data portraits for people, an information asset for big data, and the purpose of big data technology is to turn data into information that people can understand. Big data focuses on the overall and the rules. Corporate brands use data analysis to tag consumers with data tags and push tag content. This is the precise delivery of brand communication strategies. Through big data, the brand communication strategy is refined. The information dissemination will inevitably lead to the imbalance of information supply and demand. Big data simplifies people's perception of information ignoring the individualized differences and easily leading to the results of thousands of people. And you like to eat watermelon, but you like sweet, I like water, consumers are more willing to believe in similar people when purchasing opinions. Therefore, brands use labels as consumers to define the matching method of intelligent demand.

In the process, the label is produced. It is no longer the brand, but the user itself. To achieve point-to-point information balanced output. Netease Cloud's precise algorithm always pushes the song list that suits your taste. It is therefore a "paradise for independent and niche music lovers". Naturally, this is also a product of big data, but the classification of songs is difficult to put. Controlled, NetEase cloud music lacks professional classification means compared with old cool dog music and QQ music, but why is it more popular? Aside from Neteese Cloud's own marketing communication strategy, collaborative filtering algorithm is its core competitiveness. This algorithm is attributed to the invention of Amazon engineers. A customer who bought this thing may also buy another thing. The prediction of the algorithm The standard depends on the similar consumption patterns between people, the songs I like, and you like, the collaborative filtering algorithm is its core competitiveness. It is divided into user-based and project-based types. Based on the user refers to you and me, then it is possible that there is a song you like in my song list. I can push you a song list without a collection. Based on the item, the song is matched based on the image resemblance. After comparing the similarity, according to the user's historical preference, the single song is pushed for another user.

3.2   From the Recipient's Perspective, the Transition from Graphic Video Readers to Your Digital Media Users and Information Consumers

Recently, some video websites have launched idol trainees, creating 101 and other idols [4]. This group of idols does not give a distant sense of distance like traditional idols. These new idols are idols created by the masses in social media activities. In the traditional sense, the audience turns into idols behind the scenes, participate in online voting, and surrounding goods. Selling out of stock, fans in order to canvass their own idols, through fund raising, brushing the list and other ways to increase the exposure of idols, forming an unlimited economic benefits.

The convenience of the media technology gives the user a mobile phone to carry, so that the public's awareness of the demand for the media in daily life is strengthened, thus changing the media in traditional life as a tool for entertainment. Family gatherings are not fun and laugh, only the phone and the party. Nowadays, young information audiences are getting along with the media. It is true that digital progress has dramatically changed the way we communicate, from cash payments to scan code payments, from SMS to WeChat information, from traditional calls to video calls, from entities. Shopping to Taobao Tmall, from offline classroom to MOOC, from the 90 s, after the 00 digital aborigines to the digital new immigrants after 70, 80, the media is everywhere, all the time, we are used to Internet technology all the time which has changed the way we live with the media society. We used to consume media, and now we use the media to create content.

Any voice can enter the market without any threshold. Relatively anonymous identity ID, relatively free speech market, relatively equal communication channels, which is given to us by the mass media. Great rights, such a way of jumping off the system is also a kind of speech guarantee. We have more and more expression channels, and our desire to express is gradually awakened. We are eager to pay attention to our own opinions. I hope that my experience can be a reference for other people's consumption behavior, so the experience and sharing of public comments, word of mouth, etc. came into being [5]. More and more people recorded their own consumption evaluations on the APP. They were dissatisfied after the purchase, through word of mouth. Spread to relatives and friends, the impact is limited, now on the Internet, the impact of the Internet's communication effects is difficult to estimate, forcing businesses to pay attention to the feelings of each consumer, not to mention, as long as sharing travel in personal accounts and videos Attractions, small shops, and food will instantly become a gathering place for followers.

At the same time, online reviews will also greatly influence consumers' purchasing decisions. We no longer only believe in the gorgeous evaluation of advertisers, but are more willing to believe in the evaluation of experiences from strangers. According to statistics, 65% of consumers will read user reviews in the car vertical forum, and 46% hope that KOL will recommend more products worth buying. More and more brands cooperate with Net Red to promote products. However, in the rush of information, we are overwhelmed by the rush of information. Slowly also trained some of the techniques and skills to initially identify information.

3.3   Reshaping the Way Brand Content Is Produced and Disseminated

In the media scenario of the whole network convergence, the user's voice is released. The production content of users shows strong vitality. For the distribution method, the "one-to-many" point distribution mode of traditional media has gradually shifted to the "many-to-many" multi-platform distribution mode [6]. The popular short video matrix mode is often to publish the same content in multiple channels. Whether it is Weibo, Douban, Zhihu or B station, you can get similar content push, even subscribe to the video selection section directly on the singularity, watch the latest video of the vibrato every day, this is Matrix propagation mode of vibrato.

The brand orientation of the consumer center, the traditional brand communication logic follows the clear planning and ideas of the brand's own development, emphasizing that all results must be under the control of the brand owner. The online brand communication logic advocates an open brand orientation, and believes that the brand is growing up in an open source environment, and consumers are in the leading position of brand open source controls. For example, Alibaba's Brand Hub new retail platform integrates content service organizations and provides brand-name butler services to retailers in a streamlined and standardized mode [7]. It is favored by more and more new retail brands. The brand's official website is the most eye-catching segment. The presentation highlights the importance of consumer orientation to the brand in the digital age. Consumers want to be treated specifically, hoping to centralize marketing, not as a data for a thousand people.

Reverse pyramid brand communication model, traditional brand theory tells us that the steps of brand building can be constructed by brand pyramid theory, and the most representative AIDA model (attention, interest, desire, action) is top-down. The goal is to expand the base, build awareness, and retain a small number of consumers to take further action. The online brand logic is just the opposite. It is represented by a reverse pyramid model. Starting from the apex, we first find the key opinion leader KOL, using their influence, through buzzer marketing and virus transmission to form a word-of-mouth effect. External diffusion, and finally reached the grassroots. Foreign countries have formed a stable KOL culture. This culture has become increasingly effective in driving Chinese luxury goods. More and more luxury brands are adopting KOL promotion models in China, such as google boy MS-bank in the field of Weibo, which is good at Text graphics video conveys brand information and consumption concept to the general public, thus forming a word-of-mouth effect to drive public brand recognition and purchase behavior.

3.4   Remodeling the Brand Strategy

3.4.1   Focusing on the Brand Extension Strategy of the Ecosystem

Traditional brand extensions focus on sub-brands or magnify the functionality of a brand for brand extension. In the 5G era, the development of digital technology, brand extension advocates the trans-screen derivative. The extension of brands in different media needs to consider the problem of consumer adaptation. When the Internet develops to the top of the dividend, the wisdom brand logic believes that it should focus on the construction of the brand ecosystem, based on the open platform brand, and connect a large number of commercial organizations and resources beyond the industrial boundary. Co-existing value creation, forming a structured community relationship and network effect of mutual dependence, mutual coordination and reciprocal cycle. In 2018, Haier washing machine sales won the sales champion for 10 consecutive years, and its brand extension strategy through the clothing network ecosystem [8]. The upstream and downstream industries are connected together to provide users with solutions from washing, care storage, collocation, and purchase. Form a brand extension in a transformation rather than a transition. Breaking through the barrier between smart care and the Internet of Things. Realize a win-win situation for all parties in the ecosystem.

3.4.2   Emotional Connection of IP Characters

To put it simply, it is to create more emotional connections with consumers through attractive IP in content strategy, enhancing content piggybacking, content native capabilities and multi-platform content interaction capabilities. In March 2018, the number one player logged in to China and created a box office miracle in just three days [9]. What media strategy did he adopt? First of all, the number one player has gathered 138 IP characters of different sizes. Each IP has a group of ashes fans. The domestic marketing team has locked in the vertical audience of heavy gamers and senior fans, and is well-known through inviting games and animation. KOL's feature drafting and detonation of word of mouth, these media strategies have won a wide second spread of the film, and successfully achieved emotional connection with consumers.

The new era of digital technology has arrived. Internet technology has truly realized the wisdom of life scenes and brought unlimited possibilities for brand development. People have evolved from a single laundry to an intelligent management of the entire life cycle of clothing. Through the cooperation with a number of companies to achieve the brand's lasting vitality, this is inseparable from the brand-oriented brand extension strategy [10]. Say goodbye to closed category positioning and improve the ability of resource connections.

4   Conclusion

With the advancement of technology and the substantial improvement of media technology, users' personalities and needs are no longer mysterious. They can complete sophisticated user portraits through intelligent technology, and establish user activity and user viscosity through media integration strategies. Capable of creating fissile growth in brand value.

The biggest challenge is the widening of information channels, faster and farther distribution, brand survival crisis and increased competition. Brands must not only win in the fierce competition, but also be cautious in the competition to avoid the outbreak of brand crisis, because the era of media convergence is no longer like the traditional media era, can suppress the speed and scope of negative news through various controllable forces, in the era of word-of-mouth, the brand crisis can destroy any brand.

References

1. Wu, W.: Analysis of brand communication transformation in the 5G era. Mod. Mark. 85 (2019). (in Chinese)
2. Lu, P.: Moving towards the 5G era, winning in the smart brand. China Advert. 7–10 (2019). (in Chinese)
3. Sun, Y., Wang, F.: Research on the design and promotion of museum cultural creative products from the perspective of marketing communication, pp. 2–7 (2018). (in Chinese)
4. Yao, W.: On brand communication in micro-media. China Newspaper Industry, pp. 44–45 (2019). (in Chinese)
5. Xu, W.: Thinking with "central kitchen" see the brand communication in the age of media. China Advert. 30–32 (2018). (in Chinese)
6. Li, J.: An analysis of the relationship between publishing brands and audiences in the age of media and media. China Publishing Media Business Newspaper, pp. 1–2 (2019). (in Chinese)
7. Xu, X., Yu, Y.: Brand image design and promotion in digital media environment. Art Educ. Res. 32–34 (2017). (in Chinese)
8. Mano, Zhan, Q.: the shaping and dissemination of brand image in the era of new media. Western Leather. 78–79 (2018). (in Chinese)
9. Anke: the brand image in the new media environment, written and edited, pp. 22–23 (2016). (in Chinese)
10. Jie, L.: The Impact of new media communication on branding. Brand Res. 150–152 (2018). (in Chinese)

Radial Basis Function Neural Network and Prediction of Exchange Rate

Ming Fang and Chiu-Lan Chang$^{(\boxtimes)}$

Fuzhou University of International Studies and Trade, No. 28, Yuhuan Road,
Shouzhan New District, Changle, Fuzhou City, Fujian, China
Jochang76@qq.com

Abstract. This study aims to predict the exchange rat based on radial basis function neural network (RBFNN). Concerning the selected sample point, it only responds to the inputs of neighboring samples and hence has better approximation performance and overall optimization than other forwarding networks, in addition to being simple in structure and fast in training speed. RBFNN in empirical risk minimization methods has been studied, their adaptability to exchange rate prediction has been explored, and these methods have been verified by exchange rate data. The contributions are proposing an exchange rate prediction algorithm based on RBFNN and more accurately predicting short-term and long-term exchange rate variation trends.

Keywords: Radial basis function · Neural network · Prediction

1 Introduction

The purpose of machine learning must require minimization of the expectancy risk for the requisite objective function. It is not hard to imagine that the arithmetic average of the sample can be utilized to replace the ideal expectancy. This is the principle behind empirical risk minimization. The artificial neural network is the most commonly used method based upon empirical risk minimization, having achieved extremely wide application in many fields because of its powerful intrinsic adaptability and learning aptitude, solving numerous problems unsolved by traditional methods, and playing a significant role in scientific research and practical life. A connection model (neural network) consists of some simple units like neurons with weighted connections between these units. Each unit has a state which is determined by the other input units connected to it. The purpose of connection learning is to distinguish the equivalence of the input mode. Connection learning trains the network through all types of examples, and produces an internal representation of the network, whereby it can identify other input examples. Learning is mainly represented in adjusting connection rights in the network, and this type of learning is non-symbolic, with a high capacity for parallel and distributed processing, and has achieved great success and development. artificial neural network has distributed information storage, parallel information processing, self-organized learning, robustness, non-linear mapping, and other properties, and has become ubiquitous in pattern identification, artificial intelligence, signal processing, economic predictions, and other fields [1]. artificial neural network methods are

© Springer Nature Singapore Pte Ltd. 2020
M. Atiquzzaman et al. (Eds.): BDCPS 2019, AISC 1117, pp. 1287–1292, 2020.
https://doi.org/10.1007/978-981-15-2568-1_179

numerous, among which the radial basis function is an effective feed forward neural network, as it has smaller support set. Concerning the selected sample point, it only responds to the inputs of neighboring samples and hence has better approximation performance and overall optimization than other forwarding networks, in addition to being simple in structure and fast in training speed [2]. Therefore, radial basis function neural network (RBFNN) is selected as representative of the methods based upon the principle of the minimization of empirical risk.

The widely-used type of neural network, RBFNN, is introduced into the field of exchange rate prediction based upon the empirical risk minimization principle. RBFNN is a data-driven algorithm has been a computation method which can set up models in the historical input data and predict output through the models, Previous studies use data-driven algorithm such as Lewbel [3], and Swanson and White [4], Bodurtha et al. [5] proposed its application to stock markets and Gencay [6] proposed its accessibility to exchange rate prediction. Presently, RBFNN has been widely applied to every field of prediction, e.g. predicting land and water volume [7], bank deposits [8], biotoxicity [9], and the degree of air pollution [10]. In the background of all the above-mentioned applications, RBFNN takes data as the time series for prediction.

2   Radial Basis Function Neural Network System

Neural networks and other feed forward neural networks have a similar neural network structure. A neural network is composed of only three layers, each of which has unique characteristics. The input layer consists of some perception neurons, its function connecting the external input variable and the internal neurons of the network, playing the role of buffering and connecting throughout the entire network, and directly transmitting the input variables to the neurons of the hidden layer. Neural networks have only one hidden layer, and its function is to map the input variable onto the hidden layer, which is a process of non-linear conversion. During the process of transmitting the input variable to the hidden layer, a radial symmetric kernel function is selected as the activating function of the hidden layer neuron, and this function assumes the duty of being the "basis" for a set of input variables. It is a non-negative and non-linear, radial, centrosymmetric damping function. This function is sensitive to the output variable approaching the central point of the kernel function and can produce stronger output signals, and hence the neural network has the properties of simple structure, fast computation, and local function approximation. The structural objective of the entire network is ascertained when the network parameters of the hidden layer are ascertained. Under common circumstances, the more neuronal nodes the hidden layer of the neural network has, the stronger its computation and mapping capacities, and the better its function approximation, being able to approximate a complex function curve with any accuracy. However, more neurons in the hidden layer means it has higher spatial dimensions, and the performance criteria of the network are closely related to the spatial dimensions of the hidden layer. Under circumstances of higher spatial dimension, the neural network has better approximation accuracy, but the consequent price is increased complexity in the neural network. The algorithm framework of RBFNN used in this study is shown in Fig. 1.

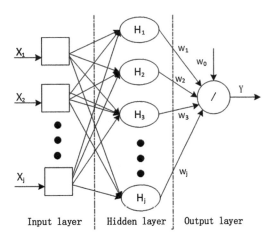

Fig. 1. Algorithm framework of RBFNN

RBFNN output layer choice function sets to $F(x) = \sum_{i=1}^{n} w_i \lambda(\|x - c_i\|)$, where c_i denotes the center of the basis function, with the Euclidean norm often chosen, and w_i is the weight vector. When the Gaussian kernel function is selected as the basis function, the ith hidden layer node output-response of the neural network is expressed as $y_i^2(x_k) = \exp\left(-\|x - c_i\|^2 \big/ 2\sigma_i^2\right)$ which defines the hidden layer of the RBF neural network. The corresponding output layer of RBFNN is $y_q^3(x_k) = \sum_{i=1}^{n} w_{iq} y_i^2(x_k) + b_q$ where the weight of the ith neuron of the hidden layer, and the qth neuron of the output layer is expressed as w_{iq}, and the threshold of the qth neuron of the output layer is expressed as b_q.

The width of the RBFNN's hidden layer is also an important factor affecting its classified prediction ability and the width of the central point determines the weight of the hidden layer and the input layer, along with the action range on input variable response, and is a key variable in realizing the ultimate mapping result of the entire network. Therefore, the difficulties of establishing an RBFNN lie in: ascertaining the relevant parameters of the hidden layer neurons (i.e. the number, center, and width of hidden layer neurons), and the connection weight of the hidden layer and the input layer.

3   Using Radial Basis Function Neural Network to Predict Exchange Rate

The RBFNN is trained by the standard of RMSE, with a training step length of 0.1–1. The next step is to set various parameters. The width of the radial basis function needs to be set first. The radial basis function is a type of locally distributed, centrally symmetric, non-negative non-linear function, which has two main parameters, the center of basis, i.e. the symmetric point, and the width of basis, i.e. in how big a region the obvious output response is produced. Between these two parameters, the width of the RBF is very important. The network is theoretically able to functionally approximate any input and output sample but cause over adaptability or non-adaptability in the functional approximation if improperly selected. Under general circumstances, the selection of width is determined by the distance between input vectors, with the requirement of being bigger than the minimum distance and smaller than the maximum distance. If the constant width is too big, a larger number of neurons are required in the network, every neuron is basically the same and network training analysis cannot be conducted. Figure 2 shows the procedures of RBFNN to predict the exchange rates.

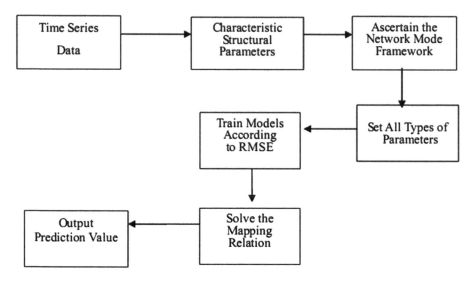

Fig. 2. Procedures of using RBFNN to predict exchange rate

4   Prediction Errors Using RBFNN

We use the EUR/USD, JPY/USD, RMB/USD and exchange rate from January, 2013 to December, 2018. Total of 55 monthly-step exchange rate data points, 40 of which are used for sample modeling and 15 of which for testing. We predict the 41st datum on the 36th–40th data points, and return the prediction result to a time series to construct the new characteristic parameters ensuring that the prediction continuously proceeds. RBFNN presents non-linearity, non-locality, non-stability, non-convexity and other characteristics, introduces the RBF kernel function, and at the same time applies three types of processing units to the neural network (input unit, output unit, and hidden unit), decreasing the complexity of calculation, and accelerating learning and convergence. RBFNN based on empirical risk minimization have a stronger ability to solve non-linear problems and may solve the problem of exchange rate prediction to a certain extent. However, the algorithm accuracy decreases, restricted by the size of data set and increasing non-linearity in the problem. Table 1 shows the prediction errors using RBFNN.

Table 1. Prediction errors using RBFNN

Exchange rate	MAE	RMSE
EUR/USD	2.415	0.0007
JPY/USD	3.972	0.0021
RMB/USD	4.531	0.0039

5 Conclusion

Along with the rapid development of financial globalization, spurred a global financial crisis and an economic recession and hence caused exchange rate prediction to evolve into an important economic issue, drawing wide attention. However, the foreign exchange market is a non-linear system with multiple variables, in which correlations between all factors are perplexing, exacerbating the difficulty of exchange rate prediction. In real applications, exchange rate fluctuations and varying trends are very complex, and the execution speed of the algorithm must surpass the variation speed of exchange rate at the same time as the exchange rate is precisely predicted. Although numerous studies pertaining to exchange rate prediction methods are currently available, the majority of the algorithms have been constrained by their complexity, and relevant research analysis has not been conducted on the applicability to data sets of the algorithms commonly used in exchange rate prediction. RBFNN is introduced into the field of exchange rate prediction based upon the empirical risk minimization principle. This method both inherits the empirical risk minimization principle and introduces the kernel functions of RBF, has a higher prediction accuracy, simple structure, fast training speed, and different from the ordinary feedforward neural networks, with the best approximation performance and overall optimization.

In subsequent further research work, more exchange rate data will be required to express information pertaining to the confidence value and confidence interval of the prediction results and other uncertainties, to supply more scientific and reasonable reference information for use in investment and international trade. Setting of algorithm parameters is required for realizing the greatest performance in the exchange rate prediction of various types of algorithms. Therefore, a more scientific identification of algorithm parameters is a rational concept for enhancing the prediction performance of algorithms.

Acknowledgments. This research was supported by the Education Department of Fujian Province, China (Grant number: JZ160491); Social Science Fund of Fujian Province, China (Grant number: FJ2018B075); Fuzhou University of International Studies and Trade (Grant number: 2018KYTD-14).

References

1. Yao, X.: Evolving artificial neural networks. Proc. IEEE **87**(9), 1423–1447 (1999)
2. Er, M.J., Wu, S., Lu, J., Toh, H.L.: Face recognition with radial basis function (RBF) neural networks. IEEE Trans. Neural Netw. **13**(3), 697–710 (2002)

3. Lewbel, A.: Comment on artificial neural networks: an econometric perspective. Econometr. Rev. **13**(1), 99–103 (1994)
4. Swanson, N.R., White, H.: Forecasting economic time series using flexible versus fixed specification and linear versus nonlinear econometric models. Int. J. Forecast. **13**(4), 439–461 (1997)
5. Bodurtha Jr., J.N., Cho, D.C., Senbet, L.W.: Economic forces and the stock market: an international perspective. Glob. Financ. J. **1**(1), 21–46 (1989)
6. Gencay, R.: Linear, non-linear and essential foreign exchange rate prediction with simple technical trading rules. J. Int. Econ. **47**(1), 91–107 (1999)
7. Carvalho, G.R., Brandão, D.N., Haddad, D.B., do Forte, V.L., Ceddia, M.B.: A RBF neural network applied to predict soil field capacity and permanent wilting point at brazilian coast. In: IEEE 2015 International Joint Conference on Neural Networks, pp.1–5 (2015)
8. Li, X., Deng, Z.: A machine learning approach to predict turning points for chaotic financial time series. In: 19th IEEE International Conference on Tools with Artificial Intelligence, vol. 2, pp. 331–335(2007)
9. Melagraki, G., Afantitis, A., Sarimveis, H., Igglessi-Markopoulou, O., Alexandridis, A.: A novel RBF neural network training methodology to predict toxicity to Vibrio fischeri. Mol. Divers. **10**(2), 213–221 (2006)
10. Lu, W.Z., Wang, W.J., Wang, X.K., Yan, S.H., Lam, J.C.: Potential assessment of a neural network model with PCA/RBF approach for forecasting pollutant trends in Mong Kok urban air, Hong Kong. Environ. Res. **96**(1), 79–87 (2004)

Information Interface of Artificial Intelligence Medical Device Information

XiYuan Wang, Ting Fu[✉], Yiyun Zhang, and Dandan Yan

Donghua University, 1882, Yan'an West Road, Shanghai, China
shbjy001@163.com

Abstract. At present the research by the Chinese scholars on intelligent electronic medical monitoring equipment is still limited to system planning and technology implementation means, while few scholars focus on research on human-machine relationship design, especially human-machine interface design, thereby resulting in the isolation phenomenon between technology and art, as well as between production and user.

Keywords: Artificial intelligence · Medical equipment · Human-machine interface

1 Three-Level Model of Emotionalized Design for Human-Machine Interface of Medical Monitoring Equipment

People have different levels of emotional experiences in product. Norman, a cognitive psychologist, divides people's emotional experiences into three categories: instinctive emotions, behavioral emotions, and reflective emotions [1]. The changes in user emotions are divided as follows: evocation, association and identification. Evocation aims to stimulate the user's emotional response via some ways, and this emotional reaction can be projected inside the products; association refers to connection with products as a result of chemical-reaction-like wonderful changes in projection after use of products. Ultimately this association will gain user's identification and acceptance on the products. Similarly, the emotionalized design of the human-machine interface (HMI) hereunder can be divided into three levels: design of sensory level, design of efficiency level and design of cognitive level, as shown in Fig. 1.

2 Analysis on the Emotionalized Factors Influencing HMI Design of Medical Monitoring Equipment

2.1 Analysis on Emotionalized Factors in Sensory Level

Sensory level emotionalized design focuses on intuitionistic visual experience that a product brings to users. Regarding HMI of medical monitoring equipment, it means users' emotional experience from visual elements of HMI. It is not only an important form of information exchange between users and equipment, but also a main factor

© Springer Nature Singapore Pte Ltd. 2020
M. Atiquzzaman et al. (Eds.): BDCPS 2019, AISC 1117, pp. 1293–1303, 2020.
https://doi.org/10.1007/978-981-15-2568-1_180

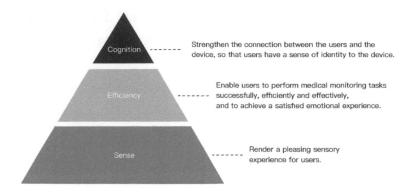

Fig. 1. The three layer design model of the emotional design of human-computer interaction interface in medical monitoring equipment

affecting the user sensory level experience. The visual elements of the human-machine interface mainly include color, graphic, text and icon, which complement each other and directly affect the user's sensory level emotion visually.

(1) Color factor

People are inherently sensitive to colors, the American Popular Color Research Center's research shows that: facing a wide variety of products, people only take 7 s to select what they are interested from these items: this is well known as 7-second law. Within the short time, the role of color accounts for 67%, as the first important factor influencing people's decision-making [2]. Therefore, in the same way, in the medical monitoring equipment HMI design, color selection and matching also directly affect the user's sensory level emotion, and determine the user's first impression to the HMI.

(2) Graphical factors

Graphics are regarded as the medium of information, and its transferring speed is much faster than text. The more artistic graphics can capture viewers' attention and quickly deliver information [3].

(3) Text factors

As an important sub-element in the design of interface visual element, the text plays a role of carrying, disseminating information and expressing thoughts and feelings. The different arrangement pattern will visually bring users different experiences.

2.2 Analysis on Emotionalized Factors in Efficiency Level

Efficiency-level emotionalized design is aimed to ensure users to complete the medical monitoring task smoothly via the HMI, and to meet the user's practical demands of the equipment. Thereby a good efficiency level design solution shall contain the interface readability and interface usability [4].

(1) Interface readability

HMI aims to stimulate user's visual experience to transmit functional information, which requires good readability of the visual elements in interface: interface graphics, text, pictures and other information, within the limited range of the interface, can be reasonably arranged [5]. It is regarded as an important channel for establishing emotional connection between users and products. Because of the specific nature of medical monitoring equipment, users require higher readability of interface: in medical monitoring equipment HMI, any and all visual information should be transmitted, expressed and understood, in a clear and accurate manner. The distribution of information shall be compliant with human eye observation habits, thereby to ensure rapid transmission of information.

(2) Interface usability

In the HMI of medical monitoring equipment, the interface usability can be subdivided into easy to understand, easy to learn and easy to operate.

2.3 Analysis on Emotional Factors in Cognitive Level

Good cognitive level emotionalized design solution should be capable to stimulate user's resonance, trigger user's identification and dependence on the interface. Therefore the efficiency level emotionalized factors can be summarized as the interface pleasure and interface resonance.

(1) Interface pleasure

Martin EP Seligman, a social psychologist, has pointed out that human control on the outer world can lead to a pleasant and positive emotional experience. On this basis, people are more willing to try new things and accept new challenge [5].

(2) Interface resonance

The HMI resonance is a strong sense of identity and emotional experience arising from effective and positive communication with the interface by users during the operation process, and such the communication is related to the feedback of interface. Reasonable and friendly negative feedback is capable of reducing the user's mis-operation probability and sense of frustration effectively, and maintaining users' good emotional experience in the operation process.

3 Emotionalized Design and Analysis on Human-Machine Interface in Medical Monitoring Equipment

3.1 Analysis on User Characteristics of Medical Monitoring Equipment

The medical monitoring equipment has a relatively complex and mixed customer group, and is characteristic of multiplicity [6]. According to the user relationship in medical service, the users of medical monitoring equipment can be generally divided into two major categories, respectively the subjective users and the objective users. The subjective users refer to the professional medical treatment staff and play a dominant and behavioral subject role; the objective users include the patients and the non-professional caregivers who are responsible for caring for the patients, and play a subordinate and behavioral follower role.

3.2 Analysis on Emotional Demands for Users of Human-Machine Interface in Medical Monitoring Equipment

3.2.1 Analysis on Emotional Demands of User Sensory Level

Considering that the emotions on sensory level are related to the initial responses of people, we mainly adopt questionnaire investigation method to study the emotional demands on sensory level and dig out the equipment preferences of users on sensory level. In addition, in-depth interview method is combined to dig out the reasons behind the preferences. The colors, image styles and text arrangement method of mainstream product interface in medical monitoring equipment are categorized, concluded and collected into investigation questionnaires. 50 questionnaires in total are issued to the subjective users and the objective users respectively in the Department of Surgery and Department of Internal Medicine in Guangdong Huizhou Center People's Hospital and 48 questionnaires are recycled (25 from the subjective users and 23 from the objective users). Thus, there are 48 effective questionnaires and the efficiency rate is 96%. The specific investigation results are as follows.

(1) Interface colors

In the questionnaire, we select the two most representative color samples in mainstream brand products to carry out user emotional preference investigation (refer to Table 1), respectively A. Single main color + auxiliary no-color system and B. Single main color + auxiliary multi-color system.

Table 1. Color classification of human-computer interaction interface in medical monitoring equipment

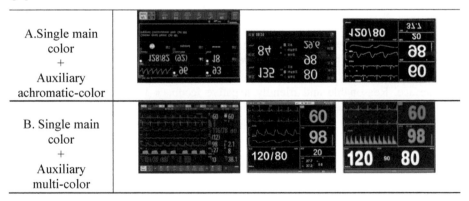

A.Single main color + Auxiliary achromatic-color	
B. Single main color + Auxiliary multi-color	

(2) Interface images

The images of human-machine interface in intelligent medical monitoring equipment include functional icons and medical images. It is found through the investigation on image design styles in market mainstream brand products as shown in Table 2 that the images can be generally concluded into two major categories, respectively realistic style and abstract style (as shown in Table 2).

Table 2. Pattern classification of human-computer interaction interface in medical monitoring equipment

A. Abstract style	
B. Realistic style	

(3) Interface text

The interface text factors influencing user emotional experience on sensory level include colors with texts, font, word size and arrangement method. Specifically, due to properties of medical monitoring equipment, the fonts and arrangement methods adopted by different brands are relatively standard. Therefore, we only analyze two variables, namely text colors and word sizes (Table 3).

Table 3. Text classification of human-computer interaction interface in medical monitoring equipment

A. the same color and word size		B. the same color but different word sizes	
C. different colors and word sizes		D. different colors but the same word size	

The users who choose C and D are in the majority, which indicates that users are inclined to arrangement of texts with different colors. In the meanwhile, there is a tiny difference in the proportion of users which choose C or D and A or B, which indicates that the users are not very sensitive to the variable, word size. It is found through in-depth interview that the subjective and objective users express that the primary demand for interface texts is clarity. Therefore, they are inclined to texts with different colors. Specifically, the subjective users indicate that texts with different colors can effectively lower the agitated feeling due to long-time data observation; the objective users indicate that different colors are more active and are easy to be distinguished and recognized, thus avoiding misoperations.

3.2.2   Analysis on Emotional Demands of Users on Efficiency Level

The emotions on the efficiency level refer to the feelings of users when they are using the products and are related to the realization of product functions. The user demands can only be specific and determined under the understanding, analysis and study of user behaviors. In other words, the user demands are satisfied through user behaviors and the user behaviors embody the user demands. [7] As a result, the study of the user emotions on efficiency level mainly adopts observation method for investigation.

(1)   Analysis on emotional demands of subjective users on efficiency level

In the case of Department of Cardiothoracic Surgery in Huizhou Center People's Hospital, there are 35 beds and 22 medical workers in total. Each work shift requires 4–6 workers. Their work nature requires them to be concentrated at every moment, respond quickly and judge accurately. Therefore, the usability of human-machine interface in monitoring equipment is the most fundamental emotional demand of medical staff on efficiency level. Including interaction effectiveness, which refers to that the interface can provide continuous operational feedback to make the users able to accurately understand current operations and how to operate next; the interaction efficiency, which refers to that the operation steps to finish the tasks are simple; interaction comfort, which refers to that the interface functions and layout satisfy user operational habits.

(2)   Analysis on emotional demands of objective users on efficiency level

As for the objective users, the utilization of medical monitoring equipment is to control or cure the diseases. Therefore, the learnability of human-machine interface in monitoring equipment is the most fundamental emotional demand for objective users on efficiency level, including understandable interface information, which refers to that the interface information can be delivered intuitively, clearly and accurately; learnable functional operations, which refer that the interface functions can be easily understood and operated.

3.2.3   Analysis on Emotional Demands of Users on Cognitive Level

The emotions on cognitive level refer to the product memory after the users use the product and the meaning the product utilization and are established on the basis of emotions on both sensory level and efficiency level. Products with good emotional design on cognitive level should be moving and unforgettable. Therefore, the emotional demand of users on cognitive level of human-machine interface in medical monitoring equipment is mainly centered on whether the interface can satisfy user psychological appeals. The difference in social individual attributes of subjective and objective users leads to different psychological appeals towards the same thing.

The hierarchy theory of needs from Maslow assumes that each individual has the need to be respected and to realize themselves [8]. This goes more for the medical staff in special work posts.

3.2.4   Construction of User Emotional Demand Model

In combination with the analysis results of emotional demands of subjective and objective users on three levels for human-machine interface in medical monitoring equipment above, the category classification kansei engineering is adopted to carry out multi-level recursive deduction on emotional demands of subjective and objective users. Their emotional demands are specified on multiple levels to finally construct the emotional demand models for both subjective and objective users of medical monitoring equipment as shown in Fig. 2.

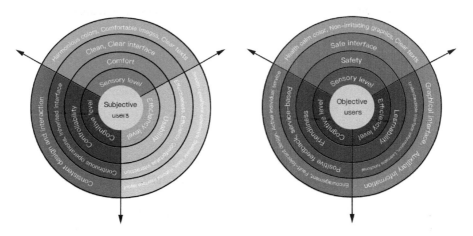

Fig. 2. Emotional needs model of the subjective user and objective user

3.3   Thinking: The Principle of Emotionalized Design for Human-Machine Interface in Medical Monitoring Equipment

3.3.1   Design Principle on Sensory Level

(1)   Rational color matching unified with interface functions
       The color matching corresponding to psychological and physical characteristics of subjective and objective users is adopted to guarantee unified color and interface functions, provide users with comfortable sensory experience and effectively lead users to operate and reduce misoperation probability.
(2)   Simple image and easy to identify
       Simple, highly summarized and understandable image forms are adopted to minimize the emotional sensation to users and facilitate user understanding and reception.
(3)   Clear text with high recognition
       Clear texts with high recognition can enable medical staff to carry out barrier-free information reading within a certain distance and provide patients with a sense of trust.

3.3.2   Design Principle of Efficiency Level

(1)   Rational interface layout and convenient operations
(2)   Foresighted and responsive operations
       Foresighted operations refer to the guidance and inspiration to the interaction operations before they are carried out.

3.3.3 Design Principle on Cognitive Level

(1) Consistent interaction form

The consistency in interaction form enables the medical staff to carry out seamlessly-connected operations in different equipment, reduce learning cost and offer them stronger confidence during diagnosis process, thus realizing their own values.

(2) Fault-tolerant design and positive feedback

This can effectively reduce the misoperations of users, especially lower the operational risk in medical care field. In addition, it can improve user confidence and satisfy the emotional demands of user controllability and friendliness while giving the user a sense of safety.

4 Emotionalized Design Practices of Human-Machine Interface in Medical Monitoring Equipment

During emotionalized design of human-machine interface in medical monitoring equipment, the contents of monitored data vary greatly and the demonstration of human-machine interface differs due to diverse medical monitoring equipment. Therefore, the emotionalized design aiming at sensory level, efficiency level and cognitive level is both unified and different. In the case of Good Friend household ECG, the human-machine interface mainly includes three major interfaces, respectively real-time ECG, record replay and local storage. The layouts of real-time ECG interface and record replay interface are basically consistent. Therefore, we only select one of them as the design object. In combination of the emotional demand model of subjective and objective users.

4.1 Emotionalized Design of Visual Elements in Human-Machine Interface

The visual elements of human-machine interface form users' first expression on medical monitoring equipment and serve as a significant channel for users to undertake information exchange with the equipment as well as a major factor influencing user emotions on sensory level. Good sensory emotional cognition can improve the trust of users in products and influence subsequent functional operations.

4.2 Emotionalized Design of Interface Layout of Human-Machine Interface

Layout of human-machine interface is the core to satisfy user emotional demands on sensory level and a major channel to correlate the users with the products emotionally. Interface layout design refers to adopting rational methods and clues to arrange and plan the interface elements and operation procedure comprehensively. The layout of human-machine interface mainly includes interface pattern, window structure and interaction function expression. The interface pattern refers to the arrangement and demonstration method of interface color, image and text by specific contents and topics. The window structure refers to the horizontal structure of interface operation procedure and is generally divided into wide structure and deep structure. The wide structure has shallower levels and its top level interface is directly connected with many different sub-modules. The access path is short and can reduce users' time spent on looking for information; the deep structure has deeper levels and multiple window transfers are needed from the top level interface to the smallest sub-module. The structure is complex and the access path is long, which requires high user skills. Therefore, simple and clear interface pattern and wide window structure can more satisfy the emotional demands of users for learnability and usability on efficiency level. The interaction function expression of human-machine interface is realized by a series of processed during interface interaction operations and is generally divided into pre operation, under-operation and post-operation stages. During user-interface interaction, they are more inclined to stable, continuous and smooth operational experience, which refers to that there are foresight before operation, feedback during operation and cancelability during operation.

4.3   Emotionalized Design of Human-Machine Interface

The interaction form of human-machine interface in medical monitoring equipment refers to the information communication method between people and equipment. This interaction from is a bilateral transmission during participation process and includes interactions on thinking, psychology, behaviors, sensory feelings and languages [9]. Based on the user emotional demands on sensory level, the emotionalized design of interaction form of human-machine interface in medical monitoring equipment is mainly embodied in following two respects: the first one is the emotionalized design of interaction context and the second is the consistency in interaction form. The interaction context describes an operation context correlated by interface behaviors and user behaviors during interaction process from the perspective of information delivery during human-machine interaction process. The design of interaction context refers to that each interface element can be dynamically correlated with the user operation behaviors, thus enabling user operations with continuous design [10]. The consistency in interaction form refers to the consistency in operations which users need to finish tasks and can improve user experience.

In the design of interaction form of Good Friend household ECG, as shown in Fig. 3, monthly health report function is added into the local storage interface, whose contents include comparison, analysis and conclusion of monthly heart rate monitoring results and relevant living health suggestions. After data collection is carried in real-time ECG, a functional entrance for sharing to professional doctors and looking up health tips is added to guide the users to carry out the next operation. This active individual service design not only improves the friendliness of equipment interface and gives the users emotional care but strengthens the connection between users and products. Good Friend household ECG supports three terminal versions, respectively cellphone, iPad and computer. Tiny adjustments to interface layout are made to different terminal versions. However, the interaction form basically remains consistent.

For instance, on iPad terminal, the function operation region of real-time ECG interface is located on the right of interface and arranged from upper to lower. On cellphone terminal, the function operation region of real-time ECG interface is likewise located on the right of interface and arranged from left to right as well as inherits the same labelization and status feedback as shown in Fig. 4.

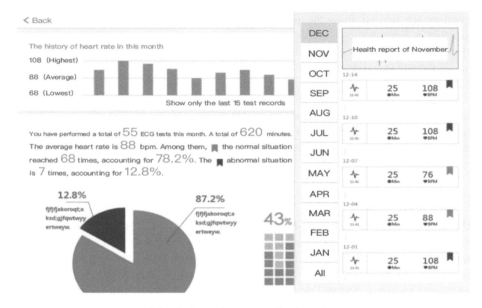

Fig. 3. Proactive personalized service

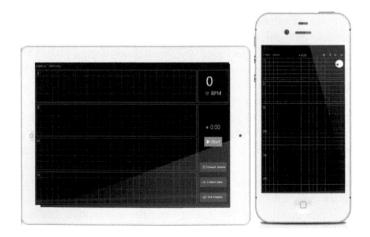

Fig. 4. Multi terminal universal

5 Conclusions

Integrating emotionalized concept into design of human-machine interface of medical monitoring equipment contributes to improving the usability and availability, thus further promoting the market competence of medical monitoring equipment. Based on this, a three-level design model for emotionalized design of human-machine interface in medical monitoring equipment is proposed. Then, based on the user investigation prospective, an emotional demand model for subjective and objective users of medical monitoring equipment is constructed and the principles for the emotionalized design are concluded, which provides certain theoretical reference for the design of human-machine interface in medical monitoring equipment.

References

1. Norman, D.A.: Design psychology 3: emotional design. In: Ou, Q., He, M. (eds.) Translate. China Citic Press, Beijing (2012)
2. Cui, Y.: The Application of color psychology in life. Dazhong Wenyi (15), 124–124 (2013)
3. Tian, F., Wang, M.: The important application of geometric figure in packaging design. Jiannan Lit. (12), 75–75 (2013)
4. Huo, D.: Analysis the graphical symbolic. Sci. Technol. Assoc. Forum (Last Half Month) (05), 91–92 (2010)
5. Zang, C.: Study on the emotional design of mobile applications interface based on information cognition. East China University of Science and Technology (2013)
6. Zhang, S., Ren, B.: Study on user interface design based on user's emotional needs. Art Panorama (6) (2016)
7. Liu, H.: Study on product design of soybean milk machine based on user behavior analysis. Shandong University (2013)
8. Zhang, X., Zhang, T.: Study on the mental health of health care provider in China. Mod. Bus. Trade Ind. (08), 254–255 (2008)
9. Xia, X., Huang, H.: Study on interactions and language conversion in digital display design. Packag. Eng. (8), 24–27 (2013)
10. Shun, P.: The gesture-based context user interface. Packag. Eng. **35**(8), 96–100 (2014)

Application Study on the Stress Test of Liquidity Risk Management of City Commercial Banks in Our Country

Fengge Yao and Meng Zhang[✉]

Harbin University of Commerce, Harbin 15001, Heilongjiang, China
Zhangmeng1@hrbb.com.cn

Abstract. In recent years, with the development of capital market and the enhancing of risk correlation, liquidity risk looms large and has become a comprehensive risk with destructive force in the highly leveraged capital market. Duo to the small scale, non-diversified service range of city commercial banks and restrained by technical conditions and human resource, their ability to manage liquidity risk remains to be further improved. Stress test of liquidity risk, as a highly effective management tool, is conducive for city commercial banks to preventing potential liquidity risk and improving their ability to withstand liquidity risk. This paper takes certain city commercial bank in Northeast China as an example and gives a cause analysis of its liquidity risk; selects stress indicators and stress-bearing indicators rationally to build a stress conduction model to evaluate the potential liquidity risk it carries and accordingly put forward corresponding countermeasures and suggestions.

Keywords: City commercial bank · Liquidity risk · Stress test

1 Introduction

In recent years, with changes occurring in management environment, business model and the structure of assets and liabilities, some city commercial banks have encountered problems such as reduction in source of fund, the decreasing of asset liquidity, lending and deposit term mismatch and the increasing of liquidity risk, etc. with the gradual deepening of financial market, the connection between financial institutions become increasingly close. A local liquidity problem could possibly cause a tight liquidity condition across the whole banking industry.

This paper gives an introduction to the significance and implementation of stress test of the liquidity risk management of city commercial banks. Combining quantitative analysis with qualitative analysis, it gives an empirical analysis of the stress test on the liquidity conditions of certain city commercial bank to create a feasible test approach theoretically and practicably applicable to city commercial banks, evaluating the potential liquidity risk it carries and putting forward suggestions to improve liquidity risk management and stress test. It is of great significance theoretically and practically to the management control of city commercial banks.

© Springer Nature Singapore Pte Ltd. 2020
M. Atiquzzaman et al. (Eds.): BDCPS 2019, AISC 1117, pp. 1304–1312, 2020.
https://doi.org/10.1007/978-981-15-2568-1_181

2   Theoretical Overview

Stress test aims to estimate banks' ability to withhold risks when facing extreme conditions of "small probability but possible". As a useful risk management tool, it can help city commercial banks to better adapt to the ever changing market environment and macroeconomic environment [1]. The emergence and development of stress test, as a supplementary to VaR model, is based on historical and hypothetic circumstances to analyze the liquidity risk of commercial banks or even the entire banking system and study on the possible impact of extreme circumstances on commercial banks. Furthermore, it quantifies the risk and improves commercial banks' ability to cope with unexpected major events.

In terms of the stress testing approach of liquidity risk, foreign scholars mainly employ dynamic analysis approach, and also introduce mathematical methods of statistics such as fuzzy theory, theory of extreme values and data envelopment analysis to improve the credibility and accuracy of risk testing [2]. While domestic scholars mostly utilize static indicators and ratio analysis to evaluate the liquidity conditions of commercial banks from the perspective of parallel comparison and combining with qualitative analysis put forward corresponding liquidity risk management measures [3].

In empirical research on stress testing, foreign institutions employ proper stress test models and approaches comprehensively to evaluating liquidity risk according to their actual conditions, which grows rapidly and is well-developed, whereas domestic scholars mainly imitate and refer to foreign experience. With liquidity gap analysis as its foundation, it runs simulations by selecting corresponding risk factors to evaluate quantitatively the liquidity conditions of commercial banks in our country under the impact of essential factors. Furthermore, since currently a uniform stress testing standard is yet to be established, there have been large disagreement among scholars [4].

3   Identification of Liquidity Risk of City Commercial Banks

3.1   Liquidity Risk of City Commercial Banks and Its Features

Liquidity risk refers to the fact that commercial banks are unable to provide financing service for customers timely, which affects their profitability and even causes payment difficulties. It mainly results from the liquidity difficulty caused by commercial banks' inability to cope with declining liabilities or increasing assets, thus affecting the operation of city commercial banks and even leading to bankruptcy [5].

Specifically, liquidity risk mainly has the following four features:

3.1.1   Small Probability
It means that it's highly unlikely to happen and in experiments it is normally overlooked. But once happening, it will be destructive.

3.1.2   Elusiveness
The elusiveness of the liquidity risk of city commercial banks is represented in two aspects: on one hand, it refers to poor risk management, with inadequate recognition of

liquidity risk. On the other hand, it refers to that liquidity risk itself is elusive and is an overall and indirect risk, easily ignored by city commercial banks.

3.1.3 Endogeneity

The main function of commercial banks is to provide savers with deposit contracts with strong liquidity while borrowers with contracts with poor liquidity, creating liquidity for society while obtaining margin interest. The essential feature of city commercial banks is their role as credit intermediary. The endogenous root of liquidity risk of city commercial banks is providing liquidity insurance for customers through liquidity transferring, meanwhile shifting liquidity risk to banks themselves.

3.1.4 Destructiveness

When emergencies happen in market, most savers will choose to withdraw money. With other factors constant, it would be difficult for city commercial banks to have sufficient money meeting the demands of savers without deficit. If not resolved timely, it is very likely that large-scale bank runs will occur and liquidity risk amplified, causing liquidity crisis from credit impaired to bankruptcy.

3.2 The Source and Manifestation of the Liquidity Risk of City Commercial Banks

3.2.1 The Source of the Liquidity Risk of City Commercial Banks

Compared with operational risk, market risk and credit risk, the reasons for the formation of liquidity risk are more extensive and complicated. It is an overall risk, and specifically, it manifests in the following aspects:

The unmatched term structure of assets and liabilities. Liquidity risk directly results from the unmatched term structure of assets and liabilities. The assets of city commercial banks mainly comprise term loans while liabilities consist of demand deposit and time deposit that can be withdrawn in advance.

Changes in Central Bank's Monetary Policy. Central Bank mainly use regulatory instruments such as deposit-reserve ratio, rediscount rate and open market operations to influence the liquidity of commercial banks. When Central Bank adopts a tighter monetary policy (such as raising deposit-reserve ratio, increasing rediscount rate and issuing bond in open market), the aggregate amount of credit will decrease and money supply is lower than demand, causing city commercial banks' inability to meet customers' demand to withdraw deposits or on loans.

The Variation of Market Rate. When market interest rises and exceeds the interest of city commercial banks, the phenomenon of "financial disintermediation" emerges. In order to obtain a higher profit, savers will transfer their money from city commercial banks to market investment, with banks losing large amount of deposits. Meanwhile, since loan rate is lower than market rate, borrowers are more willing to loan large amount of money from banks, leading to more capital outflow.

The Degree of Maturity of Financial Market. City commercial banks' ability to asset realization or other forms of raising fund is closely connected with the degree of maturity of financial market they are in. in a mature financial market, city commercial banks is able to raising fund through asset realization timely and conveniently to pay debts that are due. However, the financial market of our country is still not mature enough to provide adequate liquidity for banks within a short time. Therefore, city commercial banks are likely to face liquidity challenges.

3.2.2   The Manifestation of the Liquidity Risk of City Commercial Banks

Liquidity risk comes with the advancement of business operations of banks, which generally manifests as extremely inadequate liquidity, difficulty in raising fund and the unexpected fund outflow.

3.3   The Specific Procedures of Conducting Stress Test on the Liquidity Risk of City Commercial Banks

Article 13 of the guidelines on the stress test of commercial banks (2007) promulgated by China Banking Regulatory Commission states the steps and procedures of stress test on commercial banks: identify risk factors, design stress circumstances, select assumed condition, determine testing procedure, conduct test regularly, analyze testing results, locate potential risk points and vulnerable parts, report results according to internal business process, take emergence measures and other improvement measures and report to regulatory authorities, etc. [6].

Combining the process of liquidity risk management of city commercial banks with the stress testing steps, namely:

3.3.1   Identify Liquidity Risk and Determine Risk Factors

One of the most important parts in stress test is to determine risk factors. The selection of risk factors is a process in which the liquidity conditions of commercial banks are simplified or analyzed, which directly affects the effect of stress test. Therefore in selecting risk factors, we have to be extremely cautious.

3.3.2   Set Stress Circumstances

One of the key steps of stress test is stress circumstance setting. An improper circumstance setting will render it meaningless. In general, we can refer to historical trend or the range of variation of risk factors within a certain period and modify them according to empirical data.

3.3.3   Select Stress Testing Approach and Evaluate Risks

Stress test should select proper econometric model after take full account of the nature of risk factors and the level of difficulty in obtaining data.

3.3.4   Conduct Risk Analysis According to Stress Testing Result and Make Risk Decision

We should analyze stress testing result and give explanations, take remedial measures and address potential risks timely.

3.3.5   Couple Back Results and Draw up Risk Report

The stress testing results should be reported to higher management and board of directors timely.

4 The Implementation of Stress Test on the Liquidity Risk of Certain Commercial Bank

This paper takes a certain city commercial banks in our country as an example and conducts stress test on its liquidity risk according to the availability of information and data.

4.1 The Selection of Liquidity Risk Factors

The process of selecting risk factors in stress test is a process in which the liquidity conditions of commercial banks are simplified or analyzed, which directly affects the effect of stress test. Commercial banks should identify their main weak points in liquidity. Therefore, it is necessary to determine several important risk factors, analyze them and build a specific stress circumstance so that potential risks can be identified [7].

This paper argues that the bank's liquidity risk mainly arises from the following aspects:

4.1.1 Interest Rate
The variations of interest rate not only exert influence on assets and liabilities sensitive to interest rate change, but also on their term structure, especially assets and liabilities with options or additional options;

4.1.2 Economic Conditions
Macroeconomic conditions and the liquidity of financial market can affect the liquidity risk of banks directly or through credit risks and financing commitments;

4.1.3 Financing Capacity of Institutions
It is closely related to financing cost and credit rating and can affect liquidity risk on multiple levels;

4.1.4 The Credit Risks and Confidence of Counterparties
Lack of confidence on the part of a counterparty will cause time deposits loss and option exercise, etc. The flow resistance of counterparties' liabilities should also be taken into consideration, which is not only related to whether the liabilities are guaranteed and the information source of counterparties, but also closely connected with counterparties' relevance to banks.

According to the principles of risk factor selection, combined with the current management status and data acquisition of the bank, rate of RMB to USD, huge deposit loss within a short time and financing cost are selected as the risk factors of liquidity stress test.

4.2 Scenario Design

Scenario represents the result of all the risk factors coming together within a certain period and fully describes the uncertainty of the future. In general, we can refer to historical trend or the range of variation of risk factors within a certain period for degree of impact and modify them according to empirical data. A proper range for a market risk analysis of stress scenario is probably ten days, while multiple time frames are likely needed to evaluate stress scenario of risk exposure, for over ten days or longer. For city commercial banks, short-term and long-term market conditions should be considered simultaneously to make them focus on the short-term impact while considering its duration so that the accuracy of liquidity stress test on city commercial banks can be improved. According to the actual situation of the bank, this paper takes the following stress test scenarios (Table 1):

Table 1. Liquidity stress test scenarios of the bank

Scenario setting	Mild	Moderate	Severe
The decline of the exchange rate of RMB against the U.S. dollar	1%	1.5%	2%
Huge amount of deposit loss within a short period	7.5%	10%	12.5%
The increasing of financing cost	0.35%	0.5%	0.65%

4.3 Data Selection

According to the frequency of stress test and the characteristics of liquidity risk factors, we select the data of the bank from July, 2011 to January, 2019 as essential data, which mainly comes from internal statistics and annual reports of the bank and the statistical data of the People's Bank of China.

4.4 The Determination of Variables

This paper takes the above-mentioned liquidity risk factors as explanatory variables and an indicator that can reflects the liquidity condition of the bank as explained variable to conduct analysis of aggression and build a simple risk model. Due to the lack of statistical data, this paper can only use loan/deposit ratio as the main indicator to evaluate the liquidity risk of the bank and regard it as explained variable to conduct analysis of regression. For convenience's sake, Y represents explained variables, namely, loan/deposit ratio; X_1 represents financing cost, namely, the seven-day Shibor on the market, X_2 represents the ratio of deposit loss in the total deposit in a short run, X_3 represents the devaluation ratio of RMB against U.S. dollar.

4.5 Model Structure and Test Analysis

This paper employs a simple and direct LRM model, and the model structure is analyzed as follows:

$$Y = \alpha + \beta_1 x_1 + \beta_2 x_2 + \beta_3 x_3 \tag{1}$$

Through a simple analysis of regression with econometric software, an empirical result is reached as below:

$$Y = 0.617 + 0.364x_1 + 0.652x_2 + 0.271x_3 \tag{2}$$

According to the scenario set and through building a regression mode, liquidity risk of the bank is calculated respectively under three types of impacts [8] (Tables 2, 3 and 4):

Table 2. Liquidity conditions of the bank when the seven-day SHIBOR raised

Item	Mild	Moderate	Severe
Increasing range of the seven-day SHIBOR	0.35%	0.50%	0.65%
Single factor variation	69.64%	69.69%	69.75%
Rate-of-change	0.19%	0.26%	0.35%

Table 3. The liquidity condition of the bank when deposit loss increases within a short run

Item	Mild	Moderate	Severe
Deposit loss	7.50%	10.00%	12.50%
Single factor variation	71.45%	73.08%	74.71%
Rate-of-change	2.80%	5.14%	7.48%

Table 4. Liquidity condition of the bank when RMB is devalued

Item	Mild	Moderate	Severe
Devaluation of RMB	1.00%	1.50%	2.00%
Single factor variation	69.78%	69.92%	70.05%
Rate-of-change	0.39%	0.59%	0.78%

Through drawing comparison among the three scenarios, we can see from the above three tables that if merely a single extreme scenario is taken into consideration, large amount of deposit loss within a short run exerts huge influence on the liquidity condition of the bank. Therefore, while paying attention to the three factors, more importantly measures should be taken to prevent banks from the influence of deposit loss within a short run, which can even affect the daily operation of banks. Thus, it is necessary to conduct stress test on the liquidity risk of the bank (Table 5).

It can be seen from the above table that if the bank's loan/deposit ratio reaches up to 71.84%, it will be hit by stress; when it reaches the moderate level, loan/deposit ration could be up to 75.48%, exceeding the upper limit 75% set by regulatory authorities, and this is time for the bank to take urgent funding plan to address the situation.

After the evaluation of liquidity risk with the model, emergency measures should be taken to ease, prevent or mitigate the liquidity risk on different levels according to

Table 5. Liquidity condition of the bank under different impacts

Item	Mild	Moderate	Severe
Increasing range of the seven-day SHIBOR	0.35%	0.50%	0.65%
Deposit loss	7.50%	10.00%	12.50%
Devaluation of RMB	1.00%	1.50%	2.00%
Overall stress scenario of liquidity risk	71.84%	73.66%	75.48%
Rate-of-change relative to standard scenario	3.35%	5.97%	8.59%

the evaluation result, avoiding the huge impact brought upon banks by extreme adverse events. The stress-releasing measures of liquidity risk of different levels include bill repo, inter-bank discount, repurchase of credit assets, transfer of credit assets, due to banks and negotiated deposit, etc.

5 Strategies and Recommendations for Liquidity Risk Management of City Commercial Banks

How to optimize asset structure reasonably is an issue requiring urgent solutions for the liquidity management of domestic banks. Meanwhile, the significance of deposit management, as the impact factor of the liquidity of liabilities, cannot be overlooked. In order to reduce crisis, this part gives a detailed analysis of the liquidity risk management of banks in our country from the perspectives of asset structure, liability management and stress test and puts forward some countermeasures and recommendations to prevent and mitigate the negative influence brought upon banks by liquidity risk [9].

5.1 The Optimization of Asset Structure

Unmatched term structure of assets and liabilities is an intrinsic feature of commercial banks. Bringing terms of loans well under control is a critical way to prevent potential risks from evolving; reasonable liquid assets, improving the ability to asset realization; develop intermediary business and off-balance sheet business actively and promote asset securitization reasonably to cope with market competition and adjust asset structure.

5.2 Deposit and Liabilities Management

Liquidity risk and financial cost should be balanced to seek a reasonable deposit structure; sensitivity analysis of source of fund as needed; closely monitoring the situation of liability concentration to prevent heavy reliance on a single source.

5.3 Stress Test Management

A liquidity stress test system according with the actual situations of city commercial banks should be established to ensure the accuracy of liquidity risk data; the stress test result should be applied to the overall risk management procedure, upgrading liquidity risk and stress test to a strategic level to ensure the validity and accuracy of the test results and the balance of banks' internal capital chain and their sustained development [10].

Acknowledgements. Special item of general secretary Xi Jinping of philosophy and social sciences of Heilongjiang Province, Study on Financial Support for Heilongjiang Province's Modernized Agricultural Development under the Background of Supply-side Structural Reform. Item number: 16JYH01.

References

1. China Banking Regulatory Commission, Administrative Measures for the Liquidity Risk Management of Commercial Banks (for Trial Run) (2014). (in Chinese)
2. Zhou, H., Pan, B.: Comparative study on the management and implementation of stress test of liquidity risk. Stud. Int. Financ. (4) (2010). (in Chinese)
3. Zhu, Y.: The theory and practice of stress test of liquidity risk. Chin. Rev. Financ. Stud. (02) (2012). (in Chinese)
4. Liu, W., Yan, Y.: Measurement methods and empirical analysis of liquidity. Stat. Decis. (06) (2011). (in Chinese)
5. Chen, K.: Study on the liquidity risk management of china's state-owned commercial banks based on stress test. Southwestern University of Finance and Economics (2013)
6. Basel Committee on Banking Supervision: The Basel Committee's response to the financial crisis: report to the G20. BIS, October 2010
7. Peng, J.: Study on stress test of banking industry based on macro-prudential regulation. China Urban Financ. (10), 80–81 (2015). (in China)
8. Fang, Y.: Research on systematic risk of china's banking industry—three stress tests in the perspective of macro-pruden. Econ. Theory Bus. Manag. (2), 48–66 (2017). (in Chinese)
9. Research group of the investigation and statistics department of the People's Bank of China. Statistical and applied research on systemically important Banks from the perspective of macro-prudential management. Financ. Regul. Res. (7), 1–15 (2018). (in Chinese)
10. Wu, W., Li, X., Li, S.: Exploration and practice of liquidity stress test of local legal financial institutions. Gansu Financ. (4), 66–68 (2019). (in China)

Multi-regional Logistics Distribution Demand Forecasting Method Based on Big Data Analysis

Dongmei Lv[(✉)]

Hunan Vocational College of Railway Technology, Zhuzhou, China
cc356398632@163.com

Abstract. When the conventional logistics distribution demand forecasting method predicted the multi-region logistics distribution demand, there was a shortcoming of low analysis accuracy. To this end, a multi-region logistics distribution demand forecasting method based on big data analysis was proposed. Big data technology was introduced, a multi-regional logistics distribution demand forecasting framework was built, and the construction of multi-regional logistics distribution model was achieved; Relying on the different needs of logistics distribution between regions, the demand forecasting model was embedded to realize the forecasting and analysis of multi-regional logistics distribution demand. The test data shows that the proposed big data forecasting method is 57.23% higher than the conventional method, which is suitable for the forecast of logistics distribution demand in different regions and regions.

Keywords: Big data analysis · Multi-regional distribution · Logistics network · Demand forecasting

1 Introduction

Choosing the right forecasting model to predict logistics demand has important strategic significance for upgrading and optimizing the logistics industry. Common logistics forecasting methods are: growth rate method, moving average method, time series method, etc. Since the actual logistics forecast data often has the characteristics of multiple indicators, nonlinearity and small samples, and there are redundant indicators in the data, the prediction accuracy of most prediction methods is not high in practical applications, and it is difficult to ensure the validity. For this kind of logistics prediction problem, this paper removes the redundancy index based on the algorithm based on the difference matrix in the rough set attribute reduction. Based on the attributes of reduction, a single SVM prediction model is improved, and the input parameters of the SVM model are optimized by genetic algorithm, and a higher prediction accuracy is obtained [1]. This paper gives the specific steps of the method, and predicts the total freight volume of Guangdong Province with actual data, and verifies the effectiveness of the method.

© Springer Nature Singapore Pte Ltd. 2020
M. Atiquzzaman et al. (Eds.): BDCPS 2019, AISC 1117, pp. 1313–1321, 2020.
https://doi.org/10.1007/978-981-15-2568-1_182

2   Construction of Multi-regional Logistics Distribution Demand Forecasting Model Based on Big Data Analysis

For the door-to-door mode, considering the actual situation of the delivery, the time penalty function is constructed with the minimum total cost of the delivery activity as the target. Under the constraints of vehicle capacity limitation, time limit and number of vehicles in the distribution center, an optimal home delivery strategy model with time window constraints is established.

2.1   Introduction of Big Data Technology

China's e-commerce and the booming development of various chain stores have injected new vitality into the rapid economic development, and also brought great demand for urban community logistics and distribution. Firstly, the concepts of logistics and common distribution are defined. The development status of logistics and common distribution mode in China is analyzed, and the existing logistics distribution modes at home and abroad are summarized [2]. The connotation of the common distribution mode of logistics is analyzed. According to the different methods of customer pickup, it is divided into two modes: customer self-acquisition and door-to-door delivery. The influence of service level and logistics cost on the choice of distribution mode is analyzed. For the customer self-acquisition mode, the best service scope model of the site is constructed considering the factors such as the maximum walking time of the consumer and the shortest delivery time, delivery cost and service quality. The model was modified based on the customer's linear time satisfaction and satisfaction adjustment factors [3] (Fig. 1).

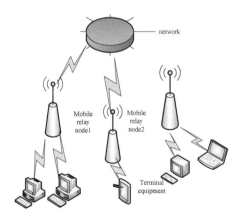

Fig. 1. Network remote control system

The main reason of big data information is from the logistics industry. Through the analysis of examples, the rationality and effectiveness of the distribution demand forecasting method and the distribution model optimization model are verified [4, 5].

2.2   Building a Logistics Distribution Database Prediction Framework

Global competition characterized by demand-driven and fast-responding market opportunities enables companies to continuously pursue efficient operations and respond to high-speed changes in demand. In the case where the profit margin becomes smaller and smaller as the "first profit source" increased by reducing the material consumption and the "second profit source" increased by saving labor consumption, modern logistics has become the third source of profit for the company by improving operational efficiency, reducing costs, and increasing the response speed to customer demand as much as possible, and becoming the third source of profit for the company [6, 7] (Fig. 2).

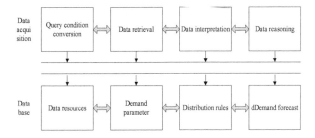

Fig. 2. Distribution demand forecasting framework under big data

Based on the introduction of big data technology, the evaluation framework of logistics distribution database was introduced to realize the construction of multi-region logistics distribution demand forecasting model [8–10].

3   Realizing the Forecast Analysis of Multi-regional Logistics Distribution Demand

3.1   Determination of Logistics Distribution Demand in a Certain Area

Many enterprises regard logistics distribution as an important part of their competitive strategy in their future planning, and some even classify their distribution capabilities as the core competence of the enterprise. Based on this, the domestic and foreign research status of the theory and application of GIS application technology and logistics distribution path optimization problem are analyzed. By summarizing the original theory, the technical route is designed on the premise of putting forward the research content. In logistics distribution, the research content and research classification of the shortest path problem, and the detailed analysis of the calculation method for solving the shortest path problem in logistics distribution. By comparing and analyzing the characteristics and applicability of various algorithms, it is decided to use genetic algorithm to calculate the shortest path of logistics distribution. The implementation method and steps of the genetic algorithm to solve the shortest path problem are given by the example of the company. First of all, the basic functions that the enterprise logistics distribution system should have are analyzed. Secondly, the process outline design and related database design for the enterprise logistics distribution system enable the company's overall information and supplier information to be managed in a unified manner; Finally, a reasonable development environment is applied to realize the integrated development of the enterprise logistics distribution system supported by the Internet of Things application technology.

As an advanced logistics method, modern logistics distribution is developing rapidly in the direction of automation, information, network and intelligence. By proposing the logistics distribution route optimization model and the Internet of Things application integration scheme, functional modules such as order forecasting function, sales volume and unified vehicle management of distribution vehicle data, and shortest path analysis are designed. It improves the ability to visualize, predict, and quickly judge the results of logistics distribution decisions. To lay the foundation for the future application of the Internet of Things in the logistics and distribution of food enterprises (Table 1).

Table 1. Multi-regional logistics distribution demand data sheet

Region	Population volume	Purchasing power	Distribution demand
Southeast	High	High	High
Central section	High	High	High
Northwest	Low	Low	Low

In the purpose of stating the connotation of logistics demand forecasting and forecasting the demand for food logistics, and on the basis of quantitative and qualitative two flow prediction methods, several major factors affecting the demand for food logistics and distribution are determined: consumption capacity, consumption time, sudden weather changes and competitor sales methods. Through the statistical analysis of the sales volume in the previous stage, the influencing factors of the sales volume of the point-of-sale products are calculated, and the sales volume of a certain product in the next stage is predicted. In turn, decision makers can more accurately formulate logistics distribution plans, reduce secondary logistics costs, and bring more benefits to enterprises (Fig. 3).

The number assigned to each distribution center can be expressed by the following formula:

$$A = (DS - I/D) \times D \tag{1}$$

Where A represents the quantity supplied, DS represents the number of days of supply, I represents the inventory of the distribution center, and D represents the daily demand of the distribution center. The different requirements for logistics distribution in different regions are shown in Table 2:

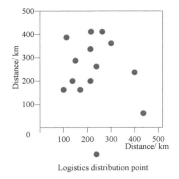

Fig. 3. Multi-regional logistics demand map

Table 2. Logistics and distribution information

Distribution mode	Time consuming	Cost	Efficiency
Walk	Long	Low	Low
Ride a bike	Short	Low	High
Robot distribution	Short	High	High

At present, China has not yet formed a sound community logistics distribution system. The distribution objects are diversified and diversified, and the degree of cooperation among distribution enterprises is low, making it difficult to achieve economies of scale. The utilization rate of the distribution center is low and the construction is repeated, resulting in high logistics costs; The distribution line lacks reasonable optimization, the vehicle has a high idling rate, low distribution efficiency, and causes social problems such as environmental pollution and traffic congestion. The trend diagram is shown in Fig. 4.

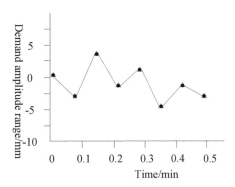

Fig. 4. Schematic diagram of distribution demand trends

Based on the above background, based on the research status at home and abroad and the analysis of the current situation of community logistics distribution in China, this paper focuses on the construction of the community logistics common distribution model. For the instantaneous economic batch, it can be expressed as the following formula:

$$Q = \sqrt{\frac{2AC}{E}} \qquad (2)$$

$$E = y - x^{(0)} \qquad (3)$$

In the formula, Q represents the economic batch, A represents the annual demand, C represents the purchase cost, and E represents the storage cost. The community logistics needs are divided into three categories: general consumer goods, fresh food, postal express, etc. According to their respective characteristics, the second exponential smoothing method, the grey forecasting method and the system dynamics model are used to predict the demand.

3.2   Embedded Multi-region Logistics Demand Forecasting Model

In recent years, logistics demand has maintained a rapid growth trend, and the logistics industry has been highly valued by the state and relevant departments, which is both an opportunity and a challenge for the modern logistics industry. At present, however, the high idle rate of logistics resources, unreasonable planning of logistics nodes and insufficient informatization have become the important factors that restrict the rapid development of China's logistics. As the pilot work of urban joint distribution continues to deepen, many cities are focusing on the theme of "accelerating the construction of urban logistics system, integrating logistics resources and optimizing logistics distribution network", and actively exploring new ideas to improve logistics efficiency and reduce costs. Its prediction model embedding process is shown in Fig. 5.

In order to solve the problem of urban distribution and improve the logistics distribution system, this paper proposes to build a three-level logistics distribution system supported by logistics park - distribution center - terminal joint distribution network with joint distribution as the core idea.

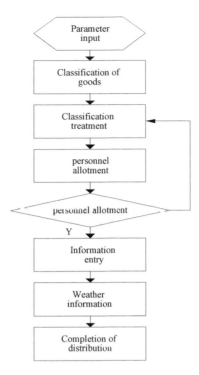

Fig. 5. Prediction model embedding process

4 Simulation Experiment

In order to ensure the effectiveness of the multi-regional logistics distribution demand forecasting method based on big data analysis proposed in this paper, with the continuous development of information technology, the flow of information and capital flow in the process of logistics and distribution can be realized in an instant, but the low level of logistics and distribution has become the bottleneck of the development of e-commerce in China. In the logistics decision-making, the choice of logistics distribution mode should be considered first. In order to verify the effectiveness of multi-regional logistics distribution demand forecasting method based on big data analysis, simulation experiments were carried out.

4.1 Test Data Preparation

Firstly, the demand and supply situation of the end logistics are analyzed separately, that is, the end logistics demand is analyzed based on the retail business change and community service development; Then, the differences, advantages and disadvantages and applicability of the various new modes in which the end logistics has been run are analyzed. The decision path of each node in the e-commerce supply chain when selecting the above mode is proposed. The parameter setting results are shown in Table 3.

Table 3. Test parameter settings

Region	Distribution time	Innovation ability	Living standard
Eastern region	1.5 h	High	High
Central region	3 h	High	Low
Western region	4 h	Low	Low

4.2 Analysis of Test Results

At present, the development of e-commerce in China is relatively fast, and the enthusiasm of consumers in online shopping is heating up. It is especially important to study logistics distribution and choose the best logistics distribution mode. The analysis and simulation curve for the demand for multi-regional logistics distribution is shown in Fig. 6.

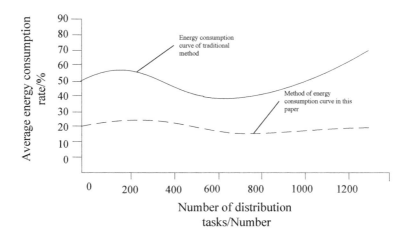

Fig. 6. Effectiveness analysis simulation curve

Based on the results of the test curve, the multi-regional logistics distribution demand forecasting method based on big data analysis is more convincing. According to the established shortest path problem model and the distance matrix between urban sales nodes, Matlab language is used to realize the application of genetic algorithm programming, and the shortest path result in the process of logistics distribution is output. Forecast the demand for food logistics. Based on the verification and analysis of the examples, the actual application system of the enterprise is developed and the implementation method of the logistics distribution process is proposed.

5   Conclusions

This paper proposes a multi-regional logistics distribution demand forecasting method based on big data analysis. Through a certain investigation, the current situation and existing problems of logistics distribution are analyzed, and multiple regression prediction and grey prediction are used to predict logistics demand. According to the predicted logistics demand, the parameter method is used to estimate the total scale of the logistics park in the region, and provide suggestions for future park planning. Summarize the four common forms of the end common distribution network, and analyze their respective advantages and disadvantages and the strategy of common distribution implementation, and finally construct the urban logistics public information platform. The construction of the city's three-level logistics system and the development of urban common distribution play an important role in solving the problem of low efficiency of urban distribution and alleviating urban traffic pressure.

Acknowlegements. Hunan Province philosophy and Social Science Fund Project: "Internet +" in the context of "Internet +" in the rural logistics and transportation service network system optimization of Hunan province "(number: 16YBG025).

References

1. Cam, M.L., Zmeureanu, R., Daoud, A.: Cascade-based short-term forecasting method of the electric demand of HVAC system. Energy **119**, 1098–1107 (2017)
2. Chen, M., Yuan, J., Liu, D., et al.: An adaption scheduling based on dynamic weighted random forests for load demand forecasting. J. Supercomput. (1),1–19 (2017)
3. Ren, S., Chan, H.L., Ram, P.: A comparative study on fashion demand forecasting models with multiple sources of uncertainty. Ann. Oper. Res. **257**(1–2), 1–21 (2017)
4. Gao, D., Wang, N., He, Z., et al.: The bullwhip effect in an online retail supply chain: a perspective of price-sensitive demand based on the price discount in e-commerce. IEEE Trans. Eng. Manag. **PP**(99), 1–15 (2017)
5. Alsaleh, M., Abdul-Rahim, A.S., Mohd-Shahwahid, H.O.: An empirical and forecasting analysis of the bioenergy market in the EU28 region: evidence from a panel data simultaneous equation model. Renew. Sustain. Energy Rev. **80**(12), 1123–1137 (2017)
6. Shaikh, F., Ji, Q., Shaikh, P.H., et al.: Forecasting China's natural gas demand based on optimised nonlinear grey models. Energy **156**, 182–196 (2017)
7. Cao, H., Li, T., Li, S., et al.: An integrated emergency response model for toxic gas release accidents based on cellular automata. Ann. Oper. Res. **255**(1–2), 617–638 (2017)
8. Ge, X.-L., Huang, Y., Tan, B.C.: Multi-stage combined city logistics distribution problem considering the traffic restrictions. Control Decis. **32**(5), 789–796 (2017)
9. Tsai, Y., Chang, K.W., Yiang, G.T., et al.: Demand forecast and multi-objective ambulance allocation. Int. J. Pattern Recognit. Artif. Intell. (5), 1859011 (2018)
10. Yan, P., Zhang, L., Feng, Z., et al.: Research on logistics demand forecast of port based on combined model. J. Phys: Conf. Ser. **1168**, 032116 (2019)

Chronic Intermittent Hypoxia Training on Rat Myocardial Mitochondria Atpase Impact Study

Qiaozhen Yan and Liping Dong$^{(\boxtimes)}$

66 Xuefu Road, Wenshan City, Yunnan Province, China
yanqiaozhen0721@163.com

Abstract. To study the chronic intermittent hypoxia training on rat myocardial line grain of ATP content and Na^+, K^+ ATPase and the influence of Ca^{2+}, magnesium 2+ ATPase. Methods: 32 male Wistar rats as the research object, according to the low oxygen training program can be divided into intermittent group, acute group and the control group, among which chronic intermittent hypoxia training, intermittent simulated altitude 3 km, 2w, then simulated altitude training 5 km, 2w training, daily training 4 h, finally simulated altitude 8 km, placed 4 h. Acute group immediately under simulated conditions of 8 km altitude 4 h; The control group are not hypoxic training. After the expiry of the low oxygen, beheaded executed after, separation of myocardial mitochondria, determination of ATP enzyme activity. Results: (1) atpase chronic group content $(9.04 \pm 4.71$ mg/100 mg of BW, $(4.96 - 1.17)$ in the acute group mg/100 mg of BW and the control group $(4.38 \pm 0.95$ mg/100 mg of BW were significantly higher, significant difference, statistically significant $(P < 0.05)$. (2) chronic group of Na^+, K^+ ATPase activity (2.55 ± 1.41) mu mol, pro - 1 mg/h, (2.66 ± 1.07) in the acute group mu mol, pro - 1 mg/h and the control group (3.08 ± 1.37) mu mol, pro - 1 mg/h had no significant difference. (3) chronic group of Ca^{2+}, magnesium 2+ ATPase activity (1.17 ± 0.34) mu mol, pro - 1 mg/h, the control group (1.28 ± 0.42) mu mol, pro - 1 mg/h no significant difference, but the acute group (0.58 ± 0.14) mu mol, mg/h significantly higher pro - 1, the difference has statistical significance $(P < 0.05)$. Conclusion: chronic intermittent hypoxia training to ensure the Ca2+, magnesium 2+ ATPase activity has a positive meaning, at the same time can significantly increase the content of ATPase, help to improve myocardial motion function, adapt the rat hypoxic environment.

Keywords: Hypoxic training · Mitochondria · Cardiac muscle · Na^+ · K^+ ATPase · Ca^{2+} · Magnesium 2+ ATPase

1 Introduction

Mitochondria are important organelles, mainly involved in cell respiration and ATP generation, and play a key role in providing energy for physiological activities during hypoxia uptake [1]. Previous studies suggest that acute hypoxia can cause the changes of myocardial mitochondrial ATP enzyme activity, but the change of the research on

© Springer Nature Singapore Pte Ltd. 2020
M. Atiquzzaman et al. (Eds.): BDCPS 2019, AISC 1117, pp. 1322–1326, 2020.
https://doi.org/10.1007/978-981-15-2568-1_183

different level differences [2, 3], and less domestic research on the project, in order to further study systematically and effectively make up the blank in the research of domestic, the author to 32 male Wistar rats as the research object, has launched a based on different hypoxia training, the rats myocardial mitochondrial ATP enzyme activity, specific report as follows.

2 Data and Methods

2.1 Animal Model

The research objects were 32 male Wistar rats, all 8w old, purchased from experimental animal center of huazhong university of science and technology. They were divided into three groups according to the random number table method. Among them, 11 patients in the chronic group received chronic intermittent hypoxia training, which was firstly placed in low-pressure oxygen chamber, simulated 3 km altitude, and lasted for 4 h every day for a total of 2w. Then the oxygen chamber was adjusted to simulate an altitude of 5 km for 4 h every day for a total of 2w. After the training, the oxygen chamber was adjusted to simulate an altitude of 8 km for 4 h. 12 patients in the acute group were given indoor air breathing. After routine feeding for 4w, they were immediately put into an oxygen chamber, simulating an altitude of 8 km for 4 h. Control group 9 cases, only routine feeding.

2.2 Detection of Myocardial Atpase

Specimen preparation after the chronic group and the acute group were exposed to hypoxia for 4 h, all the rats were put to death without head, the hearts were removed, and the tissues such as great blood vessels and atria were removed by washing in low-temperature normal saline. Homogenate buffer was prepared, the main components of which included sucrose 250 mmol/L, EDTA 5 mmol/L and tris-hci 10 mmol/L. The rat heart was cut into pieces in the homogenized buffer to form the homogenized heart. The rat heart was placed into a centrifuge for 10 min, and the upper layer was collected at night and centrifuged for 15 min. The sediment was collected, and the homogenized buffer was placed into the centrifuge for 2 times, which lasted for 15 min. The centrifugal environment is controlled in the range of (0–4) °C [4, 5].

ATPase activity determination ATPase content was determined by Lowry method, and trace inorganic phosphorus interpreted by ATPase was determined by ammonium molybdate phosphorus method designed by Reinila et al., so as to determine Na^+, K^+-atpase, Ca^{2+}, Mg^2+ -atpase activity, whose units were expressed as enzyme activity per hour (mumol •mg pro-1/h).

Statistical methods SPSS19.0 statistical software was used to process the data. The data were presented in the form of ($\pm$s), and t-test was performed. $P < 0.05$ was considered as significant difference, with statistical significance.

3 The Results

3.1 Changes of Atpase Protein Content

Statistics showed that atpase protein content was highest in the chronic group (9.04 ± 4.71) mg/100 mg BW, followed by the acute group (4.96 ± 1.17) mg/100 mg BW, and less in the control group (4.38 ± 0.95) mg/100 mg BW. The chronic group was significantly higher than the acute group and the control group, and the difference was statistically significant (P < 0.05) (Table 1).

Table 1. Changes of atpase protein content (±s)

Group (n)	Atpase protein content (mg/100 mg BW)	*t, P	&t, P
Chronic group (11)	9.04 ± 4.71		
Acute group (12)	4.96 ± 1.17	2.9101, <0.05	
Control group (9)	4.38 ± 0.95		2.9064, <0.05

Note: *represents comparison between chronic group and acute group; The chronic group was compared with the control group.

3.2 Changes of APT Enzyme Activity

Pair-pair-comparison of Na^+ and K^+-atpase activity in the three groups showed no significant difference (P > 0.05), but Ca^{2+} and Mg^{2+}-atpase activity in the chronic group was significantly higher than that in the acute group, and the difference was statistically significant (P < 0.05). Data are shown in Table 2.

Table 2. Changes of atpase activity (±s)

Group (n)	Number	Na^+, K^+-ATPase ($\mu mol \cdot mg\ pro^{-1}/h$)	Ca^{2+}, Mg^{2+}-ATPase ($\mu mol \cdot mg\ pro^{-1}/h$)
Chronic group (11)	11	2.55 ± 1.41	1.17 ± 0.34
Acute group (12)	12	2.66 ± 1.07	0.58 ± 0.14
Control group (9)	9	3.08 ± 1.37	1.28 ± 0.42
*t, P	–	−0.2119, >0.05	5.5306, <0.05
&t, P	–	−0.8469, >0.05	−0.6480, >0.05

Note: *represents comparison between chronic group and acute group; The chronic group was compared with the control group.

4 Discuss

Mitochondria are the main organelles for energy supply of eukaryotic cells, while myocardial mitochondria provide the energy basis for the normal motor function of the heart, and its main role is to participate in cellular respiration and generate ATP to store energy [6, 7]. The physiological functions of mitochondria are mainly realized by Na^+, K^+-atpase, Ca^{2+}, Mg^{2+}-atpase on the surface of the cell membrane [8, 9]. Therefore, cell uptake performance can be effectively analyzed by studying the changes of mitochondrial ATPase activity.

In this study, atpase content in the chronic group (9.04 ± 4.71) mg/100 mg BW was significantly higher than that in the acute group (4.96 ± 1.17) mg/100 mg BW and the control group (4.38 ± 0.95) mg/100 mg BW, with significant difference and statistical significance ($P < 0.05$). The activity of Na+ and K+ -atpase in chronic group (2.55 ± 1.41) mumol •mg pro-1/h was not significantly different from that in acute group (2.66 ± 1.07) and control group (3.08 ± 1.37). The activity of Ca^{2+} and Mg^{2+}-atpase in the chronic group (1.17 ± 0.34) muon mol•mg pro-1/h was not significantly different from that in the control group (1.28 ± 0.42), but significantly higher than that in the acute group (0.58 ± 0.14), with statistically significant difference ($P < 0.05$). Through this study can see that chronic intermittent hypoxia training can effectively improve the rat myocardial mitochondrial protein content, and to ensure that Ca^{2+}, magnesium 2+ ATPase activity returns to normal level, this might be due to chronic intermittent hypoxia training makes cells of rats produced adaptive change, the mitochondrial proliferation, both in volume and quantity rise, thus higher protein content. However, no significant changes in Na+ and K+ -atpase activity were observed, suggesting that hypoxia injury was unrelated to Na+ and K+ -atpase, but only related to Ca^{2+} and Mg^{2+}-atpase activity, which could provide certain theoretical basis for us to solve the problem of hypoxia ingestion [10].

In conclusion, it can be seen from the study of this case that chronic intermittent hypoxia training can make adaptive changes in rat myocardium, increase the content of ATPase, and the activity of Ca^{2+} and Mg^{2+}-atpase tends to the normal level, which can gradually adapt to the hypoxia environment.

References

1. Li, J., Xing, L.: Effects of different hypoxic training patterns of 3500 m on myocardial mitochondrial free radical metabolism and respiratory chain function after exhausting exercise in rats. J. Shanghai Inst. Phys. Educ. 36(1), 51–55 (2012)
2. Huang, L.: Effects of hypoxic exercise on the structure and function of myocardial mitochondria in rats. J. Capit. Inst. Phys. Educ. 21(4), 462–465 (2009)
3. Wu, G.: Effects of high residence, high exercise and low training on the activity of respiratory chain enzyme complex of myocardial mitochondria in rats. Northwest Normal University (2012)
4. Shen, K.: Effects of intermittent hypoxia training on myocardial morphology of obese mice. Central South University (2009)
5. Zhao, J., Li, H., Bo, H., et al.: Effects of hypoxia and treadmill exercise on mitochondrial ultrastructure of left ventricular muscle in rats. Chin. J. Rehab. Med. 26(12), 1104–1107 (2011)
6. Harred, J.F., Knight, A.R., McIntyre, J.S.: Inventors. Dow chemical campany, assignee eXpoXidation process. USPatent 3(17), 1927–1904 (2012)

7. Zhang, Y., Li, W., Yan, T., et al.: Early detection of lesions of dorsal artery of foot in patients with type 2 diabetes mellitus by high-frequency ultrasonography. J. Huazhong Univ. Sci. Technol. Med. Sci. **29**(3), 387–390 (2011)
8. Foley, R.N., Parfrey, P.S., Sarnak, M.J.: Epidemiology of cardiovasc-ular disease in chronic renal disease. J. Am. Soc. Nephrol. **9**(12Suppl), S16–23 (2013)
9. Malyszko, J.: Mechanism of endothelial dysfunction in chronic kidney disease. Clin. Chim. Acta **411**(19/20), 1412–1420 (2010)
10. Izumi, S., Muano, T., Mori, A., et al.: Common carotid artery stiffness, cardiovascular function and lipid metabolism after menopause. Life Sci. **78**(15), 1696–1701 (2012)

Visualization Research on Regional Brand Co-construction Control Policy System of Agricultural Products – Taking Sichuan Province as an Example

Xu Zu[✉], Fengqing Zeng, and Lan Yang

Business School, Si Chuan Agricultural University, Chengdu 611830, China
403008983@qq.com

Abstract. This paper analyzes the regional brand control policies of Chinese agricultural products by content analysis method and common word analysis method. Conclusions of this paper possibly not merely provides scientific reference for governments to improve the regional brand co-construction control policy system for agricultural products, also arguably provides guidance and suggestions for building and effectively maintaining brand image.

Keywords: Agricultural products · Regional brand · Control policy

1 Introduction

On June 26, 2018, "Opinions of the Ministry of Agriculture and Rural Affairs on Accelerating the Promotion of Brand Strong Farmers" highlighted that agricultural development has entered a new stage, and brand awareness should be enhanced and brand building level should be improved. Agricultural product regional brand is different from the general corporate brand or regional brand, which has the characteristics of openness, non-exclusiveness, externality and regional cultural interactivity and multi-subjectivity, and the influence factors are complicated and varied [1]. The interest disharmony and fuzzy construction subject and mechanism among multiple subjects easily lead to problems such as "tragedy of the Commons", "lemon market" effect, lack of power for brand construction and lack of power for brand competition [2, 3]. Then, how to better manage the co-construction of regional brand of agricultural products in the whole life cycle, and maintain the consistency between the high quality of regional brand products of agricultural products and the high standard of service? This is also the problem that the government and enterprises need to solve urgently.

Therefore, this paper starts with the regional brand policy of agricultural products, and provides guidance and suggestions for better maintaining the brand image, so as to help China's precision industry to alleviate poverty and realize the co-construction, sharing and co-governing sustainable development of rural revitalization.

© Springer Nature Singapore Pte Ltd. 2020
M. Atiquzzaman et al. (Eds.): BDCPS 2019, AISC 1117, pp. 1327–1332, 2020.
https://doi.org/10.1007/978-981-15-2568-1_184

2   Material Collection and Research Methods

2.1   Sample Source and Research Method

First, searching for the keywords "agricultural brand", "brand building" and "agricultural products" on official websites, which including the National Government and the Sichuan Provincial People's Government and on public websites such as Xinhua. A total of 34 policies on the construction of agricultural products brands were found. Then, two doctoral students and a research team consisting of three master students screened the policies to ensure the accuracy of policy research. Finally, there are 18 policy documents that are highly relevant to research.

This paper mainly uses the common word analysis method and content analysis method to explore the regional brand control policy system of agricultural products. The content analysis method possibly overcomes the subjective drawbacks of qualitative research and conducts "quantitative" research on the content of agricultural product brand policy, so as to more accurately grasp the content of the policy [4]. Nowadays, this method is commonly used in modern technology, such as artificial intelligence, scient metrics and information retrieval and so on. Which has achieved some significant research results [5, 6].

2.2   Standardized Coding and Credibility Checking

To better quantify and code the content of the policy system, first of all, carefully study the content of each policy. Then, the word frequency appearing in the control group of the keyword group is classified and counted, and then the coding type is kept as perfect and exhaustive as possible. Under the principle of mutual exclusion and independence [7], the two master students independently screened and processed the keyword groups based on the word frequency and the combined policy content in the keyword group to prove the accuracy of the selection results. Through reliability test statistical formula:

$$K = \frac{2M}{N_1 + N_2} \tag{1}$$

The formula can detect the credibility, where M is the number of protocols between two encoders, and N1 and N2 are the number of decisions made by each encoder. The results show that the integration consistency of agricultural product regional brand policy keywords is 84%. Nunnaly (2009) stated that previous studies were credible if the degree of agreement was above 0.7 [8]. However, the research team invited two relevant professional teachers to further modify the existing keyword integration to ensure the accuracy of the policy system research. In the end, with the help of the teachers, 32 keywords were identified for the regional brand-related policy system for agricultural products.

3 Results and Discussion

3.1 Extract Keyword Results

In this paper, the key words of regional brand co-construction Control policy system of agricultural products are extracted, and 32 key phrases of relevant policy system of agricultural products regional brand are finally determined, which could be segmented into three modules of policy targets, suitable objects and action means (see Table 1).

Table 1. Key words of agricultural product brand construction control policy system

	Key words of agricultural product brand policy					
	Keyword	Frequency	Keyword	Frequency	Keyword	Frequency
Policy goal	Competitiveness	11	Standardization	6	Time-honored brand	3
	Popularity	10	Constitutive property	5	Additional value	3
	Influence	7	Industrialization	4	Survival of the fittest	2
	Intellectual property	6	High quality	4	Go out	1
Suitable object	Agricultural department	3	Entrepreneur	3	Units concerned	2
	Artel	3	The State Council	2	Operator	1
Role means	Demonstration plot	5	Normalize	2	Marketization	1
	Innovation center	3	Product certification	2	Large-scale	1
	Exposition	3	Diversification	2	Integrated utilization	1
	Television Station	2	Professionalization	2	Internet	1
	Manufacturing industry	2	Information technology	2		

3.2 Network Mapping

By using the integration of the Ucinet software Netdraw software for processing, obtain a network map of agricultural product regional brand policy. In the network map, the denser the network line is, the more frequent the co-occurrence of the two keywords it represents [9]. Meanwhile, the closer the keywords are to the middle of the map, the higher their central position in the network map is [10]. It can be seen from Fig. 1 that the positions of "competitiveness", "influence" and "popularity" in the network map are close to the center, which indicates that the regional brand building Control policy system of agricultural products is significantly related to these several keywords.

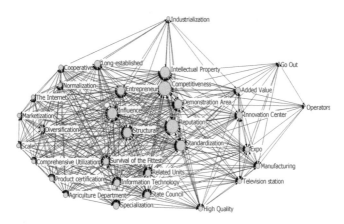

Fig. 1. Network map of key words of agricultural product brand policy

3.3 Point Degree Center Degree Analysis

The point and centrality analysis of regional agricultural product brand related policies is shown in Table 2.

Table 2. Agricultural product brand policy point degree center degree analysis results (part)

Keyword	Degree	NrmDegree
Competitiveness	78.000	31.452
Popularity	76.000	30.645
Influence	62.000	25.000
…	…	…
Industrialization	19.000	7.661
Operator	10.000	4.032
Go out	10.000	4.032

It can be seen that the absolute centrality of "competitiveness" is 78.000, and the relative centrality is 31.452, which is the highest point. The key word plays an obviously crucial role in the agricultural product brand policy. the keywords of "Reputation" and "Influence" occupy the position. The "Demonstration zone" is located at the center of absolute point and the highest degree of relative point. It can be concluded that in the agricultural product brand policy, "Competitiveness" is the main control and control objective, and the activities of enterprise brand cultivation and pilot demonstration of industrial cluster regional brand building are the main means to enhance the brand cultivation ability of enterprises.

4   Research Results and Suggestions

4.1   Research Results

By comprehensively optimizing the classification of keywords, the three main modules of policy targets, suitable objects and action means are formed, and the general framework of regional brand control policy system for agricultural products is outlined, which is as shown in Fig. 2.

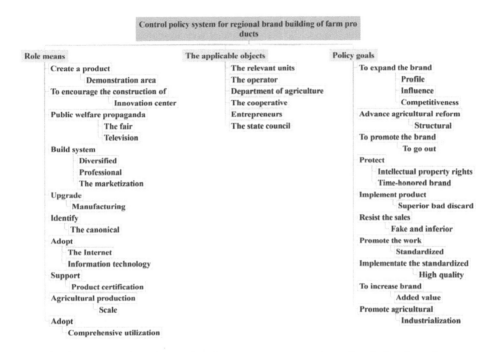

Fig. 2. Regional brand control policy system structure chart of agricultural products

4.2   Research Suggestions

Firstly, from the view of the government, the general agricultural product brand policy includes not only a variety of publicity activities such as the display of branded agricultural products, but the quality certification of the agricultural products itself is not so fancy, and the actual product quality certification of agricultural product brands has not yet been established. And the problem has been brought under control. Secondly, from the view of keyword group extraction, there are fewer laws and regulations governing agricultural product brand control by the control policy system. The next step is to start with the improvement of relevant laws and regulations, so that laws and regulations become a powerful tool for building agricultural products.

Then, from a business perspective, through the analysis of the policy system, SMEs will merge into large-scale agricultural production, giving priority to the development of "old brand" brands, which will help to expand the scale of enterprises and enhance their corporate image. Finally, from the analysis of policy social network, we can see that the "Competitiveness", "Reputation" and "Influence" of agricultural products are at the center of the network analysis chart, indicating that the government work mainly focuses on these aspects. Therefore, in the following work, we should continue to increase the promotion of agricultural product branding and further strengthen the support for the construction of agricultural product brands, which will be of great significance to the construction of agricultural product brands.

Acknowledgements. Fund Project: Sichuan Provincial Key Research Base of Philosophy and Social Sciences—Sichuan Center for Rural Development Research General Project (CR1910); Sichuan Provincial Key Research Base of Philosophy and Social Sciences—Research Center for Sichuan Liquor Industry Development General Project (CJY19-03); Sichuan Agricultural University Social Science Key Project (2018ZD04); Sichuan Province Philosophy and Social Science Key Research Base–Sichuan Agriculture Featured Brand Development and Communication Research Center General Project (CAB1810).

References

1. Shen, P.: Empirical research on influencing factors and mechanism of regional brand equity of agricultural products. J. Econ. Technol. **28**(5), 85–89 (2011). (in Chinese)
2. Huang, L.: Analysis on the main body and mechanism of regional brand building of agricultural products. Sci. Technol. Manag. Res. **28**(5), 51–52 (2018). (in Chinese)
3. Guan, C.: Research on collaborative management of regional agricultural products brand. Acad. Res. **6**(1), 74–79 (2012). (in Chinese)
4. Kassarjian, H.H.: Content analysis in consumer research. J. Consum. Res. **4**(1), 8–18 (1977). (in Chinese)
5. Chu, J., Guo, C.: The basic principle of the word analysis and EXCEL implementation. Intell. Sci. **29**(6), 931–934 (2011). (in Chinese)
6. Su, J., et al.: Comparative analysis of national and local science and technology innovation policies based on content analysis. Sci. Sci. Manag. Sci. Technol. **33**(6), 15–21 (2012). (in Chinese)
7. Li, G.: Content analysis method of public policy: theory and application. Chongqing University Press, Chongqing (2007). (in Chinese)
8. Zhao, Z., Yu, W.: Enterprises' micro-blogging interaction: based on chinese electric household industry. In: Proceedings of the Seventh International Conference on Management Science and Engineering Management, vol. 241, no. 1, pp. 135–144. Springer, Heidelberg (2014)
9. Ma, F., Wang, J., Zhang, Y.: Mapping knowledge map for domestic life cycle theoretical research-based on strategic coordinate map and conceptual network analysis. Intell. Sci. **28**(4), 481–487+506 (2010). (in Chinese)
10. Yu, W., Zu, X., Liu, Y., Zuo, R.: The visualization research on control policy system of social risk of large engineering's environment pollution. In: Proceedings of the Ninth International Conference on Management Science and Engineering Management, vol. 362, no. 1, pp. 651–662. Springer, Heidelberg (2015)

Dynamic Bayesian Model of Credit Rating for Credit Bonds

Jianhua Wu[1(✉)], Xuemei Yuan[2], and Ying Zhang[1]

[1] School of Mathematical Sciences, University of Jinan, Jinan 250022, China
wu88172968@163.com
[2] Academy of Financial Research, University of Jinan, Jinan 250002, China

Abstract. This paper proposes a dynamic Bayesian model for quantifying credit rating changes of bonds to address the lag phenomenon of rating transitions in the existing rating models. Based on the default probability process, the dynamic Bayesian method is used to obtain the best credit rating. By introducing bias parameters into the model, the judgment of investors can be integrated in the update, so to obtain a variable bond rating transitions model.

Keywords: Credit bonds · Default · Rating transitions · Bayesian

1 Introduction

In the bond credit rating, the core task is credit rating and dynamic adjustment, that is, the transfer of credit rating. The transfer of credit rating indicates the change of bond quality, reflects the change of debt issuer's ability to repay principal and interest, and also leads to a series of repricing of bonds and derivatives, thus affecting investors' investment choices. Many theoretical and empirical studies have proved the information value and risk disclosure function of credit rating and credit rating transfer, such as Rhee [1], Barnard [2] and Sajjad and Zakaria [3]. Therefore, how to conduct credit rating dynamically and construct a reasonable credit rating transfer matrix is an important research topic.

In the theoretical research and rating practice, many scholars have explored the estimation and verification of credit grade transfer matrix. Lando and Skodeberg [4] used semi-parametric regression technology to test two kinds of non-markov effects in grade transfer: temporal dependence and historical grade dependence. For example, Fuertes and Kalotychou [5] also support these results, in which it is found that there is duration dependence and degradation momentum in the rating process. McNeil and Wendin [6], a family of generalized linear mixed models are fitted for ordered multi-variate response variables, in which a markov process with time-random effects is assumed.

Kim and Sohn [7] proposed a non-markov stochastic effect polynomial model to estimate the transfer probability of credit rating. Li and Jin [8] verified the rating transfer matrix model based on markov chain with actual data. Zhou and Jiang [9] reviewed domestic research on credit rating transfer matrix. Du [10] studied how to quantify the credit grade transfer matrix by using the semi-markov chain model, and

© Springer Nature Singapore Pte Ltd. 2020
M. Atiquzzaman et al. (Eds.): BDCPS 2019, AISC 1117, pp. 1333–1339, 2020.
https://doi.org/10.1007/978-981-15-2568-1_185

compared the advantages of markov chain model and semi-markov chain model. Liu [11] investigated the application of credit rating matrix based on the rating data of real estate companies.

This paper will build a Bayesian credit rating transfer model to obtain a moderate rating transfer speed. The model is driven only by the probability of default, and requires little more than a Bayesian update. The model well fits the real process of changing the rate of credit rating.

2   Construction of Bayesian Dynamic Rating Model

2.1   Selection of Credit Rating Drivers

This paper does not adopt the multivariate statistical method based on transformation, but USES another method with more economic significance to select variables. From the practice of credit rating, it has become a common practice to analyze the credit problem by using the default probability based on the model in the bond market. Therefore, in many models, default probability is regarded as the proxy variable of a series of state variables that drive enterprise credit quality.

2.2   Dynamic Bayesian Model of Credit Rating

Suppose that the credit rating index of a bond that has not defaulted is R_t, which means the rating of the bond at time t, which is k, $k = 1, 2 \ldots, K$. In this paper, 1 is used to represent the highest credit quality, and K is used to represent the worst credit quality (which means that $K + 1$ is the default state).

In the pricing of credit debt and its derivatives, the rank transfer matrix is often used as a constant. But in fact, because of the potential macroeconomic state variables, the rank transfer matrix is non-homogeneous. Suppose that for any t, the one-year transfer probability R_t of credit rating r_t satisfies the following markov process

$$f(R_{t+1}|R_t) : r_{t+1} = r_t + M \cdot (\theta - r_t)\Delta t + \sigma \varepsilon_t, \quad \forall t \tag{1}$$

Where M is the mean recovery ratio, θ is the long-term mean of the process, ε_t is the independent normal random impact process, and σ is the time-varying impact coefficient. It should be noted that the addition of the mean response term $M(\theta - r_t)\Delta t$ in the rating is to explain A phenomenon existing in the bond rating market, that is, high-quality credit tends to drift to lower rating, while low-quality enterprises tend to drift to higher rating. Equation (1) can be estimated from the time series data of the transfer probability of historical credit rating.

In the previous setting, grade index R is positively correlated with default probability p. In other words, when the probability of default p rises, rating index will also rise.

Therefore, a conditional mapping H is set from default probability p_t to grade index R_t, that is, there exists a function $R_t = h(p_t)$. Given the sample data of credit rating R_t and default probability p_t, the probability distribution function can be visualized as the histogram of default probability in a certain level. The above analysis can be summarized as the following probability distribution: a credit rating corresponds to a series of default probabilities

$$p_t \sim f(p_t|R_t) \tag{2}$$

Given the parameter probability distribution, we can estimate the parameters by using the timing data of default probability and credit rating. This means that ratings do not change as frequently as the probability of default.

Next, we construct a dynamic Bayesian model to capture the characteristics of this hierarchical transfer velocity in the middle. Suppose the credit rating and the historical default probability sequence of a bond at time t, the credit rating transfer probability density of the bond is $f(R_{t+1}|R_t)$, and the posterior density of the bond at time t is $f(R_t|p_{1:t})$, then the prediction density of the credit rating at time $t + 1$ can be deduced as follows.

$$f(R_{t+1}|p_{1:t}) = \int f(R_{t+1}|R_t)f(R_t|p_{1:t})dR_t \tag{3}$$

If we discretize (3) above, we get the following discrete form

$$f(R_{t+1}|p_{1:t}) = \sum_{k=1}^{K} f(R_{t+1}|R_t = k)f(R_t = k|p_{1:t}) \tag{4}$$

This model can be used to predict credit rating. Combined with Eq. (4), the maximum posterior prediction of the credit rating R_{t+1} at time t to time t + 1 can be obtained, as follows

$$\tilde{R}_{t+1} = \arg\max_{k} f(R_{t+1} = k|p_{1:t}), k = 1, 2, \ldots, K \tag{5}$$

If the default probability changes from p_t to p_{t+1} from time t to time $t + 1$, the above prediction density $f(R_{t+1}|p_{1:t})$ can be further updated by using bayes' theorem to obtain the posterior density of the credit rating R_{t+1} at time $t + 1$ as follows

$$f(R_{t+1}|p_{1:t+1}) = f(p_{1:t+1}|R_{t+1})f(R_{t+1})/f(p_{1:t+1}) \tag{6}$$

This model determines whether the change in default probability is also equal to the change in rank. If we discretize (6) above, we get the following discrete form

$$f(R_{t+1} = k|p_{t+1}) = \frac{f(p_{t+1}|R_{t+1} = k)f(R_{t+1}|p_t = k)}{\sum_{k=1}^{K} f(p_{t+1}|R_{t+1} = k)f(R_{t+1} = k|p_t)} \tag{7}$$

Combined with Eq. (7), the maximum posterior estimate of the credit rating R_{t+1} at time t + 1 can be obtained, as follows

$$\hat{R}_{t+1} = \arg\max_{k} f(R_{t+1} = k|p_{t+1}), k = 1, 2, \ldots, K \tag{8}$$

The advantage of this model is that it can be used not only to predict the change of credit rating, but also to adjust the credit rating automatically. The characteristics of this model are consistent with the actual performance of grade transfer.

2.3 Bayesian Dynamic Rating Model with Subjective Intervention

By constructing the modified Bayesian model below, we can capture the impact of these biases and thus provide a more flexible system to better describe the evolution of credit ratings with default probability. To this end, we introduce K new parameters in the update rule (γ_k):

$$f_{\gamma_k}(R_{t+1} = k|p_{t+1}) = \frac{\gamma_k f(p_{t+1}|R_{t+1} = k)f(R_{t+1}|p_t = k)}{\sum_{k=1}^{K} \gamma_k f(p_{t+1}|R_{t+1} = k)f(R_{t+1} = k|p_t)}, k = 1, 2, \ldots, K \tag{9}$$

We can use these new parameters γ_k as the bias coefficient, which makes the conditional probability of each level become biased. For example, if rating agencies prefer grade k = 3, the value of A will be larger than other γ_k's. There are other benefits to introducing these parameters. By optimizing them, we can ensure that the historical system generated from the data produces a fitting transition matrix that is as close to the empirical matrix as possible. Therefore, the introduction of bias parameter γ_k makes the calibration more effective and in line with the actual situation.

2.4 Method for Checking Model—Sequential Importance Resampling

In the previous expressions (3) and (6), there is no analytic expression for the integral of the posterior distribution, so it needs to be obtained by approximate method. In this paper, the sequential importance resampling (SIBS) method will be used for approximate calculation. After obtaining the sample point set of credit rating R_t through the above algorithm, the approximate calculation of credit rating R_t estimation can be performed.

$$\hat{R}_t = E(R_t|p_{1:t}) = \frac{1}{N}\sum_{i=1}^{N} R_t^{(i)} \tilde{w}\left(R_t^{(i)}\right) \tag{10}$$

The algorithm has two advantages: first, it is very easy to implement. Second, as long as the dimensions of R_t are not very large, the approximate estimation of the posterior distribution $p(R_t|p_{1:t})$ has good properties. In fact, the algorithm is consistent and asymptotically normal, and it quickly "forgets past errors" to ensure that past errors are not accumulated.

3   Simulation Test Analysis of Model

Next, monte carlo (MC) simulation of default probability is used to verify the validity of the model. Firstly, the default probability is set in the simulation test. $p_{k,t}$ is the probability that a bond with a credit rating of k will default in a given year at time t. Suppose the default probability follows the markov process as follows:

$$p_{k,t+1} = p_{k,t} + G_k \cdot (\bar{p}_{k,t} - p_{k,t})\Delta t + \varepsilon_{k,t}, \quad \forall t \geq 0, k = 1, 2, \ldots, K \qquad (11)$$

Where G is the mean recovery ratio and $\bar{p}_{k,t}$ is the average default probability of bonds with a credit rating of k at time t, which is equivalent to the credit rating $R_t = k$. $\varepsilon_{k,t}$ is the independent normal random impulse process.

In the simulation test, The maximum number of cycles was set as N = 1000, which could be interpreted as 1000 bonds requiring rating. Finally, the following simulation sample credit rating transfer matrix is obtained, as shown in Table 1.

Table 1. Simulation experience matrix for level transfer

Initial rating	Rating after one year							Default
	Aaa	Aa	A	Baa	Ba	B	Caa-C	
Aaa	0.8901	0.0829	0.0071	0.0005	0.0012	0.0000	0.0000	0.0000
Aa	0.0069	0.8964	0.0780	0.0059	0.0006	0.0014	0.0004	0.0000
A	0.0010	0.0197	0.8905	0.0549	0.0104	0.0027	0.0010	0.0005
Baa	0.0010	0.0034	0.0607	0.8705	0.0432	0.0120	0.0098	0.0019
Ba	0.0019	0.0029	0.0079	0.0789	0.7971	0.0891	0.0821	0.0113
B	0.0018	0.0080	0.0091	0.0091	0.0696	0.7914	0.3962	0.0598
Caa-C	0.0015	0.0003	0.0005	0.0019	0.0081	0.0727	0.6045	0.1778
Default	0.0000	0.0000	0.0000	0.0000	0.0000	0.0000	0.0000	1.0000

Next, Fig. 1 below shows the contour and surface of the empirical transfer matrix.

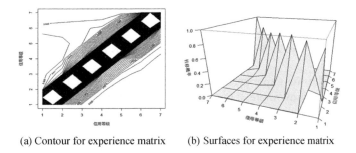

(a) Contour for experience matrix   (b) Surfaces for experience matrix

Fig. 1. Simulation experience matrix

It can be easily seen in Table 1 and Fig. 1 that the probability of bonds remaining in the same grade is higher, and the reason for this phenomenon is that the grade is returned to the mean. This can be used to describe a phenomenon in the real bond rating market, in which a bond's rating changes only if there is a sufficient number of updates to trigger a change in rating.

Throughout the period T = 60, the rank transfer matrix is calculated using the simulated rank change. The simulation test was repeated for 50 times, and the final fitting matrix was the average value of the transfer matrix in all 50 simulation tests. Table 2 below shows the results of the "fitting matrix" with bias parameter gamma $\gamma_k = 3$,

Table 2. Fitting of rating transition matrix and Bias parameters $\gamma_k = 3$

Initial rating	Rating after one year							Default
	Aaa	Aa	A	Baa	Ba	B	Caa-C	
Aaa	0.8907	0.0859	0.0029	0.0011	0.0001	0.0000	0.0000	0.0000
Aa	0.0057	0.9377	0.0499	0.0018	0.0059	0.0019	0.0004	0.0000
A	0.0499	0.0039	0.8901	0.0019	0.0250	0.0079	0.0009	0.0029
Baa	0.0703	0.0040	0.0079	0.8704	0.0029	0.0520	0.0108	0.0017
Ba	0.0129	0.1407	0.0043	0.0018	0.7699	0.0629	0.0823	0.0064
B	0.0025	0.0260	0.0260	0.0490	0.0003	0.8299	0.4062	0.0628
Caa-C	0.0014	0.0003	0.0005	0.0023	0.0083	0.0827	0.5945	0.1687
Default	0.0000	0.0000	0.0000	0.0000	0.0000	0.0000	0.0000	1.0000

Next, Fig. 2 below shows the contour and surface of the transfer matrix

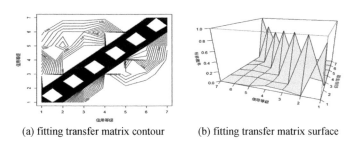

(a) fitting transfer matrix contour   (b) fitting transfer matrix surface

Fig. 2. Fitting transfer matrix with bias parameters

Compared with the experience matrix in Table 1 and Fig. 1, it can be found that Table 2 and Fig. 2 show more abundant credit grade transfer status. For example, after the introduction of bias parameters, the credit grade transfer matrix has a new feature, that is, with the decline of the initial rating, the probability of returning to the rating state after one year is gradually reduced. Of course, we can also set different bias parameters to obtain various types of transfer matrix, so as to depict various complex and changeable rating transfer situations in the rating market.

4   Conclusions

This paper proposes a dynamic Bayesian credit rating model based on the probability of default to simulate the change of credit rating. The bias parameter γ_k in the model can be appropriately adjusted to fit the historical data, so as to incorporate the subjective judgment factors of decision makers into the credit grade transition fitting. In fact, credit rating is a rough measure of default probability, and embedding bias parameter information in the model will increase the judgment of a small number of decision makers. In the model proposed in this paper, compared with the change of default probability, grade transfer tends to occur at a lower frequency, which is consistent with the fact that the credit rating in the bond market has a greater stickiness than default probability. The sequential importance resampling algorithm proposed in this paper also makes the model easier to implement.

Acknowledgments. This research was supported by National social science foundation of China (No. 15ZDB163); National social science foundation (No. 17BJY184); National natural science foundation of China (No. 11701214); Shandong natural science foundation project (No. SZR1810); Shandong university humanities and social science program general project (No. J17RA103);

References

1. Rhee, R.: Why credit rating agencies exist. Econ. Notes **44**(2), 161–176 (2015)
2. Barnard, B.: Rating migration and bond valuation: ahistorical interest rate and default probability term structures. Social Science Electronic Publishing (2017)
3. Sajjad, F., Zakaria, M.: Credit rating as a mechanism for capital structure optimization: empirical evidence from panel data analysis. Int. J. Financ. Stud. **6**(1), 1–14 (2018)
4. Lando, D., Skodeberg, T.M.: Analyzing rating transitions and rating drift with continuous observations. J. Bank. Financ. **26**, 423–444 (2002)
5. Fuertes, A.M., Kalotychou, E.: On sovereign credit migration: a study of alternative estimators and rating dynamics. Comput. Stat. Data Anal. **51**(7), 3448–3469 (2007)
6. McNeil, A.J., Wendin, J.: Dependent credit migrations. J. Credt Risk **2**, 87–114 (2006)
7. Kim, Y., Sohn, S.Y.: Random effects model for credit rating transitions. Eur. J. Oper. Res. **184**(2), 561–573 (2008)
8. Li, J., Jin, J.: Establishment of credit rating transfer matrix based on Markov process. J. Guangdong Vocat. Coll. Financ. Econ. **7**(3), 68–71 (2008). (in Chinese)
9. Zhou, L., Jiang, L.: Review of discussion on credit rating transfer. Econ. Theory Econ. Manag. V(2), 76–80 (2009). (in Chinese)
10. Du, K., Xie, Q.: Research on credit grade transfer probability based on semi-markov model. Bus. Res. **11**, 14–22 (2011). (in Chinese)
11. Liu, Q.: Dynamic change of credit rating in China's real estate industry. tianjin University of Finance and Economics, Tianjin (2014). (in Chinese)

Research and Design of Intelligent Service Platform for "Migratory Bird" Pension

Xueqin Huang[1], Ge Dong[1], and Qiang Geng[2(✉)]

[1] The College of Engineering and Technology, Hainan College of Economics and Business, Haikou 571127, China
[2] Network College, Haikou University of Economics, Haikou 571127, China
Gq_9@163.com

Abstract. Hainan Island is like spring all the year round, and the average temperature in winter is above 22 °C. Every November, a large number of the migratory birds elderly come to the island for winter, and leave the island in March next year. Many migratory birds elderly said that the original chronic disease has been significantly alleviated here, but there are many inconveniences in strange places. In order to make migratory birds elderly integrate into the local area as soon as possible, facilitate the management of local organizations and provide all kinds of old-age services for the migratory birds elderly, this paper designs a migratory bird care service platform combining Internet plus technology, integrates the resources of public pension services, optimizes allocation, and improves the quality and efficiency of the elderly care service for migratory birds.

Keywords: Hainan Island · Migratory bird elderly · Intelligent service · Pension platform · Design

1 Introduction

Hainan Island, a treasure island in the south of China, has long summer and no winter, with an average annual temperature of 22–27 °C. It has always been known as "a natural greenhouse". In addition, there is no industry in island. Green plants are perennially flourishing and the air is fine. So many old people would like to spend the winter in Hainan Island. Data shows that more than one million migrant birds people come to Hainan every winter from all over the country [1]. With the arrival of a large number of elderly migrant birds in winter every year, while bringing about the development of related industries in Hainan, new challenges have also been put forward for the migrant birds elderly in the old-age service facilities and management.

After the migratory old people enter the island, because of the different living habits, uneven personal qualities, inadequate provision of social and public resources and other issues, inevitably lead to conflicts with local residents. In addition, the original intention of migratory old people entering the island is to provide healthy old-age care. In view of the needs of migratory old people, urban management avoids conflicts between migratory old people and local residents, at the same time, it is necessary to provide services for the healthy old-age care of migratory old people [2].

© Springer Nature Singapore Pte Ltd. 2020
M. Atiquzzaman et al. (Eds.): BDCPS 2019, AISC 1117, pp. 1340–1346, 2020.
https://doi.org/10.1007/978-981-15-2568-1_186

With the deepening of the concept of intelligent support for the elderly and the continuous development of Internet and technology, the elderly who choose "migratory birds" and the elderly with strong mobility will be able to provide intelligent old-age services for migratory birds by taking advantage of information technology and related industries under the line. Aiming at the migratory birds elderly in Hainan Province, this paper designs a service platform for the migratory birds elderly, which mainly integrates the public pension service resources, optimizes the allocation, and improves the quality and efficiency of the service for the migratory birds elderly.

2   Demand Analysis

"Migratory Bird" Intelligence Service Platform for Old-age Care is mainly participated by local government, or relevant bodies of private organizations, the elderly, children of the elderly or other guardians, medical institutions, volunteers and enterprises providing old-age care services. These participants have different functional requirements (Fig. 1).

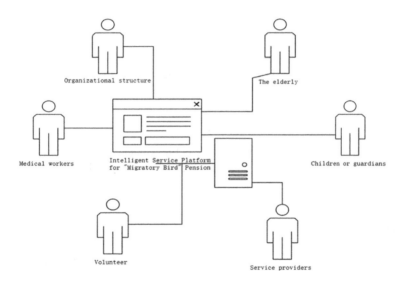

Fig. 1. The main participants

2.1   Organizational Needs

Organizations play an indirect role in service, mainly in management. In order to provide better health care services for migratory birds elderly, organizations need to audit, guide, supervise and standardize the service providers, understand the basic situation and needs of the elderly, and then provide targeted policy support, services and management.

2.2   Institutional Needs for Service Delivery

There are many types of service providers, such as residential hotels, apartment hotels, tourist attractions, household services, etc. After registration, they can display the services on the platform for the migratory birds elderly or their families to order and settle accounts.

2.3   Needs of the Elderly

Before entering the island, the migratory old people can simply understand the basic situation of Hainan Island, the surrounding environment and customs of the living area, the location of health entertainment places and so on.

2.4   Needs of Children or Supervisors

Children or supervisors who are inconvenient to accompany are always aware of the living conditions of the migratory birds elderly. When the elderly encounter an emergency, they can know at the first time, and can order various services for the elderly through the platform.

2.5   Needs of Medical Workers

Register the health data of the elderly, such as the history of common diseases and hereditary diseases. Through the platform, combined with the wishes of the elderly, can provide long-term health care services for the elderly.

2.6   Volunteer Needs

Combining with the system positioning, we can know whether the elderly in the service coverage have volunteer docking, timely understand the needs of the elderly, and provide help to the elderly.

3   Functions of Service Platform

The service platform provides various services according to the basic information and residence address of the migratory birds elderly. The residence address can be obtained by inputting or positioning system of the platform, and then targeted services will be pushed to the terminal of the elderly and their families for the elderly and their families to choose [3] (Fig. 2).

3.1   Login/Registration

Registration/login includes service providers and service recipients. Organizational managers need to audit each service provider before they can allow them to log on to their services on the platform.

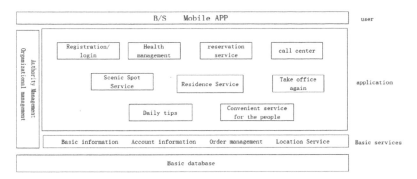

Fig. 2. Platform function architecture diagram

For service recipients, i.e. the elderly who enter Hainan Province for migratory bird-based old-age care, organizations and service providers need to know the basic information of the elderly in order to facilitate the provision of targeted services.

The platform needs to be registered before it can be used, and only the mobile phone number is needed for registration. After successful registration, the first information that needs to be entered is the name, sex, age, physical condition, place of origin, medical history, genetic diseases, occupational type, children or other guardians, date of entry/exit, and so on [4].

Basic information such as name, age, sex and place of origin will provide basic data for organizational departments and service providers, and determine service policies and content directions. Medical history, genetic disease information, in addition to the elderly emergency medical information can quickly give medical service providers, but also volunteers, community service is a guiding direction. Entering children (guardians) can facilitate children (guardians) to share data with the elderly through the platform, and can order some old-age services for the elderly. Entry/exit registration convenient platform can promptly push all kinds of service information for the elderly and guardians, and also provide data statistics for organizations and platforms, and formulate all kinds of service policies and contents.

3.2   Reservation Service

Platform can be positioned according to the mobile phone used by the elderly, pushing the peripheral services to the user interface, booking services can provide a variety of door-to-door services: sanitary cleaning, installation of various electrical appliances, carrying heavy objects, catering services, routine physical examination without large-scale equipment, etc., the migratory birds elderly in need can make reservations on their own. Appointments can also be made with the help of children far from the mainland.

3.3   Convenient Service

The migratory birds elderly can learn about local bus routes through "convenient service", locate nearby shopping malls, fitness, chess and card entertainment places and so on, so as to facilitate the elderly to familiarize themselves with local customs, make more friends, get familiar with the surrounding environment of their residence as soon as possible, and integrate into local life [5].

3.4 Residence Service

Some migratory birds elderly live in the form of renting houses or apartment hotels. Service providers can provide various housing sources and hotels for the elderly through "residential services" in a regional and price-based manner [6]. At the same time, the surrounding information should be introduced in detail, including hospitals, transportation, shopping, entertainment places, exercise places, noise situation and so on.

3.5 Health Management

"Health management" includes registration, medication reminders, counseling services, dietary recommendations, medical records management and other routine health data management. "Dietary suggestions" can push daily dietary suggestions and cooking methods according to medical records management data. According to their own situation, the migratory birds elderly can input information in the "medication reminder" to remind the elderly of timely medication by ringing bells and voice. In the "consultation service", there is a list of medical workers and volunteers, with detailed professional introduction. It is convenient for the elderly to choose according to their own situation. They can either leave voice messages or dial directly to communicate [7].

3.6 Scenic Spot Service

The "Scenic Spot Service" module mainly pushes the surrounding tourist attractions according to their location. Old people can travel to nearby scenic spots according to their distance or needs. According to the needs of the migratory birds elderly, scenic spots can provide personalized services such as pick-up and delivery services, catering services and booking lounges.

3.7 Call Center

Call centers connect the children, volunteers, community workers or hotel service providers of the migratory bird elderly. When the elderly send an emergency call, call centers automatically send alarms to the connectors at the same time or directly connect to mobile phones [8]. Volunteers, community workers or hotel service providers arrive at the elderly people's residence at the first time. After confirming the situation, we will deal with it further.

3.8   Take Office Again

In this module, the information platform is used to display the special post information provided by enterprises and institutions, and the personal information of the elderly with working ability for both sides to choose. The main purpose is to provide the elderly with the willingness to work with a platform to continue to play the residual heat, and to contribute to the construction of Hainan [9]. For example, some colleges and universities in Hainan Province have a learning mode of winter primary school for about one month. Older people with professional ability and qualifications can provide their own professional knowledge during this period to teach students in Hainan universities. As well as the professional consultation work of some enterprises and institutions and the difficult outpatient clinics in hospitals, the migratory elderly people can show their skills and provide valuable resources for the local development of Hainan.

3.9   Daily Tips

Introduce Hainan's local customs and local people's ways of dealing with people, so as to facilitate the elderly to understand the local, and avoid unnecessary disputes due to different habits [10]. There are many kinds of tropical fruits in Hainan Island, which are rich in nutrients, but there are also many calories which are too big to eat more. Moreover, individual constitution or disease are not conducive to eating some tropical fruits. Therefore, in this module, the types, nutritional value and matters needing attention of tropical fruits are introduced in detail, so as to make it convenient for migratory birds to choose by themselves.

4   Conclusion

To sum up, climate and geographical resources in Hainan will attract more and more elderly people to live here. At the same time, Hainan needs more human and intellectual experience and resources in the construction of free trade ports (areas) under the background of international tourism island. Therefore, in order to provide migratory elderly people with a more comfortable life, even for the working environment of exerting waste heat, it is particularly urgent and important to build an Internet based, intelligent and elderly care platform with all kinds of targeted and practical functions. It is believed that the application of the platform will change the living status of the "migratory birds" elderly in Hainan, so make them really feel the enthusiasm and livability of Hainan.

Acknowledgments. This research was supported by Hainan Province University Scientific Research Grant (Project No: Hnky2018-82).

References

1. Li, Y., Zeng, Y.: Study on the living situation of the "migratory bird" the old-age population in different places—taking Hainan Province as an example. J. Demogr. (01), 56–65 (2018)
2. Wang, B., Zhao, H.: Analysis of the contradiction between supply and demand of public services for the "migratory bird" pension group: taking Sanya City as an example. Adm. Forum (02), 103–109 (2019)
3. Li, X.: Design and implementation of intelligent old-age management system based on B/S architecture. Hunan University
4. Liang, M., Mao, N., Chen, W., Yu, A., Chen, J., Xu, X.: Design and implementation of service management system for old age institutions. Innov. Appl. Sci. (05), 1–2 (2017)
5. Li, C., Bi, X.: Research on intelligent old-age service system and platform construction. E-government (06), 105–113 (2018)
6. Yang, Z., Liu, N.: Design of information-based intelligent pension management system. Manag. Technol. Small Medium-Sized Enterp. (Late issue) (10), 155–156 (2016)
7. Zhi, Y.G., Ging, W.: Application of internet of things in home based elderly care service in China. J. Shanghai Technol. (9), 267–271 (2016)
8. Yu, T.L.: A study on the way of "migratory birds" supporting the aged in different place. Soc. Sci. Front (08), 276–280 (2018)
9. Yi, C., Bing, X.I.: There is no worry about "migratory birds". Rural Agric. Farm. (03), 9 (2018)
10. Godfrey, M., Johnson, O.: Digital circles of support: meeting the information needs of older people. Comput. Hum. Behav. **25**(3), 633–642 (2017)

Injury Characteristics of Different Dummies Based on Frontal Impact Test

Haiming Gu[✉], Lei Lou, and Xiongliren Jiang

China Automotive Technology and Research Center Co., Ltd.,
Tianjin 300300, China
guhaiming@catarc.ac.cn

Abstract. Crash dummies are one of the essential tools in crash tests, used to simulate injuries of real people in accident. But in fact, vehicle restraint systems are often developed according to specific dummies, and cannot cover all the population, so passengers of different sizes will show different injury characteristics in accidents. This article introduces the types of dummies and injury evaluation indexes. A single factor variable control is used to compare the injury and movement of the 5th, 50th and 95th dummies through sled impact method. The influencing factors of the difference are analyzed from four aspects.

Keywords: Injury · Dummy · Sled impact

1 Introduction

Crash dummies, also known as standard dummies or human physical models, are the most basic vehicle crash test tools. Replace real people in car crash tests, so that technicians can analyze various technical data of cars at the moment of collision. Due to the need to measure the acceleration, velocity and load and deformation of each part of the human body, the size and mass of each part of the mannequin are also very close to the real person. Most of the impact dummies are made of metal, plastic and silica gel. The chest is made of steel, the shoulder and foot bones are made of aluminum, and the pelvis is made of relatively fragile plastic. The collision dummies not only have the same appearance as real people, but also have highly anthropomorphic spines, ribs and skin. The crash dummy's body is littered with sensors that can feed up to 180 channels and refresh the data 2,000 times a second. The performance requirements of dummies used in crash tests usually include: (1) the deformation characteristics of the size, weight distribution, joint motion, chest and other parts under load are similar to those of real people; (2) it shall be able to measure the acceleration, load and other parameters of the corresponding parts of the human body; (3) small differences between individuals, good repeatability, and good durability.

Initially, crash dummies were used as aircraft seat ejection tests, and in 1960 the us developed a car crash test simulation dummy VIP. The national association of automotive engineers standard SAE is used to specify the size and weight of crash dummies. Hybrid II dummies occurred in 1972 and in 1976, the United States to further its reform and development of the closer to the human body characteristics of Hybrid III

© Springer Nature Singapore Pte Ltd. 2020
M. Atiquzzaman et al. (Eds.): BDCPS 2019, AISC 1117, pp. 1347–1354, 2020.
https://doi.org/10.1007/978-981-15-2568-1_187

dummy, so far, the Hybrid III dummies have different percentile dummies. Commonly used in frontal collisions, there are mainly 5th, 50th and 95th.

1.1   Specifications of Dummies for Frontal Impact Test

Hybrid III dummy is the most widely used the best test dummy, biological simulation performance, in terms of body characteristics, the damage index, can accurately reflect the real situation of reality, is widely used in car crash regulations at home and abroad and NCAP, at the same time in the aerospace, medical and sports equipment and other fields have a wide range of applications. In addition, the 95th percentile male large body dummies and the 5th percentile female small body dummies were developed by scaling method to evaluate the protective effect of restraint system on passengers of different sizes. The specifications of the three dummies commonly used in frontal collisions are shown in Table 1.

Table 1. Hybrid III dummy specification

Specifications	Weight (kg)	Sitting height (mm)
Hybrid III 50th	78.15	884
Hybrid III 5th	49.98	790
Hybrid III 95th	101.31	935

1.2   Human Injury Evaluation Indexes

Commonly used human injury evaluation indexes include head HIC, viscosity index VC, etc.

(1) HIC, a head injury indicator
The Wayne State Tolerance Curve (WSTC) [1] is the basis of quantitative evaluation of head collision Tolerance. Versace further studied the WSTC curve and proposed a new head injury index HIC.

$$\text{HIC} = (t_2 - t_1) \left[\frac{1}{t_2 - t_1} \int_{t_1}^{t_2} a(t) dt \right]^{2.5}$$

Where: a - head center of mass synthesis acceleration (g); t_1, t_2 - any two moments in the collision process.
    Normally, the maximum integral interval is set as 36 ms, and when HIC is limited to evaluating head contact collision injury, the integral interval is set as 15 ms.

(2) Viscosity Index VC
The crush volume index of sternum could not well reflect the possibility of injury caused by high-speed collision.

Viscosity index VC is the product of deformation velocity V(t) and relative extrusion deformation amount C(t) [2]. The units of VC are the same as the units of velocity m/s. The test showed that the tolerance level $VC_{max} = 1$ m/s with a 25% probability of causing serious injury.

Table 2 shows the base value of injury evaluation [3].

Table 2. Base value of injury evaluation

Body area injury evaluation index		5%	50%	95%
Head	HIC ($t_2 - t_1 \leq 15$ ms)	1113	1000	957
	Forward bending moment (Nm)	104	190	258
Head/neck joint surface	Backward bending moment (Nm)	31	57	78
	Axial tensile (N)	–	–	–
	Axial compression (N)	–	–	–
	Shear force (N)	–	–	–
Chest	Acceleration in the thoracic spine (g)	73	60	54
	Viscosity index (m/s)	1	1	1
Femur	Axial compression (N)	–	–	–
Knee	Leg bone to thigh bone displacement (mm)	12	15	17
	The compression load (N)	2552	4000	4920
Tibia	The axial load (N)	5104	8000	9840
	Fc- Limit of axial force (kN)	22.9	35.9	44.2

2   Effects of Body Size on Occupant Injury Risk

In a car crash, body size affects the risk of injury. Based on studies of traffic accident data and cadaver studies, there are several hypotheses about the effect of body size on injury risk.

(1) Bubble Effect Hypothesis

Arbabi [4] and Wang [5] et al. pointed out that in the process of car collision, fatty soft tissue is similar to a foam air cushion, thus providing extra protection for human internal organs. However, recent cadaveric tests have shown that abdominal injuries are more likely to occur when the cadaver in the test is larger [6]. Turkovich et al. [7] simulated and studied the injury mechanism of obese occupant in head-on vehicle collision accident and the impact of obesity on the risk of injury of male occupant by establishing a joint model of multi-rigid-body dummy model and finite element model of human abdomen

(2) Mass Change Hypothesis

Viano et al. pointed out that larger body size means that passengers have more kinetic energy in the process of car collision, which requires greater restraint to prevent passengers from contacting with the internal parts of the vehicle [8]. Kent et al. [9]

compared the dynamic response differences between obese cadavers and non-obese cadavers by carrying out frontal sled crash tests.

(3) **Hypothesis of Body Shape Change**

Considering that body shape will affect the wearing position of safety belt, Reed et al. [10] statistically analyzed the space position of safety belt worn by a large number of samples relative to bone gap structure. The results showed that the position and Angle of the abdominal seatbelt were worse for the occupant in the vehicle collision.

3   Study on Injury Difference of Dummies with Different Body Types

Based on the above analysis, different body types and genders have significant effects on occupant injury. Therefore, plan to further verify the extent of such effects through experiments. Using Hybrid III 5th, 50th, 95th dummies, three frontal impact tests were carried out, by adopting the same seat, seat belt, pulse to control the consistency of input variables, so as to analyze the influence of the size and gender differences to dummy injury.

3.1   Test Plan

In the form of dynamic crash test, three dummies, 5th, 50th and 95th, were placed in 3 tests. Seat positions are adjusted according to the dummy to ensure compliance with regulations. In the three tests, the impact pulse, seats and safety belts were consistent, and the injury and movement posture of the dummy were collected after the test.

The test adopts the frontal collision pulse of 100% rigid barrier of a certain car model, with the speed of 50 km/h, pulse peak value of 43.5 g and pulse width of 95 ms, as shown in Fig. 1.

The test adopted the form of seat dynamic test. The seat was installed on the rigid plane, and the rigid column was used to simulate the fixed point of the safety belt. The safety belt was a real car safety belt, and the burst moment of the safety belt was 10 ms.

Seat position settings in the test were as follows:

5% dummies: the seat was placed in the front and the highest position, and the backrest was the seat design angle.
50% dummy: seat was placed in the middle of front and back, middle of top and bottom, backrest was the design angle.
95% dummies: seat was placed in the last and lowest position with backrest at design angle.

3.2   Movement Contrast of Dummy

In the three experiments, there were significant differences in the movement posture of the dummy, which was reflected in the intensity of the dummy's movement. In order to quantify the difference in motion posture, we adopted the maximum head displacement and peak time of the dummy, as shown in Table 3 and Fig. 2. Female dummy head

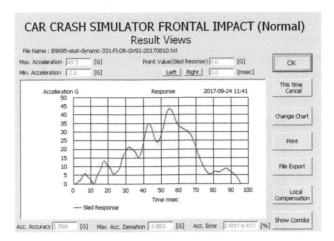

a) Acceleration

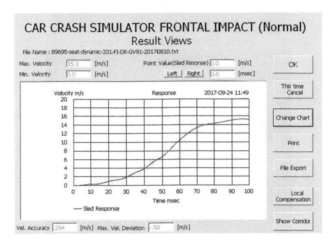

b) Velocity

Fig. 1. Crash pulse

displacement was only 360.58 mm, while 50% and dummy displacement was relatively close to 95%, almost twice as many female dummy, there exist more than 10 ms differences in peak time, it shows that: the same constraint system for larger dummy movement restrictions is poorer, and for smaller female dummy has good restriction.

Table 3. Dummy movement contrast

Parameters	5%	50%	95%
Maximum head displacement	360.58 mm	623.65 mm	726.33 mm
Peak time	116 ms	127 ms	131 ms

a) 5% female dummy movement

b) 50% male dummy movement

c) 95% male dummy movement

Fig. 2. Dummy movement contrast

3.3   Dummy Injury Contrast

The injuries of the dummy were mainly compared in terms of head acceleration, neck shear force, neck stretching force, neck torque, chest acceleration, chest compression, and thigh force, as shown in Table 4. It can be seen that under the same constraint system configuration, the three dummies have significant differences in key indicators such as head HIC, neck torque and chest compression.

Table 4. Comparison of injuries to key parts of the dummy

Index	HIC	3 ms resultant acceleration	Neck moment My	Chest compression	VC	Left thigh force	Right thigh force
5%	718	65.15	73.2	22.2	0.1	1.50	1.27
50%	467	62.66	84.49	32.7	0.18	2.45	1.94
95%	1311	90.98	51.29	44.5	0.2	2.31	2.26

4   Conclusion

Under the same restraint system conditions, completely different test results appeared in 5%, 50% and 95% dummies, mainly due to the following reasons:

(1)   Different sitting positions lead to the space size difference of the restraint system. This is reflected in the fact that 5% dummy sits in the front, 50% in the middle, and 95% in the rear. Therefore, under the same seat belt installation position, the initial gap between the seat belt and the dummy is different, and the movement time course is different, so the time and effect of restraint system intervention are different.

(2)   Different dummy masses lead to different dummy momentum. At almost the same speed, the increase of the mass of the dummy will inevitably lead to the increase of momentum, and this part of energy will eventually be absorbed by the restraint system and the dummy, so it will be reflected in the injury of the dummy.

(3)   Different heights of the dummy lead to the different position of the safety belt relative to the dummy. The change of seat belt position was directly reflected in the chest compression index of the dummy.

(4)   Different biomechanical characteristics of the dummy lead to the difference in bending amplitude of the neck, spine and other parts of the dummy in the tests. As the dummies are substitutes designed by researchers based on the biomechanical data of volunteers and cadavers to replace the real people used in car crash tests, their biomechanical properties must follow the real human characteristics. Therefore, the selection of different design prototypes inevitably leads to different structural stiffness and deflection of different types of dummies.

Based on the analysis of the above four points, different height, weight, physique and sitting position may lead to significant differences in the ultimate injury of the dummy. Therefore, it is necessary for us to adopt dummies closer to the real situation in the regulations according to the local human characteristics, so as to make the local automobile safety design closer to the national conditions.

References

1. Namjoshi, D.R., Good, C., Cheng, W.H., et al.: Towards clinical management of traumatic brain injury: a review of models and mechanisms from a biomechanical perspective. Dis. Models Mech. **6**(6), 1325–1338 (2013)
2. Petitjean, A., Baudrit, P., Trosseille, X.: Thoracic injury criterion for frontal crash applicable to all restraint systems. SAE Technical Paper (2003)
3. Sun, L., Duan, D., Liu, S., et al.: A research on the optimization of restraint system for the protection of occupants with different statures. Autom. Eng. **11**, 1312–1318 (2016)
4. Arbabi, S., Wahl, W.L., Hemmila, M.R., et al.: The cushion effect. J. Trauma Acute Care Surg. **54**(6), 1090–1093 (2003)
5. Wang, S.C., Bednarski, B., Patel, S., et al.: Increased depth of subcutaneous fat is protective against abdominal injuries in motor vehicle collisions. In: Annual Proceedings/Association for the Advancement of Automotive Medicine. Association for the Advancement of Automotive Medicine, vol. 47, p. 545 (2003)
6. Untaroiu, C.D., Bose, D., Lu, Y.C., et al.: Effect of seat belt pretensioners on human abdomen and thorax: biomechanical response and risk of injuries. J. Trauma Acute Care Surg. **72**(5), 1304–1315 (2012)
7. Turkovich, M., Hu, J., van Roosmalen, L., et al.: Computer simulations of obesity effects on occupant injury in frontal impacts. Int. J. Crashworthiness **18**(5), 502–515 (2013)
8. Viano, D.C., Parenteau, C.S., Edwards, M.L.: Crash injury risks for obese occupants using a matched-pair analysis. Traffic Injury Prevent. **9**(1), 59–64 (2008)
9. Kent, R.W., Forman, J.L., Bostrom, O.: Is there really a "cushion effect"?: a biomechanical investigation of crash injury mechanisms in the obese. Obesity **18**(4), 749–753 (2010)
10. Reed, M.P., Ebert-Hamilton, S.M., Rupp, J.D.: Effects of obesity on seat belt fit. Traffic Injury Prevent. **13**(4), 364–372 (2012)

Abnormal Behavior Detection Based on Spatio-Temporal Information Fusion for High Density Crowd

Honghua Xu, Li Li$^{(\boxtimes)}$, and Feiran Fu

Changchun University of Science and Technology,
Changchun 130000, Jilin, China
honghuax@126.com, dilihan0128@163.com,
fufeirancust@163.com

Abstract. With the strong demand of real-time detection of crowd abnormal behavior, an algorithm of abnormal behavior recognition based on video sequence for high-density crowd without any training stage is proposed, aiming at realizing online real-time detection of crowded events in intelligence video surveillance system. The problem of blurred targets and limited pixels of target expression will make it difficult to segment targets, which will affect the recognition of group behavior. In addition, severe occlusion limits the performance of traditional visual tracking methods. The use of classifier will result the dependence of scenes, and even affects the online deployment of intelligence video surveillance system. In the stage of detection process, the new method includes motion region segmentation, motion blobs extraction and crowd activity image creation, and then the information entropy of crowd activity image is used for the recognition of crowd behavior. The new method carries out experimental and quantitative data analysis on selected public data sets, and com-pares it with other online crowd event detection methods. The experimental results show that the method can detect ab-normal crowd behavior in high-density situations without training stage, and has better detection effect than the state-of-the-art technology.

Keywords: Abnormal behavior · Motion region segmentation · Blobs extraction · Crowd activity image

1 Introduce

Automatic event detection in crowded scenes is a challenging task in computer vision [1]. In order to solve the problem of group behavior detection, many detection methods have been proposed, but the online hypothesis limits the application of these methods in intelligent video surveillance system. For the analysis of crowd activities, it is necessary to develop online algorithms to achieve reliable anomaly detection. Automatic detection of abnormal events in crowded scenarios can avoid crowd-related disasters and ensure public safety.

Abnormal are usually defined as "deviations from standards, normal or expected events". This means that abnormal events can be identified as irregular situations

© Springer Nature Singapore Pte Ltd. 2020
M. Atiquzzaman et al. (Eds.): BDCPS 2019, AISC 1117, pp. 1355–1363, 2020.
https://doi.org/10.1007/978-981-15-2568-1_188

related to normal events. Traditional computer vision methods [2–4] may be ineffective in analyzing video sequences of crowded scenarios. This is because in the case of high density, blurred targets and limited target expression pixels will cause difficulties in target segmentation, and mutual occlusion of targets also limits the performance of traditional visual tracking methods.

Other factors limiting the effectiveness of existing methods for detecting abnormal events are the requirement of online computing and training phase. Online requirements limit the application of anomaly detection methods in practice. For example, in order to avoid hurting people, it is better to find panic situation as soon as possible. Due to the lack of appropriate training data, the application of methods based on classifier and training [5–7] will be limited. In fact, since it is not easy to find data about real emergencies in crowded scenarios, classifier-based methods can only be applied to the processing of specific video sequences.

In order to overcome the problems of the demand of large amount of data, scene dependence and non-real-time, a method based on statistical analysis combining feature matching and image segmentation is proposed, which uses image entropy to detect crowd abnormal behavior. By calculating the entropy difference of group activity map, the method can detect group abnormal behavior in real time without training stage.

2 Motion Region Segmentation

The purpose of motion region segmentation is to find descriptive visual features of crowd movement in observation scenes, and extract the motion regions of group behavior. The region segmentation process of visual-based is shown in Fig. 1. In this process, the feature tracker is used to detect and track the local visual features of adjacent frames, then spatial and temporal filtering is carried out to discard the static features and retain the moving characteristics of the behavior, and then the binary motion region of the video frame is obtained by calculating the probability grid in a given time window and image segmentation is carried out to get motion region mask.

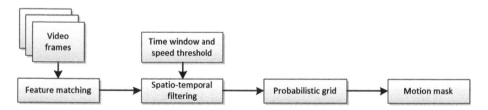

Fig. 1. Motion feature extraction based on vision.

SURF feature descriptor is used to detect and match local visual features. The distance between SURF features is very small, which shows good immunity to adjusting parameters, and can still work well under the condition of low computational requirements. The output of tracker at time t is a set of corresponding feature points captured in two consecutive frames at time T $-$ 1 and t, respectively. The set of feature points is represented as $F(t) = \{f_i^{t-1}, f_i^t | i = 1, \ldots, n\}$. After calculating $F(t)$, time filter is used to detect moving regions and create binary mask images of moving regions. The moving region contains only moving points in the scene, and the thresholds τ and γ are used to filter non-moving feature points. Among them, τ represents the time window size of the length of the historical sequence, and γ is the threshold of the speed of the pixels, which is the minimum speed value of the moving feature points.

$$v_f^t = \frac{\sqrt{(f^{t-1}(x) - f^t(x))^2 + (f^{t-1}(y) - f^t(y))^2}}{frame\ rate\ in\ a\ second} \tag{1}$$

The vector v is stored in memory. When each new frame arrives, if the feature is discarded, then we get a set of filtering features, which are recorded as. The probabilistic grid is used to weigh the moving points. The mesh is divided into several cells with the same size as the input image. Each cell corresponds to a pixel. The weighting process is shown in Fig. 2.

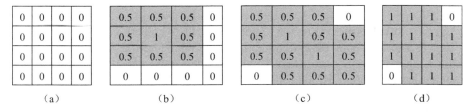

Fig. 2. The computational process of probabilistic grids

(a) All cells are initialized with a value of 0. (b) Each tracked feature point has an allocation value of 1, and its eight adjacent points have an allocation value of 0.5 (c). Considering all the feature points, there may be a point with a value of 1, and its adjacent points have a value of 1. (d) If the value of the adjacent points is 1, the value of all the adjacent points is also set to 1.

The initial value of each cell is 0, as shown in Fig. 2(a). For each feature point belonging to $F(t)^*$, the value of the corresponding cell is set to 1, and the value of the eight adjacent cells of the cell is assigned to 0.5, as shown in Fig. 2(b). If there are cells with values of 1 in the neighborhood of a cell with a value of 1, as shown in Fig. 2(c), then the values of all cells of their neighborhood are set to 1, as shown in Fig. 2(d). By further modifying the grid, the adjacent moving points can be clustered.

After setting the unit value of the probabilistic grid, the cell with the value of 1 is regarded as a white, and the other cells are regarded as black to generate a binary motion mask. The moving mask image provides the moving feature region detected in the image.

3 Motion Blobs Extraction

Motion blobs extraction is a series of operations, such as edge segmentation, triangulation and motion region aggregation, to get together the boundary regions of moving objects in a global way on video frame images, and then obtain the motion blobs images of group activities. Figure 3 shows the process of extracting motion blobs.

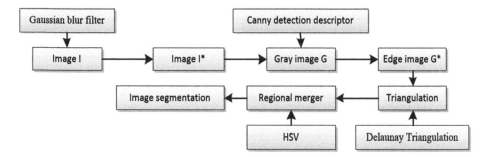

Fig. 3. The process of motion blobs segmentation

In the process of moving region segmentation, firstly, binary moving region image is used as a mask to filter the current input image I, and a new image I* is obtained by using a filter with 3*3 Gauss kernel, which is converted into gray image G. Then Canny edge extraction algorithm [8] is used to create an image G* containing edges. Then Delaunay triangulation method is used to triangulate the image. The content of edge image is used as input of Delaunay triangulation process, and the triangular subdivision surface of image is calculated. After triangulation, the merging process is used to merge the triangle regions. Considering each triangle in the graph in turn, the average value of the HSV color of all the pixels in its circumferential circle is calculated. If the mean of HSV color of a pair of triangles is less than w, they are merged into one region.

After splitting blobs on the image I*, a set F_{blob} of corresponding feature points is generated. In order to find the moving blobs, Formula 1 is used to filter the blob features, and a new set F_{blob}^* of filtering features is obtained. By projecting the set F_{blob}^* onto the segmented image, the set of moving blobs can be detected. If the region of the blob contains at least one feature point $f \in F_{blob}^*$, it is considered to be a moving blob. In this way, the moving blob image can be obtained.

4 Crowd Activity Image and Behavior Detecting

Crowd activity image is to collect data on time window w and make statistical analysis. The process of creating crowd activity image is to input a set of binary blobs images, as shown in Fig. 4(a), by using a three-dimensional mesh with the size of $m*n*w$, as shown in Fig. 4(b), and then generate crowd activity image, as shown in Fig. 4(c).

The width m and height n of the mesh are the same as the input image, while the depth w corresponds to the length of the time window. If the corresponding pixel p in the blob image is white, the value of corresponding volume pixel in the three-dimensional mesh is set to 1. The depth of pixel a is represented by a group of 1, and the number of 1 equals the number of white pixels corresponding to the blob image in the time window w. In this way, the depth of the pixel a in the 3D mesh determines the duration of each point p in the scene.

The gray value in the activity map is strictly related to the persistence of the pixels in the time window w. That is, the brighter points in the activity map indicate the higher activity intensity. In the experiment, the length w of the time window is set to the frame rate of the video sequence. Figure 4 shows the calculation process of the activity image. It can be noted that the crowd activity image only considers the part of the image that contains real motion.

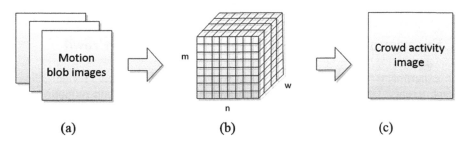

Fig. 4. Computational process of activity image for crowd motion scene. (a) Binary blob images (b) Three-dimensional grid (c) Crowd activity image

Image entropy [9–11] is a statistical method, which reflects the average amount of information in an image. The entropy of an image represents the aggregation feature of the gray distribution in the image. Let p_i denote the proportion of pixel i in the gray image, and the calculation of the gray image entropy see the definition of Formula 2.

$$H = -\sum_{i=0}^{255} p_i \log p_i \tag{2}$$

Where p_i is the probability of a value appearing in the image, and it can be obtained by gray histogram. When the image is a pure color image, there is only one gray value (white or black). At this time, the entropy is the smallest, H = 0, which means that the image does not contain any target, and the amount of information is 0. When the gray value of each pixel is different, the information entropy of the image is maximum. It can be considered that every single pixel of the image is an independent target. Therefore, the greater the entropy H of the image, the more uniform the distribution of gray values, the more objects in the image, the greater the amount of information in the image. For an image, when the gray value of each pixel is different, the entropy is the maximum.

Image entropy measures the uncertainty of image value by calculating the information of image. Because incidental events provide more information than frequent events, the occurrence of sudden conditions can be detected by monitoring the instantaneous change of entropy value. If the entropy difference of the activity map is greater than the threshold value, it indicates that an abnormal event has occurred. The calculation of the entropy difference of the activity map is shown in Formula 3.

$$Entropy(I_{t+1}) - Entropy(I_t) > e_v \qquad (3)$$

The implementation steps of the new method include motion region segmentation, motion blobs extraction, crowd activity image creation and calculating the entropy of crowd activity image (shorted as CAIE). Experiments were conducted on public video datasets, and the above methods were used to detect the interesting events in crowded scenes. In experiment, quantitative analysis and evaluation were done for the new method, and compared it with other the state-of-the-art methods.

5   Experiment of Abnormal Behavior Detection in Crowd

The experiments were done on UMN data set [13], which included 11 video clips of escape events. These videos were shot in three different indoor and outdoor scenes, including lawns, indoor corridors and outdoor squares. Each video contains about 20 people who walk in different directions, and then an unusual event causes people to escape. UMN data sets have labels of ground truth, and each frame in the video sequence is marked with "normal" or "abnormal", where "abnormal" indicates that an event of interest is occurring.

In order to evaluate the performance of CAIE method, each scene of UMN dataset was detected and drew corresponding ROC curves. The ROC curves of different scenarios are shown in Fig. 5.

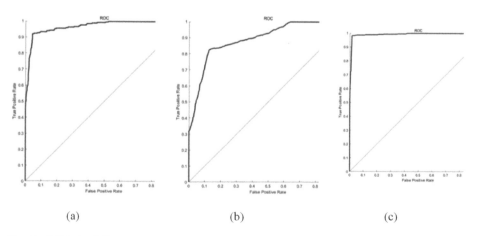

(a)   (b)   (c)

Fig. 5. ROCs in different scenes of UMN data set. (a) ROC of UMN data set scene1 (b) ROC of UMN data set scene2 (c) ROC of UMN data set scene3

From the ROC curves of different scenes, CAIE method does not need to use classifiers to detect crowd behavior, and achieves good results on different video sequences.

It should be noted that there are reports of AUC value approaching 1 in the literature on group behavior detection methods [12]. However, this performance is achieved by analyzing the entire video sequence or by training and learning, that is to say, the method takes advantage of knowledge about future information of events. This analysis is useful for obtaining models of different group behaviors, but it may be ineffective for practical online applications.

In order to make a fair comparison with the published results, the whole UMN data set is regarded as a complete video sequence, and the posed method is compared with the state-of-the-art online methods. Table 1 shows the quantitative comparisons in the UMN dataset.

Table 1. The AUC of the whole UMN data set is compared with other methods.

Method	Type	Area under ROC curve (AUC)
Pure optical flow	Online	0.84
Social force	Online	0.86
SRC	Online	0.93
Chaotic invariants	Online	0.89
CAIE (posed method)	Online	0.94

From the Table 1, we can see that the method of CAIE outperforms Pure Optical Flow [14], Social Force [14], SRC [15] and Chaotic Invariants [16]. CAIE is better than these methods because it uses moving blob sets instead of a single feature, and takes into account the change of temporal and spatial information, which makes the method more robust to sudden changes in the scene. In addition, CAIE method is not based on classifier to detect abnormal events, so the method is more flexible in different environments.

6 Conclusion

The crowd safety management in public places is a matter of concern. Using computer vision algorithm to detect the changes of abnormal crowd activities is helpful to positively respond to public safety emergencies. In the past decade, people have made great progress in related technologies, including crowd division, crowd counting and event detection. However, understanding crowd behavior is still challenging, especially in highly crowded scenarios.

A crowd behavior detection algorithm based on video sequences without any training stage is presented and introduced in this paper, which realizes online real-time detection of high-density crowd abnormal events in intelligent surveillance system. The detection process of this method includes motion region segmentation, moving blobs extraction and crowd activity map creation. Finally, crowd behavior detection is carried out by using information entropy of crowd activity image. This method effectively solves the problems of target blurring, occlusion and limited expression pixels of target in the process of group behavior recognition in high-density crowded scenes. In addition, unlike the traditional methods using classifier and training, the new method has no dependence on the environment and can realize real-time detection, which is more conducive to the online deployment of intelligent surveillance system.

The proposed method performs experiments and quantitative data analysis on public dataset of UMN. The experimental results show that the method can detect abnormal group behavior in high-density crowd scene without training stage. Compared with other online crowd event detection methods, the results show that this method has better detection effect than the online state-of-the-art technologies.

References

1. Julio Jr., C.S.J., Musse, S.R., Jung, C.R.: Crowd analysis using computer vision techniques. Signal Processing Magazine IEEE **27**(5), 66–77 (2010)
2. Duong, T.V., Bui, H.H., Phung, D.Q., Venkatesh, S.: Activity recognition and abnormality detection with the switching hidden semi-Markov model. In: Proceedings of the IEEE Computer Society Conference on Computer Vision and Pattern Recognition (CVPR), San Diego, CA, USA, 20–25 June 2005, vol. 1, pp. 838–845 (2005)
3. Yamato, J., Ohya, J., Ishii, K.: Recognizing human action in time-sequential images using hidden Markov model. In: Proceedings of the IEEE Computer Society Conference on Computer Vision and Pattern Recognition (CVPR), Champaign, IL, USA, 15–18 June 1992, pp. 379–385 (1992)
4. Brand, M., Oliver, N., Pentland, A.: Coupled hidden Markov models for complex action recognition. In: Proceedings of IEEE Computer Society Conference on Computer Vision and Pattern Recognition (CVPR), San Juan, PR, USA, 17–19 June 1997, pp. 994–999 (1997)
5. Chen, Y., et al.: Abnormal behavior detection by multi-SVM-based Bayesian network. In: International Conference on Information Acquisition 2007
6. Iosifidis, A., Tefas, A., Pitas, I.: Neural representation and learning for multi-view human action recognition. In: 2012 International Joint Conference on Neural Networks (IJCNN). IEEE, pp. 1–6 (2012)
7. Sreenu, G., Durai, M.A.S.: Intelligent video surveillance: a review through deep learning techniques for crowd analysis. J. Big Data **6**(1), 48 (2019)
8. Zhao-Yi, P., Yan-Hui, Z., Yu, Z.: Real-time facial expression recognition based on adaptive canny operator edge detection. In: Second International Conference on Multimedia & Information Technology (2010)
9. Gu, X., Cui, J., Zhu, Q.: Abnormal crowd behavior detection by using the particle entropy. Optik – Int. J. Light Electron. Opt. **125**(14), 3428–3433 (2014)
10. Márquez, J.B.: Activity recognition using a spectral entropy signature. In: ACM Conference on Ubiquitous Computing (2012)
11. Aktaruzzaman, M., Scarabottolo, N., Sassi, R.: Parametric estimation of sample entropy for physical activity recognition. In: International Conference of the IEEE Engineering in Medicine and Biology Society (2015)
12. Dalton, A., Olaighin, G.: A comparison of supervised learning techniques on the task of physical activity recognition. IEEE J. Biomed. Health Inform. **17**(1), 46–52 (2012)

13. University of Minnesota: Unusual crowd activity data set, Detection of Unusual Crowd Activity. http://mha.cs.umn.edu/proj_events.shtml
14. Mehran, R., Oyama, A., Shah, M.: Abnormal crowd behavior detection using social force model. In: 2009 IEEE Computer Society Conference on Computer Vision and Pattern Recognition (CVPR 2009), pp. 20–25 (2009)
15. Cong, Y., Yuan, J., Liu, J.: Sparse reconstruction cost for abnormal event detection. In: Computer Vision and Pattern Recognition. IEEE (2011)
16. Wu, S., Moore, B., Shah, M.: Chaotic invariants of Lagrangian particle trajectories for anomaly detection in crowded scenes. In: IEEE Conference on Computer Vision and Pattern Recognition (CVPR), pp. 2054–2060 (2010)

English Situational Teaching Assisted by Multimedia Network

Yuyan Jia[✉]

Dalian University of Science and Technology, Dalian 116052, Liaoning, China
jiayuyan@hotmail.com

Abstract. Aiming at promoting the efficiency and interesting in English teaching process, the paper studied on the English situational teaching assisted by multimedia network. The application of modern information technology in the development of college English subject-related knowledge content education and teaching activities is an inevitable development trend brought about by the continuous advancement of the development process of social education and teaching and the continuous improvement of the application level of scientific information technology. When the modern network information technology is better applied to the teaching process of college English subject-related knowledge content, the relevant on-the-job instructors must also better play their own educational leaders who can play in the education and teaching classroom. Multimedia modernization teaching tools are the most advantageous teaching tools that relevant in-service instructors can use to create teaching scenarios for students. The experiment result shows the proposed method can improve the overall performance in English teaching.

Keywords: User's feeling · English situational teaching · Multimedia network

1 Introduction

The application of modern network technology in the process of college English teaching is an inevitable trend in the development of teaching under the influence of the universal application of computer Internet technology. The application of modern Internet information technology in the process of college English teaching and education has significant advantages and advantages [1, 2]. The application of the so-called modern network technology means in the teaching process of college English subjects specifically refers to the application of modern multimedia teaching methods in the teaching process of college English subjects. Through the application of modern online education and teaching methods, relevant on-the-job instructors can present the knowledge content related to college English subjects to the students in a more three-dimensional manner as pictures, music, or video clips.

In the process of launching the modern English network teaching activities, the relevant on-the-job teaching teachers will arrange a certain class time for the students to discuss and explore the corresponding English knowledge content in groups. Flexible teaching methods and scientific teaching concepts provide powerful preconditions for the ultimate realization of the personalized teaching of relevant on-the-job instructors

© Springer Nature Singapore Pte Ltd. 2020
M. Atiquzzaman et al. (Eds.): BDCPS 2019, AISC 1117, pp. 1364–1371, 2020.
https://doi.org/10.1007/978-981-15-2568-1_189

and the improvement of students' ability to learn autonomously on English content [3]. The level of mastery of computer operating techniques by in-service instructors can sometimes directly influence or even determine the quality of their own education and teaching.

The application of modern information technology in the development of college English subject-related knowledge content education and teaching activities is an inevitable development trend brought about by the continuous advancement of the development process of social education and teaching and the continuous improvement of the application level of scientific information technology. When the modern network information technology is better applied to the teaching process of college English subject-related knowledge content, the relevant on-the-job instructors must also better play their own educational leaders who can play in the education and teaching classroom. Multimedia modernization teaching tools are the most advantageous teaching tools that relevant in-service instructors can use to create teaching scenarios for students.

Multimedia-assisted teaching tools are the main bridge and link between the campus network and teaching resources and the college English teaching classroom. Through the display and teaching of the rich and vivid network of teaching resources, students can, to a certain extent, help them better grasp and understand the use value of related English knowledge content in daily life. So as to attract students' attention to the knowledge content of English subjects to the maximum extent [4–6]. The better application of modern network technology in the process of college English teaching is an inevitable trend driven by the continuous progress in the development of social education and teaching. Only the relevant education and teaching staff truly understand the importance of the reform of the university education and teaching model, can we establish a more scientific concept of cognition. Under the guidance of the correct concept of knowledge, the relevant on-the-job teachers can adopt more reasonable teaching methods, change the traditional teaching concepts, and finally realize the orderly development of the network teaching model of college English.

2 The Basic Model and Research Design

In the network environment, teaching resources are open, teaching tools are open, and teacher-student exchanges and evaluations are open to each other [7]. The emergence of the Internet has truly provided us with an open teaching environment that has greatly enriched the form of traditional English classrooms. English teaching is for communication, and effective English learning should be an interactive process. The characteristics of the network enable the interactive teaching of English on the Internet, communication between teachers and students, and discussion among students, all of which can be conducted online [8–10]. This type of teaching is supported by the computer network conference system. Both teachers and students complete the learning task through a real-time teaching process, which is a typical simulation of the traditional teaching process.

Interactions between students and teachers include interactions between teachers and individual students, and interactions between teachers and group students. The former is suitable for individual counseling. When students encounter difficulties in learning, they are asked to ask teachers to answer the questions. Teachers use the interactive interface to ask the same questions to many students, discuss and negotiate through the communication space, and finally obtain consensus. With the development of computer network technology, the adult English teaching system can establish a new type of adult English teaching model under the support of computer technology [11]. In the campus, the combination of modern distance education and traditional education can be used to establish the network-assisted teaching system model. Through the analysis of the problems existing in the construction of adult English teaching mode in computer networks, this paper discusses the application of computer network in adult English teaching and aims to provide reference for adult English education in China. The biggest drawback of adult English teaching is short class time. Students cannot master the English knowledge they have learned in a short period of time. They are both busy with work and life, but also need to carry out correspondence learning, which has great obstacles to learning. In addition, it is difficult for teachers to communicate and communicate with students directly [12]. This also causes teachers to be in a position to prepare for each lesson. When teaching, students are taught according to the text. There is basically no communication and interaction between teachers and students. Computer network-assisted teaching can provide teachers with a platform for sharing teaching resources and strengthen communication between teachers and students.

Although many adult education schools have established a computer network teaching platform, due to the influence of funds and technology, the teaching platform does not have a wealth of teaching resources and cannot maximize the function and role of the resource sharing platform. Adult education has a certain disadvantage in competition. Although many adult education schools in China have devoted much effort to education information, deploying computers for English teachers requires teachers to use multimedia as much as possible to conduct education and teaching. However, many teaching resources are concentrated in the hands of teachers and it is difficult to share with teachers and students. Many adult education schools respond to higher levels of inspections and assessments, have a large number of paper-based books, and have relatively few links in computer network teaching resources. Although many adult education schools have established online English teaching platforms, they lack the professional network teaching resources and cannot meet the needs of students' specific operations (see Fig. 1).

3 The Mathematical Basic

Although some adult education schools have established online English teaching platforms, there are fewer servers. The storage space is seriously inadequate and the amount of teaching resources stored is limited, which cannot meet the needs of students' learning. To a certain extent, it limits the progress and development of adult English education mode. Subject to various objective conditions, the utilization of online platform teaching in adult English education is not high. The established online

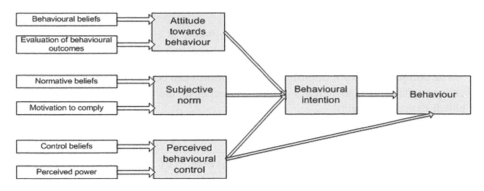

Fig. 1. Virtual environment structure.

teaching platforms often have too simple forms, unreasonable page layouts, and slow updating of teaching resources.

Build a computer network-assisted English teaching system model:

$$O_j^{(1)} = x_j \tag{1}$$

Computer network assisted adult English teaching system:

$$
\begin{aligned}
x_1 &= e(k) - e(k-1) \\
x_2 &= e(k) \\
x_3 &= e(k) - 2e(k-1) + e(k-2)
\end{aligned} \tag{2}
$$

Construct teaching curriculum video resources:

$$net_j^{(2)}(k) = \sum_{j=0}^{M} w_{i,j}^{(2)} O_i^{(1)}; \ O_i^{(2)}(k) = f\left(net_i^{(2)}(k)\right) \tag{3}$$

Teachers should pay attention to the following aspects in the construction of computer network English teaching:

$$net_l^{(3)}(k) = \sum_{i=0}^{Q} w_{li}^3 O_i^{(2)}(k); \ O_l^{(3)}(k) = g\left(net_l^{(3)}(k)\right) \tag{4}$$

$$O_1^{(3)}(k) = K_p; \ O_2^{(3)}(k) = K_i; \ O_3^{(3)}(k) = K_d$$

The computer network English teaching platform can provide teachers and students with an interactive basis:

$$\Delta w_{li}^{(3)}(k) = -\eta \frac{\partial E(k)}{\partial w_{li}^{(3)}} + \alpha \Delta w_{li}^{(3)}(k-1) \tag{5}$$

The computer network English teaching platform can combine English reading and writing organically, and it plays a direct impetus to the improvement of students' English expression ability:

$$\frac{\partial E(k)}{\partial w_{li}^{(3)}} = \frac{\partial E(k)}{\partial y(k)} \frac{\partial y(k)}{\partial u(k)} \frac{\partial u(k)}{\partial O_1^{(3)}(k)} \frac{\partial O_1^{(3)}(k)}{\partial net_1^{(3)}(k)} \frac{\partial net_1^{(3)}(k)}{\partial w_{li}^{(3)}(k)} \tag{6}$$

The construction of adult English teaching system model under computer network is a new type of adult English teaching mode:

$$\frac{\partial net_1^{(3)}(k)}{\partial w_{li}^{(3)}(k)} = O_i^{(2)}(k) \tag{7}$$

Virtual design technology is one of the most important technologies in the digital era:

$$\begin{aligned} \frac{\partial u(k)}{\partial O_1^{(3)}(k)} &= e(k) - e(k-1) = x_1 \\ \frac{\partial u(k)}{\partial O_2^{(3)}(k)} &= e(k) = x_2; \quad \frac{\partial u(k)}{\partial O_3^{(3)}(k)} = e(k) - 2e(k-1) + e(k-2) = x_3 \end{aligned} \tag{8}$$

Virtual design technology makes full use of analog simulation technology, but it is different from general simulation technology. It has features of virtual reality. Conceptual design is a preliminary stage of design. Its purpose is to obtain enough information about the style and shape of the product. It is an important stage of the design process, because 60% to 70% of the product cost is determined by this stage. In the concept design stage, people need to study a large number of feasible design schemes to determine the best design for economic efficiency (see Fig. 2).

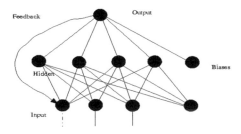

Fig. 2. Feedback neural network.

The virtual concept design uses virtual reality technology to provide designers with input methods based on speech recognition and gesture tracking. The designer can easily manipulate the products and parts in a three-dimensional virtual environment to perform various shape modeling and modification.

4   The Experiment Result and Discussion

The appearance of virtual assembly design technology brings hope to solve this problem completely. It can help designers to discover the assembly defects in the design in time. Because the current CAT systems and virtual reality systems do not integrate well, they hinder the application of virtual assembly technology. Although the virtual assembly technology has not been fully applied to product analysis and evaluation, the reliability and application value of this technology have been widely recognized (see Table 1).

Table 1. The interactive analysis of teaching process

Month	Model						Structure			
	(1)	(2)	(3)	(1)	(2)	(3)	(1)	(2)	(3)	(4)
−11	1.006	.992	16.5	1.007	.992	20.4	1.006	.989	24.1	1.000
−10	1.014	.983	17.3	1.015	.982	20.2	1.015	.972	73.4	.999
−9	1.017	.977	7.9	1.017	.977	3.7	1.018	.965	20.4	.998
−8	1.021	.971	9.5	1.022	.971	12.0	1.022	.956	9.1	.998
−7	1.026	.960	21.8	1.027	.960	27.1	1.024	.946	9.0	.995
−6	1.033	.949	42.9	1.034	.948	37.6	1.027	.937	19.4	.993
−5	1.038	.941	17.9	1.039	.941	21.3	1.032	.925	21.0	.992
−4	1.050	.930	40.0	1.050	.930	39.5	1.041	.912	41.5	.993
−3	1.059	.924	35.3	1.060	.922	33.9	1.049	.903	37.2	.995
−2	1.057	.921	1.4	1.058	.919	1.8	1.045	.903	0.1	.992
−1	1.060	.914	8.2	1.062	.912	8.2	1.046	.896	5.7	.991
0	1.071	.907	28.0	1.073	.905	28.9	1.056	.887	35.8	.993
1	1.075	.901	6.4	1.076	.899	5.5	1.057	.882	9.4	.992
2	1.076	.899	2.7	1.078	.897	1.9	1.059	.878	8.1	.992
3	1.078	.896	0.6	1.079	.895	1.2	1.059	.876	0.1	.991
4	1.078	.893	0.1	1.079	.892	0.1	1.057	.876	1.2	.990
5	1.075	.893	0.7	1.077	.891	0.1	1.055	.876	0.6	.989
6	1.072	.892	0.0	1.074	.889	0.2	1.051	.877	0.1	.987

In the case that the hardware is not powerful enough, to ensure a sufficient image refresh rate, CAT data needs to be converted to a polygon format and it is likely that further reductions will need to be implemented. Limited to the current computer capabilities, in order to complete the ideal interaction in the virtual environment, the system must compromise between the depicted realism and the necessary refresh rate. Such trade-offs should take into account the specific application environment of the system. Proper texture rendering is beneficial to enhance the realism of the graph and does not increase the number of triangles (see Fig. 3). Virtual design systems can be roughly divided into two categories: enhanced visualization systems and virtual reality-based CAT systems. With the promotion of computer technology and virtual reality technology, virtual design technology will rapidly develop. The application of this technology can not only improve design efficiency, shorten the product development and verification cycle, improve product quality and reduce costs, but also help to sprout new design ideas. Virtual design technology will bring us greater technological change.

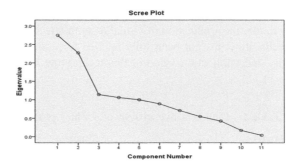

Fig. 3. Experiment result.

The results of virtual environment art design should not only possess the aesthetic beauty and color beauty of general environmental art, but also highlight the consideration of dynamic beauty and participation in beauty. In order to obtain aesthetic feeling, there are many ways of artistic design. One way is to carry out artistic design on the basis of imitation, such as imitating natural objects or industrial products and creating art. This method is suitable for design tasks such as landscape tourism, product testing, and skill training. The other is to use art to create art. Composition art breaks the traditional art of figurative depiction, using the rules of beauty to design art. It is a modeling concept that combines several units of different forms into a new unit, or splits a unit into several units and then combines them. This approach is more suitable for virtual game design or future environmental design (see Fig. 4).

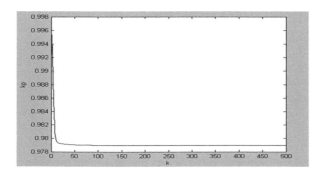

Fig. 4. Conceptual virtual environment.

The development of information technology has brought both opportunities and challenges to English education in Chinese universities. The use of multimedia technology to teach English to college students in China not only improves students' learning efficiency, but also stimulates college students' interest in learning English and develops students' autonomous learning ability. Then it may improve college students' English performance, and ultimately achieve the purpose of the society to cultivate high-quality talent. When colleges and universities conduct English education for college students, they must keep up with the times and constantly update their teaching concepts and teaching methodology.

References

1. Chan, J.Y.H.: Gender and attitudes towards English varieties: implications for teaching English as a global language. System **76**, 62–79 (2018)
2. Dybowski, C., Sehner, S., Harendza, S.: Influence of motivation, self-efficacy and situational factors on the teaching quality of clinical educators. BMC Med. Educ. **17**(1), 84 (2017)
3. Huijgen, T., Grift, W.V.D., Boxtel, C.V., et al.: Teaching historical contextualization: the construction of a reliable observation instrument. Eur. J. Psychol. Educ. **32**(2), 1–23 (2016)
4. Garn, A.C.: Multidimensional measurement of situational interest in physical education: application of bifactor exploratory structural equation modeling. J. Teach. Phys. Educ. **36**(3), 323–339 (2017)
5. İnal, D., Özdemir, E., Kıray, G., et al.: Review of doctoral research in English language teaching and learning in Turkey (2009-2013). Lang. Teach. **49**(3), 390–410 (2016)
6. Choe, H.: Identity formation of Filipino ESL teachers teaching Korean students in the Philippines. Engl. Today **32**(1), 5–11 (2016)
7. Durksen, T.L., Klassen, R.M.: The development of a situational judgement test of personal attributes for quality teaching in rural and remote Australia. Aust. Educ. Res. **45**(3), 1–22 (2017)
8. Roure, C., Pasco, D.: The impact of learning task design on students' situational interest in physical education. J. Teach. Phys. Educ. **37**(1), 1–29 (2017)
9. Short, K.G.: The 2015 NCTE presidential address: advocacy as capacity building: creating a movement through collaborative inquiry. Res. Teach. Engl. **50**, 349 (2016)
10. Sato, T., Miller, R., Delk, D.: Secondary physical educators' positioning of teaching english language learners at urban schools. Urban Education (2018)
11. Palmer, D.H., Dixon, J., Archer, J.: Identifying underlying causes of situational interest in a science course for preservice elementary teachers. Sci. Educ. **100**(6), 1039–1061 (2016)
12. Loukomies, A., Juuti, K., Lavonen, J.: Investigating situational interest in primary science lessons. Int. J. Sci. Educ. **37**(18), 1–23 (2016)

Self-balancing Car Control Based on Active Disturbance Rejection Control (ADRC)

Haibo Yu[1], Ling Zhang[2(✉)], Shengyong Yang[2], Liping Zeng[3], and Gangsheng Li[4]

[1] Fundamental Computer Department, Ocean University of China, Qingdao 266100, Shandong, China
[2] College of Engineering, Ocean University of China, Qingdao 266100, Shandong, China
zljoan@163.com
[3] College of Electrical and Electronic Engineering, Guangdong Polytechnic College, Zhaoqing 526000, Guangdong, China
[4] Department of Education, Ocean University of China, Qingdao 266100, Shandong, China

Abstract. The Active Disturbance Rejection Control (ADRC) algorithm is becoming more and more popular in the field of controller design. Compared with the traditional PID control algorithm, ADRC algorithm can achieve better performance. ADRC consists of a tracking differentiator, a feedback control law, and an extended state observer. For the self-balancing two-wheeled car control, it can be taken as a single-stage inverted pendulum model. Based on the formulated mathematical differential equation model, we apply the ADRC algorithm to control the self-balancing car. Furthermore, to show the effectiveness of ADRC, we compare its performance with PID controller by Simulink simulation tool of MATLAB. Simulation results show that the ADRC algorithm performs better than PID in smaller overshoot, lower steady-state error and faster convergence rate.

Keywords: ADRC · Self-balancing car control · PID

1 Introduction

For the control of self-balancing two-wheeled smart car, the PID control algorithm was widely used [1–4] due to its advantages of simple principle, easy realization, independent parameters and simple tuning method. But the PID control has its shortcomings, which can be summarized as the following four aspects: (1) The way of error generation is not very reasonable. (2) Differential of error has not been well handled. (3) Feedback also has some shortcomings. (4) Defects of linear combination. To overcome these defects of PID, the active disturbance rejection control (ADRC) can provide corresponding solutions, avoid some defects perfectly, and guarantee the control performance of the system [5–10].

ADRC is composed of PD (transition process), feedback control law (non-linear combination) and extended state observer (ESO), which can improve the control effect that PID cannot achieve. In this paper, the PID controller and the ADRC are built by

M. Atiquzzaman et al. (Eds.): BDCPS 2019, AISC 1117, pp. 1372–1379, 2020.
https://doi.org/10.1007/978-981-15-2568-1_190

Simulink simulation tool of MATLAB to control the self-balancing car model respectively. By comparing the simulation results, it can be seen that the performance of ADRC is better than that of PID controller.

2 Self-balancing Car Model

Balance control of the self-balancing car can be simplified to a single inverted pendulum model, as shown in Fig. 1. In Fig. 1, θ is the self-balancing car inclination; L is the center of gravity height; m is the mass; a_1 is the acceleration of the self-balancing car; a_2 is the angular acceleration caused by external force interference. The motion model is written as

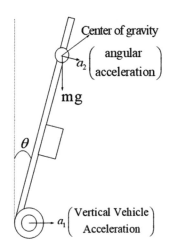

Fig. 1. Simplified model of the self-balancing car (single inverted pendulum model)

$$mL\frac{d^2\theta}{dt^2} = mg\,\sin\theta - ma_1\,\cos\theta + mLa_2 \tag{1}$$

When θ approaches zero, $\sin\theta \sim \theta$, then we have

$$L\frac{d^2\theta}{dt^2} = g\theta - a_1 + La_2 \tag{2}$$

When $a_1 = 0$, Eq. (2) can be rewritten as follows

$$L\frac{d^2\theta}{dt^2} = g\theta + La_2 \tag{3}$$

The corresponding input-output transfer function can be obtaind

$$H_{(s)} = \frac{\theta_{(s)}}{X_{(s)}} = \frac{1}{s^2 - \frac{g}{L}} \qquad (4)$$

In order to balance the vehicle, the following conditions need to be satisfied.

(1) The inclination angle and angular velocity of the vertical vehicle can be measured.
(2) The acceleration of wheels can be controlled.

3 Active Disturbance Rejection Controller

ADRC consists of the following parts: the tracking differentiator (arranging transition process), the feedback control law (non-linear combination), and the extended state observer (ESO). Its structure is shown in Fig. 2.

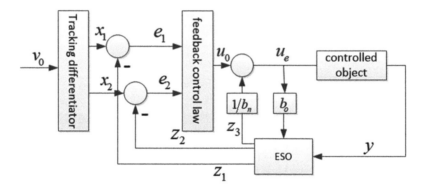

Fig. 2. Basic structure of ADRC

ADRC uses the error and the differential of error as the inputs, and then an ESO is designed to avoid the instability of the system caused by discontinuous differential variation of signals. For ADRC algorithm, jump caused by differential can be avoided and better control performance can be achieved. The corresponding second-order system model is

$$\begin{cases} \dot{x}_1 = x_1 \\ \dot{x}_2 = f(x_1, x_2, w(t), t) + bu \\ y = x_1 \end{cases} \qquad (5)$$

where $f(x_1, x_2, w(t), t)$ is completely unknown. The typical second-order ADRC algorithms are as follows.

Step 1: arranging transition process:

$$\begin{cases} \dot{v}_1 = v_1 \\ \dot{v}_2 = \text{fhan}(v_1 - v_1 v_2, r_0, h) \end{cases} \tag{6}$$

Step 2: extended state observer (ESO):

$$\begin{cases} e = Z_1 - y \\ fe_1 = \text{fal}(e, 0.5, h) \\ fe_2 = \text{fal}(e, 0.25, h) \\ \dot{Z}_1 = (Z_2 - \beta_{01} e) \\ \dot{Z}_2 = (Z_3 - \beta_{02} fe_1 + b_0 u) \\ \dot{Z}_3 = (-\beta_{03} fe_2) \end{cases} \tag{7}$$

Step 3: feedback control law (nonlinear combination):

$$\begin{cases} e_1 = v_1 - Z_1 \\ e_2 = v_2 - Z_2 \\ u_0 = \text{fhan}(e_1, ce_2, r, h_1) \end{cases} \tag{8}$$

Step 4: Disturbance compensation from control variable:

$$u = \frac{(u_0 - Z_3)}{b_0} \tag{9}$$

PD control will show the error and its first derivative. The output of ESO is the error, first order derivative of error, and the system output. The relationship between the three variables is approximately linear, and there is no large first-order derivative at the step input. Hence the ADRC has better tracking performance.

4   Simulation of Self-balancing Car Control

4.1   Simulink Simulation of Self-balancing Car Model

Using MATLAB Simulink tool to build a controlled self-balancing car model, the Simulink model of the controlled object is shown in Fig. 3.

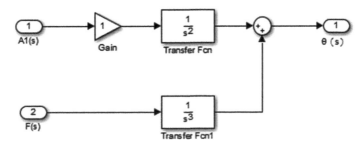

Fig. 3. The simulink model of the controlled object

4.2   Simulink Simulation of PID Control for Self-balancing Car

The controlled self-balancing vehicle model is added with a step input. A sinusoidal wave is added to the interference signal. The amplitude of the sinusoidal wave is set to 1, and the frequency is carried out according to the frequency given by the system. The simulation of the PID design is shown in Fig. 4. The adjustment of PID parameters is shown in Fig. 5.

The output waveform of PID controller is shown in Fig. 6. From the simulation result, we can see that when the amplitude of the interference signal is close to that of the input signal, there will be no stable output and have a large overshoot.

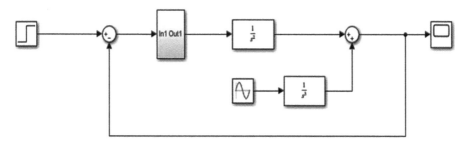

Fig. 4. PID simulation

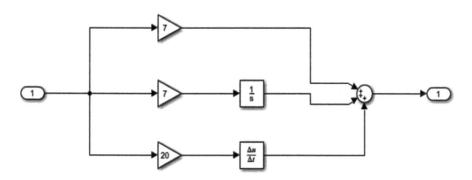

Fig. 5. PID parameters

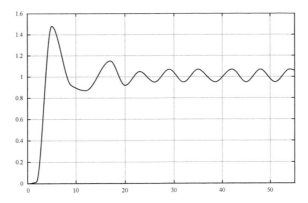

Fig. 6. PID output waveform

4.3 Simulation of ADRC for Self-balancing Car

A sinusoidal wave is added to the interference signal. The amplitude of the sinusoidal wave is set to 1, and the frequency is carried out according to the frequency given by the system. In our experiment, the linear ADRC (LADRC) is used. The simulation implementation of LADRC is shown in Fig. 7. The extended state observer (ESO) is shown in Fig. 8.

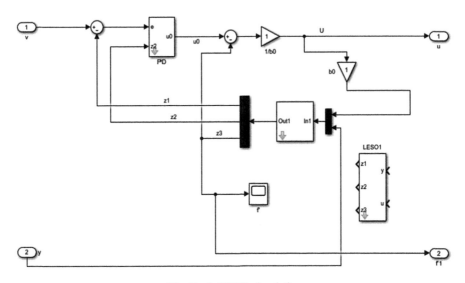

Fig. 7. LADRC simulation

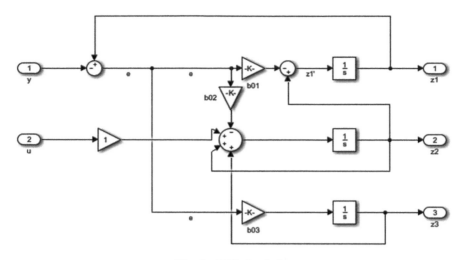

Fig. 8. ESO simulation

The ADRC has two parameters to be tuned, i.e. ω_1 and ω_2, which are set to 100 and 20 respectively. The output signal is shown in Fig. 9. Figure 10 shows the interference tracking result.

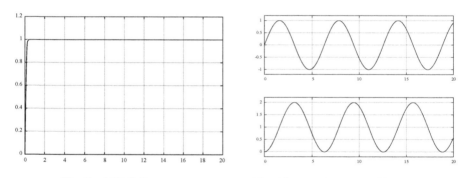

Fig. 9. ADRC Output   **Fig. 10.** Interference tracking for ADRC

For the active disturbance rejection controller, it has no overshoot, no fluctuation, and stabilizes quickly. Furthermore, it has the ability to track the interference, effectively. Depending on the position of the signal, there will be a certain delay due to the inertia, but the change of the interference can be clearly seen and handled conveniently.

5   Conclusion

ADRC is an effective method for controlling the self-balancing car, which performs better than PID in many aspects based on the simulation experiments. By comparison of control performance between PID and ADRC, we can see that ADRC has better performance in overshoot, adjustment time, steady-state error and interference tracking ability.

Basically there is no overshoot of ADRC, while for PID the overshoot is obvious. The overshoot will cause some negative effects on system. The adjustment time of PID is much longer than that of ADRC. PID control generally has a steady-state error, and ADRC has almost no steady-state error. ADRC has the ability to track interference, while PID doesn't. This is important for certain control systems, which can monitor and feedback the real-time variation of the error. Hence, ADRC algorithm is more effective in controlling the self-balancing car.

Acknowledgements. This project is partly sponsored by the National Natural Science Foundation of China (No. 51979256, No. 61603361 and No. 41506114).

References

1. Hou, Z.: Design of intelligent balancing vehicle control system based on Arduino. Int. J. Plant Eng. Manag. **23**(3), 38–41 (2018)
2. Zheng, B.G., et al.: LQR + PID control and implementation of two-wheeled self-balancing robot. Appl. Mech. Mater. **590**, 399–406 (2014)
3. Alqudah, M., Abdelfattah, M., Boiko, I., Alhammadi, K.: Dynamic modeling and control design for a self-balancing two-wheel chair, pp. 1–4 (2016). https://doi.org/10.1109/icedsa.2016.7818556
4. Zimit, A., Yap, H.J., Hamza, M., Siradjuddin, I., Hendrik, B., Herawan, T.: Modelling and experimental analysis two-wheeled self balance robot using PID controller (2018). https://doi.org/10.1007/978-3-319-95165-2_48
5. Gao, Z.: Scaling and bandwidth-parameterization based controller tuning. In: Proceedings of the American Control Conference, pp. 4989–4996 (2003)
6. Huang, Y., Xue, W.: Active disturbance rejection control: methodology and theoretical analysis. ISA Trans. **53**(4), 963–976 (2014)
7. Shao, S., Gao, Z.: On the conditions of exponential stability in active disturbance rejection control based on singular perturbation analysis. Int. J. Control 1–21 (2016)
8. Qiu, X.B., Dou, L.H., Shan, D.S., et al.: Design of active disturbance rejection controller for electro-optical tracking servo system. Opt. Precis. Eng. **18**(1), 220–226 (2010)
9. Chen, F., Xiong, H., Fu, J.: The control and simulation for the ADRC of USV. In: IEEE Chinese Automation Congress (2016)
10. Lin, S.C., Tsai, C.C., Huang, H.C.: Adaptive robust self-balancing and steering of a two-wheeled human transportation vehicle. J. Intell. Rob. Syst. **62**(1), 103–123 (2011)

Teachers' Personal Homepage Construction of Medical University Based on the Website Group

Chunlan Zhao, Hongquan Song, and Qinglian Yu[✉]

Qiqihar Medical University, Qiqihar 161006, China
yuqinglian@qmu.edu.cn

Abstract. The teachers in medical colleges lack the knowledge of dynamic webpage programming, so they cannot design complicated personal homepage by themselves, due to the heavy teaching tasks, even public course teachers majoring in computer science, they also cannot spend more energy designing personal homepage. Therefore, there is great practical significance to establishing a good operation interface personal homepage system which can customize the page content independently for medical college teachers. In this paper, we proposed how to establish a simple operation and rich content customization homepage system based on the website group platform for the professional teachers. The studies show that with this system, all professional teachers can quickly establish a personalized personal homepage according to their own needs, the basic information, teaching achievements, scientific research achievements and other important information could be shown in the personal homepage, for society and students, they could know the teachers and the level of college better. For teachers, with the rapid development of information technology and increasing demand of construction of personal homepage, it will play more important role to construct teachers' personal homepage.

Keywords: Medical university teachers · Website group · Personal homepage · Personalized customization · Informatization

1 Introduction

With the rapid development of computer network technology and upgrading of electronic products, the traditional personal teachers' information display board has been unable to meet the needs of the age. The personal homepage of college teachers is mainly used to display the personal elegant demeanour of teachers and provide an interactive interface between teachers and students. It is not only a window for the outside world to know the teachers in college, but also a platform for academic exchanges [1, 2]. In the universities of developed countries, such as Europe and America, almost every teacher has the personalized personal homepage [3, 4]. According to the survey, at present, only a few college in China have internal personal homepage, which is not good for the teachers' academic development [5–7]. In order to build a good image, college teachers need to improve their self culture, but also need to enhance online discourse power. For the teachers' personal homepage, on the one

© Springer Nature Singapore Pte Ltd. 2020
M. Atiquzzaman et al. (Eds.): BDCPS 2019, AISC 1117, pp. 1380–1385, 2020.
https://doi.org/10.1007/978-981-15-2568-1_191

hand, it plays the role of displaying personal academic achievements externally, on the other hand, it could strengthen the communication between teachers and students internally [8]. Based on the website group platform, in this paper, we design a user-centered personal homepage display system which provides communication for teachers and students [9].

The personal homepage designed in this paper is based on Website group platform, aim to provide a convenient and self-help way to build the personal homepage for colleges teachers. The teachers don't need to master professional knowledge of web page building, according to the step provided by the system only, the teachers themselves can finish a series of website construction process, including the website building, preview, publish, maintenance etc. Through choosing a different page templates, each teacher can create a personalized homepage. The teachers have more independent display space, at the same time attract students to participate, jointly create a good campus learning atmosphere, better display the elegant demeanor of university teachers.

2   Introduction of Website Group Management Platform

Website group management platform (WebPlus) is a platform tool to build information portal websites. It is convenient to integrate structured and unstructured information data and various application systems, show the website portal information. With the background functional components, the platform can display personal information flexibly, build personalized personal homepage for teachers. The teachers' basic information, teaching achievements, scientific research achievements and other important information are displayed through the personal homepage, so that the society and students can better understand teachers and the faculty of the university. Teacher personal homepage system is not only a platform to show personal elegant demeanour, but also an interactive platform for teachers and students.

2.1   Visual Background Editing

WebPlus system is based on B/S mode, [10–12] adopts J2EE architecture, and supports mainstream commercial database platform. B/S architecture is the trend of current system development. The adoption of B/S architecture enables more processing tasks to be placed on the server, reducing the pressure of the client. At the same time, users only need a browser to access the system, no additional client programs need to be installed. The background editing system of webPlus system requires no professional programming knowledge. Uploading information is as simple and convenient as using word. Background windows can upload various formats documents directly, including video files [13].

3   Personal Homepage System Designed Based on WebPlus

Designing the teachers' personal homepage, should be according to the actual work of teachers in medical college, the needs of students, tourists, teachers of college and other users, display teachers' personal basic information, scientific research achievements, teaching courses on demand and other functions. On the one hand, the teachers' personal homepage are required to fully display the characteristics of teachers' personalized style. On the other hand, teachers' personal homepage is required to be convenient to use, upgrade and maintain, operate simply. Whether or not you have a programming background, you can choose the front page template and information presentation mode according to your preference, design personalized personal homepage. Each function module of the system is shown in Fig. 1.

3.1   Introduction of the Foreground Function Modules

The teachers' personal homepage mainly show the basic personal information of teachers, including gender, age, title, education background, etc., and also show the personal teaching situation, teaching courses, teaching courseware, courses video, the teachers' scientific research achievements, teaching research projects, papers and works, patent information, etc. Teachers' personal homepage should also include interactive modules to facilitating teacher-student interaction and peer communication.

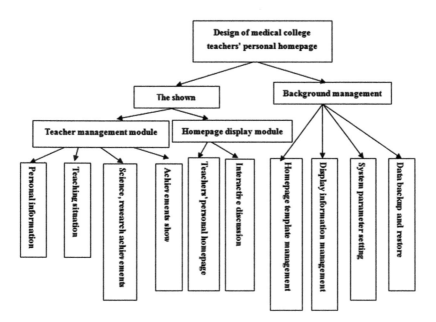

Fig. 1. System function module

3.2   Analysis of Background Function Modules

Teachers of college can upload their basic personal information in the background of the homepage system, and can also implemente the cross-system call by clicking load. Teachers can choose the page template according to their own preferences, avoid single contents of everyone's homepage, and achieve personalized page customization. Each college departments should manage the teachers' personal homepage of the department uniformly, check the information released by teachers and prohibit bad information, Backup data regularly in the background of the system, when the system failure can restore access quickly.

3.3   Teacher Personal Homepage Design

With the rapid development of information technology, the colleges' teaching is increasingly relying on information platforms. The informationization construction of colleges promote teaching quality and construction of campus culture. The teacher personal homepage system is also an important part to show the informationization construction of the colleges. The teacher's personal homepage system can not only display the teacher's elegant demeanour, publicize the teacher's image, but also promote the communication between teachers and students with more flexible way, so as to strengthen the mutual supervision between teachers and students, promote the teaching quality.

3.3.1   Information Collection of Personal Homepage

Teachers' personal homepage mainly displays the teachers' elegant demeanour, which mainly includes personal introduction, professor subject, educational experience, scientific research projects, papers, patents, editing books, etc. The information are collected in the form and displayed on the front page after checking. The teacher's personal information acquisition module is shown in Fig. 2.

3.3.2   Function Modules of Personal Homepage

On the one hand, the aim of designing the teachers' personal homepage is to show teachers' personal elegant demeanour and the campus landscape, on the other hand, is to strengthen the interaction between teachers and students, the mutual understanding among teachers, improve the teachers' strength.

3.3.3   Management and Maintenance of Personal Homepage

Because of friendly user interface of the teacher's personal homepage system, without the professional knowledge, it is easy for teachers to completing a series of website construction, such as website building, preview, releasing and so on. There is more independent display space for the teachers. The system is easy to expandable, which is convenient for the follow-up maintenance and upgrade of system functions.

In order to fully guarantee the information security of teachers' homepage system, WAF security devices are deployed in front of the Web server in the network layer, and an active defense mode is adopted to protect and monitor the server cluster of teachers' homepage system through HTTP intervention [14]. The department of the teacher is

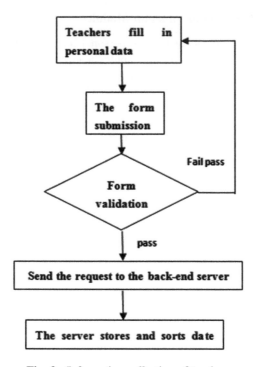

Fig. 2. Information collection of teachers

responsible for checking and maintaining the backup information released by the teacher, so as to preventing bad information of the personal homepage. Once security accident occurs, there will be the bad impact on the reputation of the teacher and the university [15]. In order to avoiding paralyzing of the teacher's personal homepage system because of server problems. The system can backup/recovery, which can backup the data and applications in the station group system, so that the system can quickly resume normal operation when it encounters security risks such as computer crimes, computer viruses and human operation errors [16].

4   Conclusion

With the rapid development of information technology, the Internet has been integrated into students' daily life, Students are accustomed to know the outside world through the Internet. Compared with the form of poster publicity, it is easier for students to understand and accept teachers' style through the system of teachers' personal homepage, which can play a good effect on promoting the communication between teachers and students and the construction of university spirits. Teachers can easily share what they have seen and felt through their personal homepage, and timely understand students' feedback, It is helpful for teachers to carry out teaching work and promote teachers' scientific research guidance to students. At the same time, there is also a platform for teachers to understand and learn from each other and improve the teaching work. The teachers' personal homepage displays teachers' elegant demeanor in a flexible way, which is beneficial for colleges and universities to publicize university spirit to the society. At the same time accept social supervision, get good social feedback.

Acknowledgement. This work was supported by guidance project of the science and technology bureau of Qiqihar, Heilongjiang Province under grant No. GYZD-2017010.

References

1. Si, J., Gao, Y., Yang, J.: Personal homepage construction exploration based on the website of university teachers'. J. Inform. Commun. (9), 283–284 (2018)
2. Jin, B., Xu, F., Wu, D.: Design of customized homepage system for university teachers. Microcomput. Appl. **32**(21), 78–80 (2013)
3. Wang, S., Zhang, G., Liu, J.: Design of an expandable website platform for quality course cluster. In: Proceedings of ICYCS 2008, Zhangjiajie, Hunan, China, pp. 2588–2591 (2008)
4. Yao, Y., Yang, W., Li, Y., Gao, F.: The design and development of computer network quality course website. In: Proceedings of ICETC 2010, Shanghai, China, 22 June 2010, pp. 413–1416 (2010)
5. Zhao, D.: Teacher file management system based on B/S model. Jilin University, Jilin (2009)
6. Liu, J.: Design and implementation teacher blog system based on PHP. Ocean University of China, Qingdao (2008)
7. Junhua, Li: Research and development of college teachers' personal homepage system based on web text mining. J. Dali Univ. **10**(4), 26–29 (2011)
8. Zhu, Y.: Research on media image of teachers in sina.com universities. Jiangxi Normal University, Nanchang (2017)
9. Zhang, W., Liu, J., Yang, Y.: Exploration and practice of personal homepage system for college teachers. China Educ. Netw. (11), 64–65 (2017)
10. Rongge, M.D.I.: A three-tier design model for ASP. Net-based web applications. Microcomput. Appl. **3**, 007 (2002)
11. Himmel, M.A., Hoffman, R.D., Mall, M.: System and method for preventing duplicate transactions in an Internet browser/Internet server environment. U.S. Patent 6,237,035. 22 May 2001
12. Zhao, X.: Design and implementation of network teaching system based on B/S structure. Comput. Mod. (2), 8–10 (2010)
13. Yu, Q., Qi, T.: Design and realization of online teaching platform based on WebPlus in medical university. In: 2018 3rd International Conference on Intelligent Computing and Cognitive Infornatics (ICICCI 2018), vol. 25, pp. 77–80 (2019)
14. Zeng, X.: Design and application of college information security based on firewall and waf security equipment. Electron. Technol. Softw. Eng. (9), 201–202 (2018)
15. Liu, Z., Zhang, X., Lu, X., Chen, P., Zhang, Y., Wang, Y.: Research on the construction and application of website clusters in Chinese universities. China Educ. Inform. (19), 6–9 (2017)
16. Li, K., Mao, W., Zhang, Y., Ding, L.: Construction of personal homepage system of college teachers based on B/S structure. China Educ. Inform. (01), 93–96 (2019)

Mechanism Analysis of Chaotic Motion of Direct-Drive Permanent Magnet Synchronous Wind Turbine

Li Yang[1,2(✉)], Tianmin Huang[1], Lei Hu[2], and Shuhua Wang[2]

[1] School of Electrical Engineering, Southwest Jiaotong University,
Chengdu 610031, Sichuan, People's Republic of China
185059568@qq.com
[2] Sichuan Engineering Technical College,
Deyang 618000, Sichuan, People's Republic of China

Abstract. In order to study direct-drive permanent magnet synchronous wind turbine (D-PMSG), affine transformation and time transformation are used to transform it into dimensionless Lorenz chaotic model. Based on Lorenz chaotic model, The Jacobian matrix eigenvalues of equilibrium points under different inputs are analyzed, and the conditions of Hpof and stability are discussed through Lyapunov stability theory. The simulation results show that D-PMSG will produce unstable Hopf bifurcation motion while different inputs or parameters change. When the wind speed changes, the D-PMSG unit will enters the chaotic state, which seriously affects its stable operation.

Keywords: D-PMSG · Lorenz model · Lyapunov stability · Chaos

1 Introduction

Compared with doubly fed machines, Permanent Magnet Synchronous Machine (PMSM) has the advantages of simple structure, high efficiency, no need for rotor excitation, and no need absorb reactive from the power grid. In recent years, PMSM has been widely used in wind turbines. Direct-drive Permanent Magnet Synchronous Generator (D-PMSG) [1–3] has become an important development direction in the field of wind power generation due to the elimination of gearboxes, low transmission loss, low maintenance cost and high reliability. However, when the parameters are in a specific area, the PMSM will have intermittent oscillation of torque and rotating speed, and chaotic phenomena such as unstable control performance [4, 5]. Therefore, many scholars at home and abroad have made many outstanding contributions to the study of the chaotic characteristics of PMSM [6–9]. D-PMSG also exhibits chaotic motion like PMSM under certain parameters or working conditions, so that intermittent oscillation of torque, rotating speed and output power occurs during power generation, which causes the output power of the D-PMSG to generate large oscillations and fluctuations, and in turn has a greater impact on the power grid, but so far there has been relatively little research on the analysis and control of chaotic motion in D-PMSG.

© Springer Nature Singapore Pte Ltd. 2020
M. Atiquzzaman et al. (Eds.): BDCPS 2019, AISC 1117, pp. 1386–1393, 2020.
https://doi.org/10.1007/978-981-15-2568-1_192

In order to study the chaotic characteristics of the D-PMSG system, the mathematical model was transformed into a dimensionless Lorenz mathematical model through linear affine transformation and time scale transformation. The Lyapunov stability theory was used to study the conditions of D-PMSG generating Hpof bifurcation and the process leading to chaotic motion under three different input conditions. The existence of chaotic motion of D-PMSG under certain condition theoretically has been confirmed. The simulation shows that the D-PMSG enters the chaotic motion state when certain parameters change or the wind speed changes under different inputs.

2   D-PMSG Mathematical Model

2.1   Wind Turbine Model

From the aerodynamics of the wind wheel, the power captured by the wind wheel can be written as follows [10]:

$$P_r = \frac{1}{2}\rho A C_p(\lambda, \beta)v^3 \tag{1}$$

Where P_r is the wind turbine output power (kw), ρ is the air density (kg/m^3), A is the wind wheel area (m^2), v is the wind speed (m/s) and $C_p = C_p(\lambda, \beta)$ is the power coefficient which is the Nonlinear function the tip speed ratio (λ) and the pitch angle (β).

The relationship between wind wheel torque and wind speed and rotating speed is given as follows:

$$T_r = \frac{P_r}{\omega_r} = \frac{1}{2}\rho A C_p(\lambda, \beta)\frac{v^3}{\omega_r} = \frac{1}{2}\rho \pi r^3 v^2 \frac{C_p(\lambda, \beta)}{\lambda} = \frac{1}{2}\rho A r v^2 C_q(\lambda, \beta) \tag{2}$$

$$C_q(\lambda, \beta) = \frac{C_p(\lambda, \beta)}{\lambda} \tag{3}$$

Where $C_q = C_q(\lambda, \beta)$ is the torque coefficient, r is the blade radius(m) and T_r is the aerodynamic torque of wind turbine.

2.2   D-PMSG Mathematical Model

Assuming that there is no motor core saturation, the inductance and mutual inductance of each winding are linear; ignore eddy current and hysteresis loss in the motor; the motor back electromotive force is sinusoidal; there is no damper winding on the rotor, and the permanent magnet has no damping effect. Then under the dq rotating shaft system, the mathematical model of PMSM [11] is:

(1) Voltage equation

$$u_d = R_s i_d + \frac{d\varphi_d}{dt} - \omega_e \varphi_q$$
$$u_q = R_s i_q + \frac{d\varphi_q}{dt} + \omega_e \varphi_d \tag{4}$$

(2) Flux linkage equation

$$\varphi_d = L_d i_d + \psi_f$$
$$\varphi_q = L_q i_q \tag{5}$$

(3) Electromagnetic torque equation

$$T_e = \frac{3}{2} n_p \psi_f i_q \tag{6}$$

(4) Equation of Mechanical Motion

$$J_{eq} \frac{d\omega_g}{dt} = (T_e - T_L - B_m \omega_g) \tag{7}$$

Taking the stator d and q axis currents i_d, i_q and ω_g as state variables, a permanent magnet synchronous motor model with a surface-mount uniform air gap (Ld = Lq = L) is as follows:

$$\frac{d i_d}{dt} = -\frac{R_s}{L} i_d + n_p i_q \omega_g + \frac{u_d}{L}$$
$$\frac{d i_q}{dt} = -\frac{R_s}{L} i_q - n_p i_d \omega_g - \frac{n_p \psi_f}{L} \omega_g + \frac{u_q}{L} \tag{8}$$
$$\frac{d\omega_g}{dt} = \frac{3 n_p \psi_f}{2 J_{eq}} i_q - \frac{B_m}{J_{eq}} \omega_g - \frac{T_r}{J_{eq}}$$

Where $u_x, \varphi_x, i_x, L_x -x \left(_{x=d,q} \right)$ is Motor terminal voltage, stator flux, stator current and synchronous inductance; T_e is Electromagnetic Torque; R_s is Stator Winding Resistance; ω_e is Angular Speed $\omega_e = n_p \omega_g$; n_p is Pole Logarithm; ω_g is Rotor Angular Speed of Generator, $\omega_g \approx \omega_r$; J_{eq} is Equivalent Rotating Inertia; ψ_0 is flux linkage; B_m is Turning Viscosity Coefficient of Generator; $T_L = T_r$ is Load Torque.

3 Chaotic Characteristics of D-PMSG

Use linear affine transformation and time scale transformation, $x = \lambda \tilde{x}$, $t = \tau \tilde{t}$, among them: $x = [i_d, i_q, \omega_g]^T$, $\tilde{x} = [\tilde{i}_d, \tilde{i}_q, \tilde{\omega}_g]^T$, $\lambda = [\lambda_1 \quad 0 \quad 0; \ 0 \quad \lambda_2 \quad 0; \ 0 \quad 0 \quad \lambda_3 \ ]$. The Lorenz chaotic model of D-PMSG is given as follows:

$$\frac{d\tilde{i}_d}{d\tilde{t}} = -\tilde{i}_d + \tilde{i}_q \tilde{\omega}_g + \tilde{u}_d$$
$$\frac{d\tilde{i}_q}{d\tilde{t}} = -\tilde{i}_q - \tilde{i}_d \tilde{\omega}_g + \gamma \tilde{\omega}_g + \tilde{u}_q \tag{9}$$
$$\frac{d\tilde{\omega}_g}{d\tilde{t}} = \sigma(\tilde{i}_q - \tilde{\omega}_g) - \tilde{T}_r$$

Among them: $\lambda_1 = \lambda_2 = B_m/\tau \psi_f n_p^2$, $\lambda_3 = 1/\tau n_p$, $\tau = L/R_s$, $\sigma = \tau B_m/J_{eq}$, $\gamma = -(3\tau n_p^2 \psi_f^2)/(2 B_m L)$, $\tilde{u}_d = \frac{\tau}{\lambda_1 L} u_d$, $\tilde{u}_q = \frac{\tau}{\lambda_1 L} u_q$, $\tilde{T}_r = \frac{\tau}{\lambda_3 J_{eq}} T_r$, σ, γ are free numbers, determined by system parameters. The balance point of the system (9) satisfies:

$$\tilde{i}_d^e = (\tilde{\omega}_g^e)^2 + \frac{\tilde{T}_L}{\sigma} \tilde{\omega}_g^e + \tilde{u}_d$$
$$\tilde{i}_q^e = \tilde{\omega}_g^e + \frac{\tilde{T}_L}{\sigma} \tag{10}$$
$$(\tilde{\omega}_g^e)^3 + \frac{\tilde{T}_L}{\sigma}(\tilde{\omega}_g^e)^2 + (\tilde{u}_d - \gamma + 1)\tilde{\omega}_g^e + \frac{\tilde{T}_L}{\sigma} - \tilde{u}_q = 0$$

Where $\left(\tilde{i}_d^e, \tilde{i}_q^e, \tilde{\omega}_g^e\right)$ in formula (10) is the equilibrium point of the system. Then the characteristic equation of the Jacobian matrix is given by:

$$D(\lambda) = |\lambda E - J| = \lambda^3 + (2 + \sigma)\lambda^2 + [1 + 2\sigma + \sigma(\tilde{i}_d^e - \gamma) + (\tilde{\omega}_g^e)^2]\lambda$$
$$+ \sigma[1 + \tilde{i}_q^e \tilde{\omega}_g^e + \tilde{i}_d^e - \gamma + (\tilde{\omega}_g^e)^2] \tag{11}$$

Studying the characteristic equation $D(\lambda)$ to judge whether the equilibrium point of the system (10) is stable in the sense of Lyapunov, and thus obtain the conditions for the Hpof bifurcation to occur. The mathematical conditions of D-PMSG from limit cycle to Hopf bifurcation and then to produce chaotic motion under certain conditions are studied theoretically.

When $\tilde{u}_q = \tilde{u}_d = \tilde{T}_L = 0$, there is no input and no load for the D-PMSG. This can be used to study the brake operation process of motor no-load and power-off. Solve-formula (10), the three balance points S_0 (0, 0, 0) and $S_{1,2}$ $(\gamma - 1, \pm\sqrt{\gamma - 1}, \pm\sqrt{\gamma - 1})$ of the system can be known. Where S_0 is the zero equilibrium point, and $S_{1,2}$ are non-trivial equilibrium points. For two non-trivial equilibrium points, the characteristic polynomial of the Jacobian matrix can be written as:

$$D(\lambda) = \lambda^3 + (2 + \sigma)\lambda^2 + (\sigma + \gamma)\lambda + 2\sigma(\gamma - 1) \tag{12}$$

From formula (12), when $\gamma = \gamma_h = \frac{\sigma(\sigma+4)}{\sigma-2}$, the eigenvalue to $S_{1,2}$ is given by:

$$\lambda_1 = -(\sigma+2), \lambda_{2,3} = \pm j\sqrt{\frac{2\sigma(\sigma+1)}{\sigma-2}} \tag{13}$$

Since λ_1 is a negative real number, $\lambda_{2,3}$ is two pure imaginary numbers. Therefore, when $\gamma = \gamma_h$, the two equilibrium points are unstable equilibrium points, so the formula (9) will produce Hpof bifurcation. From the mathematical model (9), the values of the σ and γ only depend on the PMSM parameters, cannot satisfy the conditions of the formula $\gamma = \gamma_h$. There is the conclusion: if the system parameters are unchanged, the Hpof bifurcation will not occur at these two equilibrium points.

When $\tilde{u}_q = \tilde{T}_L = 0, \tilde{u}_d \neq 0$ indicates that D-PMSG is unloaded and the stator winding is only powered by the d-axis voltage-supply. From formula (10), there are equilibrium points S_0 $(\tilde{u}_d, 0, 0)$ and $S_{1,2}$ $(\gamma-1, \pm\sqrt{(\gamma-u_d)-1}, \pm\sqrt{(\gamma-u_d)-1})$. Then the condition of Hpof bifurcation can be obtained as follows:

$$\tilde{u}_d = \tilde{u}_{dh} = \frac{\sigma^2 - \sigma\gamma + 4\sigma + 2\gamma}{2-\sigma} \tag{14}$$

When $\tilde{u}_d = \tilde{u}_{dh}$, $\tilde{i}_d, \tilde{d}_q$ and $\tilde{\omega}_g$ will produce Hpof bifurcation. When $\tilde{u}_d > \tilde{u}_{dh}$, the system will enter a chaotic state, and an unstable operation will occur. Where $\tilde{u}_d$ is determined by the system parameters and the stator d-axis voltage. So the system may enter a chaotic state as long as the system parameters change or u_d cause disturbances.

$\tilde{u}_d \neq \tilde{u}_q \neq \tilde{T}_L \neq 0$ is the general case of D-PMSG operation. Using the Kardan formula to solve the cubic Eq. (10), and the three solutions for $\tilde{\omega}_g^e$ are represented as:

$$\tilde{\omega}_{g1}^e = \sqrt{-\frac{q}{2} + \sqrt{\Delta}} + \sqrt{-\frac{q}{2} - \sqrt{\Delta}}$$

$$\tilde{\omega}_{g2}^e = \xi\sqrt{-\frac{q}{2} + \sqrt{\Delta}} + \xi^2\sqrt{-\frac{q}{2} - \sqrt{\Delta}} \tag{15}$$

$$\tilde{\omega}_{g3}^e = \xi^2\sqrt{-\frac{q}{2} + \sqrt{\Delta}} + \xi\sqrt{-\frac{q}{2} - \sqrt{\Delta}}$$

Where $\xi = \frac{-1+\sqrt{3}j}{2}$, $\Delta = \left(\frac{q}{2}\right)^2 + \left(\frac{p}{3}\right)^3$, $p = \tilde{U}_d - \gamma + 1 - (\tilde{T}_L/\sigma)^3/3q = 2(\tilde{T}_L/\sigma)^3$ $/27 - [(\tilde{T}_L/\sigma)(\tilde{U}_d - \gamma + 1)]/3 + (\tilde{T}_L/\sigma) - \tilde{U}_q$, Substituting the above formulas into the equilibrium Eq. (10), the solutions $\tilde{i}_d^e$ and $\tilde{i}_q^e$ of the other two state quantity equilibrium states can be found. When $\Delta > 0$, the equation has a real root and two complex roots; when $\Delta = 0$, there are three real roots, when $p = q = 0$, there are three repeated zero roots, two of the three real roots are equal, but when $p, q \neq 0$, two of the three real roots are equal; when $\Delta < 0$, there are three unequal real roots. Therefore, only the case of considering $\Delta \leq 0$ is of research significance. As in the above analysis, the conditions for Hopf bifurcation can be determined as follows:

$$\sigma \tilde{i}_q^e \tilde{\omega}_g^e = 2 + 4\sigma + (\sigma^2 + \sigma)(\tilde{i}_d^e - \gamma) + 2(\tilde{\omega}_g^e)^2 + 2\sigma^2 \qquad (16)$$

As can be seen from the above formula (16), when the system parameters change or the wind speed (load torque) changes, the system will enter a chaotic state.

4   D-PMSG Analysis of Chaotic Characteristics

The simulation parameters is from literature [12]. When $\tilde{u}_q = \tilde{u}_d = \tilde{T}_L = 0$, according to the system parameters, we get $\sigma = 16$, $\gamma = 45.92$, $\gamma_h = 22.86$. As is shown in Fig. 1 (a), under different initial conditions, these curves eventually tend to balance points (0, 0, 0).The five curves respectively represent the motions of five different initial states, so that the simulation shows that D-PMSG is always stable in this state, and the above analysis results are verified. But when the system parameters change so that $\gamma = 45.92 > \gamma_h$, the three equilibrium points will become unstable, and the system will enter the chaotic state, as shown in Fig. 1(b).

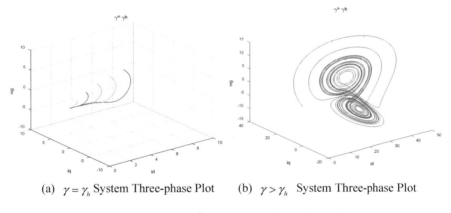

(a) $\gamma = \gamma_h$ System Three-phase Plot   (b) $\gamma > \gamma_h$ System Three-phase Plot

Fig. 1. $\tilde{u}_q = \tilde{u}_d = \tilde{T}_L = 0$ system three-phase plot

When $\tilde{u}_q = \tilde{T}_L = 0, \tilde{u}_d \neq 0$ from formula (14), we can calculate $u_d = u_{dh} = 23.06$. Figure 2(a), (b) are the state three-phase diagrams which are respectively $u_d = u_{dh} + 8$ and $u_d \approx u_{dh}$. From Fig. 2, we can find the state of i_d, i_q, ω_g moves from limit cycle to Hopf bifurcation and chaos. It indicates that D-PMSG will appear chaotic oscillation near the synchronous speed and the operation is unstable. Once the system parameters change or the disturbance occurs, the system may enter the chaotic state.

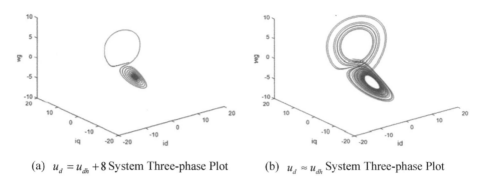

(a) $u_d = u_{dh} + 8$ System Three-phase Plot   (b) $u_d \approx u_{dh}$ System Three-phase Plot

Fig. 2. $\tilde{u}_q = \tilde{T}_L = 0$, $\tilde{u}_d \neq 0$ system three-phase plot

When $\tilde{u}_d \neq \tilde{u}_q \neq \tilde{T}_L \neq 0$, taking $\tilde{u}_d = -0.542$, $\tilde{u}_q = 0.824$ and $v = 7.89$ m/s, the three-phase plot curve is shown in Fig. 3. From the figure, the trajectory of the system is relatively slow to reach the stable equilibrium point, but after a period of time, the system gradually reaches the stable state operation. When $v = 12.6$ m/s, the system state trajectory is shown in Fig. 4. The chaotic attractor appears in the system, and the system enters the state of chaotic motion. Once $\tilde{u}_d$ and $\tilde{u}_q$ are determined, with the change of the wind speed, the wind turbine torque $\tilde{T}_r$ changes, and then $\tilde{T}_r$ also changes after linear affine transformation and time scale transformation. So for D-PMSG in the actual operation, it is of practical engineering significance to study the instability of the system with the change of wind speed.

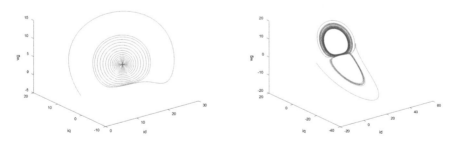

Fig. 3. $v = 7.89$ m/s system three-phase plot   **Fig. 4.** $v = 12.6$ m/s system three-phase plot

5   Conclusion

When the D-PMSG system is within a certain parameter range or the input changes, the motor speed and torque will be violently oscillated, the control performance will be unstable, and the system will generate complex nonlinear motion-chaos or limit cycles, but currently the chaotic motion analysis for D-PMSG is rare. Based on the D-PMSG mathematical model, the chaotic motion model is obtained by using linear affine transformation and time transformation. The system parameters in the conditions of D-

PMSG in steady state, limit cycle and chaotic state are analyzed by applying Lyapunov stability theory and Hpof bifurcation theory. The theoretical and simulation results show that the chaotic motion of D-PMSG will be affected by the change of input voltage under certain parameters. As the wind speed increases, the D-PMSG appears irregular oscillation and enters the chaotic running state under the condition of other parameters matching. This study confirms that under certain conditions, the D-PMSG system will enter into chaotic operation, which is not conducive to the stable operation of wind turbines, and is ready for the study of chaotic control of wind turbines.

Acknowledgments. This research was supported by National Natural Science Foundation of China (Grant No. 61473239), and Key Projects of the National Natural Science Foundation of China (Grant No. 61134002).

References

1. Liao, J., Qu, X., et al.: Optimal wind-energy tracking control of direct-driven permanent magnet synchronous generators for wind turbines. Power Syst. Technol. **32**(10), 11–15 (2008). (in Chinese)
2. Zhao, R., Wang, Y., Zhang, J.: Maximum-power point tracking control of the wind energy generation system with direct-driven permanent magnet synchronous generators. Proc. CSEE **29**(27), 106–111 (2009). (in chinese)
3. Li, J., Ren, Q.: Study on supply air temperature forecast and changing machine dew point for variable air volume system. Build. Energy Environ. **27**(4), 29–32 (2008). (in Chinese)
4. Chen, J.H., Chau, K.T., Chan, C.C.: Chaos in voltage-made controlled DC drive system. Int. J. Electron. **86**(7), 857–874 (1999)
5. Zhang, B., Li, Z., Mao, Z.: A primary study on an erratic behavior and chaotic phenomena of electric machine drive systems. Proc. CSEE **21**(7), 40–45 (2001). (in Chinese)
6. Hemati, H.: Strange attractors in brushless DC motor. IEEE Trans. Circ. Syst. **41**(1), 40–45 (1994)
7. Zhang, B., Li, Z., Mao, Z.: Anti-control of its chaos and chaos characteristics in the permanent–magnet synchronous motors. Control Theory Appl. **19**(4), 545–548 (2002). (in Chinese)
8. Jing, Z.J., Yu, C.Y., Chen, G.R.: Complex dynamics in a permanent magnet synchronous motor model. Chaos Solit. Fract. **22**(4), 831–848 (2004)
9. Cao, N., Shi, W., Zhu, C.: Simulated analysis of dynamic characteristics of permanent magnet direct-drive synchronous wind turbine generators. Electr. Mach. Control Appl. **44**(1), 104–109 (2017). (in Chinese)
10. Chen, Q., Ren, X.: Chaos modeling and real-time online prediction of permanent magnet synchronous motor based on multiple kernel least squares support vector machine. Acta Phys. sin. **59**(4), 2310–2318 (2010). (in Chinese)
11. Zheng, G., Zou, J., Xu, H.: Adaptive backstepping control of chaotic property in direct-driven permanent magnet synchronous generators for wind power. Acta Phys. Sin. **60**(6), 501–508 (2011). (in Chinese)
12. Yang, G., Li, H.: Sliding mode variable–structure control of chaos in direct-driven permanent magnet synchronous generators for wind turbines. Acta phys. Sin. **58**(11), 7552–7557 (2009). (in Chinese)

Design and Practice of Teaching Model of Measurement and Evaluation of Mathematics Education Based on the Flipped Classroom Concept

Jinming Cong[✉], Meiling Wang, and Zhiqin Huang

School of Mathematical Sciences, University of Jinan, Jinan, China
hzq-1230@163.com

Abstract. The flipped classroom is a new type of teaching organization emerging in recent years, and it's also an effective way to realize blended learning. The combination of these can realize the innovation and development of teaching methods. This article proposes the blended teaching models based on the flipped classroom concept, including traditional classroom teaching, partial flipped classroom and complete flipped classroom. Based on the model, the teaching design and practice of the course "measurement and evaluation of mathematics education" for graduate students majoring in subject teaching (math) were carried out, and suggestions are made for the effective application of the mode to solve the problem of teaching, improving teaching quality, promoting teaching reform.

Keywords: Practice of teaching model · Mixed teaching model · Mathematics education

1 Introduction

Blended learning is a new way of teaching and learning which combines face-to-face teaching with digital learning. Blended teaching can play a leading role in the teaching process, such as guiding, enlightening, monitoring and so on. Meanwhile, it can also reflect the enthusiasm, initiative and creativity of students in the learning process, and the traditional face-to-face classroom teaching combined with the modern network multimedia teaching in order to obtain better teaching results [1].

However, the flipped classroom is a new form of classroom teaching organization, that the course teacher provides learning resources in the form of teaching video, students complete the viewing and learning of teaching resources before class and teachers and students work together to answer questions, explore writing and interact in class in informationalized environment. The subversion of traditional teaching process and the "student-centered" thinking are the real essence of flipped classroom.

At present, the scholars at home and abroad generally believe that the flipped classroom can not only increase the interaction between teachers and students and the students' personalized learning time, but also is a new "blended learning style" which

© Springer Nature Singapore Pte Ltd. 2020
M. Atiquzzaman et al. (Eds.): BDCPS 2019, AISC 1117, pp. 1394–1400, 2020.
https://doi.org/10.1007/978-981-15-2568-1_193

was the result of the significant reform of classroom teaching mode under the guidance of the educational thought marked by "B-learning".

We believe that the idea of flipped classroom is throughout the blended teaching which can not only give full play to the advantages of blended teaching, but also change the single learning mode and stimulate students' learning enthusiasm and initiative. The new type of mixed teaching provides a new idea for our educational reform.

2   The Construction of Mixed Teaching Model Based on the Concept of the Flipped Classroom

2.1   The Mixture of Traditional Classroom Teaching and Modern Network Multimedia Teaching

For a long time, our school education has adopted the traditional classroom teaching, that is the teaching mode of "teachers speak and students listen". Teachers are the subject of teaching activities and the impartator of knowledge, while students are the recipients of knowledge. Although this mode has many disadvantages, its fundamental purpose is to obey the needs of subject teaching and to impart systematically and completely the cultural and scientific knowledge accumulated by human society over thousands of years. Therefore, most of the subject teaching still adopt lecturing teaching mode at present. This model imparts basic knowledge and skills such as the basic concepts, theories and principles of the discipline to students through teachers and students can acquire a large amount of knowledge quickly, intensively and systematically, and get the fastest and best development through the guidance of teachers [2].

However, the traditional way of transferring knowledge in class is single, which is not conducive to arousing students' learning initiative, depriving students of their emotional life in class teaching, causing a dull situation in class teaching, and not conducive to the development of students' innovation ability. Modern network and multimedia technology can build a friendly and realistic learning environment for teaching, and can provide rich image of teaching resources, diverse ways of knowledge acquisition and diverse ways of communication between teachers and students. Blended teaching combines the advantages of both, gives full play to the leading role of teachers and the dominant position of students, and makes learning methods more diversified, learning approaches more diversified, and learning experience more visual. It is a very practical and effective teaching model in the current teaching reform.

2.2   The Mixture of Traditional Classroom Teaching, Partial Flipped Classroom and Complete Flipped Classroom

The flipped classroom reverses the traditional teaching concept and teaching sequence, arouses the attention of the majority of educators with a brand new teaching method. More and more schools and teachers apply flipped classroom into teaching practice. However, some people questioned that the flipped classroom teaching model reduced the role of teachers, failed to mobilize students' interests and had a poor learning effect. Some studies also showed that flipped classroom did not significantly improve the teaching effect. The effective implementation of flipped classroom is influenced by many factors. If we fail to grasp it well, it will become a mere formality, only like a flipped classroom. However, in fact, students fail to complete the task of self-study after class, and the seminar in class cannot be carried out effectively, and the learning effect is even worse [3].

We believe that using the classroom lectures teaching or completely using the reverse is not desirable. There can be a cross section between the classroom lectures teaching and the flipped classroom, that teachers teach in class through choosing the key part of the teaching content, but other parts can be made of students autonomous learning, choose a particular topic or project based on the content to let the students in the form of group or student representatives are based on the theme or project-based inquiry learning, share the achievements of study, the discussion in class, teachers guide students to think and discuss, and solving puzzles and added.

This mixed approach is an effective attempt to explore the realization of flipped classroom in the course teaching. It not only focuses on the teaching of professional knowledge, but also gives students a certain degree of learning freedom, arouses the enthusiasm of students to participate in learning, and effectively avoids the similarity of flipped classroom in appearance but not in spirit caused by various factors. In blended teaching, three mixed forms can be adopted: teaching, partial flipped classroom and complete flipped classroom, which can be selected flexibly according to the characteristics of the course content and the characteristics of students.

2.3 The Mixture of Teacher Teaching Activities and Student Learning Activities

The main subjects of classroom teaching activities are teachers and students under the guidance of modern education theory, but more emphasis is placed on students' autonomy and initiative [4]. The development process of all teaching activities and the creation of teaching environment are carried out around the goal of "promoting the development of students". In the mixed teaching mode based on the concept of flipped classroom, teachers' teaching activities and students' learning activities are shown in Table 1.

Table 1. Teachers' teaching activities and students' learning activities

Teachers' activities	Classroom teaching, Prepare and publish learning tasks and resources, Teacher comments, Organization discussion, Troubleshooting, Homework correcting, Interact with students, Supplementary explanation
Students' activities	Classroom lecture, Resource based self-study activities, Complete individual or group assignments, Presentation about the projects, Question and Discussion, Interactive communication, Extra-curricular exercise

The flipped classroom upside down the traditional teaching philosophy and process that emphasizes "Teach knowledge after class and internalize knowledge between classes". Meanwhile, the students' learning and teachers' teaching activities' sequence and combination will have different change, that students become the protagonist of the classroom. In order to internalize or correct understanding of knowledge, classroom activities are mainly focused on students' problem exploration or topic report, learning results sharing, discussion and exchange, questioning and other activities. And teachers are becoming the leader of learning activities, mentors and interlocutor.

2.4   The Mixture of Formative Evaluation and Summative Evaluation

The mixed teaching mode based on the concept of the flipped classroom emphasizes the teacher's lectures, and emphasizes students autonomous learning and cooperative learning based on resource, pay attention to cultivate students' information literacy and expression ability to communicate, aroused the students' cognitive mode, the study way, and a profound change in the teacher's teaching mode and the roles between teachers and students, so in course examination also should notice to use multivariate evaluation system, pay attention to the combination of formative assessment and summative assessment, and to increase the proportion of formative assessment. Formative evaluation includes students' attendance, homework, quality and effect of individual or group study, results display, and activity of participating in communication and discussion. Summative evaluation is usually conducted in the form of a final exam, which examines whether the overall results of students' course learning meet the requirements of teaching objectives [5].

We propose a blended teaching model based on the concept of flipped classroom from the above analysis, as shown in Fig. 1.

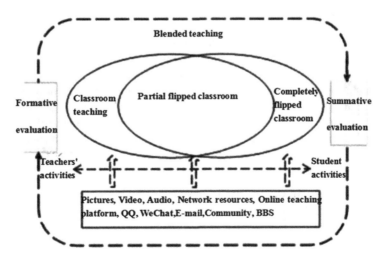

Fig. 1. Mixed teaching mode

3   The Design of the Course "Measurement and Evaluation of Mathematics Education" Under the Mixed Teaching Mode

3.1   Course Design Scheme

Measurement and Evaluation of Mathematics Education is one of the subject teaching (math) graduate elective classes and is set up in the first semester graduate record in our school. It has 32 periods and uses the textbook edited by liu ying and zeng wanting [6]. This course helps students to master the basic knowledge of measurement and evaluation of mathematics education and the method of measurement and evaluation of mathematics education. This course needs to cultivate students through the measurement of middle school mathematics teaching and make quantitative evaluation of middle school mathematics teaching effect. This course is a combination of theory and practice. We select 10 chapters of teaching content, and choose appropriate mixed teaching methods according to the requirements of teaching objectives and the characteristics of different chapters [7].

1. The basic knowledge of the basic terms, concepts and theories of the measurement and evaluation of learning education is the key content for students to master. This kind of teaching content is suitable for classroom teaching and multimedia assisted teaching. Teachers and students discuss the basic knowledge together. Extracurricular students read relevant extension resources recommended by the teacher to deepen their understanding of the basic concepts and theories of the course [8].

2. Chapter involves the content of the mathematics education measurement, such as related factors analysis, mathematics education of interval estimation and hypothesis testing of mathematical education and so on. The content of these chapters should be carried out in the form of classroom lecture and part of flipped classroom, so that students can realize it through SPSS or STATA software, so as to master the relevant factor analysis method, interval estimation method and hypothesis testing method, and realize the learning activities in the form of flipped classroom.

3. The content of chapters is distributed in a modular way. Each module focuses on one content, and each module is in a parallel relationship. Such content can be taught in the form of completely flipped classroom. Students in different study groups represent different module themes and study the content of this module. Students share, discuss and exchange their learning achievements with their classmates in class and compare and evaluate between the different groups, which is conducive to students' in-depth understanding of the teaching content [9].

According to the above discussion, the overall teaching planning and design of measurement and evaluation of mathematics education is shown in Table 2.

Table 2. Mathematics education measurement and evaluation curriculum design

Blended teaching	Teaching content arrangement	Scheduling	Learning style	Practice after class	The inspection way
Classroom teaching + expanded resources after class	1. Descriptive statistics of math achievement 7. Survey of mathematics education 9. Overview of mathematics education evaluation	8	Lectures, extracurricular reading materials, QQ and email communication	16 times	Process evaluation and summative evaluation accounted for 50% respectively
Classroom teaching + partial flipped classroom	2. The application of normal distribution in education measurement and evaluation 8. Quality analysis of math tests 10. Mathematics classroom teaching evaluation	8	Self-study before class, class lectures, individual or group reports, teachers and students, students exchange and share		
Completely flipped classroom	3. Analysis of relevant factors in mathematics education 4. Statistical inference of math achievement 5. Interval estimation in mathematics education 6. Hypothesis testing in math education	16	Self-study before class, Group report, Communication between teachers and students, resources sharing		

4 Conclusion

The teacher-oriented "teaching" changed to the student-oriented "learning" under the teaching mode of the flipped classroom [10]. The original teaching content and presentation need to be redesigned when the classroom teaching activity stretch into the teaching activity of "Pre-class → Lesson → post-class". We need to think about how to organize pre-class activities, classroom teaching and after-class evaluation. And we need to repeatedly explore what should be completed by students before class, what should be arranged for discussion and cooperation in class, and what should be taught by lecturing method. What is the basis for teachers to think about and what kind of support and help they get in the design is directly related to the success of the flipped classroom design.

Acknowledgments. This research was supported by Teaching Research Project of Qilu Institute of Technology (Grant No. JG201823).

References

1. Zhang, C., Li, H.: Teaching design of advanced mathematics micro-course based on BOPPPS model: taking "sequence limit" as an example. Western Qual. Educ. **3**(2), 163 (2017)
2. Hu, L., Zhang, B.: The flipped classroom and flipped learning: an analysis of the effectiveness of "flipped". J. Dist. Educ. **46**(09), 13–24 (2016)
3. Qu, L.: Advocating group cooperative learning and constructing efficient classroom teaching – the enlightenment of "learning whole word tower". Career Horizon (07), 101–103 (2014)
4. Luo, P., Lv, X., Li, M.: Research on the teaching mode of "flipped classroom". Educ. Teach. BBS (8), 151–152 (2017)
5. Jiang, S.: Research on mixed teaching mode based on micro-video divided classroom. Shanghai Normal University, Shanghai (2018)
6. Liu, Z.: Understanding of measurement and evaluation of mathematics learning. J. Hubei Corresp. Univ. **31**(02), 134–136 (2018)
7. Luan, Q.: Discussion on double-truth teaching model of measurement and evaluation of mathematics education. J. Hefei Norm. Univ. **36**(06), 69–71 (2018)
8. Feng, Y.: Problems and countermeasures of multiple intelligence diagnosis. Mod. Educ. Technol. (05), 27–33 (2007)
9. Yan, Z., Yan, Q., Liu, Z.: Research and practice of mobile teaching in higher vocational colleges—taking lanmoyun class as an example. Inform. Technol. Educ. (04), 173–177 (2017)
10. Abdousikur, G., Hua, L., Abdazi, G.: Some thoughts on learning the new teaching mode of "flipped classroom". Course Educ. Res. (23), 20–21 (2016)

Dual-Channel UAV Broadband Communication Network Radio Technology

Feifan Liao[1(✉)], Li Zou[1], Zhenyi Wang[1], Yanni Wang[2], and Ke Li[2]

[1] College of Information and Communication, National University of Defense Technology, Wuhan 430014, China
Liaofeifan@126.com
[2] Department of Operational Support, Rocker Sergeant School, Qingzhou 262500, China

Abstract. This paper introduces a dual-channel UAV broadband communication network radio technology design method, which can quickly construct a non-line-of-sight, multi-service broadband communication network without existing communication facilities. The method describes the overall technical architecture, workflow, dual-channel MAC layer protocol and routing protocol design method for adapting the topology dynamics of broadband communication network radio. The designed self-organizing network radio is constructed for emergency support and mobile combat conditions. The multi-node non-fixed IP broadband ad hoc network meets the requirements of data communication in line-of-sight and non-line-of-sight environments, and is especially suitable for complex terrain communication environments, and is an effective complement to military and civilian emergency communication means.

Keywords: UAV · Self-organizing network · Dual-channel · Emergency communication

1 Introduction

When the communication base station is damaged or destroyed and cannot work normally, or it is difficult to construct a fixed base station in a harsh environment, the UAV-based emergency communication self-organizing network station can play an important role as an air access point or mobile node. Conventional communication nodes have unparalleled advantages: First, deployment is rapid. When an emergency situation causes communication interruption, the UAV air node can be deployed, which is small in size, quick in construction, and responsive. The Beidou navigation system can quickly reach the communication interruption point; the second is efficient communication. The self-organizing communication system based on unmanned platform adopts dynamic networking technology to realize interconnection and intercommunication between unmanned aircrafts. It has the advantages of self-organization and self-repair, and has the advantages of efficient and rapid networking, which can meet the requirements under specific emergency conditions. Communication security requirements are an effective supplement and improvement of the existing emergency communication system [1–3].

M. Atiquzzaman et al. (Eds.): BDCPS 2019, AISC 1117, pp. 1401–1409, 2020.
https://doi.org/10.1007/978-981-15-2568-1_194

However, for airborne drone self-organizing networks, reducing signal collisions and improving the utilization of multi-hop shared wireless channels have been an important goal of researchers [4]. Therefore, the design and optimization of the Channel Access Control (MAC) protocol [5] has always been an important issue in this field. In the early research work of UAV self-organizing network, researchers generally used the IEEE 802.11 DCF [6], a typical access protocol in the field of ad hoc networks. UAV nodes use omnidirectional communication to compete and access channels [7–9], this method generally has problems of exposing the terminal and hiding the terminal, and the concealment is not high. In recent years, the stealth drone technology developed by various military powers in the world has put forward new requirements for the radio frequency stealth performance of the UAV data link [10, 11]. Under this trend, the omnidirectional communication mode between UAV nodes It must be replaced by the directional communication method with good RF stealth performance and strong anti-interference ability. Therefore, how to closely combine the characteristics of the directional communication method, and give full play to its advantages of long transmission distance and high spatial multiplexing [12]. The flexible and efficient narrow beam directional access protocol for man-machine self-organizing network design has become one of the most important problems that must be solved in this field.

This paper presents a design method of UAV broadband communication network radio based on directional channel and omnidirectional channel, which combines the advantages of directional communication and omnidirectional communication, and combines the routing algorithm under the condition of high dynamic topology. A multi-node non-stationary IP broadband ad hoc network for emergency support and mobile combat conditions is constructed, which is based on channel broadband wireless autonomous network construction technology and narrow beam directional communication technology based on ad hoc network. It satisfies both line-of-sight and non-line-of-sight. The demand for data communication in the environment effectively solves the problem of the exposed terminal and the hidden terminal of the ad hoc network, and greatly improves the concealment performance and anti-interference performance of the network. This technology has important theoretical and practical value for constructing low-altitude communication network under tactical conditions, effectively supplementing the existing emergency (tactical) communication network, and meeting the rapid opening of emergency or wartime broadband communication network.

2 System Design

2.1 Architecture

The structure of the system communication link is shown in Fig. 1. It mainly consists of four parts, namely the air data communication unit, the signaling communication unit, the ground-to-ground wired communication and the ground-to-ground wireless communication unit. The air data communication unit is mainly It is used to realize high-speed data wireless transmission. The signaling communication unit is mainly used for wireless ad hoc network protocol data exchange. The terrestrial wired communication unit is mainly used for optical fiber data access in the grounded wired access scenario, and the ground wireless communication unit. It is mainly used for network access communication of terrestrial wireless terminals in the ground access wireless access application scenario. In addition, the Beidou communication machine and the inertial navigation information interface are designed to share and distribute their own position information for each node in the air self-organizing network to support the smart active antenna beam. Forming. The digital communication unit is mainly composed of three sub-modules: COFDM digital modulation and demodulation, radio frequency transceiver, and smart active antenna. The digital modulation and demodulation sub-module based on FPGA completes COFDM digital modulation and demodulation, and the RF transceiver module adopts AD9361 integrated chip. The frequency is configurable. The center frequency works at 1.4 GHz. The baseband data modulated by the FPGA is transmitted to the RF transceiver sub-module based on the AD9361 through the FPGA intermediate layer card standard interface protocol. The RF transceiver sub-module moves the signal to the RF and passes the smart active antenna (built-in The power amplifier module is transmitted in the form of electromagnetic waves, and the receiving end repeats the above process in reverse. The receiving data is transmitted from the drone to the ground through the optical fiber, and is displayed on the terminal through the application software; the signaling communication unit is mainly modulated by GMSK digital Demodulation, RF transceiver, omnidirectional antenna consists of three sub-modules, GMSK digital modulation and demodulation is also based on FPGA implementation, RF transceiver module is built with discrete components, the center frequency works at 800 MHz, the RF architecture will be described in detail later, here is not Make a statement.

2.2 System Workflow

The system work flow chart is shown in Fig. 2. The general steps are as follows:

(a) Each node obtains time information through the Beidou device, and performs time synchronization of the entire network according to the time information;

(b) Execute the self-organizing network routing protocol through the signaling channel, establish and maintain a routing table, periodically check whether there is a node change, and if necessary, update the routing table;

(c) Each node obtains the Beidou positioning information through the Beidou device, and interacts the respective position information in the network through the signaling channel. If the node position changes, the position information is updated immediately to prepare for the orientation direction calibration;

(d) The self-organizing network is adjusted to enter the service processing state. If a user data service is accessed, the routing path required by the service is queried;

(e) completing MAC data access on the signaling channel;

(f) Data channel beam-scanning antenna performs beam alignment. If the data chain includes multiple nodes, the beam scanning antenna switches according to the MAC data access protocol to complete data interactive transmission.

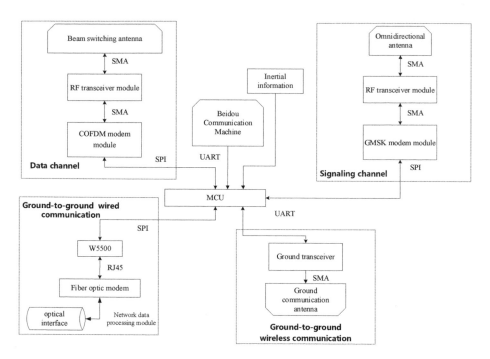

Fig. 1. System architecture

2.3 Media Access Protocol Design

The physical layer uses an omnidirectional antenna and a directional antenna. The two sides of the communication must use the directional antenna only after knowing the relative positions of each other. The MAC layer transmission mode in this case is called the directional mode. Otherwise, all MAC frames can only be transmitted through the omnidirectional antenna, which is called the MAC layer transmission mode in this case. In omnidirectional mode. The omnidirectional antenna carries a total of three channels: a common channel (800 MHz), a transmit busy channel (795 MHz), and a receive busy tone channel (805 MHz). In omni mode, only the common channel is used; in directional mode, all three channels are used, but for the same node, one of the three channels can be used at the same time, and the directional antenna carries only one channel, that is, data. Channel (1.4 GHz). The medium access mode adopts carrier sense/collision avoidance, and the transmitting node obtains the common channel use right through competition, and the node that fails the competition needs to perform backoff to reduce interference to the channel.

In the data transmission process in the omni mode, when either the sender and the receiver have no location information of the other party, the data transmission can only be performed on the common channel of the omnidirectional antenna. At this time, the operation timing of the transmitting node, the receiving node, and the neighboring node is as shown in Fig. 3.

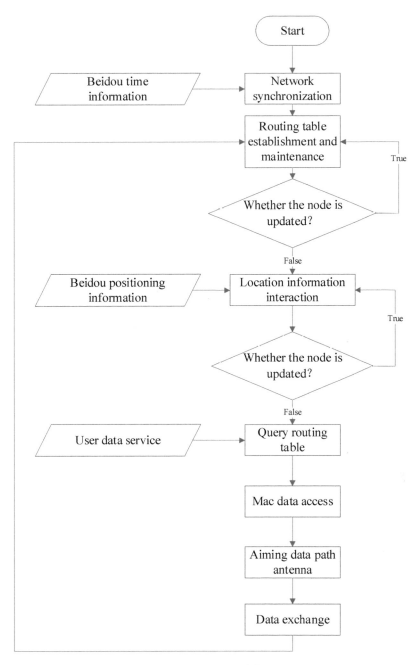

Fig. 2. System workflow

The data transmission process in the directional mode shows the action timing of the transmitting node, the receiving node, and the neighboring node as shown in Fig. 4.

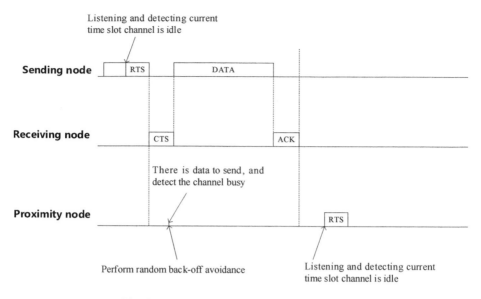

Fig. 3. Data transmission process in omni mode

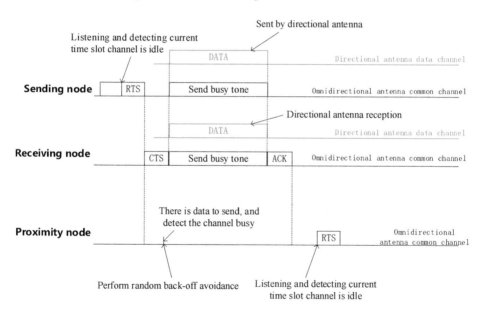

Fig. 4. Data transmission process in directional mode

Whether the omnidirectional mode or the directional mode of the data transmission process requires the communication parties to have the other party's location information negotiated on the common channel, and the negotiation process is:

If there is no receiver location in the sender's location information, the receiver is informed of the omni mode transmission through the RTS frame, and the receiver adopts the omni mode after receiving the RTS frame.

If the sender's location information holds the receiver's location, the receiver is advised to use the directional mode transmission through the RTS frame, and the receiver sees its location information after receiving the advice of the directional mode transmission.

When the receiver's location information holds the location of the transmitting party, the transmitting party is informed by the CTS frame to agree to use the directional mode transmission, and the receiving party broadcasts to receive the busy tone, ready to receive data.

If there is no transmitter location in the receiver location information, the transmitter is notified by the CTS frame to adopt omni mode transmission, and is ready to receive data. After receiving the CTS frame, the transmitter transmits data in the omni mode.

2.4   Routing Protocol Design

Based on the MAC layer, this paper uses the batman-adv routing protocol to realize the end-to-end data transmission of the UAV support node. In the introduction of the MAC layer, it is mentioned that to use a directional antenna, both the transmitting and receiving parties must know each other's orientation or the transmitting and receiving parties know each other's positions. Each node of the UAV security network uses GPS or Beidou to determine their respective geographical locations. When the key issues are concerned, how to report their respective location information to other nodes. In the Batman-adv routing protocol, each node periodically broadcasts its own OGM frame or forwards OGM frames of other nodes. As a result, each node knows the existence of other nodes. By expanding the location information of the node in which the OGM frame is added, after Batman-adv's route establishment calculation, each node can know the location of other nodes. The format of the OGM frame designed in this paper is shown in Table 1.

Table 1. OGM frame format

Type	Version	TTL	FLAG
Frame Number			
Source Address(First 4 bytes)			
Source Address (Last 2 bytes)		Last hop Address (first 2 bytes)	
Last hop Address (last 4 bytes)			
Reserved	TQ	TVL frame length	
Longitude (Unit: One in ten thousand of minute)			
Latitude (Unit: One in ten thousand of minute)			

The processing flow of each node broadcasting the OGM frame is shown in Fig. 5. The specific steps are as follows:

(a) If the last hop address in the OGM frame is the same as the address of the local node, it indicates that the OGM frame sent by the local node is directly discarded.

(b) If the source node address in the OGM frame is consistent with the local node, it indicates that the OGM frame sent by the node is broadcasted back through the neighbor node, and it can be presumed that the local node and the neighbor node are bidirectional links, and the neighbor is marked with the neighbor. The node is a bidirectional link and discards the OGM frame;

(c) If the OGM frame number is within the sliding window range, update the sliding window of the node and rebroadcast the OGM to other nodes; otherwise, discard the OGM frame.

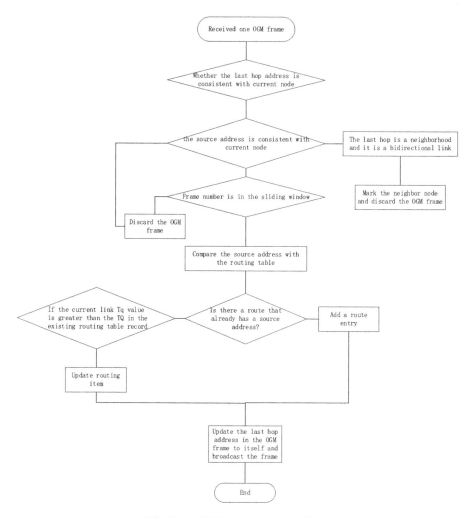

Fig. 5. OGM frame processing flow

Through the above process, routing information is established in each node, and the main content of the routing information is other arbitrary node address, the next hop address reaching the node, and the link quality TQ reaching the node.

3 Summary

This paper introduces a UAV broadband communication network radio technology based on omnidirectional channel and directional channel, which combines the routing algorithm under the condition of high dynamic dynamics, dual channel media access protocol design, narrow beam directional communication technology, etc. The dual-channel ad hoc network radio station is used for emergency support and mobile combat conditions. The omnidirectional channel is mainly used for node neighbor discovery and access control. The directional channel is mainly used for broadband data transmission. This design method can effectively solve the problem. The human-machine self-organizing network exposes the problem of the terminal and the hidden terminal, and the broadband data service in the directional channel transmission can greatly improve the anti-interference performance and the communication concealment performance of the system.

References

1. Han, B., Chen, Y.: Algorithm for dominating list of heterogeneous chains in wireless ad hoc networks. Comput. Sci. **45**(09), 135–140 (2018)
2. Chen, Y.: Comparative research and simulation analysis of wireless ad hoc network routing protocols OLSR and AODV. Comput. Knowl. Technol. **14**(08), 22–24 (2018)
3. Sharma, G., Ganesh, A., Key, P.: Performance analysis of contention based medium access control protocols. IEEE Trans. Inf. Theory **55**(4), 1665–1682 (2009)
4. Ying, J., Ming, Z., Shigang, C.: Achieving MAC-layer fairness in CSMA/CA networks. IEEE/ACM Trans. Netw. **19**(5), 1472–1484 (2011)
5. Bianchi, G.: Performance analysis of the IEEE 802.11 distributed coordination function. IEEE J. Sel. Areas Commun. **18**(3), 535–547 (2000)
6. Goddemeier, N., Rohde, S., Wietfeld, C.: Experimental validation of RSS driven UAV mobility behaviors in IEEE 802.11s networks. In: 2012 IEEE Globecom Workshops (GC Wkshps), pp. 1550–1555 (2012)
7. Alshbatat, A.I., Liang, D.: Cross layer design for mobile ad-hoc unmanned aerial vehicle communication networks. In: 2010 International Conference on Networking, Sensing and Control (ICNSC), pp. 331–336 (2010)
8. Yanmaz, E., Kuschnig, R., Bettstetter, C.: Achieving air-ground communications in 802.11 networks with three-dimensional aerial mobility. In: 2013 Proceedings IEEE INFOCOM, pp. 120–124 (2013)
9. Wang, L.: Radio frequency stealth technology and development ideas of aircraft. Telecommun. Technol. **53**(8), 973–976 (2013)
10. Chen, G.: Integrated radio frequency stealth technology for multi-function phased array system of advanced fighter aircraft. Modern Radar **29**(12), 1–4 (2007)
11. Ulinskas, T., Hughes, T., Lewis, T.: The challenge of directional networking. In: 2014 IEEE Military Communications Conference (MILCOM), pp. 1093–1098 (2014)

Passenger Ship Safety Evacuation Simulation and Validation

Jun Sun, Yuzhu Zhu$^{(\boxtimes)}$, and Pengfei Fang

Dalian Maritime University, Dalian 116026, People's Republic of China
1726724205@qq.com

Abstract. Due to the great loss caused by passenger ships' maritime accidents, the evacuation of passenger ships has received more and more attention. IMO has introduced a series of guidelines for comprehensive evacuation analysis and certification of large passenger ships during the design phase. In this paper, the personnel evacuation simulation tool Pathfinder based on the technology of Agent is used to establish the passenger ship evacuation model, and the real ship exercise data in the European "SEAFGUARD" project is used for simulation verification. The experimental results show that the model established in this paper can pass the 12 IMO tests about passenger ship evacuation simulation, meet the certification standard of the "SEAFGUARD" project, and can provide certain decision support for passenger ship safety evacuation simulation and passenger ship evacuation plan optimization.

Keywords: The safety of passenger ship · Emergency evacuation · Pathfinder · Microscopic simulation

1 Introduction

In recent years, due to the great loss and impacts caused by passenger ship accidents, it has brought great obstacles to the development of the passenger ship industry. In view of the huge loss caused by major maritime disasters and the increase in the number of passenger ships, the issue of evacuation of passengers at sea has received increasing attention. The International Maritime Organization has put forward a series of guidelines for the implementation of passenger ship comprehensive evacuation analysis, and the passenger ship is certified at the design stage [1–3]. Comprehensive evacuation drills are considered to be the most effective way to demonstrate that passenger ship's construction meets evacuation requirements, but generally the cost of full evacuation drills is high and difficult to organize. With the rapid development of modern computer technology, the use of computers for simulation research has become an important auxiliary means to improve the design, manufacturing and operation management in various industries. Modeling and simulation tools provide an effective way to help improve ship design for passenger ship evacuation studies.

© Springer Nature Singapore Pte Ltd. 2020
M. Atiquzzaman et al. (Eds.): BDCPS 2019, AISC 1117, pp. 1410–1419, 2020.
https://doi.org/10.1007/978-981-15-2568-1_195

For the study of ship evacuation model, it is mainly based on the existing personnel evacuation model, using social force model or cellular automaton method, combined with the characteristics of ship structure, to establish a ship personnel evacuation model [4–10].

Generally speaking, scholars have gradually carried out research on evacuation models of ship personnel, but the research is relatively preliminary. Although the evacuation of personnel in the local area of the ship can be analyzed from the simulation point of view, the verification of the evacuation experiment of the ship personnel in the real environment is lacking. So this paper used Pathfinder software to establish the passenger ship evacuation model and validate the established model using the actual ship evacuation data in the European "SAFEGUARD" project [11].

2   The Introduction of Pathfinder

Pathfinder is an agent-based personnel emergency evacuation simulation engineering software developed by Thunderhead Engineering of the United States. The software consists of three modules: analog computing module, graphical user interface, and 3D result display. The software motion environment is completely constructed by a three-dimensional triangular grid. In terms of building mechanism modeling, it can be used in conjunction with the actual floor construction mode. It can be used for 3D modeling of building structures such as doors, rooms, corridors, stairs, exits, etc. The model forms the walkable area of the person by rooms and connects the rooms of the same or different floors through the doors, stairs and corridors. The personnel evacuate to the exit. In terms of personnel modeling, Pathfinder can define the behavior of each person, build the size of the person's height and shoulder width, define the characteristics of the person, the walking speed and the specific walking route, and specify the evacuation exit of different personnel. Pathfinder can display the escape path and escape time of each individual in the evacuation model, as well as the number of escapes and time of each exit, and can display the group's escape path through 3D animation effects by using game and computer graphics simulation technology.

3   The Requirements and Verification of Evacuation Simulation Tools in IMO

IMO MSC Circ.1533 proposes four verification methods for the functional requirements of the evacuation simulation tool, including 12 tests, as shown in the following Table 1.

The models constructed by Pathfinder are in accordance with the expected requirements of IMO MSC Circ.1533. This paper lists the models and results of Test 6 and Test 9 for illustration.

Table 1. 12 tests of IMO on ship evacuation

Test	Description
Test 1	Maintaining set walking speed in corridor
Test 2	Maintaining set walking speed up staircase
Test 3	Maintaining set walking speed down staircase
Test 4	Exit flow rate
Test 5	Response duration
Test 6	Rounding corners
Test 7	Assignment of population demographics parameters
Test 8	Counterflow–two rooms connected via a corridor
Test 9	Exit flow: crowd dissipation from a large public room
Test 10	Exit route allocation
Test 11	Staircase
Test 12	Flow density relation

3.1 Test 6

Test 6 is a Rounding test, and the expected result is that 20 people near the left corner will successfully bypass the corner without penetrating the boundary. The simulation results are shown in Fig. 1 and are consistent with the expected results.

Fig. 1. The simulation result of Test 6

3.2 Test 9

Test 9 is a evacuation test for large public cabins. There are four exits in the public room and 1000 people are evenly distributed in the room, and personnel leave from the nearest exit. Select a group of 30–50 year old males from the IMO MSC Circ.1533 appendix and assign walking speed among the 1000 people. First open the four exits and then close the two exits. The expected result is a doubling of the time to empty the room. The simulation model is established as shown in Fig. 2. The simulation result is that the evacuation time is 210.8 s when all four outlets are open, and the evacuation time after closing the upper two outlets is 400.3 s, which is in compliance with the requirements.

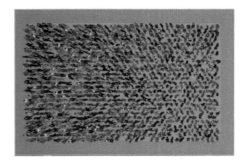

Fig. 2. The model diagram of Test 9

4 Passenger Ship Evacuation Simulation Experiment

4.1 The Layout Information of the Passenger Ship

This paper uses the three-layer passenger ship in the European "SAFEGUARD" project as the simulation and verification object. The passenger ship has a three-layer structure, including different public spaces, such as seating area, retail area and restaurant area, bar area and general area (Fig. 3). It has four assembly stations, three of which are located on deck 1 (AS A, B and C) and the other on deck 2 (AS D). The different decks are connected by several stairs.

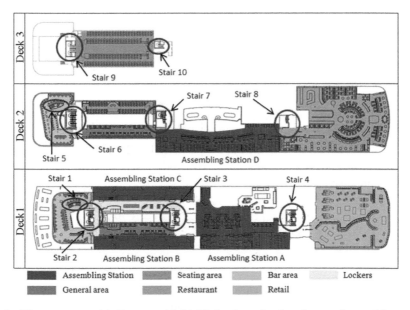

Fig. 3. The passenger ship Layout with highlighted starting locations and assembly stations

4.2   Model Construction and Numerical Simulation

According to the CAD drawing of the three-layer passenger ship given by the European "SAFEGUARD" project, the Pathfinder software is used to establish the passenger ship evacuation model.

There are 1,349 passengers on board, and 764 of them are tracked by wearing sensors during evacuation drills. Of the 764 passengers, some passengers have the same initial position and evacuation end position who are not counted. In addition, 569 passengers were not tracked, and some of these passengers participated in the evacuation drill and had a certain impact on the evacuation results. Therefore, the project assumes that 250 people of the 569 passengers participate in the evacuation drill, and the final number of simulations is determined to be 1014. The personnel are randomly distributed in each area according to the requirements in the project. The simulation result is recorded by 480 people with sensors that can be tracked. The destination is 4 Assembling stations. The speed distribution of personnel is in accordance with the requirements of IMO MSC Circ.1238, which is consistent with the requirements for personnel speed distribution in the subsequently revised IMO MSC Circ.1533. The reaction time distribution function of each region is as shown in the following equation.

Seating area:

$$y = \frac{1}{\sqrt{2\pi}(0.608)x} \exp\left[-\frac{(\ln(x) - 3.413)^2}{2(0.608)^2}\right] \tag{1}$$

Bar area:

$$y = \frac{1}{\sqrt{2\pi}(0.924)x} \exp\left[-\frac{(\ln(x) - 3.432)^2}{2(0.924)^2}\right] \tag{2}$$

General area:

$$y = \frac{1}{\sqrt{2\pi}(1.032)x} \exp\left[-\frac{(\ln(x) - 4.019)^2}{2(1.032)^2}\right] \tag{(3)}$$

Restaurant area:

$$y = \frac{1}{\sqrt{2\pi}(0.847)x} \exp\left[-\frac{(\ln(x) - 3.796)^2}{2(0.847)^2}\right] \tag{4}$$

Retail area:

$$y = \frac{1}{\sqrt{2\pi}(0.89)x} \exp\left[-\frac{(\ln(x) - 2.479)^2}{2(0.89)^2}\right] \tag{5}$$

According to the initial distribution of the above personnel, evacuation destination, speed distribution, response time distribution function, defining personnel behavior, adding 1014 passengers to the model, as shown in Fig. 4.

Fig. 4. Passenger ship evacuation simulation model after adding people

4.3   Simulation Results

The average value of the model after repeated operation for 10 times is recorded as effective data to reduce the contingency of the simulation results. The comparison between the obtained evacuation simulation data and the actual experimental data is shown in Figs. 5, 6, 7, 8 and 9.

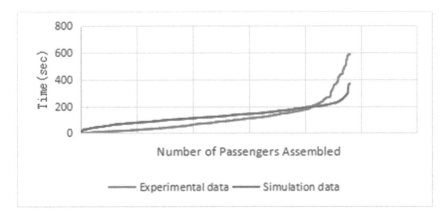

Fig. 5. The overall evacuation curve

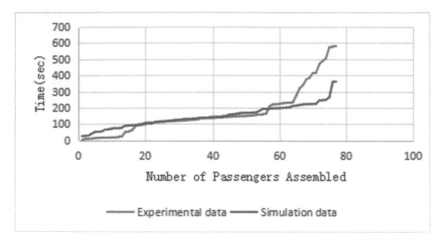

Fig. 6. The evacuation curve of Assembling Station A

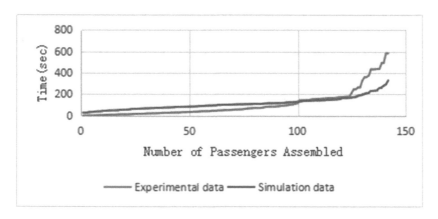

Fig. 7. The evacuation curve of Assembling Station B

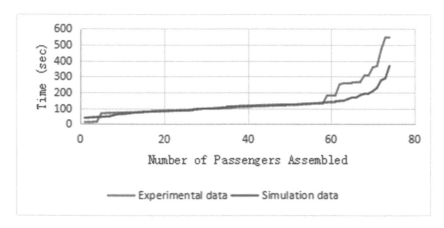

Fig. 8. The evacuation curve of Assembling Station C

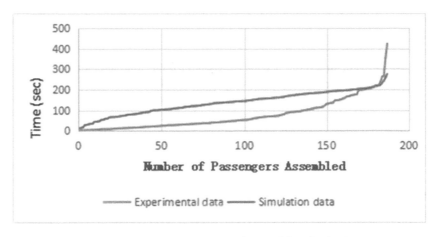

Fig. 9. The evacuation curve of Assembling Station D

The experimental data and the simulation data are different, but in general, the trend of the curve are roughly the same. There are several main reasons for these data deviations:

① During the entire evacuation process, a large number of unmarked passengers finally participated in the evacuation drill. Throughout the evacuation process, the presence of these passengers affected other marked passengers and affected their psychology and behavior during evacuation. Especially in crowded areas during evacuation, such as stairs and narrow corridors.

② The exact starting position and ending point of each marked passenger are unknown, and they are randomly distributed in a certain area. Since the starting and assembling areas on the passenger ship have a certain size, the personnel are randomly distributed, and these areas can be up to 48 m in length, so the individual assembling time may cause a large deviation, which has a certain influence on the simulation results.

③ The personnel response time given in the data is based on the region. Therefore, it is impossible to accurately assign the response time to each person in the model. For each passenger's response time, the experimental and simulation data deviation may also cause a certain influences on simulation result.

4.4   Verification of Results

Four evaluation indexes are given in the European "SAFEGUARD" project to verify the simulation data.

The recommended acceptance criteria are: (1) ERD $\leq$ 0.45; (2) 0.6 $\leq$ EPC 1.4; (3) SC $\geq$ 0.6, s/n = 0.05; (4) % diff TAT within 45%.

The verification process of the simulation results is as follows: First, calculate the four indicators and the total evacuation data. They should all meet the acceptance criteria. Then calculate the ERD, EPC and SC of the four assembling stations, a total of 12 indicator parameters. At least 9 items should meet the acceptance criteria, and at most one parameter of each assembling station should not meet the acceptance criteria. If the above two are met, the model can be proved to be suitable for passenger ship evacuation simulation research.

The above evaluation indexes were calculated for the simulation data obtained from the evacuation model established in this paper, and the results obtained are shown in Table 2. It can be proved that the model based on Pathfinder is suitable for the evacuation simulation research of passenger ships.

Table 2. Calculation of simulation results

Assembling station	Number of people	ERD	EPC	S/n = 0.05	%diff TAT
Total evacuation data	480	0.4085148	1.014339	0.84229	−37.20%
AS A	77	0.3833537	1.283747	0.6454251	
AS B	142	0.4347441	1.175923	0.9069964	
AS C	74	0.3425952	1.323503	0.8630361	
AS D	187	0.697673	0.6259437	0.7187543	

5   Conclusion

By referring to the literature on passenger ship evacuation and IMO MSC Circ.1533 for the new passenger ship and the existing passenger ship evacuation analysis guide, this paper selects the Pathfinder software for ship evacuation research. Firstly, according to the requirements of the Convention, 12 tests were carried out, and then the three-layer passenger ship in the European "SAFEGUARD" project was selected to construct the evacuation scene and the simulation data was obtained. The simulation data was tested according to the requirements of the project, and the data met the certification standards, which proved that the model based on Pathfinder was suitable for the evacuation simulation study of passenger ships.

References

1. International Maritime Organization, 2002. Interim Guidelines for Evacuation Analyses for New and Existing Passenger Ships, 6 June
2. International Maritime Organization, 2007. Guidelines for Evacuation Analysis for New and Existing Passenger Ships, 30 October
3. International Maritime Organization, 2016. Revised Guidelines for Evacuation Analysis for New and Existing Passenger Ships, 6 June
4. Shouheng, Liao: Research on Evacuation Model in Passenger Ship Under Inclination. Dalian Maritime University, Dalian (2010). (in Chinese)
5. Kanke, Wu: Research on Evacuation Model of the Stairs in Passenger Ship. Dalian Maritime University, Dalian (2010). (in Chinese)
6. Chen, M., Han, D., Yu, Y., et al.: Ship passenger evacuation model based on agent theory. Comput. Eng. Sci. **35**(04), 163–167 (2013). (in Chinese)

7. Ha, S., Ku, N.K., Roh, M.I., et al.: Cell-based evacuation simulation considering human behavior in a passenger ship. Ocean Eng. **53**, 138–152 (2012)
8. Roh, M.I., Ha, S.: Advanced ship evacuation analysis using a cell-based simulation model. Comput. Ind. **64**(1), 80–89 (2013)
9. Wang, W.L., Liu, S.B., Lo, S.M., et al.: Passenger ship evacuation simulation and validation by experimental data sets. Procedia Eng. **71**, 427–432 (2014)
10. Ni, B., Lin, Z., Li, P.: Agent-based evacuation model incorporating life jacket retrieval and counterflow avoidance behavior for passenger ships. J. Stat. Mech. Theory Exp. **2018**(12), 123405 (2018)
11. Galea, E., Deere, S., Filippidis, L.: The SAFEGUARD validation data set-SGVDS1 a guide to the data and validation procedures [EB/OL. (2012-11-30)]. http://fseg.gre.ac.uk/validation/ship_evacuation

Research and Application of Judgement Technology of Typical Medium and Low-Voltage Faults Based on Big Data Technology

Jianbing Pan[✉], Yi An, Bei Cao, Yang Liu, and Jingmin Xu

State Grid JiangXi Electric Power Research Institute, Nanchang, Jiangxi, China
283253654@qq.com

Abstract. In order to find the disconnection of medium and low-voltage lines timely and effectively, based on the big data of power supply established by the special work of marketing, distribution and regulation data sharing and integration, this paper constructs a research and adjustment model for typical medium and low-voltage faults from the perspective of electrical value (voltage, current, unbalance rate) in the low-voltage side of the distribution transformer. It analyzes the logical relationship of corresponding electrical characteristic values after faults such as line disconnection, single-phase loss and single-phase grounding of the distribution transformer when the 10 kV distribution transformer operates under different load factors, winding connection group, three-phase unbalance rate and system voltage, and applies fault warning model in the daily operation and maintenance of the distribution network, to improve the quality of active repair service, shorten the fault finding and users' power outage time, and improve the reliability of power supply.

Keywords: Line disconnection · Marketing · Distribution and regulation · Big data · Active repair · Reliability

1 Introduction

At present, the fault monitoring level of the medium and low-voltage distribution network is low. The dispatcher can only detect the faults like entire distribution line outage or single-phase grounding based on SCADA system information, but cannot discover the faults in certain part of the distribution line, the station area and the low-voltage line in time. Meanwhile, since overhead line is exposed in the wilderness, it is susceptible to external factors, resulting in various types of disconnection faults [1–7], and long-time disconnection fault may cause damage to the human body, equipment and power grid. Literature [1] proposes the use of negative sequence current to detect disconnection faults, and Literature [3] proposes a disconnected region analyzed by using the negative sequence voltage amplitude of each load point of the feeder and the minimum power supply path of the load point, which is relatively complicated. Literature [4–7] study the electrical quantity characteristics of the 10 kV side faults, but do not take the change of the electrical quantity on the low-voltage side of the distribution transformer into consideration. At the same time, with the application of distribution

M. Atiquzzaman et al. (Eds.): BDCPS 2019, AISC 1117, pp. 1420–1427, 2020.
https://doi.org/10.1007/978-981-15-2568-1_196

automation [8–10], electricity information collection and other application systems and the in-depth development of the special work of marketing, distribution and regulation, the big data center of power distribution has been basically established [11], and the network structure of station, line, distribution transformer, receiving point, meter box and user meter is constructed.

Based on the big data base of power distribution [12], this paper studies the research and judgment technology of typical medium and low-voltage faults based on the low-voltage electrical eigenvalues of distribution transformer, and realizes on-line monitoring of typical faults, such as 10 kV distribution line disconnection, low-voltage neutral line disconnection and single-phase grounding, by analyzing the logical correspondence between faults and electrical eigenvalues and positioning faults with GIS and power supply network topology, so as to take targeted repair measures to reduce the probability of power outage and personal or equipment damage, and improve active repair quality and efficiency.

2 Typical Features

Combined with existing conditions, low-voltage electrical characteristics are analyzed as follows:

2.1 One Normal Phase and Two Decreasing Phases

When the phase voltage of the low-voltage side of a Dyn11 station presents one normal phase and two decreasing phases, it indicates that there is a phase loss fault of a certain phase on the high-voltage side of the station. The reason for the phase loss may be due to the fusing or drop of fuse in the high-voltage side or the disconnection of the medium-voltage line.

As can be seen from Fig. 1, B-phase winding of the high-voltage side is subjected to the same voltage, the voltages of A-phase and C-phase are changed to 1/2 of the original voltage, and the phases are respectively shifted backward and forward by 60° to be in the same direction. According to the principle of electromagnetic induction, the b-phase output voltage of the low-voltage side is normal, and a-phase and c-phase voltage are reduced to half of the original voltage value, and the line voltage $U_{ac} = 0$.

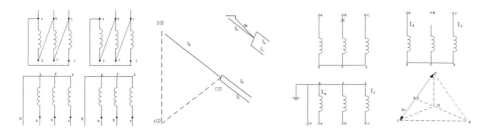

Fig. 1. Phase loss voltage vector for different connection groups

It is difficult to achieve complete three-phase balance with the load in distribution transformer. Assuming that I_a is greater than I_c, synthesis in the A-phase iron core $\Phi_A < \Phi_C$, and $U_A < U_C$. According to the magnetic flux closure principle, the magnetic flux in the three iron cores of the transformer should meet $\Phi_A + \Phi_B + \Phi_C = 0$, and since the winding currents of A-phase and C-phase are in the same direction, and the current of B-phase is in the reversed direction, the direction of the magnetic flux generated will be reversed, the magnetic flux relationship becomes $-\Phi_A + \Phi_B - \Phi_C = 0$, that is, $\Phi_A + \Phi_C = \Phi_B$. The low-voltage load is unbalanced and the system voltage remains unchanged, the faulty phase voltage with heavy load is smaller than the phase voltage with light load, and the sum of two-phase phase voltage vectors is equal to the normal phase voltage of the other phase. In the same way, characteristic values of B-phase or C-phase phase loss voltage are calculated.

2.2 One Zero Phase and Two Decreasing Phases

When the phase voltage of the low-voltage side of a Yyn0 distribution transformer station presents one zero and two decreases, it indicates that there is a phase loss fault in a phase on the high-voltage side of the station. Assuming that B-phase in the high-voltage side is out of phase, the two-phase winding of A-phase and C-phase in the high-voltage side is connected in series to the grid, causing B-phase potential is zero, the A-phase and c-phase voltage decreases to 0.866 times of the original voltage, and the voltage of a-phase and c-phase in the low-voltage side of the distribution transformer decreases to 0.866 times of the original phase voltage. The vector diagram is shown in Fig. 1.

Assuming Ia is greater than Ic, it will cause synthesis in A-phase iron core $\Phi_A < \Phi_C$, and $U_A < U_C$. According to the magnetic flux closure principle, the magnetic flux in the three columns of transformer core should meet $\Phi_A + \Phi_B + \Phi_C = 0$. Since the current directions of A-phase and C-phase winding are the same, the magnetic flux relationship changes to $\Phi_A + \Phi_B - \Phi_C = 0$ or $-\Phi_A + \Phi_B + \Phi_C = 0$, that is, $\Phi_A + \Phi_B = \Phi_C$ or $\Phi_C + \Phi_B = \Phi_A$, the magnetic flux in B-phase iron core column is not zero, that is, the b-phase voltage drop on the low-voltage side of the distribution transformer generates a small voltage, and the low-voltage c-phase voltage is equal to the sum of the voltages of a-phase and b-phase, or a-phase voltage is equal to the sum of b-phase and c-phase voltages.

2.3 One Decreasing Phase and Two Rising Phases

When the phase voltage on the low-voltage side of a distribution station is normal and the phase voltages on the user side present one decreasing phase and two rising phase, the neutral line on the low-voltage side of the station may be disconnected.

Assuming that when the neutral line of distribution transformer is disconnected (Fig. 3), the impedance value of neutral point $Y_O = 0$, the voltage of neutral point is $\dot{U}_{oo'} = \frac{\dot{U}_{ao}Y_a + \dot{U}_{bo}Y_b + \dot{U}_{co}Y_c}{Y_a + Y_b + Y_c + Y_o} \Rightarrow \dot{U}_{oo'} = \frac{\dot{U}_{ao}Y_a + \dot{U}_{bo}Y_b + \dot{U}_{co}Y_c}{Y_a + Y_b + Y_c}$. If the three-phase load is balanced, the displacement voltage of neutral point $\dot{U}_{oo'} = 0$; if the three-phase load is severely unbalanced and the neutral line is disconnected, supposing

$Z_a = 0.5R, Z_b = Z_c = R$, the admittance are respectively $Y_a = 2/R$, $Y_b = Y_c = 1/R$, the voltage of neutral point voltage is $\dot{U}_{oo'} = \frac{\dot{U}_{ao}Y_a + \dot{U}_{bo}Y_b + \dot{U}_{co}Y_c}{Y_a + Y_b + Y_c} = \frac{\dot{U}_{ao}}{4}$, the voltage of a-phase is $\dot{U}_{ao'} = 0.75\dot{U}_{ao}$, the voltage of b-phase is $\dot{U}_{bo'} = 1.145\dot{U}_{ao}e^{-j130.89°}$, and the voltage of c-phase is $\dot{U}_{co'} = 1.145\dot{U}_{ao}e^{j130.89°}$.

2.4 One Rising Phase and Two Decreasing Phases

When the phase voltage on the low-voltage side of a distribution station is normal and the phase voltages on the user side have one rise and two decreases, the neutral line on the low-voltage side of the station may be disconnected. As analyzed in Sect. 1.5, if the three-phase load is seriously unbalanced and the neutral line is disconnected, assuming $Z_a = R, Z_b = 0.5R, Z_c = R/3$, the neutral point voltage is $\dot{U}_{oo'} = 0.2887\,\dot{U}_{ao}e^{j150°}$, then a-phase voltage is $\dot{U}_{ao'} = 1.258\dot{U}_{ao}e^{-j6°35'}$, b-phase voltage is $\dot{U}_{bo'} = 1.04\dot{U}_{bo}e^{-j16°8'}$, and c-phase voltage is $\dot{U}_{co'} = 0.76\dot{U}_{ao}e^{-j10°54}$.

In the three-phase lighting circuit with unbalanced load, once the neutral line is disconnected, the neutral point displacement will be biased toward the heavy load, causing change in the voltage drop of each phase load; due to the impact of load voltage drop, the phase voltage of the phase with heavy load decreases, the bulb is relatively darker than normal, and the motor may not start; due to the impact of load voltage drop, the phase voltage of the phase with light load rises, the bulb is relatively brighter than normal, and some user equipment may burn out.

3 Simulation Analysis

To verify the accuracy of the calculation, S11-M-200/10 transformer is taken as an example, and the parameters are shown in Table 1. Matlab/Simulink is used to simulate the faults like medium and low-voltage line disconnection, single-phase phase loss of distribution transformer and neutral line disconnection under distribution transformer's operating conditions of different load ratios, winding connection groups, three-phase unbalance ratio and system voltage.

Table 1. Parameters of the three-phase transformer

High voltage (kV)	10	Low voltage (kV)	0.4
High resistance (Ω)	3.1968	Low resistance (Ω)	5.12E-03
High reactance (H)	3.12E-02	Low reactance (H)	7.99E-08
Excitation resistance (Ω)	24748.81	Excitation reactance (H)	272.911

3.1 Phase Loss of Distribution Transformer

The phase loss of distribution transformer includes faults such as fuse drop in the station area, line disconnection and jumper burnout. The simulation results are shown in Table 2.

Table 2. The simulation results of phase loss

Connection groups	Load rate	Voltage	Unbalance rate					
			0			30%		
			a	b	c	a	b	c
Dyn11	0–80%	Upper	119.9	239.9	119.9	139.6	239.9	100.6
		Standard	114.5	228.9	114.5	133.2	228.9	95.99
		Lower	98.12	196.2	98.12	114.2	196.3	82.3
	100%–120%	Upper	119.4	238.9	119.4	139.2	238.9	100.1
		Standard	114	228	114	132.8	228	95.57
		Lower	97.72	195.4	97.72	113.9	195.5	81.94
Yyn0	0–80%	Upper	0	207	119.9	0	243.1	171.8
		Standard	0	197.6	197.6	0	232	146
		Lower	0	169.4	169.4	0	198.9	140.6
	100%–120%	Upper	0	205.8	205.8	0	241.4	171.4
		Standard	0	196.4	196.4	0	230.4	163.6
		Lower	0	168.4	168.4	0	197.5	140.3

It can be seen from the data in Table 2 that after the single-phase phase loss fault occurring in the high-voltage side of Dyn11 type transformer, the low-voltage phase voltage will present one normal and two decreases, and the fault phase voltage will drop to 81.94 V; after single-phase loss of Yyn0 type high-voltage distribution transformer occurs, the low-voltage phase voltages will present two decreases or one normal and two decreases, so the fault phase will be at least 0 V.

3.2 Neutral Line Disconnection

Three-phase unbalance and distribution load ratio are adjusted by changing the three-phase load, the low-voltage voltage and the unbalance rate before and after the fault are measured, and the simulation data are shown in Table 3.

Table 3. The simulation results of neutral line break

Connection group	Load rate	Voltage	Load imbalance rate before failure							
			10%				30%			
			A	B	C	After	A	B	C	After
Dyn11	0–20%	Us	230	230	230	5.30%	230	230	230	18.02%
		Uc	238	227	227		256	219	219	
	20–80%	Us	229	229	229	5.25%	229	229	229	18.09%
		Uc	237	225	225		255	218	217	
	100–120%	Us	228	228	228	5.00%	229	229	228	17.64%
		Uc	236	225	224		254	217	216	
Yyn0	0–20%	Us	230	230	230	4.91%	230	231	230	14.54%
		Uc	235	235	222		246	246	202	
	20%–80%	Us	229	229	229	4.91%	229	230	229	14.61%
		Uc	233	233	221		245	245	200	
	100–120%	Us	228	228	228	4.68%	228	229	228	15.01%
		Uc	232	232	220		245	245	199	

Note: light load is 0–20%, normal load is 20%–80%, heavy load is 80%–100%, and overload is 100%–120%.

It can be seen from the data in Table 5 that if neutral line disconnection occurs in the distribution transformer, the voltage on the user side presents one rise and two decreases, and the voltage will increase up to 256 V, which is enough to damage the power consumption equipment; the voltage on the user side may also present one decrease and two rises, and the minimum voltage drops to 198.7 V, which affects the voltage pass rate. The three-phase unbalance rate will change greatly before and after the fault, and Yyn0 drops more than Dyn11, which is about 50% of the original.

4 Field Application

In view of the actual needs of the scene for 10 kV disconnection and phase loss fault of distribution transformer, the mining system is used to collect the abnormal conditions of public or special variation voltage for detection, power supply network topology diagram is applied to research and develop line disconnection monitoring module, and the research and judgment model is shown in Fig. 2.

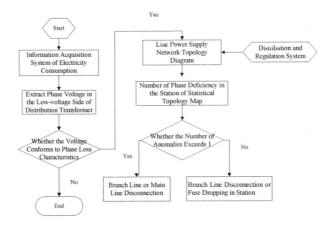

Fig. 2. Wire break test model

It can be seen from Fig. 3 that the phase loss model based on the typical characteristics of the low-voltage side voltage of distribution transformer can effectively detect line disconnection or fuse drop in the station.

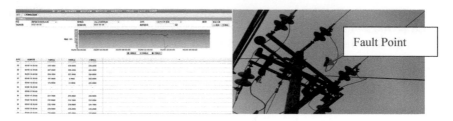

Fig. 3. Fault scene

5   Conclusion

The fault judgment model of the low-voltage electric quantity characteristic value of the electric big data only requires the arrangement of main monitoring station on the top layer, without adding any hardware devices in the distribution network, which greatly reduces the equipment investment and the daily operation and maintenance work of the grass-roots team. The monitoring station arranged can actively detect and locate the disconnection and phase loss of distribution transformer that cannot be found by the traditional dispatch, reduce the accidents caused by the disconnection, including personal injury, equipment damage, grid trip and other events, and improve the safety and reliability of power supply.

At present, as 2009 module and non-module energy meter are not qualified with active reporting function and smart meters are not completely covered in some areas, some user-side voltages can not be obtained in time. Thus, the positioning accuracy of low-voltage line grounding fault and neutral line disconnection is relatively low.

References

1. Zhu, L., Li, C., Zhang, H., et al.: Negative sequence current distributing and single-phase open-line fault protection in distribution network. Power Syst. Protect. Control **37**(9), 35–43 (2009)
2. Jia, Q., Liu, L., Yang, Y., et al.: Abrupt change detection with wavelet for small current fault relaying. In: Proceedings of the CSEE, vol. 21, no. 10, pp. 78–82 (2001)
3. You, Y., Liu, D., Li, L., et al.: Detection method based on load monitoring for 10 kV overhead line single-phase. Power Syst. Protect. Control **40**(19), 144–149 (2012)
4. Liu, Y., Meng, Q.: Study on dual characteristic of two kinds open phase running. Power Syst. Protect. Control **36**(23), 14–17 (2008)
5. Qu, G., Li, C.: Simulative study on steady-state performance of DFIG. Electric Power Autom. Equip. **25**(12), 35–38 (2005)
6. Zhang, H., Pan, Z., Sang, Z.: Analysis of complex line breaking and grounding fault in the power system with neutral point indirectly grounded. Relay **32**(18), 6–9 (2004)
7. Jia, W., Chen, J., Xu, C., et al.: Fault diagnosis for line-open with line-to-ground on radial distribution network by method of double frequency diagnosis. In: Proceedings of the CSU-EPSA, vol. 11, no. 2, pp. 20–24 (1999)
8. Wang, D., Song, Y., Zhu, Y.: Information platform of smart grid based on cloud computing. Autom. Electric Power Syst. **34**(22), 7–12 (2010)

9. Wang, G., Li, B., Hu, Z., et al.: Challenges and future evolution of control center under smart grid environment. Power Syst. Technol. **35**(8), 1–5 (2011)
10. Qu, C., Wang, L., Qu, N., et al.: Knowledge processing and visual modeling method in smart grid. Science Press, Beijing (2013)
11. Liu, K., Sheng, W., Zhang, D., et al.: Big Data application requirements and scenario analysis in smart distribution network. In: Proceedings of the CSEE, vol. 35, no. 2, pp. 287–293 (2015)
12. Mu, L., Cui, L., An, N.: Research and practice of cloud computing center for power system. Power Syst. Technol. **35**(6), 170–175 (2011)

Research and Application of Multi-level Coordination and Risk Assessment of Rural Distribution Lines

Jianbing Pan[(⊠)], Yi An, Bei Cao, Yang Liu, Zaide Xu,
and Jingmin Xu

Electric Power Research Institute of State Grid Jiangxi Electric Power Company,
Nanchang, China
283253654@qq.com

Abstract. There are some problems in the 10 kV feeder switch and line intelligent switch in the existing substations, such as over-protection or no protection range, and failure of coordination between upper and lower levels. To solve these problems, this paper proposes a method of protection setting calculation based on system parameters and two-phase short-circuit fault types. Taking the single radiation line as an example, it analyses in detail the application of this method in protection coordination, over-current setting of switches at all levels, and the risk of leapfrog trip. The results show that this method can effectively increase the protection range, optimize the switch layout, and greatly simplify the protection setting calculation while ensuring the selectivity of the protection device. Finally, based on the above calculation method, the protection setting calculation software is designed. It can automatically complete setting calculation and verify protection range according to relevant parameters of the system, and automatically generate risk assessment report of original setting value.

Keywords: Distribution line · Two-phase short circuit · Setting · Protection coordination

1 Introduction

Traditional three-stage over-current protection is applied to 10 kV outgoing line protection of substation to protect the whole feeder line. After the intelligent switch is installed in the circuit, it fails to form effective protection cooperation with it, which leads to misoperation or refusal of protection, and seriously affects the application effect of local fault isolation. Literature [1, 2] discusses the configuration of relay protection in distribution network. The quick-break protection is set by avoiding the maximum short-circuit current at the end of the line and considering the operation mode and fault type of the system. The method described in [3] is simple in configuration and can prevent the expansion of power failure range, but does not apply to instantaneous faults. Literature [4] proposes a strategy of layered setting of protection device, but does not consider the case of single-radiation power supply. According to the characteristics of short power supply radius and multiple sections of distribution

© Springer Nature Singapore Pte Ltd. 2020
M. Atiquzzaman et al. (Eds.): BDCPS 2019, AISC 1117, pp. 1428–1436, 2020.
https://doi.org/10.1007/978-981-15-2568-1_197

lines, literature [5, 6] adopts a protection scheme of distribution network circuit breaker based on time differential coordination. This scheme is not suitable for distribution lines in low load density areas with long power supply radius. Literature [7] analyses the characteristics of inrush current caused by single and multiple transformers. Literature [8] discusses the setting principle of branch protection of distribution lines, and gives the setting principle of circuit breaker protection and fuse protection. Literature [9, 10] studies the setting method of distributed power supply when connected to distribution network.

In terms of the problems of distribution line setting, protection coordination and fault location, this paper proposes a setting calculation method. This method is based on the system parameters and the types of two-phase short-circuit fault in the low load density area of distribution line protection setting. Moreover, it optimizes the position of the intelligent switch layout, and at the same time designs the calculation software for the protection setting, which can automatically calculate the setting, verify the protection range, and complete the risk assessment.

2   Calculation Method of Two-Stage Over-Current Protection Setting for Distribution Lines

To ensure the selectivity and sensitivity of relay protection of the distribution lines, the setting calculation is simplified to meet the requirements of protection coordination when three-stage over-current protection is put on the outgoing switch of substation. The section and branch switches of 10 kV distribution lines are adjusted according to two-stage over-current protection, i.e. quick-break and over-current protection [11].

2.1   Calculation Method of Quick-Break Protection Setting for Two-Phase Short Circuit

The traditional three-stage current protection sets its stage I according to the three-phase short-circuit current at the end of the line. There may be no protection range when a two-phase short-circuit occurs on the line. The data [5] show that the probability of two-phase fault is much higher than that of three-phase fault.

Therefore, when setting the quick-break protection current, the two-phase short-circuit current at the end of the feeder section is set according to the maximum operation mode of the system. The reliability coefficient is Krel = 1.1, that is:

$$I_{set}^{I} = K_{rel} \times \frac{\sqrt{3}}{2} \times \frac{55}{\frac{1}{S_{K.max}} + \frac{x_0 L}{10.5^2}} \tag{1}$$

After the quick-break protection setting is set according to formula (1), the protection range is only for the feeder section of this level, and avoids the occurrence of leapfrog trip [12].

2.2   Calculation Method of Over-Current Protection Setting Based on Delay Level Difference

Quick-break protection can not protect the full length of the line, so it is necessary to configure over-current protection as reserve protection. The fixed value of over-current protection is set according to the carrying current of the wire and the maximum load current, which can protect the whole line and have time delay. All levels of protection can meet the selectivity requirements through delay level difference ∆T, as shown in Fig. 1 [1].

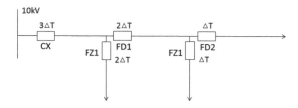

Fig. 1. Over-current protection meets selectivity requirements

According to Fig. 1, the two-stage protection coordination of outgoing switch and intelligent switch of substation is carried out. After that, when a fault occurs at any point of the distribution line, the fault isolation is completed by the only switch protection action upstream of the fault, which meets the selectivity requirements and reduces the fault section. Finally, it achieves the effect of "branch line fault does not jump trunk switch, trunk fault does not jump substation switch" [13].

2.3   Calculation Method of Avoiding Inrush Current Protection Setting

The multiple of the inrush current of multiple distribution transformers is smaller than that of a single transformer. Moreover, as the number of transformers increases, the difference between the two gradually increases. After getting the rated current of distribution transformer, the quick-break protection setting of intelligent switch is set according to the maximum inrush current avoiding the distribution. (The fundamental component of inrush current generally does not exceed 10 times of rated current of downstream distribution transformer [14]).

According to the relationship between the ratio of inrush current and its occurrence probability, the current setting value of quick-break protection can be determined by the maximum risk probability [7] of misoperation of the protection that can be accepted. After introducing inrush coefficient K_y, the quick-break protection setting method of intelligent switch is as follows:

$$I_{set}^1 \geq K_y \sum_{i=1}^{n} I_{Ni} = \frac{K_y \sum_{i=1}^{n} S_{Ni}}{\sqrt{3}U_N} \tag{2}$$

In the formula, I_N is the rated current of distribution transformer, S_N is the rated capacity of distribution transformer, and U_N is the rated voltage of distribution transformer. Take $K_y = 5.3$ (the probability of misoperation is below 5%) and set the time delay as 20–40 ms to ensure that the inrush current is lower than the quick-break protection setting after its attenuation.

3 Setting Calculation and Verification of Typical Single Radiation Line

Figure 3 is a 10 kV single-radiation distribution line. When the substation outlet switch CX is set according to the traditional setting method, the quick-break is 1940A/0 ms, over-setting of timing limit is 440 A/300 ms, setting of FD1 quick-break protection is 1190A/0 ms, setting of FD2 quick-break protection is 852A/0 ms, FD3 setting is 710A/0 ms. The fixed-time over-current protection can only form a first-level coordination due to the delay of 300 ms. According to Sects. 3.1 and 3.2 in the paper, the setting calculation method is optimized as follows.

3.1 Protection Setting Calculation Based on System Related Parameters

The short-circuit capacity is 150 MVA under the maximum operational mode of the system. The short-circuit capacity is 100 MVA under the minimum operational mode, and conductor impedance is 0.38 Ω/km. The lengths of conductors at each level are shown in Fig. 2.

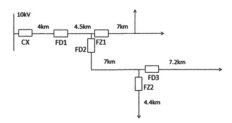

Fig. 2. Installation location diagram of switches

The protection setting of substation outgoing switch and line intelligent switch follows the methods of Sects. 3.1 and 3.2 in the paper.

Quick-break protection avoids two-phase short-circuit current at the end of feeder section under the maximum operational mode of distribution system, and presets reliability coefficient Krel = 1.1. Fixed-time over-current protection forms protection coordination through time delay, and the delay level difference ΔT takes 150 ms. The total line load is 7700 KVA and the rated current is 444 A. Therefore, the over-current setting is set according to the line distribution and the current amplitude at full load operation, Iset.CX = 444 A. The current setting of line switches decreases according to the protection coordination coefficient (1.2 times of the main line and 1.5 times of the branch line).

3.2   Verification of Protection Range

After setting the quick-break protection setting according to the above method, the protection ranges of new and old protection setting of each switch are compared as shown in Table 1.

Table 1. Circuit and three-phase short-circuit

Operation mode and short circuit type(%)	CX New/old	FD1 New/old	FD2 New/old	FD3 New/old	FZ1 New/old	FZ2 New/old
Two-phase short circuit in minimum operation mode	62.1/105.5	57.4/104.4	63.5/69.2	55.5/14.7	63.5/56.9	33.0/61.1
Two-phase short circuit in maximum operation mode	86.5/129.7	78.9/126.2	77.3/82.6	68.9/29.0	77.3/66.7	55.1/82.9
Three-phase short circuit in minimum operation mode	79.2/133.1	90.0/144.7	98.5/104.7	100.0/58.1	98.5/86.2	100.0/100.0
Three-phase short circuit in maximum operation mode	107.0/157.3	111.5/166.1	112.3/118.5	100.0/72.0	100.0/100.0	100.0/100.0

After setting the quick-break protection according to the method in Sect. 3.1: ① The protection ranges of all switches in two-phase short circuit protect the feeder section of this stage. The range of protection is large while satisfying the requirement of selectivity; ② When three-phase short circuit occurs in the maximum operational mode of CX, FD1 and FD2, the protection range extends to the lower level, exceeding 0.28 km, 0.502 km and 0.861 km respectively, which do not meet the requirement of selectivity. However, due to the extremely low probability of occurrence, it is within the acceptable range; ③ The protection range of the last stage switches: FD3, FZ1 and

FZ2 in three-phase short circuit is basically the full line length; ④ The delay of quick-break increases by 20 ms, and after verification, the quick-break protection settings are all avoiding the inrush current. From Table 3, if the protection setting calculation is optimized based on the relevant parameters of the system, the comparison with the traditional protection setting is as follows: ① When a short-circuit fault occurs in the FD1–FD2 section, the risk of FD1's leapfrog trip is reduced by 96.0%. When a short-circuit fault occurs in the FD2–FD3 section, the risk of FD2's leapfrog trip is reduced by 92.2%; ② When a two-phase short-circuit fault occurs downstream of FD3, the protection range of FD3 is increased by 40.4%; ③ The contrast between FD2 and FZ1 protection settings is close; ④ FZ2 has a better protection range under the traditional protection settings.

3.3   Optimization of Switch Distribution

The system parameters are used to calculate the protection settings of each switch. In that case, if the protection range is less than 20% when the two-phase short circuit occurs in the minimum operational mode of the system, namely, the spacing cannot meet the requirements of protection coordination, the position of the sectional switch should be moved back to increase the quick-break protection range of the upper switch. To enable CX and FD1 to have better protection range of two-phase short circuit, the minimum protection range of CX and FD1 is calculated when the spacing between CX-FD1 and FD1–FD2 is different, as shown in Table 1. It can be concluded from the results of data distribution in Table 2: ① If the minimum protection range of CX is greater than 2 km, the spacing of CX-FD1 should be greater than 3.5 km; ② The necessary condition for switches at this level to obtain an ideal protection range is that the spacing of the switches at the lower level is greater than that of this level. ③ In this case, FD3 is 7.2 km away from the end of the line due to the location determination of FZ2, so the CX-FD3 spacing is 15.5 km. If the spacing of CX-FD1 is increased step by step, the spacing of CX-FD1 should be less than 4.1 km. ④ From ① and ③, The spacing range of CX-FD1 is 3.5–4.1 km, and the minimum range of CX-FD1 is more than 60% when the first spacing is 4.0 km. Similarly, FD1–FD2 spacing is 5.0 km and FD2–FD3 spacing is 6.5 km.

4   Design of Automatic Calculation Software for Setting

To solve the problems in setting calculation of distribution network, it is necessary to consider the system parameters, protection coordination of switches at all levels, protection range, power supply topology and other factors. Based on these factors, a set of automatic calculation software is designed, which can optimize the setting design of intelligent switches.

Table 2. Minimum protection range of CX and FD1 of different spacing

CX\FD1 (%)	FD1– FD2 2 km	FD1–FD2 3.5 km	FD1– FD2 4 km	FD1–FD2 4.5 km	FD1– FD2 5 km	FD1– FD2 6 km
CX-FD1 2 km	33.6\24.2	33.6\53.2	33.6\57.5	33.6\61.5	33.6\64.4	33.6\68.8
2.4 km	43.3\22.3	43.3\52.0	43.3\56.6	43.3\60.7	43.3\63.6	43.3\68.2
2.8 km	50.0\20.1	50.0\51.0	50.0\55.7	50.0\59.9	50.0\62.9	50.0\67.6
3.2 km	55.2\19.2	55.2\49.9	55.2\54.9	55.2\59.1	55.2\62.1	55.2\67.0
3.6 km	59.1\17.6	59.1\48.8	59.1\54.0	59.1\58.2	59.1\61.4	59.1\66.3
4.0 km	62.1\15.0	62.1\47.7	62.1\53.2	62.1\57.4	62.1\60.7	62.1\65.7
4.4 km	64.9\13.9	64.9\46.5	64.9\52.4	64.9\56.6	64.9\59.9	64.9\65.1
4.8 km	67.1\11.8	67.1\45.7	67.1\51.5	67.1\55.8	67.1\59.1	67.1\64.5

4.1 Software Architecture Diagram

The automatic calculation software of setting is composed of database, man-machine interface, parameter input module, topology analysis module, setting calculation module, risk assessment module and output module. The software architecture diagram is shown in Fig. 3.

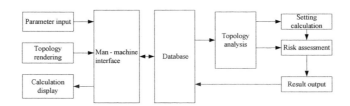

Fig. 3. Automatic setting calculation software architecture diagram

Parameter input: the input of maximum/minimum operational mode, protection setting, tower number, wire type, total load, etc. (Notes: Because there are many types and diameters of wire in 10 kV line, when the type of wire is input, the error of setting is less than 3% when the middle diameter of wire is used to calculate. This method can be used to simplify the setting calculation in rough calculation.);

Topology building: the establishment of the model position relationship of the switch;

Setting calculation: the automatic calculation of protection setting according to system parameters and methods in Sects. 3.1, 3.2 and 3.3;

Database: the storage of input and output data information.

4.2   Risk Assessment

The automatic calculation software of setting value includes risk assessment module, and can verify the protection range and sensitivity by the input of setting. If the risk assessment of FD1 in Sect. 4.1 in this paper is carried out, by calculating the inherent setting, the protection ranges of FD1 in the case of two-phase and three-phase short circuit all comes to the next level, and there is the risk of leapfrog trip. After optimizing the setting calculation through the system related parameters, the leapfrog trip probability is calculated according to the occurrence probability of two-phase and three-phase short circuit 1:9 (data in Table 1). The leapfrog trip probability under the maximum operational mode is shown in Table 3.

Table 3. The leapfrog trip probability of FD1

FD1	Two-phase	Three-phase	Comprehensive
Pre-optimization	16.8%	42.5%	19.4%
Optimization	0.0%	7.4%	0.7%

5   Conclusion

(1) The quick-break protection setting of the distribution line avoids the two-phase short-circuit current setting of the end of the feeder section of the distribution system in the maximum operational mode, and meets the selectivity requirements. At the same time, it should avoid the inrush current setting that may be generated by distribution transformer, which can distinguish three-phase and two-phase short-circuit faults, and reduce the trip risk by more than 90%.

(2) In low load density area, primary switches distribution of the line should be more than 2 km away from the switches within the station. To ensure that all levels of protection have an ideal sensitivity and protection range, the spacing should be increased step by step.

(3) The calculation method of system parameters and two-phase short-circuit fault has designed the calculation software of protection setting. Based on this, the protection setting calculation, the verification of protection range and the optimal distribution of switch position can be completed quickly so as to improve the calculation efficiency.

(4) In this paper, the calculation method of multi-level coordination and setting for practical distribution lines can ensure that the minimum protection range at all levels is more than 50%. This not only guarantees the selectivity of the protection device, but also greatly simplifies the calculation of the protection setting. However, it does not take into account the setting of small hydropower and other multi-power supply, so the relevant research needs to be further deepened.

References

1. Liu, J., Liu, C., Zhang, X., et al.: Coordination of relay protection for power distribution systems. Power Syst. Prot. Control **43**(9), 35–41 (2015)
2. Wan, S., Hu, C., Zhang, L.: Discussion on several technical problems of relay protection in distribution network. Distrib. Utilization **22**(3), 12–15 (2005)
3. Liu, H., Liu, Y.: Investigations for the use policy of 10 kV distribution network's demarcation switch. Shandong Electric Power **41**(3), 34–36 (2014)
4. Xu, Y.: Relay protection setting and use of 10 kV distribution network in urban area. Electric Power **49**(10), 49–51 (2016)
5. Liu, J., Zhang, Z., Zhang, X., et al.: Relay protection and distribution automation based fault allocation and restoration for distribution systems. Power Syst. Prot. Control **39**(16), 53–57 (2011)
6. Cui, Q., Bo, C., Li, W., et al.: Exploration on protection setting coordination of 10 kV distribution lines. Distrib. Utilization **26**(6), 32–34 (2009)
7. Liu, C., Liu, J., Zhang, Z.: Analysis and discussion on probability distribution of inrush current in distribution network. Autom. Electric Power Syst. **41**(4), 170–175 (2017)
8. Zhang, L., Xu, B.: Configuration and tuning of sub-feeder protection in distribution network. Power Syst. Technol. **41**(4), 170–175 (2017). Apparatus **40**(5), 1589–1594 (2016)
9. Cui, H., Wang, C., Ye, J., et al.: Research of interaction of distributed PV system with multiple access points and distribution network. Power Syst. Prot. Control **43**(10), 91–97 (2015)
10. Shang, J., Tai, N., Liu, Q., et al.: New protection method for distribution network with DG. Power Syst. Prot. Control **40**(12), 40–45 (2012)
11. Liu, J., Zhang, X., Zhang, Z.: Improving the performance of fault location and restoration based on relay protection for distribution grids. Power Syst. Prot. Control **43**(22), 10–16 (2015)
12. Wu, R., Zhang, Y., He, Y., et al.: Study on sub-feeder protection installation in distribution systems. High Voltage Apparatus **42**(4), 281–283 (2006)
13. Zhao, M., Liu, Z., Xu, Q.: Analysis of on-off inrush characteristics of 10 kV distribution lines. Jiangsu Electr. Eng. **25**(1), 33–34 (2006)
14. Zhong, C.: The way of lowering the bad influence of the magnetic inrush current. Power Syst. Prot. Control **37**(23), 170–171 (2009)

Belligerent Simulation Method Based on the Multi-Agent Complex Network

Jian Du[✉], Huizhen Li, and Yipen Cao

Department of Fundament, Army Academy of Armored Force,
Beijing 100072, China
jiandu2531@163.com

Abstract. Common Lanchester belligerent model, it often ignore the position information of belligerent units and the autonomy of combat units. The multi Agent modeling technology and the complex network method are applied to combat modeling, the relationship between Agent is abstracted into many kinds of multi-Agent complex networks, the structure composition of the belligerent model based on multi-Agent complex network is discussed, and the components of the belligerent model based on multi-Agent complex network are analyzed in detail. Taking the belligerent process as the evolution process of multi-Agent complex network driven by Agent autonomous behavior, a belligerent model based on multi-Agent complex network is established. It is of great significance to simulate the combat process and study the combat mechanism by combining the multi-Agent modeling technology with the complex network method.

Keywords: Agent complex network · Warfare modeling · Combat unit

1 Introduction

Using the theory of complex adaptive system as the theoretical guidance, a new method is provided for modeling the war model based on the multi-Agent modeling technology. The multi-Agent modeling technology is based on the interaction of the host with the adaptability and other subjects or the environment to study the emergence of the whole [1]. In that modern high-tech condition, the battle of the past great army will gradually change into small-scale, multi-energy, integrated combat, small-scale, high-quality, well-equipped and quick-reaction sub-unit. The capability and adaptability of the combat troops are the important factors that affect the fighting capacity. The independent operational capability of a single combat unit, especially human factors, plays an increasingly important role in the course of the war [2].

2 Modeling Method Based on Multi-Agent Complex Network

Under this new war model, how the individual perception, killing ability, defense ability and decision-making ability will affect the war form, and how the individual killing situation and perception have changed in a specific war situation are all the directions worthy of our study. If each individual is regarded as a node, the node is

© Springer Nature Singapore Pte Ltd. 2020
M. Atiquzzaman et al. (Eds.): BDCPS 2019, AISC 1117, pp. 1437–1443, 2020.
https://doi.org/10.1007/978-981-15-2568-1_198

connected with the enemy node it can sense, which constitutes a perceptual network that changes with the engagement, and the belligerent network formed by the individual to attack an enemy individual must be included in the perceptual network [3]. The simulation process is understood by the evolution of multi-Agent complex networks, as shown in Fig. 1.

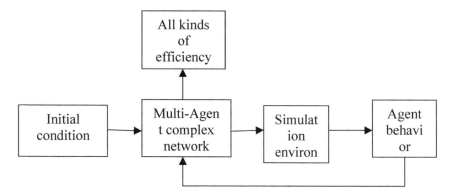

Fig. 1. Analysis method of complex network model based on multi-Agent

3   Modeling of the Engagement Structure

The definition of the engagement model based on the multi-Agent complex network is as follows [4]:

$$(\varepsilon, CA, Rn, CMP)$$

Among them, ε is a battlefield environment, Rn is a multi-Agent complex network, CA is an Agent individual, CMP is a model parameter.

3.1   Battlefield Environment

Battlefield environment ε is the space contained in the whole battlefield. It also includes objects distributed in battlefield space O, any Agent individual CA, obstacle, etc. [5]. Battlefield environment is a relative concept. The battlefield environment of a single combat unit Agent refers to the state of all other Agent except itself and the battlefield environment in which they are located.

3.2   Combat Unit of Agent

In the combat model based on multi-Agent complex network, combat unit Agent abstracts people and their weapons and equipment under operation control as the main body of combat Agent, as the actor in the combat model, and its behavior constitutes the driving force to the evolution of the network [6].

3.3 Multi-Agent Complex Networks

Multi-Agent complex networks Rn describe the interrelationship between objects O. The belligerent network $F_{R,B}$ composed of Agent belligerent relationship, the communication network $C_{R,B}$ composed of Agent communication relationship and the perceptual network $S_{R,B}$ composed of Agent perceptual relationship are subordinate to Rn each other. In addition, there are subordinate relations can be abstracted as network relations.

4 Operational Model Parameters *CMP*

The parameters of the combat model are the description of the conditions of the combat simulation and the simulation results, and which are the way for the model users to control the operation of the model in the simulation experiment and analyze the experimental results after the simulation experiment [7]. The parameters of combat model can be divided into control parameters and observation parameters.

4.1 Control Parameters

Control parameters include the model simulation of various initial conditions, simulation step size, simulation termination conditions and so on. For example, the initial value of the parameters for the combat environment model, the initial number of various combat Agents in the combat model, the initial values of the attributes and behavior rules parameters in the combat Agent model, the advance step of the simulation clock, the time conditions of the end of the simulation, the number and state conditions of the remaining combat Agent at the end of the simulation.

4.2 Observation Parameters

Observation parameters include all kinds of statistical analysis variables of the state transition process of combat Agent [8]. For example, the maneuvering trajectory of combat Agent, the types of other combat Agent engaged with it, the parameters of state analysis, the parameters of combat effect analysis, and other variables defined according to the purpose of simulation, and so on. Because the complex network method is introduced into the model, all kinds of network parameters are also included in the observation parameters. The structure of the engagement model based on multi-Agent complex networks can be described as follows Fig. 2.

The combat unit Agent obtains information from the battlefield environment and guides Agent to choose and execute its behavior. Then, after the Agent execution behavior, it has an impact on the battlefield environment. This interaction process can be represented in steps 2 and 4 based on the component diagram of multi-Agent complex network engagement model. As shown Fig. 3.

Which abstract the relationship between the Agent individuals into a multi-Agent complex network, so the Agent's influence on the battlefield environment can be expressed indirectly as the influence of the multi-Agent complex network on the

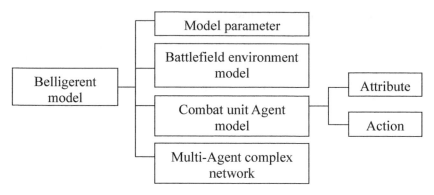

Fig. 2. Structure diagram of belligerent model based on multi-Agent complex network

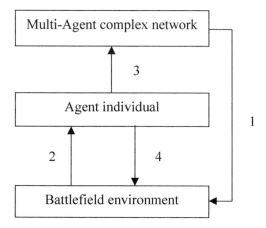

Fig. 3. Component diagram of belligerent model based on multi-Agent complex Network

battlefield environment, that is, the fourth step in the Fig. 3 can be decomposed into steps 2 and 3, Steps 1, 2 and 3 constitute a cycle. Based on the multi-Agent complex network, the engagement process is studied as the evolution of the multi-Agent complex network, which is based on the steps 1, 2 and 3 above: step 1, with the initial multi-Agent complex network as the starting point, the multi-Agent complex network under the initial conditions of different models gets different battlefield environments. Step 2, the Agent execution behavior is guided by different battlefield environments. Step 3, the performance of the Agent behavior is to change the complex network of the multi-Agent.

4.3 Battlefield Environment Model

The battlefield environment ε is the space E which is included in the whole battlefield, that is, the space of the battlefield O [9]. The system also includes the object space distributed in the battlefield space, the individual Agent of any Agent CA, the obstacle, and the like. The battlefield space E is defined in detail here. The field space can be represented by a two-dimensional coordinate or a set of points in a three-dimensional rectangular coordinate system and a number of k function values.

For two-dimensional battlefield space $V2$,

$$E = (x, y, f_1(x, y), L, f_k(x, y)),$$
$$x \in R, y \in R, k \in N$$

It's to make a difference for three-dimensional battlefield space $V3$, there are

$$E = (x, y, z, f_1(x, y, z), L, f_k(x, y, z)),$$
$$x \in R, y \in R, z \in R, k \in N$$

If E that space is a two-dimensional field, the distance between any two point $(x_1, y_1)(x_2, y_2)$ on it and the distance is the geometric distance between them:

$$L = \sqrt{(x_1 - x_2)^2 + (y_1 - y_2)^2}$$

All of the fighting units, Agent, and other objects in the war are distributed in this field. The combat unit, Agent, performs various intelligent activities in the battlefield space, and the other objects have a certain influence on the various behaviors of the Agent in the field of the battlefield.

5   Multi-Agent Complex Network

The multi-Agent complex network R_n element describes the mutual relation between the objects O. The communication network $C_{R,B}$, which is composed of the Agent, is a communication network $C_{R,B}$, which is composed of the Agent communication relationship $S_{R,B}$, and the perceived network environment composed of the Agent-aware relations is subordinate R_n to the communication network.

5.1   Perceptual Network

The perceived relationship between the opposing operational units is represented by a directed edge between the corresponding nodes, such as, for example, i discovery j a link denote $\langle i, j \rangle \in R \times R \cup B \times B \cup B \times B \cup B \times R$. All possible perceptual relationships form a sensing network, represented as a dynamically changing direction graph S, i.e.

$$S = S_{R,R} \cup S_{R,B} \cup S_{B,R} \cup S_{B,B}$$

in which

$$S_{R,R} = \{R, SE\}, SE \subseteq R \times R; S_{R,R} = \{R, B, SE\}, SE \subseteq R \times B;$$
$$S_{B,R} = \{R, B, SE\}, SE \subseteq B \times R; S_{B,R} = \{B, SE\}, SE \subseteq B \times B;$$

5.2   Communication

The communication relationship between the one-to-one operational units is represented by a directed edge between the corresponding nodes, If one of the combat units has a communication connection it is shown as a link between i and j, it is denoted $\langle i, j \rangle \in C_{R_0 \cup C_{B_0}}$ [10]. Whether there is a communication connection between the two nodes is set at the beginning of the operation of the model, and if the node is not damaged, the communication connection is always present.

When the base station has a communication connection from i direction j, it may communicate between r_i and r_j, it can be denoted $\langle i, j \rangle \in R \times R \cup B \times B$, In which $D(i, j)$ represent the distance between the two combat units, and r' is the radius of the communication for i. All possible communication relationships form a communication network which is represented as a dynamic and simple graph $C_{R,B}$, i.e. $C_{R,B} = \{R, CE\}$, $CE \subseteq R \times R; C_B = \{B, CE\}, CE \subseteq B \times B$.

6   Conclusion

The fighting relationship between the opposing combat units is represented by a directed edge between the corresponding nodes, such as the number of anti-personnel. All possible engagement forms a belligerent network, represented as a dynamic and dynamic two-part map, that is, a hand; a hand it is also possible to form a membership network by a membership relationship, etc., which are not listed here, it can think of to make a difference.

Acknowledgments. This research was supported by Army Academy of Armored Force (Grant No. 2019019).

References

1. Benowitz, K.M., Brodie, E.D., Formica, V.A.: Morphological correlates of a combat performance trait in the forked fungus beetle, Bolitotherus cornutus. PLoS One **7**(8) (2018)
2. Pyne, S., Biswas, J., Sinha, D.: A new systems approach to combat arsenic induced carcinogenesis. South Asian J. Cancer **2**(2), 82 (2013)
3. Krichevskii, G.E.: Dual-use technologies in production of XXI soldier combat uniform and civilian goods. Russian J. Gen. Chem. **83**(1), 185–192 (2013)
4. Hausen, M., Soares, P.P.: Physiological responses and external validity of a new setting for taekwondo combat simulation. PLOS One **12**(2), e0171553 (2017)
5. Kuznetsov, A.V.: A simplified combat model based on a cellular automaton. J. Comput. Syst. Sci. Int. **56**(3), 397–409 (2017)

6. Yoo, C., Park, K., Choi, S.Y.: The vulnerability assessment of ground combat vehicles using target functional modeling and FTA. Int. J. Precis. Eng. Manuf. **17**(5), 651–658 (2016)
7. Newell, N., Masouros, S.D., Pullen, A.D., et al.: The comparative behaviour of two combat boots under impact. Injury Prev. **18**(2), 109–112 (2012)
8. Matt Ritter, E., Bowyer, M.W.: Simulation for trauma and combat casualty care. Minimally Invasive Therapy Allied Technol. **14**(4–5), 224–234 (2005)
9. Seo, K.-M., Choi, C., Kim, T.G., et al.: DEVS-based combat modeling for engagement-level simulation. Simulation **90**(7), 759–781 (2014)
10. Swartz, S.M., Johnson, A.W.: A multimethod approach to the combat air forces mix and deployment problem. Math. Comput. Modelling **39**(6–8), 773–797 (2004)

Design of Power Carrier Wave Machine Based on TCP/IP Reference Model

Jun Qi$^{(\boxtimes)}$, Wenjie Zhang, and Yubo Liu

State Grid Liaoning Electric Power Supply Co., LTD., Information and
Telecommunication Branch, Shenyang 110006, Liaoning, China
qij0427@163.com

Abstract. The traditional power carrier wave machine based on digital theory
cannot meet the needs of large distance and stable transmission of actual power
signals. Therefore, it is necessary to design power carrier wave machine based
on TCP/IP model of computer network communication. According to the reference model of TCP/IP for computer network communication, it designs the
overall hardware circuit of carrier based on network interface layer, including
network layer, transport layer and application layer. The network layer solves
the problem of carrier signal attenuation and improves the effective communication distance by exploring the parent node networking algorithm and routing
strategy. It uses OFDM technology to demodulate carrier signal and improve the
stability of carrier communication. The experimental results show that the
designed carrier wave machine achieves data transmission and collection. In
addition, the networking algorithm at the network layer can efficiently construct
the network, reduce the packet loss rate, improve the communication stability,
and enhance the signal modulation ability.

Keywords: Computer network communication · TCP/IP reference model ·
Power carrier wave machine · Transmission algorithm · Data processing thread

Power carrier communication transmits power data through carrier modulation, which
has the advantages of low cost and high efficiency, and has high application value in
power system [1]. Due to the interference of the environment, power carrier communication will have the problems of signal attenuation and communication distance
shortening, and communication stability will be reduced. Seeking effective methods to
enhance the carrier communication distance and stability of power carrier wave
machine has become a hot topic in the current power field [2].

1 Power Carrier Wave Machine Based on TCP/IP Reference Model of Computer Network Communication

According to the reference model of TCP/IP of computer network communication, this
paper designs the carrier wave machine including network interface layer, network
layer, transport layer and application layer.

© Springer Nature Singapore Pte Ltd. 2020
M. Atiquzzaman et al. (Eds.): BDCPS 2019, AISC 1117, pp. 1444–1449, 2020.
https://doi.org/10.1007/978-981-15-2568-1_199

1.1   Overall Design of Network Interface Layer

The network interface layer designs the hardware circuit of the overall power carrier wave machine. The hardware structure of the designed carrier wave machine is described in Fig. 1. MCU initializes the system and operates the data through EP9315 embedded processor. Carrier signal is modulated and demodulated by the programmable communication chip, FPGA, of ProASIC3 series. MCU realizes the reading and writing processing of FPGA through different pins, and uses the carrier transmission circuit to filter and expand the modulated carrier signal. The carrier circuit performs filtering and gain adjustment on the collected signals [3].

1.1.1   Design of Network Layer

When designing the network layer of power carrier wave machine based on TCP/IP reference model of computer network communication, this paper analyzes the attenuation problem of carrier communication signal of carrier wave machine. It improves the communication distance of carrier signal by relay forwarding performance. The routing table is created while relay forwarding. By exploring the parent node networking algorithm, the networking efficiency and communication quality are improved. After the routing table is created, the relay forwarding scheme is implemented by sending processing functions and acceptance functions at the network layer [4]. If there is a problem with the routing table modification scheme in a certain path, it ensures that the carrier wave machine detects the communication terminal problem. At the same time, it connects with the network again and corrects the routing table in real time.

1.1.2   Design of Transport Layer

The signal communication process of power carrier machine is interfered by the environment, which will lead to data loss. The reliability transmission of data is realized by ACK scheme [5]. The sliding window scheme ensures the stability of data transmission in the corresponding period. Transport layer sets up algorithm rules through transfer mode control and sequence number control to implement the operation process.

1.1.3   Design of Application Layer

The upper layer of carrier is the application layer. It sets up the operation process of the overall carrier communication data, initially sets up the network layer, transport layer and hardware, and constructs the data operation thread. In addition, it completes the protocol operation of the data through the transmission of the network layer and the transport layer as well as the acceptance function.

1.2   Hardware Design of Power Carrier Wave Machine

1.2.1   Design of Transmission and Receiving Circuits of Carrier Wave Machine

The structure of the transmission and the receiving circuit of the carrier wave machine designed in this paper are described in Figs. 2 and 3 respectively. The power amplifier circuit is used to amplify the signal modulated by FPGA, so that the signal power can meet the transmission demand of power line. The switching control circuit regulates the opening and closing state of the power amplifier circuit. If the signal is transmitted normally, the power amplifier circuit runs smoothly, otherwise it is prohibited, which greatly reduces the power loss of the power amplifier circuit [6]. The filter circuit can filter out the harmonics in the carrier signal and obtain the valuable signal.

1.2.2   Module Design of Network Layer

(1) It explores the design of parent node networking algorithm, and explores the parent node networking algorithm based on a single secondary carrier node into the network. If all the carrier nodes are integrated into the network, the network will be completed [7].
The networking conditions are as follows: The number of carrier nodes in the network is n, the physical address of the primary carrier is 0, and the physical address of the secondary carrier is 1, 2, $\cdots$, n − 1; A carrier node in the network can communicate with one or more other nodes; Carrier nodes can detect SNR (Signal to Noise Ratio). When SNR is lower than P, the communication path can not transmit signals.

(2) The carrier node access network process, the single carrier node network access process is described in Fig. 4. The primary carrier and the secondary carrier $1 \sim 4$ are logical topology of the current network. The secondary carrier 5 is physically electrically connected to other carriers, but these nodes are not yet connected to the network. The access node is DY, and its integration into the current network process is as follows:

  ① DY broadcasts a signal detection packet, indicating that it is connected to the network;

  ② Signal detection packet nodes in the network count their physical addresses, regard the aggregation packet as the SNR of DY node to the current node, and transmit aggregation packet to the primary carrier.

  ③ The primary carrier obtains the node with the smallest hops in the node with the requirement of effective SNR, which is the best parent node, or takes the node with the highest SNR as the best parent node. The primary carrier collects N nodes satisfying the specification from all aggregation source nodes.

(3) DY group is used to transmit link detection packet for the host to the temporary parent node of its decision. In Fig. 4, DY transfers link detection packet from secondary carrier 2.

(4) DY' s temporary parent node collects the link detection packet and then feeds back the link detection packet to DY. If DY can collect the link detection packet of the parent node, the parent node will be integrated into the routing table.

(5) After DY has successfully implemented the detection, the decision-making operation packet is transmitted to the primary carrier. The decision-making operation stores the topology of DY and its child nodes in the node routing table through hierarchical feedback measures [8].

(6) When the primary carrier collects the decision-making operation packet, the notification packet is fed back into DY to realize DY's access to the network and integrate the carrier node into the network.

1.3   Module Design of Transport Layer

1.3.1   Design of Reliability Transfer Mode

The reliability transfer mode waits for the ACK response of the receiver after the sender transfers the data through the response transfer scheme. If no ACK response is obtained, data transfer is implemented again, otherwise subsequent packets are transferred; If the ACK response is not received after N times of continuous transfer, the target node can not transfer smoothly and needs repeated network access.

1.3.2   Basic Principles of OFDM Technology

OFDM technology can divide high-speed serial data of power carrier wave machine into large amount of low-speed parallel data. This can improve the pulse width of the symbol and enhance the anti-multipath fading ability of the carrier signal. By reducing the crosstalk between different sub-carriers by OFDM technology, sufficient frequency interval should be ensured to reduce the frequency usage.

OFDM signal is used to describe the set of orthogonal modulated sub-carriers transmitted in parallel, and its mathematical expression is:

$$x(t) = \sum_{n=-Y}^{Y} \left[\sum_{k=0}^{N-1} X_{nk} g_k (t - nT) \right] \tag{1}$$

$$g_k(\mathbf{t}) = \begin{cases} e^{2xft} & t \in [0, T_s] \\ 0 \end{cases}$$

$$f_k = f_0 + \frac{k}{T_s} k = 0, \ldots, N-1 \tag{2}$$

In the formula: Xn.k is used to describe k symbols to be transmitted in the signal stream of the n-th frame; Ts is used to describe the effective period of each OFDM symbol, and the sub-carrier quantity of OFDM is N; The central frequency of the k-th sub-carrier and the minimum central frequency of all sub-carriers are fk and f0. Sub-carriers in the frequency domain show orthogonal correlation, and carrier demodulation is completed based on this correlation, so:

$$\int_R g^k (t) g3dt = \begin{cases} T.\delta(k-1)1 = k \\ 01 \neq k \end{cases} \tag{3}$$

The obtained mathematical expression of the demodulator is:

$$X_n = \frac{1}{T_s} \int_{nT_s}^{(n-1)T_s} x(t) g_k^3(t) dt \tag{4}$$

OFDM signal can be obtained by FFT. After N modulation symbols are input and transformed by FFT at N points, data are obtained, and N time domain values of OFDM composite signal are obtained. After D/A conversion, the waveform of OFDM signal can be obtained. By multiplying the signal with the real carrier, the ODFM signal can be transformed into the required frequency band.

1.4   Module Design of Application Layer

The main program design process in the application layer module is described in Fig. 6. The main program collects the driver layer data of the configuration file [9, 10]. The parameters of driver layer are used to set the serial port and the original parameters of the FPGA device, create data processing threads for different devices, and control the communication data.

2   Experimental Analysis

The performance of power carrier is verified by experiments. By exploring the performance of parent node networking algorithm, the networking time and communication quality of this algorithm are analyzed. At the same time, the detection results of carrier communication packet loss rate are investigated to verify the reliability transfer performance of the power carrier wave machine in this paper.

2.1   Networking Method Detection

In order to detect the networking performance of this method, the time consuming of networking of this method, clustering routing method and neural network method are analyzed, as described in Fig. 7. It can be seen from the analysis of this figure that the neural network method has the highest time consuming of networking, and the method in this paper has the lowest time consuming of networking.

Communication quality is the ratio of SNR between host and slave, and the average SNR is positively correlated with communication quality. By analyzing Fig. 8, it can be concluded that, compared with the other two methods, the communication quality after networking of the method in this paper is the highest. The communication quality of the neural network method is the lowest, and there are high fluctuations.

2.2   Test and Analysis of Transport Layer Algorithms

The network server is the primary carrier and secondary carrier, and the client is host A and host B. Host A presents data from carrier color through TCP debugging assistant.

2.3   Modulation Detection

In order to detect the signal modulation of the power carrier wave machine designed in this paper, the simulation experiment is carried out. It sets the baud rate R = 50B for M signaling, and the carrier frequency is 3640 Hz and 3750 Hz respectively when data 0 and 1 are transmitted.

3 Conclusion

This paper designs a power carrier wave machine based on TCP/IP reference model of computer network communication. It consists of network interface layer, network layer, transport layer and application layer. Each layer cooperates with each other to achieve efficient transmission and collection of power carrier data. Thereby, it reduces the communication packet loss rate, improves the communication stability, and enhances the communication distance and reliability of the power carrier.

Acknowledgement. Supported by Science and Technology Project of Liaoning Province Electric Power Company Limited (2019YF-65).

References

1. Chen, X.: Design of monitoring system for communication power supply based on TCP/IP protocol. Chinese J. Power Sources **39**(8), 1760–1761 (2015)
2. Zhang, G., Xiong, X.: Interference characteristics and suppression techniques in power line communication. Telecommun. Sci. **32**(2), 182–188 (2016)
3. Deng, T.: Communication design of remote monitoring system based on secondary C/S model. Microelectron. Comput. **6**, 126–129 (2015)
4. Tao, Y., Zhao, Z.: Design of substation monitoring and control system based on streamline TCP/IP protocol. Chinese J. Power Sources **40**(5), 1127–1128 (2016)
5. Xiao, Y., Cao, M., Li, C., et al.: Adaptive impedance matching of the PLC based on PSO. Electric Power **47**(1), 133–137 (2014)
6. Hong, W., Xu, Z.: Design and development of power line carrier communication system for variable refrigerant volume air- conditioning systems. J. Comput. Appl. **36**(8), 2187–2191 (2016)
7. Xiang, M., Wen, C., Wen, C.: An IPv6 based fragmentation independent retransmission mechanism for power line communication. Power Syst. Technol. **39**(1), 169–175 (2015)
8. Xu, Z., Wang, X., Li, J., et al.: Electromagnetic compatibility influence of low voltage power line carrier communication and motor- driven household appliances. Electric Power Constr. **8**, 77–83 (2014)
9. Qian, C., Chen, L., Zhang, W.: Design and analysis of high- speed power line remote communications system based on OFDM. Electr. Measur. Instrum. **52**(24), 101–106 (2015)
10. Song, J., Tan, Y., Zhang, H., et al.: Adaptive impedance matching for power line communication based on quantum particle swarm optimization. Comput. Eng. Appl. **51**(1), 228–233 (2015)

Difference of Chinese and Western Ancient Architectural Characteristics Based on Big Data Analysis

Yuehang Zhao[✉], Lijie Yin, and Qi Wang

Jilin Engineering Normal University, Changchun 130052, Jilin, China
tougao_008@163.com

Abstract. Big data as a new information environment at present, there are few research and Application on architectural characteristics. In this context, this paper aims to analyze and mine the differences between Chinese and Western ancient buildings, find out the intrinsic relationship between Chinese and Western ancient buildings, design mining algorithms to analyze and classify the differences of ancient building features, and realize the information mining based on the large data of ancient building features. a certain amount of building data analysis is used as the driving force to carry out the experiment. The attribute classification of Chinese and Western ancient buildings is realized by using the information entropy splitting criterion in data analysis and data mining, and a decision tree is established to provide computer decision support for the difference of the characteristics of ancient buildings. The experimental results show that the classification algorithm proposed in this paper achieves more than 91% of the correct rate of classification for the characteristics of ancient buildings in China and the West, basically realizes the classification and recognition of the differences between the characteristics of ancient buildings in China and the West, and provides support for the design evaluation of the different architectural characteristics of ancient buildings in China and the West.

Keywords: Big data analysis · Data mining · Classification algorithm · Differences between Chinese and Western ancient architecture features

1 Introduction

As a record and witness of historical and cultural architecture, there are many obvious differences between Chinese and Western architecture. Ancient Chinese architecture is a prominent representative of ancient oriental architecture [1]. As a part of cultural heritage, ancient Chinese architecture is of great significance, especially in the process of digitalization and modelling of these buildings [2]. Western architecture has experienced different historical periods with different styles, and has obvious characteristics of the development period [3]. With the advent of the era of big data, the gap between the ability of data generation and the ability of big data analysis is getting bigger and bigger [4]. Traditional data analysis models can no longer meet the requirements of large data, so a new data analysis model is urgently needed to analyze and process large data in various fields [5]. The ability of big data to process large amounts of data and

M. Atiquzzaman et al. (Eds.): BDCPS 2019, AISC 1117, pp. 1450–1457, 2020.
https://doi.org/10.1007/978-981-15-2568-1_200

extract useful insights from data has radically changed society. This phenomenon has been applied in many industries including construction industry [6]. Data mining technology can be used to find features in events and predict them [7]. A large number of data mining and data analysis techniques have been applied in various fields of the construction industry. Researchers have used multiple correspondence analysis, decision tree, decision tree set and association rules to analyze the Iranian National Building Accident Database from 2007 to 2011 [8]. Cheng [9] extracts useful information from building defect database by establishing data analysis model. The information expressed in association rules can be predicted and analyzed by building defect. In order to deal with the increasing amount of building information and data generated in the life cycle of construction projects, visualization, information modeling and Simulation in large data technology have become the key to the development of capital facilities and infrastructure [10].

In this paper, data analysis and data mining theory in big data technology are introduced into the comparison of the characteristics of Chinese and Western ancient buildings, and a classification model based on data mining method is established for the differences between Chinese and Western ancient buildings.

2   Method

2.1   Architectural Characteristics of Ancient China and the West

As a witness and recorder of history and culture, ancient architecture is a kind of art that solidifies the cultures of all over the world. There are many obvious differences between Chinese and Western ancient architecture, which embodies different cultural connotations, styling styles and aesthetic concepts. This paper expounds the characteristics of ancient Chinese and Western architecture from the following perspectives:

bModeling characteristics: most of the external shapes of ancient Chinese buildings are tall, closed and axisymmetric. The courtyards of houses are inward-looking. The layout between buildings is relatively compact. Each courtyard has a clear division of labor, rich courtyard shape, and the internal courtyard is square and reasonable. The main concept of Western architecture can be obtained from any direction of the front of the western ancient buildings, but the wall is not closed or depressed.

Decoration characteristics: Chinese traditional architectural decoration is an important part of ancient architecture, in addition to having different practical values, there are strong regional cultural characteristics and artistic style of the times. The sculpture and window lattice decoration on the doorway are not only beautiful but also functional. The western ancient architectural decoration mostly embodies the Western civilization of ancient Greece.

Material Selection Features: Most ancient Chinese buildings are based on wood and soil, brick and earth, etc. The basement of palaces and walls of ordinary houses are mainly rammed earth walls or adobe walls, showing basically a yellow system, mostly the use of yellow system color; while the material of ancient western buildings is mainly stone, which is reflected ard and strong in architecture.

Architectural objectives: The ultimate goal of ancient Chinese architecture is "harmony between man and nature", which shows the high harmony between ancient architecture and nature through "artistic conception". Architectural images emphasize the spirit of traditional culture and harmony. Western ancient architecture has a stronger revolutionary tradition, and its characteristics are obvious at every stage of development.

2.2 Big Data Analysis

As the basis of big data analysis, big data is also the essence and foundation of the "Internet +" era. Big data technology is gradually changing the way people live and understand the world. Compared with the previous data, big data is a collection of the previous data, which describes the more detailed and deeper content within the reachable range, and covers the data edges that could not be reached in the past, including the previously data non-quantifiable content. Some researchers define big data as:

It is a kind of data collection, which is far beyond the capability of traditional database software tools in acquisition, storage, management and analysis.

The data scale is huge, the speed of data flow is fast, and it has various data types and low-density data value.

It can not only express direct causality, but also predict the relativity and possibility of development between things.

Big data analysis is a process of processing data, that means the data processing process of solving problems and assisting prediction and decision-making by processing and processing the collected large data to a certain extent. Big data analysis mainly includes acquiring basic data, data mining, data model analysis and visualization of data. Data mining is to select data suitable for data mining from data sources of different structures, pre-process data, i.e. data denoising and re-processing, then transform data and get data model, and finally mine data.

2.3 Data Mining Method for Chinese and Western Architectural Characteristics

(1) Classification mining method
    Classification is used to determine which predefined target class the mining object belongs to. There are many kinds of classification criteria for ancient buildings, according to a certain standard classification can be divided into many types and several sub-categories. The classification standards based on architectural features can be the use function, architectural style, architectural technology, etc.

(2) Cluster Mining Method
    Different from classification mining methods, clustering analysis belongs to unsupervised learning, and clustering analysis does not need to rely on the classification method of training set. The static data are categorized without considering the change of data. When conditions change, in order to get different

clustering results, information elements or correlation functions can be transformed.

(3) Conduction mining method

A transformation not only causes the change of the object of action, but also causes the change of the object related to it because of various correlations between things. The former transformation is called active transformation, and the conduction transformation of the former is usually the change caused by the related object.

2.4   Data Classification Algorithms

The information entropy of sample set is also called prior entropy or average uncertainty. Assuming that the data sample set of a sample number is defined as sample X, the number of samples is x, and the information entropy of X refers to the information content of sample X. The classification algorithm determines the selection of classification attributes according to the information content.

Assuming a total of K samples belong to the category set Y, then $Y = \{y_1, y_2, \ldots y_{k-1}, y_k\}$ denotes the set of categories. If the set of data samples X is further divided into k subsets, then $X = \{x_1, x_2, \ldots x_{k-1}, x_k\}$ and the number of samples in the sample set X_i are x_i, then the information entropy of the sample set is

$$E(X) = \sum_{i=1}^{k} u_i log_2 u_i. \quad (u_i = \frac{x_i}{x}) \tag{1}$$

In a given sample set X, H with L different attribute values is defined as containing $\{h_1, h_2, \ldots h_{l-1}, h_l\}$, $X = \{x_1, x_2, \ldots x_{l-1}, x_l\}$ denotes that set X is divided into subsets with L samples. x_{ij} means the numbers of $x_j(j = 1, 2, \ldots l)$ that belongs to the set $y_i(i = 1, 2, \ldots k)$; The probability that u_{ij} represents the sample subset x_j and the sample belongs to class y_i is:

$$u_{ij} = \frac{x_{ij}}{x_{1j} + x_{2j} + \ldots + x_{k-1j} + x_{kj}} \tag{2}$$

The information entropy of subset x_j is

$$E(x_j) = \sum_{=1}^{k} u_{ij} log_2 u_{ij} \tag{3}$$

The information entropy of the sample set divided into two parts by attribute H is:

$$E(H) = \sum_{j=1}^{k} \frac{x_{1j} + x_{2j} + \ldots + x_{k-1j} + x_{kj}}{x} E(x_j) \tag{4}$$

Information gain can be calculated as: $G(\frac{X}{H}) = E(X) - E(H)$; The attributes selected according to the decreasing speed of information entropy are called information gain. The idea of this classification algorithm is based on the calculated information entropy value. If the attribute value obtained is the highest information gain, the classification attribute of the node is selected as this attribute. This ensures the maximum amount of information after classification and minimizes the uncertainty of samples.

3 Experimental Simulation

3.1 Experimental Data

In this experiment, 356 pictures of famous Chinese and Western ancient buildings were selected as samples, and key feature points such as porch, window lattice decoration, roof shape, material, courtyard and color were marked. The building was an object, and the attributes were shown as relevant feature points, as shown in Table 1.

Table 1. Data set of Chinese and Western ancient architecture

No.	Porch	Window lattice	Eaves	Material quality	Courtyard	Color
1	Corridor	Carve patterns	Dipper	Wood	Four-enter courtyard	Cyan
2	Column	Woodcut	Shallow-vaulted roof	Wood and brick	Introverted enclosed space	Chinese red
3	Statue	Circular	Arch of vault	Stone	Lawn	White
4	Column type	Curves and circles	Pointed arch	Stone	Lawn	Yellow
5	Marble sculpture	Fan-shaped	Circular dome	Marble	Lawn and statues	White
6	Corridor	Woodcut modeling	Wingspan	Wood and brick	Buildings, walls, corridors	Chinese red

3.2 Establishing Classification Attributes of Chinese and Western Ancient Architectures

This paper establishes a classifier for Chinese and Western ancient building data sets, and verifies the classification results by testing data. This paper evaluates the effect of the classifier by using the indexes of correct rate and correct classification result, chooses the attributes of the classification of the ancient building data set, and then makes the judgment of "whether or not Chinese and Western architecture" according to the decision tree. Firstly, the information entropy is calculated according to the formula. According to the classification principle of this paper, the information gain value is compared and the window lattice is selected as the classification attribute. The basic steps of the algorithm are as follows:

(1) Establish the initial root node of the data set. It decides whether the sample data of the root node are all in the same class firstly. The algorithm ends and the node is marked as a leaf node when it is yes. Otherwise, select the attributes of the classification and make the next classification.

(2) Dividing samples, comparing each value of the classification attribute, and expanding it to a corresponding branch.

(3) Repeat the top-down loop call while executing steps 1–2 until the conditions are met before jumping out of the loop. The algorithm is over.

4 Experimental Results and Analysis

(1) Simulation results

According to the classification accuracy, classification time, iteration test and other indicators, this study evaluates the classification effect of the classification algorithm on the different characteristics of Chinese and Western ancient buildings. The results are shown in Table 2.

Table 2. The effect of classification algorithms on different classification of Chinese and Western ancient architectures

	Accuracy of feature point extraction	Search efficiency	
		Extraction time (s)	Search times (frequency)
Porch	93.4%	126	26
Window lattice	91.3%	265	32
Eaves	93.5%	8.51	19
Material quality	94.1%	4.1	7
Courtyard	96.2%	7.21	6
Color	91.0%	6.52	8

(2) Analysis of experimental results

The experimental results show that the classification algorithm separately classifies and detects the key feature points such as portico, window lattice decoration, roof shape, material, courtyard, color and so on, and compares them according to the classification accuracy, extraction time and extraction times. The classification accuracy of selecting six feature attributes of ancient buildings is very high, among which the accuracy of gardening is the highest, reaching 96.2%.

(3) Classification Performance Evaluation

This study evaluates the classification effect of Chinese and Western ancient building feature classifiers, and the comparison result of the accuracy rate is shown in Fig. 1.

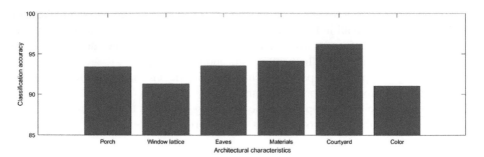

Fig. 1. Classification correctness rate of ancient architectural characteristic differences

Among the six characteristic attributes of ancient buildings, window lattice features have the most extraction time and the number of iterations has reached 32 times due to the location of feature points detected by window lattice. In the whole classification search process, the number of iterations for Chinese ancient buildings is more than that for Western ancient buildings, and the classification time for Chinese ancient buildings is also higher than that for Western ancient buildings.

5   Conclusion

Based on the analysis of the characteristics of ancient buildings in China and the West and the connotation of data analysis in the large data environment, this paper expounds the characteristics of data mining based on ancient buildings and common data mining methods. With the help of the basic theory of data mining, the theoretical framework of building feature data mining is established, and a realizable data mining method is formed which can be used to distinguish the features of ancient buildings. Finally, the classification algorithm in data mining is combined with the characteristics of China and the West ancient buildings through experiments, and the classification algorithm is used to analyze the differences between the characteristics of ancient buildings in China and the West, and the extracted features are classified into different building categories through association rules. The experimental results show that the classification algorithm can extract different characteristic parameters of ancient buildings quickly and accurately, and has a good classification effect.

This research fully utilizes data analysis and data mining technology of big data technology, and provides a new method for analyzing the characteristics of ancient buildings in China and the West. After classifying the differences of the analysis feature points of ancient buildings by the comprehensive classification algorithm, we can preliminarily understand the regular characteristics of the differences between Chinese and Western buildings, and get the specific architectural form characteristics through the obtained classification feature rules. On this basis, we can distinguish the geographical areas of ancient buildings and further predict the geographical areas of buildings. Domain. The next step will be to use association rules for classification and prediction, such as predicting the geographical division of buildings, so as to guide the selection of architectural design strategies in specific areas in the future.

References

1. Liu, J., Wu, Z.K.: Rule-based generation of ancient Chinese architecture from the Song Dynasty. J. Comput. Cult. Heritage (JOCCH) **9**(2), 7 (2016)
2. Li, L., Tang, L., Zhu, H., Zhang, H., Yang, F., Qin, W.: Semantic 3D modeling based on CityGML for ancient chinese-style architectural roofs of digital heritage. ISPRS Int. J. Geo-Inf. **6**(5), 132 (2017)
3. Amiri, N.: Modernism and postmodernism in architecture, an emphasis on the characteristics, similarities and differences. Turkish Online J. Des. Art Commun.-Tojdac 1626–1634 (2016)
4. Alyass, A., Turcotte, M., Meyre, D.: From big data analysis to personalized medicine for all: challenges and opportunities. BMC Med. Genomics **8**(1), 33 (2015)
5. Wang, D., Sun, Z.: Big data analysis and parallel load forecasting of electric power user side. Proc. CSEE **35**(3), 527–537 (2015)
6. Bilal, M., Oyedele, L.O., Qadir, J., Munir, K., Ajayi, S.O., Akinade, O.O., Owolabi, H.A., Alaka, H.A., Pasha, M.: Big Data in the construction industry: a review of present status, opportunities, and future trends. Adv. Eng. Inform. **30**(3), 500–521 (2016)
7. Goh, Y.M., Ubeynarayana, C.U.: Construction accident narrative classification: an evaluation of text mining techniques. Accid. Anal. Prev. **108**, 122–130 (2017)
8. Amiri, M., Ardeshir, A., Fazel Zarandi, M.H., Soltanaghaei, E.: Pattern extraction for high-risk accidents in the construction industry: a data-mining approach. Int. J. Injury Control Saf. Promot. **23**(3), 264–276 (2016)
9. Cheng, Y., Li, Q.: GA-based multi-level association rule mining approach for defect analysis in the construction industry. Autom. Constr. **51**, 78–91 (2015)
10. Leite, F., Cho, Y., Behzadan, A.H., Lee, S., Choe, S., Fang, Y., Akhavian, R., Hwang, S.: Visualization, information modeling, and simulation: Grand challenges in the construction industry. J. Comput. Civil Eng. **30**(6), 04016035 (2016)

Development and Application of the School Physical Education Computer Integrated Management System

Zhikai Cao[✉] and Zhongqi Mi

Jianghan University, Wuhan 430000, Hubei, China
e13685231980@126.com

Abstract. With the development and progress of the modern science and technology, the computer technology has been widely used in all walks of life, which helps to improve the work efficiency and reduce the management costs. This paper briefly analyses the development and application of the computer integrated management system of the school physical education. The application of the computer and network technologies in the school physical education and teaching management is not very high. In order to make the computer technology and the network management technology better applied and promoted in various disciplines and especially better applied in the management in the school physical education and teaching, this paper focuses on the analysis and research of the computer physical education and sports network management.

Keywords: School · Sports computer · Integrated management system · Development mechanism · Application strategy

1 Application Principle of the Computer Integrated Management System of the School Physical Education

Colleges and universities have developed a lot of projects in the field of the computer application in sports, but the establishment of the sports computer application system and management is still in the exploratory stage [1–3]. It is just at the beginning. Little is known about the problems that should be paid attention to in the establishment of the sports computer application system and the management system.

1.1 The Control Function of the Campus LAN

Building a good campus LAN, various departments can easily query and modify the information and the data, reduce the burden of the manpower and the material resources to a certain extent, and truly realize the control function of the office automation [4]. For example, linking the web pages of various education departments through the LAN can conveniently query various documents and information of the physical education and teaching, and linking the school library and the information room can be better conducive to students' learning.

M. Atiquzzaman et al. (Eds.): BDCPS 2019, AISC 1117, pp. 1458–1464, 2020.
https://doi.org/10.1007/978-981-15-2568-1_201

1.2   Remote Access and Control Function to the Internet

By accessing the Internet, we can timely understand the various developments of our major and facilitate the students to accept the new knowledge learning [5]. This function is the key to connecting the external network.

1.3   Components of the Computer Integrated Management System for School Physical Education

The system mainly consists of seven subsystems, which are installed on the server and connected with various working equipment of various departments to form a complete working system. It is convenient to control and manage the school physical education work. Seven subsystems include: the physical education and teaching, the physical fitness testing, the sports competitions, the extracurricular sports activities, the file management, the professional team training, and the venue equipment management and so on.

2   Characteristics of the Computer Application Management System of the Physical Education in Colleges and Universities

2.1   Comprehensiveness

The college sports computer application management system includes the teaching, the mass sports, the after-school training, the scientific research, the logistics management and other aspects, and each subsystem has many functional modules. For example, the management system of the physical education is composed of students' achievements in the physical education, and has the functions of statistics and analysis, which can make the sports computer management scientific and systematic.

2.2   Usualness

The data processing and analysis of the daily work and the teaching of the physical education in colleges and universities are often carried out. In a sense, the computer can provide faster, more comprehensive, more correct and timelier services for the sports managers than the manual, eliminating a large number of the manual scoring, summary, tabulation, printing and other work, so as to facilitate the inquiry, modification and storage. For example, to inquire about the students' P.E. results, the students' achievements in each semester can be displayed and printed by inputting their school number or department, class and name.

2.3   Periodicity

Many sports work in colleges and universities has periodic characteristics. For example, the track and field games, basketball, volleyball, football, table tennis, badminton and other competitions are usually carried out once a year. The manpower, material and financial resources are wasted and the time is wasted when arranging them manually. If it is done by the computer, only a few parts of the date, personnel, referees and athletes can be printed and used.

2.4 Marginality

The application of the sports computer in colleges and universities is the result of the close combination of the computer science and the sports science. The application of the sports computer is the process of the combination of two kinds of the science and technology. Therefore, we should pay attention to the organic combination of the two kinds of the science and technology, and do a good job in the research of the sports computer application.

3 Attention Problems in Implementing the College Sports Computer Application Management System

3.1 Perfecting the Leadership

This work should be composed of the main leaders of the Ministry of Physical Education and the experienced intermediate and senior teachers. The main responsibility is to formulate the computer application programs, the implementation methods, the use of funds, and the staffing and personnel training, and solve problems arising in the implementation and application process.

3.2 Training of the Multi-level Talents

A series of characteristics of the sports computer application management require a multi-level, reasonable knowledge and intelligent group. There are senior programmers (responsible for the design of the sports system, the implementation, optimization and application of the sports system, and the entire process of activities), programmers (responsible for the implementation, optimization and application of the system), and operators (responsible for the application of the sports system). From the perspective of the development and long-term consideration, the application and management of the sports computer in colleges and universities should be well done, and the knowledge and the intelligent group structure of all kinds of the personnel should be reasonable. Therefore, we must pay attention to the training of the senior and intermediate personnel responsible for the development and the daily maintenance and optimization of the sports system in the computer software personnel. In the application of the sports system, we should pay attention to the training of operators, provide necessary assistants for the middle and senior personnel, and meet the requirements of the input and the application of the daily sports information. Because of the marginality of the sports computer application, while training the above-mentioned personnel, we should also pay attention to training them with the corresponding level of the hardware and the software knowledge and all kinds of the sports knowledge, so that they can gradually develop into the high-level dual-disciplinary marginal talents at different levels.

3.3   Enhancing the Secondary Development Function of the Software

As more and more computer projects are applied in the field of the college sports, the scope is expanding, the functions are improving, and the links between the developed systems are getting closer and closer. The continuous improvement of the development and management of the school physical education will put forward higher requirements for the application of the computer in the physical education, which requires the preparation of the second development system and the enhancement of the maintainability, flexibility and portability of the software.

3.4   Reserve the Data Interface to Prepare for the Opening of the Information High-Speed Channel

The sports computer application management system has close and numerous information exchange relations with other information systems. With the wider application of the computers in various departments, the widespread use of the information networks and the opening of the information networks, and the wider interconnection between various departments and schools, the sports computer management information system cannot be isolated and closed. Besides collecting and using the data from the inside of the system, it must also collect the information and data from the outside of the system and other sports management systems, to realize the sports information and data sharing.

4   Development and Application Strategies of the School Physical Education Computer Integrated Management System

4.1   Test Settings

Under the guidance of the user's requirements, the system provides the users with the corresponding optional settings, including the assessment items, the assessment results, the functional testing and the physical examination parameters. The comprehensive management system can modify or delete all the above management items, and provide the optional setting function, which can help the specific objectives of the school sports management to be formulated reasonably. In this system, users only need to input the original score into the computer, and the system will automatically evaluate the grade of the score to achieve a mixed score.

4.2   Statistical Analysis

Whether it is the test data at different levels or at different ranges, the system can be analyzed and counted accurately. For example, first calculate the total score of a specific group, and then calculate the standard deviation of the average and percentage, and finally draw the statistical chart and tables according to the corresponding data. The analysis dimension of the system is multi-faceted. It can cover students' gender, age and class. When the single dimension is used as the analysis angle, it can also extend from this dimension to the other dimensions.

4.3   Integrated Management

The coordination management in different fields has always been a difficult problem in the school physical education management. The system can effectively achieve the comprehensive management of the objectives in different fields, run through the management objectives in the field of the physical education management and other fields, expand the scope of the resource sharing, and improve the level of the data transmission. The system basically realizes the "one-stop" management mode, covers all aspects of the school physical education management, overcomes the incompatibility drawback of the traditional management system software, and improves the overall level of the school physical education management.

4.4   Improving the Teachers' Computer Level and the Application Ability of the Technology

With the advent of the information technology, the new requirements have been put forward for the popularization and application of the computer integrated management system. In order to make good use of this technology to serve the teaching practice, on the one hand, schools should strengthen the construction of the personnel team and actively improve the teachers' computer integrated management, and especially the physical education teachers, through the regular training and the centralized training. On the other hand, schools should start from the objective needs of the physical education teaching practice, recognize the role and effect of the computer integrated management system teaching technology, and purchase enough hardware equipment for the physical education teaching, and gradually strengthen the government's investment in the physical education of the computer integrated management system.

4.5   Strengthen the Practice and Innovation of the Technology in the Application of the Physical Education and Teaching

The teaching technology of the computer integrated management system belongs to the category of the computer technology, which has certain requirements for the quality of talents. Therefore, in the process of mastering and applying the computer technology, the physical education teachers should pay attention to the renewal of the technology application and the accumulation of the experience, unable to adapt to the needs of the environment and improve their ability. Therefore, the physical education teachers can constantly sum up their experience in the application process and feedback the information through the communication and the innovative exploration of the teaching mode, so as to continuously improve their teaching abilities and effects.

4.6   Increase the Construction of the Hardware Facilities

The computer integrated management system technology needs the assistance and support of certain quantity and the quality of the hardware facilities. For this reason, after recognizing the importance of the computer integrated management system technology in the physical education and teaching practice, the relevant departments of schools should increase their investment in the computer aided technology hardware facilities, consolidate and build up the personnel team, and actively use the modern information technology to carry out the physical education teaching model in light of the actual situation of the school teaching. The innovative exploration of the computer integrated management system technology will be applied to the extreme with the equipment and facilities provided by the schools and the modern information technology, and the application effect of the computer integrated management system technology in the teaching will be excavated.

4.7   Strengthen the Development and Promotion of the Sports Assistant Teaching Software

Major in the market-oriented road, and strengthen the separate development and promotion of the sports assistant teaching software, supplemented by the independent innovation of the physical education teachers. The software companies develop the software, employ the front-line sports teachers and the relevant experts as consultants, and develop some intelligent and operable software that conforms to the level of the modern sports teachers and students' acceptance abilities. The relevant competent departments should vigorously promote the use of the developed technical software, and organize the relevant teachers to conduct the unified learning and practice, so as to train the software application talents for the grass-roots schools.

5   Conclusion

Introducing the teaching technology of the computer integrated management system in the physical education and teaching is a new way to reform the teaching mode of the physical education. It is of great practical significance to strengthen students' physical education qualities and improve the teaching efficiency. In the practice of the physical education and teaching of the computer integrated management system, the computer is the basis of the hardware and the support the guarantee for realizing the teaching goal of the computer integrated management system. The design and development of the software is the key problem for the effective application. This requires the relevant departments to carry out the professional argumentation and research on the realization of the subjects and the teaching objectives, and then to plan the basic procedures to achieve the goals according to different teaching objectives and students' actual situations. Teachers should constantly improve their professional qualities and computer application levels to lay a solid foundation for the implementation of the modern computer integrated management system and the training of new sports talents, in order to ensure the healthy and smooth development of our physical education.

References

1. Ma, Y.: Development and application of the school physical education computer integrated management system. Electron Technol. **4**, 41 (2012). 37
2. Qiu, B., Ding, W., Ma, Q.: Development and application analysis of the school sports computer integrated management system. New Technol. New Process **8**, 55–58 (2015)
3. Wu, J.: Development and research of the comprehensive management system of the school physical education and health. J. Henan Inst. Educ. (Nat. Sci. Edn.) **2**, 81–83 (2010)
4. Zhong, M.: Development of the integrated management system for college students. Comput. Program. Skills Maintenance **14**, 42–45 (2009)
5. Zhou, F., Yi, C.: Brief discussion on the construction of the computer application management system for college sports. Charming China **13**, 156 (2011)

Prevention and Control of Financial Risks by the Financial Computer System

Lianxiong Du[✉]

School of Economics, Guangdong Peizheng College, Guangzhou 510000,
Guangdong, China
jianhuazhao@yeah.net

Abstract. With the continuous development of the Internet and the computer technologies, the computer and the network risks have attracted more and more attention. Because of the emergence of various kinds of the network security problems, the development of the financial industry has been seriously affected. In this paper, starting from the development of the financial industry, the main causes of the risks of the computer systems in the financial industry are discussed. Based on the analysis, this paper puts forward some reasonable suggestions on the risk prevention of the computer system in the financial industry.

Keywords: Finance · Computer system · Financial risk · Prevention mechanism · Control strategy

1 Financial Risks in the Financial Computer System

The computer financial risk is mainly manifested in the weak security technology and measures of the financial computer system [1, 2]. Facilities are relatively backward, and the anti-virus ability of the system is poor, and the system cannot effectively prevent the "hacker" intrusion. This paper discusses the main causes of the financial computer risks caused by the computer system and the preventive measures.

1.1 The Security Technology of the Financial Computer System Is Backward

The information technology has a wide range of applications and fast changing speed, which often makes some financial enterprises have to face the application challenges of the new information technologies before fully developing and utilizing the functions of the financial computer systems currently in use [3]. This makes the security precautionary technology of some financial computing systems always in the end. The latter position cannot deal with and prevent the sudden financial security risks in time [4]. In addition, due to the backwardness of the financial computer system security technology, it also leaves some highly skilled network hackers a chance to take advantage of. Many network hackers, tempted by the huge economic interests, often take risks by forging credit cards for malicious overdraft, using the network technology to crack the bank passwords, modifying the business data and other illegal purposes such as stealing the bank funds [5].

© Springer Nature Singapore Pte Ltd. 2020
M. Atiquzzaman et al. (Eds.): BDCPS 2019, AISC 1117, pp. 1465–1470, 2020.
https://doi.org/10.1007/978-981-15-2568-1_202

1.2   Lack of the Professional Ethics of Software Developers

The software developers are the key personnel in the development of the financial computer systems. The performance and operation quality of the financial computer systems depend to a large extent on the knowledge, technology and ability levels of the software developers. The software developers can directly access the internal information of the financial enterprises. Although the current financial enterprises have high requirements for the quality of the software development personnel, due to the lack of the professional ethics of some staff members, it also greatly exacerbates the financial risks of the enterprises.

1.3   Improper Management of the Financial Computer System

The financial computer system cannot be built once and for all, but needs to be actively managed by the relevant personnel to maintain the security and stability of the computer system operation. At present, there are many problems in some financial computer systems, such as the improper management, which provide opportunities for some non-internal operators to access and use the computer system, which also brings great challenges to the security of the computer systems.

2   The Root of the Financial Risks in the Financial Computer System

The risks of the computer systems in the financial industry will directly damage the economic interests of the customers, hinder the scientific development of the financial industry, and then affect the development of the local economy and society. We should improve the risk prevention awareness of the relevant personnel in the financial industry in an all-round way. In order to improve the level of the computer system risk prevention in the financial industry, we should improve the level of the computer system risk prevention in the financial industry in an all-round way, and establish the relevant risk early warning mechanism.

In the entire process of the financial market operation, some factors such as the policies and the economic situations and so on are closely related to the overall financial activities, so these corresponding factors play a very important role in the operation of the overall financial market. And the relevant technical levels of the corresponding financial staff and the errors have a considerable impact on the operation of the financial market as a whole. For the financial industry, every financial enterprise must have a stable financial risk management and control system, and the risk control work must be dealt with with more energy. The corresponding audit work is very critical, and the level of the work is directly related to whether the risk management and control can be guaranteed.

Because the biggest platform for the financial activities is the Internet, which is widely spread, the Internet finance is essentially a model of the virtual economy. Compared with the traditional financial model, the Internet finance is completely virtualized for the object of the financial services, audiences and even the media.

Therefore, for both sides of the transaction, there are some shortcomings in the process of the transaction, such as the unclear identity information of both sides, and the asymmetric information sources of the credit qualification evaluation of both sides, which makes the transaction more like a knife-edge transaction, thus increasing the degree of the financial risks.

In the design and the technical operation of the Internet financial products, each Internet financial platform or enterprise has its own unique operation methods and modes, which make the vast number of users lose their personal interests because they are not familiar with the relevant knowledge, complicated operation process and other reasons leading to operational errors. At present, the main regulatory background of China's current financial industry laws and regulations is based on the business development of the traditional financial industry practitioners and the standardization of the transactions. However, for the Internet financial industry, which is developing so fast and radiating so widely nowadays, there are obvious loopholes in the existing laws and regulations, both in the applicable situation and in the scope of the supervision.

3   Preventive and Control Measures of the Financial Computer System for the Financial Risks

3.1   Strengthen the Safety Consciousness of the Financial Practitioners, Raise Their Ideological Awareness and Strengthen Their Ideological Education

In the process of the financial risk prevention, the corresponding accountants will play a very important role. Their moral and professional qualities are directly related to the level of the financial risk management and control. Therefore, in order to improve the level of the financial risk management and control, it is particularly crucial to improve the corresponding comprehensive quality of the accounting personnel in the financial risk management and control. This should be considered from the following aspects: to strengthen the ideological and moral training of accountants, to enable accountants to have good ideological and moral, and not to damage the financial order in order to obtain money with ignorant conscience. At the same time, in the ideological work, we should also build the concept of the risk prevention of accountants, so that the ability of accountants to control risks is constantly improved. Corresponding measures should be taken to improve the working level of accountants, enhance their professional quality and constantly expand their knowledge. Increase their areas of study, so that their ability to control the financial risks continues to improve, and improve the quality of their specific financial accounting work, so that their ability to control the financial risks continues to improve.

3.2   Strengthen the Management, Improve the Internal Control System of Safety, and Form a Strict System of Prevention

Due to the improper management, the current financial computer system is also facing greater security risks, so the financial enterprises should further improve the management of the financial computer system. In order to avoid the confusion of personnel and unclear responsibilities in the computer system management departments, the relevant management departments should improve the management system and actively build a strict prevention system. For example, the relevant management departments can institutionalize the management regulations, clearly divide the responsibilities of the staff in various positions, stipulate that the computer operators and accountants should not be mixed, and strengthen the management of the password setting, and regularly change and remove the traces of the passwords in the computer systems. The rapid development of the social economy has put forward higher requirements and expectations for the financial field, and the corresponding work should be carried out at a high level. The corresponding enterprise capital control work needs higher level requirements, which need to strengthen the macro-control of the financial enterprises, and the corresponding financial supervision should be further strengthened. In order to achieve their work, efforts should also be made in the following directions. First, efforts should be made to stabilize the exchange rate and avoid the emotional fluctuations and adverse reactions in the international market. The second is to strengthen the construction of the foreign exchange market and stabilize the flow of the related funds. In a word, the macro-control work should be carried out effectively and scientifically to stabilize the flow of funds in the international market and avoid the outbreak and intensification of the financial risks in the market.

3.3   Strengthen the Research of the Financial Computer Security Technology to Improve the Security of the System

In addition to strengthening the management of the staff, the security maintenance of the computer system also needs to seek breakthroughs from itself, and constantly improve the security performance of its application software. In view of this, the initial research and development personnel of the computer system should keep pace with the times. On the basis of the careful analysis of the security performance of their own system, various network hacker technologies should be carefully studied, and the security maintenance of the data communication should be strengthened through the encryption technology, the firewall technology, the filtering technology and the self-identification. The active application of such technologies will continuously enhance the computer system's defense against the illegal intrusion.

3.4   Perfect the Laws and Regulations and Establish the Legal Barriers for the Financial Computer Security

Law and morality are the foundation of maintaining the social harmony and the stability. Through the preventive education, the internal personnel can build up the good moral consciousness and form the good professional ethics. However, for the crimes that have already happened, we still need to rely on the laws to crack down on them severely, so as to be an effective deterrent to other bad intentions of the personnel. Therefore, the financial enterprises should also build up a strong legal consciousness while increasing the publicity of the law popularization. Once they find that the financial information has been stolen or leaked, they should report to the local public security department immediately, and solve difficult disputes and contradictions by legal means. If we want to create an effective financial risk monitoring system and constantly improve this system, we must focus on many aspects. First, we should have a sound and scientific financial accounting mechanism, which will support the overall financial risk monitoring system. Secondly, we should make a full analysis of the risk-causing mechanism, and summarize the corresponding focus according to the corresponding analysis contents, and the consideration of these focus points should also be carried out, and the corresponding mechanism of these focus points should be fully constructed in the future, and constantly improve the level of the risk prevention, which is quite important.

3.5   Strengthen the Supervision of the Computer Inspection

Strengthening the inspection of the computers is also an important measure to prevent and control the financial risks. Therefore, the relevant financial management departments should strengthen the inspection of the computer system operation and the personnel operation to ensure the stability of the system operation and standardization of the personnel operation by combining the regular inspection with the random inspection. The establishment of the Internet financial security system needs the close cooperation of the computer hardware and the software technology to ensure. At the hardware level, the primary task is to increase the investment in the network security technology, introduce the most advanced security system to enhance the financial system's ability to resist the malicious intrusion, and ensure that the financial economy can operate in a safe environment. In the data management, the digital verification, the key verification and other technologies can also effectively enhance the confidentiality of the Internet finance. In addition, we should strengthen the security login authentication system in the software aspect, and the graded authorization, the multiple authentication login and other ways can effectively eliminate the invasion of the illegal users.

4   Conclusion

The rapid development of the computer technology has brought great convenience to the human life. However, due to the influence of factors such as the form of the computer technology carrier, the network and the information characteristics, many computer systems still have great security risks in the specific application practice. The financial computer system is an information system to better serve the financial enterprises, improve the efficiency of the financial enterprises' operation and management, and forecast the accuracy of the future demand. It plays an important role in maintaining the good production and operation of the financial enterprises. At present, under the influence of the network hackers and the non-legal personnel in the financial enterprises, the financial computer system for the financial risk prevention and control needs to be further strengthened.

Acknowledgement. Empirical Analysis of the Young Innovative Talents Project (Humanities and Social Sciences), Entrepreneur Reputation and Financial Performance of Enterprises in Guangdong Science and Technology Department.

References

1. Zhang, J.: Talking about the computer system risk and prevention strategy in the financial industry. Discovering Value **16**, 53 (2012)
2. Yang, J.: On the computer system risks and the preventive strategies in the financial industry. EDN China **1**, 102–104 (2014)
3. Xiong, X.: Analysis on the main strategies of China's banking industry to prevent computer system risks. China Comput. Commun. (Theoret. Edn.) **09**, 10–11 (2012)
4. Tang, Y.: On the risk and preventive strategies of the commercial banks. China Econ. **4**, 197, 199–200 (2012)
5. Qin, L.: On the computer technology and financial risk prevention. Discovering Value **19**, 51 (2014)

Construction of the Resource Database of Piano Teaching Psychology

Jing Fan[✉]

College of Music and Dance, Fuyang Normal University, Fuyang 236000,
Anhui, China
bingnanhuo@163.com

Abstract. The piano learning and playing is a way for performers to express their understandings of the music and composer's thoughts by analyzing the composer's intentions and emotions through the information provided by the music score, combining their own music knowledge and experience. Because the music is not created by the performers, to fully understand the music, the performers need to overcome the differences in time and space to consciously explore the connotation of the music, and there are many problems to be solved in the process of learning the piano. In our teaching, it is very important for teachers to grasp students' personality characteristics. The formulation of the learning methods and plans should be in line with students' personalities. The construction of the resource database of the piano teaching psychology not only conforms to the characteristics of the times, but is also an important requirement for optimizing the teaching system at present. Therefore, it is very important to study the construction mechanisms of the resource database of the piano teaching psychology.

Keywords: Piano teaching psychology · Resources · Database · Construction research

1 Research on the Piano Teaching Psychology

Performers need to constantly overcome difficulties in the process of playing the music. There are many playing skills to learn. They need to have a high degree of control over their performances and achieve perfect performances [1–3]. Therefore, the playing requires performers to overcome difficulties, adjust their emotions and feelings, and perform purposefully. This is a necessary will process. Cognition is the precondition for the generation of wills. The characteristic of the will is that it has conscious purpose. After the performer has formed a certain understanding of the music and the performance, the purpose of the performance is produced [4]. After the purpose is determined, the performer should choose the way to achieve the goal according to the existing experience and knowledge of the performance. This is the process of understanding.

Will also has a great influence on the cognitive process. In the process of understanding the music and the performance, it is impossible for performers to overcome difficulties without will [5]. The main body of the piano performance is the performer.

M. Atiquzzaman et al. (Eds.): BDCPS 2019, AISC 1117, pp. 1471–1475, 2020.
https://doi.org/10.1007/978-981-15-2568-1_203

The emotion that the performer will produce can become the motive force and the resistance of the will action. Will is a regulator of mood. The strong will can control and regulate emotions, inhibit bad emotions, and play a positive role in promoting the piano learning and playing. The voluntary action is the conscious, purposeful and planned conscious action. The purpose of the piano performance is embodied in the performer's pursuit of the performance effect.

The psychological process of the will action is very important in the piano learning. The psychological process of the volitional action is a process of the adjustment of the consciousness to the active action. Players weigh their motivation and set goals before their action. As the volitional action, the piano learning is always accompanied by certain motivation. As far as the piano learning is concerned, the motivation and the purpose of our learning are usually not the same. The entire piano learning process is to achieve the final big goal by completing a series of the specific and indirect goals.

The role of will is reflected in the stage of implementing the decisions. The performer's subjective wishes will be transformed into the objective results through the practical actions. Only by learning a series of textbooks step by step and improving various skills can we finally begin to practice the concert repertoire. The piano learning requires the performers to study and practice a lot of skills and musical knowledge and experience. The characteristic of learning the piano as a volunteer action is that performers need to face many difficulties, which requires the performers to constantly transcend physical and psychological limitations and overcome difficulties and setbacks with strong will. As a teacher, we should pay attention to students' psychologies and give correct guidance, which will often achieve twice the result with half the effort.

2   The Contents of the Resource Database Construction of the Piano Teaching Psychology

In the piano teaching process, the stage performance is an important link. It is one of teachers' duties to instruct performers to perform successfully. At the end of the performances, teachers should pay attention to the communication with the performers. Teachers should evaluate the performances objectively in time, find out the problems and encourage the performers to make further efforts. First of all, we should fully understand the importance of the database. The importance of the database is embodied in the data support of the business system. It stores and organizes the data generated in the business process, and integrates and analyses the data through the front-end interface of the business software, which provides the most intuitive judgment support for the business development and changes. Before designing the database, we must fully understand the functional requirements, the time requirements, the platform requirements, the price requirements and the maintenance requirements of the software. In the actual development, these five aspects constitute the research contents before the software development. Only by meeting these requirements can the database application software be of high use value.

Secondly, it is very important to define the naming rules of different components of the database before designing the database. Otherwise, with the development process going on, the names of different database elements will be confused very quickly.

Generally, we can consider the following aspects: whether the table names are singular or plural, how to simplify the table aliases from the table names, and what uniform naming rules should be adopted for the column names in tables. For example, fields of the integer type can be prefixed with _N, and the fields of the character types with C_ as the prefix, and the fields of the money type with _M as the suffix, and the date columns with D_ as the prefix and so on.

Finally, create the E-R diagram. Clearly drawing the ER diagram is not only for designing the database, but its function is also similar to the engineering drawing. It can change from the designer's idea to the concrete drawing. On the one hand, it can let the designer discover some inadvertent details and make the design better. On the other hand, it can also make other people understand the database design and become a powerful tool between the designers and the users.

3 Construction Strategy of the Piano Teaching Psychology Resource Database

3.1 Determine the Construction Model

The construction of the resource database of the piano teaching psychology should be mainly based on the self-built database, supplemented by the development and introduction of the database and the cooperative co-construction of the database. The self-built database contains documents with high academic value reflecting the latest research results and the frontier trends of the disciplines. Introducing the database can not only quickly acquire the relevant scientific achievements at home and abroad, master the international leading science and technology, and promote the academic research and development, but can also learn from the advanced database construction experience. Utilize the network resource sharing, complete the cross-regional and cross-industry cooperation, use the computer network technology, maximize the co-construction and sharing of resources, avoid the waste of the personnel and the equipment acquisition funds caused by the repeated construction.

3.2 Constructing the Psychological Bank of the Piano Teaching

The music emotion comes from the life emotion. The music emotion is the concentration and sublimation of people's general emotion. It is the emotion aroused by the music works in the hearts of the performers and the appreciators. We must have experienced this kind of emotions directly or indirectly in life, thus producing resonance. Emotion in music is the concentrated expression of the life emotion, which accurately portrays the artistic images in the combination of the perception and rationality. In the huge database, finding the psychological database of the piano teaching quickly and accurately is comparable to looking for needles in a haystack. Therefore, it is necessary to tailor the piano teaching psychology to provide the personalized resource database of the piano teaching psychology. The psychological engineers of the piano teaching customize the personalized navigation for the piano teaching psychology according to users' needs and combining with the psychological products and technologies of the piano teaching, so as to facilitate the daily viewing of the database users. At the same time, the database management can set the user privileges according to the user identity, and the system will automatically judge the user's identity and privileges when logging in. After the user logs in, only the operation allowed by the permission can be performed. The psychological database of the piano teaching not only fully meets the needs of the piano teaching, but also ensures the safety.

3.3 My Favorites

For the psychological information of the piano teaching that users often need to check, if they need to search every time they log in, it will bring a lot of unnecessary troubles. In order to make it easy for users to view the psychological information of the piano teaching with high browsing frequency, the database should set up the collection function of the psychological information of the piano teaching. When the users browse the psychological information of the piano teaching, they also hope to check it again next time. They can click on the collection link of the psychological record of the piano teaching, and then save it to the designated collection catalogue. When the user logs in next time, if he needs to check the piano teaching psychology, he cannot search it, but just open it in the collection catalogue, and it is very convenient to browse the information of the piano teaching psychology.

3.4 Statistical Analysis

The resource database of the piano teaching psychology is not only enough to study the technology of the single piano teaching psychology, but also to have a comprehensive understanding of the psychological distribution and technology development of the piano teaching in industries, regions and competitors. The powerful statistical analysis system of the computer network can automatically generate hundreds of the analysis reports combining graphics, tables and texts, support the users' secondary processing, and finally realize the online exporting, intuitively showing the application of the piano teaching psychological topics, the psychological distribution of the piano teaching in the region and the development trend of the piano teaching, which are the important basis for the piano teaching psychological leaders to make scientific decisions for the survival and development of the piano teaching psychology. The musical imagination is formed on the basis of the existing musical imagery. The music imagination is a process in which the human brain forms a brand new image on the basis of the sound effect of the music. The musical imagination is produced by processing the past images in the human memory. Its content comes from the objective reality, not from the empty space. When playing the piano, the imagination produced by the performer has both unconscious imagination and reconstructed imagination. The unconscious musical imagination is the unconscious imagination caused by the stimulation of the musical signals. It is a combination of the life experiences related to the music in the past.

3.5   Teaching Comments

The teaching commentary is a kind of the explanation and annotation for the piano teaching psychology, which can further explain and excavate the piano teaching, expound the uncertainties in the piano teaching psychology, and make a judgment on its intrinsic value, so as to provide reference for the piano teaching researches. Through the recommendation function of the piano teaching, it can call on many users in colleges to comment on the psychological information of the piano teaching, so as to accumulate valuable technical resources for the psychological technology department of the piano teaching and give reasonable feedback to the educational activities of colleges and universities.

4   Conclusion

The construction of the resource database of the piano teaching psychology is a systematic and long-term work. The databases that have been built, and are under construction and need to be built should be continuously supplemented and improved according to the actual use effect and the update and development of the information. Because of its late start, local universities have many experiences and theories that can be directly learned and used for reference from the practice of the construction of the university database of the Ministry of Education. However, more importantly, it is necessary to constantly improve the construction of the database in practice in the light of the actual situations, the subject characteristics and the future development of the universities and the purpose of serving the local economic construction.

Acknowledgement. Funding Project: The Study of the Linear Thinking in Chinese Style Piano Music Works (No. SK2019A0325).

References

1. Dou, D., Shi, M., Zhao, R., et al.: Social eco-psychology: a new orientation to explore the relationship between individuals and the environment. J. Beijing Normal Univ. (Soc. Sci. Edn.) **5**, 43–54 (2014)
2. Li, H., Ma, J., Wang, N., et al.: Research progress on the organization and management of the scientific data at home and abroad. Libr. Inf. Serv. **23**, 130–136 (2013)
3. Yu, F., Peng, K., Zheng, X.: Psychology under the background of the big data: reconstruction and characteristics of the discipline system of the Chinese psychology. Chinese Sci. Bull. **5**, 520–533 (2015)
4. Wang, Y., Yu, Z., Luo, Y., et al.: Psychological research using the Internet: a survey of the West and China. Adv. Psychol. Sci. **3**, 510–519 (2015)
5. Wu, W., Wu, A., Yang, J.: The basic logic of the Open Data – Taking NYC Open Data as a sample. J. Zhejiang Univ. Technol. (Soc. Sci. Edn.) **4**, 388–393 (2014)

Video-Based Motion Human Abnormal Behavior Recognition Algorithms

Yingying Feng[(⊠)] and Linguo Li

College of Information Engineering, Fuyang Normal University,
Fuyang 236041, China
pengyuchen79@163.com

Abstract. The existing video-based algorithms for the recognition of abnormal behaviors of moving human beings have problems of low timeliness and recognition rate, so the video-based algorithm for the recognition of abnormal behaviors of moving human beings is proposed. The median filtering method is adopted for the equipment and the environment in the video produced by the influence of noise to filter out, on this basis, the background difference method is used to detect human movement accordingly, based on test results, through regional correlation method to track human movement, will get the image binarization processing, to extract of human movement behavior characteristics, based on human movement behavior characteristics, through the fuzzy algorithm to criterion of human movement behavior, achieved the identification of the abnormal behavior of human movement. Through experiments, it is found that the time efficiency of the proposed abnormal behavior recognition algorithm is reduced by 18.23% and the recognition rate is increased by 31.7%, which fully indicates that the proposed abnormal behavior recognition algorithm has better recognition effect.

Keywords: Video · Image · Motion · Abnormality · Behavior · Recognition

CLC Number: TU57 · Document identification code: A

1 Introduction

For a country, the security work is the basis of ensuring the national security. With the continuous development of science and technology, the video surveillance system has become an indispensable technical means in the security work, playing an important role in many areas of the security protection, and is widely used in all aspects of life, so that more and more people favor it. The video-based recognition of abnormal human behaviors has become one of the most important directions of our researches nowadays [1]. The video-based recognition of abnormal human behaviors can not only filter a large number of the useless information in the video surveillance system and effectively complete the safety protection work, but can also save a lot of manpower and material resources, and create great economic benefits for our society. At the same time, according to the recognition results of the abnormal human behaviors, the relevant staffs can be promptly reminded. If suspicious people are found, alarm them and

© Springer Nature Singapore Pte Ltd. 2020
M. Atiquzzaman et al. (Eds.): BDCPS 2019, AISC 1117, pp. 1476–1483, 2020.
https://doi.org/10.1007/978-981-15-2568-1_204

prevent accidents, so that the safety of life and property is greatly improved, and the aftermath of the traditional monitoring system is perfectly solved.

As far as the existing research is concerned, the research on the video-based abnormal human behavior recognition algorithm abroad is earlier and the technology is relatively more mature. It mainly uses the CAVIAR system to monitor in public places, and the KNIGHT algorithm to recognize abnormal human behaviors, which can effectively prevent and safeguard fighting and stealing. The domestic research in this field started late, but we also achieved some results [2]. For example, the Eagle Mage Intelligent Video Recognition and Monitoring System produced by Beijing TeraMage Technologies Co., Ltd. has been applied to the major metro stations in Beijing. It can monitor and identify the abnormal behaviors of the moving human body in the metro stations all the time and ensure the safety of the metro stations. However, the existing video-based abnormal human behavior recognition algorithms have problems of low timeliness and recognition rate, which cannot meet the needs of today's society. Therefore, a video-based abnormal human behavior recognition algorithm is proposed.

2 Design of the Video-Based Motion Human Abnormal Behavior Recognition Algorithms

2.1 Video Filtering

The images extracted from the video surveillance system will contain a lot of noise in the images due to the reasons of the equipment itself or the influence of the environment in the process of surveillance, which will make the images have pseudo-motion points, which has a great negative impact on the recognition of abnormal human behaviors. In order to improve the final recognition rate, it must be preprocessed and the noise contained in the images will be removed, and at the same time, we should enhance the quality and clarity of the images, facilitate the recognition of abnormal human behaviors in the subsequent movements [3]. The specific image processing process is shown below.

The noise in images mainly refers to the factors that hinder people's understanding of the information. It belongs to an unpredictable random error and can only be analyzed by the probability and statistics method. Nowadays, the most widely used noise elimination method is the filtering. The filtering methods are mainly divided into the Gauss filtering method, the mean filtering method and the median filtering method. After reading the literature, it is found that the median filtering method can suppress the random noise distribution and protect the edge information of the images while removing the noise. Therefore, the median filtering method is used to remove the image noise.

The basic idea of the median filtering method is as follows. Firstly, the pixels in the images are sorted according to the gray value. If there are bright spots or dark spots in the images, they are arranged at the edge. The noise points are removed by the median filtering method, and the median of the gray value of the pixels in the field is used to replace the pixel value of this point.

If the pixel value of a point in a video image is set as $p(x, y)$, and the filter window is set as M, and its specification is $5 * 5$, and then the median filter in a video image is expressed as:

$$m(x, y) = median\{p(x, y)\} \tag{1}$$

Through the above process, the video image filtering process is completed to prepare for the following human motion detection.

2.2   Motion Human Detection

On the basis of the above filtered videos, the corresponding detection of the moving human body is carried out. That is to say, the moving change area in the video is extracted completely. Effective moving human body detection is the key technology of the abnormal behavior recognition. The effect of the moving human body detection directly affects the result of the abnormal behavior recognition [4].

There are three main methods for the human motion detection: the frame difference method, the background difference method and the optical flow method. According to the relevant literature, the background subtraction method has moderate complexity and good robustness, and can detect the whole motion area.

Firstly, the weighted average background model is used to model the background to avoid the phenomena of "holes" and "double shadows" in the images. The image sequence is set as $I_k(k = 1, 2, \cdots, n)$, in which $I_k(x, y)$ represents the pixel value of k frame image at the point (x, y). $I_1(x, y)$ is represented as the initial background $B_1(x, y)$. Each frame is iteratively added and the background is accumulated. The background image is obtained as follows.

$$B_k(x, y) = \sigma I_k(x, y) + (1 - \sigma)B_{k-1}(x, y) \tag{2}$$

Among them, $B_k(x, y)$ represents the background image after the cumulative k frame, and σ represents the weight factor, with the range of $(0, 1)$. The value of σ has a great impact on the effect of the differential image, so it is necessary to set its value appropriately. $B_{k-1}(x, y)$ represents the background image after the cumulative $k - 1$ frame.

Secondly, based on the constructed background, the background difference image is partitioned. Set the k frame difference image to:

$$D_k(x, y) = |I_k(x, y) - B_k(x, y)| \tag{3}$$

$D_k(x, y)$ represents the difference image between the k frame image $I_k(x, y)$ and the background image $B_k(x, y)$.

$D_k(x, y)$ is divided into n sub-blocks on average, and each sub-block is represented as $D_{ij}^k(x, y), i = 1, 2, \cdots, m; j = 1, 2, \cdots, n$. The total pixels of each sub-block are calculated and compared with the threshold. If the pixel size is larger than the threshold value, the sub-block is considered to be the foreground sub-block, and the background sub-block is considered to be the background sub-block [5]. The set threshold is expressed as:

$$T = L \times W \times \theta \times 255 \tag{4}$$

Among them, L represents the length of the sub-block, and W is the width of the sub-block, while θ is the threshold coefficient and 255 is the maximum gray value of the pixel.

Finally, the background is updated. With the changing environment, if only one background is used, the detection effect will be adversely affected. Therefore, it is necessary to update the background in the real time according to the real situation. The process of the background image updating is as follows:

$$B_k(x, y) = \begin{cases} B_{k+1}(x, y) & k = 1 \\ \sigma I_k(x, y) + (1 - \sigma)B_{k-1}(x, y) & k > 1, Q < T \\ B_{k-1}(x, y) & k > 1, Q \geq T \end{cases} \tag{5}$$

Among them, Q represents the sum of the sub-blocks' pixels, and T represents the set threshold.

At the same time, the binarization method is used to extract the moving human body, and the corresponding results of the moving human body detection are obtained. It provides support for the following motion human tracking.

2.3   Motion Human Tracking

On the basis of the above results of the human motion detection, the human motion is tracked. Simply put, it is to search the position of the same moving human body in different video frames. The region correlation method is mainly used to track the moving human body [6].

Firstly, the corresponding relationship between two the adjacent frames of the moving human body is established. Based on this relationship, the corresponding detection of the moving human body in each frame image is carried out, and the outer rectangle is extracted. It can be found that the outer rectangle of the two adjacent frames overlaps each other, and the overlapping area is large. Its overlap area is the most matching rate of the moving human body, which can be represented as $A_{i,j}$. The larger the matching rate is, the more successful the matching is. Its expression is as follows:

$$A_{i,j} = \max \left(\frac{area(OL_{k-1,i} \cap OL_{k,j})}{area(OL_{k-1,j})}, \frac{area(OL_{k-1,i} \cap OL_{k,j})}{area(OL_{k,j})} \right) \tag{6}$$

Among them, $OL_{k-1,i}$ and $OL_{k,j}$ represent the outer rectangle of the two adjacent frames of the moving human body separately, and $area(OL_{k-1,i} \cap OL_{k,j})$ is the overlapping area.

Through the above process, the moving human body is matched and the tracking of the moving human body is completed.

2.4 Motion Human Behavior Feature Extraction

On the basis of the above results of the human motion tracking, the binarization method is used to extract the characteristics of the human motion behaviors, which provides the data support for the following recognition of the abnormal human motion behaviors [7]. The simple behavioral characteristics cannot accurately express the human behaviors, so it is necessary to mention the trajectory and velocity of the center of the mass to express the behavioral characteristics of the human body. The specific process is shown below.

The process of extracting the trajectory of the centroid motion is: to binarize the video image and set the size of the binarized video image to be $P * Q$, and the centroid coordinate (x, y) of the moving human body is represented as:

$$x = \frac{P_{10}}{P_{\infty}}, y = \frac{P_{01}}{P_{\infty}} \tag{7}$$

Among them, P_{ij} represents the value corresponding to the pixel value of the binary video image.

The main formulas for calculating the velocity of the motion are as follows:

$$V = \frac{S_h}{S_v} \tag{8}$$

Among them, S_h expresses the displacement of the center of the mass of the moving body in the horizontal direction, and S_v expresses the displacement of the center of the mass of the moving body in the vertical direction.

The above process completes the extraction of the human behavior characteristics and provides the data support for the recognition of the abnormal human behaviors.

2.5 Recognition of the Abnormal Behaviors of the Motion Human Body

Based on the above extracted human behavioral characteristics, the abnormal behavior of the human body is identified by using the fuzzy theory [8].

In the process of recognizing the abnormal behaviors of the moving human body, the behaviors of the identified objects are the complicated non-linear process with the time-space changes, and it is difficult to obtain an accurate mathematical model [9]. The fuzzy theory mainly imitates human's discriminant thoughts, and judges the behaviors through the processing of the fuzzy information.

The process of the fuzzification is to map the eigenvalues of the human motion to the fuzzy variables in the corresponding universe, mainly to determine their membership functions. The membership function is expressed by the Taylor expansion.

$$F(\phi) = \frac{m(x, y)}{B_k(x, y)} \otimes \prod_i A_{i,j} \wp \frac{(x, y)}{V} \tag{9}$$

The rule knowledge base is mainly composed of the abnormal behavior database and the rule base, which provides the database support for the abnormal behavior recognition of the human motion.

The process of the fuzzy recognition is as follows. Input the fuzzy value of the human motion characteristics and recognition rules into the fuzzy recognition algorithm, and get the corresponding recognition results. The fuzzy value for the recognition of the abnormal behaviors of the motion human body:

$$\bar{F} = \sum_{i=1}^{n} (F(\phi) \times k_n) \tag{10}$$

Among them, k_n represents the weight factor.

The main purpose of defuzzification is to transform the above results into the exact quantities. The range of the exact quantities is $[0, 1]$, and 0 represents the normal behaviors while 1 represents the abnormal behaviors. The rule of defuzzification is:

$$U = \begin{cases} \bar{F} > 1 & 0 \\ \bar{F} \leq 1 & 1 \end{cases} \tag{11}$$

If the abnormal behaviors of the human body are recognized, the real-time alarm is given [10].

Through the above process, the recognition of the abnormal human behaviors is realized, which provides more advanced technical support for the security work.

3 Experiments and Result Analysis

The above process realizes the design of the recognition algorithm for abnormal human behaviors, but whether it can solve the existing problems still needs further verification. Therefore, the simulation experiments are designed to verify the performance of the proposed algorithm.

In the process of designing the simulation contrast experiment, the performance of the algorithm is mainly reflected by the timeliness and the recognition rate. The detailed experimental results are analyzed as follows.

3.1 Contrastive Analysis of Timeliness

The timeliness refers to the time between the abnormal behavior recognition and the alarm. The shorter the timeliness is, the better the performance of the algorithm will be.

The timeliness of the proposed algorithm is much lower than that of the existing algorithm, and the average value of the proposed algorithm is 18.23% lower than that of the existing algorithm.

3.2   Comparative Analysis of the Recognition Rate

The comparison of the recognition rates obtained by experiments is shown in Table 1.

Table 1. Recognition rate contrast table

Number of experiments	Existing algorithms	Propose an algorithm
10	45%	80%
20	62%	81%
30	32%	86%
40	35%	79%
50	66%	78%
60	51%	69%
70	55%	91%
80	49%	89%
90	56%	93%
100	72%	94%

As is shown in Table 1, the recognition rate of the proposed algorithm is much higher than that of the existing algorithm. The maximum recognition rate of the proposed algorithm is 94%, and the average is 84%. The maximum recognition rate of the existing algorithm is 72%, and the average is 52.3%. The average recognition rate of the proposed algorithm is 31.7%.

The experimental results show that the proposed algorithm reduces the timeliness by 18.23% and improves the recognition rate by 31.7%. It fully demonstrates that the proposed algorithm has the better recognition effect.

4   Conclusion

The proposed algorithm reduces the timeliness by 18.23% and improves the recognition rate by 31.7%. It can provide more effective protection for the human safety. However, in the process of the experiment, the experimental results are deviated from the actual results due to the setting of the parameters, but the overall trend is unchanged. In order to recognize the abnormal human behaviors more accurately, it is necessary to further study and to optimize the proposed algorithm, and provide more advanced technological support for the video surveillance system.

Acknowledgements. Anhui province outstanding young talents support program "Research on multi-target recognition and tracking technology based on information fusion in intelligent monitor in", project number (gxyq2017157); Major teaching reform project in anhui province "Research and practice of computer applied innovative talents training mode based on EPT-CDIO", project number (2016jyxm0777); Anhui provincial science and technology department fund project "Research on fuzzy multi-threshold tumor image segmentation and recognition based on swarm intelligence optimization", project number (1908085MF207).

References

1. Du, G., Chen, M.: Research on the motion object abnormal behavior detection algorithms based on the intelligent video analysis. Video Eng. **12**(12), 23–26 (2018)
2. Zhang, J., Zhang, Z., Yang, K., et al.: Human behavior recognition based on the local features of the time dimension. J. Xi'an Univ. Technol. **33**(2), 169–174 (2017)
3. Han, F., Ying, J.: Detection and recognition of the human abnormal behaviors based on the Bayesian network. Softw. Guide **17**(189(07)), 13–17 (2018)
4. Zhang, R., Wang, S.: Target behavior recognition algorithm based on the convex template. Radio Commun. Technol. **43**(4), 75–79 (2017)
5. Li, S., Rao, W.: Video-based detection of human motion areas in mines. Comput. Sci. **45**(4), 291–295 (2018)
6. Zhao, X., Wang, H., Ji, X.: Research on the human behavior recognition based on the new projection strategy. Comput. Eng. Sci. **40**(9), 95–101 (2018)
7. Nie, Y., Zhang, P., Feng, H., et al.: Human motion recognition in the 3D video based on the motion standard sequence. J. Terahertz Sci. Electron. Inf. Technol. **15**(5), 841–848 (2017)
8. Wang, Y., Li, J., Tian, S.: Abnormal behavior detection based on the fuzzy ISODATA clustering and the histogram entropy value algorithms. Mod. Electron. Technol. **40**(12), 120–123 + 127 (2017)
9. Ke, G.: Manifold learning-based human abnormal behavior recognition method. Pioneering Sci. Technol. Monthly **30**(04), 118–120 (2017)
10. Li, W., Wang, L., Yao, D., Yu, H.: Fuzzy recognition of the human abnormal behaviors based on the motion characteristics analysis. Acta Medicinae Universitatis Scientiae et Technologiae Huazhong (Nat. Sci. Ed.), **42**(07), 87–91 (2014)

Teaching Practice of "Computer Aided Design" for Clothing and Textile Major Based on the Three-Dimensional Technology

Shan Fu$^{(\boxtimes)}$

Academy of Fine Arts, Jiangxi Normal University of Science and Technology,
Nanchang 330000, Jiangxi, China
bopingzhang@yeah.net

Abstract. With the development of the human society, the fashion design has deepened its practical, aesthetic and economic awareness, and constantly enriching the styles, colors, materials and decorations. The computer aided design, such as color matching, volume measurement and tailoring drawing, opens up a new way of the garment design. Under the background of the mature development of the computer technology, this technology has been widely penetrated into every field of our society. The course of fashion design in colleges and universities aims to cultivate students' cognitive and application abilities in the fashion forms and colors. The combination of the computer technology and the fashion design in colleges and universities is more conducive to the improvement of students' professional abilities. The increasingly strong social demand for the garment CAD requires us to constantly update the teaching contents, select the advanced garment CAD software, explore the advanced and reasonable teaching methods and means, and train the required talents for the society.

Keywords: Three-dimensional technology · Clothing and textile specialty · Computer-aided design · Teaching practice

1 Teaching Innovation Background of "Computer Aided Design" for Clothing and Textile Major Based on the Three-Dimensional Technology

The course of the computer aided design is mainly for the major of the fashion art design. Taking the current student training program as an example, it can be seen from the curriculum setting that the teaching thinking mode of the computer-aided design course is that the freshmen in Grade One are mainly enrolled in the general education courses, while in Grade Two they are mainly given priority to the basic courses of the fashion design, learning the aesthetic principles of the design, capturing popular elements and colors, realizing from the abstract design to the concrete process, and completing the entire design process [1, 2]. In the third grade, the students began to integrate the basic knowledge of the fashion design into the comprehensive application, learning the project design of men's, women's, knitting and children's wear by

© Springer Nature Singapore Pte Ltd. 2020
M. Atiquzzaman et al. (Eds.): BDCPS 2019, AISC 1117, pp. 1484–1489, 2020.
https://doi.org/10.1007/978-981-15-2568-1_205

modules, and contacting the assistant design software and the CAD system to gradually integrate into the industrial production process [3].

1.1 The Curriculum Arrangement Is Not Closely Related to the Basic Theoretical Curriculum

Although the computer aided design curriculum is an optional course, the software involved in the teaching belongs to the basic tools that must be mastered by the clothing design specialty. In the long-term teaching process, it is found that all students will take this course, which shows the importance of the course [4]. At present, the curriculum of the computer-aided design is arranged in the fifth semester. There are two drawbacks. First, the theoretical courses related to this course are all concentrated in the second grade. In the teaching process, it is often necessary to review the previous theoretical knowledge in order to achieve the purpose of the knowledge point convergence [5]. Second, in order to participate in more practical exercises, design students are enthusiastic about participating in all kinds of the national fashion design competitions. In the information age, almost all competitions choose more convenient electronic design draft, so the teaching arrangement of this course is too backward.

1.2 The Teaching Contents Are Distributed in Blocks and the Knowledge Points Are Not Effectively Integrated

The process of the colors, the effect design, fabrics, the process analysis, the plate making, placing and layout in the course of the computer aided design is the embodiment of the complete design process. Its teaching purpose is not only to achieve the software operation, but also to effectively reflect the design thinking from the multiple perspectives. However, at present, the teaching contents of this course are composed of two parts and six main knowledge points, which are distributed in blocks, and the integration of the knowledge points is only reflected in two aspects. One is the embodiment of the fabric pattern design in the garment effect drawing, and the other is a series of the teaching practice from the plate making, grading and discharging based on the plotter. This makes the knowledge points unable to be effectively integrated. In addition, the course involves three kinds of the software, but the teaching contents among the software are self-contained and there is no cross-teaching.

1.3 Positioning Mistakes and Insufficient Intensity of the Practical Teaching

Some teachers and students are wrong about the role orientation of the major of the fashion art design. They think that the future development is just to create relevant works in the office directly with the help of computers and other tools. In the process of the teaching and learning, they face the awkward situation of making a car behind closed doors and relying solely on the subjective imagination, which directly affects the final teaching quality.

1.4   Lack of Teaching Experience and Lack of Enthusiasms for Learning

For some teachers majoring in the fashion art design, they have not really participated in the fashion design. Without their own practical experience, they will naturally subconsciously avoid the practical teaching forms in the teaching process. At the same time, some students are not interested in the major of the fashion art and design because of many factors, so they are inevitably not enthusiastic in the learning process, which will affect the learning effect in the future.

1.5   Enterprise Support Is Not Enough to Apply What They Learn

Compared with other disciplines and enterprises, the major of the fashion art design is more closely related. However, in the actual teaching process, it is not closely related to some colleges and enterprises in enterprises. It cannot provide students with an opportunity to learn and apply, and to exercise themselves in practice. Therefore, some courses that should be practiced have become paper talkers, unable to play its application role.

2   The Practical Value of "Computer Aided Design" Teaching in Clothing and Textile Major Based on the Three-Dimensional Technology

2.1   Enrichment of the Teaching Demonstration

The teaching process of the apparel structure is as follows. Through the digitizer, CAD and other software, the classroom teachers explain the formulaic theoretical knowledge of the structure, while demonstrating the setting before the plate making, the drawing of basic lines, the line trimming and measurement, the generation of cutting, the evolution of cutting, the addition of seam, and the adjustment of yarn and other process of the operation. According to the demonstration, the students make the corresponding records, and then carry out exercises. Finally, the teachers make questions, tour guidance and homework comments. Not only is the operation very simple and fast, but also helps teachers to understand the classroom effect and students' knowledge points and the project mastery in a timely manner.

2.2   Simplification of the Operation Process

The formulation of the clothing tailoring scheme is an indispensable follow-up knowledge structural system in the course of the clothing structure. However, in the normal teaching, it is very difficult for us to see the teachers demonstrating the layout operation on the blackboard. The main reason is that explaining the layout method and changing principle on the blackboard is too complicated and tedious. We should not only prepare the industrial template in advance, but also draw the template on the blackboard once. Some styles cannot be drawn on a blackboard. The students are also difficult to understand because of its abstraction. This problem can be effectively solved by using the garment CAD software. With the help of the corresponding CAD software, the layout system can quickly realize the movement, replication and rotation of the template, demonstrate the layout methods of various styles, and calculate the material consumption. This kind of the teaching is not only intuitive and visual, but also easy to operate. It is easy for students to master.

2.3   Dynamization of the Static Graphics

The dynamic process of drawing can be demonstrated by using the garment CAD software (PowerPoint2000 and Fuyi garment CAD are used in the courseware making). The drawing examples are novel and the structure explanation is clear and accurate. It can also show the entire operation process from the style design to the structure design to students through the multimedia teaching equipment. The data recording can also be recorded step by step according to the drawing process, which is convenient for the learning and consolidation in the future reference, and can effectively improve students' practical and innovative abilities.

2.4   Convenience of the Structural Design

The plate-making system in the garment CAD is the most efficient and characteristic module in the garment enterprises. Through CAD, we can operate the knowledge system of the industrial pattern-making, seam-laying, multi-size garment layout and plate-laying, which not only saves time, but also carries out the training of the enterprise simulation teaching.

3   Teaching Mechanisms of "Computer Aided Design" for the Clothing and Textile Major Based on the Three-Dimensional Technology

3.1   Exploration of the Teaching Scheme

Firstly, through the virtual stitching of the two-dimensional CAD patterns and introducing students' creative design patterns, the virtual display of the clothing colors, patterns and fabric textures can be realized, which provides a platform for students' creative expression in their design. In addition, the clothing three-dimensional virtual fitting technology can also present the dynamic effect of the virtual models. By setting the background, simulating the lighting contrast and loading the model actions, the virtual models can be generated, which can directly feedback the design effect of students, and increase the interest of the courses and the human-computer interaction.

Secondly, the layout is the key to determine the comfort of the ready-made clothes. The three-dimensional fitting system has the simulation calculation of the clothing interior space, the human body pressure state and the fabric tension state. The loose or tight degree of the layout design is clear at a glance. Moreover, the two-dimensional pattern modification can be feedback to the three-dimensional platform in the real time, and the structure adjustment of the three-dimensional virtual clothing piece can also be reflected in the two-dimensional pattern, realizing the linkage modification between the two-dimensional pattern and the three-dimensional virtual ready-to-wear.

Thirdly, the three-dimensional fitting system has its own fabric attribute database. According to the adjustment of the longitude and the latitude density, the weight, the thickness, the drapability and the elasticity parameters, different types of the fabric simulation results can be obtained. The new fabrics function of the system can input the measured data of the real fabrics through the instrument to form a self-built fabric library. The simulation results are very close to the real fabrics, and meet the needs of students to design the clothing with special fabrics, such as the down clothing.

Fourthly, the three-dimensional fitting system can obtain different shapes of the human body, such as the fat body and the pregnant woman and so on, by adjusting nearly 30 size parameters of the virtual models, such as height, chest circumference, waist circumference and hip circumference. It can help students design and modify the clothing patterns under different shapes, cooperate with the optical three-dimensional scanning system, and build the human parameters. The database has increased the diversity of the student model design.

3.2   Teaching Reform Practice of the Computer Aided Design Curriculum

The computer aided design (AutoCAD) is a practical and applied course. In the process of the teaching, teachers should flexibly apply various teaching methods to their teaching according to the actual situations, and constantly improve the effectiveness and quality of their teaching, so that students can draw lessons from one another and use them flexibly, so as to cultivate students' creative thinking. In the constant exploration and trial, we should improve the effectiveness and the quality of the teaching and cultivate the high-quality comprehensive application-oriented talents with innovative consciousness for the society. According to the teaching reform plan, the computer-aided design course is arranged in the first part of the fifth semester without changing the training plan, so as to prevent it from divorcing from the theory. In the case of the unchanged total class hours, the teaching contents of the course are adjusted, and the three-dimensional fitting module is added, and the teaching method is reformed. The software tool explanation was combined with the professional characteristics and reorganized into the case teaching. The Photoshop application forms nearly 10 teaching cases of the pattern design, the fabric performance and the jewelry drawing, and the Illustrator application forms nearly 8 teaching cases of the accessory drawing, the seam expression and the process details. The application of ET forms 6 teaching cases around the prototype punching, the dart transfer, and the fold segmentation and so on. CLO-3D forms 5 teaching cases along the operation process of the virtual sewing, the fabric property setting and the pattern modification, totally 29 teaching cases for the purpose of the tool explanation. Finally, the project creative design method is adopted, focusing on the specific themes, from the entire process of the clothing effect drawing design, the style drawing and the industrial model design to the final virtual fitting, so that students can master the comprehensive application of the software tools in practice and achieve the teaching purposes.

3.3   Constructing the Practical Mechanisms of the Comprehensive Teaching

As the front-end research and development link, the garment design will combine the digital construction with the virtual reality technology. The curriculum teaching reform of "Computer Aided Design" for the major of the fashion art design based on the three-dimensional technology is an exploratory experimental teaching project. By integrating the CLO-3D three-dimensional fitting system, the links of the two-dimensional pattern virtual stitching and the three-dimensional dynamic display are added, and the design effect of students is presented more intuitively, and the knowledge structure of the computer aided design teaching has been improved and a closed-loop teaching system has been formed. Especially the implementation of the project creative design makes students become the main body of the classrooms, and students' creative practical ability has been greatly improved, and the good teaching effect has been achieved. At the same time, the teaching reform also provides a direction for the teaching practice of the information technology courses in the garment specialty.

4   Conclusion

In the course of the several years' teaching, focusing on the employment characteristics of the clothing enterprises, we have made great achievements in the classroom teaching by constantly adjusting the curriculum system, enhancing students' practical abilities, improving the teaching means and methods, and optimizing the assessment methods. Students are more interested in learning this course, more active in the classrooms than before, and pay more attention to and study the clothing materials in peacetime, which has played a certain role in enhancing the teaching effectiveness of this course. We should keep pace with the times, teach students in accordance with their aptitude, and cultivate the applied talents suitable for the development of the modern clothing enterprises.

References

1. Zhao, L., Zhang, R., Liu, H.: Classified and stratified teaching of the textile students based on the professional characteristics. Text. Apparel Educ. **4**, 294–296 (2015)
2. Li, Z., Zhang, H.: Research on the reform of strengthening the practice teaching in the computer aided machinery design. Educ. Teach. Forum **5**, 204–205 (2012)
3. Yu, J.: Research on the construction of the integrated course system for the fashion design major in secondary and higher vocational schools. Way Success **25**, 60 (2016)
4. Wu, Y.: Discussion on the construction of the course group system of the fashion design specialty. Vocat. Tech. Educ. Forum **54**(32), 50–51 (2011)
5. Yang, H.: New thoughts on the teaching reform of the fashion design specialty. Text. Rep. **4**, 54–56 (2016)

Application of the Big Data Technology in the Enterprise Management

Yumei Guan[✉]

Sanya University, Sanya 572000, Hainan, China
yingguocugb@126.com

Abstract. With the rapid development of the information technology, especially in the new situation of the deep implementation of the "Internet +" strategy, the big data technology has been applied to all aspects and all walks of life. This will have a certain impact on the management mode of enterprises. Enterprises should have a clear understanding of this, and in a specific process of its implementation, we should combine the big data technology with the enterprise management closely. We should not only give full play to the positive role of the big data technology, but also make great efforts to innovate the enterprise management mode, so that the big data can play a greater role in promoting the enterprise management reform and development.

Keywords: Big data technology · Enterprise management · Applied research · Value significance

The era of the big data has had a tremendous impact on the management of enterprises. From the impact on the decision-making environment of managers to the impact of managers themselves, all of them affect the management of the contemporary enterprises. In the era of the big data, whether the value of the big data can be brought into play and make best use of the advantages and bypass the disadvantages in the use of the big data will make a positive judgment for the current enterprises to make management decisions.

1 Characteristics of the Big Data

The big data refers to the data sets that cannot be captured, managed and processed by the conventional software tools in a certain time range. It is the huge and the diversified information asset with the high growth rate that requires a new processing mode to have the stronger decision-making power, insight and the process optimization ability.

1.1 High-Capacity

In today's era, data, such as the movement of everyone, are generated regularly from one or more fixed sources. According to the relevant data forecast, by 2020, the amount of the data recorded by the data system can reach 30ZB, so it is not difficult to see that the data of today's era has the characteristics of the large capacity.

© Springer Nature Singapore Pte Ltd. 2020
M. Atiquzzaman et al. (Eds.): BDCPS 2019, AISC 1117, pp. 1490–1495, 2020.
https://doi.org/10.1007/978-981-15-2568-1_206

1.2 Diversification

The types of the data are mainly divided into the structured data and the unstructured data. The structured data is the database, such as the hospital database, the enterprise financial system and the education card. The unstructured data refers to some data, including videos, pictures and networks. With the rapid development of Alibaba e-commerce platform initiated by Ma Yun, the unstructured data in China has shown explosive structural growth in recent years.

2 Necessity of the Enterprise Management Innovation in the Big Data Era

The reason for the enterprise management innovation is that the era of the big data brings both opportunities and challenges to the growth of enterprises. Opportunities mainly refer to the great economic value contained in the information data in the era of the big data. Enterprises can apply the collected information to the strategic decision-making and the operation management in accordance with the innovation-driven development strategy, which can effectively improve the economic efficiency of enterprises and obtain higher profits, thus laying a solid foundation for enterprises to expand their scale. Basically, this is the intrinsic realistic need for the development and growth of enterprises. Challenge refers to many problems in the enterprise management in the era of the big data. To solve these problems, we must strengthen the innovation in the management.

2.1 Internal Needs of the Enterprise Development

In the context of the big data era, the enterprise management innovation can bring huge benefits to enterprises. The author mainly analyses the necessity of the internal enterprise management innovation from three aspects: optimizing the enterprise operation, improving the enterprise competitiveness and allocating the human resources scientifically. Optimize the operation of enterprises. The essence of the enterprise operation is to develop as many customers as possible to improve the economic benefits of enterprises. The most important way to develop a large number of customers is to understand the diversified needs of customers and put forward strategies to meet them. The big data can make us know consumers' personal information and needs, the consumption tendency and other important data information, and the diversified needs of customers. Therefore, only by innovating the diversification and precision means and the mining data efficiency can we continuously optimize the operation of the enterprises and take the initiatives in the market competitions.

Further improve the competitiveness of the enterprises. In the past market competitions, the government's policy support and the location advantages play a more important role in enhancing the competitiveness of enterprises. However, in the context of the big data era, the data information constitutes one of the elements of the enterprise competitions. The scientific analysis of the collected data can not only know the actual needs of the customers and seize the market share, but also accurately understand the market situations and formulate more scientific, reasonable and valuable enterprise strategies. Based on this, the enterprise management innovation can improve the decision-making efficiency, show the potential values, and improve the core competitiveness of the enterprises while saving costs.

Scientifically allocate the enterprise human resources. Regardless of the environment, talent is the premise and foundation for the development and growth of enterprises. Enterprises can make use of the resources provided by the big data information to set up the personal file database of the employees, understand their demands and meet their needs, so as to improve their loyalty to enterprises and avoid the brain drain. At the same time, the establishment of the employee database can also attract more useful talents and lay a talent foundation for the development and growth of enterprises.

2.2 Challenges of the Big Data for Enterprises

Large scale, strong timeliness and complex composition are the basic characteristics of the big data. These basic characteristics bring tremendous difficulties and challenges to the data processing and the data analysis of enterprises. Only through the enterprise management innovation can these difficulties be gradually overcome and can the economic benefits of enterprises be continuously improved.

The enterprise data analysis emphasizes the real-time. At present, the social and economic development is exceptionally rapid, and the environment of the enterprise development is also in a dynamic change. Enterprises need to collect, collate, screen and analyze the data on an exceptionally large scale. If enterprises want to extract the valuable information from the massive data information, they must ensure the efficiency of the data parsing. In fact, it is very difficult for enterprises to guarantee the timeliness of the data information processing.

The integration of the enterprise data information is difficult. With the development of the Internet and the electronic information technology, there are many kinds of channels for the existence of the data information, such as the social networking platform and the e-commerce platform. The data information collected from these channels is various, diverse and complex, such as video, audio, pictures, geographical locations, and telephone information and so on. In fact, it is easier for enterprises to process the structured information with relatively fixed formats, and there is still a lack of the efficient means for the unstructured or semi-structured data processing.

The decision-making concepts need to be changed urgently. Business operators usually have the relatively sufficient basis for making specific decisions. The data information has the natural attribute of assisting the decision-making. The data information can enhance the rationality and scientificalness of the decision-making, reduce the management risks, and improve the quality and efficiency of the decision-making. In the context of the big data era, enterprises can no longer make analysis and decisions only through the simple information data, but must thoroughly analyze the data and information, and compare the external competitive environment, so as to improve their own problems more effectively and obtain the development opportunities. Enterprises must change their decision-making concepts and consciousness in the real time, and constantly improve the efficiency of the enterprise management.

It is very important to ensure the data security. In the current context of the big data, the data information acquired by enterprises has a lot of personal privacy, and its storage is carried out through computers and networks, so there are huge security risks. In such a complex information management environment, strengthening the management innovation is the only way to ensure that the data information is not lost and not leaked, to ensure the data security.

3 Application Strategy of the Big Data Technology in the Enterprise Management

3.1 Improving Managers' Awareness of the Data Business Values

Enterprise staffs must correctly understand the commercial values of the data, so that they can guide the staff to pay attention to the collected data, improve the reliability of the data, and lay a good foundation for the future data analysis. Managers must correctly judge the reliability of the data, so as to improve the commercial value of the data. Enterprise managers, in particular, must attach great importance to the data, and guide grassroots staffs to pay attention to the data collection. At present, there is a very fierce competition among enterprises, and the market environment is also one of the factors affecting the development of the enterprises. Whether an enterprise can develop smoothly or not must be able to adapt to the market environment, improve the customer trust, and enable customers to purchase products manufactured by the enterprises. The enterprise staffs must have a targeted understanding of the customer needs, in order to enable the staffs to carry out better product services, and increase the communication with customers, so that every customer can experience the services of the enterprises. When a company's marketing department collects the data, it should give it to the professionals for the comprehensive analysis, so the enterprise must employ professionals for the data analysis, in order to better collect the data for the professional analysis, which has a certain role in promoting the future planning of enterprises. Enterprise leaders should actively absorb talents in the human resources, through the systematic professional training, so that every staff of the company can play their maximum value.

3.2 Using the Big Data to Adjust the Environment of the Management Decision

The popularity of the Internet makes the contemporary data grow at a rate of 50% every year. The data in the market affects the decision-making of most enterprises. Under the background of the big data, how to collect and distribute the effective data has an impact on the plan and the final decision-making of enterprises. The big data urges the decision-makers to have a reasonable expectation for the future, and there are differences in the final decisions made in different expectations. When using the data, enterprises need to analyze and screen out the available and effective data for their own enterprises. Most of the management decisions of the contemporary enterprises need to analyze and integrate the data to get their main development direction in the future.

However, different ways of the data processing in different enterprises result in the decision-making errors. How to make use of the big data has become the key link and the growth basis for the enterprises in the market. Enterprises need to improve their competitiveness through the big data. In the era of the big data, the use of the data is not only a digital technology, but also related to the survival and development of enterprises. Having a set of the data processing methods suitable for their own enterprise development is the core element that the contemporary enterprises must master. Under the circumstances that the decision-making background has changed, more factors need to be taken into consideration when the decision-makers choose the management mode suitable for the development of enterprises. The basic way for the sustainable development of enterprises is to transform the new challenges into the advantages of their own development.

3.3   Enriching the Management Data with the Big Data

In the era of the big data, the decision-makers' choice of the enterprise management mode can better reflect the decision-making ability and the technical levels of managers. Finding the data needed by the company in a large amount of the data is the key link in the process of the enterprise's choice of the management mode. The data in the era of the big data is large and abundant, and the data contained are also pluralistic. Screening and selecting the data is the basic quality that the enterprise management must possess. The value of the big data mainly comes from the integration and analysis of the debris data. The collection of the historical data for the business operation is the main basis for predicting the future development direction. The collection of the historical data requires the business managers to constantly mine the debris data in order to obtain the effective data. In the era of the big data, the data acquisition and analysis not only has a certain guiding role in the choice of the enterprise management mode, but also is very helpful to enhance the ability of the managers themselves. When the decision-maker is under great pressure, it is impossible to make a correct judgment simply by relying on the subjective choice of the human beings. This is the advantage of the big data. The analysis and decision-making of the big data has many advantages, but this does not mean that the choice of the enterprise management mode needs to rely entirely on the judgment of the data. It is an assistant tool in the decision-making process of the enterprise managers, and the comprehensive analysis of the decision-makers should be the real decision-making mode of the enterprise management.

3.4   Using the Big Data to Improve the Decision-Making Level of the Policymakers

In the traditional choice of the enterprise management mode, the decision-makers often make choice of the enterprise management mode through their personal experience. The use of the big data improves the decision-making system of the decision-makers, and the focus of the decision-making is the collection of the data itself. In the past, most of the problems found in the enterprise management are the frontline employees at the grass-roots level. In the era of the big data, decision makers can get the current situation of the enterprise development through the analysis of the data, and find out the existing problems. While the decision-making ability has been greatly improved, they will also rely more on the interaction between the frontline employees and the enterprise management layers. At present, the advantages and disadvantages of the enterprise management will be exposed to the public to a large extent. It can no longer lead people to the trend of the enterprise management status and make the market of the enterprise development transparent. This requires the managers of enterprises to combine the actual situations of the enterprises and avoid making the wrong decisions when choosing the management mode of enterprises. The analysis of the big data is also the main content that the contemporary entrepreneurs need to focus on. Only by constantly improving the ability and the decision-making level of managers themselves and choosing the right business management model can they find their own position in the market which is gradually moving towards transparency and develop continuously.

4   Conclusion

The big data technology brings challenges and opportunities to the enterprise management and promotes the enterprise management to change. In many aspects of the enterprise management, it is affected by the big data. Enterprises should take the initiatives to raise their awareness, change their ideas and innovate their ways to adapt to the new changes. Even though the times are changing constantly, managers have the ultimate decision-making power in choosing the enterprise management mode. The external assistant environment can only help decision-makers reduce the probability of mistakes where they are prone to make mistakes. Improving the management decision-making levels of managers is the core of the survival and development of enterprises.

References

1. Wen, M.: On the application of the data mining technology in the electric power enterprises in the big data era. Mod. SOE Res. (24) (2015)
2. Zhang, J.: Brief analysis of the application of the big data technology in the enterprise strategic management. Comput. Knowl. Technol. **20**, 17–20 (2016)
3. Zhao, M.: Discussion on the corporate financial strategic management in the big data era. Money China **23**, 258 (2015)
4. Wang, H.: Research on the strategy of the human resource management changes in the big data era. Sci. Technol. Innov. **8**, 55–56 (2016)
5. He, H.: Application of the big data technology in the enterprise financial management. Enterp. Reform Manag. **4**, 154 (2016)

Design of the Ideological and Political Education Platform for the Teacher Education Public Course in Teachers' Colleges and Universities Based on the Video on Demand Technology

Hongyu Hu[✉] and Guoping Liang

Mianyang Teachers' College, Mianyang 621000, Sichuan, China
guoxiwu11@163.com

Abstract. The teacher education public course is a compulsory basic course for non-pedagogical normal students in normal universities, and it is an important course for the formation of teachers' professional identity and educational ideal. Based on the connotation and significance of the curriculum ideological and political education and the current situations, the curriculum characteristics and the educational situation of the implementation of the teachers' education public courses, the design of the platform for the ideological and political education of teachers' education public courses in teachers' colleges and universities based on the video-on-demand technology is not only in line with the trend of the times, but also more in line with the comprehensive requirements of the current ideological and political education. Through the combination of the data compression technology, the network technology and the database technology, we can make full use of the campus network to serve teachers' education public courses in normal universities and improve the teaching quality of schools.

Keywords: VOD technology · Curriculum ideological and political education · Teacher education public course · Curriculum reform

The course of the ideological and political theory should be strengthened in the course of improvement, enhance the affinity and pertinence of the ideological and political education, meet the needs and expectations of students' growth and development, make all kinds of courses and the ideological and political theoretical courses go in the same direction and form synergistic effect, which is a centralized summary and scientific explanation of the course of the ideological and political education. Through the VOD system, we can combine rich and colorful multimedia teaching resources such as the online teaching multimedia courseware and the scientific and technological lectures, and realize a real-time and interactive learning environment, so as to reform the traditional educational means and teaching mode, improve the teaching efficiency and optimize the teaching process.

© Springer Nature Singapore Pte Ltd. 2020
M. Atiquzzaman et al. (Eds.): BDCPS 2019, AISC 1117, pp. 1496–1502, 2020.
https://doi.org/10.1007/978-981-15-2568-1_207

1   Analysis of the Principle and Characteristics of the VOD Technology

1.1   Working Principle of Video on Demand

The interactive VOD system is composed of the front-end processing system, the broadband conversion network, the user access network, and the user terminal equipment (set-top box plus TV or computers) and so on. The VOD system can adopt not only the centralized processing structure, but also the centralized management and distributed processing. It can flexibly select HFC, FTTB, FTTC, ADSL, VDSL and other access modes. The user set-top boxes can display and communicate the video/audio/data through TV and PC through the interfaces matched with access mode according to their needs. The user initiates the first communication call to the nearest VOD service access point through his VOD terminal, requests the use of the VOD service, and sends requests to the video server through the VOD service upstream channel (such as the computer network, the telecommunications network, and the cable television network and so on). The system responds quickly, displays the VOD on the user's TV screen, and audits the user's information to determine the user's identity. Users make choices according to the VOD and request to play a certain program, while the system decides whether to provide the corresponding services according to the results of the auditing.

1.2   Characteristics of the VOD (Video on Demand) System

VOD is a new media mode in recent years. It is the product of the integration of the computer technology, the network communication technology, the multimedia technology, the television technology and the digital compression technology. The VOD technology enables people to broadcast video programs and information in the library freely on the computers or TVs without resorting to the video recorders, DVDs and cable TVs. It is an interactive system that can freely select the contents of the video programs. The essence of VOD is that users of the information actively acquire the multimedia information according to their own needs. It differs from the information publishing in the biggest difference. One is initiative, while the other is selectivity. This is a way for the information recipients to self-improve and self-develop according to their own needs. This way will more and more meet the deep needs of the information resource consumers in today's information society. It can be said that VOD is the future mainstream way of the information acquisition in the multimedia video and audio.

2 Design Background of the Ideological and Political Education Platform for Teachers' Education Public Course in Teachers' Colleges and Universities Based on the Video on Demand Technology

2.1 Current Situation of the Implementation of the Ideological and Political Education in Teachers' Education Public Courses in Normal Universities

The teacher education public course refers to a series of the public compulsory courses for non-education majors in normal universities, aiming at broadening students' knowledge of education and teaching, helping to form their correct educational concepts and cultivating their basic skills of teaching and researches. At present, there are the following problems. The first is the marginalization of their status. Students generally attach importance to their professional courses. The teacher education public courses have become the "disaster areas" where students skip classes. Secondly, the teaching methods of most teachers are old and single, which cannot stimulate students' interests. Due to the large proportion of students in public classes, it is difficult to implement "teaching according to aptitude" and the interaction between teachers and students objectively, which also causes some difficulties in the classroom management, resulting in the poor teaching effect, the low efficiency and the loose classrooms. The teacher education public course is a general course for teachers' major. Practice has proved that this kind of courses can achieve good educational effect. While teaching knowledge, it should also elaborate the logic, spirit, value, thought, art and philosophy behind the knowledge, and effectively transmit the correct value pursuit and the ideal belief to students in the form of "moistening things silently". This kind of curriculums contributes to the formation of students' educational ideas and ideals, and is a good soil for the development of the curriculum ideological and political education. The audiences of the public courses of the teacher education are quasi-teachers. They will become the backbones of the national basic education causes. The education ideal, the education ideas and the teacher morality training of normal students are particularly important. In recent years, the media exposed endless incidents against teachers' morality. From the pre-school education to the higher education, all of them are contrary to the training objectives of normal education. It is necessary to rectify and guide normal students through "curriculum ideological and political education".

2.2 Design Significance of the Ideological and Political Education Platform for Teachers' Education Public Course in Teachers' Colleges and Universities Based on the Video on Demand Technology

VOD can publish the school audio and video news on the Internet, and can broadcast activities and meetings live on the Internet. Breaking through the limitation of the teaching time and space, the teaching activities can occur in all areas of the network connection. At the same time, the teaching time has become more flexible, unlike the traditional classroom teaching, which can only be carried out in a well-arranged teaching time, so that the individual teaching, the collaborative learning and other forms can be strengthened. It can broadcast the teaching process live and automatically generate the multimedia courseware materials. The teaching materials in the form of the multimedia, which are scientific, interesting and artistic, have good flexibility. Provide the perfect multimedia on demand system and the convenient and effective content retrieval function, change the status of the learners from the recipients to the main body of the teaching, and make the media play a role as a cognitive tool for learners to learn, master and use the knowledge rather than a simple demonstration tool. The audio and video teaching materials can be converted into the digital format of the streaming media and uploaded to the VOD library. Teachers use this function to change their role from the knowledge instructor to the learning instructor and information organizer. The streaming media integrated management system can support most of the mainstream database servers, and has the strong versatility, which makes the VOD video system more convenient and fast in the construction of the information campus. The system can manage and maintain remotely in the way of the Web, support the classified and graded management of programs and users, realize the remote monitoring and the maintenance management, and save a lot of manpower and material resources in the management of the traditional teaching materials. The live programs have the flexible functions of the timing recording and the automatic publishing, which can set the working schedule and the program schedule of the live channels, and the station can automatically start or stop the live broadcasting according to the schedule. It can set the automatic recording program schedule in the process of the live broadcasting and generate files, which can be automatically uploaded to the classified catalogue designated by the on-demand server through the network. The recording table is flexible in the use and convenient in the management.

3 Design Strategy of the Ideological and Political Education Platform for the Teachers' Education Public Courses in Normal Universities Based on the Video on Demand Technology

3.1 Based on the Ideological and Political Concept of the Course, Do a Good Job of the "Top-Level Design"

First of all, we should improve the curriculum system of the ideological and political education in colleges and universities. "He who does not seek the world for a time is not enough, and he who does not seek the overall situation is not enough to seek a region." The implementation of the curriculum ideological and political education needs to build a three-dimensional education system of all staff, all-round and the entire process. This system includes not only the ideological and political theoretical courses, but also the comprehensive literacy courses and the professional courses of various departments and disciplines. The public course of the teacher education is a general course for normal students. According to the connotation and requirements of the curriculum ideological and political education, the general education system should be revised around such aspects as the system mechanism, the training objectives, the curriculum design, the teacher selection, the textbook selection and the teaching methods.

3.2   Innovate the Teaching Mode and Teaching Carrier, and Highlight the Orientation of Educating People

The curriculum ideological and political education should ultimately return to serving the growth and development of the students. Therefore, the curriculum reform should embody the "student-based" concept, change the traditional "teacher-centered" teaching mode, fully mobilize the enthusiasms of the students, and infiltrate the ideological and political education into the teaching mode, the teaching methods and the teaching carrier to reflect the goal of educating people. Under the background of the "Internet+" and the big data era, the reform of the teaching methods has a new support and carrier. The "flipped classroom" brings us a learning mode combining the "online" and the "offline" learning. In order to gain more time for students to think, discuss, ask questions and answer questions, we can designate or make unified teaching videos for students to study before class, and organize group discussions, group work or case analysis in class.

3.3   Design of the School Video on Demand System

The school VOD system generally consists of three parts: the front-end system, the network system and the user system. The front-end system is generally composed of the server group and the network management group. It is used to store and manage the data information and the user information, process the user's interactive command information and send the video data. The network system is divided into the backbone network and the local network backbone network, which requires the high bandwidth to connect the front-end system. The local network, as the next level network of the backbone network, transmits the information to each client devices. The client is a PC terminal with the display device. Ultimately, it is the client that controls the VOD and realizes the VOD. The video on demand system is based on the Ethernet network. It is used to play the video data, provide the user interface and realize the interactive VOD. The principle of the system function realization is as follows. Firstly, the client sends a request to the video server. After the video server receives the user's request, the scheduler decides when and on which channel to transmit the required video stream according to the specific scheduling scheme, and sends this information back to the client.

The SSH framework is a set of the frameworks, which is a popular development framework for the business layer Spring, the presentation layer Struts and the persistence layer Hibernate. It is an open source integration framework for the web applications. The responsibilities of the SSH framework are divided into four layers: the business logic layer, the presentation layer, the domain module layer and the data persistence layer, which can help developers build the Web applications with good reusability, clear structures and more convenient maintenance in a short time. Using the SSH development model, not only the data of the controller, the view and the model are separated, but also the management and maintenance of the code are more convenient. The advantages of the high cohesion and the low coupling are realized. Using the SSH framework can not only greatly improve the reuse rate of the code, but also benefit the cooperation between the development teams, improve their working efficiency, and save costs.

VOD server: It is the core component of the VOD system. In the VOD system, the load is the largest, so the VOD server uses the multi-machine cluster. The server operating system is Windows 2003 Server, and needs to install 9.0 or more streaming media service software. This is a cross-platform streaming media server with the strong network management capabilities, supporting a wide range of the media formats. At the same time, the client needs to install the corresponding playback software. User's VOD page is mainly composed of "home page", "classified VOD", and "program query" and so on. Users can easily and quickly understand all kinds of the information, select the required programs, and obtain the required video files through the "program query". On the VOD page, users can watch the program by clicking on the mouse. Pages are mainly written by scripts such as ASP, JSP and .NET, while the database access uses the ADO access technology to query the video data database. The management page is mainly composed of "list", "add", and "edit" and so on. Administrators can easily publish and edit the video file information through the WED pages. They can select categories through menus, and at the same time, they can select the corresponding video files in the list of categories to modify and delete. After the above work is completed, the simple campus VOD system is basically realized.

4   Conclusion

In the era of the "Internet+", we use the Internet to collect useful and educational materials. It is far from enough to explain the basic concepts and principles in the public education and teaching courses, and to present the knowledge according to the texts alone. A large number of cases should be used to assist the understanding. By adopting the B/S three-tier structure technology and some experience accumulated by myself, these technologies are deeply applied in the process of the system development, which improves the performances of the entire system. Under the background of the common development of the television technology and the information technology, the video-on-demand technology has emerged. This technology has created a new teaching mode and has gradually become an indispensable and important technical means in the field of our education.

Acknowledgement. This paper is the result of the social science research of Mianyang Federation of Social Sciences in 2018, which is "the special subject of school-local co-construction", "Exploration and practice of the curriculum ideological and political development in teachers' education public courses in normal universities" (MYSY2018QN03).

References

1. Wang, M.: On the application of marxist methodology in education. Intelligence **17**, 109 (2014)
2. Lei, Z.: Innovation of the ideological and political work in colleges and universities from the perspective of the new media. J. Fujian Radio TV Univ. **40**(1), 26–29 (2014)

3. Qu, S.: Challenges and countermeasures of the ideological and political work in colleges and universities in the micro-era. Econ. Prospects Bohai Sea **11**, 128 (2017)
4. Ding, H., Yuan, G., Zhang, Z., et al.: Realizing the web page layout with CSS+DIV development technology. Exp. Sci. Technol. **4**, 39–41 (2012)
5. Liu, W., Yan, W.: Application of the task-driven method based on the micro-video in our teaching – advanced application of office automation course as an example. China Mark. **211** (28), 207–208 (2016)

Application of the Computer Simulation Technology in the College Football Teaching

Jun Jin[⊠]

University of Shanghai for Science and Technology, Shanghai 200093, China
88279110@qq.com

Abstract. With the development of our society and the wide application of the computer technologies, the traditional teaching methods in college football teaching cannot meet the needs of the social development, so it is very helpful and necessary to add the computer simulation technology to the college football teaching to train new talents. The application of the computer simulation technology in the college football teaching is not only a bold breakthrough and innovation in the football teaching mode, but also an important demand for optimizing students' understandings and cognitions. Compared with other computer technologies, the simulation technology has higher requirements for the computer hardware, more rigorous implementation steps and more standardized process.

Keywords: University · Football teaching field · Computer simulation technology · Application thinking

In the 21st century, the computer technology has become a rapidly developing industry, and through the combination with other industries, it has also led to the development of other industries, with a very considerable market capacity [1]. The application of the computer simulation technology in the field of the football teaching in colleges and universities will provide effective help for students to better understand the movement posture and the skill requirements of football.

1   The Present Situation and Problems of the Football Teaching in Colleges and Universities in China

Through the interpretation of the development trend of football in the world today, the team relies on the overall skills and tactics more than the individual stars to win [2]. The fast pace of attack and defense and the fierce confrontation require the higher physical fitness of students. The competition strategies, tactics and football concepts are diversified. The characteristics of the current world football are the outstanding integrity, the comprehensive technology, the high-intensity confrontation, the abundant physical strength, and the formation tends to be more conducive to the balance of attack and defense [3]. The football course is one of the constituent elements of the physical education curriculum in colleges and universities. The teaching goal of the football course is to train students who master the professional football theoretical knowledge and have the perfect personality and the lifelong learning concept.

© Springer Nature Singapore Pte Ltd. 2020
M. Atiquzzaman et al. (Eds.): BDCPS 2019, AISC 1117, pp. 1503–1508, 2020.
https://doi.org/10.1007/978-981-15-2568-1_208

1.1   The Teaching Objectives Are Superficial and There Is a Lack of Substantive Teaching Guidance

At present, there are still some problems in the football teaching, such as the single teaching mode, the outdated teaching methods, teaching contents unable to attract students' interests, and being unable to meet the needs of students' development [4]. The results show that 56% of the football teachers still focus on cultivating students' knowledge and abilities, and 27% of them regard improving students' interests in learning and making students enjoy the fun of the football learning as their teaching objectives, while only 17% of them will train students' consciousness of lifelong exercise as the goal of their teaching. It is not difficult to see that the goal of the football teaching in colleges and universities has not reached the goal of cultivating students' innovative consciousness and improving students' creative abilities.

1.2   Formula-Based Teaching Mode and a Lack of Excavation of Students' Characteristics

The survey results show that teachers have the low recognition of the teaching contents and the teaching contents are obsolete and rigid. For example, the football skill training courses are mere formality, and the role of improving students' football skills is negligible [5]. The formula-based teaching mode affects students' enthusiasms for learning and makes students bored and dull, and they cannot really feel the positive significance of football. At present, the football teaching in colleges and universities is mainly divided into the following steps. Firstly, teachers teach the theoretical knowledge related to football, and secondly, teachers demonstrate the football movements and students practice under the guidance of the teachers. Lastly, teachers comment on the students. This teaching mode places students of different physical qualities and different sports levels on the same starting line, and students' actual needs and levels are not taken into account, and students' individual differences and acceptance abilities are not fully respected, which is not conducive to the formation of the lifelong exercise awareness, and nor to the improvement of their sports abilities.

1.3   Overemphasizing the Promotion of Technologies in Teaching

The football teaching in colleges and universities usually combines the football techniques and tactics to teach. Although this form of teaching can improve students' football skills to a certain extent, the improvement of technologies only stays at the surface. Teachers overemphasize the promotion of technologies, fail to fully tap the potential of students, fail to cultivate students' sense of cooperation and competition, and ultimately fail to achieve the established teaching objectives.

1.4   Fewer Competitions and a Lack of Communication

It is very difficult to organize football matches in colleges and universities, with a long cycle and a high demand for venues, which ultimately hinders the football exchanges between colleges and universities. The football teaching should attach importance to the individualized development of students and help them establish the lifelong sports consciousness. Teachers should take the individualized development of students and the improvement of students' sports consciousness as the teaching objectives and basis, formulate the relevant teaching plans, innovate the traditional teaching methods, formulate various evaluation systems, and focus on training students' innovative abilities and the spirits of unity and cooperation. Colleges and universities should provide facilities and equipment for the football teaching in terms of the hardware construction to meet the hardware needs of the football teaching.

2 Application Feasibility of the Virtual Simulation Technology in the College Football Teaching

The virtual technology emerged in the United States, which is a very comprehensive computer technology. It is an interdisciplinary subject of the computer graphics, the three-dimensional tracking technology, the pattern recognition, the simulation technology and other related technologies. "Virtual" means that there is no real existence, which is created by the computer to simulate the physical things or the environment. The virtual reality is characterized by immersion, interaction and imagination. Immersion refers to the degree of the authenticity of the user's existence in the simulated environment. Interactivity refers to the user's operability of the virtual objects in the simulated environment and the feedback timeliness of the obtained environment. Imagination mainly reflects the vastness of the imagination space that the virtual technology can provide for the users. The virtual technology can greatly expand the human understanding, which is not limited to the real environment. What's more, users can realize the recognition of the impossible environment or things on the premise of the free imagination.

The application of the virtual simulation technology in the modern education has greatly promoted the development of the teaching technology and realized the leap-forward development of the modern educational technology. The virtual simulation technology achieves a breakthrough in the inherent mode of the traditional teaching methods. On the basis of the virtual simulation technology, the virtual learning environment can be created independently. In this virtual learning environment, learners can acquire more knowledge through interaction with the virtual learning environment, and at the same time can greatly improve their personal learning abilities, and constantly innovate their own learning methods. The virtual simulation technology refers to using part of the simulation software skillfully on the basis of the construction of the experimental model, the circuit environment of the real simulation test model, and a series of the concrete practical operations under the action of the experimental model, which are mainly embodied in the two levels of the virtual simulation and the virtual reality.

The virtual simulation technology is the combination of the computer simulation technology and the system simulation technology. The virtual simulation technology is necessary, feasible and superior in the modern teaching, especially in the college football teaching. The computer simulation is a comprehensive technical subject which is practical and has the strong application characteristics. It has been widely used in the field of education. It has been widely used in physics, chemistry, electronic circuit and other experimental courses. Relevant simulation software has appeared one after another, and has played a very good role. However, the computer simulation technology is seldom used in the field of the physical education. Therefore, the computer simulation technology is introduced into the physical education. In our teaching, it will play an active role in promoting the future development of the physical education and teaching.

In the training of students' movements, the computer can simulate the process of the sports virtually. Without the actual training, the students can get training with the comparable or better results. Or the computer can describe the students' participation in the football process digitally, and make various complicated analysis, and use the computer digital information input, fast processing, intelligent processing and other functions to deal with student's movement process, with the characteristics of large capacity, fast data processing, accurate, complete and rapid information feedback, easy access, and scientific analysis and so on. It provides a scientific basis for coaches and scientific researchers to make a comprehensive analysis of students' football training. It is a powerful promotion for the scientific training and matches. Or it can analyze and display the details of excellent students' football sports process, so that the unnecessary detours can be avoided in the ordinary training, so as to quickly assist students in making their movements more standardized.

3 Specific Strategies of Applying the Computer Simulation Technology in the College Football Teaching

The teaching contents related to the college football teaching contents, which are difficult for students to grasp or are interesting to students, are displayed and released through the panoramic technology of the computer simulation technology. Students can choose the teaching points which are interesting to individuals or doubtful in college football teaching classes at any time after class for repeated observation, to promote students' grasp of the college football teaching contents, integrate into a good football atmosphere, and effectively enhance the mastery of the sports skills and their interest in learning.

3.1 All-Round Simulation of the Football Teaching Environment

The computer simulation technology, in a simple understanding, is that to meet people's actual needs, through the computer simulation of a variety of the realistic scene environment, in the course of learning football, because of the different equipment of each school, students are afraid of the injuries in practice, so that students in the learning process do not like this sport. If the virtual reality environment is simulated, the virtual scene can avoid students' injuries in practice, and improve the teaching equipment and the teaching environment.

3.2   Multi-type Teaching

The physical education and teaching has a very high demand for the sports venues, equipment and environment, and if restricted by one of them, we cannot carry out certain courses. Football, as a result of various factors such as the venue, equipment and environment and so on, is limited to the promotion of the school courses, so it cannot meet the diversification of the physical education. Through the computer simulation technology, we can solve these courses which cannot be carried out because of the external factors. By constructing the simulation scenarios, students can practice in the virtual reality environment, which can not only achieve the goal of the football sports teaching, but can also ensure the safety of the students.

3.3   Shooting and Making the Football Videos

Pre-preparation stage: The Premiere and the DetuStitch software are used to synthesize the panoramic camera and the balance vehicle. After adding the panoramic video hopping, the voice explanation and the text annotation, the system is tested and used through the panoramic publishing platform. Make detailed project development plans, including the schedule, equipment borrowing arrangement and the working content arrangement. The teaching practice of the panoramic sports course based on the computer simulation technology in the teaching links focuses on ensuring the clarity and authenticity of the panoramic videos so as to give full play to their intuitive and immersive characteristics.

Taking the football upside-down training as an example, eight viewpoints are selected for shooting on the playground, and the viewpoints are set around the demonstrators of the football upside-down. Another method of the shooting can add a balance car to assist the shooting, and the panoramic camera is placed on the balance car, with the demonstrator of the football upside-down as the center, and the distance of the radius 2 meters is the circular trajectory, to shoot while moving at a uniform speed. Image synthesis and post-production: Classify the panoramic videos and organize the effective video materials into a file. The panoramic video is adjusted by the Premiere and the After Effects software, such as the saturation of the position, the brightness and other illumination conditions and so on. The panoramic video is stitched into the seamless panoramic video. Write a program to make the panoramic video viewpoint jump transfer, the text annotation box and the voice explanation and other functions, and finally conduct the internal testing.

3.4   Specific Application in the Teaching and Training Mechanism

In the application of the computer simulation technology and the football teaching videos in colleges and universities, 720 degree scenes are displayed. The viewers can choose any angle to watch according to their personal wishes and needs. Because of the different vividness, authenticity and arbitrariness of the free choice of angles between the traditional videos, it is easier to impress the viewers and make them authentically and conveniently integrated into the sports. The interactive learning experience can stimulate students to participate in the sports and the teaching activities to the greatest extent. The application of the computer simulation technology in the football physical education teaching is a new experience and feeling for the college football teaching, teachers and students. It is not only a new way to widen the teaching means of football in colleges and universities, but can also inject new creativity and innovative consciousness into the teaching and learning process of teachers and students. It will promote teachers and students to better accomplish and realize the goals and tasks of the football teaching in colleges and universities, stimulate the research and exploration of sports, and improve the positive effect of the football teaching in colleges and universities.

4   Conclusion

With the continuous development of the science and technology, the field of the computer technology has evolved from the simple data processing to the human-computer interaction, and eventually to the integration of the simulation technology, which is gradually applied to all aspects of our society. The simulating technology is the abbreviation of the computer simulation technology. It combines many fields of the technology, provides the technical support for many fields in our country, and plays a vital role in the development of many fields in our country. In the college football teaching, the training of students' movement is influenced by many factors. Many sports techniques not only have the complex structures, but also need to complete a series of the complex technical actions in an instant. Moreover, many movements themselves are very dangerous, which brings challenges to students' training.

References

1. Li, H.: The practical application of the virtual simulation technology in the computer courses. Digital Commun. World **6**, 296 (2015)
2. Zhou, J.: Application and analysis of the computer virtual simulation technology in the NC teaching. China Sci. Technol. Panor. Mag. **20**, 190 (2012)
3. Meng, Q.: Brief analysis of the computer virtual simulation technology and its applications. Comput. CD Softw. Appl. **22**, 36 (2011)
4. Xu, K.: Analysis and application of the computer virtualization technology. Digital Technol. Appl. **3**, 117 (2016)
5. Song, J., Xu, W.: Application of the multivariate simulation technology in the computer network teaching. China Sci. Technol. Inf. **6**, 169–170 (2013)

Design and Implementation of the Big Data Management Decision System Based on the Hadoop Technology

Deli Kong[✉]

Nanjing Institute of Mechatronic Technology, Nanjing 211135, Jiangsu, China
1138452255@qq.com

Abstract. With the rapid development of the mobile Internet, the amount of data is increasing at the level of TB or even PB every day. People's requirements for the efficiency and security of the data access are also constantly improving. The traditional data storage technology is unable to deal with the massive data. How to store and read the data efficiently has become a hot issue of research and concern. Hadoop is a mature solution in the big data storage. It has many advantages, such as high reliability, high scalability, high fault tolerance and high efficiency. It is also open source and free of charge. It is very suitable for the scientific research. Therefore, this paper chooses the Hadoop platform to build the big data storage system.

Keywords: Hadoop technology · Big data · Management decision · System design · Implementation

Hadoop is a framework that can efficiently distribute the massive data. It mainly consists of two core technologies: HDFS (Distributed File System) and Map/Reduce (Distributed Computing Framework) [1]. This project adopts the new technologies such as Hadoop distributed architecture and the ecological components, plans to integrate and rationally utilize the existing infrastructure resources such as servers, networks and storage in higher vocational colleges, and builds a big data cloud platform center to meet the requirements of the future information development.

1 Technical Support of the Big Data Management Decision System Based on the Hadoop Technology

1.1 Analysis of the Working Principle of the Hadoop Framework

Hadoop is an open source framework, and its essence is the large-scale data that can be used to write and run the distributed application processing. Compared with other frameworks, Hadoop has the characteristics of convenience, expansibility and easy operation. Especially the convenience of Hadoop makes it a dominant advantage in the process of writing and running the large distributed programs [2]. With the help of Hadoop, users can appreciate the advantages of the distributed computing rules to a large extent. Hadoop can solve the problem of the time-consuming data transmission

© Springer Nature Singapore Pte Ltd. 2020
M. Atiquzzaman et al. (Eds.): BDCPS 2019, AISC 1117, pp. 1509–1513, 2020.
https://doi.org/10.1007/978-981-15-2568-1_209

very well in the process of the big data processing by using the distributed storage, the migration code and other technologies [3]. More importantly, the data redundancy mechanism enables Hadoop to recover gradually from the single point failure.

The basic structure of the Hadoop framework includes the distributed file system HDFS and MapReduce. HDFS mainly uses the Master/Slave architecture. An HDFS cluster contains the NameNode nodes and the DataNode nodes. NameNode belongs to the central server. Its main function is to manage the namespace of the file system and to be responsible for the file access [4]. In the cluster system, usually a DataNode is run in a node, which mainly manages the data information in the node, handles the file read and writes the requests sent by the client, and creates and replicates the data module under the scheduling of NameNode. In addition, Hadoop can also complete the MapReduce distributed computing. MapReduce can divide the total task into several sub-tasks, and each sub-task can be processed in any cluster node.

1.2 Map Reduce Programming Model

The MapReduce (Map-Merge Algorithms) model is an abstract model of the higher-order parallel functions proposed by Google. According to the related reports, millions of MapReduces are executed every hour in the Google cluster [5]. It relies on the idea of the functional programming to abstract the general operations of the massive data sets into two sets of operations, Map and Reduce, which greatly reduces the difficulty of the distributed parallel computing programs. In such a computing model, there are two key links: the mapping Map and the aggregating Reduce. So programmers must be required to implement the above two functions, the Map function and the Reduce function, to calculate a set of the input keys, so as to obtain another pair of the output keys.

On the Hadoop platform, the MapReduce applications consist of a Mapper class, a Reducer class, and a driver function for creating JobConf. Sometimes you can also implement a Combiner class as needed, which is actually an implementation of the Reduce function. Input: The application automatically provides the Map and Reduce functions, and specifies the specific location of the input/output and the specific parameters required for other operations. This process divides the large files in the directory into separate data blocks. Map: This model can treat the user job input as a set of the key/value pairs. The MapReduce model can automatically call the Map function to process one of the key-value pairs, thus forming a new key-value pair. Shuffle and Sort: In the Shuffle period, the network provides all Reduces with matching key pairs for all the Map outputs, and in the Sort period, the Reduce input is grouped according to the Key value. Generally speaking, Shuffle and Sort are executed together. Reduce: For each Key, the user-defined Reduce function is executed to get a new key-value pair. Output: Write the results of Reduce into the output directory.

1.3 Application Characteristics of the Hadoop Big Data Analysis

As a typical representative of the distributed computing, Hadoop has more advantages than other distributed frameworks. Scalability: Hadoop can distribute the data and complete the computation among available computer clusters without stopping the cluster services, which can easily extend to thousands of nodes. Simplicity: Hadoop implements a simple parallel programming mode. Users can write and run the distributed applications without knowing the underlying details of the distributed storage and computing. They can process the large-scale data sets on clusters, so users using Hadoop can easily build their own distributed platform.

2 Design of the Big Data Management Decision System Based on the Hadoop Technology

Through the data center of the big data cloud platform, we gather all kinds of the structured and unstructured multi-source heterogeneous data to form a unified data center resource pool. We use the big data mining algorithm and the data visualization technology to form all kinds of the data service models to provide the educational and teaching decision support for school leaders.

2.1 System Design Objectives

This project mainly designs and implements the big data center system based on the Hadoop College. Firstly, the system development environment is built on the Linux system. Aiming at various business information systems of the college, the seamless integration of the existing information systems and the Hadoop-based data center is realized, including the unstructured data and the semi-structured data storage, the structured data storage and the big data mining platform. The structure of the project is as follows: the related theories and technologies. Many latest theories and technologies of the computer science have been used in the construction of the data center, including Hadoop, the HDFS distributed file system, the MapReduce batch computing model, the Spark memory computing framework, the Mahout machine learning and data mining database, the HBase distributed non-relational database and the Hive data warehouse tools.

2.2 System Design of the Big Data Cloud Platform

Firstly, the overall framework of the system is designed, and the core design issues are elaborated. Next is the implementation of the structured data storage, which is elaborated from two aspects: the system storage design and the database table structure. The unstructured data storage part mainly elaborates the design of the data storage system and the structure of the data storage table. Finally, the design of the big data machine learning platform is described. A data cleaning method based on the functional dependencies between the schema attributes is proposed. In order to improve the data quality of the query result records from the multiple Web databases, this method can effectively repair the incomplete, inaccurate and incorrect attribute values in the query result records by using the entity recognition technology and the functional dependencies among the pattern attributes. At the same time, an incremental data integration method is proposed. That is, the order of the integration is determined by the result of the data quality evaluation of the record set, which effectively improves the efficiency of the data integration.

3   Implementation of the Big Data Management Decision System Based on the Hadoop Technology

3.1   Environment Construction and Implementation of the Big Data Cloud Platform

According to the design scheme of the platform, the function of the platform is developed. Firstly, the data center platform is built, and after the system is completed, the structured data storage is developed by taking the card communication information system as an example. Secondly, the video teaching system is taken as an example to develop and realize the storage of the unstructured data. Finally, the realization of the machine learning platform is expounded. The system consists of four parts: the user layer, the system management layer, the data storage layer and the infrastructure layer. User layer: The user sends the Ajax requests through the WebApp server, and the server responds to the corresponding JSON data to the browser, and the front end calls the response function REST interface through the Angular JS framework, and finally, the page data is rendered automatically through the Angular JS bidirectional data binding. The system management layer: It is responsible for connecting the application program with the underlying data operations, and using the JFinal MVC framework, a large number of files and user operations are encapsulated as the REST interface, providing the data return of the front-end Web pages interface, and sending the data results required by users back to the client in the JSON format. It mainly provides the user management, the directory management, the file management and other services. Storage data layer: Hadoop cluster the works in this layer, which is composed of HDFS and Jetty. HDFS provides the distributed storage support. One Name Node stores the metadata of files. The multiple Data Nodes provide the storage implementation for files. The Jetty server provides the external REST interface call function. At the same time, this layer also encapsulates a large number of the HDFS operations.

3.2   Data Acquisition and Query

HDFS is responsible for the data partitioning. HDFS automatically divides the data into the data blocks. The default size of the blocks is 64M. This user can set it up on his own. When the data is partitioned into blocks, the data block processing job is divided into several Map tasks that can run independently, which are assigned to different cluster nodes to execute, generate intermediate files in a certain format, and then create indexes by merging these intermediate files by several Reduce tasks. When users query, the global index cluster forwards the query requests to satisfying the data blocks to achieve the routing lookup function, which reduces the pressure of the local index cluster and improves the query speed. The global index block is realized by the data block summary and the local index summary. The global index cluster is built by the DHT technology. The nodes of the local index cluster establish the local index table of "key value – data block summary" for each data block, and then the global index cluster node establishes the global index table of "key value – data block". This reduces the length of the routing list for each index item and organizes the global index items through a consistency hash. When customers query, they first hash the key value of the query, then find the global index table of the data block that maintains the key value through the consistency protocol, and then forward the request to the local index table corresponding to the data block, and finally get the query result.

3.3 Design and Test of the Big Data Mining Algorithms

According to the application requirement of the college business system, the ETL tools are used to import the data from the core databases of various business systems into the big data cloud platform, and the machine learning algorithms such as the classification and clustering are used for the iterative learning to generate the data service model and the continuous feedback optimization to guide the teaching practice. An effective query result record extraction technique is proposed. In order to avoid the semantic matching of a large number of pages and ensure the efficiency of the data extraction, this paper first determines the exact search result page based on the method of the URL matching, then locates the query result record by using the pattern attribute path identified in the query result pattern extraction, and implementing the extraction and annotation of the query result records. The wrapper based on the attribute path can effectively improve the execution efficiency of the record extraction in the continuous query result pages.

4 Conclusion

With the continuous development of the information age, the data generated by the Internet is growing explosively. The traditional single-chip computer architecture is unable to cope with such a large amount of data. The cloud computing provides a new solution for eh large-scale data processing. Hadoop is one of the open source projects of the Apache foundation, which collects Mapreduce, HDFS, HBase, pig-sub-projects in one, showing the excellent computing, processing and scheduling capabilities.

References

1. Gao, H., Yang, Q., Huang, Z.: Standardization of the key technologies for the big data analysis based on the Hadoop platform. Inf. Technol. Stand. **5**, 27–30 (2013)
2. Wang, T., Shen, J., Yin, R., Ma, Z.: Application of the big data analysis technology in the power industry based on the Hadoop platform. Electron. Technol. Softw. Eng. **024**, 177 (2016)
3. Song, W.: Research on the data mining technology based on the Hadoop platform. Pract. Electron. **12**, 76–77 (2014)
4. Yuan, C.: Data mining and analysis based on the Hadoop cloud computing platform. China Comput. Commun. **15**, 58–59 (2015)
5. Dai, Z., Sheng, H., Wang, L.: Big data analysis and processing based on the Hadoop platform. Telecommunications **6**, 59–60 (2015)

Multimedia Teaching Reform of Skiing Based on the VR Technology

Xiaofeng Li[(✉)]

Hulunbuir University, Hulunbuir 021008, Inner Mongolia, China
kingguogang@sohu.com

Abstract. In view of the unfavorable factors of the unreasonable allocation of the class hours, the restriction of the site conditions and the high difficulty of the action technologies in the traditional skiing teaching methods, this paper puts forward the idea of using the modern education and multimedia technologies to overcome the shortcomings of the traditional skiing teaching methods, aiming to break the traditional teaching methods and effectively improve the quality of the skiing teaching. Through the analysis of the traditional skiing teaching mode and teaching methods, this paper expounds the positive role of the computer and multimedia courseware in the skiing teaching with examples, probes into the characteristics and advantages of using the multimedia courseware in the skiing teaching, and puts forward concrete opinions and suggestions.

Keywords: VR technology · Skiing · Multimedia teaching · Reform strategy

The VR technology is a technology in recent years. With the development of computers, it combines the hardware with the corresponding software to realize a more real human-computer interaction. Similarly, the application of the VR technology in the multimedia teaching can also provide students with a real learning environment, provide more real virtual space to exercise students' practical operation abilities, have the stronger interaction, and let students immerse themselves in the learning environment, which is of great help to improve the current teaching quality.

1 Factors Influencing the Traditional Skiing Teaching Quality

Under the impetus of "ice and snow sports", the skiing teaching is gradually systematized [1]. However, because students have different growth environments, personal hobbies and interests, they have different mentality towards the skiing learning, which makes skiing teaching quality unable to improve rapidly as a whole.

1.1 On the Student Side

According to the relevant survey and data, the influencing factors of students are mainly embodied in the following three aspects. The first is the students' fear and inferiority complex. Usually, students will have the phenomenon of fear and inferiority [2]. There are many reasons. According to the learning situations of using skills,

M. Atiquzzaman et al. (Eds.): BDCPS 2019, AISC 1117, pp. 1514–1520, 2020.
https://doi.org/10.1007/978-981-15-2568-1_210

students will show different functional responses in different cognitive states, which make the students' brain automatically produce the fear reaction when they perceive the danger of skiing [3]. At the same time, when practicing the skiing movements, some of them are more difficult. If they are hindered, they will make students feel inferior, which will lead to students not dare to learn more difficult movements, and ultimately reduce the quality of the skiing teaching [4]. The second is the impact of injuries. Due to various injuries in the skiing learning, and the injury rate is very high, some students will form psychological shadow after the injury, which will generate the psychological moods of fear and terror.

1.2   On the Teacher Side

At present, the influencing factors of the teachers mainly refer to the incompleteness of the skiing related theoretical knowledge and the unscientific teaching methods of textbooks and skills [5]. Among them, the theoretical knowledge of skiing is too much about the problems that should be paid attention to in the practical practice, but not the close combination of each kind of the theoretical knowledge and the practical teaching links. As a result, students have some problems, such as the incomplete knowledge of safety and the inadequate knowledge of the skill theory. At the same time, teachers did not make the teaching plans according to students' actual learning situations, did not set the teaching objectives and choose the teaching contents reasonably, which resulted in the inappropriate teaching methods, teaching content levels beyond the students' acceptance range and other problems, greatly reducing the quality of the skiing teaching.

1.3   In the Aspect of the Teaching Matching

Because the skiing teaching is the first time that many students are exposed to skiing, if the slope of the site is too steep and there are too many bends, it will enhance the psychological fear of some students, thus affecting the quality of the skiing teaching. Therefore, the teaching matching is one of the important factors affecting the quality of the skiing teaching. We must pay attention to the reasonable setting of the height, the slope and the length, and choose the type of ski according to the actual learning situations of students, so as to give full play to the auxiliary teaching role of the teaching venues and equipment. Generally speaking, the ski shoes or the skis are made of the ABS material or the plastic. They have the high hardness. They can protect and fix the feet and the ankles in time in the course of the practice. To a certain extent, they can improve the safety of the skiing teaching. However, students will feel heavy after wearing the ski shoes, which can make them walk inflexibly and inconveniently. Especially when standing and taxiing, it is easy to break, which eventually leads to students' fear and terror of difficulties. Therefore, the skiing teaching facilities are an important factor affecting the quality of the teaching.

2   Application Strategy of the VR Technology in the Multimedia Teaching Reform of Skiing

How can we apply the virtual reality technology in the skiing multimedia teaching? On the premise of using various technologies, the virtual teaching platform of the multimedia experiment with imagination and interaction is established by means of constructing the virtual space and other methods. The research shows that the VR technology is mainly realized by virtue of the hardware and the software resources of the virtual reality, so as to realize the human-computer interaction. By virtue of the virtual space, we can provide the users with the visual, auditory and tactile experience, and the users can interact and perceive the three-dimensional objects. The application of this technology in the practical teaching can enhance students' operational abilities, optimize the experimental teaching environment, and enhance the level of the teaching management.

In the process of our country's educational reform, the skiing teaching has become an important part of the curriculum reform in some colleges and universities. It is also an important manifestation of the continuous promotion of the quality education. Only by constantly improving the quality of the skiing education can we better promote the overall development of students' comprehensive qualities and abilities. According to the actual situations of the skiing teaching, there are many factors affecting the quality of the skiing teaching, which need to be paid great attention to in order to truly improve the current situation of the skiing teaching.

Because the skiing teaching is relatively difficult, how to accomplish the task of the skiing teaching with the high quality is a long-term research topic for the physical education teachers in the northern universities. The multimedia application in the skiing teaching has many teaching advantages, such as stimulating students' enthusiasms for learning, enriching the teaching contents of skiing, improving students' ability to explore independently, and accelerating students' correct mastery of the skiing skills and so on. At the same time, it also shows that it cannot combine closely with the sports practice on the teaching site in time. Teachers' ability to use the multimedia influences and limits the overall level of the multimedia teaching, the lack of the multimedia teaching hardware facilities and the single courseware. Some countermeasures are put forward to improve the skiing teaching quality, such as changing the ideas, renewing the teaching mode, strengthening the classroom interaction, mobilizing students' learning enthusiasms, improving teachers' comprehensive qualities, enhancing the scientific level of the multimedia teaching, increasing the investment in the funds and improving the hardware environment of the multimedia teaching.

3   Theory of the Multimedia Teaching Reform of Skiing Based on the VR Technology

3.1   VR Technology

The virtual reality (VR) is an advanced human-computer interaction technology which can effectively simulate human's behaviors of seeing, hearing and moving in the natural environment. It uses computers to generate the realistic vision, hearing and smell to perceive the world and make users feel immersed in it. The virtual reality is developed on the basis of many related technologies. It combines the latest achievements of the computer graphics and images, the simulation technology, the sensing technology, the display technology, the artificial intelligence, the language recognition, the network technology and other disciplines. The virtual reality is the intersection and integration of the computer science and the information science. Generally speaking, the virtual reality has four characteristics: the multi-perception. In the virtual environment, users not only have the visual perception of the general computers, but also have many perceptions, such as hearing, touch, force, motion, and even smell and taste. Immersion: It refers to the degree of the reality that users exist in the virtual environment as protagonists. The virtual reality system can temporarily separate the users from the external environment and integrate them into the generated virtual world wholeheartedly. Interactivity: It refers to the degree of the user's manipulation of objects in the virtual environment and the natural degree of the feedback from the environment. Users can talk and interact with the virtual environment in the real time through the three-dimensional interactive devices from inside to outside and from outside to inside. Autonomy: It means that objects in the virtual environment can move autonomously according to various models and rules.

3.2   Constructivist Teaching Theory

Constructivism holds that learners' knowledge is acquired through the construction of the meanings under certain circumstances, with the help of others, such as cooperation, communication and the use of the necessary information between people. The ideal learning environment should include four parts: situation, cooperation, communication and meaning construction. Situations in the learning environment must be conducive to the meaning construction of the learners' learning contents. In the teaching design, creating a situation conducive to learners' construction of meanings is the most important link or aspect. Collaboration: It should run through the entire learning process, including the cooperation between teachers and students, and between students and students. Communication: Communication is the most basic way or link in the process of the cooperation. In fact, the process of the collaborative learning is the process of communication, in which each learner's ideas are shared by the whole learning group. Communication is a vital means to promote the learning process of every learner, and the construction of meanings is the ultimate goal of the teaching activities, and everything should be carried out around this ultimate goal. In the design of the teaching process, constructivists put forward that if our teaching is too simple to divorce from the situation, it should not be from simple to complex. In this process, students should find out the sub-tasks needed to complete the overall task and the knowledge and skills needed to complete the tasks at all levels.

4 Multimedia Skiing Teaching Reform Strategy Based on the VR Technology

The application of the multimedia technology in the skiing technology teaching in colleges and universities has a very important practical significance. At the same time, it has a series of effects on the skiing technological teaching. It mainly embodies in the selection of the teaching contents, the change of the teaching organization forms, the new demand for teachers and the regulation of students' learning process. The definition of the key and difficult points in the skiing technological teaching is an important prerequisite for the successful production of the multimedia courseware and the improvement of the teaching quality. According to the different teaching stages of the skiing technology, the emphasis of our teaching is to adapt to the skiing, and the difficulty is to improve the skiing speed and ability of different difficulty. The VR learning solution is divided into three parts: the virtual learning, the intelligent learning and the wise learning.

4.1 VR Skiing Learning Scene Management Platform

The virtual learning builds the virtual reality scenes to realize the immersion teaching mode for learners. On the basis of the virtual reality, the relevant information of the learners is collected and the learning effect is analyzed intelligently, so that the learning scenes can be refined and the better teaching effect can be achieved. Using the computer technologies, the virtual reality technology and the network technology, we can solve the problems of the shortage of the teachers, the non-standard teaching contents and the single teaching mode for the traditional schools, realize the transformation of the teaching concept, the innovation of the teaching means and the reform of the teaching mode, realize the simplification, flexibility and high efficiency of the teaching work, and create the VR virtualized scenarios, to enable the students to acquire practical working skills in the virtual scenes. The training scenarios can be divided into the daily learning scenarios, the real-life examination scenarios, and the knowledge breakthrough scenarios and so on. In each category, the multiple scenarios can be set according to the actual situations of the daily work.

4.2 VR Skiing Examination Management Platform and the Learning Information Management System

By setting the exam questions in the learning scenarios, and then assessing students' learning situations through the real-life exercises, the platform can make each exam into a drill, thus comprehensively examining the effect of students' acceptance of their learning. Collect the user information of the participants in the VR skiing learning scenarios, and the behavior information they take in the learning process, such as the student attribute information, the student participation information, the learning focus information and other information related to students.

4.3 Skiing Learning Effect Analysis System

Through the above-mentioned learning information, through the decision analysis system and various statistical analysis techniques, an integrated and business-independent data analysis platform is established to realize the application of the report statistics and the training process monitoring and analysis, which can provide the fast, scientific and rational reliability for evaluating the learning effects and improving the learning scenarios.

4.4 Multi-person Cooperative Training System

The system can solve the problem of the collaborative training. In addition to realizing the three-dimensional real-time interaction, the system also includes the collaborative communication and the information management functions. The multi-person collaborative training platform is a real-time three-dimensional scene in which many students use the computer networks to cooperate with each other through the network communication to complete a task. At the same time, the system provides a knowledge base of the relevant information and records the entire operation process to provide the expert evaluation system. By using the artificial intelligence technology and the virtual simulation technology, the multi-person collaborative virtual environment can be joined with the interactive mechanism of the expert evaluation, and the learning process and the teaching scheme can be adjusted and evaluated at any time. Thus, the evaluation of the learning results can include not only the objective quantitative evaluation information, but also the active empirical judgment information.

5 Conclusion

The skiing technology teaching has many characteristics, such as more basic technical links, less teaching hours, difficult to control the center of gravity of the body on the snow, fast moving speed, limited protection function of teachers, and prone to injury accidents. Using the VR technology to carry out the skiing teaching activities can effectively turn boredom into fun, enhance the teaching interests and values, make teachers more ski-proficient in the skiing teaching class, make students more interested in the skiing class, and make students' creative quality lively and active development.

References

1. Zhang, J.: On the development and construction of the photographic simulation laboratory based on the VR technology. China Educ. Light Indus. **1**, 91–93 (2014)
2. Liu, Z., Liang, X., Huang, M.: Construction of the network virtual simulation laboratory platform based on the cloud computing technology. China Comput. Commun. **9**, 245–246 (2016)
3. Fan, X.: Deep immersion: development of the VR technology. Internet Econ. **11**, 30–35 (2016)

4. Liu, T., Chen, Q.: Research on the integrative development of the VR technology and the Internet education. Inf. Constr. **7**, 197 (2016)
5. Lei, Q.: Application of the VR technology in the English teaching. China Educ. Technol. Equip. **23**, 131–132 (2016)

Big Data Forecast on the Network Economy

Xiaoxuan Li[1(✉)] and Qi Wu[2]

[1] School of Economics, Fuyang Normal University,
Fuyang 236000, Anhui, China
xianghongxue1@yeah.net
[2] School of Physics and Electronic Engineering, Fuyang Normal University,
Fuyang 236000, Anhui, China

Abstract. With the increasing amount of the data information, the value of the data information is constantly highlighted. Under this circumstance, the big data technology for the Internet data collection, collation and analysis emerges as the times require, and plays an important role in the social and economic development. The organic combination of the big data analysis and the cloud computing technology is very in line with the background of the times. Under the traditional data analysis framework, the deep mining ability of the statistical data is obviously insufficient, and the efficiency of the data economic statistics and the quality of the economic statistics cannot be effectively improved.

Keywords: Big data · Prediction · Network economy · Analysis · Application research

China's network economy has stepped into the fast lane of the accelerating development, which has a strong supporting role for China's economy. However, according to higher standards, there are still many weak links, especially in promoting the informatization, digitalization and intellectualization of the network economy [1]. In order to promote the development of the network economy, we must play a positive role of the big data, and actively explore the innovative way of the network economy development, so that the big data can be widely used in the network economy system, forming a strong support force [2].

1 Analysis of the Application Advantage of the Big Data Forecasting to the Development of the Network Economy

The big data means that the size of the data has exceeded the ability of the typical database software tools to capture, store, manage and analyze [3]. The data processing, communication, aggregation, storage and analysis have penetrated into every link of the social management and become an important part of it. The ubiquitous sensors and microprocessors, as well as the Internet and the social media, have produced the big data [4]. Because the most important part of the network economy is openness, integration, cross-border and personalization, it is very important to apply the big data to the network economy. The application of the big data to the network economy can further enhance the coordination of the network economy. The most important thing is

© Springer Nature Singapore Pte Ltd. 2020
M. Atiquzzaman et al. (Eds.): BDCPS 2019, AISC 1117, pp. 1521–1527, 2020.
https://doi.org/10.1007/978-981-15-2568-1_211

that the big data has a strong function of the information collection. Because the network economy mainly adopts the way of "on-line", and the interconnection among the economic entities, enterprises and consumers should be realized [5]. The big data technology can coordinate all aspects.

The big data, represented by the accumulation of the Internet platforms, has its unique advantages in forecasting the network economy. Timeliness: The data accumulated through the Internet platform is stored in the network space. When the transaction happens, all the information such as the transaction data and price and so on will leave traces of record in the network immediately. It can be extracted by certain methods and technologies for processing and analyzing problems without the time lag. Precision: The data provided by the network platform records the information according to the actual occurrence when the event occurs, which reduces the human operation and provides relatively more original data, rather than the data information collected by the humans after processing, which is more accurate, with relatively low cost.

In order to reduce the costs, the traditional data collection process will try to collect the overall data rather than the detailed data information. In the era of the network big data, there is little difference between extracting the general data information and collecting a certain kind of the data separately. It can provide more detailed and meaningful data information without significantly increasing the cost. Large sample size: By using the big data information of the Internet, we can get the sample information of the whole or close to the whole, but not by statistical sampling to get the sample information to infer the overall information.

2 Problems of the Network Economy Under the Background of the Big Data

Although the application of the big data in the main body of the network economy, especially for some large-scale and high-quality main body of the network economy, is more prominent in this respect, the vast majority of the main body of the network economy still have a lot of the ambiguous understanding in the application of the big data, and there are many constraints in the application process, so that the network economy has not yet adapted to the era of the big data, and even restricted the scientific and healthy development of the network economy, which needs to be improved seriously. To carry out the in-depth analysis of the problems of the network economy under the background of the big data, the most prominent are the following three aspects.

First, the network economy lacks innovative ideas. The idea is the precursor of the action. For the development of the network economy, we must always go on the road of innovation. In spite of the overall development of China's network economy, not only has it made progress compared with the past, but also with the joint efforts of the country and the main body of the network economy, China's network economy has entered the fast lane of the accelerated development. Some main bodies of the network economy have not yet deeply realized the role of the big data in the development of the network economy, so they have not paid much attention to its application. Therefore, they lack effective collection, analysis and collation of the enterprise information, the business information and the consumer information, and their production and operation activities lack pertinence and systematicness, which is an important bottleneck restricting the development of the network economy.

Second, the network economy lacks the field expansion. With the emergence of the big data technology, the development model of the network economy has undergone profound changes. The traditional e-commerce model has not adapted to the needs of the development of the situation. It is necessary to reform the existing model, especially in the field of the network economic development. However, in view of the current operation of China's network economy, the application of the big data is very limited, which directly leads to the lack of the effective expansion in the field of the network economy, and still remains at a shallow level in many cases.

Third, the network economy lacks the systematic integration. For the network economy, in order to make the functions and roles of the big data play an effective role, the most fundamental thing is to further improve and perfect the network economy system, so that it has a strong integration function, which depends on the innovative application of the big data. For example, in the process of its development, some main bodies of the network economy also integrate the big data system with other systems organically. The financial system, the management system and the ERP system are relatively weak. The big data has not penetrated into each management system, resulting in the weak integration of the system.

3 Application of the Big Data Forecasting in the Network Economic Analysis and Research

It is not difficult to find that there is still great potential and space for the domestic research and application in this field after reviewing and combing the researches on monitoring and forecasting of the network economy by using the big data of the Internet at home and abroad. It is embodied in the following three aspects:

3.1 Change from the Traditional Network Economic Statistics to the Internet Non-uniform Data

The traditional network economic data depend on the survey and statistics to a great extent. In the trade-off between the accuracy and timeliness, the official statistical departments usually sacrifice the timeliness to ensure the accuracy. This inevitably leads to the lag of the data publishing time. And abandoning accuracy for timeliness seems to do more harm to monitoring and forecasting the network economy. The "big" of the big data is reflected in the fact that it no longer depends on various statistical data. All kinds of the non-uniform data and the unstructured data can be used as resources. Search the data, the social data, Micro-blog, We-Chat and forums and so on can be used to monitor and predict the network economy. In this regard, many useful attempts have been made abroad. The big data collection channels are no longer limited to the statistical surveys, because the diversity of the data types expands the channels and scopes of the data collection. Various texts, images, videos and broadcasts can become the objects and channels of the information acquisition through the big data technology and methods.

3.2 Change from the Total Forecast of the Network Economy to the Forecast of the Leading Indicators of the Network Economy

Monitoring and forecasting the total amount of the network economy has been the focus of the domestic research. But compared with the foreign countries, we still have a big gap. On the one hand, foreign countries have applied the big data methods and technologies to the real estate, the stock market, the automobile, the tourism, the medical treatment and the unemployment, which are closely related to the network economy. However, there are few studies on these areas in China. On the other hand, although the big data has not entered China for a long time, it has attracted great attention in the international arena. Many countries have raised the big data to the national development strategy, hoping to be the world leader in the big data field.

3.3 Change from the Medium and Long Term Monitoring and Forecasting to the Real-Time Monitoring and Forecasting

The current network economy monitoring and forecasting ability is not enough to fully meet the needs of the economic and social development. The retail search index established by Vosen and others not only successfully predicts the individual consumption in the United States, but also is more accurate than the Consumer Confidence Index of the American Conference Committee and the University of Michigan Consumer Confidence Index. The latter two indices are calculated on the basis of the statistical data from the social surveys.

4 Analysis and Research Method of the Big Data Forecast on the Network Economy

Because of the unique characteristics of the big data, it is difficult for the traditional processing methods to deal with the analysis of the big data directly. New technical methods are needed to process and analyze the big economic data. At present, the use of the big data analysis of the network economy has achieved some results, but it is far from enough and suffers from great limitations.

4.1 Data Acquisition

The traditional statistical and measurement models use the data obtained through statistics and research, which are the structured data information directly used for processing and analysis. Use the data provided by the Baidu Index and the Internet companies such as Google. This method mainly takes advantage of the advantages of the search engines, and provides free search data for users. This method is also the most widely used data source at present. It uses Google Trends search volume to do the network economic prediction and analysis. This method is simple and convenient to obtain the data, with the low technical difficulty and the low cost, and the direct query can be obtained. Customize the data information by using the "web crawler" technology. The web crawler is a program or script that automatically crawls the information of the World Wide Web according to certain rules. Commonly used programs or languages for the data crawlers are: Python, Java, C++, and C+ and so on. Big data for enterprises: At present, the most abundant data resources are the Internet companies. Because they have great advantages in their respective fields and invest more resources to build the databases, they have great data advantages. Compared with the former two methods, the internal data of these companies are more complete and the data quality is better.

4.2 Dimensional Disaster

In the network economic model, due to the low value density of the big data, it is often necessary to obtain enough data information through the big data mining technology, which includes the data information on the multi-dimension, so the explanatory variables will increase greatly. Therefore, there will be a problem of the high-dimensional data in the research, that is, too much available information often. There will be too many explanatory variables related to the explanatory variables, resulting in the so-called "dimension disaster". In order to solve this problem, we need to reduce the dimensions of the variables and get the most relevant variables for the macro-prediction purposes. There are two ways of the thinking reduction: the feature selection and the feature extraction. The feature selection refers to replacing the original feature set with the feature subset, while the feature extraction refers to projecting the high-dimensional data into the low-dimensional space.

4.3 Unstructured Data

There are various sources of the big data information, such as the network logs, audio, pictures, URLs, texts, and geographic location information and so on. Only about 10% of them belong to the structured data, and the remaining 90% of the data information is unstructured data. The prediction and analysis of the network economy cannot be separated from the processing of the unstructured data. The Internet public opinion information and other information are more text information, which cannot be handled without a unique processing method. At present, the mature technology is the text information classification processing. The common text classification algorithms are the Naive Bayesian classification, the center nearest distance discrimination, the k nearest neighbor algorithm, the decision tree algorithm, and the artificial neural network and so on.

4.4 The Change of the Research Problem Paradigm

In most cases, the traditional econometric models assume that there is a definite functional relationship between the explanatory variables and the explanatory variables. However, it is difficult to directly assume that there is a definite functional relationship between the explanatory variables and the explanatory variables. In the actual process, the large number of the explanatory variables cannot be directly assumed to be a definite functional relationship with the network economic indicators. According to the explanatory variables and the network economic variables, there is only a correlation, there is no causal relationship. This is more difficult to deal with by using the traditional econometric models.

4.5 Data Noise Problem

The noise data refers to the information acquired that is not valuable for the research purposes. The existence of the noise data problem leads to two serious consequences: the increased costs. Because of the existence of a large number of the noise data in the data, the process and difficulty of the data processing are increased, and the corresponding research and analysis costs are greatly increased. Increase the error of the research conclusion. Because the noise data are not recognized, a large number of the invaluable data information will be added to the research process, which will bring uncertainty to the research conclusions and increase the error of the results, which will lead to the worthless of the research conclusions and even the wrong conclusions. In the process of the big data network economy research, the scope of the data sources is becoming wider and wider, and the means of the data collection is expanding, and the ability of the data collection is increasing. However, the large amount of the data collected is not directly related to the purpose of the research. Therefore, in the actual research process, it is very important to identify the most relevant data information for the research purposes.

5 Conclusion

With the rapid development of the big data, the international academia and the network economic policy makers have realized the revolutionary impact of the big data on the network economic analysis, and gradually tried to combine the concepts, methods and technologies with the network economic analysis of the big data. This paper makes a preliminary discussion on the research and application of the Internet big data in the network economic monitoring and analysis, hoping to provide useful reference for the future government and academic research.

Acknowledgement. Anhui Natural Science Foundation Youth Project: Signal extraction and noise processing of search engine data under the big data strategy (No. 1908085QG305).

References

1. Han, M.: Talking about the application of the big data in the economic field. Netw. Secur. Technol. Appl. **1**, 80–81 (2015)
2. Li, P.: Further improve the application of the big data analysis. Netw. Econ. Manag. **6**, 9–10 (2016)

3. Li, H.: Problems and countermeasures of the big data applied in the economic field. Res, Technol. Econ. Manag. **10**, 79–84 (2015)
4. Jie, J.: The impact of the big data on commercial banks. Stat. Manag. **5**, 73–74 (2014)
5. Shi, J.: An analysis of the impact of the big data on the logistics management. Charming China **23**, 51 (2016)

Problems and Countermeasures in the Interaction Between Teachers and Students in Computer Intelligence Classrooms

Shenghong Liu$^{(\boxtimes)}$

Hankou University, Wuhan 430212, Hubei, China
guanjuchen@126.com

Abstract. With the rapid development of the Internet, the popularization of the mobile technology and the information technology, the smart phones, as a new learning tool based on the network technology, have been widely used in more and more courses. The advent of the new generation learning mode of "Internet +smart phones" has changed students' learning location and learning space, and students' learning is more casual, intelligent and personalized. How to maximize the positive role of the smart phones in their learning and how to develop their strengths and avoid their weaknesses is a problem that teachers engaged in the front-line teaching should ponder. In the computer intelligence classrooms, how to effectively solve the problems existing in the process of the teacher-student interaction and put forward the targeted and reasonable countermeasures is of vital importance.

Keywords: Computer intelligent classroom · Teacher-student interaction · Problems · Research countermeasures

The Internet technology and the artificial intelligence technology have become the catalytic factors of the educational reform. It is worth exploring how to promote the transformation of students' training from the knowledge-based education to the ability-based and accomplishment-based education on the basis of the deep integration with the classroom teaching.

1 Construction of the Computer Intelligence Classrooms

In order to achieve the goal of the future-oriented personnel training, we must adopt different training methods and paths from the traditional education. Without the strong support of the information technology, it is difficult to achieve in the actual scale and conditions [1]. The "Future School Innovation Plan" of the Chinese Academy of Educational Sciences is to explore the effective path of the future education from the aspects of the curriculum, the learning style, the information technology, the learning space and the school organizational structures under the new goal of the personnel cultivation [2]. Among them, the information technology is an all-round penetrating factor. The "materialized" elements of the intelligent technology and the information

© Springer Nature Singapore Pte Ltd. 2020
M. Atiquzzaman et al. (Eds.): BDCPS 2019, AISC 1117, pp. 1528–1533, 2020.
https://doi.org/10.1007/978-981-15-2568-1_212

technology must be combined with the "human" elements of the teachers and the students to play a role [3]. Only under the guidance of a new educational concept and the teaching practice can they play a role.

The classroom practice exhibition reflects the great catalytic effect of the intelligent information technology on the classroom change from the real teaching scene of the front-line teachers [4]. In many demonstration schools, some schools integrate the cloud-end big data platform with the STEM teaching and the PBL project-based learning to design the integrated courses, take the problem as the guidance, record and analyze the learning process intelligently and dynamically from the aspects of the design, implementation, evaluation and reflection, and rely on the big data to accurately train students' innovative thinking and practical abilities [5]. Some schools use the holographic cloud-end platform to promote the integration of the inquiry and the collaborative learning in mathematics, and to accurately grasp the dynamic learning situation and the practical application of the visualization of the teaching contents.

How to change the AI from "opponent" to "helper" to better promote the human learning and growth is a widespread concern in today's society. We will implement the National Intelligence Education Project, set up courses related to the artificial intelligence, gradually promote the programming education, and encourage the social forces to participate in the development and promotion of the programming teaching software and games that are fun to teach. Education is not accomplished overnight, and nor is AI + education. It will be a long process for schools, teachers and students to adapt to AI. In a certain period of time, if we cannot see the positive changes brought by AI but we should also bear the burden of using and learning it, this AI + education product may soon be abandoned. Therefore, in order to make AI + educational products really fall into the practical application, the products must be easy to use. When the education meets AI, the education will be innovated and AI will meet the challenge of the market. The artificial intelligence has the characteristics of comprehensiveness, cross-cutting and strong application, and colleges and universities are still exploring the specialty and the curriculum setting. At present, we need to further improve the curriculum system and explore the training mode of the compound talents. Besides the professional basic courses, the ethics related courses should also be covered to help students establish their correct ethical value orientation.

2 Problems in the Interaction Between Teachers and Students in the Computer Intelligence Classrooms

The intelligent classroom is the organic combination of the artificial intelligence and the classroom teaching. Some people call it the intelligent classroom for short. Now the artificial intelligence is not only a learning content, but also a learning tool and a learning means. If there is no AI content in the students' knowledge structure, it is a major defect in the knowledge structure and the ability structure.

2.1   Problems Existing in the Concept of the Interactive Teaching Between Teachers and Students

The intelligent classroom is a real learning space and a community of teachers and students. The learning community embodies the new educational concept, and also promotes the traditional Chinese cultures in which teaching is mutually beneficial and there is no class. There is a historical inheritance relationship between the ancient teaching and the present learning community. In the past, we used to make no social distinctions in our teaching, but we did not teach students in accordance with their aptitude. Now the learning community reflects the teacher-student relationship of the mutual benefit in our teaching. There are three main problems in the teaching concept of the effective interaction between teachers and students. First, in the teacher-student interaction, many teachers think that the teacher-student interaction is a form. Most teachers think that the teacher-student interaction means eye or language communication between teachers and students in class, such as questioning in class, or some students do some small actions in class. Teachers stop and understand through actions or eyes. It's one-sided. Secondly, the importance of the effective interaction between teachers and students is not enough. Many school leaders think that students and teachers are under great pressure. Because of the college entrance examinations, there is not much time to organize the teacher-student interaction, so the interaction between teachers and students is put on hold. This is a problem of the insufficient attention. Finally, the teacher-student interaction is unbalanced, and some subjects are paid more attention to. It is possible that there will be more effective interaction between teachers and students.

2.2   Problems in the Design of the Interactive Teaching Between Teachers and Students

One of the most remarkable features of the intelligent classroom is the organic combination of AI and the classroom teaching. AI is not only the content of our learning, but also a tool of our learning, and also a means of our learning. From the overall trend, education cannot go to the future if it only depends on the offline, does not intervene in the online, and does not intervene in the network. The problems of the effective teacher-student interaction in the teaching design are mainly reflected in two aspects. On the one hand, some teachers think that the effective teacher-student interaction is unnecessary in the course of preparing lessons, as long as they have a good idea, and some teachers think that the effective teacher-student interaction is difficult to grasp in class, and this part is omitted. Most of these two problems are reflected in the older teachers and fewer in the new teachers. This is also a kind of the empiricism. On the other hand, the effective interaction design between teachers and students in the course of preparing lessons is unscientific and inaccurate. That is to say, when preparing lessons, it is very clear that the interaction should be carried out when teaching goes to this stage, but there are no specific measures and plans for the interaction.

2.3 Problems in the Implementation of the Interactive Teaching Between Teachers and Students

The application of the big data technology in our education has been strengthened in depth and breadth. At present, the whole information architecture is changing. The "cloud + end" is the structure we advocated before, but the future is the cloud plus the edge, that is, the small individuals gathered at the edge gradually. There are teachers' and students' platforms in the classrooms. Teachers' and students' platforms will stay and send the necessary ones to the clouds instead of staying locally. Of course, teachers 'and students' platforms will form a small closed loop. There are two problems in the implementation of the interactive teaching between teachers and students. First, the effective interaction between teachers and students is easily neglected in the process of the teaching implementation. This is due to the long-term examination-oriented education, which makes the interaction between teachers and students not so important in our teaching. All teaching is centered on examinations, teaching according to the requirements of examinations. It is not the ultimate goal of our teaching to improve students' comprehensive qualities. Secondly, the effective teacher-student interaction tends to be formalized. That is to say, many teachers do not pay attention to the effect of the teacher-student interaction, but take the interaction as a buffer for our teaching, so that students can relax slightly in the long learning process.

3 Solutions to the Interaction Between Teachers and Students in the Computer Intelligence Classrooms

The structural optimization refers to the optimization of the relationship between the learning contents and the learning styles after the breaking of the classroom teaching contents and the boundaries, including the updating of the learning contents, the personalized arrangement of the learning styles, the flexible arrangement of the learning contents and learning styles and so on. The formation of the good teacher-student interaction in the classroom teaching requires the efforts of teachers, students and the school management departments to work together to form effective synergies.

3.1 Positioning the Teachers' Roles in the Good Interaction Between Teachers and Students in the Classroom Teaching

In the computer intelligence classroom teaching, teachers play the role of guides from beginning to end. The traditional teacher-student relationship is a relationship of "I say and you listen". In order to form a good interaction between teachers and students in the classroom teaching, we must break this mode of thinking. First, teachers need to lower their posture, guide students with a parallel relationship between students and non-guidance relationship, and carry out the classroom teaching. Therefore, teachers must change their roles to be friends. Teachers must believe that students have enough wisdom and ability to effectively communicate and have dialogues with the teachers in class and form an effective learning situation. Secondly, teachers should study the syllabus and the textbooks carefully, grasp the essence of the textbooks holistically and macroscopically, and form a knowledge atmosphere for the effective communication with the students. Because the classroom time is limited and precious, teachers are required to prepare lessons adequately to avoid failing to grasp the key and difficult points of the classroom teaching and the phenomenon of how large or inadequate classroom teaching capacity will not occur. Only in this way can the goal of the classroom teaching be clear, and the organizational process orderly and efficient, and the purpose of the teaching effect remarkable.

3.2   Forming a Harmonious Atmosphere of the Good Teacher-Student Interaction in the Classroom Teaching

Teachers are the guides and pioneers of students' knowledge learning. This requires teachers to regard students as friends and equal partners on the road to learning, rather than commanders. Teachers should fully respect the dignity of students, let students have a sufficient sense of acquisition, and have a very strong internal driving force for learning. In this way, students can form a kind of ability and consciousness of active learning and effective interaction with teachers. Only then can they have the intention to improve the efficiency of the effective interaction with teachers in the classroom teaching.

3.3   Designing the Effective Interactive Discussion Topics for the Classroom Teaching

The computer intelligence classroom teaching runs through the whole semester. The themes of each class teaching are different. Therefore, teachers must grasp the courses taught macroscopically. In the previous lesson preparation links, according to the themes of the classroom teaching, we should carefully design the effective interactive discussion topics among students. In this way, in the classroom teaching, teachers can easily control the classroom, form effective interaction between teachers and students around this topic, and guide students to carry out the active and fruitful topic discussion. This discussion must be conducted in accordance with a certain procedure, rather than the students' rambling and disorderly speech. And this is a very test of teachers' classroom organizational ability. To this end, we should clearly discuss the role of the team and the individual play, and clearly discuss the theoretical basis of the topic construction, so that the discussion has a solid theoretical foundation. Grasp the scale and the temperature of the discussion, so that the discussion is not only active and warm, but also we cannot make it become the verbal attacks, or even the personal attacks, thus losing the value and the original intention of the topic discussion.

3.4   System Design and Guarantee of the Teacher-Student Interaction in the Classroom Teaching

The necessity and importance of the teacher-student interaction in the computer intelligence classroom teaching has been generally accepted by the educational circles. However, how to achieve the effective teacher-student interaction in the classroom teaching is a topic worthy of study, which involves not only the teachers and the students, but also the school management. On how to effectively promote the subject teaching of the school management level, one is to enhance the awareness of the relevant management departments. Let the relevant functional departments fully realize that the interaction between teachers and students in the classroom teaching is not only an important part of the modern classroom teaching, but also an important measure to improve the quality of the classroom teaching. The second is to improve the system design of ensuring the interaction between teachers and students in the classroom teaching. The relevant management departments of schools should promulgate the relevant rules and regulations to ensure the interaction between teachers and students in the classroom teaching as soon as possible. Let the teaching mode of the teacher-student interaction in the classroom teaching be able to follow the rules, thus forming the institutionalization of the teacher-student interaction in the computer intelligence classroom teaching. Thirdly, we should form an incentive mechanism for teachers to interact with the students in the classroom teaching. The performance management of the classroom teaching is a difficult job to quantify. Therefore, we should establish and improve the incentive mechanisms of the teacher-student interaction in the classroom teaching to stimulate teachers' interests and enthusiasms in the teacher-student interaction in the classroom teaching, so as to improve the level of the classroom teaching.

4 Conclusion

The path of the classroom reform in the Internet era can be expressed in "Process reengineering and structural optimization". The teaching process of the modern curriculums is basically the five-stage teaching method in Herbart's era, on the basis of which some innovations have been made. The intelligent classroom teaching process will be more flexible, diverse and elastic, and the teaching and learning, the lecture and practice, the in-class and out-of-class, the self-study and guidance, the individual learning and the collective learning, and the formal courses and the micro-courses have the multiple arrangements, which can change depending on the changes of the teaching contents, the teaching objectives, the teaching environment and the teaching conditions and so on.

References

1. Miao, X.: Some experiences on training the practical abilities in the classroom teaching of artificial intelligence. Course Educ. Res. **10**, 250 (2016)
2. Zhou, M.: Application and exploration of smartphone-assisted mobile English translation learning. J. Hubei Corresp. Univ. (7) (2015)
3. Wang, L., Qi, X.: Exploration of the smartphone-assisted classroom teaching. Indu. Sci. Tribu. **22**, 170–171 (2015)
4. Yang, L., Li, C., Hou, X.: Design of the voice navigation teaching assistant system based on smart phones. J. Wuhan Univ. Technol. **32**(5), 157–160 (2010)
5. Wang, G.: Research on the innovation of the ideological and political teaching in the higher vocational colleges in the era of the popularization of smart-phones. Univ. Educ. (11) (2016)

Multidimensional University Students' Ideological and Political Education Model Based on the ASP Technology

Xin Liu[✉]

Zhengzhou University of Science and Technology,
Zhengzhou 450064, Henan, China
hailanqi@aliyun.com

Abstract. The ideological and political educational system in colleges and universities is an important part of the current college students' training system. It bears the important responsibility of systematically educating college students in Marxist theory and helping them form correct ideological and moral concepts. With the development of the times, the importance and the evaluation methods of the ideological and political education in colleges and universities have been deeply studied. Because of the importance of the ideological and political education, the role of the evaluation methods in the entire education system cannot be underestimated. The multi-dimensional ideological and political education model of college students based on the ASP technology has a solid theoretical foundation, which is conducive to the scientific evaluation of the researches.

Keywords: ASP technology · Multi-dimensional · College students · Ideological and political education · Model research

With the popularization and application of the mobile Internet in recent years, the educational information platform is migrating and developing to the mobile platform. The educational evaluation platform included is no exception [1]. In order to meet the needs of the educational evaluation application in the mobile scene, the design of a unified server-side information processing center is particularly urgent [2]. It not only meets the needs of the mobile interface access, but also reduces the cost of access to various devices and accesses the information at will.

1 Problems in the Current Ideological and Political Education Activities of College Students

In view of the maturity of the modern network technology and the hotspot sharing technology, it has become a learning trend to use the mobile phones to watch videos for learning [3]. This paper proposes the design of the mobile learning system for the ideological and political education in colleges and universities based on the micro-video, thus achieving the effective design of the mobile learning system with the ASP technology and the video coding compression technology [4]. Finally, the micro-video

© Springer Nature Singapore Pte Ltd. 2020
M. Atiquzzaman et al. (Eds.): BDCPS 2019, AISC 1117, pp. 1534–1539, 2020.
https://doi.org/10.1007/978-981-15-2568-1_213

function in the system is implemented effectively. Through the system implementation, students can use the mobile phones to study the ideological and political education anytime and anywhere, effectively realize the mobile learning and optimize the effect of the ideological and political education.

1.1   Lack of Attention to the Ideal and Belief Education

The core of the ideological and political education in colleges and universities is the education of college students' ideals and beliefs. Ideals and beliefs are the basis of helping college students to succeed [5]. In today's society, with the popularity of the Internet, college students are in a more complex society, which is full of diverse cultures and values. When pluralistic cultures and values bring them the wider horizons and more choices, "de-ideologization" and "universal value" theory, the historical nihilism and other erroneous trends of thoughts are also filled with college students, which will make them more likely to fall into a state of perplexity if they are not mature. If they are not properly guided at this stage of the university, it will inevitably lead to some confused college students' escaping from the reality, thus doubting the socialist ideal. Some college students choose to sacrifice their collective interests for their own selfish desires, and some college students blindly escape from the present situation. In fact, they waste their time and are unwilling to strive for their own goals. Some even turn to criticism and mockery of Marxism as a gimmick to besiege and support Marxists in order to broaden their horizons.

1.2   Emphasize the Theoretical Education and Neglect the Practical Training

The ideological and political education is a highly theoretical subject, which requires us to study systematically. The traditional ideological and political course is mainly about explaining the theoretical knowledge of books, which is abstract and full of political preaching. There is a widespread problem of emphasizing the theories over the practice, which makes it difficult to arouse the emotions of college students and therefore there is a lack of attraction and appeal. In addition, some ideological and political educators cannot scientifically analyze the reform. The impact of the economic globalization on the traditional ideology, the impact of the network on the traditional ideology, the lack of scientific analysis of the hot issues such as the moral decline, the ideological westernization, the corruption, and the environmental pollution and so on, which have emerged in recent years, make it impossible to give convincing answers to college students. As a result, the ideological and political education cannot get the favor of college students, with a lack of attraction.

1.3   The Educational Ways and Means Face Challenges

The traditional ideological and political education in colleges and universities emphasizes the "inculcation theory" and adopts the closed education. With the development of the economic globalization and the information network, college students cannot be educated in the prescribed time and place in the traditional way. As long as there is an Internet computer, students can acquire the required knowledge at any time, and quickly understand the information of politics, economy and social life happening at home and abroad. The concept of "wall" in universities will gradually disappear, and the relatively narrow educational space previously closed will become an open and global educational space. How to adapt to the situation of the network and how to modernize the traditional ways and means of education are the problems that must be solved immediately in the ideological and political education of college students.

2   Analysis of the Technical Characteristics of ASP

ASP is the abbreviation of Active Server Page, meaning "Active Server Page". ASP is an application developed by Microsoft instead of the CGI scripting program. It can interact with the database and other programs. It is a simple and convenient programming tool. The ASP web page file format is.asp, now commonly used in a variety of dynamic websites. ASP is a server-side scripting environment that can be used to create and operate the dynamic web pages or the web applications. The ASP web pages can include HTML tags, plain texts, script commands, and COM components and so on. Using ASP, you can add interactive contents (such as online forms) to a web page, or you can create a web application that uses the HTML pages as the user interfaces.

At the same time, the ASP technology can combine components and HTM and so on, integrate them into corresponding applications, and make professional web pages according to customer's customization requirements. Different from the scripting languages such as JavaScript and VBS, the ASP technology needs to execute the command scripting language in the server, and push the final generated web interface into the browser. When developing the software with the ASP technology, developers do not need to consider the adaptability of generating the software. In the process of using the computers, users can access the network or the special servers to meet the operating conditions of the software. The application of the ASP technology in the process of the maintenance software can avoid the decentralized maintenance of each terminal and reduce the workload of the maintenance server.

The software developed by the ASP technology is directly generated in the Web server. PWS and IIS and so on can support the operation of the ASP software. When users use the software and files generated by ASP, the Web server in Microsoft can directly respond to the user's application requirements and use the ASP engine to further run the application files. In addition, when the user has the script compatibility problems in the process of using, the ASP engine will solve the incompatibility problems in the process of running the software by calling scripts.

3   The Significance of Designing the Multidimensional College Students' Ideological and Political Education Model Based on the ASP Technology

With the development of the information technology, the phenomenon of using the network to learn has become very common. With its advantages of digitalization, high speed, globalization, interaction and openness, the network has brought new opportunities for the development of our university teaching, and also provided a learning environment for the contemporary college students to share resources. However, when people apply the network to the distance learning, they find that the disadvantages of this new learning method are also obvious.

The quickness of the network improves the efficiency of the ideological and political work for college students. The ideological and political work of college students is of pertinence and timeliness. Only when it is timely and fast can we grasp the initiative of the work. The network is known as the "expressway" of the information. Even if it is far away, it can also "face-to-face" discuss problems with students, exchange ideas and exchange information, overcome the obstacles caused by the natural and geographical conditions, shorten the space-time distance, and greatly improve the efficiency of the ideological and political work of college students. Through the network "group" type of the active and rapid dissemination of new ideas, theories and policies, the work stage, educational methods and means have made great breakthroughs.

In order to adapt to the development of the times and realize the leap in the teaching quality of the ideological and political theory course in colleges and universities, the only way is to combine the network learning with the face-to-face learning in the traditional classroom. The blended learning is characterized by the organic combination of the traditional classroom learning and the online learning. It aims to complement each other's strengths and complement each other's weaknesses in order to achieve the best teaching effect. The mixed learning platform of the ideological and political courses in colleges and universities is the core of the network education based on the mixed learning mode. The construction quality of the mixed learning platform will directly affect the teaching effect of the ideological and political courses in colleges and universities.

With the emergence of the Internet cloud platform, many colleges and universities across the country take it as a carrier for our education and teaching. In order to adapt to this information situation, the ideological and political education has also established corresponding space resource courses to carry out the ideological and political education and teaching. However, the construction of the cyberspace resource curriculum is only the basis of using this network learning platform for our education and teaching. In order to improve its quality, it is necessary to carry out the evaluation research of the application effect of the cyberspace resource curriculum. As the most important course for college students' ideological and moral cultivation, the ideological and political course plays a very prominent role in the college teaching. Therefore, the construction of a scientific and reasonable evaluation system of the ideological and political curriculum teaching is of great help to improve the current teachers' teaching.

4   Multidimensional University Students' Ideological and Political Education Model Based on the ASP Technology

The popularization of the network has a great impact on the teaching of the ideological and political course. Therefore, how to use the network platform to expand the classroom space, stimulate students' learning initiatives and enthusiasms, and enhance the effectiveness of the ideological and political curriculum teaching has become an urgent problem to be solved. The main function of the ideological and political education website is to provide a platform for teachers and students to exchange ideas and learn. Therefore, the construction of the ideological and political education website should fully reflect the latest important events, especially the timely reflection of the major events and the important speeches. In addition, the ideological and political teachers can also interact with students on the Internet and conduct the appropriate guidance of public opinions.

4.1   Multidimensional University Students' Ideological and Political Education Model Structure Based on the ASP Technology

The development environment of this model is the Asp+Access mode. ASP (Active Server Pages) is a developing non-compiled application environment. It can combine Html, script and reusable ActiveX server components to build the dynamic and powerful Web-based commercial applications. It supports the server-side scripts based on the IIS and supports VBScipt and JScipt completely. It has high efficiency and short development cycle. It has the strong expansion ability.

4.2   Multidimensional Ideological and Political Education Model for College Students Based on the ASP Technology

The ideological and political educational model should not only start with the appearance of the web page, but also attract the attention of the teachers and the students by using the vivid pages. More importantly, it should prepare rich and colorful contents for the web sites. The author sets the main sections of the website as: the current affairs and politics section, the theoretical study section, the notice and announcement section, the Party membership navigation section, the advanced deeds section, the online communication section, the film and television space and other sections.

4.3   Comment Management Section

The news management system includes several modules, such as the news management, the news addition, and the news comment management and so on. In the background news add page, the administrator chooses the news classification according to his authority. The system administrator can add news in all categories. The news administrator can only add news in his authority classification. In the news management page, administrators can modify or delete all their published news. In the news comment management page, administrator users can access the news comment page according to their rights, and manage the news comments.

4.4   Security Management Module

The verification code is a popular way in many websites nowadays. The main function of the verification code is to prevent the batch registration. It can effectively prevent this problem from continuously trying to log on to a certain registered user by violent cracking of a specific program. Although it is a bit troublesome to log on with the verification code, it is still necessary for the community to use this function and it's also important. Set the blacklist type of the uploaded files to ensure that it is not possible to upload these files directly. At the same time, the system imposes strict restrictions on the extension of the uploaded files, allowing only a number of letters in English. The file names uploaded to the host are generated by the system to ensure that hackers cannot upload any executable files in various ways. After uploading files using the system upload function, the system will scan the current upload directory, delete the files that appear in the blacklist type immediately, and lay down the last impossible breakthrough to ensure that hackers cannot invade the website system by uploading.

5   Conclusion

At present, the network has become the most important way for college students to obtain the information after school, and its advantages and disadvantages have begun to gradually highlight. It has become an urgent and important task for the ideological and political education to enter the network, use the network cultures to educate and guide college students. The construction and development of the ideological and political education website has opened up a new teaching platform for the ideological and political education in colleges and universities. The ASP technology is used to realize the main business functions of the ideological and political education model, which enriches the training space for students.

Acknowledgement. Foreign Languages Evaluation Research of National Vocational Colleges and Universities: Study on the Critical Thinking Evaluation of Non-English Major in Vocational Colleges [FLAB005].

References

1. Zhang, L.: Website security research based on ASP. Technol. Wind **15**, 244 (2010)
2. Dong, L.: Design and implementation of the professional teaching resources base based on the ASP technology. J. Liaoning Teachers Coll. (Nat. Sci. Ed.) **18**(4), 55–58 (2016)
3. Lu, L., Hou, C., Wang, J., Huang, P.: Design and implementation of a literature information service platform for the primary and secondary school teachers' research based on the ASP technology. J. Ningbo Inst. Educ. **13**(4), 69–72 (2011)
4. Xu, L.: A brief talk on the website security research and preventive methods based on ASP. Electron. World **8**, 118 (2016)
5. Shi, X.: Design and implementation of the teacher information management system based on Asp. Intelligence **28**, 58 (2012)

Life Expectancy and Its Reliability Analysis of Hydraulic Hose Based on Impulse Test

Yin Lv$^{(\boxtimes)}$ and Keli Xing

School of Electromechanical Engineering and Automation, Shanghai University,
Shanghai 200444, China
hongqi25642100@126.com

Abstract. In this paper, a hydraulic hose impulse test bench has been developed. The impulse test of the same specification hydraulic hose has been carried out using the hydraulic hose impulse test bench and the test data have been collected. Based on Weibull distribution, the life expectancy and confidence interval of the specification hose are obtained by using Matlab to calculate the hose impulse test data. Finally, Kolmogorov–Smirnov test of the hose impulse test data is performed using Kolmogorov test table, thus confirming the reliability of the test data. According to research, it is proposed that Weibull distribution can be used to analyze the life expectancy of the hose, thereby reducing the number of trials and energy consumption and protect the environment.

Keywords: Hydraulic hose impulse test bench · Life expectancy · Weibull distribution · Kolmogorov–Smirnov test

1 Introduction

Hydraulic hoses are important accessories for the entire hydraulic system. The performance of hydraulic hoses also greatly affects the safety, reliability and service life of the entire hydraulic system. Hydraulic hose pulse test benches have been researched in China since 1976. In the 40 years since then, various research institutes and companies have continuously improved the behaviour of the test benches. However, the analysis of experimental data is rare.

Because when the company produces a new hose and connectors, it should be tested for quality and provide test data to ensure its reliability. Therefore, the impulse test of the hose with the same specifications is performed on the hydraulic hose pulse test bench and the test data is collected. However, each hose impulse test consumes about 600 kWh and takes about 60 h. For the hoses of the same specifications, the company needs to perform 40 sets of hose impulse tests and collect data each time. Finally, the reliability of the hose is determined by the probability of passing the test. It cause a very high cost of both energy and time, result in a low economic benefit. Since Weibull distribution is most accurate for the life expectancy of mechanical and electrical products and can be analyzed with small sample test data, in this paper, a life expectancy and confidence interval of the 40 groups of hose impulse test data by Weibull distribution has been presented. Finally, Kolmogorov–Smirnov test has been

M. Atiquzzaman et al. (Eds.): BDCPS 2019, AISC 1117, pp. 1540–1546, 2020.
https://doi.org/10.1007/978-981-15-2568-1_214

used to verify the data, which proves that the hose impulse test data of the hose impulse test bench is reliable and can be calculated with Weibull distribution.

The predicted fatigue life obtained from the Weibull distribution for the five experimental data showing a strong similarity with the estimated fatigue life for a large sample of 200 fatigue test data [1]. Therefore, it has been conferred that fewer hose impulse tests can be performed, and accurate data on hose life expectancy can still be obtained after Weibull distribution in this paper. Companies can get a reduce of the number of tests to 5–10 times by using Weibull distribution, which is also accompanied by about 2,000 h time reduce, and 20,000 KWh electricity reduce.

2 Hydraulic Hose Impulse Test Bench Principle

The most important function of the self-developed and manufactured hydraulic hose pulse test bench is how to provide the high temperature and high pressure hydraulic pressure impulse curves that meets the experimental requirements [2–5]. According to the national standard, the test bench should enable the liquid fluid to circulate through the sample, apply the internal pulse pressure to the sample with a rating of (1 ± 0.25) HZ, and ensure the rate of pressure rise [6].

Component 1 is a frequency conversion motor and a constant rate pump. Component 2 is an electromagnetic relief valve for regulating system pressure. Component 4 is a booster cylinder. The boost ratio is 1:4. Therefore, the boost pressure of the booster cylinder, (which also is the test pressure of the tested hose), is 4 times the set value of the electromagnetic relief valve [7–9].

Component 14 is an accumulator, which will add hydraulic oil when the system is pressurizes, to make sure the whole system is more stable and avoid overshoot for a long time, so that the pressure can be set up faster. Component 9 is a slippage pump. And component 10 is are mote pressure regulating valve. Both of them are the components of the auxiliary hydraulic system, to supplement the hydraulic oil when the system releases pressure [10].

Although hydraulic hose can be tested by hydraulic hose pulse test bench, test data can not directly explain the life characteristics of the hydraulic hose. Calculation and analysis of the data of the hydraulic hose pulse impulse test carried out on the test bench has been shown in the following sections.

3 Data Collection of Hose Impulse Test

Data of impulse test in 2016, which contains 40 hoses with same specification from a certain company has been collected. Among them, 19 hoses reached 220,000 pulse tests and 21 hoses didn't. The number of failures of these hoses are 6000, 14000, 18000, 20000, 30000, 31000, 37000, 60000, 69000, 76000, 82000, 87000, 103000, 110000, 110000, 140000, 160000, 160000, 178000, 184000, 218000, respectively. In accordance with national standards, pulse frequency of the test bench is 1 Hz. When number reaches 220,000, it takes about 60 h and 600 kWh. This company rules that the hydraulic hoses that reached 220,000 pulses can met the requirements and will be regarded as qualified products. Therefore, truncated test method was adopted in this paper, which means the hose that has not failed after 220,000 pulse tests is no longer tested.

In the past, after collecting the data of the hose impulse test, the company determines whether the hose is qualified and can be put into production by determining the number of the standard hose as the qualified product, but it can not accurately predict the life expectancy of this type of hydraulic hose. Therefore, in the following text, the life expectancy of the hydraulic hose will be obtained by Weibull distribution of the experimental data of the hose, which can calculate the reliability of the hose more accurately.

4   The Weibull Distribution of the Impulse Test of the Hose

4.1   The Theoretical Basis of Weibull Distribution

Weibull distribution is the theoretical basis for reliability analysis and impulse test, which is widely used in data processing of impulse test because of the feasibility to estimate the distribution parameters of its life by means of probability.

Weibull distribution, especially of three-parameter distribution is the most adaptable distribution model in reliability analysis. With the change of three parameter values, it can be completely equivalent or similar to other kinds of distributions. Weibull distribution can be used to explain many different types of problems, especially in the study of life phenomena of components or parts.

Weibull distribution can predict important life characteristics such as life expectancy, reliability, and failure rate at a certain moment. And it can provide simple and effective charts when needed, which is easy for researchers to understand if they have insufficient data. The results obtained by Weibull distribution are also very accurate under small sample data.

Based on the Weibull distribution mentioned above, it has been used to calculate the data of the hydraulic hose pulse impulse test in this paper.

4.2   Weibull Distribution Formula

The three-parameter Weibull distribution of the fault density function is:

$$f(t) = \frac{m}{\eta - t_0} \left(\frac{t - t_0}{\eta - t_0} \right)^{m-1} \cdot \exp\left[-\left(\frac{t - t_0}{\eta - t_0} \right)^m \right]$$

Among them, $t \geq t_0, \eta > 0, m > 0$. m is the shape parameter, it determines the shape of the curve; η is a scale parameter, also called a characteristic life, which can be understood as the life expectancy. t_0 is a positional parameter, that is, the minimum life span, which indicates that the product will not fail until this time, and will not affect the shape of the curve and will only shift the entire curve.

In this paper, what the company wants to obtain is the life expectancy of hydraulic hoses of the same specifications. It does not require the addition of two additional parameters. Under this condition, Matlab can achieve this purpose more conveniently. Therefore, Matlab has been used to calculate the pulse impulse test data of the hydraulic hose.

4.3   Weibull Distribution Based on Matlab

For the Weibull distribution, Matlab comes with this function library. Therefore, using Matlab for Weibull distribution can be more efficient, and the estimated life of this size hose can be more accurately.

Assuming that the above hydraulic hose impulse test data obey Weibull distribution, Matlab is used to calculate the confidence interval of its life expectancy and average life.

Running instructions of Matlab shows as follows:

data = [6000, 14000, 18000, 20000, 30000, 31000, 37000, 60000, 69000, 76000, 82000, 87000, 103000, 110000, 110000, 140000, 160000, 160000, 178000, 184000, 218000, 220000, 220000, 220000, 220000, 220000, 220000, 220000, 220000, 220000, 220000, 220000, 220000, 220000, 220000, 220000, 220000, 220000, 220000, 220000] (Input experimental data)

censoring = [0, 0, 0, 0, 0, 0, 0, 0, 0, 0, 0, 0, 0, 0, 0, 0, 0, 0, 0, 0, 0, 1, 1, 1, 1, 1, 1, 1, 1, 1, 1, 1, 1, 1, 1, 1, 1, 1, 1, 1] (Taking the 220000 pulse as the truncation, the data of the pulse lifetime not reaching the 220000 pulse is recorded as 0, others as 1.)

[a, b] = wblfit (data, α, censoring) (Weibull distribution of hose impulse test data)

Among them, [a, b] is the result of Weibull distribution, a is the life expectancy, b is the confidence interval for life expectancy, $100 \times (1 - \alpha)$ is the confidence interval of the test data, take α as 0.05, Confidence interval is 95%.

After substituting the above data and censoring, the hose life expectancy is a = 294074 times, and the confidence interval b is 181,338 times to 476,897 times.

Through the above calculation using Weibull distribution, it has obtained the life expectancy of this type of hydraulic hose. However, the nonparametric test is required for whether the hose impulse test data conforms to the Weibull distribution. Among the nonparametric test, Kolmogorov–Smirnov test is more accurate. Therefore, Kolmogorov–Smirnov test has been selected to test the impulse test data of the hydraulic hose, and determines whether it is in accordance with the standard.

5   Data from Kolmogorov–Smirnov Test

5.1   Theoretical Basis of Kolmogorov–Smirnov Test

Since the characteristics of the hose impulse test sample data are poorly known, nonparametric tests are required. The commonly used nonparametric test methods include binomial distribution test, chi-square test, and Kolmogorov–Smirnov test. Binomial distribution test is more often used in binary problems, such as the positive and negative aspects of coins. There are similarities between Chi square test and Kolmogorov–Smirnov test. Both of them compare the difference between the actual data and the theoretical data, but Kolmogorov–Smirnov test is more accurate.

Therefore, this paper selects Kolmogorov–Smirnov test method to test the pulse life data of the hydraulic hose.

Kolmogorov–Smirnov test can use original sample data to infer whether the original sample obeys a theoretical distribution and is suitable for exploring the distribution of the random variable of continuous type.

The Kolmogorov–Smirnov test is a test of goodness of fit. Goodness of fit test is used to test whether the overall distribution of a batch of classified data is consistent with a certain theoretical distribution. By studying whether the distribution of the sample observation is consistent with the theoretical distribution of the set, and by analysing the differences between the two distributions, it determines whether there is a reason to think that the observation results of the sample are derived from the assumed overall distribution of the theory.

Kolmogorov–Smirnov test is more accurate than the Chi square test. It compares the data that needs to be statistically analyzed with another set of standard data, and finds the deviation between them. Therefore, Kolmogorov–Smirnov test to test the data has been used in this paper.

The basic idea of using K-S test in this paper is: First, the impulse test data of the hose are arranged according to the number of lifes from small to large, and the distribution is assumed. Then the failure probability is calculated according to the assumed formula, and the calculated failure probability is compared with the actual failure probability in the experimental data, and their maximum deviation value D is obtained. From the truncated impulse test $D_{n,T}$ limit distribution table, determine whether the deviation value D is located within the specified confidence interval. If the deviation value D value is within the prescribed confidence interval, the detected data is satisfied, and if the deviation value D value exceeds the prescribed confidence interval, the detected data is unbelievable.

5.2   Kolmogorov–Smirnov Test Based on Kolmogorov Test Table

Assume that the hose impulse test data obeys the exponential distribution H: $Fo(t) = 1 - e^{-\frac{x}{294074}}$. Among them, 294074 is the life expectancy of the hose calculated from Weibull distribution above.

Establish maximum deviation: $Dn_t = Sup\{|Fo(t) - Fn(t)|\}$. Among them, $Fn(t) = \frac{n_{loseefficacy}}{n_{totalnumber}}$, $n_{loseefficacy}$ is the actual number of failures before the life pulse number. For example, before the 20000 pulse number, the actual failure time is 4, which is 6000, 14000, 16000, 20000, that is, $n_{loseefficacy} = 4$, $n_{totalnumber}$ is 40 of the total sample in the previous article, that is, $n_{totalnumber} = 40$.

$\alpha = 0.05$ is set in the above. Then $1 - \alpha = G_T(d) = 0.95$.

The distribution of the hypothesis of the hose pulse hypothesis in this paper is $Fo(t) = 1 - e^{-\frac{220000}{294074}} = 0.527$, Check the limit distribution table of $D_{n,T}$ in truncated impulse test in Kolmogorov test table. Obtain $Fo(t) = 0.5$, $D_T, a = d = 1.2731$.

Calculate the maximum deviation Dn_t, as shown in Table 1.

Table 1. Dn_t operation table

t_i	Cumulative failure number	$Fo(t_i)$	$Fn(t_i)$	$|Fo(t_i) - Fn(t_i)|$
20000	4	0.0657	0.1	0.0343
40000	7	0.1272	0.175	0.0478
60000	8	0.1846	0.2	0.0154
80000	10	0.2382	0.25	0.0118
100000	12	0.2883	0.3	0.0117
120000	15	0.3351	0.375	0.0399
140000	16	0.3788	0.4	0.0212
160000	18	0.4196	0.45	0.0304
180000	19	0.4578	0.475	0.0172
200000	20	0.4934	0.5	0.0066
218000	21	0.5235	0.525	0.0015

$$Dn_t = Sup\{|Fo(t) - Fn(t)|\} = 0.0478$$

$$\sqrt{n} \cdot Dn_t = \sqrt{40} \times 0.0478 = 0.3023$$

Because $0.3203 = \sqrt{n} \cdot Dn_t < D_{n,T} = 1.2731$, so accept hypothesis Fe (T), that is, hose impulse test data obeys the exponential distribution of $Q = 294074$.

6 Conclusion

In this paper, the test data of the self produced hydraulic hose impulse test bench are collected, and the Weibull distribution is carried out to the hose impulse test data of the test bench test of the company by Matlab, thus the life expectancy of the modified hose is obtained. After the Kolmogorov–Smirnov test of the hose impulse test data, it is concluded that the test data of the hydraulic hose pulse test bench are reliable. In the future, the impulse test data of the hose can be carried out by Weibull distribution.

This conclusion can greatly help the hydraulic hose test after the company, and will greatly reduce the manpower and material resources. Facilitate about 80% of the cost saving and energy consumption for the company in one year, to meet the requirements of green environmental protection.

It is hoped that people in need can get guidance to use Weibull distribution to process the life expectancy of mechanical and electrical products from this paper.

References

1. Lv, Z., Yao, W.: Determining fatigue life distribution with small samples. Chin. J. Appl. Mech. **25**(2), 323–326 (2017)

2. Cho, J.R., Yoon, Y.H., Seo, C.W., Kim, Y.G.: Fatigue life assessment of fabric braided composite rubber hose in complicated large deformation cyclic motion. Finite Elem. Anal. Des. **100**, 65–76 (2015)
3. Drinkaus, P., Armstrong, T., Foulke, J.: Investigation of flexible hose insertion forces and selected factors. Appl. Ergon. **40**, 39–46 (2009)
4. Pavlov, A.I., Polyanin, I.A., Kozlov, K.E.: Improving the reliability of hydraulic drives components. Procedia Eng. **206**, 1629–1635 (2017)
5. Fang, Z., Gao, L.: Estimation of parameters of three-parameter Weibull distribution in life analysis. J. Armored Force Eng. Inst. **13**(1), 70–74 (1999)
6. Moebus, F., Pardal, J.M., dos Santos, C.A., Taraves, S.S.M., Louzada, P.A., Barbosa, C., dos Santos, D.S.: Failure analysis in high pressure thermoplastic hose fittings submitted to cold forming by swaging process. Eng. Fail. Anal. **74**, 150–158 (2017)
7. Huang, R., Carriere, K.C.: Comparison of methods for incomplete repeated measures data analysis in small samples. J. Stat. Plan. Inference **136**, 235–247 (2006)
8. Kwak, S.-B., Choi, N.-S.: Micro-damage formation of a rubber hose assembly for automotive hydraulic brakes under a durability test. Eng. Fail. Anal. **16**, 1262–1269 (2009)
9. Zhang, Z.: Failure analysis of φ10 mm, φ14 mm high-pressure polytetrafluoroethylene hose assemblies during pulse test and the improvement. Shanghai Chem. Ind. **139**(6), 18–23 (2014)
10. Zhu, F., Yi, M., Ma, L., Du, J.: The reliability and life index of hydraulic components with grey prediction methods. Huazhong Univ. Sci. Tech. **24**(12), 24–25 (1996)

Multi-interactive Teaching Model of German in Colleges and Universities Under the Information Technology Environment

Lei Ma[(✉)]

School of Foreign Studies, Northwestern Polytechnical University,
Xi'an 710129, Shaanxi, China
zhilin87445667@126.com

Abstract. The integration and application of the network multimedia technology and the German curriculum in the information age represents the development trend of the German teaching reform nowadays. The teaching mode can promote the teaching effect twice with half the effort. The multi-interactive teaching mode based on the constructivism theory is a new teaching mode supported by the information technology. The multi-interactive German teaching mode in universities under the information technology environment has changed the relationship between teachers and students in the course of the classroom teaching and between teachers and students and the teaching contents. It can improve the quality of the German teaching, effectively improve the teaching effect and improve students' abilities to use the German language. A new teaching mode supported by the information technology fully embodies the principle of learning as the main body. It focuses on the connotation and theoretical basis of the multi-interactive teaching mode to improve the quality of the German teaching in universities.

Keywords: Information technology environment · University German ·
Multiple interaction · Teaching mode · Applied research

As a small language, German is developing slowly compared with the mainstream foreign language teaching. Compared with the active and in-depth study of the teaching ideas pursued in the mainstream language teaching, the German teaching lays more emphasis on the development of the basic education [1]. How to improve the teaching quality and enable students to have the proficient language application abilities has become the goal that educators in this field have been pursuing.

1 Analysis of the Application Background of the Multi-interactive Teaching Model of German in Colleges and Universities Under the Information Technology Environment

Most students like to use the multimedia (videos and pictures and so on) in the classroom for the audio-visual teaching, while some students think that the application of the multimedia in the German classrooms improves the quality of our teaching [2].

© Springer Nature Singapore Pte Ltd. 2020
M. Atiquzzaman et al. (Eds.): BDCPS 2019, AISC 1117, pp. 1547–1553, 2020.
https://doi.org/10.1007/978-981-15-2568-1_215

The audiovisual teaching plays an important role in the German teaching, but there are many problems in this lively teaching form.

1.1 The Concept and Materials of the Audio-Visual Education Are Outdated

At present, the German teaching theory has been introduced, studied and used for reference by the foreign teaching theories, and especially after the 21st century, it has made more reference and imitation to the foreign teaching theories. Although we have also extracted some experience and given guidance to our teaching, the teaching effect has not been changed in expectation [3]. Especially in the audio-visual teaching, there is a lack of the corresponding theoretical guidance, guiding the teaching practice and standardizing the teaching activities. The lack of the theoretical depth can be attributed to the following reasons: The scarcity of the research institutions for the teaching theories and practice, the lack of the platform for the accumulation of the practical experience, and the lack of attention to the audio-visual teaching and so on [4]. It is necessary to give more in-depth analysis and reflection on the way of the learning and imitating for many years, carefully analyze the differences between our foreign language teaching environment and foreign countries, clearly and profoundly understand our unique personalities, recognize the teaching characteristics of different languages and different periods, and recognize the differences between their contents and ideas. In the teaching activities, we constantly compare with the specific ideas and contents of the relevant teaching theories, and constantly sublimate our unique foreign language teaching theories, so as to promote the improvement of the audio-visual teaching [5]. At present, the main audio-visual textbooks used by the German majors are few in the quantity and outdated in contents. Although they play a gradual role in the training of students' oral and listening skills, they are systematic and scientific. However, due to their early publication and outdated contents, some contents of the video textbooks are not related to the social environment in which students are now living and cannot stimulate students' enthusiasms for learning.

1.2 Lack of the Teachers' Role Awareness

The interweaving of the "teaching" and "learning" runs through the main line of the teaching activities. Teachers usually occupy the main position. A large number of studies have shown that teachers' personal qualities have become the short board of the cask effect in the teaching activities. The language audio-visual teaching is closely linked with the modern information. It not only needs the knowledge structure of the general foreign language teachers, but also needs the profound modern information. In addition, the modern teaching theoretical knowledge such as pedagogy and psychology should also be possessed by teachers. Only by understanding the changing law of students' psychological activities in the process of learning and flexibly applying the teaching methods to different students and different situations can students improve. A large number of the research and practice in the audio-visual teaching also require teachers to have the characteristics of "directors", command various factors in the teaching, mobilize students, lead students, and make the classroom activities around the teaching purpose. However, in the real German audio-visual classrooms, due to the small number of the professional German teachers and the heavy teaching tasks, there are less innovative reforms in the research field of the teaching methods, and there is a lack of reasonable and effective teaching experience to assist the improvement of the teaching efficiency. Among the students, the interaction with the body language is relatively less. This reflects that there are fewer interactive links in the classroom design, and most of the time teachers are personally explaining the audio-visual contents. In the limited audio-visual classes, the teaching usually centers on the "audio-visual", with fewer opportunities and short time for the language output, the insufficient practice in the language use such as the situational conversation, and the communication and interaction between teachers and students. Limited to the traditional questions and answers, students tend to have a low ability to use the language in practice and will soon forget it.

1.3   Students' Self-confidence in Their Basic Language Ability

Because of the practical characteristics of the audio-visual teaching, students must participate in the classrooms, and show their audio-visual abilities in the classrooms. Of course, it is necessary for the audio-visual courses to have a certain language foundation. The current German teaching classes are similar to the English teaching classes. Emphasizing the training of students' grammar and reading abilities while neglecting the cultivation of the oral and listening abilities results in students' lack of confidence in their foreign language abilities and their enthusiasm for learning will be frustrated once they encounter something they do not understand in the process of the audio-visual teaching. From the survey results, it can be seen that in the German learning, some students think that their weaker link is the learning of the basic language ability, and students think that they spend less time on the basic language ability learning, and the training distance between listening and speaking is far from being skilled in using the basic foreign language. And some students think that their basic foreign language ability is the weaker link. Without the self-confidence in the speech ability, once you encounter words or contents that you can't understand in the audio-visual classroom activities, it will affect your mood, and then lead to fear and anxiety, which will directly affect the interaction effect in the subsequent activities.

2   Requirements for the Construction of the Multivariate Interactive Teaching Model of German in Universities Under the Information Technology Environment

2.1   Hardware Requirements

Under the information technology environment, the university German teaching mode cannot be separated from the corresponding network hardware and the network resources, which requires the school to allocate good network equipment and teaching resources. Especially for the listening materials, it is better to have the original German listening materials, which are from shallow to deep and from easy to difficult, and can improve students' interests in these listening materials and cultivate their good cultural literacy in the light of the current hot politics and the latest news.

2.2 Software Requirements

Schools need to develop the relevant teaching evaluation software, which not only enables students to learn by themselves, but also evaluates students' learning and mastery of the German listening through the fixed teaching software. Statistics show students' listening practice time and listening materials through the network, evaluates students' listening level in different time periods, and gives them timely information. Teachers' teaching feedback can help teachers understand the progress of students' listening and the problems they encounter, and help students solve the related problems.

2.3 Requirements for Teachers

Under the information technology environment, the teaching mode of the listening requires teachers to change from the simple listening training to the self-directed classroom teaching methods, answering questions and solving puzzles, and organizing the classroom discussions, so as to help students learn more effectively.

2.4 Requirements for Students

The multi-interactive teaching of German in universities under the information technology environment requires students to have a better understanding of the network-related knowledge and operation, and to be able to independently complete the browsing and downloading steps. In addition, students also need to understand the teachers' teaching objectives, formulate their own learning plans, make clear arrangements for their own listening-related training, regularly evaluate and feedback their listening level, and promote the improvement of the listening level through the continuous summary.

3 Strategies for Constructing the Multivariate Interactive Teaching Model of German in Colleges and Universities Under the Information Technology Environment

Under the information technology environment, the multi-interactive teaching of German in universities is a new attempt. Proper application can not only effectively improve students' listening levels, but also exercise students' self-study abilities. Only when teachers and students work together and cooperate with each other can the teaching quality of the German listening be steadily improved.

3.1   Identifying Students' Targets in the German Learning

In the standard sense of the autonomous learning, the determination of the learning objectives should be made by students, but the current German teaching mode and all aspects determine that students are unlikely to freely choose learning contents and determine the learning objectives. Therefore, in most cases, the learning objectives are determined by the teachers, and of course, it can also be said that the curriculum or the examination instructions are determined. Determining the learning objectives cannot simply list the contents according to the curriculum standards, which is not helpful for students to learn independently. For students who have just started to learn independently, they can help them through the "preview plan". College German preview programs should include several contents: the basic knowledge, and the basic skills and standards of our learning, suggestions on the learning methods and steps, and tests after the self-study. Writing should be based on students' knowledge and characteristics, in accordance with the logical order, so as to guide students to study independently. In our daily teaching, the preview plan is compiled and sent to the group. Students are required to check the relevant preview plans when they study independently, so that students will not be unable to learn independently because they forget to carry the paper materials.

3.2   Student Self-study

After completing the first step, students can be required to study independently according to their learning objectives and preview plans. In the process of the autonomous learning, we should pay attention to the following points: to ensure the autonomous learning time. This point should be allocated according to the actual situation. At the beginning of the autonomous learning mode, the time is longer and the later period is shorter. The complex extra-curricular time and the intra-curricular time and the simple contents can be arranged directly in the classrooms. The mode of self-study: Generally, follow the mode of "teaching → learning guidance → self-study". At the beginning, teachers should take students to read together and tell them how to learn when they read. After the students have mastered the basic methods, the teachers should tell the students what the key point of reading is, and mark the key contents and difficult problems of the textbooks with fixed symbols, which are convenient for the teachers to explain with questions, and also for reference in future review. Once you have formed a habit, you can let go. Guidance in the learning process: In the past, the learning process guidance was only through the self-study syllabus. Now with the network, we can play the convenience of the network, so we can guide through QQ groups, and guide students through texts and video dialogues. It can also send the courseware and video files to students through the file sharing, so that students can download them according to their needs, assist them in their learning and ensure the quality. They must be required to master what they can master first through their self-study, and they can be asked to complete it through the random network checks.

3.3   Teachers' Key Point Explanations

Through the self-study, students will still be unable to understand and master some problems, or the depth of their self-study is not enough, which requires teachers to explain. At this time, the explanation is a high-level explanation, which cannot be comprehensive, so that we should highlight the key points and solve difficulties. In short, help students achieve this goal as a benchmark. Since students have already had the foundation of the self-study in German, we can use the PPT courseware, Geometric Sketchpad and FLASH animations to assist our teaching, without worrying about the side effects of the too large classroom capacity brought by the courseware.

3.4   Practice Consolidation

If the first several links can be well completed, most students should be able to understand and master the objectives of the German teaching in universities. But at this stage, it is impossible for students to master and apply the knowledge and skills they have learned thoroughly. Even many students seem to have mastered them, but in fact they have only learned to imitate. Many students reflect that they understand in class, understand the examples and still can't do the exercises after class. This is the case in fact. Therefore, we should consolidate our knowledge and skills through practice. When designing the test questions, we should pay attention to the design variant training, guide students to generalize and transfer the knowledge, and pay attention to the principle of easy problem design before difficult problem design. In the course of the practice, students should be given the individual counseling, and especially those with learning difficulties. This can be achieved through the attention in German classes at universities, and through the Internet in the after-school time.

3.5   Class Summary

The systematization of knowledge is the only way for students to improve their German accomplishments. Only through the systematization of the knowledge can students really understand the subject contents and can they memorize and apply it well. At the beginning, it is not easy for students to learn these, so they can do it according to the following points: to include all the contents of this section, to embody the logical connection and the knowledge structure to make it a whole, to highlight the key points, and to be concise and terse.

4   Conclusion

At present, the traditional teaching mode is mostly used in the German teaching in colleges and universities in China. This paper integrates the information technology environment into the multi-interactive teaching mode of German in universities. By making full use of the network teaching resources and technologies, it can stimulate students' enthusiasms and abilities to learn independently, which has a pioneering significance for the development of the German teaching in colleges and universities.

Acknowledgement. Research on spatial narrative of Novalis' poetic novel. supported by "the Fundamental Research Funds for the Central Universities".

References

1. Chen, Y., Zhang, Y.: Exploration of the multiple interactive teaching model of college English. J. Chongqing Univ. Sci. Technol. (Soc. Sci. Ed.) (8), 198–199 (2010)
2. Guo, Y.: Learning-orientation and multi-interaction – a study of college English teaching model in the information technology environment. J. Heilongjiang Coll. Educ. (7), 165–167 (2013)
3. Wang, C.: Research on the multi-interactive teaching model of college English in the multimedia environment. China Educ. Technol. Equip (14), 123–124 (2016)
4. Liu, Y., Sun, F.: Consideration and practice of the effective college English teaching in the context of the information technology integration. J. Heilongjiang Coll. Educ. (2), 150–152 (2015)
5. Yang, X.: An empirical study on the multivariate interactive teaching model of college English under the network environment. J. High. Educ. (19), 53–54 (2015)

Application of the Computer Vision Technology in the Agricultural Economy

Yue Meng and Wenkuan Chen[✉]

Sichuan Agricultural University, Chengdu 610000, Sichuan, China
lbkgw0@yeah.net

Abstract. The computer vision is a science that studies how to make the machines "see". It refers to the machine vision that uses the camera and the computer instead of the human eyes to recognize, track and measure the targets, and further do the graphic processing to make the computer processing more suitable for the human eyes to observe or transmit to the instrument to detect the images. With the rapid development of the computer technology, the computer vision technology has been widely used in the agricultural machinery, such as the field machinery, the agricultural products processing machinery and the agricultural products sorting machinery, which can realize the automation and intelligence of the agricultural machinery and promote the sustainable development of the agriculture.

Keywords: Computer vision technology · Agricultural economy · Application strategy · Value advantage

The computer vision mainly refers to the use of computers to analyze images, so as to control certain actions or obtain the data describing the scenery. It is an important field of the artificial intelligence and the pattern recognition [1]. The computer vision technology has a very wide range of applications and value advantages in the agricultural economy, and it is also an important breakthrough point in the development of the modern agricultural economy [2].

1 Application Connotation Analysis of the Computer Vision Technology

The computer vision emerged in the 1970s, which involves a wide range of disciplines, including vision, the CCD technology, automation, artificial intelligence, pattern recognition, digital image processing and computers. At present, the computer vision technology is mainly based on the image processing technology. It simulates the human eyes through the computer vision, takes the close-range photographs of the crops using the spectra, and uses the digital image processing and the artificial intelligence technology to analyze and study the image information [3]. The main steps of the computer vision technology include the image acquisition, the image segmentation, the preprocessing, the feature extraction, and the processing and analysis of the extracted features [4].

© Springer Nature Singapore Pte Ltd. 2020
M. Atiquzzaman et al. (Eds.): BDCPS 2019, AISC 1117, pp. 1554–1559, 2020.
https://doi.org/10.1007/978-981-15-2568-1_216

As a scientific discipline, the computer vision technology studies the related theories and technologies, trying to build an artificial intelligence system that can obtain 'information' from the images or the multi-dimensional data [5]. The information referred to here refers to the information defined by Shannon that can be used to help make a "decision". Because the perception can be seen as extracting the information from the sensory signals, the computer vision can also be seen as the science of how to make the artificial systems "perceive" from images or the multi-dimensional data.

The computer vision is not only an engineering field, but also a challenging and important research field in the field of science. The computer vision is a comprehensive subject, which has attracted researchers from various disciplines to participate in its research. These include the computer science and engineering, the signal processing, physics, Applied Mathematics and statistics, neurophysiology and cognitive science and so on.

Vision is an integral part of various intelligent and autonomous systems in various application fields, such as manufacturing, inspection, document analysis, medical diagnosis, and military. Because of its importance, some advanced countries, such as the United States, have listed the study of the computer vision as a major fundamental issue in science and engineering, which has a wide impact on the economy and science, namely the so-called grand challenge. The challenge of the computer vision is to develop the visual capabilities for computers and robots that are comparable to the human level. The machine vision requires the image signals, textures and color modeling, the geometric processing and reasoning, and the object modeling. A competent visual system should integrate all these processes closely. As a discipline, the computer vision began in the early 1960s, but many important advances in the basic research of the computer vision were made in the 1980s. Now the computer vision has become a mature subject different from the artificial intelligence, the image processing, the pattern recognition and other related fields. The computer vision is closely related to the human vision. A correct understanding of the human vision will be very helpful to the study of the computer vision.

2 Feasibility Analysis of the Application of the Computer Vision Technology in the Agricultural Economy

The computer vision is to use various imaging systems instead of the visual organs as the input sensitive means, and the computers instead of the brain to complete the processing and interpretation. The ultimate goal of the computer vision research is to enable computers to observe and understand the world through the vision, like the human beings, and have the ability to adapt to the environment independently. It takes a long time to achieve the goal. Therefore, before achieving the ultimate goal, the medium-term goal of people's efforts is to establish a visual system, which can accomplish certain tasks according to a certain degree of the intelligence of the visual sensitivity and feedback. Therefore, the goal of the current research efforts is to achieve a visual-assisted driving system with the road tracking ability on the expressway, which can avoid collision with vehicles in front of it. In the computer vision system, the computer plays the role of replacing the human brain, but it does not mean that the computer must complete the processing of the visual information according to the method of the human vision. The computer vision can and should process the visual information according to the characteristics of the computer systems.

However, the human visual system is the most powerful and perfect visual system known so far. As can be seen in the following chapters, the study of the human visual processing mechanism will provide inspiration and guidance for the study of the computer vision. Therefore, it is also a very important and interesting research field to study the mechanism of the human vision and establish the computational theory of the human vision by means of the computer information processing. This research is called the computational vision. The computational vision can be considered as a research field of the computer vision. In the new era, the computer technology has been developed in a long-term way, and its application in the field of the agriculture has created great social benefits.

With the progress of the science and technology, the modern agriculture has experienced a variety of the production modes from the initial stage of the development to the process of gradually approaching maturity. The traditional production models are relatively backward in the production tools and low in the production efficiency. Nowadays, with the wide application of the computer vision technology in agriculture, the level of the technology and the equipment has been greatly improved. Through the selection of the fittest, the informationization is more prominent in front of people. It can be said that the symbol of the modern agriculture is the agricultural information-ization. The development of the agricultural economy needs to keep up with the modernization of the agriculture, and the modernization of the agriculture must rely on the informatization of the agriculture.

3   Application of the Computer Vision Technology in the Agricultural Economy

3.1   Application in the Field Work

In the field work, the application of the computer vision technology is late. In recent years, due to the environmental protection policies put forward, the application of sowing, plant protection and fertilizer application machinery in the farm operations has become more and more widespread. The computer vision technology is mainly used in the seedling grafting, the field weeding, the pesticide spraying, and the fertilization and sowing. In order to identify weeds effectively and spray herbicides accurately, the morphological characteristics of the binary images of soybeans, maize and weeds commonly seen in the Midwest of the United States were analyzed. It was found that plants can effectively distinguish dicotyledons from monocotyledons within 14 to 23 days after growing, with the highest accuracy of 90%. The system mainly uses the machine vision guidance system to align the sprinkler with the top of each row of crops, and automatically adjusts the sprinkler according to the width of the crops, so as to ensure the consistency between the width of the crops and the width of the fog droplet distribution, thus effectively saving pesticides.

3.2   Application in the Agricultural Products Processing

With the rapid development of the computer vision technology and the computer technology, the computer vision technology is widely used in the automation of the agricultural products processing. For example, Jia P and others put forward the image processing algorithm. This algorithm is mainly based on the horizontal direction of catfish and the centroid position and the angle of the main axis. It detects the orientation of the catfish and the position of the dorsal fin, the pelvic fin, and the head and tail, so as to determine the best cutting position. In addition, Huang Xingyi et al. analyzed the color characteristics of the germ rice without dyeing when studying the production process of the germ rice. The parameters of the color characteristics of the germ rice were obtained as the saturation S. At the same time, the computer vision system was used to automatically and nondestructively detect the embryo retention rate of the germ milled rice, and the results were basically consistent with the results of the artificial evaluation.

3.3   Application in Sorting the Agricultural Products

When classifying and identifying the quality of the agricultural products, the computer vision technology can be used for the non-destructive testing. Generally, the computer vision technology does not need to contact the measured objects. It can directly use the surface images of the agricultural products to classify and evaluate their quality. It has the advantages of the unified standards, the high recognition rate and the high efficiency. The computer vision technology in the detection of the agricultural products mainly concentrated in cereals, vegetables and fruits. The computer vision technology was used to evaluate the quality of the broad bean. The theory distinguishes stones, different fava beans, being too small and damaged and qualified by different discrete methods, and classifies their characteristic parameters by using the shadow images. The final classification results are in good agreement with the statistical classification results.

3.4   Monitoring Growth

The infiltration of the computer technology in the field of the modern agriculture has realized the real-time monitoring of the plant growth. It can not only accurately understand the growth parameters such as the leaf area, the leaf circumference and the petiole angle of plants in monitoring facilities, but also judge their maturity according to the color and size of fruits, so as to take the targeted supplements. Water, fertilization and other measures play an important role in improving the plant growth efficiency and quality. In addition, the computer vision technology can be used to accurately locate the growth and distribution of weeds in the crop and soil background, which provides a strong basis for the efficient and variable spraying of the herbicide reagents. At present, the methods of the acoustic measurement, trapping and near infrared are mainly used to analyze the situation of pests and diseases in the crop growth at home and abroad, but the research on the application of the computer vision technology in the agricultural insect identification is still relatively scarce. Zheng Yongjun, Mao Wenhua and other scholars have put forward the idea of using the locust morphological feature factors to classify the moving areas of detecting the sky and grassland sub-images in the paper "Recognition method of grassland locusts based on the machine vision", which will greatly improve the recognition rate of the jumping locusts. The combination of the computer vision technology and the insect mathematical morphology theory will also play a greater role in the agricultural pest monitoring, with broad application prospects and development prospects in the future.

3.5   Quality Grading

With the development of the modern agriculture, the computer technology has been widely used in the germplasm detection and the quality analysis. Especially represented by the computer image processing technology, it refers to the acquisition of the image signals through the CCD camera into the digital signals, and the computer processing technology. The application of the computer recognition technology in the agricultural germplasm detection not only saves a lot of the manual investment, but also improves the efficiency and accuracy of the germplasm detection. In addition, the non-destructive testing of the agricultural products can also be carried out by using the computer image processing technology. According to the physical parameters such as the surface color, shape, brightness and size, the quality grade of the agricultural products can be evaluated.

3.6   Information Dissemination

Under the background of the information and the network era, people's life and working methods have changed tremendously, and their dependence on the computer technology is increasing. Similarly, the development of the modern agriculture is inseparable from the computer technology. The greatest value of the agricultural products lies in meeting people's living needs. Only when the quality is recognized by the society can we get good benefits, and thus achieve the stable and sustainable development. The agricultural product from planting to consumption is a complicated and systematic process, which contains the abundant information. Usually, the agricultural products are affected by the seasons, regions and other factors, and if the promotion is not effective, it is easy to affect its later sales, and then there is a backlog of the unsalable phenomena, resulting in the immeasurable waste of resources and economic losses and so on. The application of the computer technology in the modern agriculture has established an agricultural industry system, which closely links the agricultural development with the market demand and ensures the sustainability of the agricultural development to the greatest extent. Specifically, the computer technology can be used to create a network information service platform. While timely understanding of the economic market and making the industrial rationalization adjustments, consumers can also intuitively grasp the information of the crop growth and processing, thereby enhancing the consumer confidence and interest. The application of the computer technology in the modern agriculture has accelerated the dissemination of the agricultural information and played a positive role in adjusting and optimizing the agricultural industry system.

4   Conclusion

With the wide application of the computer technology in various fields of the social economy, many intelligent agricultural machinery systems have emerged in the agricultural production, such as the agricultural information system, the crop simulation system, the agricultural expert system, the agricultural mechanization visual system, the traceability information system, and the plant growth monitoring system and so on. The computer technology has become an important means to promote the agricultural technology innovation and improve the labor efficiency, but there are also some problems. At present, the application of the computer vision technology in the agricultural economy in China has a certain gap compared with the developed countries, but its development potential and application prospects are broad, and it will occupy more and more components in the future agricultural production.

References

1. Wang, C., Zheng, H., Zhang, H.: Application prospect of the internet technology in agriculture. Fujian Agric. Sci. Technol. (12) (2015)
2. Zhou, F.: Application of the computer technology in the agricultural technology. Consume Guide (10) (2015)
3. Ren, Z.: Research on the agricultural information collection system of Heilongjiang reclamation area based on the mobile message. Mod. Agric. (1) (2016)
4. Hao, Z., Zhao, A., Jin, X., et al.: Using the computer vision to study the color change in white tea processing. J. Fujian Agric. For. Univ. (Nat. Sci. Ed.) (3), 325–329 (2010)
5. Li, W., Tang, X., Tang, Y., et al.: Study on the relationship between the color change and the quality of green tea based on the visual technology. Food Res. Dev. (5), 1–4 (2015)

Information Collection of the Agricultural Database in the Agricultural Economy

Yue Meng and Wenkuan Chen[✉]

Sichuan Agricultural University, Chengdu 610000, Sichuan, China
lbkgw0@yeah.net

Abstract. The modern agriculture usually refers to that with the application of the modern agricultural technology and equipment and the science and technology in the production activities under the implementation of the modern management mode. The establishment of the basic agricultural database and the special agricultural information service system has provided a powerful grasp for combing and integrating the agricultural information resources, provided a powerful data support for the management decision-making, improved the management levels, and well met the current needs of the innovative management of the agricultural economy.

Keywords: Agricultural economy · Agricultural database · Information collection · Research model

With the development of the informationization, the agricultural informationization has gradually infiltrated into China's agricultural industry, and the development of the agricultural informationization today has its distinct characteristics. These characteristics are mainly manifested in the application of the information technology. At the same time, it can effectively change the mode of the production and integrate the production information resources through the application of the information technology, so as to achieve the effect of promoting the sustainable development of the agriculture.

1 Value Analysis of the Agricultural Database Information Collection in the Agricultural Economy

The application of the agricultural database information collection is conducive to the scientific agricultural production. By collecting the necessary information in the production process, the land and the climate in the production process can be accurately controlled, and the production work can be carried out scientifically and reasonably, and the speed and the quality of the agricultural production can be effectively improved.

1.1 Improving the Speed and the Quality of the Agricultural Production

The development history of the agriculture is generally divided into three stages from the perspective of the productivity: the primitive, the traditional and the modern

© Springer Nature Singapore Pte Ltd. 2020
M. Atiquzzaman et al. (Eds.): BDCPS 2019, AISC 1117, pp. 1560–1566, 2020.
https://doi.org/10.1007/978-981-15-2568-1_217

agriculture. The basic characteristics of the primitive agriculture are nomadism and collection, and the simple agricultural activities are carried out through the use of the stone tools, while the traditional agriculture has learned some experience. Farmers use the manpower, the animal power, the natural force and the hand-made iron tools to produce. The agricultural production efficiency is not high, and the self-sufficiency was the universal nature at that time. In contrast, the modern agriculture pays attention to the informationization, has the strong scientific and technological support in the production tools, and pays attention to the protection of the natural ecological environment, which at the same time can vigorously promote the growth of the agricultural economy, so the modernization and the informationization have become the trend of the agricultural development.

1.2 Effective Promotion of the Agricultural Economic Growth

The development of the informationization has greatly promoted the growth of China's agricultural economy. In the process of the agricultural production, the informationization production can not only improve the utilization rate of the resources through the rational utilization of the resources, but can also reduce costs and time, thus increasing the benefits of the agricultural producers. Through the establishment of the agricultural information network system, it can effectively manage the problems of the untimely regulation and the low management efficiency, improve the circulation efficiency of the agricultural market and reduce the transaction risk. The informationization technology can effectively transform the development mode of the agriculture in China, and gradually replace the extensive development mode with the refined development mode, so that the economic growth of the agriculture can be effectively promoted, which also has a positive impact on the overall economic growth of China.

2 Analysis of the Information Collection Principles of the Agricultural Database in the Agricultural Economy

2.1 Principles of Initiative and Timeliness

As far as the agricultural extension is concerned, the information related to the agricultural development should be collected actively, and the useful information should be collected, collated, analyzed, processed and promoted in time. For the agricultural extension technicians, they should have a high degree of the information awareness and the ability to collect, collate, analyze, process and promote the information. In the work of the agricultural extension, we should achieve "I have what other people have not", "I am excellent in what others have" and "I am expert in what others are excellent". We can timely grasp and promote the key points, difficulties and hot spots in the process of the local agricultural development, and promote the development of the local agriculture.

2.2   Principles of Authenticity and Validity of the Information

For the agricultural science and technology information, the authenticity and the validity of the information are directly related to the fundamental interests of the agriculture, the rural areas and the farmers. Therefore, in the process of collecting the information, the gold content, practicability and reliability of the information must be taken as the primary consideration.

2.3   Principles of Practicality and Efficiency

For the personnel of the agricultural technology extension, all kinds of the scientific and technological information released should be the key, difficult and hot information related to the development of the agriculture, the rural areas and the farmers in a certain region. Therefore, the practicability of the agricultural science and technology information is very important. Only the information that meets the needs of the agriculture, the rural areas and the farmers is the valuable information. At the same time, after the release of the information, the information with the good economic and social benefits is the good information.

2.4   Principles of Systematicness and Comprehensiveness

For the agricultural science and technology information, the information providers should conduct the in-depth analysis and processing of the information after collecting and collating the information, so as to systematically analyze and comprehensively analyze the sources, purposes, significance, implementation, effect and evaluation of the information, so as to ensure the value of the agricultural science and technology information.

2.5   Normative and Planning Principles

As far as the collection of the agricultural science and technology information is concerned, it should not only embody its attributes of the value, scientificity and authenticity, but should also consider the standardization of the information accumulation and collection. Through the system standardization and the perfect management of the collection of the agricultural science and technology information, the classification database of the agricultural science and technology information should be formed as the reference basis for the decision-making of the agriculture, the rural areas and the farmers. At the same time, the agricultural extension services and the agricultural economic management departments should also formulate plans for the agricultural science and technology information collection to promote the standardized and scientific management of the information collection through plans.

2.6   Principles of Prediction and Dynamics

For the agricultural science and technology information, there should be a certain degree of foresight in the process of the information collection, and a certain degree of forecasting for the development trend and the demand of the agriculture, the rural areas and the farmers, to ensure that the information provided is forward-looking and practical. At the same time, we should grasp the dynamic changes in the fields of the economic development, the social development and the scientific and technological development to ensure the guiding role of the information.

3 Design of the Agricultural Database in the Agricultural Economy

3.1 System Architecture Design

The basic agricultural database and the basic agricultural information service system are deployed on the government information resource integration and utilization service platform. The multi-tier software architecture based on the B/S structure is mainly divided into six parts: the network, the data management and service, the service component, the application service, the standard specification and security management operation and the maintenance guarantee.

3.2 Data Contents and Organizations

Through the in-depth analysis and combined with the data resources, the agricultural information resources accumulated over the years were sorted out, including the annual agricultural statistics information of China and monthly agricultural trade information. The annual information includes the agricultural comprehensive information, planting, animal husbandry, fishery, township enterprises, farming, agricultural machinery, rural energy, rural management and agricultural natural disasters information. The monthly information includes food, wheat, rice, corn, soybean, cotton, oil, sugar, meat, pigs, cattle, sheep, and poultry. The daily data include the wholesale price information and the wholesale price index information of vegetables, fruits, livestock products and aquatic products.

3.3 Functional Module

The basic agricultural database focuses on the management of the basic agricultural data, the catalogue services, the visual display, the sharing services and other functions. Based on more than 1000 indicators provided by the main data, the data resource sharing services were established, including 12 kinds of the data indicator sharing services, including the comprehension, planting, animal husbandry, fisheries, township enterprises, agricultural reclamation, agricultural machinery, rural energy, rural management, agricultural natural disasters, agricultural product prices and agricultural product trade. After the completion of the database, as an important part of the dynamic database of the basic national conditions, it provides users with the agricultural information services and sharing.

The thematic data are extracted and stored. Through collecting the thematic data from various data sources and classifying the management according to the thematic

---

information, the collection, organization and storage of the basic agricultural thematic data can be realized. The main functions include the index item extraction, the data acquisition and storage, and the directory management. Basic farming information display: The basic farming information display module realizes the visualization of the basic farming index data, mainly using the statistical maps, tables, the statistical maps and the knowledge structure maps. Basic agricultural information retrieval: Basic agricultural databases provide the information retrieval functions, and support the single-condition retrieval and the combination-condition retrieval. Single conditional retrieval: The single conditional retrieval supports the information query according to the single fields such as keywords, time and category of the basic agricultural information. The combination conditional retrieval: The information is queried by the multi-field combination, and pictures, tables and text display of the query results are given quickly. Basic agricultural data service: The basic farming data service provides a uniform format of the data service interface. Users can obtain the data information through the data service interface, thus realizing the basic farming online data sharing service.

3.4 Special Information Service System for the Basic Agricultural Conditions

According to the demand of the leaders for the agricultural information and through the development of the customized functions, the special information service system of the basic agricultural conditions constructs the columns such as the daily market price and the price index of the agricultural products, the monthly trade situation of the agricultural products and the means of the production, and the annual agricultural economic information, so as to provide thematic agricultural information letters for the leaders and the relevant business departments. The main contents include the daily market price and the price index of the agricultural products. This column realizes the visualization display of the market price and the price index of the agricultural products every day. The agricultural products include the animal products, the aquatic products, vegetables and fruits. The visual display supports many ways, including the statistical chart display, the information retrieval, the statistical map display, and the data download and print output. The monthly trade in the agricultural products and the means of production: The monthly trade column of the agricultural products and the means of production focus on the import and the export information of planting, breeding, means of production and other important indicators at a monthly frequency. The main functions include the statistical chart display, the information retrieval, the statistical map display, and the data download and print output.

3.5 Data Resource Sharing

The data resource sharing is realized by the online service and the unified retrieval. Based on the E-government network, it integrates with the central government information resources portal, provides the online services of the basic agricultural information for the leaders of the State Council, the business departments and other authorized users, and realizes the user-oriented special information services by utilizing the basic agricultural information service system based on the government affairs. The basic agricultural database can achieve the basic agricultural data information sharing. In addition, through the establishment of a unified information resource index, all pages can be retrieved and located by a unified search method. The data resources provide the interface access calls and so on.

4 Information Collection Procedure of the Agricultural Database in the Agricultural Economy

In the collection of the agricultural science and technology information, certain procedures should be followed to ensure the compliance and scientificity of the agricultural science and technology information collection. Generally, the collection of the agricultural science and technology information should be carried out according to the determination of the collection objectives, the formulation of the collection plans, the implementation of the collection work and the evaluation of the collection effect.

4.1 Determine the Acquisition Target

The contents, levels and depths of the information provided by different target groups should be treated differently, and the collection of the agricultural science and technology information in different regions should also be treated differently. After defining the information demanders, we should collect the information pertinently, so that the collected information contents can serve the information demanders. After analyzing the needs of the information demanders, the scope of the information acquisition should be defined. This range can be collected from the two dimensions. One is the horizontal acquisition, which collects the same, relevant or similar information as the information demanders, and conducts the horizontal comparative research to provide the valuable information. The second is the vertical acquisition. We can collect and analyze the continuous data and information in a certain area, space and time, and make a comparative study of the data, so as to provide the objective, reliable and referential information for the information demanders. According to the purpose of the information users, the quantity of the information collection can be determined. For example, for the analysis of the agricultural output, the information collected in the past five or ten years can be analyzed. However, the information collected in a longer period of time has no objective reference value because of the influence of the climate, the scientific and technological environment and the labor environment changes. The quantity should be mastered according to the information demanders and the actual needs of the agricultural science and technology extension.

4.2 Make the Acquisition Plans

The collection plan is the arrangement of the information collection work in a certain time node according to the actual needs of the agricultural science and technology extension. For the information collection plan, it should be standardized, careful, reasonable and operable.

4.3   Implementation of the Collection Work

In accordance with the acquisition plan, we should work in an orderly manner to achieve the scientific, rational and practical nature of the information acquisition.

4.4   Assessment of the Collection Effect

The objective analysis and evaluation of the economic and social benefits of the local agriculture, the rural areas and the farmers after the information collection and publication can provide reference for the quality and efficiency of the information collection in the future.

5   Conclusion

In order to collect and integrate the existing agricultural and rural economic data, the agricultural database should provide the information services for the users as one of the information resources of the dynamic database of the basic national conditions of the governments. The basic agricultural special information service system should be customized and developed according to the needs of the leading comrades for the agricultural information. Through the development of the function customization, the daily market price and the price index of the agricultural products, the monthly trade situation of the agricultural products and the means of the production, the annual agricultural and economic information and other columns will be realized. The completed system will provide the thematic, authoritative, reliable and time-sensitive agricultural and rural economic information services for the relevant departments.

References

1. Jiang, X., Zhang, M.: Land intensive use information collection terminal system based on the mobile GIS. Geomat. Spat. Inf. Technol. (11), 118–120, 123 (2014)
2. Chen, J., Liu, J., Tan, J.: Design and implementation of the tobacco farmland information collection system based on the mobile GIS. Guangdong Agric. Sci. (11), 184–187 (2009)
3. Yu, L.: Research on the geological disaster information acquisition system of the mobile GIS. Geospatial Inf. 10(2), 113–115 (2012)
4. Meng, W.: Design and implementation of the urban greening information acquisition system based on the mobile GIS. Stand. Surv. Mapp. 26(1), 26–28 (2010)
5. Liu, H.: Impact of the agricultural informatization on the agricultural economic growth in China. Mod. Econ. Inf. (19), 325–326 (2016)

Construction of the Computer Network Courses in College Physical Education Distance Video Teaching

Zhongqi Mi$^{(\boxtimes)}$ and Zhikai Cao

Jianghan University, Wuhan 430000, Hubei, China
1959278975@qq.com

Abstract. With the development of the computer technology, the multimedia technology and the communication technology, a new educational model, represented by the educational technology, has emerged, in which the distance education is a prominent part. Physical education is a bilateral activity of the teaching and learning with strong practicality and participation. At present, the resources on the network are fewer than those of other disciplines, and there is a lack of systematicness and integrity. For the physical education and teaching, the demonstration and the imitation of actions are very suitable to be presented in the form of the multimedia contents. Therefore, in this context, based on the mature campus network of colleges and universities, and based on the Internet technology, the computer technology, the multimedia technology and the modern communication technology, it is very necessary to design and develop a long-distance network educational system and platform for physical education.

Keywords: College physical education · Distance video teaching · Computer network course · Construction strategy

At present, all countries in the world are vigorously promoting the distance education. The lifelong, globalized and efficient education makes the distance education develop rapidly all over the world [1]. Due to various ways of realizing the distance education platform, according to the characteristics of the physical education discipline, this research adopts the web-based teaching system, the real-time interaction and the computer network courses in college physical education remote video teaching based on the on-demand.

1 The Construction Background of the Computer Network Courses in College Physical Education Distance Video Teaching

Colleges and universities can make better use of all kinds of the physical education and teaching resources, make full use of and share the limited sports equipment, sports venues, teachers and other educational resources, improve the quality of the classroom teaching, enrich students' extracurricular sports activities, and break through the time and space limitations of the traditional physical education to a certain extent. The new

© Springer Nature Singapore Pte Ltd. 2020
M. Atiquzzaman et al. (Eds.): BDCPS 2019, AISC 1117, pp. 1567–1572, 2020.
https://doi.org/10.1007/978-981-15-2568-1_218

form, which supplements and promotes students' physical education curriculum learning in and out of class, provides new learning space and time for college students [2]. It will greatly improve students' initiatives and enthusiasms in learning the physical education, enhance the popularization of the sports theoretical knowledge, and improve students' physical education quality more scientifically and systematically [3].

In recent years, college students' physical education and health education have been paid more and more attention by our society and schools. However, the traditional classroom teaching methods of the physical education theoretical courses in general colleges and universities cannot meet the needs of the current physical education and teaching [4]. At the same time, with the rapid development of the network technology, the computer technology and the multimedia technology, the modern teaching modes such as the network teaching and the multimedia teaching emerge as the times require [5]. With the rise of MOOCs, the distance video teaching has become a trend of development.

The distance video teaching of the physical education in colleges and universities based on the network platform is a kind of the distance teaching mode, which combines the video technology with the computer network technologies. Compared with the traditional physical education theoretical courses, the online video theoretical courses have the advantages of flexible time, rich contents, high learner autonomy and strong interaction between teachers and students. In this study, the methods of literature, questionnaire, mathematical statistics and logical analysis are used to analyze the theories, technologies and applications of the physical education theories and the fitness video courses in colleges and universities under the network platform.

2 Application Advantages of the College Sports Distance Video Teaching

The distance video teaching system generally refers to a basic support system suitable for the characteristics of the distance education, aiming at providing a powerful online virtual audio-visual teaching environment for the distance video education and teaching platform. Using this system, the teaching units can easily organize experts, professors and students to carry out the remote video education and teaching tasks. The distance video education is the collective name of the educational form relying on the distance video teaching system to carry out the educational activities. It is a method of providing synchronous or asynchronous video education and training relying on the contemporary video communication network. Compared with the traditional education, the distance video education is a new and brand-new mode of our education. It fundamentally breaks the time and space constraints, enabling people to learn at anytime and anywhere and also enabling more learners to access the excellent educational resources, while having a sense of the on-site education that the traditional distance education does not have. Usually, the remote video education has the functions of sharing, virtualization, interaction, assistance, retrieval and service. The sharing function ensures that learners can access various fixed and advanced learning resources more efficiently. The virtual function breaks the restriction of the region and time, realizes the education at any time, the lifelong education, and makes good use of the fragmentary time.

The interactive function, as one of the cores of the distance video education, is embodied in the two-way communication of the learning between educators and educatees through the network, so that students can get the same feeling as the traditional educational methods with the on-site education. The collaboration function is embodied in that learners can communicate and cooperate with each other and participate in the completion of the tasks together. In the retrieval function, learners can quickly access the desired resources through the retrieval function, so as to achieve the purpose of improving the learning efficiency. The service function fundamentally determines the difference between the subject-object relationship and the service object of the traditional and contemporary education and teaching.

The distance video education, while breaking through the time and space constraints, also determines the irregularity and self-discipline requirements of the educational activities. The process of the sports learning and the self-practice should be people-centered and give learners full freedom. Therefore, learners are required to have the higher self-discipline and learning concepts, and have the stronger self-planning and self-learning abilities. In the case of the insufficient and regular learning time, we should maintain the quality of the learning. Otherwise, the result of the distance video education is divorced from the essence of the education which enables the educatees to acquire the knowledge. To solve this problem, we can further strengthen the interaction of the distance video education and the targeted physical education practice to improve the quality of our teaching and the different levels of the physical education for different groups of people, avoid the waste of time caused by the repeated learning of the known contents, and make full use of the time in the process of strengthening and learning the unknown knowledge.

3   Contents of the Computer Network Courses in College Physical Education Distance Video Teaching

Through the theoretical analysis of the relevant contents of the sports remote video teaching, the characteristics and requirements of the sports remote video teaching are summarized. That is, the sports remote video teaching should improve students' learning interests and self-study abilities, master the sports exercise knowledge and sports information through the teaching of the sports health care, sports fitness, sports leisure and sports information and sports health knowledge, while developing the lifelong exercise and the healthy life habits. Through the research on the actual construction of the distance sports video teaching based on the network platform, the results are as follows. The curriculum framework needs to include the multiple modules such as the course contents, the resource center, the communication platform and the online testing, and the web page construction and the video production can adopt the ASP.NET MVC, LINQ to SQL, SQL SERVER 2005, FMS, AJAX, and caching and so on. In order to better realize the network video courses, we need to establish six levels of the system: the hardware layer, the basic platform layer, the data layer, the business support layer, the application layer and the user layer.

Through the research on the practical application of the sports remote video teaching based on the network platform, it is proved that it is feasible and practical to promote the sports and health online video courses under the current conditions of the network hardware, software, teachers, and curriculum resources and so on. Because the university sports remote video teaching based on the network platform is still in its infancy, there are still many imperfect and immature places. But it provides learners and teachers with the flexibility and diversity that the traditional classroom physical education theoretical curriculum education cannot meet, and provides a platform for the real-time education, the self-learning and the lifelong exercise for college students.

In order to construct a more reasonable future model and direction of the sports theoretical education, this paper puts forward the following suggestions. Colleges and universities should pay more attention to students' sports and health theoretical education, combine with the requirements of the modern society for students' quality, carefully analyze the content requirements of students' sports theory teaching, and develop purposefully and pertinently the sports theoretical teaching and the teaching methods, to enable students to grasp the skills of their healthy life.

4 Strategies of Constructing the Computer Network Courses in College Physical Education Distance Video Teaching

The purpose of the computer network course design in the physical education remote video teaching in colleges and universities is to provide users with an Internet-based network learning environment. The functions of the system are accomplished through the interaction between the application server and the user browser. There are various kinds of the information and data stored in the computer network courses of sports remote video teaching in sports colleges and universities, including students' information, sports competition pictures, demonstration videos of sports action technology, sports teaching text materials, students' interactive messages, and sports network examination scores and so on. These data are stored in the database server in different formats, and are organized and maintained by the database server.

All the remote objects in the distance physical education and teaching, including teachers, students and the system administrators, should be connected to the Internet. These active subjects send requests to the server through their own browsers. The server processes these requests and responds to them in time, and gives feedback to the user's requests. The information retrieved from the database is presented to the users. And for the teaching system, different subjects have different rights, so the channels used are different. That is, the browsing interface is different, in order to meet the different needs of different users.

The communication between the system layers is mainly realized by the modules between the hardware and the software. In fact, the interaction between the client and the server is the interaction process between the users and the modules. The remote system of the physical education and teaching mentioned in this paper is also divided into two parts: the server and the client. The main way for users to obtain the information is to access the server resources from the application interface. Through learning, communicating, practicing and examining, we can achieve the goal of the physical education and teaching.

A complete computer network course in the college sports remote video teaching should consist of three parts: the backbone system, the management system and the teaching resource database. Portal sub-module: This module is mainly responsible for the entry of the learning module and the background management module. In addition, it also has the functions of publishing the sports-related information, news and some enrollment information, the online registration, the online course selection and the course inquiry. The portal module mainly faces the users of the external network. It serves as an external window to connect the public sports or the extensive sports with the campus physical education. It also meets the needs of the teachers and the students for visiting other module portals.

The learning sub-module mainly includes the following functions: the real-time teaching through the network, the courseware learning, the homework submission, the discussion, question answering, counseling, teaching and research, consultation and the performance query. The system will automatically record the students' learning status and progress in the real time. Therefore, the main service objects of the system are the teachers and the students. Through the technical support, we can ensure that the intranet users (teachers and students) can complete the sports learning under the network environment. This is the basis of realizing the distance physical education. Only through a reasonable interactive learning module can users learn the knowledge through the distance physical education and teaching system.

In the literal sense, the management sub-module is mainly used to complete the management of the distance physical education in the system. The main contents of the management include students, notifications, teaching activities, sports elective courses and teaching points, examinations and achievements. The management module is mainly for the management teachers and the teaching teachers in the distance education. The content of the module is used to manage the teaching activities and all the information generated in the process of the activities, and to share the information. As the three sub-modules of the computer network course in the physical education remote video teaching in physical education colleges and universities, each module is independent of each other, and at the same time, they have some auxiliary functions and are inseparable. The three modules in the data flow are most closely related. In the management of the user-based database, the management of the user name and password is the foundation of the portal module. In the learning module, the query of students' scores, notification, viewing and information publishing are the core of the module. And in the management module, it is responsible for the management of the relevant information of students and teachers in the daily network teaching, and provides reliable information for the curriculum performance queries and the examination qualifications.

Because of the close relationship between the modules, it is necessary to establish the database with all the modules in order to ensure the uniformity of the database and facilitate the synchronous management of the information sources in the database. In addition, it should be noted that the data in the learning module will be managed by the management module. There is a very close relationship between the two, and they play an important substantive role in the long-distance teaching of our physical education.

5 Conclusion

In the construction of the long-distance video teaching of physical education in colleges and universities, we should constantly improve the structure of the curriculum system, so that the curriculum system can continuously improve and perfect with the requirements of the teachers and students, and strive to achieve a more convenient operation experience, while meeting the richer functional requirements. In a larger scope, the more colleges and universities use the online video courses, the richer the course resources will be. At the same time, more feedback from the teachers and students can also help developers improve the online video course system.

References

1. Wang, H., Ye, F.: Overall design of the computer network course in college physical education distance video teaching. Youth Lit. (14), 103 (2012)
2. Zhang, T., Wang, L.: Design and implementation of the computer network course in the college sports distance video teaching based on the cloud computing. Contin. Educ. Res. (7), 59–61 (2011)
3. Wang, X., Wang, J., Li, B., et al.: Design of the computer network course in the physical education distance video teaching in colleges and universities. Sci. Technol. Innov. Her. (3), 180 (2014)
4. Yu, Z.: Design and implementation of the computer network courses in college physical education distance video teaching based on the streaming media technology. Softw. Guide (4), 195–196 (2011)
5. Li, Y.: Theoretical research on optimizing the physical education and teaching in universities with the modern educational technology. Bull. Sport Sci. Technol. (7), 53–54, 92 (2011)

The Effective Transformation Path of Scientific Research Achievements in Applied Universities Based on the Data Warehouse

Jinli Sun(✉)

Zhengzhou University of Science and Technology, Zhengzhou 450064, China
huangmingbucea@126.com

Abstract. Compared with traditional universities, one of the more obvious characteristics of applied universities is that they pay attention to the production of applied knowledge and the promotion of the transformation of achievements so as to highlight their unique advantages in serving the society. The institutional reform in universities with the subject adjustment as the breakthrough point is the second strategic step for application-oriented universities to promote the transformation of their achievements. Building a platform for the cooperation between "government, industry, education and research" is the organizational guarantee for application-oriented universities to achieve the transformation of their achievements, and recruit the teachers with academic application value recognition from the development strategy height, and ultimately achieve the pragmatism culture of running a university for practical use of application-oriented universities. The simulation study of the effective transformation path of scientific research achievements in applied universities based on the data warehouse has brought into full play the application advantages of the modern science and technology, and better promoted the application-oriented universities to transform their own scientific research achievements into comprehensive advantages in a timely manner.

Keywords: Data warehouse · Applied university · Scientific research achievements · Effective transformation path · Simulation research

As one of the main contributors of scientific research achievements, applied universities have three characteristics: short transformation cycle, small adaptation range and low value creation [1]. How to solve these problems and promote the timely and efficient transformation of scientific research achievements in applied universities requires exploring appropriate paths in the light of the current era environment.

1 The Connotation and Characteristics of the Data Warehouse

The data warehouse is produced in order to further mine the data resources and make decisions under the condition that there are a large number of the databases [2]. It is not a so-called "large database". The purpose of the data warehouse project construction is to provide the basis for the front-end query and analysis. Because of the large redundancy, the storage required is also large.

© Springer Nature Singapore Pte Ltd. 2020
M. Atiquzzaman et al. (Eds.): BDCPS 2019, AISC 1117, pp. 1573–1579, 2020.
https://doi.org/10.1007/978-981-15-2568-1_219

1.1 The Connotation of the Data Warehouse

Data warehouse can be abbreviated as DW or DWH. Data warehouse is a strategic set that provides all types of the data support for the decision-making process at all levels of an enterprise [3]. It is a single data store created for the analytical reporting and decision support purposes. To provide guidance for the business process improvement, monitoring time, cost, quality and control for enterprises requiring business intelligence.

1.2 Characteristics of the Data Warehouse

In order to better serve the front-end applications, data warehouse often has the following characteristics: high efficiency. Data warehouse analysis data are generally divided into days, weeks, months, seasons and years and so on [4]. It can be seen that the efficiency of the daily data cycle is the highest, requiring 24 h or even 12 h, and customers can see yesterday's data analysis. Because of the large amount of data in some enterprises every day, the data warehouse which is not well designed often has problems. It is obviously not feasible to delay 1–3 days to give the data [5]. All kinds of the information provided by the data warehouse must have the accurate data, but because the data warehouse process is usually divided into many steps, including the data cleaning, loading, querying and displaying and so on, the complex architecture will be more multi-level, so because the data source has dirty data or the code is not rigorous, it can lead to the data distortion. The false information may lead to the wrong decision-making and the loss rather than benefits. The reason why some large-scale data warehouse system architecture design is complex is that considering the scalability of the next 3–5 years, in this way, the future can be very stable without spending too much money to rebuild the data warehouse system. Mainly reflected in the rationality of the data modeling, the data warehouse solution has some more intermediate layers, so that the massive data stream has enough buffer, so as not to have a lot of data and cannot run. The data warehouse technology can wake up the data accumulated by enterprises for many years, not only to manage these massive data well, but also to excavate the potential value of the data, thus becoming one of the highlights of the operation and maintenance system.

2 Outstanding Problems in the Process of the Transforming Scientific Research Achievements in Applied Universities

2.1 The Imperfect System of the Transformation of the Scientific Research Achievements Frustrates Teachers' Enthusiasms for the Transformation of Their Scientific Research Achievements

More than 75% of applied universities in China still use the quantity and quality of the scientific research projects and papers as the main reference indicators. Under this guidance, teachers conduct the scientific researches, do projects and write papers only to prepare for the evaluation of their professional titles. As long as the projects are finalized or appraised, the whole scientific research work came to an end. Teachers will not deeply consider whether the scientific research results can be commercialized and industrialized, and nor will they care whether they can produce economic benefits and social benefits. Therefore, teachers always attach more importance to topics and papers but less to applications and transformation. More than 50% of the scientific research management departments of applied universities in China do not have a set of the scientific evaluation criteria for the transformation of the research results. Statistics and awards of scientific research results at the end of the year only depend on papers and projects. Whether or not the transformation of the scientific research results cannot be regarded as the scope of the assessment, and there is no matching reward method for the transformation of the scientific research results. In the long run, the hard-earned scientific research achievements can only be put on the shelf, which greatly frustrates the enthusiasms of teachers in the transformation of their scientific research achievements.

2.2 Scientific Research Results Are Out of Touch with the Market Demand and There Is a Lack of Adaptability

At present, nearly 80% of the teachers in applied universities in China are from graduated universities to the employment universities. After years of training, their theoretical knowledge is gradually enriched. However, most professional teachers have no industry experience, with the lack of practical skills, and the lack of recognition of enterprises, new technologies, new materials and other application levels, so that the scientific research results are out of the market demand and there is a lack of market adaptability. In addition, few market personnel and professional engineers and technicians are involved in the university research team, and even fewer leaders of technological innovation. Over the years, many subjects in applied universities have been directly assigned by higher authorities or selected from literature. They are theoretical, academic, with inadequate pertinence, and there is the serious phenomenon of the closed-door car-making. The product design has not been benign to the local enterprises and markets. More than 90% of the research results remain in the laboratory stage or small trial stage. Due to the lack of the interdisciplinary collaboration, the comprehensive development capacity is obviously insufficient, and the technical matching is poor, and the maturity and applicability are not strong.

2.3 It Is Difficult for Scientific Research Achievements to Fall to the Ground Because of Less Investment in the Transformation of Their Scientific Research Achievements

The 211 and 985 series of universities in China have relatively sufficient funds for scientific researches, while other public universities have relatively tight funds for scientific researches. The funds of applied universities are more used for the school development, the specialty construction and the curriculum construction. The survey found that private colleges and universities in applied universities are self-financed, and their research funds are even more stretched. The transformation of scientific and technological achievements in scientific researches requires a lot of financial support.

Generally speaking, there are four stages in the transformation of scientific and technological achievements from R & D to the product delivery. The first stage is the laboratory stage, and at present, the support of the scientific research funds in most schools is basically enough to complete the first stage of the experimental expenditure. The second stage is the pilot stage. There are about 2,300 horizontal projects in Tsinghua University every year, with a research fund of more than one billion Yuan, but less money is invested in the pilot test. The survey results show that nearly 60% of the applied universities in China are often forced by the financial pressure to simplify the steps or even omit them. The third stage is the production stage of enterprises. And without the basic support of the second stage, the third stage often presents unexpected problems, which restrict the production. The fourth stage is the stage of putting them into the market. Because of the lack of the professional integration of applied university teachers, many of them do not understand the market and know nothing about the investment and financing, which leads to good product creativity not being accepted by the market.

2.4   Information Asymmetry in the Transformation of Scientific Research Achievements and the Lack of Information Exchange Platform for the Transformation of Scientific Research Achievements

Almost everyone of university teachers do scientific researches and do projects every year, but less than 10% of them can really transform their scientific research achievements. There exists the information asymmetry in the transformation of scientific research achievements, which is mainly the information asymmetry between the suppliers and the demanders. Nearly 94% of teachers often stop applying for patents. The disconnection between university teachers and the society and the enterprises often leads to the lack of access to the promotion and application of the patented technology. Every year, the number of the high-tech enterprises in China is constantly increasing. These enterprises need a lot of new technologies. However, enterprises do not know what valuable scientific and technological achievements are in university laboratories. Therefore, more than 90% of the scientific research achievements can only be left in laboratories, and the market transformation cannot be achieved. At present, there are few mature scientific research and technology intermediaries in China. Even if there is, then realizing the docking between the supply and the demand through intermediary will invisibly increase the cost of the communication between the two sides.

3   Simulation Research on the Effective Transformation Path of the Scientific Research Achievements in Applied Universities Based on the Data Warehouse

The scientific research achievements of applied universities should first have the market adaptability, and then solve the supply-demand docking problem through the interactive platform of the network information. In addition, the scientific research team of universities should not only have technical talents, but should also cultivate comprehensive talents who are familiar with production, management, finance, market and other important factors. Only in this way can we make a scientific judgment and reasonable evaluation on the transformation of their achievements. Research universities are the center of the output of scientific and technological achievements, and the benefits of the transformation of scientific and technological achievements expand the sources of the funding for research universities, and the transformation of scientific and technological achievements is closely related to the three functions of research universities.

3.1 The Characteristics of the Transformation of the Scientific and Technological Achievements in Research Universities

With the heavy task of the transformation, the diversity of the transformation modes, the interaction of the transformation process, the systematicness of the transformation work, the remarkable economic and social benefits, and the irreplaceability of meeting the country's major strategic needs, the factors affecting the transformation of the scientific and technological achievements in research universities in China are: the cultural factors at the macro level, the regional economic vitality factors, the institutional factors, the legal and policy factors, factors of the scientific and technological achievements at the micro level. The talent team needed for the transformation of scientific and technological achievements cannot meet the needs, and the development of the scientific and technological intermediaries is imperfect, and the development of the scientific and technological intermediaries is medium. The technology market demand of small enterprises is insufficient, and the funds needed for the transformation of the scientific and technological achievements are insufficient.

3.2 Research on the Policy of the Effective Transformation Path of the Scientific Research Achievements in Applied Universities Based on the Data Warehouse

Data warehouse is a structured data environment for the decision support system (DSS) and the on-line analysis application data source. Data warehouse studies and solves the problem of obtaining information from database. Data warehouse is characterized by the subject-orientation, integration, stability and time-varying. Generally speaking, the decision support system based on the data warehouse consists of three parts: the data warehouse technology, the on-line analytical processing technology and the data mining technology. The data warehouse technology is the core of the system. In the following articles in this series, the main technologies of the modern data warehouse will be introduced around the data warehouse technology. The main steps of the data processing are discussed. How to use these technologies in the communication operation and the maintenance system is helpful for the operation and maintenance.

The data organization of the operational database is oriented to the transaction processing task, and each business system is separated, while the data in the data warehouse is organized according to a certain subject area. The theme is an abstract concept corresponding to the application-orientation of the traditional database, which integrates, classifies and analyses the data in the enterprise information system at a higher level. Each topic corresponds to a macro analysis area. Data warehouse eliminates the data that is not useful for the decision-making and provides concise views of specific topics.

To promote the transformation of the scientific and technological achievements of research universities in China, policies and countermeasures should be put forward from the three levels of the government, research universities and enterprises. The government functional departments should reduce the examination and approval items and links, simplify the examination and approval process, establish a coordinated policy system, accelerate the establishment and improvement of the technology market, actively cultivate the science and technology intermediaries, and increase the financial, tax and economic policies to support the transformation of the scientific and technological achievements. Leaders should attach importance to the transformation of the scientific and technological achievements, and formulate the detailed rules and regulations and implement them in practice. When setting up the scientific and technological projects, they should consider their future application and the market factors to ensure the transformation of the scientific and technological achievements from the source. They should fully understand the important role of the university science and technology parks in the development of universities and build the large-scale science and technology parks. The establishment and development should be integrated into the overall construction and development plans of the schools, and the construction of talents for scientific and technological achievement management should be strengthened. Enterprises should have a long-term strategic vision, and their leaders should have a keen sense of high and new technologies, and they should attach importance to the construction of their own scientific and technological teams and strengthen their abilities to absorb and create the high and new technologies. They should increase investment in scientific and technological development funds and guarantee the special funds, and they should establish incentive mechanisms, organizational structures and organizational cultures to promote the transformation of scientific and technological achievements. Therefore, improving the interoperability and reusability of modeling and simulation is the core of the current research on the modeling and simulation technology, and realizing the sharing and reuse of the simulation resources is one of the most important tasks.

4 Conclusion

The scientific research management departments of applied universities should, in accordance with the state regulations and the actual situations of the universities, formulate a scientific and reasonable system of the income distribution and reward for the transformation of scientific research achievements on the premise of listening to the suggestions of the people who have completed the achievements and the departments that have transformed the achievements, and publicize it publicly within the scope of the schools so as to enable the teachers involved in the scientific researches to fully understand and correctly interpret the knowledge value-oriented income distribution policy, to fully mobilize the enthusiasms of the scientific research personnel in the transformation of scientific research results under the guidance of this policy.

Acknowledgement. This paper is the result of the phased research of the soft science project of Henan Science and Technology Department in 2018, "Research on the innovation model of scientific research achievements of universities based on the needs of small and medium-sized enterprises in Henan" (project number: 182400410139, project host: Sun Jinli).

References

1. Yue, H., Li, S., Ren, Z.: Significance and development countermeasures of the transformation of scientific and technological achievements in colleges and universities for college students' entrepreneurship. J. Agric. Univ. Hebei: Agric. For. Educ. Ed. (3), 1–3, 6 (2012)
2. Fu, B.: Research on the promotion and transformation of the educational and teaching achievements in applied universities. Course Educ. Res.: Res. Learn. Law Teach. Method (24), 11 (2015)
3. Bai, Y., Wang, Y., Qu, B., Pan, G., Hu, T.: Strategic research on the promotion and transformation of independent innovation achievements in the new period. Manag. Res. Sci. Technol. Achiev. (9), 17–19 (2012)
4. Zhong, W.: Performance evaluation of industry-university-research cooperation of research universities in China based on the data representation. Sci. Technol. Prog. Policy (14), 118–123 (2016)
5. Liu, J., Liu, Y., Hao, X.: Feasibility study on the transformation of scientific and technological innovation achievements of college students. China Educ. Innov. Her. (7), 4–5 (2014)

Ways of the Computer Network Serving the Spiritual and Cultural Life of Migrant Workers

Congcong Wang[✉]

College of Biology and Food Engineering,
Fuyang Normal University, Fuyang 236000, Anhui, China
weixiong619@126.com

Abstract. The 21st century has entered the information society. With the popularization of the computers and the networks, people's production and life are undergoing earth-shaking changes, and has brought great convenience. Entering the network society, the computer network penetrates into all aspects of people's life and changes the rhythm of people's life. From the online games to the online classes, from the online shopping to the online stock speculation, all can be completed through computers. The deep-seated impact of the Internet on people's languages, values, behavioral patterns and cultures is also increasing. At present, the spiritual and cultural life of migrant workers is relatively low-key. Both the projects they can participate in and their consumption capacities are affecting their daily consumption contents. Considering the background of the information age, giving full play to the advantages of the computer network will provide important help for the rich contents and the system construction of the spiritual and cultural life of migrant workers.

Keywords: Computer · Network services · Migrant workers · Spiritual and cultural life · Specific ways

Since the beginning of the 21st century, Chinese academia has formally studied the cultural life of migrant workers. At present, the most frequently used words to describe this problem are "islanding", "desertification", "marginalization", "passing passengers", "absence" and "scarcity". Some scholars focus on the negative impact of the lack of spiritual and cultural life of migrant workers from the perspective of the social stability through investigations and researches and other empirical analysis [1]. Some scholars argue from the angle of the cultural equity and the social harmony that the protection of the cultural life of the migrant workers by the governments and the society lags behind the urbanization process of our country [2]. Others analyze the reasons for the lack of spiritual and cultural life of migrant workers from the perspective of migrant workers themselves. For example, the local cultures and the small-scale peasant consciousness and so on restrain the cultural consumption consciousness of migrant workers, so as to restrain the development of their spiritual culture [3].

© Springer Nature Singapore Pte Ltd. 2020
M. Atiquzzaman et al. (Eds.): BDCPS 2019, AISC 1117, pp. 1580–1586, 2020.
https://doi.org/10.1007/978-981-15-2568-1_220

1 Analysis of the Current Situation of the Spiritual and Cultural Life of Migrant Workers

At present, the number of migrant workers in China has reached 200 million. While they make contributions to the urban construction, they also bring their traditional virtues of simplicity, kindness, diligence and bravery to the cities [4]. They don't pocket the money they have picked up and they are willing to help others. They are good to their neighbors and they are brave to do justice. They are honest, trustworthy and self-sacrificing. They have become models of the urban civilization and the social civilization. The migrant workers' educational level has generally improved, and their ideological concept is more positive. Their consciousness of rights is constantly increasing, and they can correctly understand their identity and social status [5]. They have lofty ideals, attach importance to the dual needs of materials and spirits, and attach importance to their career development prospects, and they are no longer satisfied with being the pure physical front-line worker. Although the spiritual and cultural life of migrant workers has been improved compared with the past, there is still a big gap between them and the urban citizens in terms of the activity contents, ideas and beliefs, social psychologies and emotional life.

1.1   The Activity Contents Are Monotonous, Poor and Vulgar

At present, the spiritual and cultural life of migrant workers is almost out of touch with the developed urban civilization. A survey of the leisure time allocation of migrant workers shows that the top six leisure ways are playing cards, watching TV, listening to the radio, chatting (35%), shopping, and reading books and newspapers, and other leisure ways are sleeping, doing housework, going to the dance hall and looking for fellow townsmen.

1.2   Conservative and Backward Ideology and Weak Legal Consciousness

As soon as they enter the cities, most of the migrant workers hold a wait-and-see attitude towards the new life in the cities. They often reject and resist the new things and the fast rhythm of the city life. In addition, due to the small-scale peasants' thought for thousands of years and the inherent selfish and narrow mentality of the peasant class, some of them excessively pursue the material interests and they take food and clothing as the highest goal, neglecting the spiritual and cultural life. In addition, many migrant workers know little about the law. When their own interests are infringed, they often console themselves with the mentality of "loss is happiness", endure silently, and do not resist defending their rights. Some migrant workers use extreme ways to defend their rights, such as suicide, self-mutilation and kidnapping their bosses, which violate the legitimate rights and interests of others and disrupt the social order. Seriously, it also constitutes a crime and they ruin their own future.

1.3   There Is the Inferiority Complex, with the Dissatisfaction with the Society

Because of the differences in the languages, the living habits and the behavioral patterns, the group of migrant workers unconsciously classifies themselves and the urban residents as people of two worlds. In addition, some urban residents alienate migrant workers in their lives and suspect their moral qualities. They regard migrant workers as objects of prevention, which puts more emphasis on the original self-reliance of migrant workers. The inferiority and autism make them confine themselves to the social circles based on their professions, geography and kinship relationships. They dare not do the recreational activities which only the urban residents seem qualified to do, and they cannot integrate into the urban life well. The resulting loneliness will breed discontent and resentment after being depressed for a long time, and some even evolve into the social hatred, and then retaliate against the society.

2   Summary of the Background of the Computer Network Serving the Spiritual and Cultural Life of Migrant Workers

Strengthen the leadership and recognize the importance of strengthening the cultural construction of migrant workers from a strategic perspective. Historically, the floating population has been a weak point in the social management and an important factor in the social stability. Improve the working mechanism to lay a mechanism guarantee for the cultural construction of the migrant workers. Through integrating the functions of propaganda, culture, education, trade unions, health, labor security, civil affairs and other departments, we should coordinate the establishment of the functional institutions for the cultural construction of migrant workers at all levels, and clarify the responsibilities of the cultural construction of migrant workers.

We will increase the infrastructure construction. The construction of the cultural services for migrant workers should be included in the scope of the public financial security, and the cultural facilities should be planned scientifically and comprehensively. At the same time, under the leadership of the propaganda and cultural department, according to the specific situations of various cities, the key project bank of the cultural infrastructure construction is established, and the related projects of the cultural infrastructure construction of migrant workers are included in the key project bank, and the funds are consciously invested in the grass-roots communities and the industrial areas where the migrant workers are concentrated. Construct the community cultural service platform. We should build a community-based public cultural service platform for migrant workers, incorporate the cultural needs of migrant workers into the community planning, and improve their right to use the community cultural facilities. In the areas where migrant workers are relatively concentrated, we should explore a new mode of the public cultural service, which integrates the social and urban cultures, instead of the original "campus" management mode of industrial zones.

Produce and supply the cultural products and services that migrant workers are willing to buy, afford and use. Guide the market to "tailor-made" appropriate cultural products for the vast group of migrant workers, to produce more cultural products that migrant workers can afford, understand and use, and to organize more cultural and sports activities that migrant workers are willing to participate in. Further purify the cultural market. The cultural administrative departments at all levels should enhance their sense of responsibility, intensify the inspection of the cultural markets in the rural migrant workers' gathering areas, severely crack down on the illegal Internet cafes, entertainment places and illegal lottery spots, and intensify the purification of the cultural living environment of the rural migrant workers. Strengthen the research and exploration of the cultural needs of the migrant workers. Relying on the research institutes, colleges and universities and other related research forces, we carried out the in-depth social and cultural surveys on the migrant workers, deeply analyzed the lifestyle, the living conditions, the psychological awareness, the ideological perception and values of migrant workers, and found out the characteristics of the cultural needs of migrant workers.

3 The Possibility of the Computer Network Serving the Spiritual and Cultural Life of Migrant Workers

The computer network service is a new concept gradually derived with the development of the network. It is a very important subject to scientifically define the concept of the computer network service, objectively understand the characteristics of the computer network service and its function and role in the spiritual and cultural life of migrant workers.

3.1 The Meaning of the Computer Network Service

Culture can be divided into four levels: knowledge, emotion, ethics and belief. Knowledge and emotion belong to the category of the spiritual culture. It is a dynamic concept which is constantly enriched and accumulated in practice. The computer network service is literally understood as the spiritual culture in the network virtual environment. It refers to the social psychology and social ideology based on the network virtual environment, through various cultural resources, cultural environment and cultural exchanges externalized through the network, including all spiritual life and spiritual production process in the network environment. It includes two meanings – the network culture and the spiritual culture.

3.2 Basic Characteristics of the Computer Network Services

The characteristics of the computer network services are determined by the network as a cultural carrier. There are three main aspects. Communication and innovation: The computer network service is a kind of the communication culture, with the characteristics of the two-way communication, which makes it richer and more complex. Different consciousnesses or understandings communicate with each other, and different viewpoints collide with each other and produce new viewpoints. This is the communication and innovation of the computer network services. Comprehensiveness: Now the network has become the carrier of all kinds of the cultures in the real life. It spreads and embraces all cultures, and makes them become a blending comprehensive culture. The main manifestations are as follows. First, the degree of the cultural freedom is very high, and all kinds of the cultural information can be disseminated and gathered on the Internet. Secondly, the culture on the Internet is pluralistic. The multicultural competition opens up a broader prospect for the computer network services. Openness: The openness of the computer network service comes from the openness of the network. "The network is an open structure, which can expand indefinitely. As long as it can communicate in the network, that is, as long as it can share the same communication code, it can be integrated into the new nodes. A network-based social structure is a highly dynamic open system that can innovate without threatening its balance. This characteristic of the network enables people to enter any network at will without restriction.

4   Ways of the Computer Network Serving the Spiritual and Cultural Life of Migrant Workers

According to the characteristics of the network and the requirements of the spiritual and cultural life of migrant workers, the cultivation of the computer network services actively develops and protects the excellent ones by means of downloading, transferring, guiding and shielding, while abandoning all kinds of the decadent and backward computer network service products. There are still some deficiencies in the training of the computer network services. The following are some suggestions for strengthening the training of the computer network services:

4.1   Vigorously Develop the Computer Network Services with a Positive Attitude

At present, in order to form a positive interaction between the computer network services and the spiritual and cultural life of migrant workers, the positive changes in the spiritual and cultural life of migrant workers resulting from the development of the computer network services should be explained and developed, so as to make people realize that the active computer network services will inevitably form fairness, justice, freedom, equality and collectivism. At the same time, the correct spiritual and cultural life of migrant workers will inevitably become the soul of the computer network services, thus establishing the computer network services with Chinese characteristics and making them an important part of the advanced culture.

4.2   Strengthening the Absorption of the Active Computer Network Services by Migrant Workers

The spiritual and cultural integration of migrant workers is related to the overall situation of the sustained and healthy development of China's economy and the construction of a harmonious society. We should advocate the humanistic care, and effectively protect the legitimate rights and interests of migrant workers, so that migrant workers can truly benefit, and create a good atmosphere for the entire society to pay attention to the spiritual life of migrant workers. The solution to the spiritual life of migrant workers is not what they can do by themselves, which requires the joint efforts of the entire society to fundamentally solve the problems facing the spiritual and cultural life of migrant workers. Strengthen the awareness of active computer network services. We should strengthen the psychological quality education of peasant workers and use various ways to cultivate their good psychological qualities so as to distinguish the excellent computer network services more accurately. Finally, we should vigorously strengthen the cultivation of the migrant workers' comprehensive qualities and abilities to adapt to the society and enhance their sense of social responsibilities.

4.3   Deepening the Rational Understanding of the Spiritual and Cultural Life of Migrant Workers

Our Party and our country have always attached great importance to the spiritual and cultural life of migrant workers, and the development of the computer network services has an important impact on the spiritual and cultural life of migrant workers. Therefore, we need to conduct pioneering researches and deepen the understandings of these issues in order to make the spiritual and cultural life of migrant workers get scientific and rational support. It lays the foundation for the spiritual and cultural life of migrant workers in the computer network service. The development of the computer network service has brought more opportunities and challenges to the spiritual and cultural life of migrant workers. We should take the active computer network service seriously and combine it with the spiritual and cultural life of migrant workers effectively, so as to promote the overall growth of migrant workers.

5   Conclusion

Migrant workers' migration into cities is not only a spatial migration of the rural population to cities, but also a modern sense of "cultural migration", and a process of the individual transformation from the rural people to the urban people. However, at present, the academic circles pay more attention to the economic and social aspects of migrant workers, but there is a lack of research on their spiritual and cultural needs. We hope to study the problems in the spiritual and cultural life of the new generation of migrant workers through the analysis of the correlation between their educational level and basic characteristics, life experience and social development, so as to help them overcome the "island effect", improve their enthusiasms for participating in the urban construction and jointly promote the construction of a harmonious society. The computer network serves the spiritual and cultural life of migrant workers, which not only adapts to the development trend of the information age, but can also serve the vast number of migrant workers in a more cost-effective way.

Acknowledgement. Project Fund: Key Research Project of Humanities and Social Sciences in Colleges and Universities of Anhui Province in 2019. A Research on the Development of Spiritual and Cultural Life of Migrant Workers in Northern Anhui Province in the Process of New Urbanization (No. SK2019A0311).

References

1. Li, X.: Analysis of the current situations of the spiritual and cultural life of migrant workers and its countermeasures. J. Zhengzhou Univ. Light Ind. (Soc. Sci. Ed.) **14**(4), 46–50 (2013)
2. Lu, D., Huo, X.: The current situation of the spiritual and cultural life of the new generation of migrant workers and the countermeasures and suggestions – a survey from Anding District, Dingxi City, Gansu Province. Tianjin Agric. Sci. **18**(4), 81–84 (2012)
3. Long, Y.: Current situation and countermeasures of the spiritual and cultural life of the new generation of migrant workers. Decis.-Mak. Consult. (2), 70–73 (2012)
4. Wang, X.: Problems in the spiritual and cultural life of the new generation of migrant workers and their countermeasures. Theory Res. (31), 122–124 (2013)
5. Li, W.: Research on the public cultural service of migrant workers in small and medium-sized cities in Western China – taking Shaanxi Province as an example. J. Shaanxi Univ. Technol. (Soc. Sci. Ed.) (3), 89–93 (2016)

Fuzzy-PID Control Technology in Glaze Firing Process Control System of Roller Kiln for Bone China

Wang He[✉]

Tangshan Polytechnic College, Tangshan 063299, Hebei, China
977010177@qq.com

Abstract. The precision of roller kiln temperature control is an important factor to control the lead content in the glaze of color ceramic. This paper studies the temperature control of the glaze firing process in the roller kiln and investigates the application of Fuzzy-PID control technology in the glaze firing process control system of roller kiln for bone china to achieve the "green ceramic" process requirements. Taking the single-chip microcomputer as the control core, based on fuzzy control, the temperature of the electric heating roller kiln is precisely controlled. At the same time, the GPRS network is used to realize real-time data monitoring. Finally, the fuzzy-PID improvement strategy is proposed. Hope that this research can promote the progress of China's color-burning process development.

Keywords: The progress of color-burning process · Roller kiln for bone china · Fuzzy-PID control technology

1 Introduction

In the process of producing bone porcelain, the temperature directly determines the quality of the product. Based on the complexity of temperature control in the roller kiln, it is difficult to accurately and steadily control the temperature, which often leads to excessive lead content in the glaze, although the problem of lead content in the white porcelain glaze has been solved. However, the color picture firing system, due to many factors such as flower paper pigments, the problem of high lead content in the glaze has always been difficult to overcome. The color ceramic glaze has zero lead content, ensuring that the roller kiln maintains a high temperature of 1050 °C and is continuously and accurately fired. This process has extremely high requirements for accuracy, and traditional PID control can not meet this accuracy. Therefore, using Fuzy-PID control technology to control the temperature of bone porcelain roller kiln is of great significance for the "green ceramic" manufacturing with zero lead content.

© Springer Nature Singapore Pte Ltd. 2020
M. Atiquzzaman et al. (Eds.): BDCPS 2019, AISC 1117, pp. 1587–1593, 2020.
https://doi.org/10.1007/978-981-15-2568-1_221

2 PID and Fuzy-PID Temperature Control

In modern industrial kilns during the process of ceramic firing, most enterprises use PID to control the temperature of roller kilns, and the temperature control of industrial kilns involves many factors, such as non-linear factors, variable factors, asymmetric factors, etc., even if PID is used. Based on the signal acquisition and measurement, it is impossible to satisfy the requirement of precise temperature control. Fuzy-PID control technology, using fuzzy control combined with PID control, can meet the requirements of the system control accuracy under different operating conditions. Based on fuzzy control combined with PID, the size deviation is quickly adjusted, and the high temperature and stability of the bone porcelain roller kiln can be well controlled. Solve the problem of high lead content in glazed surface under the color burning process and truly realize the "green ceramic" manufacturing process [1].

3 Application Scheme of Fuzy-PID in Process Control System of Color Firing in Bone Porcelain-Roller Kiln Glaze

3.1 Control System Design Based on Single Chip Computer

Observe Fig. 1, kiln firing stage E is firing temperature deviation, indicating error threshold, Fuzy-PID control mode, such as system judgment, enable Fuzzy control mode, using Fuzy-PID control algorithm to generate corresponding control variables, Control the output voltage of the thyristor output circuit, change the heat output of the high-temperature resistance wire in the roller kiln, and achieve the accurate and stable temperature control of the color burning system. Under the operating conditions, the PID control mode is used. The quantization factor in Fig. 1 contains and, as a proportion factor, Based on the principle of fuzzy control, the temperature reference value selects the temperature corresponding to the reference value of the heat value, the input variable selects the temperature deviation change and the temperature deviation, the output quantity takes the voltage change of the electric heating furnace, and the language variable level. Taking into account the temperature deviation caused by various disturbances such as the change of heat value, set up a basic theory field, set a basic theory field of change, and quantify the temperature deviation domain, and calculate and obtain the quantization factor and the difference, but also for the output language variable quantization domain. In the firing stage, the Ku ratio factor value needs to be adjusted several times. Too large a value may produce overshoot production shocks. If the value is too small, the adjustment time will be extended. After repeated debugging, the Ku value will be determined at 0.0001, and the three variables E, EC, and U will be determined. That is, the temperature deviation, the temperature bias change, and the voltage change of the electric heating furnace are simultaneously converted into the quantization domain change discrete quantities, that is, the three language variables E, EC, and U. Each language variable is divided into 7 files and 7 language values are

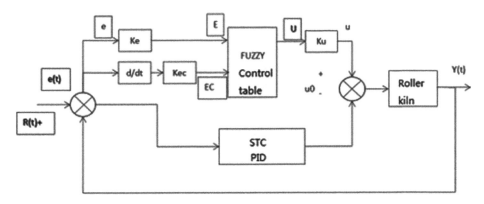

Fig. 1. Principle circuit of measurement and control system for bone porcelain roller kiln with Fuzy-PID

taken [2], that is, PB/PM/PS/ZE/NS/NM/NB, representing Zhengda/mid/small/zero/ negative/negative/negative/large, the 7-file fuzzy subsets are in the same domain as Formula 1 and formula 2 respectively:

$$E = EC = U = \{PB, PM, PS, ZE, NS, NM, NB\} \tag{1}$$

$$E = EC = U = \{-6, -5, -4, -3, -2, -1, 0, 1, 2, 3, 4, 5, 6\} \tag{2}$$

3.2 Establishment of Fuzzy Rules

Fuzy-PID control technology is used in the process control system of color firing in bone porcelain roller kiln glaze. The basic requirements include fast dynamic response speed, and automatically strengthen the proportional effect if the deviation is too large, such as formula 3:

$$IF \ E \text{iand } EC \ i \ then \ U_{ij}(i = 1, 2, A, m; \ j = 1, 2, A, n) \tag{3}$$

In order to meet the requirements, it is necessary to adjust the temperature deviation weight, and if the deviation is large, the weight value will increase. A fuzzy controller containing multiple adjustable factors is used, such as formula 4:

$$J = -[\alpha E + (1 - \alpha)E \ C] \tag{4}$$

Formula 4, [ ] For consolidation, it represents the adjustment factor under the controller, and the deviation is represented by E. Fuzy's specific control rule is as follows:

$$U = \begin{cases} -[0.9E + 0.1EC] \cdots E = \pm 6 \\ -[0.8E + 0.2EC] \cdots E = \pm 5 \\ -[0.7E + 0.3EC] \cdots E = \pm 4 \\ -[0.6E + 0.4EC] \cdots E = \pm 3 \\ -[0.5E + 0.5EC] \cdots E = \pm 2 \\ -[0.45E + 0.55EC] \cdots E = \pm 1, E = 0 \end{cases} \tag{5}$$

Follow formula 5 control rules to repeatedly debug and revise to obtain the actual query table.

3.3   Transmission Letter of Temperature Control System of Electric Heating Roller Kiln

Multiply the exact values of E and EC with their respective quantization factors, IE multiply by the same acquisition. E represents the fuzzy quantization value of temperature deviation in the quantization of temperature deviation, and at the same time obtains the fuzzy quantization value in the quantization of temperature variation. The query control is based on the rules, and the fuzzy variable quantization domain quantization value U can be obtained, and then the value is multiplied by the proportional factor to obtain the exact value of the output control amount. In the firing stage, once the temperature control is performed with the PID, the output is incremental PID control. Such as formula 6:

$$u_0(n) = K_p[e(n) - e(n-1)] + K_1 e(n) + K_D[e(n) - 2e(n-1) + e(n-2)] = K_p \Delta e(n) + K_D[\Delta e(n) + \Delta e(n-1)] \tag{6}$$

Form 6, reduce the deviation by the proportional coefficient, achieve a small overtone in the system response, eliminate static errors by the integral coefficient, improve the steady-state performance of the system, use the differential coefficient to reflect the error trend to reduce the adjustment consumption time, and realize the system's rapid response, system operation phase, The PID differential link only causes a shadow of the system deviation EC. Its role is to suppress the deviation from any direction in the response phase, brake the deviation in advance, and reduce the overtone in advance. In Type 6, it is an incremental PID control calculation [3].

In order to demonstrate the effectiveness of Fuzy-PID control technology on the temperature control of bone porcelain roller kiln, using Simulink as a tool, the mathematical model simulation calculation was carried out. The Fuzy-PID control algorithm was used to generate corresponding control variables and control the output voltage of the thyristor output circuit. Therefore, the calorific value of high temperature resistance wire in roller kiln firing is changed. The valve value is 5 °C. In the simulated bone porcelain roller kiln firing color picture stage, if the temperature deviation is > 5 °C, the temperature control is achieved in the Fuzy fuzzy control mode. If the temperature deviation is 5 °C, the temperature control is performed with the conventional PID., Simulink simulation Fig. 2 is the normal PID temperature control curve, and Fig. 3 is the Fuzy-PID temperature control curve:

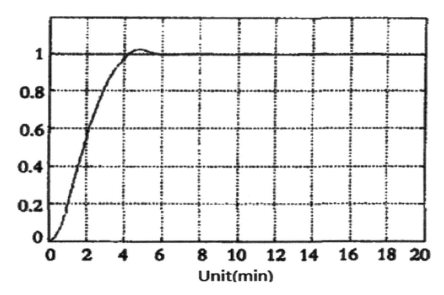

Fig. 2. General PID quality control curve

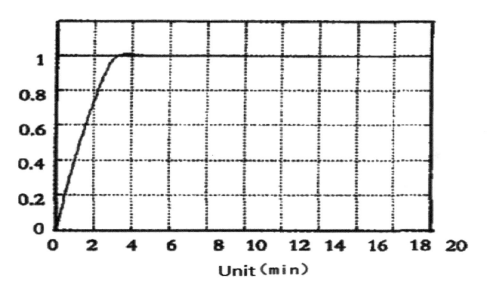

Fig. 3. Fuzy-PID temperature control curve

The observation of the conventional PID temperature control curve and the Fuzy-PID temperature control curve can be demonstrated. The conventional PID carries out the temperature control phase of the bone porcelain roller, and the control overshoot is 25 °C. The overshoot is only 12 °C under the Fuzy-PID temperature control mode based on fuzzy control. And the response speed has been improved, so the steady state accuracy has been effectively improved to meet the requirements of steady state accuracy of the color firing process temperature control in the bone porcelain roller kiln glaze.

4   GPRS Bone Porcelain Roller Track Remote Monitoring

Bone porcelain roller kiln glaze in the color burning process stage, the temperature control accuracy requirements are extremely high, effective real-time monitoring is to ensure that the system will not make mistakes, ensure the quality of products, in the design of the Fuzy-PID control system stage, the use of a single chip Smart display control instrument, The device has its own correction function. Based on the thermocouple temperature data collection, GPRS is used as the temperature data transmission method. GPRS is used to transmit the temperature data of the bone porcelain roller kiln collected by the thermocouple to the Central station platform in real time. After the platform has processed the data, Then use the mobile network to transmit to the cloud computing platform for storage, and provide users with real-time remote data acquisition, to achieve real-time management monitoring of bone porcelain roller kiln temperature [4] And ... The remote monitoring function structure of GPRS bone porcelain roller is shown in Fig. 4:

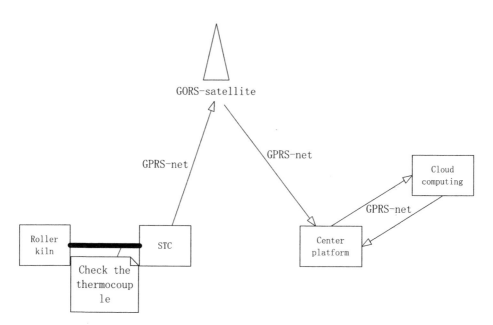

Fig. 4. Functional structure of GPRS remote monitoring system for bone china roller

5   Fuzzy-PID Improvement Strategy

The Fuzzy-PID temperature control method proposed in this study has strong robustness, good dynamic response performance and small overshoot. However, the fuzzy control has the disadvantage of poor steady-state. Because of the existence of deviation rate, the control form of PID or Fuzzy-PID algorithm based on deviation lacks a certain steady-state. Therefore, the temperature basis of Fuzzy-PID control for skeletal porcelain roller kiln is lacking. In this paper, a kind of Fuzzy-PID compound control form is proposed. Under the temperature control system of bone china roller kiln, the PID is used as the main temperature controller of the system, and the output of the fuzzy control Fuzzy controller is used as the proportional coefficient of the output of the PID control. After confirming the control form and control rules, the fast response of the Fuzzy is realized based on the fuzzy control rules with self-adjusting factors, and the work of the PID control is exerted steadily. Yes, good avoidance of defects.

Conclusion: In order to control the temperature of the roller kiln of Bone China accurately and reduce the lead content of the glaze of color ceramics effectively, a Fuzzy-PID temperature control strategy is proposed in this study. The temperature deviation is the standard. When the firing temperature deviation is greater than a certain threshold, the Fuzzy control form is automatically activated. If the temperature deviation is smaller than the threshold, the PID calculation method is automatically adopted. The simulation curve of Simulink demonstrates the fuzzy-PID control method. The application value of the method and the real-time monitoring of the temperature of bone china roller kiln based on GPRS are introduced. Combining with the disadvantage of poor steady-state of the fuzzy control, a composite adaptive improvement strategy of Fuzzy-PID is proposed, which can effectively improve the stability of the temperature control of bone china roller kiln.

Acknowledgment. Study on the application of Fuzy-PID control technology in the process control system of color firing in the glaze of bone porcelain roller kiln; Project number: 17110231a; School Zip Code: 063299.

References

1. Zhang, W., Chen, L.: Research on temperature control of electric heating furnace based on fuzzy-PID. Innov. Sci. Technol. (10), 81–82 (2014)
2. Liu, Y., Xi, X., Yan, P.: Design of electric boiler control system based on fuzzy PID control. Autom. Instrum. (10), 59–60 (2016)
3. Xiong, Y., Fang, K.: Design of temperature Fuzzy-PID controller for variable water volume air conditioning based on fuzzy rule switching. Comput. Meas. Control (2), 436–438 (2015)
4. Zhang, M., Zhao, B., Zhou, H.: Heat treatment temperature control of forging spindle based on fuzzy-PID. Foundry Technol. (10), 2562–2564 (2015)

Construction Path of the National Educational Enrollment Examination Cadre Database Based on the Big Data Analysis

Zhongxue Wang[✉]

Henan Provincial Admissions Office, Zhengzhou 450064, Henan, China
2593792052@qq.com

Abstract. Based on the computer system and the modern information technology, the big data mining and analysis technology extracts the most valuable information from the massive data, which provides an effective basis for the decision-making of the information demanders. The mining technology is a common technology in the modern data analysis. It is often used in combination with the genetic algorithm in the field of the computer research. The data mining is more scientific and reasonable, which provides the technical support for the construction of cadres in the national education enrollment examinations.

Keywords: Big data analysis · National educational enrollment examination cadres · Database · Construction path

Establishing a cadre database of the national education entrance examinations, and especially an innovative database with the reasonable structure and the complete information, is an important guarantee for building a large-scale, well-structured, scientifically developed and internationally competitive personnel team, and it is of great significance for improving the working abilities of the cadre team of the national education enrollment examinations.

1 Construction Background of the National Educational Enrollment Examination Cadre Database Based on the Big Data Analysis

With the progress of our society and the development of the science and technology, the amount of the data people get has increased dramatically. Because of the huge amount of the data, the data transfer and processing become very difficult, so the new technology is needed to manage the data.

1.1 Overview of the Big Data

The development of the big data era mainly shows the following basic characteristics: the large-scale, which mainly refers to the relatively large amount of the index data information, which generally involves all the data information of the application object,

M. Atiquzzaman et al. (Eds.): BDCPS 2019, AISC 1117, pp. 1594–1599, 2020.
https://doi.org/10.1007/978-981-15-2568-1_222

with obvious integrity. Diversity: The development of the big data era also shows the diversity of the data information, which involves many types of the data information, and the forms of the data information are also diverse, and especially the unstructured data, which accounts for an increasing proportion at present. Rapidness: The development and application of the current era of the big data also show a relatively obvious high-speed requirement. Although the amount of the data information it needs to process is large and relatively complex, it also faces the higher speed requirements, which require the strong application effectiveness. High quality: As far as the development of the era of the big data, because it involves a large amount of the data information, it will inevitably show a relatively obvious low quality density, but this does not mean that there are no high quality requirements for the application of the big data related technologies. On the contrary, in order to ensure that the application of the big data is more efficient and appropriate, it is also necessary to focus on the quality of strict checks to ensure that it can play a strong role in the decision-making process.

1.2   Data Mining Technology

As a new data processing method, the data mining technology has shown great value in the development of the current big data era. It can better realize the effective analysis of a large number of the complex data and extract some useful high-quality data, so it can be adapted to the data processing with the low-quality density. From the point of view of the specific application of the data mining technology, it shows obvious cyclic characteristics. It can make efficient use of the data information and deal with it pertinently according to different data application requirements. Finally, it can achieve the ideal benefits. With the development of the current AI technology, the data mining technology also presents many new application characteristics, and especially the application of the neural network and the genetic algorithm, which embodies the strong practical benefits and has been applied ideally in many fields. The application of the data mining technology can work well on the relational databases, the text data sources, the temporal databases, the data warehouses and the heterogeneous databases.

1.3   The Big Data Analysis Technology

The big data analysis refers to the analysis of the large-scale data. Big data can be summarized as five Vs of volume, velocity, variety, value, and veracity. As the hottest IT vocabulary nowadays, the use of the big data, such as the data warehouse, the data security, the data analysis, and the data mining and so on, around the commercial value of the big data has gradually become the focus of profits that industrial people are competing for. In the big data analysis of the asynchronous processing, the process of capturing, storing and analyzing is followed. In the process, the data is acquired by the sensors, the web servers, the sales terminals, and the mobile devices and so on, and then stored on the corresponding devices, and then analyzed. Because these types of the analyses are carried out through the traditional relational database management system (RDBMS), the data forms need to be transformed into the structural types that RDBMS can use, such as rows or columns, and need to be connected with other data. The process of the processing is called extraction, transfer, loading or ETL. Firstly, the data is extracted and processed from the source system, and then the data is standardized and sent to the corresponding data warehouse for further analysis. In the traditional database environment, this ETL step is relatively direct, because the object of the analysis is often known as the financial reports, sales or market reports, and enterprise resource planning and so on.

2 Requirements for the Construction of the National Educational Enrollment Examination Cadre Database

According to the new requirements of the National Education Congress, the personnel of the admission system should base themselves on the overall situation, pay attention to the leading role of the educational reform, grasp the primary problems of the current educational reform, analyze the specific problems, and further promote the reform of the admission system of the college entrance examinations. The cadres of the education and recruitment system should strengthen their own capacity building and strictly require themselves. Adhering to the general tone of steadily striving for progress, adhering to the new development concept, earnestly implementing the deployment of the personnel and social work conferences, taking the examination safety as the core, taking the prevention and resolution of the personnel examination risks as the grasp, adhering to problem-orientation, and building a professional team of the admission examination with political reliability, strong thinking, excellent business and serving the public are the important guarantee to promote the scientific development of the enrollment and examination undertakings.

First, we should strengthen the training and coaching. Ideologically, we should firmly abandon the narrow notion that the recruitment examination work does not require the training and is naturally familiar with it for a long time. Standing at the height of maintaining the educational equity, we should incorporate the training of the recruitment team into the cadre training plan of the education system, and carry out various types of the training at all levels in a planned, hierarchical and multi-form manner. We should strengthen the study of the political theories, insist on doing "virtual" and "real" work, learn to understand the new deployment and requirements of the Party committees and the governments at all levels on the education, do "understanding" in politics, and improve the ability of recruiters to grasp and deal with the problems politically and in the overall situation.

Second, we should improve and perfect the working system. The core of strengthening the construction of the cadre contingent in the enrollment examinations is to exert the rigid restraint function of the system. It is necessary to establish and improve the responsibility system of the post objective management, clarify the duties, requirements and standards of each functional department and its staff, monitor and evaluate the progress and the quality of the work in the entire process, and use it as an important basis for the evaluation, selection and promotion. At the same time, we should improve various working mechanisms, establish a standardized management system in the aspects of the candidates' registration, examination, admission, information management, and financial and logistics management, and strengthen the implementation of the system so as to ensure that everyone has tasks, responsibilities, requirements and systems in all aspects.

Third, we should improve our work style. We should take the implementation of the Party's mass line education reform and the ongoing "three stringent and three solid" thematic education as an opportunity to strive for the people, pragmatism and honesty. We need to improve the service workflow, ensure that the candidates and the masses have channels to respond to questions and have places to consult. We must conscientiously "answer with questions, answer with letters, answer with complaints, and investigation with mistakes".

Fourth, we should enhance the vitality of the cadres. We should strengthen the construction of the leading group of the admission examination center, select cadres who are familiar with the education, have certain level of the policies and theories, adhere to principles, deal with affairs decently and have strong sense of responsibility to serve in the admission examination institutions, so as to avoid the embarrassment and discomfort of "layman leading professionals".

3 Construction Path of the National Educational Enrollment Examination Cadre Database Based on the Big Data Analysis

With the advent of the era of the big data, the big data analysis emerges as the times require. When the analysis begins, the data is first extracted from the data warehouse and put into RDBMS to generate the required reports or support the corresponding business intelligence applications. In the process of the big data analysis, the bare data and the converted data are mostly saved, because it may need to be converted again later. The construction of the cadre database of the national education enrollment examination is an important measure and guarantees to promote the reform of the educational system, and it is also an inevitable experience based on the era of the big data.

3.1 Constructing a Government-Led Platform

From the point of view of the service portal construction and the maintenance subject, the construction of the cadre database of the national education enrollment examinations is a unified management mode. In order to improve the utilization rate of the talents and guarantee the authority, a unified deployment led by the governments should be made. This requires that the main body of the construction should be a non-profit organization as well as a government organization, with the strong information service capabilities and experience, and a wide range of the social recognition. Therefore, it is necessary to choose a national information service institution as the support of the platform construction. To optimize the structure of the recruitment cadres, on the one hand, it is necessary to be "stable". The professional backbone of the recruitment and the examination institutions cannot be changed at will. We should ensure the stability and continuity of the recruitment cadres through the emotional retention, the career retention and the treatment retention. On the other hand, we should actively strive for the attention and support of the competent educational departments and the Party committees at all levels, implement the system of the regular rotation and exchange of the cadres, and combine "weeding out the old and bringing forth the new" with "taking the old and adopting the new", so as to continuously inject fresh vitality into the cause of the enrollment and examinations.

3.2 Perfecting the Service Contents

In terms of the service contents, the existing open databases should be integrated into one platform. Among them, we should focus on integrating the social resources, achieving sharing, avoiding duplication of the construction and saving the social resources. At the same time, it pays attention to the government resources and the social resources, and provides the database services in all fields. Through facing experts, social supervision and candidates, we can realize the greatest value of the construction of the national education recruitment examination cadre database. We should enhance the service awareness, make extensive use of various means to publicize the enrollment policies and regulations, popularize the common sense in examinations, guide all sectors of our society to carry out the public welfare activities to serve the candidates, and timely carry out the consultation and guidance before, after and during the admission period, in order to think what candidates want and what candidates need urgently, and constantly improve our service qualities and working levels. It is necessary to educate and guide the recruiting cadres to build a firm ideological defense line against corruption and change, so as to achieve the self-respect, self-reflection, self-policing and self-discipline, accept the supervision on their own initiatives, and shape a good image of the recruiting cadres.

3.3 Pay Attention to the Stable Operation of the Database System

There are two problems to be paid attention to when constructing the cadre database of the national education entrance examination. The first is to establish a scientific, rational, subjective and objective service evaluation system for the service organizations, to comprehensively evaluate the contents, the retrieval system and its functions of the cadre database, the use of the database and the service of the database, and to establish the common principles, so as to ensure timely supervision of the relevant cadres. Second, attention should be paid to the issue of the information property rights. That is, attention should be paid to the use of the links resources in the case of full access to the licenses and decrees of the commercial service organizations. Broader and flexible access standards should be adopted to make every important service institution present on the platform as equally as possible. We should strengthen the professional training, adhere to the principle of "main responsibility" and "main business", and adopt a combination of "inviting in" and "going out", so that every recruiter can familiarize himself with and master the policy stipulations of the national admission examinations, clarify his duty, abide by the working rules, and be a faithful defender and recruiter of the examinee's interests, and the faithful executor of the examination policies. We should pay attention to the training of the "new recruits". We should formulate specific and feasible training plans for the new recruits, and actively play the role of "transmission and guidance", so that they can become familiar with their posts, environments and policies as soon as possible, and become "specialists", so as to ensure that the recruitment team can have qualified successors.

3.4 Integration of the Database Resources

The unified data norms and standards are the basis of the data exchange and sharing, and the scientific classification is the key to effective development and management. Under the leadership of the special leading group, the talents are scientifically classified. According to their own characteristics, a set of the scientific information classification and coding scheme for the scientific and technological talents is established, and the coding system and the basic standards for the storage are unified. At present, there are some problems in the construction of the national education recruitment examination cadre database, such as the scattered distribution, the repeated construction and the low utilization, which have not yet formed the due scale effect. Taking the field of the national education and enrollment as the catalogue, we integrate the important database resources that have been built in this region into a portal. It not only helps the users to search, select and customize the appropriate database, but also facilitates the cooperation and exchange among the database builders. It also improves the level of construction, saves the social resources and improves the use efficiency of the database.

4 Conclusion

The era of the big data is an important trend in the development of the Internet information technology. Research shows that the amount of the Internet data is increasing year by year, and the characteristics of the "big data" are becoming more and more prominent. In order to ensure the unification of the professional personnel database system structures and the seamless connection of the human resources information managed by various functional departments of the national education recruitment examination cadre database, according to the principle of "unified planning and resource sharing", the information system is constructed by the combination of the separate system and the open platform.

References

1. Wang, T., Shao, G.: Big data analysis based on the cloud computing. Fujian Comput. (7), 26–27 (2014)
2. Chen, Q., Zhang, Y., Chen, C.: Big data analysis in the cloud computing environment. Des. Tech. Posts Telecommun. (5), 1–4 (2015)
3. Wang, S.: Big data analysis based on the cloud computing. J. Jilin Radio Telev. Univ. (7), 37–38 (2015)
4. Wang, W.: Research on the customer loss based on the big data analysis. Mod. State-Own. Enterp. Res. (12) (2015)
5. Li, X.: A survey of the common software tools for the big data analysis. Digit. Technol. Appl. (11), 241 (2015)

The Patrol Inspection Technology for UHV Transmission Lines Based on the UAV Technology

Chengjia Wen(✉), Wenjiang Li, Tie Chen, Jing Chen, and Jun Liu

State Grid Chongqing Electric Power Company Overhaul Branch,
Chongqing 400039, China
xinguo7274209@126.com

Abstract. With the development of the UAV technology, it will become a normal trend to use UAV for the line patrol inspection. However, there are some problems in the UAV patrol inspection, such as the manual control, the inadequate exploitation and utilization of the image data and the extensive data management and so on. To solve these problems, combined with the UHV transmission line project, the methods of the UAV patrol inspection data acquisition under different patrol operation modes are presented. The UAV patrol inspection data are scientifically managed and applied based on the geographic information system.

Keywords: UAV technology · UHV · Transmission line · Patrol inspection technology

With the rapid development of China's economy and the transformation of the energy structures, the construction of the trans-regional transmission projects focusing on UHV has been carried out in an all-round way. This is an important choice for ensuring the national energy security, improving the energy efficiency, serving the development of the clean energy and promoting the construction of the ecological civilization. It can effectively enhance the transfer of the long-distance and large-capacity energy in China, and has a far-reaching impact on transforming the mode of the economic development and adjusting the energy structures.

1 Background Analysis of the Unmanned Aerial Vehicle Technology in the Patrol Inspection of UHV Transmission Lines

Electricity plays an important role in the development of our modern society and is the guarantee of the healthy and stable development of our national economy [1]. With the increase of the coverage area and length of the transmission lines, the pressure of the daily operation and maintenance of the power grid operation companies is constantly increasing, and especially of the construction and operation of the UHV transmission lines, which has brought great challenges to the power grid operation companies. The patrol inspection of the transmission lines is a necessary and basic work to find and

© Springer Nature Singapore Pte Ltd. 2020
M. Atiquzzaman et al. (Eds.): BDCPS 2019, AISC 1117, pp. 1600–1606, 2020.
https://doi.org/10.1007/978-981-15-2568-1_223

eliminate the hidden dangers in time to ensure the safe and reliable operation of the transmission lines [2]. In the long-term operation and maintenance of the transmission lines, various patrol methods are summarized. In addition to the traditional manual patrol inspection, new technologies are actively applied to the line patrol inspection, among which the UAV patrol inspection is a technology widely promoted in recent years, which has achieved remarkable results in improving the quality and efficiency of the patrol inspection.

In recent years, as a new inspection tool, UAV has the advantages of flexibility, low cost, easy to carry and strong environmental adaptability. For example, it is not limited by the terrain and whether the lines are live or not [3]. It is more suitable for the inspection work in the steep mountainous areas and the multi-river landforms. Because of the above advantages, UAV has been widely used in the daily inspection and the fine inspection of the transmission lines, and become an indispensable tool for the inspection personnel in the power transmission and substation.

Even in the context of the construction of the intelligent power grids in recent years, the UAV patrol inspection has begun to introduce more intelligent applications in order to achieve the full-automatic operation and further improve the efficiency [4]. The higher the voltage level of the power line, the greater the interference to UAV in its vicinity will be. The UAV affected by the electromagnetic interference will be unable to obtain the geomagnetic signals, with the remote control/map transmission signal loss and so on in flight. Driven by the science and technology, the application of UAV in the transmission line inspection can greatly improve the efficiency and quality of the transmission line operation, and further enhance its reliability and stability.

2   Problems in the Patrol Inspection of the UHV Transmission Lines

2.1   Waste of Resources

In the operation and maintenance of the modern overhead transmission lines, it is not advisable to repeatedly check a certain problem in order to confirm it. But too many times of the checking will inevitably lead to the greater resource consumption. Therefore, it shows that there is a waste of resources in this mode. Specifically, in the modern overhead transmission lines, when the operation and maintenance system finds a fault in an overhead transmission line, the manual inspection will go to the field, and the inspection scope is very wide, and usually the entire section is checked. However, due to the lack of the in-depth means of the manual inspection, it cannot detect the specific failures in the first few inspections, so repeated inspections will result in the waste of resources. In addition, there are other manifestations of the resource waste. For example, under some conditions of the overhead transmission lines with long service time, the failure rate is relatively high, which brings a burden to the artificial or technical system because of the aging of the equipment and other phenomena.

2.2   Insufficient Working System

With the increasing coverage of the overhead lines in China, the residential electricity consumption is also increasing rapidly, which makes the overhead line operation and maintenance workload increase tens of times. However, the means of operation, maintenance and repair are still very backward. It is impossible to determine the fault lines quickly after the overhead line breaks down. The precise location of the fault location can only be carried out by the traditional manual barrier removal method step by step to find the fault line. Secondly, because the working system is not perfect enough, it is easy to be affected by other external factors when carrying out the live maintenance work. Reflected on the level of the specific work, it will deviate from the expected goals, and the resulting problems will even lead to the expansion of the failure.

2.3   Poor Construction Quality

With the increase of people's demand for the electric power resources and the increase of the electric power engineering quantity, the construction time is more urgent because of the influence of the wide construction area and the large amount of engineering. At the same time, in the process of the construction, many constructors' overall quality is poor, which leads to the inadequate construction. After the construction of the entire line frame, it is prone to the operation and maintenance problems, affecting the effect of the power transmission and the quality of the electricity consumption of residents. The construction problems are manifested in the following aspects. First, the jumper clamp is not tightened, which makes the sag of the line lower and the vertical distance between the conductor and the ground insufficient. Secondly, the area through which the transmission lines pass is seriously affected by trees. The distance between trees and the transmission lines is too small, which increases the hidden danger of the safe operation of the power grid. Thirdly, the construction of the new circuit, such as the iron tower, is widely used, but in the process of the assembly, if there are non-standard problems, it will lead to the decline of its stability, which will affect the safe and stable operation of the lines.

3   Advantages and Disadvantages of the UHV Transmission Line Patrol Inspection Technology Based on the UAV Technology

3.1   Advantages of the Unmanned Aerial Vehicle in the Transmission Line Patrol Inspection

The autonomous navigation flight technology and the autonomous hovering technology are the basic functions of the UAV technology, so they can maintain a certain safety distance with the transmission lines. Their advantages are embodied in the analysis of the transmission line disasters in the disastrous weather, patrolling, the detecting transmission lines, and finding line tripping fault points and so on. The two working modes of the UAV technology are the flight control and the autonomous hovering, but it also flies by itself and manually. In the manual mode, the UAV flies according to the predetermined route under the autonomous navigation, and patrols the lines. When UAV is in the autonomous hovering state, it can take the high-definition pictures of the transmission lines, and carry out the transmission, and the ground station staff can check the fault point through the surveillance videos.

3.2   Disadvantages of UAV

Although UAV plays a very important role in the inspection of the transmission lines, it also has some shortcomings in some aspects. Because of the large size of the UAV and the relatively small battery capacity, the time to cruise with UAV is very limited. In addition, the effective control range that UAV can receive is 1000 m, and the remote control range is also limited, and UAV cannot automatically analyze the image fault points. In the aspect of the power inspection, we should strengthen the supervision of the UAV flight, give the real-time warning of the UAV flight trajectory, give early warnings when UAV is about to fly out of the legitimate airspace, and timely inform the UAV to remind the UAV pilots to change courses.

3.3   Application Trends

The core technology of the UAV operation should be popularized in an all-round way. The UAV patrol system covers 100% of all the transmission lines, and the UAV patrol management and control system of the transmission lines covers 100% of all the operation and maintenance units. We should continuously improve the UAV operation management and the technical standard systems for the transmission lines, improve the UAV sub-module in the operation inspection management and control system of the headquarters, expand the data application center capacity and functions, and carry out the rolling update of the inspection and maintenance operation mode in line with the development trend of the new technology, and achieve the full automatic and intelligent line operation management and control. Accelerate the cultivation of the compound transportation and the inspection personnel. To promote the continuous improvement of the UAV training system, on the one hand, in accordance with the national laws and regulations and the relevant policies, we should promote the basic skill training of UAV, lay the foundation for the legitimate and compliance to carry out the UAV patrol operations. On the other hand, based on the needs of the power grid patrol, deepen the training of the UAV post skills, and effectively improve the comprehensive skill levels of the patrol personnel, which will promote the transformation of the traditional inspection team to the compound inspection team.

4   Application of the Patrol Inspection Technology for UHV Transmission Lines Based on the UAV Technology

4.1   Technical Elements

Special remote control helicopters (UH) and four-rotor unmanned aerial vehicles (M4R) are often used for the patrol inspection of the transmission lines. The components of the UH include: the remote control helicopter body, the vibration absorption suspension device, the image acquisition and transmission equipment, and the ground video surveillance and control system and so on. The backstage staffs mainly use the wireless image transmission equipment to transmit the target video images of UH collected in flight to the transmission lines, so as to make timely judgment on the fault of the transmission lines. The components of M4R include the ground station and the M4R UAV, the wireless image transmission equipment and the wireless micro-high resolution image acquisition equipment in M4R. Therefore, the specific target of the collected transmission lines has the high definition and can be transmitted remotely. At the same time, M4R applies the symmetrical distribution method of four rotors to further shape the aerodynamic layout of the system, which has the good performance in taking off and landing and hovering.

4.2   Functional Overview

During the patrol inspection, UH UAV can fly to the safe distance of the transmission lines by the manual remote control, hover near the transmission lines, collect images, and transmit them to the background through the wireless image transmission equipment, while the ground station will issue instructions to guide M4R UAV to fly along the transmission lines and away from the transmission lines. The position hovering of the safe distance is to complete the function of the autonomous hovering of M4R UAV in the safe space position, so as to ensure the clarity of each angle when taking pictures. In addition, if the staffs want to adjust the image acquisition equipment and the detected equipment, they only need to adjust the heading and the M4R vibration absorption platform. The M4R ground station not only has the functions of guiding the M4R UAV to its predetermined positions and maintaining the remote control and the telemetry communication with the M4R UAV, but also has many functions, such as controlling the flight state, receiving the UAV transmission and detection images, and controlling the UAV platform states and so on.

4.3   Analysis of the UAV Flight Trajectory Prediction Technology

As a maneuverable and flexible moving target, the UAV flight trajectory prediction has the high technical difficulty. When predicting the UAV's flight trajectory, it is necessary to predict the overall movement trend of the UAV as a moving target, not just the UAV's motion state at a certain point. Therefore, it is necessary to predict whether the UAV will fly out of the operating area through the trajectory curve of the UAV in a certain period of time. The least squares curve fitting algorithm can be used to predict the flight trajectory of UAV based on the trajectory information of UAV in a period of time and the appropriate polynomial function can be used to fit the trajectory of UAV. Then, through a large number of experiments and the judgment of the trajectory type, the trajectory prediction effect of UAV can be improved by using the fitting polynomial function of the appropriate order.

4.4   Monitoring and Early Warning Technology in the Unmanned Aerial Vehicle Power Patrol Inspection

After completing the prediction technology of the UAV flight trajectory, it is necessary to solve the early warning of the UAV flying away from the operation area. For the electric patrol UAV, we should predict whether the UAV will fly out of the operation area, and send out the timely warning to the UAV pilots, in order to remind the UAV to change the flight directions or speeds, so as to avoid the UAV flying from the operation area. The least square curve fitting algorithm can be used to predict the flight trajectory curve of the UAV, and then the tangent equation of the flight trajectory curve of the UAV can be used to predict whether the flight trajectory curve of the UAV intersects with the boundary of the operation area or not. However, the area of the electric power patrol inspection operation is basically the irregular shape. How to judge whether the straight line intersects the irregular figure is an urgent problem to be solved. The problem of how to judge whether the UAV will enter the no-fly zone is solved by the monitoring and early warning technology. If the tangent of the UAV's flight trajectory curve at the current position intersects with the smallest inner rectangle of the operation area, the nearest intersection point with the current point is obtained, and the distance between the UAV and the boundary of the operation area at the current time is calculated. The alarm information, such as the distance between the UAV and the boundary of the working area at the current moment, is sent to the ground station of the UAV pilot by UDP to inform the pilot to change the flight speeds or directions of the UAV so as to avoid flying away from the legitimate flight area and ensure the flight safety. If the UAV does not intersect, the early warning of the next time will be carried out.

5   Conclusion

The purpose of the UAV patrol inspection is to ensure the reliability of the power supply efficiently. In order to complete the autonomous flight and the close shooting, the most important thing to be solved is the reliability and the anti-jamming ability of the aircraft. With the wide application of the RTK technology in the power grid patrol inspection, the above difficulties will be solved quickly and meet the development needs of the informatization and intellectualization of the power grids. Practice has proved that the improved method can provide more intelligent and efficient data acquisition mode for the UAV patrol inspection for the UHV transmission lines.

References

1. He, J., Luo, Y., Fan, C., et al.: Preliminary study on the manual joint inspection mode of the overhead transmission line UAV. Electron. Test (21), 99–101, 48 (2015)
2. Zhang, Y., Xiao, C., Guo, X.: Research on the application of the UAV patrol inspection for the high voltage transmission line technology. Shandong Ind. Technol. (18), 249 (2016)
3. Liu, K., Huang, J.: Research and application of the redundant avionics system for UAV. Comput. Meas. Control (5), 1340–1343 (2012)
4. Li, Y., Huang, H., Zhang, H., et al.: Ground nominal electric field calculation of the UHVDC transmission lines based on the radial basis function method. J. Chongqing Univ. (1), 75–80 (2013)

The Architecture of China's Cross-Border E-Commerce Platform Based on the Data Mining Technology

Xianghui Xu[(⊠)]

Institute of Finance and Economics of Zhengzhou University of Science and Technology, Zhengzhou 450064, Henan, China
wangfei843684@126.com

Abstract. With the development of the economic globalization and the information globalization, the cross-border e-commerce among countries is growing rapidly. However, due to the requirements of its own characteristics, the existing logistics distribution model cannot keep up with the needs of the cross-border business development, which has become a bottleneck restricting its development. Based on the background of the cross-border e-commerce, this paper analyses the existing logistics model. Combining the advantages and disadvantages of the model, the problems of the unbalanced development of the logistics system, the low level of the logistics informationization, the lagging level of the systematic management of the customs and the lack of the financial risk prevention mechanisms in the development of the cross-border e-commerce logistics in China are discussed. Suggestions for the development and construction of the integrated logistics platform are given, and the cohesion between the cross-border e-commerce logistics and the e-commerce platform is also analyzed in details.

Keywords: Data mining · China · Cross-border e-commerce platform · Architecture research

In order to meet the needs of the cross-border e-commerce development for the customs clearance, inspection, settlement and the tax refund business with the short cycle, low cost and high efficiency, an integrated service mode of the cross-border e-commerce is proposed, which includes the convergence management channel, the automatic data collection and the optimization of the business process [1]. The basic information service platform and the basic industries of the cross-border e-commerce are planned and designed. The overall structure and functions of the cross-border e-commerce integrated service platform composed of the service platform have achieved preliminary results.

1 The Construction Background of China's Cross-Border E-Commerce Platform Based on the Data Mining Technology

At present, the popular forecasting methods of the foreign trade product sales simply study the forecasting problem from the perspective of the third-party platform or the big data, but they do not take into account the application of the Internet platform, the

© Springer Nature Singapore Pte Ltd. 2020
M. Atiquzzaman et al. (Eds.): BDCPS 2019, AISC 1117, pp. 1607–1613, 2020.
https://doi.org/10.1007/978-981-15-2568-1_224

cross-border e-commerce and the big data fusion in the dynamic evolution forecasting of the product sales [2]. In order to improve the prediction effect of the sales volume of the export products and achieve the scalability and dynamic evolution of the prediction system, based on the research of the "Internet+" environment, the corresponding algorithm of the large volume data mining, the personalized prediction mechanism and the intelligent prediction algorithm for the cross-border e-commerce export sales volume is studied, and the corresponding algorithms such as the distributed quantitative and the centralized qualitative calculation are improved [3]. In this paper, a dynamic forecasting model of the export product sales based on the large controllable correlation data of the cross-border e-commerce is proposed, and the application experiments are carried out, and the experimental results of various models are compared and analyzed.

1.1 Cross-Border E-Commerce Market Has Great Potential for Development

Although the cross-border e-commerce enterprises have developed relatively early, even after the watt networking giants began to take up positions in the mature market, they basically completed the overall layout in the first two years, but there are still large dividends in the market, and the scale of the cross-border e-commerce platform transactions and the scale of Haitao users still have a larger room for growth [4]. At the same time, the cross-border e-commerce market is developing rapidly. With the increasingly close communication between China and the international communities and the escalation of the consumer consumption concept, the cross-border e-commerce market will continue to expand in the future, and the growth rate will remain at a relatively high level [5].

1.2 Independent Cross-Border E-Commerce Platform Starting from Sub-divisional Areas

Although the current integrated e-commerce platform backed by the Internet giant has more traffic resources, there are still no enterprises in the market that can monopolize the market or threaten to emerge. The independent cross-border e-commerce platform still has a larger space for development in the attitude of trotting. The independent cross-border e-commerce is easier to carry out the different attempts than the comprehensive cross-border e-commerce, expand its own business system, covering more complete commodity categories, and providing more brand style choices and more personalized after-sales services and other aspects, to meet the increasingly diverse needs of the users.

1.3 The Cross-Border E-Commerce Platform Should Follow the Policy Guidance

The cross-border e-commerce industry is still in the initial stage of development, and the introduction and guidance of the domestic policies have a very beneficial impact on the standardized development of the industry. For the cross-border e-commerce platform, the market regulation will break the industry bubble, and the rest of the market must follow the central policy guidance and adjust the business forms in time to adapt to more intense competitions.

2 Analysis on the Architecture Design of the Cross-Border E-Commerce Platform in China

2.1 Insufficient Understanding of the Big Data Technology and the Cross-Border E-Commerce

Many enterprises only have a superficial understanding of the online product display, the sales settlement services and the market grabbing on the cross-border e-commerce platforms. They think that only products need to be displayed on the page. The cognitive bias of the big data technology and the cross-border e-commerce makes many small and medium-sized enterprises fail to tap the greater potential in the development of the cross-border e-commerce. At the same time, in the marketing, we cannot "adjust measures to local conditions" or "vary from person to person". We cannot consider and analyze the foreign customers' consumption psychology, consumption preferences, consumption needs, and online shopping habits and so on. We can only reduce prices and promote sales singly. The cross-border e-commerce is mainly the small fragmented orders. Whether we can accurately grasp the consumer psychology and demand of the foreign consumers is directly related to the business performances of enterprises.

2.2 Operation Models Lag Behind

The ability of the big data application is insufficient. Due to various uncertainties in the foreign market faced by the cross-border e-commerce, such as the cultural differences, the differences in the customer consumption habits, and the consumer preferences and so on, it is difficult for enterprises to grasp the specific situation of the target market. The big data technology can solve this problem smoothly. In the era of the big data, the inadequate ability and technology of small and medium-sized enterprises to acquire, analyze, utilize and dig the big data will directly affect their overseas market expansion, the new customer development, and the store operation effect, as well as their product and service innovation. Most small and medium-sized enterprises engaged in the cross-border e-commerce still stay in the publication of the product information, or simply determine the product positioning based on the user reviews, or select products based on the online sales ranking, which is only the primary data analysis. If we can make full use of the big data analysis, we can achieve the precise marketing by collecting and analyzing the information about users' occupation, consumption habits, preferences and shopping evaluation.

The marketing methods are backward. The modern marketing is relationship-oriented, through the multimedia advertising or the network platforms and the media, to achieve the interaction with the consumers. However, many small and medium-sized enterprises do not know how to do a good job of marketing through the cross-border e-commerce platforms. At the same time, the phenomenon of the homogeneity of the cross-border e-commerce export products is very serious. In order to attract the consumers more quickly, most enterprises can only attract the customers through the constant price reduction and promotion, reduce their expected profits, and make enterprises fall into a vicious circle. The long-term development of enterprises is harmful but not beneficial. Therefore, in the era of the big data, the small and medium-sized enterprises need to avoid the traditional price war, and target the overseas customers for the product improvement and promotion, and implement the product differentiation through the big data technology and the modern marketing methods.

2.3 Imperfect After-Sales Services

The cross-border e-commerce and the logistics interact and go hand in hand. The logistics cost of most small and medium-sized enterprises carrying out the cross-border e-commerce business is the biggest expenditure besides the commodity costs. At the same time, the efficiency of the logistics will directly affect the consumption experience of the foreign consumers, thus indirectly affecting the sales of the enterprises' products. In the overseas warehousing and the localized operation, SMEs are unable to achieve the financial capacity, so they can only cooperate with the third-party logistics, while the efficiency of the commodity distribution is restricted by the third-party logistics, which lacks competitiveness in the logistics cost and time. At present, the cross-border logistics is mainly through three ways: the international postal package, the four international express deliveries and the freight transportation. According to statistics, 70% of the parcels in China's cross-border e-commerce export business are delivered through the international postal system. But because the postal delivery is sent out by private parcels, it is impossible to refund the tax, and the risk is high, and the speed is slow, and the loss rate is high. Therefore, the cross-border e-commerce distribution of small and medium-sized enterprises is restricted by logistics, which is not conducive to the expansion of the overseas markets. There are various disputes in the cross-border e-commerce shopping due to the material inconsistency, the freight bearing and the language misunderstanding. These disputes are difficult to solve because of the lack of the public power and the reasonable dispute resolution mechanisms. This is a big crisis for small and medium-sized enterprises. If there is a situation of returning and exchanging goods, they need to can only admit your bad luck in order to avoid trouble and lose both money and goods. If he is complained by customers, he may be fined or even sealed by the platform party. Therefore, in order to fundamentally reduce losses, avoid disputes and reasonably resolve disputes, SMEs in the process of the cross-border e-commerce business need to solve the problem urgently.

3 Research on the Architecture of China's Cross-Border E-Commerce Platform Based on the Data Mining Technology

China's cross-border e-commerce platform based on the data mining technology is a specification designed for the large-scale applications. It is usually suitable for the software programming of the server network and the client network.

3.1   Platform Layout

According to the characteristics of the model, the platform is mainly composed of four parts. The first is the platform operation management side, and the second is the consumer system. The third is the business store system, and the fourth is the online customer service system. The four systems coordinate with each other to form a multi-store mall platform to provide more convenience for transactions. The first is the client layer. This level is mainly used to realize the operation interface and display layer of the large-scale application system. Clients can be subdivided into the application clients, the Applet clients and the browser clients. The second is the Web layer. This layer mainly provides the Web services for enterprises. The Web is mainly composed of the Web components. The Web layer is mainly used to process the customer requests, call corresponding modules, feedback the information processing results to the clients, and realize the interactive use of the information. The third is the business layer. The business layer is also called the application layer or the EJB layer, which is mainly composed of the EJB components and the EJB servers. Usually, the EJB server and the Web server products are combined together. The combination of them is called the application server. The EJB layer is mainly used to implement all business logic in the information system. Therefore, the EJB layer is regarded as the core contents of the large-scale applications, which needs to be implemented by the EJB components placed in the business layer.

3.2   Mobile End of the Website

Because the data of the mobile terminal is synchronized with that of the PC terminal, the content of the mobile terminal is briefly introduced, taking the mobile terminal as an example. The mobile terminal mainly consists of two parts: the seller's center and the buyer's center. Among them, the main contents of the seller center include shops, orders, logistics, receipts, passwords, withdrawal, points, and other related functions. Buyer centers mainly include orders, receipts, logistics, passwords, coupons and applications for opening stores and other functions.

3.3   Platform Technology Architecture

The platform technology architecture is mainly based on the "OSGI" design, which can play a good dynamic loading function, and form a corresponding relationship with Bundle to facilitate the communication. At the same time, ORM can hide the data access of the system, realize the interaction, make the communication easier, and increase the stability of the platform itself. Because this platform uses the latest memory computing technology, it can speed up the data processing, and a large number of applications of the caching technology, making the system more powerful, effectively preventing the hacker attacks.

3.4   Platform Expansion

After the application of the B2B2C mode, the platform of the multi-stores mall has certain extensibility, mainly reflected in the following three aspects. The first is the security access strategy. Under the B2B2C mode, using the HTTPS SSL authentication system, set up the front-end access function, using the blacklist function, which can help to block the malicious attacks, and its technology can provide more secure access channels for the businesses and the consumers. The second is the authorization function. Through this function, we can accurately record each user's access behaviors, and automatically generate directories, provide the retrieval function, and then quickly locate the specific access behaviors. Thirdly, the anti-attack strategy is set up. For the information and the data submitted by the users, two security processes are carried out, which are the client-side security processing and the server-side security filtering, to ensure the data security to the greatest extent. Focus on exploring the underlying logic of the entrepreneurship. Choose different objects and focus, and inspect the platform design framework of the cross-border e-commerce vertical platform enterprises, the management level of the platform operation enterprises, the control ability of the supply chain management system Inspect the capability of the service entrepreneurs of the park and the base operation enterprises, and whether the supporting service system is perfect, and inspect the application of the enterprise platform rules of the cross-border retail business, the intellectual property management, and the industry competitiveness.

3.5   Platform Core Businesses

The core businesses include the business placement, the consumer transformation and the financial management and so on. To recruit the offline businesses for the society, form a multi-store mall platform, and use the professional online platform mechanism to operate. In this process, the buyers and the sellers can also carry out certain supervision and play a third-party control role. Before the realization of the consumer transformation, it is necessary to do a good job of the consumer contacts, and analyze the online transactions and the loyalty of the consumers, so as to urge businesses to improve their service behaviors, and provide more targeted and personalized services for the consumers. After the transformation of the consumers, through the effective improvement of the platform shopping process, the order and payment can be well transformed to facilitate the profit of the sellers. The financial management module achieves the regular settlement through the unified operation mode. On the one hand, it creates a good shopping environment and experience for buyers, and on the other hand, it also ensures the authenticity and traceability of the financial data.

4   Conclusion

The experimental results show that the model fully integrates the openness, the extensibility of the "Internet+" and the advantage of the dynamic prediction of the big data, and realizes the dynamic, intelligent and quantitative qualitative prediction of the export volume of the export products based on the cross-linked electricity supplier's controllable connected big data under the environment of the "Internet + foreign trade". The comprehensive prediction effect of the model is obviously better than that of the traditional model, and it has the strong dynamic evolution and the high practical value.

Acknowledgements. The study was supported by New century education and teaching reform and talent training project of local colleges and universities in Zhengzhou: Research on curriculum reform and innovation of cross-border e-commerce of International trade major based on "Integration of industry and education" [ZZJG-B8020]; Zhengzhou University of Science and Technology teaching quality project [2017JGZX10].

References

1. Zhong, W., Wu, J., Mei, S.: E-commerce application ability: concept, theory composition and empirical testing. J. Syst. Manag. (1), 47–55 (2011)
2. Wu, F., Wang, Y.: Difficulties and countermeasures of the cross-border e-commerce development in China. Mod. Bus. (7), 39–40 (2016)
3. Tong, B.: Brief discussion on the status quo of the cross-border e-commerce and the choice of the international logistics mode. Brand (4), 37–37 (2015)
4. Cao, S., Li, Z.: Research on the third - party logistics model of the cross-border E-commerce. E-Commerce (3), 23–25 (2013)
5. Dong, Z.: Problems and countermeasures of the cross-border e-commerce in China. Co-Op. Econ. Sci. (24), 72–73 (2015)

The Integrated Language Representation System of Illustrations Based on the Streaming Media Technology

Yi Yuan[1(✉)] and Di Lu[2]

[1] Wuhan Technology and Business University, Wuhan 430065, Hubei, China
linrong111786@163.com
[2] Wuhan College, Wuhan 430065, Hubei, China

Abstract. Illustration is the use of the graphic images, in line with the principle of the unity of aesthetics and practice, trying to make lines and shapes clear and lucid, easy to produce. Illustration is a universal language in the world. Its design is usually divided into characters, animals and commodity images in the commercial applications. In recent years, with the popularization of computers and the network, the modern illustration creation in China has been greatly developed. Illustration has gradually been widely used in various fields of our life and work. The diversified characteristics of the modern illustration development can no longer cover the application scope of the traditional illustration. With the development of the media technology, the application of the streaming media technology in the comprehensive language representation system of illustrations has developed into the technical connotation of the illustration art display.

Keywords: Streaming media technology · Illustration · Comprehensive language · Presentation system

With the popularity of the Internet, the demand of using the network to transmit the voice and video signals is also growing. After broadcasting and television and other media onto the Internet, they also hope to publish their own audio and video programs through the Internet. However, the volume of the audio and video files in storage is usually very large [1]. The streaming media technology improves the difficulty of transmitting the audio and video over the Internet. Using this technology, the language form of the illustration art can be effectively interpreted and displayed, and the best display effect of the art can be achieved.

1 The Connotation and Characteristics of the Streaming Media Technology

With the development of the modern technology, the network brings people a variety of forms of the information. From the first picture appeared on the network to various forms of the network video and three-dimensional animations, the network makes people's audio-visions feel very satisfied [2]. However, before the emergence of the streaming media technology, people have to download these multimedia contents to the

M. Atiquzzaman et al. (Eds.): BDCPS 2019, AISC 1117, pp. 1614–1619, 2020.
https://doi.org/10.1007/978-981-15-2568-1_225

local computers first, after a long wait (because of the bandwidth constraints, the downloading usually takes a long time) before they can see or hear the information conveyed by the media.

The streaming media refers to the continuous time-based media that uses the streaming transmission technology in the network, that is, the media that publishes the audio and video multimedia contents in the real time by means of the data stream on the Internet [3]. The audio, video, animation or other forms of the multimedia files belong to the streaming media [4]. The streaming media is a kind of the multimedia file that can be viewed and listened to by the viewers while downloading, without waiting for the whole multimedia file to be downloaded.

The traditional way of transmitting the multimedia information such as the audio and the video over the network is to download them completely and then play them back [5]. It often takes several minutes or even hours to download them. By using the streaming media technology, the streaming transmission can be realized. Sounds, images or animations can be transmitted continuously and uninterruptedly from the server to the user's computers. The user does not have to wait until the entire file is downloaded, but can watch only after several seconds or more start-up delay. When the audio and the video are played on the user's machine, the rest of the file will continue to be downloaded from the server.

The streaming media technology makes it possible for the traditional media to open up a broader space on the Internet. The radio and television media programs are more convenient to access the Internet, and the on-demand programs of audiences are simpler. The online live audio and video broadcasting will also be widely used. The streaming media technology will change the traditional media's "push" transmission into the audience's "pull" transmission. The audience will no longer passively accept the programs from the radio and the television, but receive the information they need at their own convenient time. This will improve the status of the audiences to a certain extent, enable them to take the initiative in the news communication, and also make their needs have a more direct impact on the activities of the news media.

The wide application of the streaming media technology will blur the boundaries between broadcasting, television and network. The network is not only the assistant and extender of the broadcasting and television, but also their powerful competitor. Using the streaming media technology, the network will provide the new audio and video program style, and will also form a new mode of operation. Bringing the advantages of the traditional media into full play, utilizing the advantages of the network media and maintaining the good competition and cooperation among the media are the way to develop the network in the future, and also the way to develop the traditional media in the future.

2   Research on the Development Trend of the Illustration Art

Today, with the rapid development of the economy, the illustration art still has great potential and broad market prospects because of its incomparable charm. The illustration art, also known as the demonstration art, refers to an art form that is inserted in the middle of the texts to explain and illustrate the texts. The illustration art has a long history and has been loved by people. Until today, with the rapid development of our economy, the illustration art still has great potential and broad market prospects because of its unparalleled charm. It is not only limited to explaining words, but also shows its own advantages in the graphic design such as packaging and advertising.

In the graphic design, there are great differences between the application of the illustration art and the traditional style of the graphic realism. For the traditional realistic style, it is accomplished by the careful construction and ingenious combination of patterns. It gives people a static beauty and lacks the visual dynamic feeling. The application of the illustration art in the graphic design is to combine the realistic style of photography with the traditional realistic style, to insert pictures with the real color and texture in the design, to show people the real things, to stimulate the audiences' visual nerve strongly, and to give people a real and more intuitive dynamic feeling.

With the progress of our society, the illustration art is also changing, and it has more new connotations, and also has a more unique side in the form of expression. The illustration art not only retains the traditional realistic style, but also injects new expressive techniques such as the exaggeration and generalization, which has a richer presentation force and expressive force. In addition, the illustration art can be freely combined after the graphic design is disturbed, which is also difficult to achieve by other media.

In addition to retaining the realistic traditional style, the illustration art is often used in many graphic designs nowadays to achieve the desired effect of the designers. Compared with the traditional realistic style, the abstract way of expression pays more attention to the idea of people's inner subconscious, which cannot be embodied by the realistic method. The abstract technique of expression can give people more imagi-nation space and bring people's imagination into full play. It is not specific to convey a certain message to people, but to let people understand a certain idea, and is an ideological, subconscious internal spiritual exchange, which is more expressive than the realistic style. The illustration art of the abstract style requires people to understand the essence of works by expressing and associating. This is the focus of the works of the abstract style.

The abstract style has obvious characteristics of the times and is good at repre-senting a new and fashionable product. In the graphic design, the illustration of the abstract style will leave room for people to imagine, which is called "1000 people have 1000 Hamlets". Different people will read different meanings from the abstract graphic design, which is the charm of the works of the abstract style.

3   Comprehensive Language Performance Requirements of Illustrations

Today in the new era, the modern illustration art has perfected the shortcomings of the traditional illustration art design and brought new ideas of the illustration art creation. Its scope of application goes beyond the boundaries of many traditional publishing designs. Today, it has become an art that integrates practicability and innovation.

3.1   Expressions of the Comprehensive Painting Language Forms

Modeling, colors and materials are the three basic elements of the painting language. The language of the comprehensive painting has its unique language form. Its characteristics depend on the material attributes of the material itself, including the appearance, colors and materials of the materials, as well as the relevance and directivity it brings. This kind of correlation and directivity has become a unique way of expression of the comprehensive material painting. The artist's creation is to fully tap the aesthetic value of the material attributes, and make use of the material attributes to inspire and hint people with the visual language. In addition, the comprehensive painting has strong emotional characteristics. The language form of the comprehensive painting is the same as that of other paintings. It is a way of expressing the painter's perception of objects and inner feelings. Finding a suitable material for the emotional expression is a prerequisite for the artists to express their emotions in a comprehensive form of the painting language.

3.2   Material Application of the Comprehensive Painting

From the birth and development of painting to the present, the materials of the comprehensive painting are constantly changing. In the field of the contemporary painting art, the objects and subjects of the artistic expression have changed fundamentally. The comprehensive painting has become a special form of expression. It is freer in the techniques and styles, and more in line with the trend of the modern pluralistic art. The use of the painting materials in the modern comprehensive painting has gone far beyond the meaning of the material itself, and the range of the material selection is constantly expanding. To some extent, as long as the purpose of expressing the artist's feelings and ideas can be achieved, all materials can be used for the artistic creation. Of course, the application of materials is related to artists' personal hobbies and concepts. They often look for the most suitable materials and artistic expressions according to their own conditions. The comprehensive painting materials emphasize the use of the multiple forms of expression, and pay more attention to the understanding of the essence of the painting. Generally speaking, the comprehensive painting brings different visual feelings to the contemporary painting art, has a profound impact on the artists' artistic creation, and has greatly changed the aesthetic habits of the public.

4   Research on the Integrated Language Representation System of Illustrations Based on the Streaming Media Technology

The streaming media expression is relatively simple in the form. You can see the pure video mode, and there are video + picture + schematic illustration mode, and the picture + sound mode. The streaming media refers to the use of the streaming transmission technology on the network for the continuous real-time broadcast of the media formats, such as the audio, video or multimedia files. The streaming media technology is also called the streaming media technology. The so-called streaming media technology is the network transmission technology to compress the continuous image and sound information and put it on the Web server. The video server transmits the compressed packages sequentially or in the real time to the user's computers so that the users can watch and listen while downloading, instead of waiting for the whole compressed file to be downloaded on their own computers. This technology first creates a buffer on the user-side computers and uses the next segment of the data as a buffer before playing. When the actual connection speed of the network is less than the playing speed, the player will use a small segment of the data in the buffer, which can avoid the interruption of playing and ensure the playing quality.

The color of the illustration art is more uniform, and the lines are also scattered, and the shape is simple and generous, and the color of illustrations is very rich. The illustration has formed a unique academic system of knowledge because of its diversity and complexity of colors. The color aesthetics is a theoretical system for people to explore and study the attributes, perceptions, methods and principles of the color matching. People have different preferences for colors, which is influenced not only by their nationality and region, but also by their age, interest and personality. The difference between people's preferences and dislikes for colors is reflected in all aspects of life. There are also strong differences in colors in some periods, which are closely related to the aesthetic trends, fashion trends, and social development and so on. The expressive force of the lines comes from the emotional contents of the lines, and the change of the lines reflects the change of their emotions. In the modern illustration, lines are simple and concise and fashionable, which is also the manifestation of the pursuit of the modern illustration art.

As a special way of the visual communication combined with the design and the painting, the illustration has been widely used in various fields of the social life with vivid images, intuitive expressions, strong rendering power and unique artistic aesthetic feelings, bringing colors to people's lives. The illustration itself has the value of the social aesthetics. It improves the public's aesthetics and promotes the prosperity and development of the visual culture. The most important feature of the illustration is that it has beautiful pictures and visual impact. Through this most intuitive and simple impact, it conveys to people the ideas the author wants to express. The illustration art has various forms of expression. In the past, the illustration mostly used the hand-painted watercolor, but now it mostly uses the computer drawing or the hand-painted manuscripts and the computer coloring, and presents in the form of the single or series. From the contents of the pictures, illustrations are mostly composed of characters, the scenery or the combination of the two, using the monochrome or colors. Of course, there are also some illustrations in accordance with the contents of the subject matters or social functions, purposes and other ways of expression. In short, no matter what kind of contents or forms of expression, the illustration art is an irreplaceable art form in our social life.

5   Conclusion

The painting language includes the painting skills, forms and contents, and is an organic entity of the mutual influence and infiltration. Painters integrate skills, knowledge, emotions and other painting materials to convey their emotions to people through the painting language. The illustration art and the comprehensive painting have the advantages of combining the traditional and the modern science and technology. The illustration design is not only an important part of the modern social art, but also a very important art form in the modern art. It is widely used in various fields. With the rapid development of the science and technology in China, the field of the illustration design has also ushered in opportunities for innovation and expansion. The emergence of the dynamic illustrations promotes the application of the illustration art in various fields, and promotes the development and innovation of the illustration design.

Acknowledgement. Achievement of Scientific Research Project (Project No. A2019013).

References

1. Zong, X.: On the innovation and development of the dynamic illustration in the illustration design. Art Sci. Technol. (10), 215–284 (2015)
2. Xu, D.: On the beauty of colors in the illustration art. Educ. Chin. After-Sch. (11), 153 (2011)
3. Yuan, Z.: A brief talk on the study of the color aesthetics in the illustration art. Sci. Wealth **7** (z1), 274–275 (2015)
4. Yang, S.: On the contemporary character of illustrations in the visual communication design. Art Sci. Technol. (2), 283 (2016)
5. Yang, F.: The dynamic characteristics of modern illustrations in the visual communication. Art Educ. Res. (9), 100 (2012)

The Visual Elements in Film Poster Design Based on Photoshop

Liqun Zhang[✉]

University of Shanghai for Science and Technology, Shanghai 200093, China
binglei7418887@163.com

Abstract. The movie posters are the accessories and derivatives of films, and have become an indispensable part of the film industry. The movie poster is a kind of the poster design, but it has more plot than the ordinary poster. It has the function of transmitting the movie information, attracting the audience's attention and stimulating the movie box office income. So it is also called the "business card" of movies. The movie poster design is a kind of the self-promotion, and also a direct expression of the soul of the films. This intuitive information transmission is sometimes even more convincing than the film. The Photoshop-based image processing software is full of our daily life, changing our past image processing and poster design methods.

Keywords: Photoshop · Movie poster · Design · Visual elements · Art aesthetics

The movie posters convey the film cultures and information through relatively complete visual images and picture forms. Therefore, designers should determine the contents and forms of the films by whether the graphic implication can express the audience's demand for watching the film, that is, whether the film form is persuasive or not [1]. Applying the Photoshop software to the graphic design of the movie posters can not only simplify the processing means, but also give full play to the best design effect, so that the ordinary people can also operate well to complete the picture modification [2]. The film posters can not only achieve the expected publicity effect, but also win more box office revenue for the film.

1 Photoshop's Functional Cognition

The abbreviated form of the Adobe Photoshop is PS, which is an image processing software developed and distributed by Adobe Systems. Photoshop mainly processes the digital images composed of pixels. With its numerous editing and drawing tools, it can effectively edit pictures. PS has many functions, including images, graphics, texts, videos, and publishing and so on [3]. In 2003, Adobe Photoshop 8 was renamed Adobe Photoshop CS. In July 2013, Adobe launched a new version of Photoshop CC. Since then, Photoshop CS6 has been replaced by the new CC series as the last version of the Adobe CS series.

© Springer Nature Singapore Pte Ltd. 2020
M. Atiquzzaman et al. (Eds.): BDCPS 2019, AISC 1117, pp. 1620–1625, 2020.
https://doi.org/10.1007/978-981-15-2568-1_226

Photoshop specializes in the image processing, not the graphic creation. The image processing is to edit and process the bitmap images and use some special effects. Its focus is on the image processing. The graphics creation software is to design the graphics according to its own ideas, using the vector graphics and so on [4]. The graphic design is the most widely used field of Photoshop. Whether it is the book cover, the poster or the playbill, these print products usually need the Photoshop software to process the images. The image creativity is the specialty of Photoshop [5]. Different objects can be combined by the Photoshop processing to change the images.

Functionally, the software can be divided into the image editing, the image synthesis, the tone correction and the functional color effect production. The image editing is the basis of the image processing. It can transform the images such as enlargement, reduction, rotation, tilt, mirror image, and perspective and so on. It can also copy, remove speckles, and repair and modify the damaged images. The image synthesis is to synthesize several images through the layer operation and the tool application, which is the only way for the art design. The drawing tools provided by this software make the foreign images and ideas blend well. The tone correction color can adjust and correct the colors of the images quickly and conveniently. It can also switch between different colors to meet the application of the images in different fields such as the web page design, printing and multimedia and so on.

2 Analysis of the Visual Elements in the Film Poster Design Based on Photoshop

The visual symbols have the characteristics of communication, symbolism and polysemy. First of all, the visual symbols are communicative. The meanings of the visual symbols are to convey the meanings to the audience by means of the multiple combinations of words, graphics and colors. Communicativeness is the most important characteristic and mission of the visual symbols. If the poster design cannot convey the information to others without knowing what it is, it will be a failure. The visual symbols are symbolic. The visual symbols convey the information through the symbolic devices. Through the interaction of various factors, the visual symbols need the viewer's understanding more. Because the visual symbols are symbolic, the interpretation of the information transmitted by the visual symbols has the characteristics of polysemy.

2.1 Character Symbols

The character symbols mainly refer to the visual symbols with the language as the main content. They have the strong expressive ability. They can express the contents to be conveyed directly, convey the designer's intentions and various information to the public, and give people a clear visual feeling. The proper words in the design works often play a key role of adding the touch that brings a work of art to life.

Nowadays, graphic characters have been more and more widely used in the poster design, and show many advantages with its novel and unique design methods and expressive power. The character graphicalization breaks the traditional design thinking mode, using exaggeration, deformation and other creative techniques to cleverly combine with the graphics to produce a new visual image, enhance the visual impact of the posters, and attract the attention of the audience. In the current fast-paced and information-intensive society, simple words can no longer meet the requirements of people's fast reading, which is especially easy to be ignored.

2.2 Graphic Symbols

The image symbols can be the abstract graphics or the concrete images, rich and diverse. German designer Horgo Matisse once said, "A good poster should be spoken in the graphical language rather than in the text annotations". Matisse clearly pointed out that the key to the poster design is to use the vivid graphics to attract people's attention. For example, the poster design of WWF movies can be seen at first sight by viewers. First of all, in the visual perception, the whole picture is bright, showing a unique sense of the visual art. With the deepening of the cognitive level, the brain processes these images more deeply, and the meaning behind these seemingly aesthetic graphic symbols.

2.3 Color Symbols

The color symbols refer to the means of expressing the information that needs to be disseminated through colors. The commonly used methods include: using colors to foil the overall atmosphere and highlighting the key points with the color contrast and so on. The appropriate color selection is to better display the theme, attract the public attention, and ultimately achieve the role of propaganda and warning. According to the research of the color psychology, people can feel different hues, brightness and purity of different colors. The color has the characteristics of cold and warm, light and heavy, swelling and shrinking, and advancing and retreating and so on. In the long-term social life, people accumulated a lot of emotional color psychological experience, which resulted in different color associations and color symbols, thus rising to a rational understanding and becoming a specific concept.

3 Design Strategy of the Visual Elements in the Film Poster Design Based on Photoshop

Through the conscious design, we can create the graphic pictures with the psychological association and the induction effect, and then influence the viewers and produce the emotional "collision" with them. For this reason, designers should pay attention to:

3.1 The Theme Is Clear and the Idea Is Unique

Almost every film has its own unique aspect. Before designing the posters, a relatively clear theme should be defined. After viewers see it, they can clearly identify the essential characteristics of the film and the main differences with similar products. For this reason, before deciding on the theme of the poster, the designer should have a thorough understanding of the film. If a designer can integrate novel ideas into a movie poster and organically combine the "shape" and the "meaning", then the poster design must be fresh, energetic and curious in the end. Color plays an important role in the visual communication and the strong intuition in the poster design. The unique attributes, values and functions of colors are irreplaceable by other design elements. The successful poster design usually uses its colors appropriately and meets the needs of the contents and forms. Sometimes, it even promotes a product or a service-oriented design activity to achieve success, while the wrong color design will make people pay a high price.

3.2   Image Introduction and Language Creativity

Color plays an important role in the visual communication and the strong intuition in the poster design. Graphics with the distinct features are easy to attract consumers' attention, and there are striking differences in the visual features in highlighting the propaganda strength of the individual posters, so they are easy to identify. The visual features of this kind of graphics are simple and distinct, original and novel. It is easy, natural and intuitive to watch, so it can make a deep impression on the viewers. In order to ensure the timeliness of the graphics and the themes in the process of the visual communication, the positioning of the graphics language should be creative and not complicated, and the method of using less to win more should be adopted. The so-called combination of complexity and simplicity means that only "fewer" pictures can achieve "more". The color elements cannot be ignored in the process of the poster design and the cultural information transmission. This is because, first of all, color is the most direct "language" communication, and it can attract the audience's attention in the first time, so that the viewer's "eyes" are relaxed or vigilant. Secondly, color is the most sensitive visual factor. It can let the audience appreciate the specific emotional infection, and change their psychology, even the view of the film, and then affect their desire to watch and their consumption behaviors.

3.3   Strong Sense of Form and Being Very Attractive

The aesthetic poster design usually has a strong sense of form and is easy to arouse the audience's visual appreciation taste. Designers need to break away from the conventional thinking and apply the new aesthetic elements to the poster design to create more novel, beautiful and durable forms to attract the viewers. It is worth mentioning that the simpler the form is, the stronger the poster effect will be. However, in the process of choosing and displaying the film and television images, we must be honest and not false, because the honest film and television images have the strong affinity and can attract the attention of many people. The different cultural traditions and the folk customs, people of different regions and nationalities have different perceptions and understandings of colors. As a poster designer, we need to understand the differences of the national colors, the regional colors and the custom colors, and orientate the main colors in the posters reasonably according to the different color preferences of the actual consumer groups, so as to meet the visual appreciation needs of different consumer groups, and finally achieve the commercial promotion of the film industry. Only in this way can designers design more perfect and more contemporary excellent poster design works.

3.4 Application of Photoshop in the Film Poster Design

Imitating the stamp tools: This function is the most commonly used for the modification and replacement of the details, and in the modification of small things such as portraits and LOGO and so on, it can be repeated use. The magic wand matting, the Lasso matting, the channel matting and other functions are also reflected in Word and Excel. These tools can adjust the overall tone of the picture by increasing or reducing the range of the color range selected and giving a new sense of freshness. The feathering and liquefaction tools can make the edges and diagonals of posters more soft and beautiful. Sponge tools and dilution tools and so on can enhance or reduce the saturation of the posters, according to the specific requirements to adjust the brightness. There are many posters in the design that we may need to improve the color contrast between the local areas and the regions to form a sharp contrast, and then we can use the function of retaining the high contrast value, in order to make the outline of the edge lines of the screen more delicate. In order to make the plane graphics become the three-dimensional dynamic and reliable, we can use the duplicate layer, through a series of processes such as the free change, the position adjustment, the layer style adjustment, and the transparency adjustment and so on. In order to highlight the main body or the theme of the picture, the four corners can be darkened by making a layer mask for the background layer, so as to achieve the prominent purpose. The layer mask is often used for the color gradient, which can play a very good transitional role. Although the posters are mainly painted, they also lack the cooperation of words. In addition, the effect of the font adjustment can also play the same role. Projection and relief effect is the most commonly used font prominence function in the poster design. This effect can better attract the eyes, deepen the impression and make the font more profound, and have a strong impact on the vision.

4 Conclusion

As a poster designer, we need to understand the differences of the national color, the regional color and the custom color, and orientate the main color in the poster reasonably according to the different color preferences of the actual consumer groups, so as to meet the visual appreciation needs of different consumer groups, and finally achieve the commercial promotion of the film industry. Only in this way can designers design more perfect and more contemporary excellent poster design works. Combining the visual advantages of Photoshop, we should pay attention to the use of Photoshop to maximize its functional values in the process of the picture design of movie posters.

References

1. Wang, R.: A brief analysis of the visual representation techniques in the film poster design in the commercial film era. Art Fash. (10), 49 (2014)
2. Ding, Y.: Application of the art design in film and television works. Theory Horiz. (2), 186–210 (2011)
3. Wen, Q.: An analysis of the design performance of film posters. North. Lit. (3), 121–122 (2016)
4. Huo, J.: Visual performance of the film poster design. Movie Lit. (12), 163–164 (2011)
5. Niu, L.: On the design of the visual expressiveness of film posters. Intelligence (23), 228 (2013)

Application of Streaming Media on Demand Technology in the Visual Classroom Teaching of College English

Fanghui Zhao[⊠]

Zhengzhou University of Science and Technology, Zhengzhou 450064, China
mouxue8179871@163.com

Abstract. With the continuous improvement of the campus network infrastructure, the daily learning of teachers and students in schools has become prominent. Therefore, how to build a practical campus video on demand system based on the streaming media video technology has become an integral part of the campus information construction. In the transmission of the information in the visual classroom teaching of College English, the streaming media technology, which plays while downloading, can effectively change the current situation of the bandwidth limitation of the low-speed access to the Internet, overcome the shortcomings of the traditional downloading and playing methods, realize the streaming transmission of the real-time interactive video and audio resources, and enable learners to use the streaming media technology in the courses. With more freedom in choosing the learning time and place, teachers and learners can truly realize the online real-time interaction and provide learners with a learner-centered online learning environment.

Keywords: Streaming media on demand technology · College English · Visualization · Classroom teaching

With the rapid development of the computer technology and the network technology, the way of information transmission is also changing. The new network technology provides more choices and methods for the transmission of the latest teaching information and resources [1]. The teaching method has also developed from the traditional face-to-face classroom teaching in large classes to today's remote network teaching and the multimedia teaching and other new teaching methods.

1 Application Principle of the Streaming Media on Demand Technology

The streaming media refers to the continuous media using the streaming technology in the Internet/Intranet, such as audio, video or multimedia files. The streaming media does not download the entire file before playback, but stores the beginning part of the content in memory [2]. The data stream of the streaming media is transmitted and played at any time. The implementation of the streaming transmission requires caching. Because the Internet carries out the intermittent asynchronous transmission on the basis

© Springer Nature Singapore Pte Ltd. 2020
M. Atiquzzaman et al. (Eds.): BDCPS 2019, AISC 1117, pp. 1626–1631, 2020.
https://doi.org/10.1007/978-981-15-2568-1_227

of the packet transmission, for a real-time A/V source or the stored A/V file, they will be decomposed into many packets in their transmission. Because the network is dynamic, the routes selected by each packet may be different, so the time delay to reach the client will vary, and even the first packet may arrive later [3].

For this reason, the buffer system is used to compensate for the effects of the delay and jitter, and ensure the correct sequence of the data packets, so that the media data can be output continuously without the temporary network congestion causing the playback to stop [4]. Usually, caching does not require much capacity, because caching uses a ring linked list structure to store the data. By discarding the played contents, streams can reuse the free cache space to cache the subsequent unplayed contents.

The process of the streaming transmission: After a user chooses a streaming media service, HTTP/TCP is used to exchange the control information between the Web browser and the Web server in order to retrieve the real-time data needed to be transmitted from the original information [5]. Then the Web browser on the client starts the A/VHelper program and retrieves the relevant parameters from the Web server using HTTP. The number pairs initialize the Helper program. These parameters may include the directory information, the encoding type of the A/V data, or the server address associated with the A/V retrieval.

2   The Value of the Streaming Media on Demand Technology in the Visual Classroom Teaching of College English

The streaming media data is the carrier of the media information. The commonly used streaming media data formats are .ASF and .RM and so on. The server is to store the media data. In order to store the large-capacity film and the television data, the system must be equipped with the large-capacity disk arrays, with the high-performance data reading and writing capabilities, the high-speed transmission of the external request data, and the high scalability and compatibility, which can support the standard interfaces. This system configuration can satisfy thousands of hours of the video data storage and realize the mass storage of the chip source. Network is the network suitable for the multimedia transmission protocol or even the real-time transmission protocol. The streaming media technology develops with the development of the Internet technology. It adds the multimedia service platform on the basis of the existing Internet.

2.1   Change the Traditional Teaching Concept and Strengthen the Application of the Knowledge Visualization Teaching

Today, with the rapid development of the society and the economy, education will enter the road of the rapid innovation. The quality of College English teaching also needs to be greatly improved. Strengthen the construction of the knowledge visualization teaching mode, attract students to study, enhance students' concentration, and realize the real integration of teaching and learning through strengthening the intuitive interaction, so as to cultivate students' interest in learning, achieve the purpose of the interesting learning, and achieve the promotion of the College English teaching concept. Different combinations and application methods of words, phrases and sentences in our teaching are presented in the three-dimensional form. Through illustration, the multiple meanings of words and sentences are vividly described. This multi-media teaching method with abundant pictures and texts makes students deeply impressed with the classroom knowledge, thus improving the accumulation of the classroom knowledge by college students.

2.2   Promote the Teachers' Level to Master the Teaching Means of the New Educational Ways

In most primary school English teaching, there is too traditional and single education model, and especially in some slightly backward areas, teachers are weak, and many teachers do not have the knowledge of science and technology in the new era. However, the current development of education inevitably requires teachers to upgrade themselves, enrich their teaching means, and effectively use the knowledge visualization to improve their teaching quality and efficiency. This requires teachers to actively change the traditional educational concepts, improve their own quality, expand their teaching methods and learn to use the science and technology to carry out their teaching. Educational institutions and schools should pay attention to the training and guidance of teachers, so that teachers can quickly learn and master the knowledge visualization teaching methods, and have sufficient theoretical basis and practical abilities for the new equipment, the new technology and the new concepts, so as to apply them to the teaching of the new era. In addition, in the knowledge visualization teaching, teachers can encourage students to take learning notes actively, so as to form their own learning knowledge system, which can be used in their later learning, become their learning experience, effectively improve their learning efficiency and learning initiatives, and promote students' learning enthusiasms in a virtuous circle. During the period of the teaching interaction, the use of the classroom teaching methods in which students discuss problems with each other, actively mobilize students' learning moods, so that the classroom atmosphere is lively and will not lack focus.

2.3   Increase the Investment in the Educational Innovation and Provide the Equipment Support

The application of the knowledge visualization teaching method needs to rely on a large number of the multimedia and various kinds of the scientific and technological equipment. Just like the development of the production and people's life in today's society, the scientific and technological teaching has shown a trend of development in the educational system. From the point of view of cultivating all aspects of students' learning attributes, the advanced science and technologies are indispensable. Under the circumstances of combining students' learning, life and social development, it comprehensively promotes college students to adapt to the society and enhance their various abilities. Therefore, it is not enough to strengthen the innovative ideas of teachers' educational concepts and apply the knowledge visualization teaching model. All educational departments also need to improve the support of the educational funds, provide enough scientific and technological teaching equipment for schools, and also need the strength of all sectors of our society to make practical contributions to college

English education and try their best to support and help them. Help the slightly backward areas to achieve the scientific and technological teaching. Through the use of the modern teaching and educational equipment such as the large screen video, students' learning efficiency has been successfully improved.

3 Implementation Strategy of the Streaming Media on Demand Technology in the Visual Classroom Teaching of College English

With the rapid development of the computer technology as the main body of the long-distance network teaching, a large number of College English visual classroom teaching resources are presented to learners through the network transmission. Through the network transmission of College English visual classroom teaching information, the audience of the network teaching is expanded, which is conducive to the realization of the new educational goals of digitalization, informatization and finalization.

In the era of the knowledge economy, people need to constantly learn new knowledge and skills to keep up with the pace of the times. Learning must be transformed into a life-long process. The online education breaks through the limitations of the traditional "face-to-face" teaching, and provides a new way of teaching for knowledge seekers, which is time-dispersed, resource-sharing, vast geographical area and interactive mode. Therefore, it is widely noticed by people. In view of the far-reaching significance of the distance education, the functions of our distance multimedia teaching system are to realize VOD of the teaching courseware, the live broadcast of our teaching, and the network classroom and so on. In order to carry out the distance education activities on the IP network, two basic problems need to be solved: the transmission of the audio and video stream information and the synchronization between them and the data. Because of the wide bandwidth of the audio and video information, it is impossible for students to download all the programs to the local computers and play them again. Advanced network playing technology must be adopted to realize transmitting while playing. In addition, in the process of the teaching, teachers often use the electronic teaching plans to assist their teaching, such as PowerPoint, and there is a strict time synchronization relationship between the display of the electronic teaching plans and the audio and video streams, which requires us to maintain the synchronization relationship between them in the transmission process.

The working process of the system is as follows. The input video and audio signals will be sent to the encoder of MPEG4 for encoding. The program stream output by the encoder can be stored in the storage device or sent directly to MediaServer. The main function of the MediaServer is to complete the broadcast of the program stream. MediaServer broadcasts programs from three sources. It may be the ASF files stored in the storage devices or the real-time programs transmitted by the encoders. The programs it broadcasts can also be obtained from other MediaServers. Ordinary users can access the system through LAN or the wireless network.

The streaming media technology is a technology that uses the streaming to transmit the continuous time-based media. The streaming transmission is to compress the video, audio and other media into one compression package, which is transmitted continuously and in the real time from the video server to the user's computers. It only needs to cache enough playable video capacity at the user end to start playing. The streaming media system consists of the coding tools, the streaming media data, servers, networks and players. Coding tools: It is used to create, capture and edit the multimedia data to form the streaming media format. Use the media acquisition equipment to produce the streaming media. It includes a series of tools, ranging from the independent videos, sounds, pictures, and text combinations to the production of rich streaming media. The streaming media files generated by these tools can be stored in a fixed format for use by the publisher.

When teaching reading, teachers should sort out the reading materials in advance and make a scientific and comprehensive prediction, which will be very helpful for students to follow-up English reading. Let students participate in the English reading preparation activities, which can make students curious about the English reading contents, and can also help students to further understand the English reading contents, to correctly grasp the ideas to be conveyed by the reading contents. The teaching of English reading is the key point in the teaching of reading. It requires students to sort out the ideas of the articles independently and have a clear understanding of the reading contents. The application of the text clouds in College English teaching is the key to improve the teaching effect of our English reading. Therefore, in the teaching of reading, teachers can combine the reading contents and make them into a text cloud map, so that the difficult contents of the article can be fully displayed in the form of the visual teaching, and the key points in the article can be marked in a striking way, so that students can intuitively learn.

Our English teaching is a comprehensive knowledge integration learning process, which can organically integrate the English vocabulary and grammar and so on. When teaching English reading, teachers should not only improve students' English learning levels, but also expand students' English learning scopes, so that students can understand diverse cultures and cultivate students' quality education. In the process of learning English reading, students need to correctly interpret the English words and find the internal rules between the grammar rules. The use of the mind mapping in the reading teaching can enable students to extract information, cultivate the ability to mine information and improve the learning ability of the English reading. For example, in the teaching of the English story-based reading, the article can be divided into different parts by using the clues of the story beginning, occurrence and ending in a structured way, and the article can be sorted out comprehensively, so that the students can learn from different parts and combine the reading contents into a complete story to guide their thinking. The form of the picture is displayed to the students, so that the students can have a clear and intuitive impression and better understand the contents of the article, so as to cultivate the students' English reading abilities.

4 Conclusion

To sum up, the visual teaching has become the main way of College English teaching. The integration of the streaming media technology, the visual classroom and the college English teaching activities can help students form a clear and intuitive impression of the contents of English teaching, and on this basis, carry out the in-depth study to improve students' language expression. And the ability to use English plays a vital role. The visualization teaching mode is an innovative teaching mode. In College English teaching, it makes college English teaching more scientific and standardized, promotes the comprehensive improvement of students' English abilities in the process of the continuous application, and then promotes the development and perfection of our English teaching.

References

1. Wang, Y.: Interactive teaching strategies under the reversal classroom teaching model – a case study of college English course. Data Cult. Educ. (11), 179–181 (2015)
2. Li, Y., Ma, S., Huang, R.: Learning analysis technology: design and optimization of the service learning process. Open Educ. Res. (05), 18–24 (2012)
3. Wei, S.: Learning analysis technology: mining the value of the educational data in the age of big data. Mod. Educ. Technol. (02), 5–11 (2013)
4. Zhang, Y.: Application of the streaming media technology in video on demand. Comput. CD Softw. Appl. (01), 33–34 (2010)
5. Li, W.: Application of the streaming media technology in the video on demand. Video Game Softw. (14), 54–56 (2012)

The Network System of Scientific Research Management in Local Undergraduate Colleges and Universities Based on the Web Platform

Yawen Zhou[✉]

Mianyang Teachers' College, Mianyang 621000, Sichuan, China
lihai204927@163.com

Abstract. Scientific research, as the basic function of universities, has become an important criterion for evaluating the quality of universities, and also a powerful indicator for influencing the reputation of universities. How to do a good job in the management of scientific researches is a topic that all kinds of colleges and universities discuss together. To improve the scientific research abilities and qualities of colleges and universities requires that the scientific research management of colleges and universities should be combined with the modern management science and absorb the advanced management ideas and concepts. It has become an urgent and important task for the scientific research management departments of colleges and universities to use the advanced information technologies to build a systematic platform for sharing the scientific research resources and an overall operating scientific research management network to realize the whole-process management of scientific research projects.

Keywords: Web Platform · Locality · Undergraduate colleges and universities · Scientific research management · Network system

Through the detailed requirement analysis of the scientific research management informationization in colleges and universities, the work flow of each branch in the scientific research management is sorted out, and the business flow diagram of the system is abstracted. The system follows the design principles of openness, practicability and advancement, and the data structure system and the function application system of the platform are constructed. By using the object-oriented design method, the design of the human-computer interaction interface of the system has been fully considered. A multi-user, efficient and networked information platform system for the scientific research management in colleges and universities has been preliminarily designed and constructed, and some application results have been achieved.

1 Design Requirements of the Scientific Research Management Network System in Local Undergraduate Colleges and Universities Based on the Web Platform

The scientific research management system (SRM system) has a wide range of objectives. Some SRM systems focus on the management of the scientific research funds and declared projects, while some SRM systems focus on the preservation and

© Springer Nature Singapore Pte Ltd. 2020
M. Atiquzzaman et al. (Eds.): BDCPS 2019, AISC 1117, pp. 1632–1638, 2020.
https://doi.org/10.1007/978-981-15-2568-1_228

update of the scientific research data. As a management system, the development of the SRM system must also have the operation and management of the corresponding database and the front desk for the customer service.

Before the system development, the user needs are investigated and analyzed. For university teachers, they hope to report their scientific research results, apply for the acceptance of the scientific research projects and inquire about the scientific research contents of other teachers through the Internet. For university-level scientific research managers, they hope to obtain the scientific research information of their teachers online, and to verify their authenticity in time, while grasping the overall scientific research work in an all-round way. For the scientific research managers at the school level, their work is to master and supervise the scientific research work of the entire school in general, and to formulate the corresponding policies and issue notifications accordingly.

The system administrators need to monitor and manage the operation of the system comprehensively. On the basis of a clear understanding of the functional requirements of various users for the system, the detailed requirements for the system are put forward from the business point of view, including the addition, verification, modification and deletion of the scientific research information, the user registration and the personal information maintenance, the check-up and termination of the application for the school-level projects, the issuance of the news announcements, the combination of queries, statistics, reports and the authority matching. In the system architecture design stage, the same or the similar functional requirements of different users are merged into the same functional module. The system platform is structurally divided into the research portal and the research management system, and the sub-modules of the research management system are partitioned in details.

2   The Design Background of the Scientific Research Management Network System in Local Undergraduate Colleges Based on the Web Platform

At present, in the establishment of the local undergraduate colleges and universities, the management of scientific researches is under the overall responsibility of the specialized scientific research management departments, which are the main body of implementing the scientific research management, mainly responsible for the organization, coordination and management service functions of the scientific research work in schools. Some universities can meet the increasing demands for the scientific research project management and the transformation of the scientific and technological achievements, further promote the integration of the industry, university and research, and effectively improve the service levels of the scientific research management by optimizing the establishment of institutions and the resource allocation and setting up the independent achievement transformation centers and the science and technology parks.

In addition, most of the local undergraduate colleges adopt the two-level management mode. That is, the scientific research management departments at the school level (such as the science and technology department, and the social science department and so on) are in charge of the overall scientific research work of the school as a whole, and the second-level colleges (departments) (such as the scientific research leaders, and the scientific research secretaries) are in charge of the scientific research management as "upload and delivery", which is mainly responsible for the specific affairs such as conveying notifications, organizing and declaring scientific research projects, and reviewing and awarding achievements. Unlike the school's scientific research management departments, such as the neat personnel, and the sound institutions and so on, the secondary colleges (departments) are not equipped with the specialized office space, and the personnel are extremely unstable, and the scientific research secretaries are also set up as the teaching secretaries, not to mention the professionalization of the scientific research management personnel.

Most of the traditional scientific research management modes of local undergraduate colleges adopt the two-level management. The second-level colleges generally rely too much on the scientific research management departments at the school level. They have not established their own scientific research management system, or even are only responsible for the upload and delivery of the work. They have not really carried out the work of the form review, the project supervision and the result identification. On the other hand, there are many kinds of projects that the school undertakes and participates in, which require the scientific research managers to devote a lot of energy to carrying out a series of the chain work, such as the project declaration mobilization, the mid-term inspection, the final acceptance, the result appraisal, the scientific research awards and the result transformation. The scientific research managers are tired of doing the above-mentioned complicated and trivial transactional work, and it is difficult to achieve the meticulous management.

Some colleges and universities have not carried out the scientific and systematic training for the scientific research managers, which makes the management level unable to be effectively improved. On the other hand, the managers should not only face the professional teachers in schools, but also involve various kinds of the science and technology. Therefore, they should understand not only the direction of the scientific research management, but also the discipline and professional knowledge. However, most of the scientific research management personnel are not from the management background, and because of the heavy workload, they cannot systematically learn the management and professional knowledge, and have no time to think about the practical problems such as the unit development planning, which leads to the failure of the proper role of the managers to effectively reflect and play.

3 The Design Ideas of the Scientific Research Management Network System in Local Undergraduate Colleges Based on the Web Platform

Although the scientific research management is an important part of the work of local undergraduate colleges and universities, there are still many problems and shortcomings, which require colleges and universities to fully explore the innovative institutional mechanisms and strengthen the reform of the evaluation system. The managers should enhance their learning awareness, establish their responsibility and service awareness, and broaden their management ideas, and the scientific researchers should set up the correct concepts, deal with the relationship between teaching and scientific researches, avoid short-sightedness in scientific researches and take the road of the sustainable development in their scientific researches.

3.1   System Technology Architecture

The overall architecture of the system belongs to the typical J2EE multi-tier structure. It follows the standard three-tier system architecture, adopts JAVA and the related component technologies, adopts the Oracle database as the underlying data storage platform, and adopts the Weblogic application server as the system operation supporting platform in the middle layer of the system. The system applies and integrates the corresponding intermediate components and the customized development of the system applications to achieve the overall functions of the system. The client layer of the system is implemented by the IE browser. WBM has two implementations, which develop in a balanced way and do not interfere with each other. The first is the proxy scheme, which adds a Web server to a network management workstation (proxy). This workstation communicates with the end devices in turn. The browser users communicate with the agents through the HTTP protocol, while the agents communicate with the end devices through the SNMP protocol. The common methods are CGI, JAVA, and the ACTIVE-X technology. It can be operated conveniently and efficiently with various devices, and is more suitable for the management of the large-scale network systems. The second way to implement WBM is embedding. Web functions are really embedded in the network devices. Each device has its own Web address. Managers can easily access the device through browsers and manage it. This technology mainly implements the Web embedding of each device to distribute the data in MIB to browsers and execute the corresponding network management commands such as SET and GET, so as to configure the equipment and collect the management data. The main technology involved in implementing WBM is HTML, which is a language used to generate pages that users see when they browse the WWW home page. HTML is used to construct the expressive information and provide the hyperlinks to other pages.

3.2   Application Process of Synchro FLOW in the Scientific Research Management System

The scientific research management system is established under the environment of the unified identity authentication system, which realizes the information sharing, the data interaction and the resource unification, and realizes the overall management and planning. The scientific research management system provides a unified way of the information access for the application system, so as to thoroughly solve the problems in the informatization of the scientific research institutes. The SynchroFLOW is based on the most popular and potential J2EE platform, and is developed with WFMC as the standard. It has a graphical interactive user interface, good stability, expansibility, security, distributed transaction management and flexible business process change functions. It can easily integrate the existing systems, the middleware and the e-commerce applications of the scientific research institutes. It is an ideal platform for developing, deploying and managing the key business. The workflow system has a strong flexibility in dealing with the business processes. It can quickly and arbitrarily compose the suitable processes according to the actual needs. These processes can basically solve all the scientific research approval businesses. When dealing with the business data, the workflow can easily extract the business data, so that the business data can also collaborate with each other, thus improving the flexibility and the scope of the application of the workflow system in dealing with the business processes. The project application process of the scientific research system design can be an independent process, and the generated business data can also be linked with the follow-up process, which can realize the approval and circulation of the scientific research projects.

3.3   Functional Description

Users can determine whether a project is established or not through the project approval. After the project approval, the project teams can be formed and the project implementation plans can be drawn up. Through the implementation plan, the project progress control can be made. In the process of the implementation, the project progress and the personnel arrangement can be adjusted according to actual conditions, which can reflect the current project progress and the progress analysis, the cost analysis, the revenue and expenditure, and the project inquiry, statistics, acceptance and delivery, including the project establishment, the project approval, the task allocation, the task reporting, the project reporting, and the query and statistics modules.

Project Declaration
Pre-establish the projects and establish the project archives.

Project Approval
If the superior department confirms that the project can be applied for, it shall be submitted to the superior department for the examination and approval.

Project Members
After the project establishment, input the project member information to prepare for the task assignment.

Project Reporting
In the process of the project progress monitoring and the execution institutionalization, the phased progress of the project is evaluated and analyzed by means of the project reporting, and various project analysis reports are generated according to the relevant data.

Task Allocation
After the project approval, the project managers or the project managers can assign their respective tasks and phased work arrangements to the project personnel through this module, preset the progress and objectives of the tasks, and use this module as the basis for the task monitoring.

Mission Reporting

After receiving the project tasks, the project personnel regularly report the project implementation to the project managers and the supervisors according to the schedule. The system generates the task execution report according to the report, and evaluates the execution abilities of the project personnel according to the project task assignment indexes.

Project Funding

During the execution of the project tasks, the project funds are recorded in the system and controlled by the project budget in the real time. Project leaders and supervisors check the expenditure situations at any time, and evaluate the expenditure situation of the project funds according to the allocation indexes of the project tasks.

Project Review

The project evaluation module includes three functions: the project milestone review, the project change review and the project acceptance review.

Project Results

After implementing the project tasks, the project results are archived. The system provides modules such as the document catalogue management, the result center, and the result query and so on. The document catalogue management module can make the document administrator set and maintain all document catalogues conveniently, so that the document system can strictly follow the unit's knowledge management system. Other users can maintain the corresponding document catalogues after authorization in the result management, and add and delete documents independently. All document catalogues will become the user's available document center after they are published. Users can access the result documents with the access rights. The system provides powerful mining tools for users to customize the text retrieval methods, which can quickly locate the required information.

4   Conclusion

In the process of developing this system, firstly, the data table is established in the SQL database management tool, and then the corresponding classes and the method classes to complete the basic operation are created. The elements of the page are designed according to the user's requirements. The specific functions are implemented in the aspx.cs page. This paper submits the scientific research information and the approval of the scientific research information online. The implementation process of the main functions such as the scientific research information query, the statistical report and the user authentication is explained in details. The design idea of the user interface of the system is also briefly introduced in this paper. Then it summarizes the key technologies used in the process of the system implementation, such as the master page, the RDLC report, and the Ajax technology and so on, and describes these key technologies in details, and gives an application example of the system.

Acknowledgement. Higher Education School-based Research Project, No. XBYJ201906.

References

1. Hu, K., Liao, X.: Records of the construction of the science and technology innovation platform of Hunan Agricultural University. Exp. Technol. Manag. (3), 27–30 (2013)
2. Diao, S.: Practice and exploration of the laboratory construction and management in colleges and universities. Exp. Technol. Manag. (6), 233–235 (2015)
3. Yang, D., Guo, H.: Design and implementation of the scientific research management system. Microprocessor (1), 43–45, 49 (2016)
4. Liu, Z., Wang, X., Yan, S.: Design and implementation of the network management system for the scientific research and achievement – taking Mudanjiang Medical College as an example. J. Mudanjiang Med. Univ. (6), 113–114, 115 (2014)
5. Yuan, G., Liu, Z.: Design and implementation of the scientific research information management system in universities. Stat. Manag. (6), 131–132 (2014)

The Teaching Quality Assessment System for Application-Oriented Undergraduate Colleges and Universities Based on the WEB Platform

Yanfang Zong[✉]

Zhengzhou University of Science and Technology, Zhengzhou 450064, China
liyi011169@163.com

Abstract. With the development of the higher education in China, many new problems and challenges arise in the teaching and management of colleges and universities. With the increase of the enrollment in colleges and universities, the quality of graduates decreases and their ability to work is questioned. The teaching quality of colleges and universities has become the focus of our social attention, so it is particularly important to effectively evaluate the quality of our teaching. The teaching quality assessment is one of the important means to check the teaching quality of colleges and universities, which is a routine assessment work carried out every semester in colleges and universities. The evaluation of the teaching quality is of great significance to schools, teachers and students. The teaching quality evaluation system based on the web platform developed in this paper is a convenient and efficient teaching quality evaluation system adapted to the current new situation.

Keywords: WEB platform · Application-oriented undergraduate course · University teaching · Quality assessment · System design

At present, the new technological revolution relying on the development of the Internet technologies has realized the extension of the mental labor and the enhancement of the intellectual capital. Faced with the unprecedented changes, our need for higher education is more urgent than ever, and our thirst for scientific knowledge and outstanding talents is stronger than ever [1]. The development and construction of the applied undergraduates with Chinese characteristics, the times characteristics and the distinct characteristics will make Chinese higher education unique in the field of the higher education in the world.

1 Design Background of the Teaching Quality Evaluation System for the Application-Oriented Undergraduate Colleges and Universities Based on the WEB Platform

1.1 Application Status of the WEB Platform

From the beginning of programming, it is unavoidable to deal with methods, classes, interfaces and other things. Over the time, this will naturally be summarized, resulting

M. Atiquzzaman et al. (Eds.): BDCPS 2019, AISC 1117, pp. 1639–1645, 2020.
https://doi.org/10.1007/978-981-15-2568-1_229

in the development platform. With the popularity of the Internet, it is naturally required to use the Internet as the basis for the networking to realize the sharing of the network resources. This stimulates the creativity of the software developers and forms a web development platform [2]. The Web development platform provides the design and development tools to support the layout of the Web interface, which greatly improves the development efficiency. It provides the base class system and the interface template library of the basic business document development, as well as a large number of the basic components [3]. At the same time, it integrates all kinds of services, making the functions of the business development easy to use these common services for the collaborative work, and making the integration and deployment of the business functions more convenient and easier.

Based on Web 2.0 and the software on the basis of the Internet B/S architecture, it is also the software itself, but not the final software product, but the software for the secondary development. Used for the development of various business systems, such as CRM, MIS, ERP, HIS, and OA and so on, the web development platform itself is a secondary development platform [4]. It is different from other development tools such as eclipse. It needs to write a lot of codes to complete various business modules. The secondary development platform only needs to design its own data structure and do some simple configuration and drag-and-drop operations to complete the opening of the business system. It can quickly realize the design and production of various reports, including the complex charts and reports [5]. The traditional coding development needs to compile all kinds of reports by itself. It is much simpler to use the development platform, which can easily configure all kinds of reports. The purpose of using the web development platform is to realize the resource sharing, save the development cost, improve the development efficiency and shorten the development cycle.

1.2 Background of the Teaching Quality Assessment in Applied Undergraduate Colleges and Universities

To build a first-class applied undergraduate course, we must lay a solid foundation, base on the connotation and requirements of the undergraduate education, follow the laws of the development of the higher education and the undergraduate teaching, and firmly establish the belief of "great plan for higher education, undergraduate education as the foundation, unstable undergraduate course, shaking the hills" in order to realize the innovation of the undergraduate education, the teaching innovation, the personnel training innovation and the practical training innovation. Applied undergraduates must follow the laws of the development of the higher education, and make clear their orientation by reflecting, summarizing, refining and explaining the development process of schools and the concept of running schools. Emphasis should be placed on the application, and we should vigorously promote the integration of the industry and our education, do a good job in the school-school, school-enterprise, school-site and school-institute cooperation, promote the organic link between the industrial chain, the innovation chain, the education chain and the personnel chain, and effectively improve the ability to serve the local economic and social development.

From this we can see that the current situation of the evaluation of the quality of the applied undergraduate personnel training in China is really worrying. Therefore, one of the most urgent issues in the field of the higher education is to re-examine the current paradigm of the teaching quality assessment in China's colleges and universities, and to scientifically evaluate the quality, so as to form a scientific and efficient paradigm of the "education quality". Reviewing the existing information at home and abroad, there are abundant research results on the three themes of "International Higher Education Quality Assessment Model", "Subject of Higher Education Assessment" and "China Higher Education Quality Assessment System". However, there are few studies on the paradigm of the undergraduate teaching quality, let alone the "student-centered" paradigm of the teaching quality assessment.

2 Design Contents of the Teaching Quality Evaluation System for Application-Oriented Undergraduate Colleges and Universities Based on the WEB Platform

The application-oriented undergraduate teaching quality evaluation system based on the WEB platform can collect the data of the teachers' teaching work conveniently and comprehensively, provide the results of the teachers' and the students' online evaluation of the teaching, and collect all kinds of the evaluation information quickly and centrally, so that the educational administration department can timely understand the teaching dynamics and the teachers' situation, and provide reference for teachers of the educational administration. It provides the scientific basis for the teaching quality and reduces the workload of the teaching teachers.

The system users are divided into three categories: students, teachers and administrators. Students can choose the courses on the Internet and mark and leave the messages for teachers and their courses. Teachers can check students' evaluation results and messages and evaluate the teaching quality of their peers. Managers can query and count the information of students and teachers' peer evaluation of their teaching. At the same time, teachers can view teacher list and print. Queries can display different query results according to different query conditions, and different results can also be displayed according to the statistical requirements.

In recent years, the enrollment scale of colleges and universities has been expanding, and the quality of our teaching has become an important issue of concern to the leadership of colleges and universities. In order to improve the quality of our teaching, colleges and universities have begun to develop the teaching quality evaluation system, but these systems still have problems such as the single function, the backward technology, and the lack of security and so on. Combining the characteristics and advantages of the web technology and the UML modeling technology, a teaching quality evaluation system based on the Web platform is constructed.

Combined with the basic business of the teaching quality evaluation in colleges and universities, the functions of the system are analyzed and set up. The core functions of the system are divided into the system management, the quality evaluation, the resource management and the evaluation query modules. Then, the business needs, the use case analysis and the design objectives of the system are combined. Finally, some codes and interfaces of the core functions of the system are implemented, and the basic methods of the software testing are used to test the functions and performances of the system to verify the feasibility and practicability of the system.

The teaching quality evaluation system designed in this paper mainly uses the modern technology to realize the automation and informationization of the resource management, the quality evaluation and the evaluation query, to realize the networking, automation and systematization of the management of the teaching quality evaluation system, and to improve the efficiency of the teaching quality evaluation in colleges and universities. At the same time, it strengthens the communication among teachers, experts and students, realizes the network transmission and processing of the teaching quality evaluation, and provides scientific basis for the teaching decision-making in colleges and universities.

3 Operational Authority of the Teaching Quality Assessment System for the Applied Undergraduate Colleges and Universities Based on the WEB Platform

The system is divided into the student operating interface, the teacher operating interface and the administrator operating interface.

3.1 The Student Interface Can Perform the Following Tasks

Personal information: Through this function module, students can view their school number and class information. Online teaching: Through this function module, students can select courses and evaluate teachers of their selected courses. Others: Through this function module, students can modify their passwords, ensure the user security, and be able to exit the system.

3.2 The Teacher Interface Can Perform the Following Tasks

Personal information: Through this function module, teachers can view their own teacher number and the information of their department. Information management: Through this function module, teachers can check students' scores and messages on their courses. Others: Through this function module, teachers can modify their passwords and exit the system.

3.3 The Administrator Interface Can Do the Following

Student management: Through this function module, administrators can view the student information and add, update and delete students. Teacher management: Through this function module, administrators can view teachers' information and add, update and delete teachers. Course management: Through this function module, administrators can view the course information and add, update and delete the courses. Statistics: Through this function module, administrators can check the scores of students and their teachers, as well as the ranking of the evaluated teachers in the department. Others: Through this function module, administrators can modify the passwords of students, teachers and administrators, and can exit the system.

4 Design Mechanism of the Teaching Quality Evaluation System for the Application-Oriented Undergraduate Colleges and Universities Based on the WEB Platform

The quantitative evaluation of the teaching quality is a necessary prerequisite for the effective data mining of the teaching quality evaluation. Because there are many qualitative elements and few quantitative elements involved in the evaluation of the teaching quality, the qualitative criteria can only be given in general, and the flexibility of the criteria is relatively large. In addition, the biases of the evaluators in grasping the criteria and their subjective perceptions increase the difficulty of the quantitative analysis of the teaching quality evaluation. The main characteristic of the analytic hierarchy process (AHP) is that it combines the qualitative analysis with the quantitative analysis, expresses the human subjective judgment in the quantitative form and deals with it scientifically. It fits well with the characteristics of the teaching quality evaluation and reflects the main problems it faces more accurately. So firstly, the evaluation index system of the teaching quality is established by using the analytic hierarchy process, and a multi-level evaluation model is put forward. The weight of each index is calculated by using the analytic hierarchy process to make the evaluation result more scientific and effective.

The Web development platform can be realized in two ways: the positive and the negative generation. Taking Hongtian EST-BPM platform as an example, the reverse generation is adopted. The specific operations are: drawing E-R pictures well, configuring parameters of the generated code, executing the Ant script, generating the configuration files and the corresponding pages from the DAO layer, the business layer, and the control layer to the display layer. Pages include CRUD, pagination, combination, query and other functions. A good web development platform will generally include the existing ones such as the drop-down boxes, the pop-up dictionaries, the date selection boxes, framesets, and tabs and so on.

Aiming at the quantitative evaluation index of the teaching quality based on the analytic hierarchy process (AHP) and the fuzzy comprehensive evaluation method, the improved decision tree algorithm and the Apriori algorithm are used to mine the data set of the teaching quality evaluation including the evaluation data of the teaching quality and the quantitative evaluation index obtained by the above-mentioned analytic hierarchy process (AHP). Specifically, firstly, the quantitative index system is discretized into four intervals, corresponding to the four subjective evaluation indicators of the teaching quality, namely "unqualified", "general", "good" and "excellent". The main goal of using the decision tree technology is to obtain an effective decision tree through the data mining and the learning of the sample data sets. Through the discrimination of the decision tree, the evaluation data of teachers' teaching quality input into the teaching quality evaluation system are classified into one of the four subjective evaluation index categories, so as to complete the effective evaluation of the teaching quality. The goal of using the Apriori algorithm is to extract the representative association rules from the teaching quality evaluation data sets, including the teaching quality evaluation data and the discrete quantitative indicators. These association rules represent two types of the implicit information in the data sets. The first is the implicit relationship between the teaching quality evaluation data, and the second is the implicit relationship between many teaching quality evaluation data and the discrete quantitative indicators. No matter what kind of the implicit information it is, it can be used as the result of mining the data of the teaching quality evaluation and play an auxiliary role in the decision-making of improving the teaching quality of the relevant departments.

The hierarchical analysis method is helpful to make more effective use of the advantages of the data mining technology to mine the information of the teaching quality data samples. For colleges and universities, the establishment of a scientific teaching quality evaluation system is an effective mechanism for the management of colleges and universities, as well as the main platform for the collection, processing and analysis of the teaching information. The problems reflected in the scientific teaching quality evaluation system can help guide universities to deepen the comprehensive reform of the higher education, such as the adjustment of teachers and the teaching reform. The evaluation system is generally a hierarchical and multi-objective complex system, and the commonly used method to establish such an evaluation system is the analytic hierarchy process (AHP).

5   Conclusion

As an important part of the higher education, the application-oriented undergraduates must actively adapt to and respond to the changes in the external environment, enhance their ability to predict the changes in the external environment, and constantly adjust their structure so as to make the orientation adapt to the development of the regional economy and the society, the personnel training system to the social changes, and the scientific research to the needs of the enterprises in the industry, give full play to their own advantages and realize their self-worth.

Acknowledgements. The study was supported by Ministry of Education Vocational College foreign language teaching reform project: vocational college non-foreign language student's speculative ability evaluation research [FLAB005]; Zhengzhou high educational English teaching team construction project [2016344]; Zhengzhou University of Science and Technology teaching quality project [201667]; Study on the application of cultivating college English teaching environment under the information technology in the internet era [ZZJG-B7019]; Exploration of "New Engineering" talent cultivation mode for Foreign Language Majors under the background of "Belt and Road" [ZZJG-B8021].

References

1. Zheng, L.: A divided and conquered Apriori algorithm for the direct generation of frequent itemsets. Comput. Appl. Softw. (4), 297–301, 326 (2014)

2. Lin, J., Huang, Z.: Improvement of the Apriori algorithm based on the array vectors. Comput. Appl. Softw. (5), 268–271 (2011)
3. Li, K., Chang, Z.: Design and implementation of the parallel data mining system based on the cloud computing. Microcomput. Inf. (6), 121–123 (2011)
4. Li, F.: Design and implementation of the business intelligence system based on the data mining. Mod. Electron. Technol. (11), 152–155 (2016)
5. Qi, H.: Application of the Apriori algorithms in the data mining of the user behavior monitoring in ACViS. Netw. Secur. Technol. Appl. (5), 47, 49 (2016)

The Design of English Reading Resource Library Based on WeChat Public Platform

Xiong Ying[(⊠)]

Wuhan College of Foreign Languages and Foreign Affairs, Wuhan 430000,
Hubei, China
aasriisc@163.com

Abstract. Reading is an important learning goal for college students, which can get information and improve their quality, it is particularly important in English learning. The construction of WeChat public platform's English reading resource library can expand the actual time and space dimensions of campus. This paper uses WeChat public platform to build an English reading resource database, and studies how to apply WeChat public platform to extracurricular online education and teaching effectively, so as to promote the development of College English education.

Keywords: WeChat public platform · English reading · Resource library

1 Introduction

With the rapid development of new media, new communication technology and mobile Internet, the popularity of electronic products such as computers, tablets and smart phones has increased sharply, the number of Internet users is increasing fast. As of December 2017, the forty-first China Internet development statistics report. The scale of our country's netizens is 772 million, the popularity rate is 55.8%, exceeding the global average (51.7%) 4.1% points, exceeding the Asian average (46.7%) 9.1% points, and students are the largest group of netizens, accounting for 25.1% of the total number of netizens in the country [1]. The total number of new netizens increased by 40 million 740 thousand over the whole year, the growth rate is 5.6%. The number of netizens in China continues to maintain steady growth. Among them, the number of mobile phone users in China is 753 million, and the proportion of netizens in mobile Internet users has increased from 95.1% in 2016 to 97.5%. In the 2017 WeChat data report, we learned that the average daily user of WeChat has reached more than 700 million, compared with the same period last year, it has increased by 35%. It can be seen that more and more people turn to WeChat, and WeChat's user resources are expanding. According to the results, all the students in my class are using WeChat, more than 1/3 of the students are online for 24 h, and more than 65% of the students have more than 8 h per day online [2, 3]. The data prove that the building of a reading resource library based on the WeChat public platform has a good basis for use.

© Springer Nature Singapore Pte Ltd. 2020
M. Atiquzzaman et al. (Eds.): BDCPS 2019, AISC 1117, pp. 1646–1651, 2020.
https://doi.org/10.1007/978-981-15-2568-1_230

2   Overview of the WeChat Public Platform

2.1   The Development and Application of the WeChat Public Platform

WeChat is a free chat software similar to Kik's free instant messaging service, which is launched by the Tencent Inc. Users can quickly send voice, video, picture and text through mobile, tablet and web pages, and share the wonderful content to the WeChat circle [4]. Users can also add friends and pay attention to public platform by shaking, searching numbers, nearby people, scanning two-dimensional code. In order to further expand the exchange learning, entertainment and work. The Tencent Inc has added a WeChat public platform on the basis of WeChat. Through this platform, individuals and enterprises can create a WeChat public number and can communicate and interact with a specific group of words, pictures and voice. Teachers can apply WeChat public platform to course teaching, use its powerful functions to integrate and timely release related teaching resources effectively for students to learn and use. Teachers can also use WeChat public platform to send group messages, push new micro learning resources to students every day, and students can also send problems to the platform, so as to achieve one to one interaction between teachers and students [5]. In addition, after the students log on the platform, each student's learning process can be retained in the database of the platform backstage, which is convenient for teachers to collect and analyze the students, and the first time to understand the students' learning situation. In a word, the application of WeChat public platform in teaching can alter or enrich learners' knowledge structure imperceptibly. It can be used as an effective supplement to the teaching of teachers [6].

2.2   The Features of the WeChat Public Platform

WeChat users are not restricted by time and space. They can receive messages, ask questions, leave messages, and consult at any time and place. WeChat public platform provides one to one dialogue and answer service for users' special questions. WeChat's publication and reply are immediate, so when the user submits the problem, they can immediately solve the user's problem, as long as WeChat online, there is no need to be limited by time and space.

WeChat public account can achieve precise push of messages through user grouping and geographical control. Compared with micro-blog, WeChat is the dissemination of point to point, which more precise and efficient, thus WeChat users will not miss any information. When a WeChat public account releases a message.

All concerned users will receive immediately, and will be classified according to different concerns. They will not be missed due to information flooding. In addition, WeChat's one to one communication mode is confidential highly. users can be expressed by WeChat when they can't express publicly the problems or ideas. This mode greatly reduces the possibility of information leakage, and helps to protect user privacy.

3 Construction of English Reading Resource Library Based on WeChat Platform

3.1 Collate the Materials Needed for the Construction of Reading Resource Library

At present, the College English application ability test is divided into 2 levels: A and B. A is the standard level, covering all the contents of the basic requirements. B is an excessive level, which English proficiency requirements is slightly lower than A. the author collects the material of two modules of A and B level according to these two exam levels. The type of reading questions was selected from the "four choice one" question, which was commonly seen by students. Each level selected 30 reading articles and 5 selected questions for each article, according to the proportion of scores and the shortcomings of students in reading questions. Each reading is equipped with short video and additional reading materials in order to enhance students' interest in reading and help students understand background knowledge while reading. The selection of video and additional information should consider not only the relevance between reading and understanding of articles, but also the difficulty and interest. In addition, all the materials are taken from foreign video websites and articles, which ensure the purity and accuracy of language.

3.2 The Construction and Design of the WeChat Public Platform

The account of WeChat public platform is divided into subscription number, service number and enterprise number. I applied for a subscription number suitable for individual users, according to the actual situation of the applicant. Its main functions include: Group sending: the official number initiatively pushes important notice or interesting content to users. Automatic reply: users can extract regular messages from the public number on the basis of the key specified by the platform number. One to one communication: the official number provides users with 1 to 1 dialogue service. Set up a teaching resource library on the WeChat public platform, we need to be the developer of WeChat platform first, and developers must apply for private servers for background development. Only by applying a private server to build a bridge of mutual information between public platforms and WeChat users can the WeChat public platform be created, in order to make WeChat public platform more complete, and more in line with the conditions and requirements of the implementation platform of teaching resource base.

It can be concluded that WeChat users receive and send information through the WeChat backend server. The information connection between WeChat users and the WeChat public platform is erected by the server, WeChat users are not directly related to WeChat public platform. Building private servers is the foundation and key step of developing WeChat public platform. To build WeChat backstage server for teaching resource platform, we need to rent private servers. After applying for the WeChat backend server, we need to rent private servers to build WeChat backstage server for teaching resource platform to achieve the "handshake" of WeChat backstage server and WeChat public platform. The above steps realize the timely interaction and relationship between WeChat backstage server and WeChat public platform. The development of WeChat public platform is completed, which can be used as a teaching resource platform and applied in practical teaching practice. Learners can access the platform, accept and feedback corresponding learning information only needs to use a mobile intelligent terminal to open WeChat, and scan the corresponding two-dimensional code on the WeChat public platform.

I invite relevant professional and technical personnel to deploy the first level menu and the two level menus on the server in the form of multiple dynamic web pages, and carry out pertinence re-development. At the same time, the students' answer situations (including student's name, class, student answer, standard answer, and time used) are recorded on the server for the teachers' inspection and analysis. Teachers need to edit text, audio, video and other learning resources into material and deposit the platform material library according to the needs of teaching in order to carry out information push. In addition, we can provide relevant network links for students to access the source of the original network directly. Teachers need to focus on users' group management, so as to achieve targeted information push. With the help of automatic reply, the platform can provide users with key information navigation. The automatic response function is divided into three kinds: Be added to automatic reply, Automatic message reply and Keyword auto reply. Be added to automatic reply means that when students pay attention to the public account number, the system will send message messages to students. Teachers can use this function to set up welcome information and briefs on the platform. Automatic message reply means that when a student sends information to the platform, the system sends the message to the student automatically. Keyword auto reply means that when a student sends key messages to the platform, the system returns the matching message to the student.

4   The Application of the WeChat Platform Resource Library

After the completion of the resource library, the author tried out a trial in five classes. Firstly, students log on to the mobile WeChat, scan the platform two-dimensional code and pay attention to the public number, and send their WeChat name, name and class information to the platform for the teacher to group the students according to the class. Secondly, students choose their own columns according to the menu hints on the platform. In the menu of "skill", students can know common types of reading comprehension and problem-solving skills by watching the micro Lesson Videos and text materials commonly used by teachers in college reading skills. The menu of "actual practicing", students enter B level and A-level menu respectively according to their reading ability, and then enter names, classes and student numbers, so that teachers can collect students' answer situations in the background. Thirdly, students choose an actual battle from 30 reading comprehension. After clicking "submission", students can see their answers and correct answers and understand their own problems. Students can also watch additional video and text materials of every reading comprehension, understand the relevant background knowledge and expand reading while practicing. The author investigated the application effect of the platform after nearly 3 months' trial. At present, the number of platform concerns is 273, 151 of them were students in my class. Based on the analysis of 139 valid questionnaires, the author has the following findings.

(1) Students' interest in the WeChat platform
In response to this new form of reading comprehension, 75.1% students expressed interest in reading through WeChat platform. Most of the students think that reading comprehension through the WeChat platform is more convenient and more able to interact with the teachers. At the same time, it is also convenient for teachers to expand the resources and track the students' reading situation.

(2) The use frequency of the WeChat platform
23.5% of the students said they were using it every day, and 41.7% of the students said that every 2–3 angels used it once. More than half of the students said they had completed one level (30) of reading, and 81.5% of the students thought the WeChat platform could effectively increase their reading. The construction of platform as a supplement to offline teaching is undoubtedly very effective for increasing students' reading.

(3) Resource utilization of WeChat platform
For the video resources provided in the platform and the supplementary reading materials, 76.3% of the students watched the video data, and 73.9% of the students believed that these resources were helpful for reading.

(4) Student's advice
Students also put forward a lot of good suggestion for the existing WeChat platform resource library, such as add more online interaction, increase the explanation for the problem of error concentration, add the function of accumulative integral, set the reading time limit, and expand the platform to other aspects of English, and so on.

5 A Summary

WeChat public platform has the advantages of convenience, effectiveness and integration of resources, and its application is more and more extensive. It will also be the help in the field of education and teaching. Teachers and students can interact and communicate with each other with the help of the platform, and teaching is centered around the interaction between teachers and students. With the help of WeChat platform, teachers can effectively integrate the resources of pictures, texts, voice and videos in the form of students' interest. In addition, the large-scale coverage of mobile Internet and the improvement of campus wireless network have greatly promoted students' Extracurricular online learning. With the continuous improvement of WeChat public platform technology, its application prospect in the field of university education will be broader. Therefore, teachers should actively explore and practice, make full use of the WeChat public platform to develop more suitable curriculum resources for students, enhance teaching effect and improve teaching management.

References

1. Zhang, L., Gao, W.: The construction and development of College English WeChat mobile learning platform. J. Shaanxi Youth Career Acad. (01) (2017)
2. Jia, L., Zhang, G., Shi, C.: Design and practice of the medical English reading flipping course based on the WeChat public platform and micro community. Comput.-Assist. Foreign Lang. Educ. (02) (2016)
3. Qu, Y.Q.: The construction of WeChat mobile learning platform in higher vocational colleges. J. Changsha Telecommun. Technol. Vocat. Coll. (04) (2017)
4. Liu, Z.: Feasibility study of introducing authentic materials into College English teaching under WeChat public platform. Chin. Lang. J. (01) (2016)
5. Wang, Q.: Research on the teaching mode of College English inversion based on WeChat platform. China Educ. Technol. Equip. (02) (2016)
6. Bao, W., Zhou, H., Xie, F.: An optimal design of web based online English teaching by using struts framework. Bol. Tec./Tech. Bull. **55**(10), 58–63 (2017)

Construction of Online Micro-curriculum Group of Instrumental Music Based on Streaming Media Technology

Jia Fan[✉]

Faculty of Music, Jiangxi Science and Technology Normal University,
Nanchang 330038, Jiangxi, China
ipaper2015@qq.com

Abstract. The instrumental music course is the main course of the music teaching. To promote the construction of the network micro-course group of instrumental music reasonably plays an irreplaceable role in improving students' art practice abilities, music perception abilities, innovation abilities and music aesthetic abilities in an all-round way. As a new intelligent technology, the streaming media technology will give full play to its advantages, facilitate students to learn the relevant professional knowledge and build a vivid and flexible teaching mechanism. The teaching of the instrumental music is an important course in the music teaching. The traditional teaching mode has some disadvantages which affect the students' enthusiasms and effects. By giving full play to the advantages of the streaming media technology, such as real-time and interaction and so on, it can make up for the shortcomings of the traditional teaching.

Keywords: Streaming media technology · Instrumental music · Network micro-curriculum group · Construction research

Nowadays, with the rapid development of the science and technology, the application related to the information technology has gradually penetrated into people's daily life. With the support of the network and the communication technology, the data processing and the multimedia technology has gradually penetrated into various fields [1]. Especially with the development and progress of the network facilities in our country, especially with the general development and progress of businesses, multimedia has a strong market power in this context.

1 Overview of the Streaming Media Technology

1.1 Streaming Media Technology

The so-called streaming media technology refers to a media format which adopts the streaming mode in the transmission process of the network. The main mode of streaming is the transmission of audio and video on the network [2]. The main method of the streaming media technology is to compress the multimedia information and put the final results on the server for customers to download or listen to. The biggest

M. Atiquzzaman et al. (Eds.): BDCPS 2019, AISC 1117, pp. 1652–1657, 2020.
https://doi.org/10.1007/978-981-15-2568-1_231

advantage of the streaming media technology is that users can listen or watch at any time without downloading all the information. In the process of the traditional audio transmission, the entire file is needed to be downloaded before viewing [3]. The streaming media technology solves this problem, realizes the efficient transmission and opens a new era of the information media. The application of the streaming media technology in the micro-course teaching can give full play to the advantages of the dissemination and application of the teaching resources, which not only does not need to change the current network mode, but also can receive the high-quality audio information in the context of the streaming media, fully ensuring the user experience [4]. The adoption of the streaming media technology can further reduce the load of the servers, save the investment of the hardware, and ensure the use of the broadband [5]. The streaming media technology has great advantages and utilization value in terms of the technology and the market application value.

1.2   The Realization Way and the Process of the Streaming Transmission

In the process of the streaming media transmission, it mainly consists of the following three steps. First, in a short period of time, if we cannot support a considerable number of the audio files or the multimedia transmission needs, we need to use the streaming media technology to preprocess the audio data to ensure that the preprocessing can solve the quality problem of the video. Second, the traditional media needs to download all the files before they can be used. However, the streaming media does not require the high capacity of the cache, so it can be listened and played at the same time without cache. In the process of playing, if there is the limitation of the network broadband, it will lead to the interruption of the information, and the transmission of the data will also lead to the decrease of the speed due to the interruption of the network. Adding the streaming media technology will have a certain buffering effect on the data, and the new mode will not make the video data smoother. The data buffering refers to the process that the streaming media needs to process the data before playing the file and leave it in the cache system, which is a preset process. To cache the data and play it in advance, in such a background, once there are factors such as the quality of the network transmission and so on, in the process of the file playing, the file will run normally because of the data retained. Third, there needs to be a suitable transport protocol to ensure the transport of the documents. Both the WWW technology and the TCP protocol are the hypertext protocols, but the implementation of the system needs a lot of overhead and is not suitable for the data transmission. Therefore, when controlling the streaming media information, HTTP/TCP protocol is needed to ensure the real-time transmission.

1.3   Broadcast Mode of the Streaming Media

The so-called streaming media unicast mode mainly refers to the creation of a separate data channel between the client and the media server to transmit each data transmitted by the server. By the unicast transmission, the client can query the media server, and then through the data copy, the media server can send directly to the user according to the data channel. The unicast mode will bring a certain burden to the server, which requires a long response time, and then leads to the network congestion, being unable to play. Only when the service quality reaches a certain standard can the system managers buy more matching hardware or broadband. The so-called multicast method of the streaming media refers to using the IP multicast technology to copy routers to the multiple channels at one time, which can not only promote the multicast method of the system, but also prevent the situation of the delay. If a large amount of the data is transmitted to the client at the same time, only a single data packet needs to be transmitted, so that all the clients can be transmitted to the server, and then as long as they are connected to the data stream, in this way, the broadband can greatly improve the communication efficiency.

2 Requirements for the Construction of the Online Micro-curriculum Group of Instrumental Music Based on the Streaming Media Technology

As a new teaching mode, the micro class mainly focuses on 5 to 10 min to tell the contents of a specific topic. In a busy society, this kind of mode has been widely welcomed on the network with rich forms, wide span and short time. Different from other teaching methods, combining with streaming media technology for system research and development, the micro class can achieve a win-win situation in the teaching and make students fully feel the charm of the micro classes. The network teaching has changed the traditional teaching mode, the teaching contents and the teaching methods have been greatly innovative.

The synchronous teaching is what we call the real-time teaching. Students can choose their own teachers. Education has been promoted and reform has been upgraded. Some famous teachers will have a lot of students. Even in the largest classroom, the number of students is limited. The synchronous teaching can solve this problem. The specific implementation process is to record the teacher's courses into the video and audio by using the camera and the loudspeaker after the teaching starts, input them by the encoder, then encode them into the ASF stream, send them to the server, start the station service module through the video server, and then publish the advanced streaming format (ASF) through the multicast technology, so that students can receive and learn at the terminals.

Although there are some differences in the structure of the streaming media courseware, the whole structure includes the video area, the title area, the handout area and the subtitle area. In this way, learners can listen to the lessons through the Internet, and at the same time, they can also see the key points and contents taught by teachers, and they can also interact with teachers. Students can also cooperate with each other to learn, learn independently anytime and anywhere, and improve the development space of the online teaching.

The interactive teaching refers to that teachers use the long-distance teaching method in teaching, which can be realized by some long-distance teaching equipment, including the camera, microphone, sound card, and handwriting board and so on. These kinds of the teaching equipment connect teachers and students together to form a network class. The teaching video can also be collected through the acquisition card and uploaded to the server, so that students can learn and communicate, and teachers can guide students' learning online, which can improve students' learning effect and the teaching efficiency.

The courseware on-demand teaching is to upload the PPT made by teachers through the network, making it a resource for students to learn. Students can learn the teaching contents of the PPT at any time and place, not limited to a certain time and classroom, and can freely start, pause and carry out other operations when watching the PPT. Therefore, this kind of the teaching method is richer in contents and more flexible in forms. Students' learning plans can be arranged freely. They can not only learn what they are interested in, but also speed up or postpone their learning progress without being consistent with other students.

3   Construction of the Online Micro-curriculum Group of the Instrumental Music Based on the Streaming Media Technology

The micro-course group is a series of the short and concise video courses aiming at the key and difficult points of the course core knowledge. Being short and small refers to the control of the video duration of each micro course within 5 to 10 min, and being shrewd refers to the small incision in the content selection, the refinement and the accuracy of teachers' language, and the refinement of the video production technology. A series refers to the internal connection between the selected knowledge points, which can constitute the basic knowledge backbone of a course.

3.1   Strengthen the Overall Planning of the Resource Construction

For a school, the content of the teaching resources is relatively rich, including the personnel training program, the curriculum planning, the teaching plan, the question bank, the recording and broadcasting, the micro courses, and the electronic materials and so on, so we should make a reasonable plan for many kinds of the information and carry out the construction and sorting. For many schools, the development and utilization of resources should be carried out jointly, and the standards should be unified. The repeated development of resources should be reduced, and the utilization efficiency of the talents and resources should be improved. Gather strength to develop and integrate the high-quality teaching resources, open the credit mutual recognition among different schools, and truly realize the sharing of the educational resources.

3.2   Improve Teachers' Information Technology

We should strengthen and improve the information technology education for teachers, train teachers in the application of the teaching information by levels and batches, and train the multimedia materials and the courseware development technology, the courseware software application and the video production technology. Strengthen the direct communication between the professional teachers and technicians, so that professional teachers' thoughts and ideas can be expressed completely.

3.3   Improve the Online Teaching Resources of Instrumental Music

The complete instrumental teaching resources should include the pre-class handouts, courseware, after-class exercises, exercise solutions and other supporting resources, so that learners can learn at any place without the paper learning materials. Arrange the teaching time according to the knowledge points, fragment the knowledge points, and master a knowledge point without taking too long. According to the different objects, the teaching resources are divided, and the learners choose the learning contents according to their own conditions, and combine the learning contents freely. As long as the corresponding teaching objectives are achieved, the learners with different needs can learn them.

3.4   Strengthen the Maintenance of the Instrumental Teaching Resources

Regularly maintain and update the instrumental teaching resources, and timely add the new technologies into the system. Strengthen the online communication between teachers and learners, answer the questions raised by learners in time, and optimize the teaching resources according to the user feedback. In terms of the school system, teachers are encouraged to use the online courses in the teaching process, and students' credits of using the online courses are recognized, and the teaching resources of the online courses are fully utilized. This way is normalized, so that teachers and students are allowed to adapt to different teaching modes step by step, and the online music teaching resources are the important teaching resources for teachers. Meanwhile, they are constantly enriched and improved, so that students can develop a habit of being diligent in thinking.

3.5   Construction Mechanism of the Online Micro-curriculum Group of Instrumental Music Based on the Streaming Media Technology

The mixed cloud platform construction mode is selected to build the instrumental teaching resource platform in colleges and universities. Promote the effective integration of the public cloud and the private cloud, and achieve the construction of the cloud computing. This is also an ideal model to establish a platform for the professional teaching resources. The resource platform includes the shared data and the sensitive data, and the hybrid cloud model is more suitable. The auxiliary shared data is stored in the public cloud and the sensitive data is stored in the private cloud, so as to achieve a reliable model of building a professional teaching resource platform relying on the cloud computing.

The combination of the mobile and the web modules: The combination of the micro-course modules and the basic teaching courses are suitable for the mobile end and the web end respectively, with the characteristics of "fragmentation". The mode of the effective integration of the college basic teaching and the students' mobile end "fragmentation" learning is selected, which is consistent with the characteristics of college students' learning, and the mode is established. The teaching environment of colleges and universities is composed of the basic teaching environment and the students' micro learning environment. The micro-courses not only have the characteristics of the micro-learning, but also have the characteristics of the mobile learning. According to the corresponding laws of the knowledge development and the characteristics of the educational activities in colleges and universities, the learning activities are consistent with the teaching activities in colleges and universities. The modular structure of the components is selected, which has the characteristics of personalization and reorganization. The service-oriented structural system has the characteristics of being component-based, convenient for the disassembly, multiple and free establishment mode, which is consistent with the reconfigurable instrumental needs of the learners, and its configurable, situational and personalized teaching environment.

4 Conclusion

The streaming media technology also realizes the characteristics of the multimedia, breaking through the limitations of the network. In the process of the instrumental music teaching, the distance teaching can be realized. Students are very convenient in their learning. They can watch the educational resources they need anytime and anywhere. They can also review and do exercises on the network. Through the network, students' homework can be answered, corrected and evaluated, so that students can gradually improve. The use of the streaming media enhances the role of the network. Combined with the campus network of the school, it creates a simultaneous teaching, interactive teaching and on-demand teaching platform for students.

Acknowledgement. School-level teaching reform project of Jiangxi Science and Technology Normal University (Project No.: JGYB-17-78-76).

References

1. Xu, Q.: Reflections on the transformation of the ordinary undergraduate institutions of higher learning. J. Natl. Acad. Educ. Adm. (3), 38–43 (2015)
2. Tang, S.: Analysis of advantages and disadvantages of implementing the large-scale enrollment in colleges and universities. China High. Educ. Res. (1), 88–89 (2009)
3. Liu, J.: Application and practice of the hybrid teaching in the software training course. J. Nanyang Norm. Univ. (3), 61–65 (2016)
4. Wang, M., He, B., Zhu, Z.: Micro video course: evolution, positioning and application fields. China Educ. Technol. (4), 88–94 (2013)
5. Guo, G., Zhang, X.: Study on the observation methods of the community interaction ecological problems in online courses. Distance Educ. J. (1), 71–79 (2014)

The Practical Resources of English Writing in Higher Vocational Colleges Based on the SQL Database

Zhen Fang$^{(\boxtimes)}$

Sichuan Vocational and Technical College, Suining 629000, Sichuan, China
e13685231980@126.com

Abstract. Writing is the core of the English teaching. English teaching aims at the quality education and promoting students' development. It focuses on training students' application levels and abilities of English knowledge, and effectively improves the English writing levels of vocational college students especially through giving full play to the application advantages of the educational resources. The SQL database is a kind of the database software with the function of the data query and update. Through reasonable operations, it can effectively solve the problems in the process of the English writing and improve the writing levels.

Keywords: SQL database · Higher vocational college · English writing · Practical resources

With the development of the information age, the single classroom teaching is difficult to meet the learning needs of students. The differences in the learning bases, purposes and learning styles make students have different choices and uses of learning resources [1]. The English SQL database is rich in resources and forms, which can stimulate learners' interests in learning and improve their learning efficiency.

1 Features and Functions of the SQL Database

SQL is the abbreviation of the structured query language. SQL is a set of operation commands specially established for the database, and it is a database language with complete functions. When using it, we only need to issue "what to do" command, and "how to do" is not considered by users [2]. SQL is powerful, easy to learn and easy to use, which has become the basis of the database operation, and now almost all the databases support SQL [3]. The data architecture of the SQL database is basically a three-level structure, but the terms used are different from the traditional relational model terms.

SQL includes all operations on the database, mainly composed of four parts: the data definition: This part is also called "SQL DDL", which defines the logical structures of the database, including four parts: the definition database, the basic table, the view and the index. Data manipulation: This part is also called "SQL DML", which includes two kinds of operations: the data query and the data update. The data update

© Springer Nature Singapore Pte Ltd. 2020
M. Atiquzzaman et al. (Eds.): BDCPS 2019, AISC 1117, pp. 1658–1663, 2020.
https://doi.org/10.1007/978-981-15-2568-1_232

includes three kinds of operations: insert, delete and update. The data control: The control of the user access data includes the authorization of the basic tables and views, the description of the integrity rules, and the transaction control statements and so on [4]. Rules for the use of the embedded SQL language: Stipulate the use of the SQL statements in programs in the host language.

The SQL database is a collection of tables defined by one or more SQL patterns [5]. An SQL table consists of a row set. A row is a sequence (set) of columns. Each column corresponds to a data item with a row. A table is either a basic table or a view. A base table is a table that is actually stored in a database, while a view is a definition of a table that consists of several base tables or other views. A basic table can span one or more storage files, and a storage file can also store one or more basic tables. Each storage file corresponds to the last physical file on the external storage. Users can query views and basic tables with the SQL statements. From the user's point of view, views and basic tables are the same. There is no difference. They are all relationships (tables). SQL users can be applications or end users. SQL statements can be embedded in programs of host languages, such as FORTRAN, COBOL, Pascal, PL/I, C and ADA. SQL users can also be used as independent user interfaces for end users in the interactive environments.

2 English Writing Requirements of Higher Vocational College Based on the SQL Database

Vocational college students' English writing mainly includes the ability of students to cover practical short essays, fill in English forms or translate intermittent practical texts. Although English writing in vocational colleges is not so high in terms of the vocabulary expressions, the grammar structures and the writing requirements, it is a big challenge for many vocational college students. When many vocational college students write in English, they are often afraid and at a loss, or they don't know how to write when they write. When they write, the structure of the text is often confused, and the language expression is not authentic, and there are many grammatical errors, with poor coherence, and the length of their writing is short. Chinglish can be seen everywhere. On the one hand, it is based on the contents. According to the historical, cultural, educational, sports, scientific and technological themes and 10% of the applied stylistics, the corpus is collected. The entire SQL database has collected more than 1000 compositions of various themes and genres, and has increased at any time according to the teaching needs. The source of the corpus mainly includes simulated questions and VOA English broadcast materials. On the other hand, it is established through the writing of students' proposition compositions. Under the unified regulations, the students who participated in the experiment handed in the electronic documents of their compositions and sent them to the correction network. After many times of correction, they formed the SQL database materials.

The construction of the higher vocational English SQL database extends the English learning time and place from the traditional classrooms and teaching materials to network and anytime, anywhere, making full use of the fragmented time for the fragmented learning. At the same time, the extension of the extracurricular learning time also makes up for the lack of the class hours, so that students can reasonably use their time to learn according to their own needs. The construction of the SQL database provides convenience for students and teachers. For students, the network resource base provides the courseware and the lesson plans for teachers to teach, which is convenient for deepening learning and review. At the same time, it also provides the online testing, so that students can check their learning situations and adjust their learning strategies.

The extended learning area in the resource base can meet the individual learning needs of students, such as the in-depth learning of a certain skill, or preparing for a certain exam, and so on. For teachers, we can use the resource base platform to carry out the in-depth curriculum development, upload and save materials, realize the sharing, avoid a lot of the repeated labor, and improve the efficiency of the lesson preparation and teaching. The construction of the resource base can encourage learners to use the information means to study actively and autonomously, meet learners' individual learning needs, and promote the construction of the learning campus. Various forms of the learning and teaching methods are also the trend of the teaching reform and development.

3   The Practical Mode of English Writing in Higher Vocational College Based on the SQL Database

The pre construction work of English writing teaching platform based on the SQL database mainly includes the collection and arrangement of the writing corpus, and the development and installation of the system software. In the aspect of the SQL database construction, the research group uses the relevant ideas and software provided by the network for reference, and adopts the computer programming technology and the database technology to build a student interactive higher vocational English teaching platform system.

Writing analysis: First of all, students log in to the teaching platform of the English writing based on the SQL database with the student number and the real name as the user name. After finishing the writing task assigned by the teacher, the composition will be sent to the SQL database platform in the form of the electronic manuscript. Comprehensive evaluation is carried out from the basic information, the word frequency statistics, the wrong words, the sentence number, the average sentence length, the average word length and grammar, and the evaluation results are given. At the same time, teachers are required to make further analysis and modification of the results to ensure the objectivity of the final scoring results as much as possible.

Independent learning: Students can access their own composition analysis and evaluation reports on the teaching platform, and with the help of the SQL database retrieval technology, view articles of the same subject and genre, observe and analyze the language rules in the writing, complete the self-study by combining analysis and evaluation reports, and improve their writing levels and self-study abilities. In this process, students improve their interests in learning and can learn English writing with their own free will and full enthusiasms.

Teaching mode feedback: Based on the SQL database platform of the higher vocational English teaching, combined with the traditional classroom teaching, this paper explores the teaching mode of the higher vocational English writing based on the SQL database platform. The teaching of the English writing in the higher vocational colleges gets rid of the shackles of the time and space, and at the same time, it promotes the learning and emotional communication between teachers and students, and between students and students. With the help of the network, students enrich the learning resources and develop students' independent learning abilities through the detailed and objective personal composition. Teachers can also understand students' real writing situations through students' records, which can provide help for teachers to carry out the targeted teaching.

4 The Application of the Practical Resources of English Writing in the Higher Vocational Colleges Based on the SQL Database

4.1 Retrieve the SQL Database

Compare the original contents of the students' writing with the historical data. Today, with the development of various communication channels, it is necessary to establish a data platform that is easy for students to use, so that students can master the writing skills and vocabularies in time, and improve their core competitiveness. In addition, we can collect some external data as the expansion part of the SQL database, such as the contents of the media and the contents created by users from the Internet platform, and improve the data storage, the retrieval query and the copyright management of the SQL database, so as to lay a good foundation for the further transformation and utilization. Because the database management system is a multi-user system, in order to control the access rights of the users to the data and maintain the sharing and completeness of the data, the SQL language provides a series of the data control functions, including the security control, the integrity control, the transaction control and the concurrency control. SQL is a language with the strong query function. As long as the data exists in the database, it can always be found out from the database by the appropriate methods.

4.2 Establish the SQL Database of English Writing in Higher Vocational Education

According to the requirements of the principle of "practicality first, applicability for degree", the sources of the writing information for vocational college students include the English required by their daily communication and the English skills required by their future work. According to the different themes of economy, education, history, culture, sports, and public health and so on, the SQL database which can be retrieved at any time is established. Data update includes data insertion, deletion and modification. They are completed by insert the statement, delete the statement and update the statement respectively. These operations can be performed on any basic table, but are limited on views. When a view is exported from a single basic table, we can insert and modify it, but not delete it. When a view is exported from the multiple basic tables, none of the above three operations can be performed. Data blog SQL database: Based on the social platforms such as the student micro-blogs, blogs and We-chat, as well as the contents and data of blogs and micro-blogs from grassroots, English lovers and professionals, it can also become a clue source of students' English writing. Through the collection and preliminary structural processing, it can be transformed into the SQL database of the English writing information of vocational college students which can be inquired and retrieved at any time for students' learning needs.

4.3   Case Base System

As an independent module of the database system, the case base system can be installed on the server, the workstation and the PC separately. The case database data of the database mainly includes texts, numbers, and pictures and so on. The main databases in the case base include the basic database and the case database. The database stores different data corresponding to different languages. On the one hand, it is used for the case database to manage the memory, query, addition, deletion and modification of the specific language information. On the other hand, it provides the data support for the translation materials. The basic functional requirements are as follows: file function: Browse the information existing in the database (In the browse window, we can also add, modify, delete and carry out other editing operations), file saving, print preview, printing, and exiting editing. The database records can be cut, copied, pasted, added and deleted. With the continuous emergence of new words, the data in the case database is increasing or updating. Therefore, these editing operations must be provided in order to track the current latest data changes, ensure the database update, and ensure the reliability of the design. The help system provides the detailed functions, the operation process, the parameter input requirements, and the functions of various menus and buttons in the database system. The help system should have a general description of the design process and a detailed description of how to use the database system.

4.4   Database System Management

The database system platform is set with the three-level management system of "user registration" and "password protection". The system user levels include the ordinary users, the chief engineer users, and the system administrators. Different users have different use rights. When the case base system is installed separately on the server and the PC and so on, corresponding security measures shall also be taken. General level users: Only the system can be used. Translation and review level users: The database data of the system can be modified, deleted and supplemented. System administrators: The system administrators have the highest authority and can perform all operations on the database. In addition, the database systems can automatically and manually backup the data, and can expand the field and the capacity according to the demand.

5 Conclusion

Under the background of the big data, the vocational college students' English writing is given more choices and more ways. Teachers can make full use of the information brought by the data to polish the teaching process and design more efficiently. Students can enhance their learning abilities and awareness, and strengthen their learning effect. Of course, teachers and students should take a correct view of the data, rather than use it directly without analysis. They should pay attention to the differences in the writing objects and contents, and then make full use of it. It is of great significance to improve the quality of the personnel training, cultivate the compound talents and improve the overall competitiveness of students.

References

1. Guan, H.: Application of the small written language corpus in the writing teaching. J. Jinan Vocat. Coll. (1), 22–29 (2015)
2. Zhu, L.: Training and analysis of listening and speaking abilities in English teaching in the higher vocational education. J. Liaoning High. Vocat. (8), 15–19 (2015)
3. Xi, K.: The optimization method of the database query statement. Digit. Technol. Appl. (11), 29–35 (2016)
4. Ye, C.: Application of the SQL database in the college English writing teaching. J. Changchun Norm. Univ. (Humanit. Soc. Sci. Ed.) (3), 16–21 (2014)
5. Ge, L., Li, G., Liu, B.: A corpus-based empirical study of the college English hierarchical teaching mode. J. Sichuan Int. Stud. Univ. (3), 15–19 (2014)

The Sports Cooperative Learning Education Model Based on the Computer Platform

Chuntao Jia[1(✉)], Hong Liu[2], and Hao Yu[3]

[1] College of Physical Education, Zunyi Normal University,
Zunyi 563000, Guizhou, China
jianhuazhao@yeah.net
[2] School of Education and Center of Mental Health and Psychological
Counseling, Zunyi Normal University, Zunyi 563000, Guizhou, China
[3] Department of Public Basic Education, Moutai Institute,
Renhuai 564507, Guizhou, China

Abstract. With the further implementation of the physical educational reform and the new curriculum standards, building a new teaching mode, giving full play to students' principal position and improving the quality of our physical education and teaching in an all-round way has become the focus of the current innovative development of educational activities. At present, the integration of the information technologies and our teaching activities is deepening. With the help of the computer platform, the teaching space can be effectively extended, and students' learning vitality can be stimulated, and the real innovation of our teaching activities can be realized. For the physical education and teaching, we should use the computer platform to let students fully experience and feel the vivid diversity of sports, so that students love sports and develop good sports habits and the sense of participation at the same time.

Keywords: Computer platform · Sports cooperation · Learning education · Model research

In our school physical education and teaching, the computer multimedia technology plays an important role. The majority of the physical education teachers should renew their concepts, clarify the relationship between the multimedia teaching and the traditional physical education, rationally use the multimedia to stimulate students' thinking, and mobilize students' initiatives to participate in the exercises [1]. However, the multimedia assisted instruction is only an auxiliary means, which cannot be relied on excessively. We should accept the advanced multi-media teaching concepts and combine the traditional sports teaching methods, so that the two can be combined into one. The computer multimedia technology not only changes the monotonous sports teaching mode, and provides the diversified teaching methods, but also improves teachers' teaching abilities, stimulates students' learning initiatives and enthusiasms, and lays a solid foundation for the future sports teaching [2].

© Springer Nature Singapore Pte Ltd. 2020
M. Atiquzzaman et al. (Eds.): BDCPS 2019, AISC 1117, pp. 1664–1669, 2020.
https://doi.org/10.1007/978-981-15-2568-1_233

1 Application Value of the Sports Cooperative Learning Education Model Based on the Computer Platform

With the development of the information technology, many high-tech achievements have been directly introduced into the field of education, and the multimedia technology is one of them. This creates favorable conditions for the implementation of the quality education in the physical education classroom teaching. The multimedia technology is a process of transmitting the information to students in the form of images, sounds and words [3]. Because of its vivid and interactive characteristics, it can greatly mobilize students' learning initiatives and enthusiasms, promote our teaching quality and improve students' abilities. However, with the rapid change of knowledge, the traditional physical education can no longer meet the requirements of the modern physical education [4]. How to combine various modern teaching media with the traditional teaching media organically, and especially how to introduce the multimedia computers into the physical education classrooms to meet the requirements of the modern physical education, is an important issue facing our sports workers [5].

Collaborative learning is a new learning method advocated by our modern education, which achieves the win-win goal through the cooperation, communication and sharing of participants. It is a mutual learning method with clear division of labor for a team to accomplish a common task and experience the process of the hands-on practice, the independent exploration and the cooperative communication. Starting from the collective nature of the teaching process, the cooperative learning and teaching mode focuses on the changes of the interaction between students in view of the drawbacks of neglecting the peer interaction in the traditional teaching, incorporates the cooperative group structures into the goal of the classroom teaching, and promotes the coordinated development of students' individuality and generality. Influenced by the traditional educational thoughts, the former sports teaching emphasized too much on the teacher-centered and the teaching method of imparting knowledge and skills, but seldom considered the students' principal position and their learning methods in their learning activities. The students' principal role and the interaction between students could not be brought into full play, which limited the rational development of students' personalities and qualities.

The cooperative learning and teaching model can maximize the development of students' physiological skills and psychological qualities. The "small group" psychological teaching in the physical education classroom teaching can maximize the development of students' physiological skills and psychological qualities, improve their social adaptability and creativity, and lay the foundation for students to go to the society. From the perspective of the development of the modern educational reform, the focus of the classroom teaching should be on students' own exploration and research. The collaborative learning can better meet the individual needs, abilities and interests of students, so that students can develop in the autonomous learning activities. In the process of the collaborative learning, students can master certain technical skills according to their abilities and levels, improve their abilities to detect and correct errors, cultivate students' cooperative spirits of understanding different opinions and learning to respect and cooperate with each other, and develop students' social communication abilities, self-evaluation abilities, the ability to distinguish right from wrong, and the ability to deal with the relationships between teachers and students, between students and students, between collectives and individuals, and between individuals and the society in the atmosphere of the collaborative learning.

Give students more free space for activities, so that they can discover, explore and innovate in their independent learning. The basic form of the collaborative learning is a collaborative learning group composed of 5–7 people, which makes use of the inter-action of dynamic factors to promote learning and achieves the goal of the physical education based on their group performances. Therefore, the physical education and teaching not only emphasizes the interaction between teachers and students, but also emphasizes the cooperation and interaction between students. Teachers are required to give students more free space for activities and more opportunities for communication. In the process of the collaborative learning, students' practice activities are mainly cooperative, while the competition is dominant among groups. Competition and cooperation go hand in hand, so that students' autonomy and collaboration can be fully integrated. Through the result of their learning and mastering skills among students, autonomy and cooperation are organically combined, and the individual learning makes a unique contribution to the group learning. The group learning promotes students' individual self-study, mutual learning, self-evaluation and mutual evaluation. The group learning results are integrated into the individual efforts, which reflects the collaboration among students. The interaction between students in their collaborative learning can not only improve students' physical qualities, academic performances and social skills, but can also improve their interpersonal relationships. It can also cultivate their innovative consciousness, improve their vocational skills, improve their person-ality development, and form good willpower qualities.

2   Connotation and Value of Sports Cooperative Learning Education Model Based on the Computer Platform

The so-called cooperative teaching is a brand-new educational concept. It is a teaching mode in which teachers cooperate with other relevant personnel (researchers, experi-menters, teaching assistants, and other professionals) on a specific teaching goal. The significance of the group collaborative learning is to teach in groups, that is to say, to organize students and teachers into groups or teams for the collaborative teaching. The group collaborative learning and teaching according to one's aptitude and the hierar-chical teaching methods are the harmonious and unified teaching methods. The pre-condition of practicing the teaching method of the group cooperative learning is the class teaching system. Only by ensuring that the teaching method of the group coop-erative learning is practiced under such premise can the advantages of the teaching method of the group cooperative learning be effectively reflected and the purpose of the innovation and cooperation be achieved. To effectively practice the teaching method of the group cooperative learning, we must take students as the center. Only the imple-mentation of the group cooperative learning teaching method with students as the center can embody the students' subjectivity and then effectively play the students' development, creativity and enthusiasms.

The teaching method of the group cooperative learning not only has the characteristics of cooperation, interaction, spontaneity and high efficiency, but can also effectively reflect and integrate the significance of the cooperation and competition, constantly promote students' abilities to develop and innovate, and actively research and explore the contents of the computer teaching. The teaching method of the group cooperative learning helps to improve the relationship between teachers and students, and also improves the efficiency of the cooperation between teachers and students. Under the simultaneous and concerted efforts of teachers and students, the efficiency of the computer classroom teaching can be continuously improved.

The use of the computer assisted instruction in the physical education teaching, such as the track and field, gymnastics, martial arts and ball games, has obvious advantages. Through the visual images, it can display the standardized movements, and reveal the mechanics principles of sports through animations. Students can hear and see, understand principles, and memorize key points of actions, plus teachers' guidance, help and their diligent practice, so that their motion technologies are bound to make great progress.

3 Application of the Sports Cooperative Learning Education Model Based on the Computer Platform

Because of the differences in the intelligence, physical qualities and sports basis, students often have "insufficient to eat" and "unable to eat" in the traditional teaching. A good computer-assisted instruction can achieve the man-machine conversation. Students can choose different contents and speeds according to their own circumstances, and some can also analyze their own mistakes and shortcomings in technologies. This fully reflects the individualized teaching theory of teaching students in accordance with their aptitudes.

One of the main tasks of the physical education and teaching is to enable students to master certain sports skills, and on this basis, their flexible application and creation of new sports skills. The formation of a new action must be through the process of listening and perceiving the movements of technologies, and it is the main source of information for students. Because of the new and organic combination of their visual, auditory, audio-visual, static and dynamic aspects, the computer-aided instruction has turned the original boring technical classes into the vivid and intuitive images, enabling students to show positive desire to learn and study diligently, and fully reflecting the main role of students in the re-teaching. Classroom practice has proved that students are generally interested in the computer assisted instruction.

In sports, many sports techniques are not only complex in structures, but also need to complete a series of the complex technical actions in a moment. For example, in the final force sequence of shot put of the throwing project, and in the jumping project, it is difficult for students to see these instantaneous completed movements clearly, and it is also difficult to quickly establish a complete action representation. Teachers can use the computer multimedia technologies to optimize the teaching process with its vivid images, vivid pictures, and flexible animation and music effects. Using the animation or images in the courseware, the teaching methods such as the slow action, the stop mirror and the replay are combined with the explanation and demonstration. This can help students see the technical details of each instant action clearly. Teachers can explain the essentials of each decomposition action, demonstrate the entire process of the action, and then grasp the key parts of the action, and highlight the key points and difficulties. It is faster and more complete to establish the action representation, which improves the teaching efficiency of the action learning in the cognitive stage and shortens the teaching process.

When using the computers to teach, the teacher explains and demonstrates the action, and then plays it directly to the students to watch. The students who learn can practice, but the students who don't can repeat the explanation and demonstration of the action by using the control button. And display the specific sounds and icons of the technical difficulties, key points, common erroneous actions, to remind students to pay attention to the teacher's inter-line guidance, analyze and compare them with students, ask questions and answer questions, so that students can grasp the actions intuitively, actively and stereoscopically, improve our teaching efficiency, and promote students' abilities to analyze and solve problems. For example, we can give several groups of simulation lenses of takeoff (including good, better, general and poor jumping) so that students can carry out their cooperative learning. First, observe, compare, analyze, communicate and discuss it in groups, and finally get the technical requirements and precautions of running-up, taking-off, soaring and landing, and then report to each group in groups. Finally, the teacher summarizes the main points of the techniques of the stretch-up long jump. In this way, students not only learn the knowledge and skills, but also master the learning methods. For example, it is difficult for students to grasp the feeling of "jump" when teaching "roll before jump". At the same time, it is easy to collapse when rolling over, and the teachers can only complete the demonstration at one go. Once slowing down, the action is easy to make mistakes. Therefore, using the multimedia production is much simpler, and it can turn the actions into the static, the abstraction into the image, and enhance the students' perceptual awareness.

In the process of our teaching, teachers can design some sports activities similar to the TV programs. Students can perform the group activities according to their own way, perform in turn, and request each student to join in the performances. The group teachers who perform better will give some awards. Teachers can also organize the knowledge contests on the sports knowledge, through which students' enthusiasms for learning can be improved, and students' awareness of self-learning can be enhanced, so as to achieve the effect of enhancing students' self-confidence in learning. Moreover, another advantage of the computer aided instruction system is that teachers need to cooperate with each other in the process of the courseware making. Every physical education worker should participate in the production of the teaching courseware. This greatly improves the working efficiency of the physical education, strengthens the teaching ability of teachers, and can also make up for the shortage of the teachers in teaching, and effectively stimulate the subjective initiatives of PE teachers.

4   Conclusion

In the process of the physical education, the use of the computer-assisted instruction system will stimulate the initiatives of teachers. Therefore, in the computer assisted instruction, the physical education teachers can choose materials used in the teaching through the teamwork. Therefore, in the innovative teaching of the physical education, teachers must take the initiatives to understand the different needs of students for physical education. At the same time, comprehensively consider the development factors of students' overall qualities, start from the current situation of our teaching to develop a reasonable teaching courseware, from which the enthusiasms and initiatives of physical education teachers can be mobilized.

References

1. Li, F.: The application effect of the independent teaching method in the physical education theory teaching. J. Qufu Norm. Univ. (Nat. Sci. Ed.) (2), 109–113 (2016)
2. Xu, F.: Preliminary study on the auxiliary teaching effect of micro courses in basketball elective courses in tourism secondary vocational schools. Guangdong Educ. (Vocat. Educ. Ed.) (2), 118–119 (2016)
3. Wu, X.: Micro-strategies to enhance the humanistic flavor of the basketball club activities in rural junior high schools. New Curric. (11), 260–261 (2015)
4. Zhou, L.: An analysis of the effective teaching measures of basketball passing and catching techniques in secondary vocational schools. New Educ. Era (Teach. Ed.) (36), 145 (2015)
5. Yang, Z., Dai, H.: Application of the sports education mode in football elective courses in colleges and universities. J. Tianshui Norm. Univ. (2), 116–119 (2014)

The Passive UHF RFID Temperature Sensor Chip and Temperature Measurement System for the On-line Temperature Monitoring of Power Equipment

Renyun Jin[1(✉)], Rongjie Han[2], Hua Fan[2], and Liguo Weng[1]

[1] State Grid Zhejiang Hangzhou Xiaoshan District Power Supply Co., Ltd.,
Hangzhou 311202, Zhejiang, China
weixiong619@126.com
[2] Zhejiang Zhongxin Electric Power Engineering Construction Co., Ltd.,
Hangzhou 311202, Zhejiang, China

Abstract. In the long-term work, affected by the power equipment foundation, temperature and humidity, serious overload and many other reasons, it is extremely easy to have problems such as not tight power equipment crimping and the changes of the contact parts and so on, which eventually make the contact resistance of the power equipment increase, resulting in huge risks. Through the use of the wireless temperature measurement, the on-line monitoring of the passive UHF RFID temperature sensor chip and the temperature measurement system, the abnormal information of the power equipment is collected and fed back, effectively ensuring the safe operation of the power equipment.

Keywords: Power equipment temperature · Online monitoring · Sensor chip · Temperature measurement system

The power passive wireless temperature monitoring system is mainly composed of the passive RFID temperature sensor and the reader writer, which can realize the wireless monitoring of the equipment temperature in a large range and the remote monitoring of the equipment operation status.

1 The Necessity of the On-line Temperature Monitoring of Power Equipment

Temperature is one of the most important parameters in the condition monitoring of the power equipment. Through the temperature monitoring, the operation status and the fault information of the power equipment can be known accurately and timely [1]. Monitoring the operating temperature of the power equipment, such as the oil temperature of transformers and the conductor temperature of the transmission line (overhead line and power cable) can calculate its load limit capacity and the equipment aging degree, so as to provide basis for the dynamic capacity increase or the

© Springer Nature Singapore Pte Ltd. 2020
M. Atiquzzaman et al. (Eds.): BDCPS 2019, AISC 1117, pp. 1670–1675, 2020.
https://doi.org/10.1007/978-981-15-2568-1_234

maintenance and update of the power equipment [2]. Monitoring the temperature of the stator and rotor, the high-voltage switch cabinet, the bus joint, the outdoor switch, the breaker contact, capacitor, reactor, the high-voltage cable, and transformer and so on of the generator can timely find the local or the overall overheat or the relatively abnormal temperature distribution accompanying the abnormal situation or fault, and also provide the historical data for the fault analysis.

In the long-term operation of the high-voltage switch cabinet, the ring network cabinet, the knife switch and other important equipment in the power system, there may be problems such as the equipment aging, the surface oxidation, corrosion, and the loose fastening bolts and so on. In addition, much power equipment will have the overload operation for a long time, easy to cause the high temperature anomaly [3]. If not found and handled in time, it may cause the safety accidents such as melting, burning and even explosion [4]. The electrical characteristics of the power equipment require the monitoring system to collect the temperature in a passive wireless way, and the monitoring system can monitor and record the temperature of key nodes of the equipment in the real time and carry out the safety warning based on the big data analysis [5].

The passive UHF RFID temperature sensor chip is a high precision temperature sensor tag. Based on the passive UHF RFID technology, the tag has the characteristics of the wireless temperature measurement without battery. Based on the special antenna design and the substrate composition, the tag can work well in most application scenarios, including the metal surface, and has the temperature resistance characteristics of 250 °C ambient temperature. On the premise of not affecting the insulation and safety of the power equipment, the real-time temperature monitoring of the key nodes of the power equipment is realized.

2 Passive UHF RFID Temperature Sensor Chip and the Temperature Measurement System for the Online Temperature Monitoring of Power Equipment

2.1 UHF RFID Technology

The hardware components of the temperature monitoring system are mainly composed of three parts: the temperature sensor label, the reader writer, and the background server. Among them, the background server is connected to the reader writer through the RS485 bus or the network cable, and the reader writer is connected to its antenna through the feeder, and the tag antenna is integrated on the tag chip. The basic work flow of the wireless communication system is as follows by applying the RFID technology to the tag reader writer. First of all, the reader generates a carrier signal and transmits it through its antenna. When the sensor tag enters the effective coverage area of the electromagnetic wave transmitted by the reader, the sensor tag is activated. The activated tag transmits the identification information stored in the chip to the reader antenna through its built-in antenna. The high frequency signal is transmitted to the reader antenna through the antenna regulator for demodulation and decoding, and then sent to the upper computer for the data processing.

2.2   Research on the Key Technologies of the System

Label and Antenna Selection

Ltu27-wx2 is a passive wireless and low-power temperature sensor. The chip uses the advanced UHF radio wave energy collection. The energy is obtained through the RF electromagnetic wave of 840 MHz to 960 MHz. A 512-bit erasable nonvolatile data storage unit is built in the chip to store the user information and other data. The RF chip communication interface supports the EPC global C1G2 v1.2 communication interface, which can be used with various types of the UHF RFID reading and writing equipment to build a passive wireless temperature sensor system.

Metal Resistant Design of the Sensor Label

Since the label is applied to the distribution network equipment, the influence of the metal on the label must be considered. This paper adopts a relatively low-cost and easy-to-use anti-metal design method, using ABS (acrylonitrile butadiene styrene plastic) package shell to pad up the label and the AMC structure at the bottom of the shell. The AMC structure is composed of three parts. The top layer is the ideal electrical conductor floor, and the bottom is the periodic arrangement of the metal patches, while the media is filled between the two, and the metal patches and the floor are connected by a metal via. The main functions of the ABS packaging shell are as follows. The temperature of the RF tag is measured by the wired thermistor, and the thermistor is installed near the key point. As the high-voltage environment does not allow the connecting wire to be exposed, the ABS packaging shell plays the role of the insulation protection. The AMC structure is adopted in the packaging shell to reduce the interference of the metal on the label and improve the reading rate of the label. Because the bottom of the packaging shell is laid with a metal layer, it has a good thermal conductivity for the thermistor temperature measurement.

Communication Distance Estimation

Recognition distance, that is, the maximum distance R that the RFID reader can detect the backscattered signal of the tag, is an important performance index of the system. It is determined by the minimum threshold power Pth of the wake-up tag chip and the receiver sensitivity Pmin of the reader writer. According to the Friis equation, the received energy of the tag at the distance r is calculated. Among them, R1 is determined by the minimum threshold power Pth of the wake-up tag chip, and R2 is determined by the receiver sensitivity Pmin, while the final communication distance is estimated to be a small value.

Anti-collision Mechanism

In the RFID system, when more than one tag is in the scope of the reader, there will be communication conflict, that is, collision. There are two types of collisions in this online temperature monitoring system. One is the collision caused by the multiple tags responding to the reader at the same time, and the other is the interference of the non temperature tags on the RFID system within the scope of the reader in the system. For the on-line temperature monitoring system of the switch cabinet, the number of the temperature tags is limited. In this paper, the packet polling mechanism is introduced based on the original Aloha algorithm based on the dynamic frame time slot of the reader writer, which improves the recognition efficiency.

First, the reader sends a query command to the tag, and the tag receiving the command gets energy and is activated. The tag randomly selects a time slot from the frame length 1-f to transmit the identification information, and stores the time slot number in the register SN. If the data is sent successfully, the tag will enter the sleep state and will no longer be active in the subsequent timeslot. If there is a conflict, the tag will enter the waiting state and reselect the timeslot to send the data in the next frame. The reader and the writer verify the identification information of the data sent by the tag. According to the EPC, the tag is divided into the temperature group and the non temperature group. The uploaded successful temperature tag goes to sleep, and this frame will not be queried again. The non temperature tag will be added to the blacklist, and then it will not be queried again.

Equipment Installation

The temperature vulnerable points of the high-voltage switchgear are distributed at the bus connection, the cable connection and the circuit breaker connection. The system temperature sensor can be installed at the above temperature key points, and the label can be installed at the bus connection. The antenna of the reader writer is installed on the metal door of each function room of the switch cabinet, which is located in the switch cabinet, and the antenna wire is led out to the reader writer by drilling holes on the door. Due to the RF connection between the antenna and the tag, the installation position of the reader has little influence on the communication distance. The reader can be installed outside the switch cabinet through the antenna feeder. Considering the interference of the metal to the passive tag and the distribution of the temperature nodes in different gas chambers, the method of adding redundant antenna can expand the communication range.

Background Software Development

The temperature online monitoring software is based on the C# programming language of the Microsoft.NET platform. The system software has the functions of connecting the reader and the writer, the online real-time temperature measurement, the temperature data storage, the real-time alarm, and the temperature curve analysis and so on. The main contents of the interface display are the IP address of the reader writer, the tag EPC within the antenna range, the number of the tag reads, the real-time temperature and the installation address information set according to the tag EPC. The temperature data is drawn into the two-dimensional curve, and the curve coordinates change in the real time. As is shown in the figure, "cabinet phase 1A" label temperature shows 29.26 °C (green). When the temperature exceeds the set warning threshold (75 °C, which can be set), the line turns red to realize the temperature alarm. The temperature information will be saved in the History.log text file every 30 s (which can be set), which is convenient for the monitoring personnel to query the temperature history data and print the report. The above functions well realize the online real-time monitoring of the temperature value of the key points at the operation time, and the human-computer interface is convenient for the unified monitoring and management.

3 Experiment and Feasibility Analysis

The wireless temperature sensor based on the surface acoustic wave (SAW) technology uses the piezoelectric material, which has the pure passive and wireless characteristics, and does not need to consider the sensor power supply, the high voltage insulation, the equipment rotation and other issues. It can withstand the high and the low temperature r-20(Ti00(rc), does not involve the electronic migration process in the semiconductor materials, with the long life, the resistance to the discharge shock and the electric field. The sensors are small in size, light in weight and easy to design and install.

3.1 Sensitivity Test of the Sensor Label

The sensitivity of the RFID tag chip is the minimum energy needed for the chip to be activated. The sensitivity is the most important performance index of the tag chip. The size directly affects the tag performance, such as the read-write distance. In a certain frequency band, most chip manufacturers only give a sensitivity value of the chip, but do not identify the change of the chip sensitivity with frequency. The sensitivity of the tested label tends to be stable in the frequency range of 860 MHz to 960 MHz, which is maintained at about −4 dBM and the highest in 950 MHz. Corresponding to the RFID frequency band in China, the sensitivity of the tested tag is −4.1 dBM.

3.2 Sensor Tag Reading Rate Test

Considering the influence of the metal of the switch cabinet on the tag communication, in the range of 2 m of the tag standard communication, put the reader antenna at 0–2 m of the tag, and attach the tag to the 20 cm × 20 cm metal plate. The direction of the tag and the metal plate are parallel to the reader antenna to achieve the best RF coupling. According to the experimental data, at 1 m, it can be seen that the metal partition will reflect and shield the field of the reader and writer, reducing the reading rate of the tag, but not completely unable to read. According to the experimental data of 2 m, when there is a metal diaphragm, the absorbed RF energy of the metal is converted into the electric field energy, which weakens the total energy of the original RF field strength, resulting in the label being unable to work normally. The interference of the metal plate reduces the communication distance of the tag, which cannot reach the standard of 2 m, but the reading and writing distance of 1.5 m is enough to meet the installation and temperature monitoring of the equipment.

3.3 Temperature Measurement Performance Test

In order to test the temperature measurement performance of the temperature tag, different ambient temperatures were measured at the same time and compared with the mercury thermometer. The result of the temperature measurement performance experiment shows that the temperature measurement result of the label is slightly higher than that of the mercury thermometer, but it is very close. The difference between the label and the thermometer is less than 0.5 °C. According to the experience of the daily operation and the maintenance personnel of the switch cabinet, the normal temperature of the electrical joint is 30 °C–60 °C. If overheating occurs, the temperature can reach over 75 °C, and the deviation value of the wireless temperature measurement is 0.5 °C.

3.4   Temperature Measurement Test of the Switch Cabinet

The experiment is carried out in the 10 kV high-voltage switchgear in the school high-voltage laboratory. The sensor tag is installed at the connection of the A-phase contact of the switchgear breaker. In this paper, the 24-hour temperature record data is selected to reflect the temperature change of the switchgear throughout the day. Through the analysis of the 24-hour contact temperature record, it can be seen that the RFID temperature online monitoring system can operate normally without affecting the work of the switchgear, and the recorded data correctly reflect the relationship between the contact temperature and the ambient temperature, which shows that the temperature monitoring system is feasible.

4   Conclusion

The temperature monitoring of the distribution network equipment is of great significance to the safe and stable operation of the equipment. The RFID temperature online monitoring scheme uses the passive wireless sensor tags to collect the temperature, and the sensor nodes do not need the power supply. Through the wireless data transmission, the multi-node temperature online monitoring is realized. In the process of the monitoring, the system has the following advantages: small equipment volume, easy to install; low cost and no maintenance cost; no impact on the operation of the distribution network equipment, not easily affected by the environmental factors, online real-time monitoring, and the PC provides a good human-computer interface, which is easy to operate, and has a good application prospect.

References

1. Yu, B., Xu, X.: Design and optimization of the passive wireless temperature sensor system. Comput. Knowl. Technol. (2), 253–255 (2015)
2. Wang, W., Zhang, G., Zhang, H., Liu, C., Wang, F.: A new temperature monitoring system for the substation power equipment. Urban Constr. Theory Res. (Electron. Ed.) (29), 2605–2606 (2015)
3. Li, X., Huang, X., Chen, S., Gan, J., Liu, Q., Pan, Z.: Integrated temperature monitoring system of the intelligent substation equipment based on the ZigBee network. High Volt. Electr. Appl. **47**(8), 18–21 (2011)
4. Cao, L., Zeng, H., Le, Y.: Design of the wireless temperature detection intelligent compensation system for the CNC machine tools. Modular Mach. Tool Autom. Manuf. Tech. (9), 63–65 (2012)
5. Zhang, T.: Application of the wireless temperature measurement technology in Tonggang power system. Jilin Metall. (3), 53–55 (2013)

Application of the Watershed Algorithm in the Image Segmentation

Caihong Li[1(✉)], Junjie Huang[2], Xinzhi Tian[1], and Liang Tang[1]

[1] School of Electronics and Information Engineering, Xi'an Siyuan University,
Xi'an 710038, China
huangmingbucea@126.com
[2] Foundation Department, Xi'an Siyuan University, Xi'an 710038, China

Abstract. The image segmentation is the basic prerequisite step of the image recognition and image understanding. There are many existing image segmentation methods. This paper mainly introduces the use of the watershed algorithm in some specific image segmentation and its advantages over other segmentation methods, which is illustrated through the image segmentation experiment of the number of the sticky rice grains.

Keywords: Image segmentation · Watershed algorithm · Conglutinated rice grain

The image segmentation is the technology and process of dividing the image into several specific and unique regions and extracting interested objects. The image segmentation is the basic prerequisite step of the image recognition and image understanding. The quality of the image segmentation directly affects the effect of the subsequent image processing. The existing image segmentation methods are mainly divided into the following categories: the threshold-based segmentation method, the region-based segmentation method, the edge-based segmentation method and the specific theory based segmentation method [3].

This paper mainly introduces the usage and advantages of the watershed algorithm in some specific image segmentation, and compares it with the edge segmentation.

1 Watershed Algorithm

The watershed algorithm is a region based segmentation method in the image segmentation. Different from those image segmentation methods which aim at finding the boundaries between the regions, the watershed algorithm constructs the region directly to achieve the image segmentation. [4] Generally speaking, any gray-scale image can be regarded as a terrain surface, in which the high-intensity represents the peaks and hills, while the low-intensity represents the valleys. Fill each isolated valley (the local minimum) with different colors of water (labels). When the water rises, according to the nearby peaks (gradients), water of different colors in different valleys will obviously

© Springer Nature Singapore Pte Ltd. 2020
M. Atiquzzaman et al. (Eds.): BDCPS 2019, AISC 1117, pp. 1676–1680, 2020.
https://doi.org/10.1007/978-981-15-2568-1_235

start to merge. To avoid this situation, barriers (heightening dams) need to be added in a timely manner where the water is about to merge. Continue to fill the water and build barriers until all the mountains are submerged. The created barrier then gives the segmentation results.

2   Common Operation Steps of the Watershed Algorithm

The watershed segmentation algorithm regards the image as a "topographic map", in which the pixel value of the area with the strong brightness is larger, while the pixel value of the area with the dark brightness is smaller. By looking for the "catchment basin" and the "watershed boundary", the image is segmented.

Steps:

(1) Read the image and grayscale the color image;
(2) Using the gradient algorithm to find the gradient graph;
(3) The foreground and the background of the image are marked, in which the foreground pixels in each object are connected, and each pixel value in the background does not belong to any target object;
(4) The segmentation function is calculated, and the edge line of the segmented image obtained by the watershed segmentation algorithm is applied to realize the image segmentation.

Note: it is not good to use the watershed segmentation algorithm directly. If the foreground and the background are marked differently in the image, then the watershed algorithm will achieve better segmentation effect.

The watershed transform is the input image of the basin, and the boundary point between the basins is the watershed. Clearly, the watershed represents the maximum of the input image. Therefore, in order to get the edge information of the image, the gradient image is usually used as the input image, that is, grad $(f(x, y)) = ((f(x-1, y) - f(x + 1, y))^2 + (f(x, y-1) - f(x, y + 1))^2)^{0.5}$

In this equation, $f(x, y)$ represents the original image and grad () represents the gradient operation [1].

3   Application Examples and Comparison of the Watershed Algorithm

If the objects in the image are connected, it will be more difficult to segment them. The watershed algorithm is often used to deal with this kind of problems, and usually has better results [2].

For example, we segment and count the typical conglutinated rice image.

The matlab program of the related experiments and analysis of results.

(1) Without the watershed algorithm, the code is as follows:

```
I=imread('rice3.jpg');
J=imtophat(I,strel('disk',15));
bw=im2bw(J,graythresh(J));
bw_opened=bwareaopen(bw,5);
 bw1= edge(bw_opened,'canny');
  k = imfill(bw1,'holes');
  SE = strel('disk',3);
    L = imopen(k,SE);
L1=bwlabel(L,8);
[L1,num]=bwlabel(L,8)
rgb1=label2rgb(L1, 'spring', 'c','shuffle');
figure
subplot(1,2,1),imshow(I),title('original image')
subplot(1,2,2),imshow(rgb1),title('Segmentation   results   without
watershed')
```

It is concluded that the number of the rice grains after the segmentation is num = 69. In this program, the high hat filter image enhancement operator, candy edge detection operator and morphological open operation are used to segment the image [5], and the effect is different from the actual number of the rice grains.

(2) Further segmentation by the watershed algorithm

Matlab code is continued as follows:

```
g=imgradient(I);
figure
subplot(1,2,1),imshow(I),title('original image')
subplot(1,2,2),imshow(g,[]),title('Gradient of the original image')
im=imerode(bw,strel('rectangle',[3 3]));
figure
subplot(1,2,1),imshow(I),title('original image')
```

```
subplot(1,2,2),imshow(im),title('Foreground marker')
Lim=watershed(bwdist(bw));
em=Lim==0;
figure,imshow(em),title('Background marker')
g2=imimposemin(g,im | em);
figure,imshow(g2),title('Modified gradient image')
L2=watershed(g2);
L2 = imclearborder(L2, 8);
L3=bwlabel(L2,8);
[L3,num]=bwlabel(L2,8)
rgb2=label2rgb(L2, 'hot', 'c', 'shuffle');
figure
subplot(1,3,1),imshow(I),title('original image')
subplot(1,3,2),imshow(rgb1),title('Segmentation     results     without
watershed')
subplot(1,3,3),imshow(rgb2),title('Use watershed segmentation re-
sults')
```

First, calculate the gradient of the image. Then mark the foreground of the image. The total number of the rice grains segmented by the watershed algorithm is num = 73, which is consistent with the actual number of the rice grains.

4   Conclusion

The watershed algorithm is one of the most important algorithms in the image segmentation. It has been widely used in the image processing because of its good edge detection ability and the ability to get a relatively concentrated basin. Through the above-mentioned experiments, it is found that the watershed algorithm is more suitable for the segmentation of some conglutinated images, and less suitable for the segmentation of images with complex backgrounds. At the same time, the watershed algorithm sometimes faces the problem of the image over segmentation while obtaining good edges. In this case, we can use the improved algorithm such as the marked watershed to study the segmentation [6]. In a word, the watershed algorithm has a wide range of applications in the image segmentation, which can only need in-depth studies.

Acknowledgments. The "13th Five Year Plan" Project of Shaanxi Province (SGH18h526); the "13th Five Year Plan" Project of Shaanxi Province (SGH18h530); the research and practice project of the new Engineering Course in Xi'an Siyuan University (17SYXGK004).

References

1. Qiutang, L.: Image processing algorithm based on the mathematical morphology. Electron. Technol. Softw. Eng. (06), 80–81 (2016)
2. Hongyue, C., Guoqing, Y.: Optimization method of the road extraction from the high-resolution remote sensing image based on the watershed algorithm. Remote Sens. Land Resour. **25**(3), 25–29 (2013)
3. Tianhua, C.: Digital Image Processing and Application. Tsinghua University Press, Beijing (2019)
4. Gonzalez, R.C.: Digital Image Processing. Publishing House of Electronics Industry, Beijing (2011)
5. Limei, C.: Digital Image Processing – Analysis and Implementation Using MATLAB. Tsinghua University Press, Beijing (2019)
6. Zhengjia, L.: Color image segmentation method based on the improved watershed algorithm. Bull. Sci. Technol. **9**, 172–174 (2018)

Application of the Genetic Algorithm in Water Resource Management

Minxin Li, Jiarui Zhang, Xin Cheng, and Yun Bao[(✉)]

Hohai University, Nanjing 210098, Jiangsu, China
lbkgw0@yeah.net

Abstract. With the continuous development of the social economy, the specific demand for water resources is increasing. How to use the limited water resources effectively becomes a practical problem to be solved effectively. Combined with the current situation of the water resource management in China, combined with the relevant water resource planning model and solution method, this paper analyzes the application of the genetic algorithm in the process of the water resource management, and simulates with specific cases, to verify the feasibility of the model and algorithm. The specific research results also provide an effective decision-making scheme for the current water resource management.

Keywords: Genetic algorithm · Water resources · Management mechanism · Application research

With the rapid development of our economy and the continuous growth of the population, the contradiction between the supply and the demand caused by the shortage of water resources is increasing. The shortage of water resources will seriously affect and restrict the development of the social economy [1]. How to develop and use the limited water resources scientifically and reasonably to meet the needs of the social and economic development has become the current research hotspot. The optimal allocation of water resources refers to the unified allocation of water resources or other related resources through the engineering measures and non-engineering measures in a specific basin or region under the guidance of the sustainable development strategy, using the system analysis theory and the optimization technology, and carrying out the scientific and reasonable distribution at different times and between different beneficiaries.

1 Current Situations of Water Resources in China

At present, the total amount of the water shortage in China is about 40 billion cubic meters, and the annual drought area is 2–2.6 million square kilometers, which affects the grain output of 15–20 billion kilograms and the industrial output value of more than 200 billion Yuan. There are nearly 70 million people in China who are facing the difficulty of drinking water. From the distribution statistics of the population and the water resources, we can see that the difference of the water resource distribution between the north and the south is very obvious [2, 3]. The population of the Yangtze

© Springer Nature Singapore Pte Ltd. 2020
M. Atiquzzaman et al. (Eds.): BDCPS 2019, AISC 1117, pp. 1681–1686, 2020.
https://doi.org/10.1007/978-981-15-2568-1_236

River Basin and the area to the South accounts for 54% of the whole country, but the water resources account for 81%, while the population of the North accounts for 46%, and the water resources are only 19%.

With the problems exposed by water resources becoming more and more prominent, the shortage of water resources has become one of the key problems that our country will face [4]. The situation of water resources in China is not only the shortage of water resources, but also the increasingly serious pollution and waste of water resources [5]. At the same time, due to the differences in the spatial distribution of water resources in China, the management of water resources has become the top priority of the work of our government departments.

2 Problems in the Water Resource Management

2.1 The Management System of Water Resources Is Not Perfect

In the current process of the water resource management, there is a lack of a certain interest coordination mechanism. China's "Water Law" stipulates that "the State implements a unified management system for water resources combined with the hierarchical and the sub-sector management." However, in some areas, the management of water resources is not carried out in strict accordance with the rules and regulations. The loose management is not conducive to the rational use of water resources, and also makes the contradiction between the supply and the demand of water resources further increased.

2.2 Serious Waste of Water Resources

The waste of water resources in China is mainly manifested in agriculture, industry and people's daily life. In the aspect of agriculture, irrigation is the main way. In the process of the agricultural production and labor in most areas of our country, the flood irrigation is mainly used to irrigate crops. The water resources that this kind of method plays a practical role account for nearly 1/3 of the total irrigation, and the rest 2/3 of them are leaked and evaporated in the process of the transportation. However, the phenomenon of the industrial waste is mainly the unreasonable utilization of water resources. With the adjustment of China's industrial structure, although the utilization measures of water resources have been improved to a certain extent, there are still serious differences compared with the developed countries, and the reuse rate of water resources is very low. The main problems in our daily life are the unclear distinction of the water use, the improper sewage treatment and the low quality of the drinking water.

2.3 Nonstandard Development and Exploitation of Water Resources

Most of the water resources people use in the production and life refer to the fresh water resources, but the fresh water resources only account for a small part of the global water resources, a large part of which is the sea water, which cannot be directly used. Therefore, people mostly take the way of developing and using the groundwater resources in the utilization of the fresh water resources. Because of the increase of our population, the development of the industry and the pollution and the waste of the surface water, the development and utilization of the groundwater resources has become the main method for people to obtain water resources. However, due to the imperfection of the relevant policies and regulations, people are more casual in the exploitation of the groundwater resources, resulting in excessive exploitation of the groundwater resources, which leads to serious shortage of the groundwater resources.

3 Definition and Characteristics of the Genetic Algorithm

In the modern computational intelligence methods, the genetic algorithm has been widely used in the modern water resource system problems because of its strong adaptability, global optimization, probability search, implicit parallelism and simple generality. However, in the application, it is found that the genetic algorithm still has many shortcomings to be improved, such as the solution space search strategy, the problem convergence, and the control parameter setting and so on, which are also the reason why it has long become the research hotspot of the computational intelligence at home and abroad. The cross integration is one of the main ways of the modern scientific and technological innovation, and it is also an important way of the population derivation of the genetic algorithm. It is an important way to combine the traditional, conventional or modern, intelligent mathematical methods with the genetic algorithm to improve the performances of the latter.

The genetic algorithm (GA) is a computational model simulating the natural selection and the genetic mechanism of Darwinian biological evolution, and it is a method to search the optimal solution by simulating the natural evolution. The genetic algorithm starts from a population which represents the potential solution set of the problem, and a population is composed of a certain number of individuals encoded by genes. Each individual is actually a chromosome with characteristics. As the main carrier of the genetic material, the chromosome is a collection of the multiple genes. Its internal expression (i.e. genotype) is a combination of genes, which determines the external expression of the individual shape. At the beginning, we need to realize the mapping from phenotype to genotype, that is, coding. Because the work of imitating the gene coding is very complex, we often simplify it, such as the binary coding. After the generation of the first generation of the population, according to the survival of the fittest, the generation evolves generation by generation to produce better and better approximate solutions. In each generation, individuals are selected according to the fitness of individuals in the problem domain, and carry out the cross and mutation with the help of the nature Genetic operators in genetics, to produce populations representing the new solution sets.

The genetic algorithm is also a search heuristic algorithm used to solve the optimization in the field of the computer science artificial intelligence. It is also a kind of the evolutionary algorithm. This heuristic is often used to generate the useful solutions to optimize and search for problems. The evolutionary algorithms developed from some phenomena in the evolutionary biology, including heredity, mutation, natural selection and hybridization. The genetic algorithm may converge to the local optimum when the fitness function is not selected properly, but cannot reach the global optimum.

4 Application of the Genetic Algorithm in the Water Resource Management

In some form, the optimization problem can be described as: finding a certain value combination of the components of the optimization variable house, so that the objective function can reach the optimal or approximate optimal under given constraints. The method to solve this kind of problems is called the optimization method. In the process of intersecting, infiltrating and integrating all kinds of the science and technologies, optimization has become the trend of the development of the system and even the whole world. The optimization criteria reflected in the system of hydrology, water resources and water environment have increasingly become a balance scale of people's analysis system, evaluation system, transformation system and utilization system due to the influence of astronomy, climate, meteorology, underlying surface, and humanity and so on. With the comprehensive influence of many factors, the optimization problems in the water science often show the complex characteristics of the high dimension, multi-peak, non-linear, discontinuous, non-convex, and with noise and so on.

The theoretical background of the combination of the experimental design and the genetic algorithm is the generalized experimental method. The application the foundation of the integration of the experimental design and the genetic algorithm is their strong complementary advantages. Based on this theory, the specific way and the operation method of the two-way integration of the two are put forward for the first time: the experimental design method based on the genetic algorithm (the genetic orthogonal design, and the genetic uniform design), and the improved genetic algorithm based on the experimental design. Among them, the specific implementation steps of the experimental design embedding genetic algorithm to form the experimental genetic algorithm include the following. According to the uniform design table, the initial population of the genetic algorithm is uniformly distributed. In order to improve the diversity of the population, several uniform design tables are used to combine the variables at different levels to generate a new population. It is called the deterministic uniform optimization operation to search for a certain range of the deterministic uniform distribution around elite individuals. In the search of the random normal distribution, a random variable which obeys the normal distribution is added around some excellent individuals to generate a new generation group.

The optimal design model of the penstock structure of the hydropower station is established, which takes the economy as the goal and the structural safety and the technical feasibility as the constraints. The prediction of the water resource system is a subject with the high technical and artistic requirements. It not only requires the forecasters to master a variety of the system prediction methods and technologies, but also requires the forecasters to have the ability to use these technical methods flexibly and comprehensively. Due to its strong nonlinear mapping ability and the fault tolerance, the artificial neural network has become one of the common modeling methods in the modern water resource system engineering.

BP-ANN is used to study the nonlinear combination prediction method, which can effectively avoid the tedious calculation of the weight of the traditional combination prediction model. Aiming at the problem that the weight of each model is difficult to be determined scientifically in the combination prediction, the combination problem of the prediction model is transformed into the pattern recognition problem of 0, 1 exclusive or for the first time according to the principle of "selecting the best and using it". The improved BP-ANN method is used to solve the problem and the satisfactory results are obtained. This method of determining the variable weight is essentially a process of the model optimization. Because every sample is taken from the best of each prediction model, it can ensure that the model is "always the best" at the existing prediction level, and has the advantages of being clear and easy to understand, simple and easy to operate. As a special case of the variable weight combination forecasting method, it has the high practical value in the field of the combination forecasting. The key to the evaluation of the water resource system is the reasonable construction and the effective optimization of the evaluation model. The traditional methods based on the conventional modeling and optimization methods are not competent for the comprehensive evaluation of the complex water resource system involving multi-attribute, multi-level and multi-factor.

The traditional attribute recognition model of the linear attribute measure function has a large error in the evaluation results of the random sampling. Therefore, an improved attribute recognition model based on the nonlinear attribute measure function is proposed for the first time. The evaluation experiments of the uniform random sampling and the orthogonal design show that the accuracy of the evaluation results of the improved attribute recognition model is significantly better than that of the traditional model, indicating that the attribute measure function of the index has an important influence on the comprehensive evaluation results of the attribute recognition model. Because the nonlinear measure function can better describe the actual membership degree of the evaluation index than the linear measure function, the improved attribute recognition model has the higher evaluation reliability.

The model not only fully considers the influence of the uncertainty factors on the system income, but also can weigh the economic benefit and the punishment risk and give the optimization results in the form of the interval to obtain the most scientific decision-making scheme, which can be effectively applied to the optimal allocation and management of the surface water resources. Aiming at the problem of the rational utilization of the groundwater, a non linear programming model is constructed, which takes the minimum cost of the system as the objective function, the number of wells in the study area (x) and the pumping capacity of each well (Q) as the decision variables, and takes the demand of water resources, the drawdown of the water level and the buried depth of the groundwater as the constraints. The model is solved by the genetic algorithm, and the optimal number of wells is obtained. Then, aiming at the optimization of the spatial layout of the pumping wells, the Kriging interpolation method is used to simulate the groundwater depth of the existing monitoring data, and then, based on the entropy weight method, the rationality of the spatial layout is comprehensively evaluated. Finally, the optimal well number and the evaluation results of the spatial layout are combined to simulate the optimized well location distribution map based on the GIS data.

5 Conclusion

With the expansion of the scope and the depth of the research system, the traditional methods have been increasingly limited in dealing with the high-dimensional, non-convex, non-linear and other complex problems of the modern water resource system. In recent years, with the rapid development of the modern applied mathematics and the computer technology, the artificial intelligence computing theory and the analysis methods, such as the genetic algorithm, the artificial neural network model, and the fuzzy set and so on, have been put forward for the complex system problems. The introduction of these methods has greatly promoted the development of the system analysis technology and injected new vitality into the research of the modern water resource system problems.

Acknowledgement. Foundation Projects: Self-help project of the brand specialty construction in Jiangsu universities (PPZY2015A043), Construction project of the dominant disciplines in Jiangsu universities (discipline of the water conservancy engineering).

References

1. Wang, Z., Tian, J.: Study on the optimal allocation of regional water resources based on the particle swarm optimization. China Rural Water Hydropower (1), 7–10 (2013)
2. Qiong, X.: Evaluation of the regional water resource carrying capacity based on the improved firefly algorithm and the projection pursuit model Hydropower New Energy (7), 71–75 (2015)
3. Feng, Ma., Qian, W., Wenjing, L., Guiling, W.: Evaluation of the water resource carrying capacity based on the projection pursuit model of the index system – taking Shijiazhuang as an example. S. N. Water Divers. Water Conserv. Technol. **10**(3), 62–66 (2012)
4. Xu, J.: Research on the cash flow evaluation of the listed real estate companies in China based on the projection pursuit model of the genetic algorithm. Ind. Sci. Tribune (9), 56–57 (2012)
5. Jiachao, R., Fazhu, J.: Evaluation of the financial solvency of the agricultural listed companies in China based on the projection pursuit clustering model of the genetic algorithm. J. Heilongjiang Bayi Agric. Univ. **25**(1), 95–98 (2013)

Application of the VR Technology in the Interior Design

Weiwei Li[✉]

Wuhan Technology and Business University, Wuhan 430065, Hubei, China
zhang2015mail@foxmail.com

Abstract. At present, the VR technology has experienced the entire process from scratch to development in the application process. At present, the VR technology is very mature. The virtual scene produced by this technology is exactly the same as the real design scene. Achieve the essential breakthrough of the interior design. With the help of the VR technology, users can experience the relevant indoor conditions at different times and in different environments, and fundamentally achieve a comprehensive breakthrough in the decoration design. This paper will combine the characteristics of the VR technology, and explore the application strategy of the VR technology in the interior design combined with the requirements of the interior design.

Keywords: VR technology · Interior design · Application path · Technical requirements

With the improvement of the science and technology, there are more and more high-tech elements in our life, and people are more and more yearning for a more comfortable, fast and efficient living environment, which put forward higher requirements for the interior design industry. In some emerging technologies, the VR technology has a very promising future, which mainly uses the computer hardware and various sensors to form the three-dimensional information, bringing different visual effects to people, to improve the level of the interior design.

1 Characteristics of the VR Technology

1.1 Authenticity

The VR technology can create people's sensory and perceptual abilities to things, as well as the ability of sports intensity analysis and the physiological commonness synchronization [1, 2]. The simulation scene is used to strengthen the user's perception of the surrounding space and things, and create a sense of reality.

1.2 Immersion

The VR technology creates an interactive three-dimensional dynamic scene through the modeling and simulation. With the help of handles, glasses and other devices, the

© Springer Nature Singapore Pte Ltd. 2020
M. Atiquzzaman et al. (Eds.): BDCPS 2019, AISC 1117, pp. 1687–1692, 2020.
https://doi.org/10.1007/978-981-15-2568-1_237

simulated scene is displayed in front of users [3]. Users will continue to inspire ideas according to their immersive feelings and perfect interaction with the environment. Through the communication with the user experience, the efficiency of the interior design can be improved [4].

1.3 Interactivity

The VR technology can create a real on-the-spot experience for users. With the help of the hardware systems such as handles, trackers, force feedback systems, and sensing gloves, as well as the software systems such as the model driven software and the function programming software, users can manipulate the experience, and change scenes and colors and so on, meet the requirements of the user experience, and show the interactive application effect of the VR technology.

1.4 Imaginative

The VR technology can create the undeveloped virtual things in the real world [5]. After the user experience, the new knowledge can be acquired, and the perceptual cognitive ability and the association ability can be improved, and the creative thinking can be stimulated, which is conducive to the development and creation of new things.

2 Application Status of the VR Technology in the Interior Design

The virtual reality (VR) technology is a kind of the computer simulation technology that can create and experience the virtual world. It is considered to be one of the high-end products of the new high-tech industry. The application of the VR technology in the practice of the interior architecture can mobilize the user's vision, hearing and touch in all aspects. Through the three-dimensional dynamic vision, we have a full understanding of the interior design. In the current interior design, give full play to the VR technology. The advantages and application of the system provide the basis for the stable development of the interior design industry.

With the change of the drawing methods, technologies bring high efficiency. In the early interior decoration, the space planning and design are mostly by hand drawing, which is the only way to communicate with the customers and determine the plans. The hand drawing not only needs a lot of time, but also needs one to three years of the painting experience. Drawings may not be able to be understood by customers, and customers may not understand the designer's design intent. But the emergence of CAD, sketch master and 3Dsmax software can be said to be the gospel of the interior decoration industry. CAD is used for drawing plans, elevations, and sections and so on in the interior decoration. It is also very convenient to operate. Just remember the usage of the shortcut keys, and it doesn't take long for a set of plan layout plans to be displayed in front of customers, which is both intuitive and easy to communicate. The 3Dsmax modeling and the Vray rendering software let the renderings achieve the same effect as the real photos. The emergence of these software not only allows designers to reduce the design time, but also more importantly allows customers to understand the designer's ideas and the effect of the space scene after the design more directly, which greatly improves the transaction rate. The traditional design process is carried out on paper. CAD, 3Dsmax and sketch masters and so on are finally presented on paper. This method has its own advantages. For example, it can improve the logic thinking and the hand drawing ability of designers, but it also has limitations. Because of the limitation of the space, designers can't design every tiny place completely, and the design intention and design styles expressed on paper can't be completely communicated to customers, which lead to a common problem in the design of the home decoration. The design intention expressed by the designers cannot be intuitively felt by the customers, and the cycle of the modification after modification also increases the design time and consumes a lot of energy of the designers. The application of the VR technology is not only a major breakthrough for the interior decoration, but can also reduce a lot of time needed to modify the design for designers, and customers can understand the designer's design intention, and understand the effect of the design more intuitively.

The application of the VR technology brings new visual experience to viewers. Now many design and home decoration companies are also preliminarily applying the VR technology to change the traditional design performances. The essence of this composite reality technology is to use the harmonious fusion of the virtual scene and the real scene, and to use the image processing technology to form a fusion space combining the virtual scene and the real scene, so as to provide customers with a richer and realistic visual virtual three-dimensional scene with high sense of presence. With the rapid development of the VR technology and the emergence of the VR panorama, viewers can use software to watch a 360° and 720° effect picture on their mobile phones or computers. It is more viewing than the ordinary photos, not only seeing a wider range of perspectives, but also having a sense of generation. The VR panorama is a new understanding and breakthrough of the visual images. Many interior design companies have applied the panoramic effect of VR to the design, because the traditional renderings are rendered by the 3D modeling, and from the perspective of view, they just render the model exactly the same as the actual photos. The visual immersion effect of the VR panoramic renderings is particularly excellent. It only needs to cooperate with the head display (head display device) to achieve the visual immersive effect.

3 Application Advantages of the VR Technology in the Interior Design

3.1 Strong Ability to Show Space

At present, the decoration designers use the 3DMAX for the interior design. The movement of lights and mapping materials is a primary factor for the texture and appearance of the 3D simulation environment. The renderers used the include scan lines, radiance, and rendering plug-ins and so on, which is the basis for creating a relatively high-definition image quality. For example, in a set indoor environment, walking with the mouse up, down, left and right can be free to add animations, so that consumers can participate in the experience. At the same time, through the virtual reality technology, we can quickly realize the construction of the virtual space, so that the whole space can be divided into the primary and the secondary ones, so that users can feel the environment they want to live in from different directions.

3.2   Convenient and Accurate Virtual Detection Method

The virtual reality technology can be based on the activity data of the experimenter to complete the design of the incomprehensible and more complex design relationship along the specified objects, without affecting the next process. For example, in the details of the hollowed objects and the irregular objects such as stairs and railings, the virtual scene experiencer experiences a strong sense of hierarchy of stairs, and checks the simulation environment collision test in the unidirectional realization of the complex objects through other accurate functions, and the whole scene stereo accurately detects the design scheme.

3.3   Interactive Experience

The virtual technology supports the experiencers to experience the interactive experience of the design scheme simulation, and sets the experience simulation areas on the limited areas, so that the experiencers can experience the interactive experience of all senses and the virtual technology. For example, if an experiencer wants to understand the details of the room design with a virtual device, he can run a specific software program to make the built-in camera of the software quickly reach the area limited by the experiencer, enlarge or shrink the observed object according to the needs, and observe and interact with the form and location information of the objects until he or she reaches the desired effect.

4   Application of the VR Technology in the Interior Design

4.1   Dynamic Comparison and Analysis of Various Elements in the Architectural Design

Designers use the virtual technology to build a virtual building. In the virtual network, they design the building in proportion to the real life needs. The process is very simple, similar to the process of the building blocks. On the network, they can also obtain some required material resources, enrich and improve, and take care of many details. They can observe the whole picture of the building from a comprehensive perspective. Further modifications can be made in terms of materials and colors, and the real-time switching can be realized between different buildings until the requirements are met. The final scheme can be used to redraw the structure chart, and the real-time editing and comparison can be made in terms of the element selection. The continuous modification can be made, and the timely communication with customers or more communication with other designers and we can achieve satisfactory final design. Especially in the comparison of elements, the transformation between various elements can meet the construction conditions under the specific environment state, and can also make the customer's needs and design concept to present to the maximum extent. Not only is the designer's work more convenient, but also the customer's participation will be higher.

4.2   The Advantage of the Virtual Technology in the Interior Design

The traditional interior design mode generally relies on the two-dimensional and three-dimensional imaging of drawings or computers, and the user's participation is not high enough, but the people who ultimately want to use this scheme are still customers, and their satisfaction is very important. Using the VR technology, designers can invite customers to participate in the simulation experience and understand their needs in the real time. Without a lot of professional knowledge, customers can also communicate with designers without obstacles and put forward their own suggestions. Designers can modify and perfect the schemes on the basis of some basis. In addition, in the design stage, some complex structures and scenes can be assembled by using the virtual technology, and the interaction of the VR technology can be constantly adjusted.

4.3   Space Experience Analysis of the Virtual Architecture

The ordinary computer technology has limitations for the two-dimensional and three-dimensional image display, and it cannot display all the information comprehensively. For designers, it is difficult to observe directly only some parameters, and some problems are easy to occur. The emergence of the VR technology can make up for this problem to a large extent. The virtual reality technology can feel the building from a three-dimensional perspective, and the distance can also be controlled. If the location of the construction unit and the designer is far away, the current situation of the construction can also be transmitted to the designer, so that the designer can understand the process and the state of the construction in the real time. The designers can not only put forward some constructive suggestions, and improve the quality of buildings, but also save a lot of time wasted in traffic, which can be more convenient and efficient.

One of the outstanding points of the VR technology is to provide a more real sense of movement and a more comprehensive experience mode. In the past, it is difficult for designers to achieve such an effect by the hand drawing or the computer drawing. Limited by the technical level, the analysis of the space angle by designers is also prone to the deviation, with great limitations. The VR technology can meet the all-round feelings based on hearing, vision, and touch and so on. Designers can observe from all angles, including the support of the walls, the conversion of the space perspective, and the adjustment of the unreasonable parameters will be more convenient, and the designed schemes will be more close to nature, more in line with the science, and the degree of comfort is also very high.

Customers can use the VR technology to feel some details of the scheme when accepting the scheme and all things will bring the real feelings to people, just like in the real scene. For example, some designs that are very in line with the physiological structure of the human body need this intuitive physical experience to feel the advantages, seats, mattresses that are suitable for the body structures. This type of the furniture may not be brilliant when you look at it, but you can feel its mystery when you feel it personally.

4.4   Analysis of the Physical Environment in the Architectural Design

The design of the buildings is greatly affected by the physical environment, and especially for the high-rise buildings, the middle and the high-rise buildings are greatly affected by the air flow, which will have a certain impact on the living environment and the daily life of the residents. If the building is close to each other and the air flow between the two buildings will increase the wind speed and even cause noise pollution. It will not only affect the quality of life of the residents, but also cause wear on the surface of the building, and the roof and the external decoration may be damaged to a certain extent. The VR technology can be used to simulate some situations in advance, regulate the air flow and observe the building model, improve the ventilation conditions and optimize the building structures. And use computers to make statistics of some data and indexes, which is convenient to debug the parameters and take into account more aspects, so as to get the most scientific and reasonable design schemes.

5   Conclusion

The VR technology has a great advantage in the interior decoration. There is no space limitation. Moreover, designers can introduce all design intentions to customers in the virtual space. Customers can also understand the designer's ideas at a glance and even participate in the modification of the design schemes. With the immersion of the VR technology, customers can feel whether they have achieved the desired effect and shuttle between their own virtual and real scenes.

Acknowledgement. Provincial teaching reform research project of Hubei Province: "Reform and practice of the teaching mode based on "project driven" – Core courses and examples of the environmental design major", Project No.: 2017480.

References

1. Lin, P.: Reflections on the curriculum construction of the interior design major in higher vocational colleges. J. Jiujiang Vocat. Tech. Coll. (1), 38–39 (2013)
2. Wei, Z., Liu, Q.: Application of the VR technology in the civil engineering disaster prevention. J. Eng. Manag. (2), 81–85 (2016)
3. Wang, M.: Artistic conception creation of the Chinese traditional garden based on the VR technology. Manag. Technol. SME (26), 171–172 (2014)
4. Shiming, Z.: Curriculum construction of the digital media art based on the VR technology. Technol. Wind (19), 59 (2016)
5. Wei, J.: Application of the VR technology in the interior design. Create Living (8), 54–55 (2016)

The Identity Authentication Mechanism of Human Testimony Based on the Mobile APP

Congjun Liu[✉] and Longgen Sun

School of Computer Science, Jiangsu University of Science and Technology,
Zhenjiang 212003, Jiangsu, China
hongqi25642100@126.com

Abstract. The OCR technology is becoming more and more mature. Its application in the document recognition greatly improves the efficiency of the digitization of various kinds of the document information. In the field of the mobile APP applications, more and more developers are developing the applications. The mobile APP applications bring more convenience to users. However, in many practical applications, besides completing the information registration, the APPs or the WEB system also need the data collection and verification to verify the authenticity and validity of the data information. Based on the existing OCR identification technology, combined with the face recognition and other technologies, this paper compares the face photo information data on the identity card with the living face photos, and provides the customers with the identity authentication mechanism services based on the mobile APP.

Keywords: Mobile APP · OCR technology · Human testimony · Face recognition

1 Introduction

With the development of the mobile terminals and the mobile communications (5G), the usage rate of mobile phones far exceeds that of computers. 80% or more businesses will be transferred to mobile intelligent terminal systems gradually in a few years. At present, the download application of mobile apps in major application markets is very popular. It can be solved by mobile applications regardless of clothing, food, housing and transportation, such as the current popular mode of O2O, the mobile payment online, mobile finance, and mobile online ordering offline distribution and so on [1–3]. The identity authentication app based on the mobile terminal effectively solves the need for the real-name authentication, makes the application of the real-name system more efficient and convenient, and improves the user experience, and solves the problem of the user's cumbersome input [4–6]. Based on the existing OCR (Optical Character Recognition) identification technology and OpenCV-based face recognition method, this paper compares the face photo information data on the ID card with the living face photos, and provides the identity authentication mechanism service based on the mobile APPs for customers [7, 8]. This authentication method has the advantages of convenience, rapidness, security and reliability, and can fully meet the needs of mobile APP applications in the identity authentication of human testimony [9, 10].

© Springer Nature Singapore Pte Ltd. 2020
M. Atiquzzaman et al. (Eds.): BDCPS 2019, AISC 1117, pp. 1693–1700, 2020.
https://doi.org/10.1007/978-981-15-2568-1_238

2   Face Recognition Based on OpenCV

The face recognition technology has a wide range of applications, and the algorithms of the face recognition technology are also diverse. Research directions include the template-based, the holistic face-based, the neural network-based and several other features. At present, the common recognition methods are: the geometric feature based recognition method, the line segment distance based recognition method, and the feature face based recognition method and so on.

In this paper, the OpenCV-based face recognition method is applied to verify the authentication of human testimony in mobile APPs. The flow chart of the authentication of human testimony is as follows: face detection and location, face preprocessing, face feature extraction and feature comparison. In order to ensure the accuracy of the face recognition, in the process of the image acquisition, it is necessary to ensure the full appearance of the front face, and it is best not to wear glasses, to ensure that the light is sufficient and the performance of the camera is good. Figure 1 is a flow chart of the authentication of human testimony.

Fig. 1. Authentication of human testimony flow chart

Face detection and location: Accurately locate the position of the face and determine the size of the face in the image to be recognized. By detecting the location and size, the face can be accurately segmented from the image to be recognized, which provides effective input for more accurate recognition. This paper combines the skin color detection and the AdaBoost algorithm for the face detection. The reason for choosing this method is that if we use the single skin color detection to do this work, although the efficiency is ideal, the detection range is not accurate enough, and the result image will always contain other parts including the human skin color of the face. But only when using the AdaBoost algorithm to detect, it happens to be on the contrary. Therefore, this paper chooses a combination of the two methods for the face detection. The face candidate regions are obtained by building a Gauss model in the YCbCr space, and then detected by using the AdaBoost algorithm in the candidate regions, which can accurately locate the face position in the image.Face image preprocessing: The most important step of the face recognition is to perform the face image preprocessing. Due to the influence of various external factors, such as the surrounding environment, the facial feature, the illumination and the hardware performance of the equipment, images often have various problems, such as the inconsistent size, the unclearness and the noise. The purpose of the image preprocessing is to remove or reduce the interference of the external factors (such as the illumination, the background and the equipment performance) to the greatest extent before the face recognition, remove the irrelevant interference information in the image to be recognized, improve the detectability of the effective information in the image to be recognized, and simplify the operation of the data. In the process of the face image preprocessing, this paper does two things: the image graying and the image enhancement. The image graying is to convert the RGB values of three different components of the color image input by the user into the same values. The original color image is transformed into the gray image. The calculation amount of the transformed image in the recognition is greatly reduced. In this paper, the weighted average method is used to accomplish the image graying. The purpose of the image enhancement is to improve the quality of the image, make the image visual clear, and be more conducive to the recognition and processing. The image enhancement is mainly divided into three parts: the histogram equalization, the image denoising and the image sharpening. First, the gray image is processed by the histogram equalization method, so that the gray area of the original gray image is more uniform, and the gray distance of the image pixels is enlarged, and the contrast of the image is increased, and the image quality is enhanced. Secondly, the image denoising is to filter the image in order to minimize the noise interference of the image. Finally, the image sharpening is done by using the Laplacian operator. The image sharpening can compensate the contour of the image, enhance the edge and gray jump of the image, and make the image clearer. Figure 2 shows the effect of the face image preprocessing.

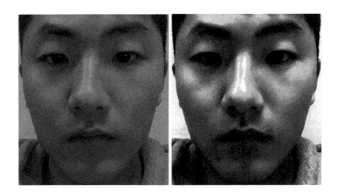

Fig. 2. Face preprocessing impression drawing

The image feature extraction: The feature extraction is based on the improved local directive pattern (LDP). LDP is a variant based on LBP (Local Binary Patterns). The LBP algorithm is very stable for the monotone linear illumination processing, but it is not ideal for images with the random noise and the non-linear illumination changes. In order to remedy this defect, Jabid and others proposed a robust feature extraction method LDP in 2010.

The basic principle of the LDP feature extraction is as follows. Each pixel is taken as a neighborhood in the center of its 3×3 range, and the convolution operation is carried out with eight adjacent pixels and eight Kirch templates (M0 $\sim$ M7). The edge gradient value corresponding to the direction of the template is obtained and recorded as mi (i = 0, 1, …, 7). Take the absolute value of the edge gradient value and sort it. Choose the maximum number of k in |mi| as the main feature, assign k bit to 1, and (8-k) bit to 0, and encode these 8 bits sequentially to produce an 8-bit binary number, and calculate the decimal value of the number. The decimal value is the LDP eigen-value of the center point of the 3×3 matrix. Namely:

$$x_{LDP}(k) = \sum_{i=0}^{7} 2^i b(|m_i| - |m_k|), \ b(x) = \begin{cases} 0, x < 0 \\ 1, x \geq 0 \end{cases}$$

Among them, $|mK|$ is the largest k value in $|mi|$ and b(x) is a symbolic function. Kirch templates M0 to M7 are shown in Fig. 3:

$$M_0 = \begin{bmatrix} -3 & -3 & 5 \\ -3 & 0 & 5 \\ -3 & -3 & 5 \end{bmatrix} M_1 = \begin{bmatrix} -3 & 5 & 5 \\ -3 & 0 & 5 \\ -3 & -3 & -3 \end{bmatrix} M_2 = \begin{bmatrix} 5 & 5 & 5 \\ -3 & 0 & -3 \\ -3 & -3 & -3 \end{bmatrix} M_3 = \begin{bmatrix} 5 & 5 & -3 \\ 5 & 0 & -3 \\ -3 & -3 & -3 \end{bmatrix}$$

$$M_4 = \begin{bmatrix} 5 & -3 & -3 \\ 5 & 0 & -3 \\ 5 & -3 & -3 \end{bmatrix} M_5 = \begin{bmatrix} -3 & -3 & -3 \\ 5 & 0 & -3 \\ 5 & 5 & -3 \end{bmatrix} M_6 = \begin{bmatrix} -3 & -3 & -3 \\ -3 & 0 & -3 \\ 5 & 5 & 5 \end{bmatrix} M_7 = \begin{bmatrix} -3 & -3 & -3 \\ -3 & 0 & 5 \\ -3 & 5 & 5 \end{bmatrix}$$

Fig. 3. Kirch template

The directions from M0 to M7 are east, northeast, north, northwest, west, south-west, south and southeast.
The principle of the LDP coding is as follows:

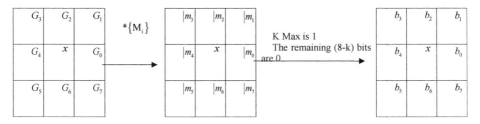

The encoding order of the 8-bit binary numbers from low to high is b0 to b7.
By consulting the relevant data, we know that in the LDP algorithm, the k value is 3, and when the k value is 3, the local feature information can be fully preserved.
In this paper, the basic LDP mode is improved. After getting the neighborhood edge gradient, the operation of taking the absolute value is cancelled. At this point, the eigenvalue calculation process is as follows:

$$x_{LDP}(k) = \sum_{i=0}^{7} 2^i b(m_i - m_k), \ b(x) = \begin{cases} 0, x < 0 \\ 1, x \geq 0 \end{cases}$$

Among them, m_k is the largest k value in m_i and b(x) is a symbolic function. It can be seen from the experiment that the improved LDP algorithm can more clearly represent the local feature information of the face image. The contrast difference map of the feature extraction between the basic LDP model and the improved LDP model is shown in Fig. 4. In this feature extraction algorithm, the feature histogram of the entire image is used to represent the local feature vectors of the face. There is no distinction between the high-contrast structure and the low-contrast structure of the image. The difference of the face area has no significant impact on the recognition effect, so the gray image needs to be partitioned.

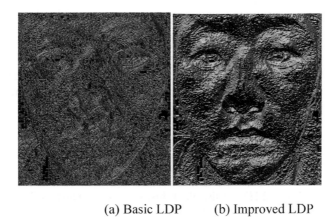

(a) Basic LDP   (b) Improved LDP

Fig. 4. Feature extraction impression drawing

The gray scale image is pre-processed by dividing n blocks of M × N. Each sub-region is marked as Ri (i = 0, 1, …, n). For each different block image, the block weighting operation is performed with its structure comparison information. The process is shown in Fig. 5. The calculation process is as follows:

$$w_{rc}(x_{LDP}(r,c)) = \frac{1}{8}\sum_{i=0}^{7}(m_i - \bar{m})^2, \bar{m} = \frac{1}{8}\sum_{\eta=1}^{8}m_\eta, w_i = \frac{1}{M \times N}\sum_{r=1}^{M}\sum_{c=1}^{N}w_{rc}(x_{LDP}(r,c))$$

Among them, xLDP(r, c) is the improved LDP eigenvalue of column c in row r, and wrc(xLDP(r,c)) is the contrast structure information of the corresponding blocks, and wi is the weight value of the corresponding blocks. Finally, the improved LDP histogram features of each block are extracted. The calculation process is as follows:

$$H_n(\tau) = \sum_{r=1}^{M}\sum_{c=1}^{N}f(x_{LDP}(r,c),\tau), f(x,y) = \begin{cases} 1, x = y \\ 0, x \neq y \end{cases}$$

Among them, Hn(τ) is the column height in the histogram when the gray value is τ, and f(x, y) is the judgment function. The final face image is represented by integrating the histogram features of each block.

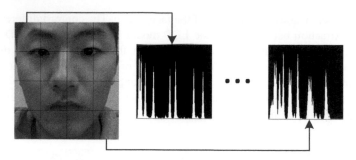

Fig. 5. Feature extraction impression drawing

In the phase of the feature comparison, considering the efficiency and accuracy of the feature matching, this paper chooses the chi-square distance method to calculate the face image similarity. The calculation process is as follows:

$$x^2(H^1, H^2) = \sum_{i,j} w_i \frac{(H_{i,j}^1 - H_{i,j}^2)^2}{(H_{i,j}^1 + H_{i,j}^2)}$$

Among them, H1 and H2 are the block histograms of images to be matched, and i denotes the block number, while j denotes the number of the improved LDP feature histograms contained in the block area (j takes 1 in this paper).

3   Experimental Results and Analysis

There are two face contrast modes in the face recognition. One is 1:1 and the other is 1:N. 1:1 contrast is to compare a pair of the different single-sample face images to determine whether they are the same person or not. 1:N contrast is to compare a specific single-sample face image with everyone in a given face database. The face information is calculated and compared, and the smallest difference is the final result of the recognition. The face recognition in this paper is used for the comparison of the human witnesses, so the comparison mode used in this paper is 1:1 comparison. Before the face comparison, a large number of the sample data are needed to determine the recognition threshold. Then, the threshold is applied to the system to judge whether it is the same person or not, and the smaller the threshold is, the smaller the gap between the face images is, that is, the greater the possibility of being the same person.

In this paper, the Yale Face Database is used to train a large number of the sample data to determine the size of the threshold. The final threshold is 10500.

In order to verify the effect of the face recognition algorithm, that is, the recognition rate, this experiment compares the face recognition effect of the basic LBP, the basic LDP and the improved LDP on the Yale face database and the real face image database respectively. In the Yale face database, including five volunteers, there are 165 face images. In the actual face image database, there are 100 face images of 10 temporary workers, in which the illumination level and their gender are different. The comparison of the algorithms focuses on their recognition rate and the time required for the successful recognition.

The experimental environment of this paper is the notebook computer, Interl i5-4200 M 2.5 GHz, 4 GB memory, Microsoft Win7 operating system. The experimental results are shown in Table 1.

Table 1. Comparative statistical results of the human testimony

Methods	Correct recognition rate/%		Average time use (ms)
	Actual face database	Yale face database	
LBP	84.1%	82.7%	32.7
LDP	90.6%	89.3%	25.8
Improved LDP	94.9%	93.3%	27.3

As can be seen from the table above, the LBP algorithm consumes the most time, which is due to the higher dimension (256) of the LBP feature histogram, and its computational complexity, while the LDP and the improved LDP have a significant reduction in time, because the k value in this paper is 3. At this time, the dimension of the feature histogram is only 56, which greatly reduces the computational complexity and improves the recognition efficiency. However, in terms of the recognition time, the improved LDP is slightly higher than LDP, because compared with LDP, the improved LDP has more steps of partitioning and calculating weights in the recognition process. It is acceptable to take into account the advantages of the improved LDP in the recognition rate and a slight increase in the recognition time.

Mobile experimental environment:

(1) IPhone 7, 32G, 2G memory, quad-core A10 processor, IOS 10 operating system;
(2) Glory 8X, 64G, 4G memory, six-core Kirin 710 processor, EMUI 8.2 (based on Android 8.1) operating system.

In the course of the experiment, the lack of the light and the wearing of the accessories were not considered. Ten experiments were carried out on 10 people in different real-time environments, including different illuminations and genders. The correct recognition rate can reach more than 95%. The average recognition time in the environment of iPhone 7 is 43.2 ms and that in the environment of glory 8X is 48.7 ms, which can fully meet the needs of this subject.

References

1. Yimin, T.: OCR-based Identity Card Recognition System. D. Huazhong University of Science and Technology (2014)
2. Yuxiu, S.: Research on the Business Card Recognition Method Based on the OCR Technology. D. Harbin University of Science and Technology (2015)
3. Ziyi, Y., Qingyong, Z., Xueqi, L.: A face detection method combining the improved AdaBoost and the skin model. J. Comput. Digit. Eng. **4**, 743–749 (2018)

 4. Li, J., Xu, T., Li, H.: Face recognition based on the improved LDP. J. Comput. Appl. Softw. **5**, 148–150 (2014)
 5. Hui, Z., Xin, L.: Single-sample face recognition algorithm based on the improved LBP. J. Comput. Appl. Softw. **12**, 220–223 (2017)
 6. Juan, T., Zhao, X.: Analysis and consideration of the MVC design mode based on J2EE. J. Comput. Modern. **10**, 54–58 (2010)
 7. Lei, G., Xing, S.: Methods and devices for configuring the timing tasks based on the Quartz framework P. Beijing Weiku Electronic Commerce Technology Co. Ltd.: CN101887381A (2010)
 8. Xu, Z., Jiao, W.: RabbitMQ message confirmation mechanism optimization. J. Comput. Syst. Appl. **3**, 252–257 (2018)
 9. Fan, Z., Luo, F.: JWT authentication technology and its application in WEB. J. Digit. Technol. Appl. **2**, 114 (2016)
10. Luyang, L., Bing, G.: Android app-based secure login authentication solution. J. Mod. Comput. (Prof. Ed.) **35**, 9–12 (2016)

Application of the Internet Technology in the Visual Communication Design

Xiang Liu[✉]

Faculty of Fine Art, Jiangxi Science and Technology Normal University,
Nanchang 330038, Jiangxi, China
guoxiwull@163.com

Abstract. In the digital era, the development and application of science and technologies are becoming more and more mature. On the premise of the information technologies and the network technologies, digital, information and media technologies are gradually integrated into public life. In the current visual communication design, the use of the Internet technology can enrich the display form of works and optimize the viewing effect. Therefore, we should integrate the development trend of the visual communication design activities and use the Internet technology reasonably.

Keywords: Internet technology · Visual communication design · Applied research · Digital media

The wide application of the digital media has a great influence on people's aesthetic and visual feelings. The change of the reading mode and the design concept also promotes the innovation of the media mode and the innovation of the modern visual communication design. Because people's life is gradually affected by digitalization, the visual communication design should be more digitalized, life-oriented and popular, so as to realize the high-quality development of the visual communication design in the digital era.

1 The Development Trend of the Visual Communication Design Activities

With the popularization of the Internet and the development and innovation of the digital media technologies, the new media technology has been involved in the visual communication design to expand the ways of the design creation thinking and the visual expression means [1]. In the era of the new media, the new design concepts inject new vitality into the visual communication design, and advanced technological innovation has opened a new creative space for the visual communication design [2]. The new media technology forms a visual innovation in the visual communication design with the dynamic graphic and the image design, and realizes interactive experience with the integration of the multiple senses [4]. The new media technology provides the new development opportunities for the art design, but there are also disadvantages, which require the design practitioners to adhere to the concept of

© Springer Nature Singapore Pte Ltd. 2020
M. Atiquzzaman et al. (Eds.): BDCPS 2019, AISC 1117, pp. 1701–1706, 2020.
https://doi.org/10.1007/978-981-15-2568-1_239

"people-orientation", continue the traditional design concept and form of expression, in order to make the biggest breakthrough in the limited innovation.

The visual communication design has a clear purpose, whose essence is to pass the specific visual information to the public most accurately and quickly through the design [5]. The visual communication design covers a wide range of contents, including the general graphic design, such as the advertising design, the packaging design, the public logo design, CI, and the logo design and so on. The visual information is the essence of its communication, or the visual communication design can be called an art of information processing. From this point of view, the visual communication design and the Internet have something in common, both of which are the human efforts and the exploration around the dissemination of the information.

Compared with the traditional media, the visual communication design based on the network platform has the characteristics of timeliness, which can reduce the time spent in the printing, storage and transportation. Moreover, because the Internet has the characteristics of the audio, video and text fusion and the strong interactive functions, the period of the information acquisition and transmission is obviously shorter than that of the traditional media, and the contents covered by the Internet are wider than that of the traditional media, and the contents and forms of the visual communication design based on network platform are richer. This kind of the enrichment has broken through the simple words, pictures and videos, but through the use of the link technology to combine words, pictures and video audio to present in front of the audience.

The visual communication design based on the network platform also has the characteristics of relying on the network transmission. Although the web page presented on the network can be viewed by the audience, this is based on that fact that the audience has a certain speed of the network transmission. When a large number of pictures and videos are designed on the web page, if the network transmission speed of the audience is slow, it will take a lot of time to receive the information. To some extent, this requires that the visual communication design with the help of the network platform cannot rely on a large number of the pictures and videos. At the same time, we should realize that the visual communication design with the help of the network media needs to integrate various elements such as vision, pages, interaction and navigation. Among them, the visual design analysis is due to the fact that when people operate in front of the computers, they are often between half a meter and one meter away from the computer display. This determines that people's vision is basically to pay attention to the contents in the center of the computer screen, and then observe the surrounding contents. If there is the strong stimulating information in a certain position of the computer screen, people's attention will be attracted by this information. It is based on the characteristics that people's vision and attention will be attracted by the stimulating information. In the visual design, people can use pictures or words with shocking senses to attract people's attention. Page design refers to the design and creation with the help of the Internet platform.

2 The Application Principle Analysis of the Internet Technology in the Visual Communication Design

2.1 Aesthetic Principles

In the visual communication design of the network advertisement, we need to enhance the beauty of the design creation through the change of graphics, and adopt the ways of the transmission and exaggeration. At the same time, we need to be able to optimize the composition of advertisements to make the whole advertisement present a kind of beauty of colors, so as to improve the appreciation of the advertisement, in order to better meet the public's aesthetic.

2.2 The Principle of Humanization

In the design of the network advertisements, we can design the ultimate goal of the advertisements is to let the audience accept the ideas of the advertisement communication. The visual communication design is to better achieve the purpose of the advertisement communication. The design of the network advertisement must adhere to the principle of humanization to better meet the aesthetic needs of the audience. At the same time, we should also be able to respect the individuals, so that we can meet the needs of different audiences as much as possible.

2.3 Principle of Innovation

In the visual communication design, we must be able to adhere to the principle of innovation, and the network advertisers must be able to dare to break through the traditional design mode, in order to better give new ideas and creativity to advertising, so that advertising works have a more unique style. The color and color selection of advertising graphics must be close to the main purpose of the advertisement, so as to make the work more infectious and strengthen the expression of the work.

2.4 Comprehensive Principle

In the design of the visual communication, we must be able to fully use a variety of ways to achieve the design goals. In the design of the online advertising, we must be able to consider the relationship between various design elements and grasp the visual elements scientifically, so as to better show the charm of the online advertising. At the same time, we must be able to flexibly use various design methods to make the whole design method effectively integrate in the network advertising, further play the comprehensive role of the network advertising, and effectively improve their own design quality and effect.

3 Application Requirements of the Internet Technologies in the Visual Communication Design

The visual communication design beyond the vision will be popular. Now the visual design we mentioned does not only include the design of the visual elements in the traditional sense, but also adds elements such as hearing. We can guess that, in the future development process, whether the visual design can add olfactory and tactile elements, we don't know. In today's society, it is in the era of the rapid development of various high-end technologies and constant renewal of innovative ideas. In such a favorable environment, the visual design is bound to break its own sky. According to the reality, when designers are creating, they not only use one aspect of the knowledge, but also cross and combine many kinds of the knowledge, which requires designers to study and explore different knowledge and master the basic knowledge of various disciplines.

The visual communication design will be more and more permeated with the spirit of the humanistic care. Now the world is moving towards the form of integration, and the development speed is faster and faster. The original single way of the information dissemination and the backward media have not met the needs of people in today's society for the multiple information. Nowadays, people pursue innovation, which has a certain impact on the previous design concept. In order to meet people's needs, necessary improvements must be made in the design, and thus, some new designs emerge. At the same time, compared with the past, the material and living standards of today's society have changed dramatically. People pay more attention to the enjoyment of their spiritual aspects, and the requirements for the design are gradually improving. Most of them want to see new design breaking through the traditional constraints.

The visual communication design will absorb more nutrients from the traditional Chinese culture, and the requirements to get inspirations from each person's vision are different. It can be said that there are great differences. The main reason for this situation is that each person's living environment, cultural background, education level, their own personality characteristics are different. Different people will have different understandings of the same visual image, so they will have different psychological reactions. These differences urge the design to be constantly updated in order to adapt to various differences. As a new subject, the visual communication design has great particularity. It is not only the traditional education content such as class, teaching, asking questions, doing homework and reviewing homework, but also the educational means of the modern science and technology to stimulate the enthusiasms and creativity of designers.

The era of the design requires that the design education must follow the choice of the times. The teaching mode in the classroom is far from meeting the needs of today's designers' learning. The creative thinking of these designs in the network media will be easier to guide designers to play their own talents and create excellent works. Designers can not only solve their doubts on the Internet, but also make personal comments on the excellent works on the Internet, from which they can understand the opinions of other designers. When the author uploads his works on the Internet, his peers timely express many opinions on the creation concepts, expression forms, color application, font change, and layout composition and so on, of the works, and put forward praise and criticism from various aspects. The discussion process is a learning process, which increases the interaction between designers. The most important thing is to push the visual communication education mode to another new space, which greatly improves the defects of the traditional design education.

4 Application Strategy of the Internet Technologies in the Visual Communication Design

In the Internet era, with the promotion of the digital media technology, the visual communication design has realized the span of the time and space, brought people the virtual aesthetic experience, and successfully brought all kinds of the visual communication design products into an innovative society.

4.1 Pay Attention to Dynamic Image Design in the Visual Communication

The typesetting design in the visual communication includes the design of texts and images. In the design of images, the four elements of displacement, reference, speed and direction should be used to dynamically express the images based on the 4D space, so as to unify the dynamic expression of the image and plane design. In the context of the digital media, the visual communication design should make full use of the modern network information technology, the multimedia technology, and the design and so on, and use the diversified design styles to design the image effects to the extent of the impact on people's vision, so as to meet people's needs for the visual experience. For example, at the new product launch, designers can dynamically design the graphics and images, integrate the appearance and functions of new products and express them again, highlight the advantages of products, handle the details and vividness of the appearance of products, present the best images of products, and show their functions to the audience from ordinary to special. On the basis of the innovation, the traditional graphic design is transformed to make the traditional graphic dynamic. For example, in the design of the dynamic posters, it can effectively connect the dynamic of the plane with the image science, and realize the perfect combination of the plane and the three-dimensional, which is more conducive to the development trend of the visual communication design in the context of the digital media, more conducive to the promotion of products with values, in line with the aesthetic and demand of the audience, rich in personality and characteristics, and more conducive to the introduction of the digital media for the visual communication design concept to provide a more creative platform.

4.2 Design of the Dynamic Characters in the Visual Communication

Text has the comprehensive attributes such as space, time and movement state. In the process of designing the dynamic texts, in order to ensure the regular movement of texts, we must pay attention to grasping its comprehensive attributes. Especially in the context of the digital media, if the dynamic texts can stay in the movement for a short time and give the audience enough time to read the texts, it will greatly improve the product attractiveness. Once any design arouses the resonance of the audience, it has played a certain effect on the subsequent publicity. In the process of designing the dynamic characters, we must abide by the principle of "diversification of the design, popularization of aesthetics, and accuracy of the information transmission", and carry out effective design based on this to meet the needs of the society. For example, when making a promotional video of an activity, the dynamic text design should be properly introduced to make the text in the promotional video present a wandering state in a fancy way. When it runs to a certain extent, it will form a complete text from the fragments of the fancy, so as to attract the audience's attention and deepen the audience's impression of the text, so as to achieve the best publicity effect.

5   Conclusion

The relationship between art and technology is widely existed in various design fields. In the traditional visual communication field based on the plane carrier, the relationship between art and technology certainly exists. However, in the new visual communication design based on the network carrier, there are some different characteristics that we need to understand. The basic elements of the visual communication design, such as words, graphics and images, are more limited than the traditional media in the network, which is mainly because the amount of the information transmission on the network is limited by hardware. Therefore, the design elements of the visual communication based on the network should be as small and precise as possible.

References

1. Binbin, S.: Analysis of the innovative design concept of the visual communication art design. Art Educ. Res. (9), 11–19 (2015)
2. Meng, X.: Teaching discussion on the innovative design concept of the visual communication art. J. Inner Mongolia Normal Univ. (Educ. Sci. Ed.) (5), 21–28 (2014)
3. Xueqin, L.: Visual communication design in the digital age. Art Sci. Technol. (11), 12–19 (2016)
4. Hui, C.: New concept of the visual communication design in the digital age. J. Southeast Univ. **12**(2), 90–93 (2010)
5. Xing, Y., Zhang, H.: Visual interaction in the visual communication design. Art Sci. Technol. (3), 101–122 (2013)

The Design of the Model Education Platform for College Students Based on the Characteristics of the Network Behaviors

Yu Liu[✉]

China West Normal University, Nanchong 637000, Sichuan, China
xianghongxue1@yeah.net

Abstract. In the process of college students' cultivation, by choosing models and propagating the stories of the learning models, the learning process is the power of recognition, dissemination and recognition of the models, and interprets, deepens and internalizes the spirits of the models. Under the Internet environment, the training methods of college students also need to be adjusted and optimized. Especially from the characteristics of college students' network behaviors, we should design a model education platform, so as to better disseminate the spirit of the models, tell the stories of the models, and achieve the best effect of the model education of college students.

Keywords: Network behavior characteristics · College students · Model education · Platform design

Among all the groups affected by the Internet, college students are the most influential and the large number of special groups. At the same time, they enrich the model education in contents and forms [1]. However, in today's context, there are many problems in the model education. To solve these problems, educators need no delay.

1 Design Background of the Model Education Platform for College Students Based on the Network Behavior Characteristics

The reason why we want to design the platform of model education for college students is that we should pay attention to the following reasons [2]. Firstly, the selection of college students' role models plays a positive role in promoting the socialist core values. Firstly, the selection of college students' role models is helpful to carry and transmit the socialist core values. As the value symbol of the vivid images, the images, advanced deeds and noble spirits of an example can carry and transmit the values of the socialist core values and make them concrete and vivid. Second, the model selection helps to enhance the recognition of the socialist core values. College students realize the recognition of the socialist core values through the rational cognition, the emotional

© Springer Nature Singapore Pte Ltd. 2020
M. Atiquzzaman et al. (Eds.): BDCPS 2019, AISC 1117, pp. 1707–1712, 2020.
https://doi.org/10.1007/978-981-15-2568-1_240

identification and the behavioral identification of the role models [3]. Thirdly, the model selection helps to enrich the connotation of the socialist core values. The enrichment and expansion of the model selection and the educational connotations in different times are in line with the mainstream values of our society and enrich and perfect the connotations of the socialist core values.

Secondly, college students' model recognition is consistent with the connotation of the socialist core values [4]. First, they have a high degree of recognition of the role models and their incentives. According to the survey of 10 different types of college students, most students want to grow up to be students' role models. They expect to be successful in their own growth, and recognize the inspiration and the leading roles of the role models. The second is that the characteristics of identifying with examples conform to the connotation of the socialist core values. College students' choice of the model traits is in turn "the love of our motherland", "honesty and trustworthiness", "diligent learning", "scientific and technological innovation" and "self-improvement". This is basically consistent with the socialist core values at the individual level of the values of "patriotism, dedication, integrity, and friendliness". Thirdly, we should give full play to the enthusiasms and initiatives of students in the selection of the models. In the model selection, college students think that their right to know and participate is very important [5]. Fourthly, to be an example, we must rely on our own efforts and the cooperation of schools and families. Schools should further improve their work on the selection of the targets, the cultivation of the focus points, the selection mechanisms and the influence of the publicity.

In addition, the characteristics of college students' role models conform to the pursuit of the socialist core values. One is the "typicality" and "generalization" of the model figures. The role models are changing from the "traditional heroes" to the "typical civilians", and more and more "grass-roots heroes" are recognized and loved by the society. The second is "diversity" and "unification" of the model types. To cultivate the models of college students' identity, we should grasp the yardstick between "diversity and richness" and "unified dominance" to truly realize the lasting vitality of the models. The third is the "visible" and "learned" propaganda of the models. More attention should be paid to the reconstruction of the model image in the model propaganda, so as to avoid the tendency of "tall and complete" in the model propaganda, and to make the public understanding of the models more stereoscopic and the images of the models more real.

2 Design of the Model Education Platform for College Students Based on the Network Behavior Characteristics

2.1 System Overall Design

The Internet user behavior analysis system designed in this paper consists of the data capture module, the data analysis module, the data chart generation module and the system interface framework module. The process of the system operation is as follows. First, the system obtains the network data packet by the data capture module, intercepts its packet header information, extracts the key feature information such as the source IP address, the destination IP address, and the port number and so on in the packet header, stores it in the database and displays the real-time captured data information in the system interface. The data analysis module makes the statistics and the analysis of the feature information captured in the database, and then obtains the data that can reflect the user's behavior characteristics. The chart module generates the intuitive, easy-to-understand and better reflecting charts based on the data from the data analysis module and displays the charts in the system interface. The system interface framework module realizes the user's operation and information display.

2.2   Database Design

This system uses the MySql database system to store the user behavior data in the running process. MySql is an open source database system with the small size, the fast speed and the low cost. It can run on various operating system platforms. Because of its open source and free characteristics, a large number of users develop MySql as a database management system. The database designed by this system includes three data tables: the capture information storage table data, the data analysis result table, and the registration port table tblregisteredports.

2.3   Introduction of the System Module

System Interface Framework Module
The system interface is mainly used for the good interaction with users to realize the operation of the system. The system interface is divided into three parts. The first part is the menu bar, which is used to implement the operation of the system, including the setup of the capturer, the start-up of the data capture, and the analysis of the captured data and so on. The second part is the real-time display of the tables to realize the display of the real-time captured feature information. The last part is the total size label, which shows the total amount of the captured data in the real time. The concrete implementation method of using the Java language is as follows: create a JFrame class object as the main interface framework, and the main interface framework is divided into three parts: the menu bar part, the table part, and the label part. The main interface framework uses the boundary layout manager to layout, adding a JPanel panel at the NORTH, CENTER and SOUTH locations of the main interface framework, adding JMenuBar objects to the panel at the NORTH locations, and adding a series of the menu bar components to realize the menu bar function. In the CENTER location, JTable objects are added to the panel JScroll Pane class objects, and then JScroll Pane class objects are added to the Jpanel panel. Finally, the real-time capture feature information is displayed in the SOUTH location, and the Jlabel class objects are added to the panel JScroll Pane class objects. The size of the captured data is displayed by constantly updating the data on the label.

Data Capture Module
The data capture module is mainly used to capture the data generated by user's network behaviors from the hosts. The capture interface shows that the "Select Network Card" drop-down box is used to select the network card to capture. The selected network card information is displayed in the text field of "Network Card Information". The "Hybrid Mode" radio button is used to realize the capture. Choose "Yes" to capture all packets

passing through the network card, including those not sent to the local computer, and "No" to capture only the packets sent to the local computer. The "Data Length" text box is used to set the size of the data captured at one time, and the "Timeout" text box is used to set the timeout. The implementation of the data capture module is as follows. First, get Device List method in the Jpcaptor class object provided by the Jpcap is used to get the list information of the network cards, and then open the network cards specified to be captured through the open Device method provided by Jpcap, and set the relevant capturer information. Then, get Packet method is used. Active capture is used to capture the data packets, and then filter the captured data, obtain the UDP and TCP data packets, intercept the header information, extract the source IP address, the destination IP address, the port number and other key feature information in the header, and finally store these data in the database.

Data Analysis Module
The data analysis module can process the original captured data and obtain the required data. The implementation method is as follows. In the process of the information exchange between the host and the server it visits, the type of the network behaviors can be obtained by acquiring the port number information in the process of the information exchange. Based on this point, the port used for the local machine is meaningless, and only the port number of the server has the corresponding meaning. The data analysis method of this module is as follows. If the information source IP is the local IP address, the corresponding destination port number is the server port number. If the information source IP is the local IP address, the source port number is several server port numbers. We process the raw data in this way, and then get the data information we need. Finally, the processed data are analyzed through the traffic activities, the protocol types, the traffic trends and other aspects to get the final data.

Chart Display Module
The main function of the chart display module is to display the processed data in a chart way. View different data analysis charts by selecting buttons in the interface. The implementation of this module is as follows. Use JFreeChart from the JAVA open source project to implement the chart. Firstly, the data set of charts is constructed, and then the methods of createBarChart, createPieChart and createLineChart provided by the abstract class ChartFactory are used to implement such methods as the bar chart, the pie chart, and the polyline chart and so on. By referring to the third-party external plug-ins, the interface can be beautified and friendly.

3   Notices for Application of the Model Education Platform for College Students Based on the Network Behavior Characteristics

At present, there are many outstanding problems in the cultivation of college students' role models. Firstly, the accelerated pace of the social change and the spread of the cultural diversity have caused new uncertainties to the model education. The proliferation of different value choices on the Internet has challenged the themes and the mainstream values unprecedentedly. Second, college students' subjective conscious-ness is strengthened, which cannot resonate with the model education at the same frequency. It is difficult for them to play the role of the model education because of the inconsistency between the selected model and their physical and mental needs. Thirdly, the opaque selection procedures and the unscientific selection conditions reduce the effectiveness of the model education.

Therefore, we should strengthen the cultivation of university students' role models from three aspects. First, in terms of the education guidance, the cultivation of college students' role models should focus on "knowing". We should deepen the education of ideals and beliefs, identify with examples, constantly strengthen the construction of the ideological and political work system, and give full play to the educational function of all courses. Second, in practice, the cultivation of college students' role models should focus on "doing". Colleges and universities actively build different types of the moral education practice platforms, establish the dynamic model banks with different themes, and use different youth models for the classification and guidance. At the same time, colleges and universities should organize students to go out of the campus and accept the social baptism. Thirdly, in the aspect of the public opinion propaganda, the culti-vation of college students' role models should focus on "micro". We should not only adhere to the correct guidance of public opinions, but also make full use of the advantages of the integration and development of the media in the new era, such as the network, television, radio, newspapers and other means, and adopt the ways and channels that college students are willing to accept to do a good job of propaganda.

There are five main ways to improve the effectiveness of the cultivation of college students' role models. One is to standardize the type of the models according to the contents of the socialist core values. In the example education of colleges and uni-versities, we should catch the students' attention, pay attention to the guidance of the mainstream values, and set an example for all or some groups of students. The second is to actively tap the internal model education resources in colleges and universities. The advanced deeds of the models around us are close to the daily study and life of college students, and they are easier to be accepted and imitated. Thirdly, we should establish a model selection and propaganda mechanism for students' universal par-ticipation. Adhere to the student-center, and fully listen to students' opinions, so that the models have the profound student foundation. Fourthly, we should actively build the campus cultures and persist in introducing literary works to educate people. We should create an environment of the cultural consciousness and pay attention to the promotion and inheritance of the model figures and advanced national spirits. The humanistic quality courses should be set up to cultivate college students' humanistic qualities and create a humanistic atmosphere in the campus environment. Fifthly, we should optimize the model cultivation environment for the integration of the campus, our families and our society. Schools should build a wide range of the exemplar training platforms, including the theoretical learning, the technological innovation, the peer assistance, and the public welfare activities and so on.

4   Conclusion

The model education is an indispensable educational method in the ideological and political education from the ancient times to the present. It plays an important role in both the information-blocked ancient times and the rapid development of the science and technology. As a new type of talents receiving the higher education, college students always lead the trend of the times. Therefore, starting from the characteristics of college students' network behaviors, the best effect of college students' model education will be achieved.

Acknowledgement. This project is supported by the Fundamental Research Funds of China West Normal University (Project number: 18E011).

References

1. Li, Y., Fan, Y., Deng, X.: A study on the effectiveness of the model education for college students under the network environment. Grand Chapter **36**, 168 (2011)
2. Zhang, J.: On the characteristics and countermeasures of college students' network behaviors. Time Educ. **15**, 58, 60 (2014)
3. Li, M.: A brief analysis of the application of the red culture in the cultivation of the socialist core values of contemporary college students – taking baise red culture in Guangxi as an example. Market Forum **10**, 12–15 (2015)
4. Zhao, H., Wang, S.: Analysis of college students' network psychological characteristics and guidance methods of their network behaviors based on the survey of Guangdong university of technology. Intelligence **34**, 48–49 (2015)
5. Zhou, S., Liu, F.: Analysis of subcultural characteristics of college students' internet behaviors. Contemp. Youth Res. **5**, 77–81 (2013)

Design of the Automobile Marketing System Based on the Big Data

Sijin Lv[✉]

Sichuan Vocational and Technical College, Suining 629000, Sichuan, China
hailanqi@aliyun.com

Abstract. With the continuous maturity of the application of the big data technology, now the technology has been widely used in the automobile industry. In many fields such as the automobile production, sales and after-sales and so on, it has given full play to the application advantages of the big data. The core of the big data marketing is to make use of the technical advantages, and accurately connect and understand the actual needs of consumers, so as to carry out the psychological intervention on the purchase behavior of the target consumers. For this reason, this paper aims at the advantages of the big data technology, and designs an automobile marketing system with application advantages.

Keywords: Big data · Automobile · Marketing system · Design ideas

After the emergence of the Web2.0 technology, it is rapidly promoting the consumption of the Internet applications of the Internet users in China. As one of the pillar industries in China, the automobile industry will inevitably take the network marketing as a necessary means. For China's automobile industry, the construction of the network marketing system is in its infancy [1]. The existing automobile network marketing system cannot really play the role of the network marketing, and whether it is the function realization or the safety consideration, the design of all aspects is not perfect [2]. The design of the automobile marketing system based on the big data is of great practical significance.

1 Design Background of the Automobile Marketing System Based on the Big Data

At present, the huge dealer system with the high operating costs and low profits has seriously affected the development of enterprises, and restricted the relationship between enterprises and users, resulting in the low brand awareness [3]. In the face of this situation, it is imperative to design and implement the automobile marketing system based on the big data application by referring to the relevant researches of the existing automobile marketing websites. The main purpose of this paper is to design and implement an automobile network marketing system based on the big data [4].

Starting from the system demand analysis, according to the marketing characteristics of the automobile products, combined with the needs of all parties, determine the system functional architecture, and then form the logical architecture according to the

M. Atiquzzaman et al. (Eds.): BDCPS 2019, AISC 1117, pp. 1713–1719, 2020.
https://doi.org/10.1007/978-981-15-2568-1_241

business levels [5]. And on the basis of the demand analysis, we use the method of the functional module division to design the system. At the same time, we use the business flow chart to explain the main functional modules such as the panoramic car watching, the appointment test drive and the registration login, and use the relevant design method of the database to complete the design of the data base.

After the implementation of the system, it mainly expounds the practicability from the aspects of the vehicle model display, the panoramic car watching, the customer relationship management and the user behavior analysis, and proves the safety and reliability of the system through the function test and the performance test, and further explains the application value of the system through usability evaluation. The system uses a uniform style of the page design to achieve the user-friendly interface, which further makes the automobile network marketing system feasible and more efficient. According to the characteristics of the automobile network marketing system, the overall framework of the system adopts the B/S architecture, and adopts the modular design to realize the automobile network marketing system based on the big data application.

2 Advantages of the Automobile Marketing System Based on the Big Data

The big data can help the automobile enterprises to provide more services for users through the sales database management and the data mining. It can change the business models such as the automobile production and marketing, and use the data insight to mine the maximum value of the automobile marketing. The task of the big data is to liberate the front-end users from trivia and create a simple, direct and convenient car life through the power of its technology background.

2.1 Quickly Predict the Market and Guide the Upper Level to Make Decisions

The big data of the automobile can understand the real demand of the market through the information analysis of the customers' consumption and use preferences collected by the network and the Internet of the automobile, and then formulate more effective product strategies to quickly adjust development, production and resource allocation. In terms of the network, automobile enterprises can analyze the Internet search volumes in a certain period of time through a third-party company to determine the market popularity, vehicle types, regional attention and other information, and can also judge the popularity of this car in the market through the analysis of the social media such as the micro-blog and We-chat, so as to take corresponding market strategies. In terms of the Internet of vehicles, enterprises can monitor the distribution and the use of their products in the market in the real time, such as collecting vehicle bus information, and analyze the driving age, driving habits, regional road conditions and vehicle conditions of users who buy products in various regions, and adjust the market strategies of the product models and regions based on the geographical locations of the products.

2.2 Precision Marketing to Improve the Transaction Rate

The massive data sources can provide the greater support for the marketing work. Enterprises can get users' behavior tendency through the network data analysis. For example, according to the time when users stay in a certain model of the web page and the number of times that the same user browses a web page, they can get the users' purchase intention. Then, they can improve the users' transaction possibility by pushing the promotion information. In the same way, through the analysis of the social media users' attention and forwarding, we can determine the purchase possibility of the user, so as to develop the accurate promotion strategies. The big data of the Internet of vehicles, through the analysis of the driving routes of the users in a region, can get the travel habits of the users in the region, and the automobile enterprises can design more accurate advertising and exhibition work. At the same time, through the analysis of the vehicle parking place, we can get the economic situations and the consumption abilities of the users, and even analyze the consumption habits of the users. On the other hand, through the analysis of the driving data, the automobile enterprises can analyze the current vehicle conditions of the existing customers, and can get the users who need to change cars, so as to carry out the secondary purchase marketing work.

2.3 Significantly Reduce Costs

Through the analysis of the big data, enterprises can improve the product quality, improve the production process and simplify the business process. For example, through the collection and analysis of the vehicle operation and the maintenance information, we can understand the quality performance of each product in the market. These data are of great value to the R & D department, the production department, the purchasing department and the marketing department. The big data analysis in the Internet of vehicles plays a positive role in reducing the costs of the marketing services. For example, through the analysis of the geographic location information of the products to be sold, the logistics situation of the enterprise can be evaluated, so as to improve the logistics levels, shorten the delivery time, reduce the inventory costs, and greatly reduce the cost of the enterprise combined with the industry's advanced TD mode.

3 Design Requirements of the Automobile Marketing System Based on the Big Data

According to the analysis of the development environment and the current situations of China's automobile data marketing, combined with the car purchase behaviors and media behavior preferences of China's automobile users, the following requirements should be met in the design process of the automobile marketing system based on the big data:

Focus on the target users, innovate the user insight methods, and pay deep attention to the post-90s emerging consumers. The post-90s have become one of the main forces of the social consumption. Studying the characteristics of the post-90s in terms of the car purchase decision-making, the life behaviors, and the consumption concepts and so on, will help the car enterprises adjust their products and marketing strategies in time, improve their brand awareness and reputation in the post-90s, and bring more sales transformation. Integrate the high-quality media, integrate and exchange resources, and realize the full scene marketing throughout the entire process of the automobile marketing. The user's life trajectory is divided into the online and offline categories by the Internet. At the same time, there are many segmentation scenarios in the offline and online scenes. The automobile digital marketing needs to integrate and exchange resources to cover as many user contacts as possible to solve the fragmentation and decentralization problems in the current marketing activities.

Innovate the marketing concepts, and carry out the value creation and the effectively deliver it to the customers with the customers in the center. The change of the real environment promotes the innovation of the automobile digital marketing concept. For the automobile advertisers, the essence of the marketing is to create values for customers and transfer the values to the customers effectively. Strengthen the effect assessment, and all the automobile marketing effects are to improve the sales as the ultimate goal. In the past, the problem solved by the digital marketing was more about how to effectively transfer the marketing contents to users. Today, the connotation of the digital marketing has extended to a wider range of fields. The marketing content reaches users and becomes the starting point rather than the end point of the digital marketing in the new era. Enterprises invest more budgets in the digital marketing of automobiles, not only to expose and click, but also to improve the sales volumes as the final goal.

Improve the marketing efficiency, find users quickly and communicate efficiently, and save users' communication costs. With the growth of the data volumes, the improvement of the data processing capacity and the spread of the data marketing scenarios, the importance of the data in the field of the digital marketing have never been highlighted. For the automobile digital marketing, using the big data technology, through the multiple directional means, quickly find the target users among a large number of users, and under the right scenario, select the right media, transfer the appropriate contents to users, and achieve the efficient communication with users, which will greatly save the communication costs, and achieve the double promotion of the marketing efficiency and effect. Create better experience, use the surreal technology such as the immersive experience, and comprehensively improve the sense of the marketing technology. The application of the surreal technology in the digital marketing can comprehensively enhance the sense of the marketing technology and bring the immersive experience to users. Through the CG image marketing, we can show the unique selling points of the product in a more three-dimensional way. Through the VR interactive driving experience, we can show the performances and scenes of the products in a more real way, attracting more audience attention. The surreal technology will not only improve the user experience of the marketing, but also help the growth of the car brand volumes and product sales.

4 Design of the Automobile Marketing System Based on the Big Data

In the era of the big data, the massive data brings the powerful decision support to the enterprise marketing. Using the relevant technical tools to analyze the network data, we can get the user's purchase intention and consumption tendency. On this basis, through the promotion of the promotional information, we can improve the possibility of the transaction. At the same time, we can use We-Chat and other social media for the data mining. By analyzing the user's attention and forwarding in the social media, we can further determine the user's purchase intention and consumption tendency, and the promotion strategy for their purchase intention is more accurate.

4.1 Optimization of the Marketing Mode

In the era of the big data, micro-blog, We-chat, communities, forums and other new media become the platforms of the enterprise marketing. The automobile marketing enterprises must adapt to the situation, develop the marketing mode suitable for the new media environment, and change from the traditional one-way transmission mode to the new media interactive marketing mode. Under the interactive marketing mode, the automobile marketing enterprises have the following characteristics: The enterprises and the users form an information community. The enterprise is no longer a single marketing subject, and the user is no longer a single information receiver. The marketing contents focus on the information sharing. In the interactive marketing mode, the contents of the marketing ideas are very important for enterprises to get the feedback from users. Therefore, it is necessary to fully understand the user psychology, create creative contents, truly let users like it, and then generate resonance, form the word-of-mouth effect, and speed up the spread of the marketing information. Focus on the user personalized experience. There is a certain gap between the traditional marketing promotion and the personalized communication needs of consumers. In the era of the big data, the automobile marketing enterprises can obtain the user data in the real time, select suitable media according to their characteristics, improve user's personal experience and realize the precise marketing through the creative marketing promotion contents.

4.2 Pay Attention to the Construction of the Data Management Platform

In the era of the big data, the massive data is the competitiveness of enterprises. Understanding the main characteristics of the potential users before the product production will effectively enhance the competitiveness of enterprises. Therefore, enterprises must attach great importance to the data collection strategically. First, the automobile marketing enterprises should collect the unilateral data. The unilateral data is a large amount of the online and offline data produced by enterprises in the marketing process, which should be managed in an aggregate way. Second, enterprises should try to realize the data sharing. On the premise of ensuring the security of the enterprise's unilateral data, the connection with the third-party data should be made to ensure the data of both sides can be interconnected. Third, focus on the information collection of competitors. Knowing the marketing situation of competitors in time can effectively improve the success rate of the enterprise marketing. The marketing information analysis of competitors mainly includes the brand communication trend analysis, the word-of-mouth category analysis, and the product attribute distribution and so on. Through the collection and analysis of these kinds of the information, we can timely monitor and grasp the status of competitors, and then protect the reputation of the enterprises and the products.

4.3   Build the Collaborative Management Structures

In the era of the big data, the accuracy and timeliness of the information determine the quality of the enterprise decision-making. Therefore, in order to improve the efficiency, automobile marketing enterprises need the fast response ability. This requires the automobile marketing enterprises to adopt the flexible ways to strengthen the cross-border cooperation with the IT companies and the e-commerce companies. Using the online and the offline management systems, we can make statistics and analysis of the customer information, with the data information as the core and the customer value promotion as the purpose, and the data decision-making results are quickly implemented in all aspects of the enterprise's entire value chain systematic lean management. In the era of the big data, the focus of the competitions among the automobile marketing enterprises has shifted from the product competition to the customer resource competition. The marketing of the automobile marketing enterprises should conform to the trend of the times, know how to use the massive data, and gain advantages in the increasingly fierce competitions.

5   Conclusion

In the era of the big data, the precision marketing has become the norm of the auto industry, and the auto insurance and the auto finance will be more accurate. The UBI auto insurance based on the analysis of the driving behaviors of car owners can help the insurance companies to minimize the loss ratio. The financial data and the credit records of car owners' banks and the abnormal driving behavior monitoring based on the Internet of vehicles can help banks and financial institutions to determine to ensure the safety of the mortgaged vehicles. The customer data will become an important intangible asset of an enterprise, and the data mining and analysis becomes the magic weapon for an enterprise to win the competitions.

References

1. Jin, M.: Research on the problems and countermeasures of China's automobile marketing mode. Utomobile Technol. **9**, 10–12 (2016)
2. Wang, H.: China's automobile marketing model and prospect. Bus. Cult. **6**, 185–189 (2011)

3. Lv, X.: Problems and countermeasures in the development of the automobile network marketing in China. J. Taiyuan Urban Vocat. Coll. **3**, 73–74 (2012)
4. Mei, Y.: Automotive marketing under the "Internet +" approach. Manag. Technol. SME **28**, 152–153 (2015)
5. Gu, W.: Development status and countermeasures of the automobile marketing mode in China. Manag. Technol. SME **10**, 49–51 (2015)

The Practice of the Multimedia Courseware for College Foreign Language Teaching Based on the Network Resources

Bingjun Ma[✉]

Foreign Languages School, Fuyang Normal University, Fuyang 236000,
Anhui, China
yingguocugb@126.com

Abstract. With the mature development of the information technology, the network multimedia teaching is gradually integrated into the practice of the foreign language teaching in colleges and universities. In contrast, the traditional teaching mode has been unable to meet the comprehensive needs of the foreign language teaching. Therefore, making full use of the advantages of the network resources and the multimedia technology to make the multimedia courseware for college foreign language teaching will essentially promote the innovation of the college foreign language teaching practice. This paper explores the design of the college foreign language multimedia courseware based on the network resources, and makes a detailed study.

Keywords: Network resources · College foreign language teaching · Multimedia courseware · Practical research

The multimedia provides the technical basis for the information age, and at the same time, it also causes great changes for the foreign language teaching in colleges and universities. The application of the multimedia teaching in the foreign language teaching is conducive to helping students to build the knowledge structure, stimulate students' interests in learning, enrich the contents of the classroom teaching, and ultimately improve the efficiency of the classroom teaching [1]. In the process of introducing the multimedia into the classrooms, it is necessary to select the appropriate way of making the courseware, the design and resources of the courseware, the multimedia teaching and the teaching environment, and strengthen the training of teachers' computer knowledge and skills [2].

1 The Practice Background of the Multimedia Courseware for the College Foreign Language Teaching Based on the Network Resources

1.1 Multimedia Foreign Language Teaching

In a broad sense, the multimedia foreign language teaching refers to a teaching method dominated by the computer technology and covering a variety of media [3]. On the one

M. Atiquzzaman et al. (Eds.): BDCPS 2019, AISC 1117, pp. 1720–1726, 2020.
https://doi.org/10.1007/978-981-15-2568-1_242

hand, the teaching subject obtains the learning contents with the help of the multimedia discs and the network teaching, and on the other hand, the teaching activity will also absorb and play the characteristics and advantages of a variety of the media, including books, tapes, slide films, electronic whiteboards, and CDs and so on, forming a joint effort to build a truly subjective foreign language teaching system [4].

1.2   Network Environment

The network environment refers to the physical connection of many multimedia computers in different places, and the realization of the network culture, and the software and hardware sharing system according to a certain protocol communication [5]. The environment always has an important influence on people and things. From the perspective of the teaching, the network environment can not only make the network resources and tools work, but also include learners' motivation, learning atmosphere, teaching strategies and interpersonal relationships. The network environment provides favorable conditions for college English to develop and update the teaching and learning methods and to implement the flexible and open teaching strategies.

1.3   Teaching Strategies

The teaching strategy is a systematic decision-making activity for teachers to achieve the teaching objectives according to the characteristics of the teaching situations. The teaching strategies include not only the adjustment and control of the teaching contents, teaching methods and teaching means by teachers, but also the adjustment and control of students' learning activities and learning methods, and should be based on the learning strategies.

2   Problems in the Multimedia Practice of the Foreign Language Teaching in Colleges and Universities

2.1   The Lack of Contents in the Multimedia Courseware

At present, in many college English classes, the multimedia courseware used by teachers is still deficient in contents. The main reason is that the college English teachers' own multimedia network technology ability is limited, and the teaching task is heavy, and it is difficult to make a detailed and targeted teaching courseware in a limited time. In the actual teaching process, English teachers often rely too much on the courseware, use the mouse and the keyboard to display the courseware mechanically and rarely modify and supplement the courseware. The teaching is too limited to the courseware itself, and the interaction between teachers and students is insufficient.

2.2 Multimedia Teaching Takes up Too Much Time in Our English Teaching

Some college English teachers give students too much time to watch the audio-visual materials, but there is not enough time to practice English. The interaction between teachers and students and the communication between students and students are reduced, and students' English application ability is difficult to enhance, and the teaching effect is not good.

2.3 The Massive Application of the Multimedia Resources Has Seriously Distracted the Students' Attention

In our teaching, in order to show the effect of the multimedia network and its related technology, teachers often neglect the guidance of the educational psychology. The extensive use of sounds, background patterns and animation effects, which are not closely related to the English classroom teaching knowledge in the teaching practice, has led to the vague themes and less emphasis in the college English teaching.

3 The Practical Advantages of the Multimedia Courseware in the College Foreign Language Teaching Based on the Network Resources

In our English teaching under the multimedia network environment, the multimedia technology integrates the multimedia information such as pictures, texts, sounds, and images and so on, which fully combines the scientificity, artistry and interactivity of the teaching materials, and greatly enhances learners' feelings and the cognition of the language and cultures. Because of the powerful function of the multimedia computer, the modularization, intelligence and network of the multimedia courseware, the teaching mode creates a good and vivid language learning environment for students to learn the language, which enables the communication between teachers and students to be carried out on a richer level, and builds an organic platform for the reality of the interaction, cooperation and autonomy of the English teaching and learning.

The network classroom teaching has a relatively scientific and comprehensive theoretical basis, takes the advantages of the above two theories of Ausubel and constructivism, and discards the short ones. It is a new modern classroom teaching mode which is formed by the integration of the information technology and the traditional classroom teaching. Its teaching activities are carried out under the dual environment of the classroom and the network, with students as the main body and the teachers as the leading idea. In the networked classroom environment, the role of the knowledge transmitter and the information source mainly includes the multimedia teaching network. As the organizer of the teaching activities, teachers play a leading role in our teaching. They mainly focus on students, correctly guide students to use the teaching network to obtain the useful information, properly guide students to arrange the learning progress and check the progress of each student, and organize the group discussion and the mutual learning. Students are not only the objects of our teaching, but also the main body of the learning. Under the guidance of the teachers, they control the learning process by themselves according to the progress of the teaching requirements. They use the interactive multimedia teaching terminals to carry out the self-study, make-up, preview, review and various simulation training, and actively solve the problems in learning.

The network classroom teaching absorbs the advantages of the traditional classroom teaching and the network teaching. First, it has the rich multimedia learning resources and provides an open and good environment for the independent learning. Second, the meaning construction of the two-way interaction and the use of various means of communication promote the problem-solving and emotional communication between teachers and students. Third, because of the direct participation of teachers, we can teach students according to their aptitude and avoid students' deviation from the goal. Fourth, it is convenient to carry out the teaching activities, so that students can actively complete the meaning construction according to their cognitive characteristics in the classroom teaching activities, and form students' personalized and creative thinking in the environment of the autonomous learning and the collaborative learning. The application of the modern educational information technology makes the network teaching break through the constraints of the time and space, and has the incomparable advantages of the traditional classroom teaching. The classroom teaching is still the main channel for the implementation of the quality education, which has the irreplaceable role of the network teaching.

4 The Practice of the Multimedia Courseware in College Foreign Language Teaching Based on the Network Resources

4.1 Application in the Audio Visual and Speaking Teaching

The multimedia teaching method combines the visual and vivid images with the sounds, and the dynamic and static image fusion text appears interactively, so as to improve the students' attention and stimulate their interest in learning. The cultivation of listening and speaking abilities has always been an insurmountable obstacle for most students. One of the important reasons is that they lack the correct pronunciation training methods and the effective listening training methods. In the case of a large number of the ineffective practice, students either greatly reduce their interests in the English listening and speaking, or simply give up. In the network and multimedia assisted teaching mode, teachers can demonstrate and teach students correct and effective pronunciation and listening methods in class, so as to fundamentally help students find out the crux of the problems and rebuild their confidence in the English listening and speaking learning. The rich network resources are convenient for teachers to search and select the images, and the real and vivid multimedia resources that conform to the contents of the teaching materials, and bring the real scenes of life to the stage. This kind of the teaching mode of the scenario representation can not only mobilize the classroom atmosphere, but also help students to further understand and master the learning contents. The multimedia teaching materials enable the "human computer" dialogue to be realized. In the process of solving the English learning problems, computers can give timely feedback to students' understanding and mastery. For example, in the audio-visual and oral teaching of New Horizon College English and New Standard College English, every listening question has a voice prompt after the input of the answer whether the answer is correct. This kind of timely feedback can stimulate the students' sense of challenge and guide them to think in the right directions. In the entire teaching process, teachers only need to conduct the computer operation and thinking guidance to make students become the real participants and speakers in the classrooms.

4.2   Application in the Teaching of Comprehensive Courses

In the course of comprehensive courses, the teaching method of combining the vision and hearing can teach knowledge more effectively. Through the sensory stimulation of the vision and hearing, students can get more information than the single listening to teachers. With the help of the multimedia courseware and videos, the boring texts in the textbook become vivid and intuitive images and pictures, which deepen the students' understandings and cognition of the textbook contents. On this basis, teachers can choose the relevant expanding contents to extend when preparing lessons, so as to broaden the students' vision, and guide students to conduct the divergent thinking and the innovative thinking. In this teaching mode, students' interests and enthusiasms in learning have been significantly improved, from the state of the passive teaching to the state of the active participation in thinking, and the teaching results have been improved on the premise of realizing the teaching purpose. College English teaching is mainly in English, which has certain basic requirements for students' English level. For students with the poor English foundation, if there is no clear instruction or guidance in the classroom, they are likely to be ignorant and easy to make a small difference. The multimedia courseware can help teachers to demonstrate the purpose of the classroom teaching, teaching contents and teaching arrangement, clarify the contents and the specific operation mode of each teaching link, and help students who have questions or don't understand and follow the teacher's teaching, so as to overcome the problems that students feel difficult to listen to in the class.

4.3   Application in the Cultivation and Improvement of Students' Initiatives in Learning

Mastering and using the network and the multimedia is not only the requirement of college teachers, but also the requirement of a new independent learning mode for modern students. Teachers should recommend and guide students to use and master the network and multimedia resources in class, and let students use and practice in the future learning process. For example, in the week before each unit, teachers should encourage all students to consult a large number of knowledge related to or derived from the unit contents through the Internet, organize and collect the collected data, edit the video files, and make the corresponding multimedia courseware. In the classrooms, we will provide students with the opportunity to show their learning achievements and let them share and exchange their learning experience on the stage. This not only greatly improves students' enthusiasms and participation in learning, but also realizes the "student-center" in the real sense, and gradually improves students' comprehensive English abilities, and especially the ability of the information understanding and analysis and the pronunciation and language expression, and stimulates students' innovative thinking abilities. As the saying goes, "Interest is the greatest teacher." The same is true for the English learning. In addition to the relatively tedious and boring college English classroom learning, students should be encouraged to use the network and the multimedia resources to cultivate their interests and expand their cultural knowledge. For example, English songs, movies, TV plays, English novels, newspapers and magazines are the very easy and practical self-study materials. However, because the resources are too large, if the selected resources have higher requirements for the English level, they will not play a role in cultivating students' interests, but will hit the enthusiasms of students' self-study. Therefore, according to the average level of students, teachers should recommend some resources with the same or higher abilities to students and guide them to study independently.

4.4 Applications in the Promotion of the Construction of the Teachers' Team

The brand-new teaching methods and models require teachers to constantly learn and keep up with the new understanding and application of the existing teaching resources, encourage and advocate teachers to continuously improve and perfect the teaching methods and models through training or academic exchanges, and constantly explore and discover their advantages and disadvantages through the teaching practice and the scientific research, so that they can play a greater role in the future teaching process. In view of the richness of the network and the multimedia teaching resources, in order to improve the working efficiency of teachers and avoid the repeated construction, a large number of the excellent multimedia courseware can be shared through the network resources. However, different teachers have similar or different understandings and expansion of the same teaching contents, so teachers can modify and improve the downloaded multimedia courseware according to the actual teaching situations.

5 Conclusion

In order to make better use of the advanced modern teaching methods, teachers should design the teaching materials and activities scientifically and reasonably under the guidance of the modern educational theories and cognitive theories, in combination with the laws of the language learning, use computers and the multimedia courseware to create an optimized learning environment and a good teaching situation, strive to realize the matching and serialization of the multimedia courseware, and introduce, self-make, and cooperate to develop the teaching software and constantly enrich and update the multimedia teaching environment according to the actual situations.

Acknowledgements. This Article is funded by Anhui smart class pilot project: English classroom teaching skill practices in smart ways (Project No.: 2017zhkt345) and 2018Anhui provincial colleges and universities major scientific research project: Theories and practices: SLA theories and foreign language teaching (SK2018ZD026).

References

1. Fu, M.: Thinking and discussion on the classroom teaching of the self-made computer multimedia courseware for college teachers. Fujian Comput. **26**(7), 189–190 (2010)
2. Hong, W.: Analysis and reflection on the problem of the PPT courseware dependence in college foreign language classroom. Comput.-Assist. Foreign Lang. Educ. **6**, 76–80 (2012)
3. Shi, A.: Problems in the use of the college multimedia courseware and improvement measures. Academy **15**, 112–114 (2015)
4. Wang, Y., Zhao, H., Peng, N.: Study on the integration and optimization of the foreign language curriculum and information resources in private colleges and universities. Sci. Technol. Inf. **13**(6), 174–180 (2015)
5. Pan, J.: Design and research of making foreign language multimedia courseware based on the Internet. Heilongjiang Sci. **7**(4), 118–119 (2016)

Design and Implementation of Big Data Asset Management System

Wen Ma[✉] and Xinyang Zhang

Yunnan Power Grid Co., Ltd. Information Center, Kunming 650011, China
zhilin87445667@126.com

Abstract. In order to understand the design and implementation of large data asset management system, this paper has carried out research work. In this study, the design and implementation process of the system are analyzed, and the framework that supports the system is described. The main contents are: data source management, data processing management, data asset management, data supply management and system user management. According to the test, it is confirmed that the system is easy to expand and compatible. Under the application of this system, we can strengthen the information management of tax department, upgrade the level of asset management, and divide the process of data capitalization into several stages, such as collection, cleaning and output. Through abundant system functions, we can greatly improve the utilization rate of data assets.

Keywords: Big data · Remote monitoring · Assets · Operation and maintenance system

1 Introduction

With the arrival of the era of big data, there are tens of thousands of data stored in the cloud. And these hundreds of thousands of data, in the context of big data, its value has been re-positioned, it can be said that under the correct management, data is the intangible most valuable asset [1–4]. However, the data itself does not have asset attributes, it needs an effective management system to complete a series of processes such as automated collection, processing, clarity, product to be of value.

At present, whether it is IT, finance or other departments, there are a lot of data generated from time to time [5–8]. However, with the increase of data and the handover of personnel, data resources can not be used, reuse and other issues have greatly increased the cost of resource management, causing many problems to management departments. In order to solve these problems, we need to design an asset management system. Like many resource management systems, this system takes computer as the operating platform, uses B/S structure and distribution as the whole life cycle of resources (generation, collection, collation, inventory, scrapping) of the database [9–11].

Through the establishment of asset catalogue of data and the governance of data quality, data can be applied by managers, and the later operation of data assets is designed and implemented, which can support the realization of data grafting such as

M. Atiquzzaman et al. (Eds.): BDCPS 2019, AISC 1117, pp. 1727–1732, 2020.
https://doi.org/10.1007/978-981-15-2568-1_243

distribution, opening and transaction of enterprise data assets, thus promoting the value realization of data assets.

2 Summary of Related Theories and Technologies

2.1 UML Unified Modeling Language

Unified Modeling Language (UML) is a graphical language with complete functions and easy to learn and use. It generally supports transaction modelling and software system development [12–17]. It can model and abstract the whole stage of software development, and display it, including requirement analysis, specification, and system construction and configuration. Nowadays, UML has been listed as the standard general design language by Object Management Organization (OMG). From different dimensions of software development, UML defines nine kinds of diagrams, such as system use case diagrams, class diagrams, object diagrams, state diagrams, activity diagrams, sequence diagrams, collaboration diagrams, component diagrams and deployment diagrams. These model diagrams can describe the system from different angles and different time nodes in an all-round way [18–20]. Then the system model merges these different focus model diagrams into a whole, which is convenient for system developers to analyze and construct the software. In the requirement analysis stage, the tax electronic data management system designed in this paper will use UML unified modeling language to model the system in multi-dimension, and help us to realize the system.

2.2 RBAC Role Access Control Technology

In short, role-based access control (RBAC) simplifies system authorization and security management by separating users from permissions by adding the concept of roles. Because the change frequency between roles and privileges of the system is less than that between roles and users. Therefore, only through the scientific and reasonable setting of role permissions, the system can easily achieve user authorization management, reduce the difficulty of system authorization management, and reduce the cost of system management. In addition, RBAC is also an extensible access control model. It can support the security policy of the system very flexibly, and the change of system privileges can be relaxed and retractable, and has greater scalability. The so-called RBAC model contains five basic elements: users, roles, resources, control and authorization. The relationship between them is: first set certain permissions to roles; then assign roles to certain users; finally, through the session to see whether the user and the role set are activated. Thus the relationship among users, roles and permissions is separated. Control object (Resource): mainly refers to the management function of the system, such as module management, system management, etc. Operation: mainly refers to the specific operation of management, such as add, delete, change, check, etc. As shown in Fig. 1.

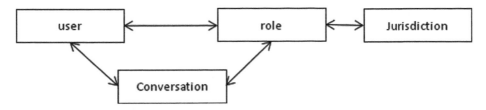

Fig. 1. Role access control model

3   System Design

3.1   Architecture Design

The gradual transition to the goal of asset operation. The system is roughly divided into data catalogue management module, which is used to solve the data problems and data assets operation module, which is used to facilitate later operation and maintenance. Users according to specific asset audit rules, the core design idea of the data assets management system is to start with the management of large data assets and give consideration to data application. Monitor and evaluate the resources in the system, ensure the quality of data in operation, and facilitate the production department to further process the data into products. The general architecture design of the system is shown in Fig. 2.

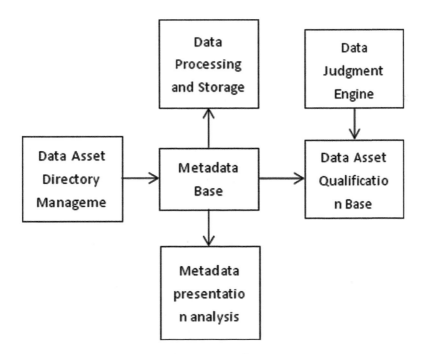

Fig. 2. Data asset catalogue and quality management structure

3.2 Functional Analysis

(1) Metadata Base: The original data in the system is stored in the database, and all kinds of data are stored in the metadata base in a unified format through the catalogue management module. At this time, the data is only data, and it does not have the property of assets, nor can it become a product. Only after subsequent processing can it become a product. As a storage module of raw data, only a unified interface is provided.

(2) Data catalogue management: The first sub-module of the system establishes and maintains a data asset catalogue. The purpose is to integrate complex data resources and send them to the source database after agreeing on a unified interface.

(3) Data asset quality database: Data in metadata database is processed through a series of processes such as interface collection, product loading (collection, verification, cleaning, desensitization) and then into data quality database. The data here can be called data asset. It has a unified format, forms corresponding views, and needs to belong to it. Long-term monitoring of sex

(4) Data quality audit: Because the system reports a large amount of data every day, it is necessary to analyze the data quality in the quality database, that is, the later maintenance function of data assets. It determines whether the system can find and solve the problems of data and equipment failure in the first time, so as to ensure the normal operation of the system. Key.

In addition, we need to configure subsidiary functions, such as data asset acquisition, data asset link analysis and management, and data service environment. Data assets acquisition: acquisition involves the management of acquisition sources, the definition of acquisition methods, and the formulation of acquisition rules. After collection, the data assets of the whole province can be displayed intuitively in the form of catalogues. 2. Analysis and management of data assets links: By modeling and displaying the whole processing link from data source to final result table, we can see the origin and development of a data object and analyze which data objects a certain index is calculated from. It can analyze the impact of the change of a certain index on the follow-up data objects. Third, the data service environment builds a mixed multi-tenant data environment, gradually establishes a system-wide business data service resource pool, and provides basic data services and wide-table data services to the outside world.

4 System Testing

4.1 Basic Process of System

The workflow of the system is roughly as follows: Data Asset Catalog module is responsible for sorting out the data in the cloud and storing it in the metadata database of the system. However, the data of the metadata database, as the original data, does not have the property of assets. After processing and corresponding inventory, it enters the database of data assets, and proceeds with the status of these assets. Analysis becomes a big data asset, which can be used by the corresponding departments.

4.2 Test

In the basic process of testing, testing is mainly aimed at the performance of the system in the actual asset management. The results show that the system first reads the administrator's audit rules, and then users need to log in to start the functional system to extract the value of data audit points and get the results. According to this result, the data quality table can be automatically generated, and the data quality can be scored. Of course, the scoring rules can also be configured. The system can determine the need for alarm according to the pre-configured alarm rules, and send warnings. Table 1 is the system login test record.

Table 1. System logon test records

Test user type	Login name	Login password	Yes/no logon successful	Test result
Advanced users	320800001	123	Yes	Adopt
General user	320800002	123	Yes	Adopt
General user	320800003	123	Yes	Adopt
General user	320800004	123	No	No passage

The data import function test is to use the data import function in the data source management module to import external data in batches according to the table structure set in the system. Firstly, we need to create a temporary table of taxpayer information test in the background database of the system, which is named DJ_NSRXX_TEST_TEMP. Table fields include: taxpayer name, identification number, registration authority, entry date. The table structure construction statements are: CREATE TABLE DJ_NSRXX_TEST_TEMP(NSRMC VARCHAER2(20), NSRSBH VARCHAER2 (20), DJG VARCHAER2(20), LRRQ DATE). According to this table structure, 49 test data are prepared and saved as EXCEL document. The EXCEL table is imported through the import function of the data source management module.

5 Conclusion

This paper gives a design scheme of data asset management system and completes the implementation and later test according to this idea. Through the development of the system and the later practice, the system can check and check a large amount of data, and collect the large amount of data through this system, as to whether the equipment is running well or not. A good detection standard is of great practical significance. In addition, the use of the system is flexible, and it can manage all the monitored equipment well. Once a device fails, the first time it can send an alarm.

References

1. Campos, J., Sharma, P., Gabiria, U.G., et al.: A big data analytical architecture for the asset management. Procedia CIRP **64**, 369–374 (2017)
2. Whyte, J., Stasis, A., Lindkvist, C.: Managing change in the delivery of complex projects: configuration management, asset information and 'big data'. Int. J. Project Manag. **34**(2), 339–351 (2016)
3. Kobayashi, K., Kaito, K.: Big data-based deterioration prediction models and infrastructure management: towards assetmetrics. Struct. Infrastruct. Eng. **13**(1), 1–10 (2016)
4. Heng, F., Zhang, K., Goyal, A., et al.: Integrated analytics system for electric industry asset management. IBM J. Res. Dev. **60**(1), 2: 1-2: 12 (2016)
5. Marzieh, H.Z., Reza, A.M., Suleiman, A., et al.: Knowledge creation in nursing education. Global J. Health Sci. **7**(2), 44–55 (2015)
6. Wang, N., Sha, J.H.: The fact research of the equipment and facilities integrated management IT system for Chinese oil company. Adv. Mater. Res. **1044–1045**, 1839–1842 (2014)
7. Hanchate, D.B., Bichkar, R.S.: Software project contacts by GRGA scheduling and EVM. Int. J. Comput. Appl. **97**(97), 1–26 (2014)
8. Santoso, F., Redmond, S.J.: Indoor location-aware medical systems for smart homecare and telehealth monitoring: state-of-the-art. Physiol. Measur. **36**(10), R53 (2015)
9. Brous, P., Overtoom, I., Herder, P., et al.: Data infrastructures for asset management viewed as complex adaptive systems. Procedia Comput. Sci. **36**, 124–130 (2014)
10. Sinha, K.C., Labi, S., Agbelie, B.R.D.K.: Transportation infrastructure asset management in the new millennium: continuing issues, and emerging challenges and opportunities. Transportmetrica A: Transp. Sci. **13**(7), 591–606 (2017)
11. Ginardi, R.V.H., Gunawan, W., Wardana, S.R.: WebGIS for asset management of land and building of Madiun city government. Procedia Comput. Sci. **124**, 437–443 (2017)
12. Zhou, K., Fu, C., Yang, S.: Big data driven smart energy management: from big data to big insights. Renew. Sustain. Energy Rev. **56**, 215–225 (2016)
13. Stein, V., Wiedemann, A.: Risk governance: conceptualization, tasks, and research agenda. J. Bus. Econ. **86**(8), 813–836 (2016)
14. Thorisson, H., Lambert, J.H.: Multiscale identification of emergent and future conditions along corridors of transportation networks. Reliab. Eng. Syst. Saf. **167**, 255–263 (2017)
15. de Montjoye, Y.A., Shmueli, E., Wang, S.S., et al.: openPDS: protecting the privacy of metadata through SafeAnswers. Plos One **9**(7), e98790 (2014)
16. Park, K.H.: What happened to efficiency and competition after bank mergers and consolidation in Korea? Soc. Sci. Electron. Publ. **33**(3), 33–55 (2016)
17. Xiang, W.S.M.R.: The development of elite Rugby Union officiating in Wales: a critical analysis. Cardiff Metrop. Univ. **59**(2), 91–115 (2014)
18. Craig, B., Peter, G.V.: Interbank tiering and money center banks. J. Financ. Intermediation **23**(3), 322–347 (2014)
19. Frandsen, B., Rebitzer, J.B.: Structuring incentives within accountable care organizations. J. Law Econ. Organ. **31**(Suppl. 1), i77–i103 (2015)
20. Berman, P., Dasgupta, B., Kaligounder, L., et al.: On the computational complexity of measuring global stability of banking networks. Algorithmica **70**(4), 595–647 (2014)

Design of the Management System for Salary Diagnosis and Optimization of Medical Staff in Shaanxi Public Hospitals of Traditional Chinese Medicine

Baoning Qi[✉], Jiaxin Wang, Mingrui Ji, Shouzhu Xu, Juan Li,
and Chuandao Shi

Department of Public Health, Shaanxi University of Chinese Medicine,
Xianyang 712046, Shaanxi, China
kingguogang@sohu.com

Abstract. The reward management is the focus of the human resource management, and the salary system is the basis of the reward management. As the reward management is very sensitive, which not only involves the interests of employees, but also affects the operation efficiency and development of the unit, therefore, we should standardize the design of the reward management from a scientific perspective, so as to achieve the best results of the staff compensation and welfare. The salary management and optimization of the medical staff in Shaanxi public hospitals of traditional Chinese medicine is of great significance to the medical work of the relevant personnel. Therefore, it is very important to give full play to the advantages of the information technology and innovate the salary management system.

Keywords: Shaanxi · Public hospitals of traditional Chinese medicine · Medical staff · Salary diagnosis and optimization · Management system

1 Background of the Salary Diagnosis and Optimization of the Medical Staff in Shaanxi Public Hospitals of Traditional Chinese Medicine

1.1 The Unreasonable Salary Distribution with the Tendency of Equalitarianism

At present, the average monthly income of the medical staff in public hospitals is not much different. According to the relevant data, in some secondary hospitals, the gap between the average annual incomes of the medical staff with high education, long working hours and professional titles is not big. The difference between the annual salaries of the medical staff with high professional title and that with the intermediate professional title is no more than 20%, while the difference between the salaries of the medical staff with the intermediate professional title and that with the junior professional title or that without professional titles is about 40%. In this case, the salary incentive system has insufficient incentive effects on the technical backbones, and has

M. Atiquzzaman et al. (Eds.): BDCPS 2019, AISC 1117, pp. 1733–1739, 2020.
https://doi.org/10.1007/978-981-15-2568-1_244

obvious equalitarianism tendency. The high quality working level can't be replaced by the high remuneration. In the long run, it will inevitably lead to the decline of the working enthusiasms of the medical staff and the decline of the medical service quality, and then affect the image of the hospitals, and hinder the development of the hospitals.

1.2   Lack of Justice in the Performance Appraisal System

The performance appraisal system is a supporting system of the salary incentive system. The purpose of the performance appraisal in modern hospitals is to evaluate the working quality, efficiency, achievements and other aspects of the medical staff, provide basis for the salary formulation, enable the hospital managers to understand the advantages and disadvantages of the medical staff in time, deepen their original advantages, and correct their shortcomings and improve the professional levels of the medical staff. At present, the performance appraisal system of public hospitals in our country is lacking in justice. Its form is superficial, emphasizing seniority and neglecting contributions, leading to some employees losing the right to participate, which is not conducive to the long-term development of hospitals. From the above analysis, it can be seen that the income of the medical staff does not conform to the labor they have paid, which leads to the lack of initiatives of the staff to work and the phenomenon of slack work to the patients. In order to avoid this situation, our public hospitals should make corresponding rectification.

2   The Value of the Salary Diagnosis and Management System of the Medical Staff in Shaanxi Public Hospitals of Traditional Chinese Medicine

In order to optimize the performance appraisal system of public hospitals in our country, we should make a performance appraisal system suitable for our hospitals according to the actual situations of the hospitals and the successful experience and development directions of the performance appraisal system at home and abroad. The hospital managers should establish scientific and reasonable performance appraisal system according to the working characteristics of different posts, with the reference data of the post value of the medical personnel, their contributions to the hospital, their own working efficiency, and their professional and technical levels and so on, and in accordance with the principles of fairness and justice, so as to promote the long-term development of the hospitals.

2.1   Improve the Working Efficiency and Reduce the Workload of the Business Personnel

The effective combination of the background calculation function and other modules enables the ERP human resource system to adjust the salary structures of the company according to the actual needs of the company, and effectively process and count the changes of the salary data information, which facilitates the management of the salary information by the business personnel. The ERP human resource management system not only reduces the difficulty of the salary information management and reduces the workload of the business personnel, but also improves the efficiency, quality and level of our work. The business personnel can not only complete the work efficiently and correctly and reduce the cost of the work, but also effectively query the data information to ensure the consistency of the information.

2.2 Meet the Needs of Processing the Salary Information of Various Structures

The ERP human resource management system is connected with the time management mode. Its characteristics of cross-region, cross-type of work and cross-department make the ERP human resource management system query and update the information in the real time according to the dynamic change of the salary information of employees. The system can effectively record and query the information, clearly show the unit information, and improve the scientificity of the salary information management. For the unit with the shift system, the ERP human resource management system can design the information processing method suitable for the situation according to the needs of the unit. After inputting the corresponding operational code, search and associate according to the background configuration, calculate the salary of the employees, improve the accuracy of the salary information calculation, further improve the salary structure of the unit, and realize the management of the unit's information.

2.3 Effectively Control the Unit's Payroll

The ERP human resource management system has the total amount management functions such as the salary warning, adjustment, and distribution and so on. Through this function, managers can master the total amount of the unit's wages and the payment of the wages, better understand the work status of the unit, and realize its control and management of the unit. The ERP system can provide reference for the managers by analyzing and controlling the indicators of the human resources, so that they can find out the irrationality of the unit in time and accurately, find out the problems of the unit, provide the scientific basis for the decision-making of the managers of the unit, enhance the competitiveness of the unit, and make the unit survive and develop better in the fierce competition.

3 Contents of the Design of the Management System for the Salary Diagnosis and Optimization of the Medical Staff in Shaanxi Public Hospitals of Traditional Chinese Medicine

When designing the reward system, the hospitals should first formulate the salary strategy through the analysis of the unit strategy, comprehensively consider the hospital factors (strategic objectives, performances and cultures), the staff's own factors (including personal qualifications, working experience, and personal potential and so on), the post factors and the labor market price and other factors, and use all kinds of the possible salary elements – the fixed salary, the variable salary, the indirect salary, the flexible welfare and the non-economic salary, to make the hospital's salary fair internally and competitive externally, and play the supporting effect of the salary on the unit strategy. In order to design a strategic oriented comprehensive salary system, there are several key steps.

The hospitals should carry out the analysis from the perspective of the unit strategy (including the understanding of the nature of employees, the core values of the unit and the development goals of the unit), position the salary, formulate the salary policy and make choices on the salary models, salary structures and salary levels. This step is critical. The formulation of the compensation strategy includes the level strategy, the structure strategy and the mode strategy. The salary level strategies that hospitals can choose include the market leading strategy, the market following strategy, the cost oriented strategy or the mixed salary strategy. The salary structure strategy has the high flexibility, high stability and harmonious model. The salary model has three models: position centered, ability centered and performance centered. The hospitals should analyze the specific situations of which strategy mode to choose in the specific implementation.

The job analysis is mainly to redesign the work, eliminate the unnecessary positions, formulate the job specifications, and form the organization chart and job description. The job evaluation (job assessment) focuses on solving the internal fairness of the salary. It has two purposes. One is to compare the relative importance of each position within the unit and get the position rank sequence, and the other is to establish a unified position evaluation standard for the salary survey, to eliminate the difference of the position difficulty caused by the different position names among different hospitals, or even if the position names are the same, but the actual work requirements and the work contents are different, so that different positions can be compared, to lay the foundation for the fairness of the post wages. It is the natural result of the position analysis and is based on the position description. Most of the relative value of the evaluation posts adopts the point value factor method. That is, the number of points is determined according to the evaluation elements, and then it is summarized, and then the post grade is determined according to the total number of the points.

The salary structure refers to the relationship between the labor value or the importance of various jobs in the hospitals and the corresponding wages payable. The salary structure design is to convert the order, grade, score or symbolic monetary value of the work value or importance into the actual salary value, so that the salary of all the work in the hospital is determined according to the same principle of the contribution law, so as to ensure the internal fairness of the hospital salary system. Then, many types of the salaries are combined into several grades to form a salary grade series, and then the specific salary range of each position in the hospital is determined.

4 Design Features of the Management System for the Salary Diagnosis and Optimization of the Medical Staff in Shaanxi Public Hospitals of Traditional Chinese Medicine

The salary management system is mainly composed of six functions: the daily input, data query, report statistics, record management, system management and data backup. Its functional structure chart, and the composition and functions of each functional module are as follows.

4.1 Daily Entry

Salary entry: ① Personal entry: Enter the number and press enter to select the salary date. If the number exists in the current month, the record will be displayed and can be modified. If it does not exist, the new record will be entered. ② Department entry: Select the salary date and the work item to be modified in the department, and then press OK. If the department record exists in the current month, the data of the month will be displayed. Otherwise, the salary of the previous month will be called. The salary record of the month will be generated after the modification and saving. (2) Personnel data entry: Similar to the salary personal entry. (3) Public accumulation funds entry. ① Modify the public accumulation funds: Enter the year and the corresponding number of the provident fund, and if any, enter the new value of provident fund and save it. ② Generate the public accumulation funds: Select the department and the year to generate the public accumulation funds and press OK. If the public accumulation fund of the department exists in the current year, ask whether to regenerate the public accumulation fund. If yes, call the salary item of last December and generate the public accumulation fund automatically. If the value of the fund is displayed and can be modified and saved, the new value of the fund will overwrite the old value of the public accumulation fund. If the public accumulation fund of the Department does not exist in the current year, the rest are the same except for the query window. (4) Generate the tax: Select the tax date to generate and press OK. The correct operation sequence is: salary entry → generate the public accumulation funds → generate tax. When the public accumulation funds of a certain number or the salary item of the month (except the only child subsidy and the prepaid medical expenses) changes, the tax should be generated again, to ensure that taxes are updated.

4.2 Data Query

Include the salary item, the personnel information, the public accumulation funds and the pay-slip query. The operation is basically the same. Just select the date to query and the query method (number, name, and department) and press OK.

4.3 Report Statistics

(1) Report by department: We can print the department group statistical report of the whole hospital in a certain month, and also print a page of report. (2) Report by

department type: We can print the department statistical report of a certain type (such as the medical treatment) in a certain month. (3) Report of the whole hospital: We can print the group summary report of all types of the departments in the whole hospital. (4) Pay-slip: We can print it in three ways: by the whole hospital, by the department and by the number.

4.4 Record Management

(1) Transfer processing: Enter the number of the person to be transferred, and then enter the number of the new department and press OK. If the person transfers before generating the salary of the current month, the personnel code of the monthly salary is the new code after transferring. If the person transfers after generating the salary of the current month, the personnel code of the monthly salary is the original code. The change of the account number is the same. (2) Modify the current month's number and the account number: When the number or the account number of a person in a certain month is wrong, we can modify it here. (3) History query: the operation is similar to the query.

4.5 System Management

(1) Department management: The number and the name of the department can be added and modified, and the department type can be generated automatically. (2) Number management: The number, name and account number of the person can be added and deleted, and the name and account number can also be modified. The coding format of the number: department code – person code. After the number of a person is deleted, all its data will be transferred into the history record.

4.6 Backups

(1) Backup basic data: There are two types of the backup: the backup of the salary and the personnel data, and the backup of the number and the department data. The former is backed up once a month, and the latter is changed when the data changes. (2) Backup bank data: Insert the U disk and press the generate key after selecting the date, a copy of the salary data will be generated and sent to the U disk (bank data) and a copy of the same data will be sent to the hard disk.

5 Conclusion

To sum up, in the case of the private hospitals and the foreign hospitals flooding into the market, Shaanxi public hospitals of traditional Chinese medicine must rectify their own shortcomings, constantly strengthen their own construction, and establish and improve the salary incentive system and performance appraisal system, so that the medical staff can get income from their work and have a basis for their work, so as to improve the enthusiasms of the medical staff, improve their working efficiency and create a benign competitive environment for them. In addition, the establishment and improvement of the salary incentive system and the performance appraisal system has accelerated the return of public hospitals from "market" to "public welfare" in the period of the medical and health reform, which, to a certain extent, promoted the medical personnel to pay more attention to the medical ethics, avoid some adverse phenomena, and then improve the overall image of the hospitals, and promote the healthy development of the hospitals. In addition, due to various reasons, there are still deficiencies in this paper, and the exploration of the salary incentive system in public hospitals is not detailed and thorough, and it is expected to be further explored in the future research.

Acknowledgement. Scientific research project of Shaanxi Administration of Traditional Chinese Medicine (2019-ZZ-ZC012).

References

1. Chunyu, X., Xinyi, Y.: Diagnosis and strategy of the compensation management in private enterprises – a case study of HB company. Reform Openning (12), 126128 (2015)
2. Wucheng, Z.: Research and optimization strategy of the salary management in small and medium-sized private enterprises. Friends Acc. (17), 57–60 (2013)
3. Ziang, Z.: How to adjust the salaries reasonably. Manag. Sci. (8), 134 (2012)
4. Chen, X., Xie, Y., Geng, Y.: British civil servant compensation system and its enlightenment. Reform Openning (5), 51–53 (2016)
5. Linda, L.: Salary adjustment method based on the nature of occurrence. J. China Youth Univ. Polit. Sci. **26**(1), 98–101 (2007)

Adjacent Car Interaction Network Based on Reliable Neighbor Broadcast Protocol

Juan Shao[✉]

Suzhou Industrial Park Institute of Vocational Technology, Suzhou 215000,
Jiangsu, China
linrong111786@163.com

Abstract. Based on Reliable Neighborcast Protocol (RNP), this paper constructs an interactive network of adjacent vehicles. RNP provides reliable message interaction for adjacent vehicles at a distance, while enabling rapid response between all adjacent vehicles. On the basis of RNP, a road capacity estimation model based on the three types of vehicles, which are manual, self-driving and cooperating car, is presented. The model estimates the road capacity by analyzing the average safety distance between the three types of vehicles, and proves that increasing the proportion of self-driving and cooperating car can greatly improve the road capacity.

Keywords: Self-driving · RNP · Interaction network

Road capacity estimation model that the three models with different probabilities in the road, the safety distance between different models estimates are also different. The safety distance of the manual vehicle depends on the safety clearance between the vehicles; the safety distance of the self-driving driving vehicle depends on the deceleration and the vehicle reaction time interval which accord with the distributed distribution; the safety of the cooperating vehicle Distance situation is complex, the model before and after it is considered to be interactive or non-interactive situation. Front and rear vehicles are not interactive, that is equivalent to the case of self-driving vehicles; vehicle in front can not be interactive, after the car can, its safe distance depends on the negotiation through the RNP deceleration, the time interval and the vehicle response time interval. Both the front and rear cars can interact, and its safe distance depends on the deceleration and negotiation time interval negotiated through the RNP. These three cases calculate the average safety distance according to their probability of occurrence. Finally, the road capacity is estimated from the average vehicle length and the overall average safety distance.

The road capacity estimation model has drawn the conclusion that unmanned and unmanned vehicles with interactive functions are as much as possible. In this paper, the following constraints are considered: First, the road capacity needs to meet at least the daily average road traffic; Secondly, by Peak Hour Factor (PHF) this factor is usually set between 0.80 and 0.95, Finally, the LOS and the vehicle density are estimated, and the LOS calculated from the combination of the three vehicle proportions is as high as possible and the vehicle density is as low as possible. In addition, the dedicated lane setting corresponds to the ratio of fixed unmanned and interactive unmanned vehicles, which can be transformed into the above models for estimation.

© Springer Nature Singapore Pte Ltd. 2020
M. Atiquzzaman et al. (Eds.): BDCPS 2019, AISC 1117, pp. 1740–1745, 2020.
https://doi.org/10.1007/978-981-15-2568-1_245

Based on the assumption of certain conditions, the safety distance, road capacity, PHF factor, LOS and road vehicle density of three kinds of motor vehicles at different proportions are simulated by computer, which is sensitivity analysis.

1 Overview

Thomson Reuters latest intellectual property and science and technology report shows that between 2010 and 2015, and vehicle unmanned technology related to more than 22,000 patents for inventions. Not just a Google company, basically the United Kingdom, the United States, France, Japan, China, Germany, Russia, Korea and other countries have a higher level of technology companies in this technology to some extent, practical. As of March 2016, Google had test-driven their fleet of driverless cars in autonomous mode a total of 1,500,000 mi (2,400,000 km). [1] Google conducted a number of research and experiments, and has achieved good results, unmanned technology has also been more and more people understand. This technology allows all drivers to lose their jobs - driving without the need for people to operate, replaced by a faster response, less susceptible to other factors affect the computer. Not even a single driver can match them, they will remember all the traffic laws, not drunk driving, fatigue driving, part of them can also interact with other vehicles, passing road information [1, 2]. These seemingly unrelated to solve the problem of traffic congestion, but it is our best breakthrough key point. Driving on the road even if there is no accident, but as long as a sudden brakes, an unreasonable lane acceleration, the overall situation will have a ripple effect of traffic.

2 Vehicle-to-Vehicle Communicating Technology

Reliable Neighborcast is a new communication paradigm, and there is a reliable neighborcast protocol (RNP) associated with it. This protocol is used in a vehicle-to-vehicle (V-2-V) communication network, which communicates the speed, location and status of nearby vehicles and provides a coordinated approach to traffic management. This example of communication matches the above-mentioned unmanned vehicle interaction, so we apply it to the real problem we are facing [1, 3, 4].

In the RNP, each vehicle is in contact with a group of surrounding vehicles—the neighborhood vehicle to which it is connected is called its neighborhood. The neighborhood of each vehicle will also have its own neighborhood, so the vehicle can communicate with more vehicles that are not in its own neighborhood. The process of communicating information is done over a wireless link, which is much smaller than the bandwidth of a wired network and has a much higher error rate. The nature of the radio, the bandwidth constraints imposed on us in the neighborhood between the use of broadcast to transmit information, rather than point to point link [3, 4].

In this paper, we chose to build RNP over the mobile reliable broadcast protocol (M-RBP). But not the only one, RNP can be built on any reliable broadcast protocol. M-RBP can provide reliable transmission for all members of the broadcast group, but also to achieve a consistent message sort. It is efficient in terms of the number of controls required for each broadcast message. M-RBP has a dynamically changing broadcast group, which is different from the earlier reliable broadcast protocol. This property makes M-RBP very suitable for mobile applications.

3 Road Capacity Model

It is necessary to solve some roads capacity problems, because the road capacity is one of the important factors. If the road capacity is large, the roads will be unobstructed. If the road capacity is small, the roads will be crowded. As for determine the size of the road capacity factor is the vehicle.

The purpose of this part is to determine what kind of car can narrow the safety distance between the preceding vehicle and the following vehicle. All the vehicles can be divided into three kinds of cars, each kind has different driving methods, they also influences the capacity of the road in different ways: Manual vehicles: the vehicle driving by person; Self-driving-vehicles: the vehicle use computer systems to driving; Cooperation vehicles: the vehicle that can send messages to other vehicles. Since the self-driving-vehicles use a system known as adaptive cruise control (ACC) that understands the traffic conditions around it, it can improve safety and increase road capacity.

4 Highway Assessment Model

In this section, the model is further extended to take into account the factors such as peak flow, road grade, and dedicated traffic lane, so that the model is closer to the actual situation.

4.1 Highway Level of Service

This section discusses the Level of Service (LOS), which is used to assess road traffic. A to F are used to define six different service levels, where A represents the best and F represents the worst.

4.2 Peak Flow

Peak hourly traffic expresses the maximum capacity requirements of the road. Assuming that this study only examines urban roads, this type of road peak hourly flow rate is small, unlike tourist routes that may explode with the arrival of holidays. This article uses the Peak Hourly Factor (PHF), which is often used to design hourly traffic.

The peak flow rate is the peak hourly flow. For urban areas, PHF is typically between 0.80 and 0.98. PHF value is relatively low, that in this period of traffic flow with a large variability; high value means that in this period of time traffic flow is only a small variability. A PHF of more than 0.95 usually indicates a high traffic flow and can be used to constrain traffic flow during peak periods.

4.3   Adjust Flow

Hourly capacity needs to be translated into the equivalent passenger flow, which takes into account the heavy truck (bus, truck and recreational vehicle) factors as well as the driver characteristics.

4.4   Highway Service Level and Vehicle Density

The road service level can be determined from the calculations above and from the tables in the first sub-section.

The road capacity model in the third section of this paper yields the hourly volume per lane, the Hourly volume in the PHF. Thus, in order to satisfy the condition of PHF of 0.80 to 0.98, a range of three percentages of manual, self-driving and cooperating car and corresponding PHFs can be calculated. In addition, the hourly capacity calculated from the road capacity model in conjunction with other parameters can determine the road service level and vehicle density, from which to determine the higher level of service, while the smaller vehicle density of manual, self-driving and cooperating car three types as a percentage.

4.5   Dedicated Lane

Assume that a lane in the road is set to self-driving lane, the manual vehicle is not allowed to enter the lane, and self-driving vehicles will not enter the other lanes. It is assumed that the vehicles in the dedicated lane are self-driving and cooperating vehicles, while the vehicles in other lanes are manual vehicles.

Through the road capacity model in the third part of this paper, we can calculate the total road capacity after setting n (one-way) lanes, and calculate the service grade of the road through the above evaluation model, so we can choose the service grade scheme Design of dedicated lanes.

5   Simulation and Conclusion

In this section, simulation steps and results are given in conjunction with the second, third and fourth part models and the relevant data, and the corresponding conclusions are given.

Self-driving and cooperating car will reduce the safety distance between vehicles, where cooperating car reduces the separation of safety vehicles. Figures 1 and 2 show the safety distances and estimated road traffic capacity (vehicles/hour/lane) for different percentages of manual driving, automatic driving and auto-driving with communication interaction, respectively. Can be seen when 100% are man-driven vehicles, the safety distance of 30.556 m, road traffic capacity of 2868.983 vehicles/hour/lane. When 100% are auto-driven, the safety distance is reduced to 19.908 m, road traffic capacity of 4130.900 vehicles/hour/lane. When 100% are auto-driving with communication interaction, the safety distance is 5.028 m and the road traffic capacity is 10720.667

vehicles/hour/lane. It is clear that autopilot and autopilot with communication inter-action can significantly increase road capacity.

In summary, without taking into account other factors under the premise of self-driving and cooperating car as possible, can greatly improve the capacity of the road to solve the problem of peak vehicle traffic problems. In addition, the road capacity at least to meet the road average daily traffic.

To sum up, the number of self-driving and cooperating car is the bigger the better, but considering the factors such as road utilization, PHF should give priority to the range of 0.80 to 0.95 combination ratio. In addition, because the dedicated lane is equivalent to a fixed proportion of autopilot vehicles and automatic driving with communication inputs, the road capacity effect is significant. Finally, due to the particularity of autopilot vehicles, the original road service level is not applicable, the corresponding data in the level of the road capacity is proportional to increase again after the assessment.

6 Conclusion

This section lists the Strengths and weaknesses of our model.
   Strengths:

(1) Using RNP to Improve the Reliability and Performance of Interaction between Unmanned Vehicles.
(2) The interaction of each type of vehicle is analyzed and modeled.
(3) Based on PHF factor, the basic scheme of increasing road capacity is put forward.
(4) A new evaluation scheme is proposed for road service grade.

   Weaknesses:

(1) Lack of data

   We have only a small amount of data, especially road data and driver characteristics of data is almost no. Thus, the estimated results may not have the ability to normalize.

(2) Ignoring the car accident or other unexpected factors

   In our model, no accident or other unexpected factors, such as weather, are considered. This means that the estimated road capacity of the model may be low.

(3) No complete consideration of traffic road traffic

   In the actual traffic road design, such as green isolation and other factors need to be considered comprehensive; and our model only highlights the unmanned vehicles encountered in the process of driving the situation.

References

1. Willke, T.L., Maxemchuk, N.F.: Reliable collaborative decision making in mobile ad hoc networks. In: Proceedings of 7th IFIP/IEEE International Conference on MMNS, San Diego, CA, 3–6 October 2004, pp. 88–101
2. Willke, T.L., Maxemchuk, N.F.: Coordinated interaction using reliable broadcast in mobile wireless networks. In: Proceedings of Networking, 2–6 May 2005, pp. 1168–1179
3. Maxemchuk, N.F., Tientrakool, P., Willke, T.L.: Reliable neighborcast. IEEE Trans. Veh. Technol. **56**(6), 3278–3288 (2007)
4. Chang, T., Lai, I.: Analysis of characteristics of mixed traffic flow of autopilot vehicles and manual vehicles. Transp. Res. Part C **5**(6), 333–348 (1997)

The Collation and Dissemination of Jilin Folktale Data Based on the Network Resource Mining

He Wang[(✉)]

College of Humanities and Science of Northeast, Normal University,
Changchun 130011, Jilin, China
359533680@qq.com

Abstract. The structures of Jilin folktales are diversified. The typical structures of the folktale are: the single line, progressive, chain and compound. It not only conveys the northeast people's attention to the vividness, touching, novelty, and ingenious ups and downs of the stories, but also contains people's aesthetic pursuit of the rhythms of the stories, includes grasping and using the story program flexibly. By giving full play to the technical advantages of the network resource mining, Jilin folk stories have the new breakthrough in the form of the data sorting and dissemination, presenting to the public in a more modern and international form, and better serving the society.

Keywords: Network resource mining · Jilin · Folktale · Data arrangement · Communication research

For a long time, Jilin is short of the fertile soil which can be rooted in Jilin culture, and the folk story materials can provide a platform. As an important carrier of the culture and education, folktale materials, with their strong impact and refined strokes, are bound to deepen the public's understandings of Jilin cultures, establish the public's awareness of Jilin regional cultures, and enhance the sense of identity and pride of the regional cultures.

1 Network Resource Mining Technology

For the network resources, it not only contains the important data and information such as the logical association, the performance and the fault, but also is affected by its complex data and the large scale, and lacks the scientific and effective management and utilization [1]. Based on this, a data mining method is developed, which is based on the data information of the network resources. It includes three aspects: the data horizontal association analysis technology, the data temporal association analysis technology and the data index technology [2]. Through the establishment of the data temporal association characteristics and the data management, and the data collation and dissemination and so on, it can effectively provide reference for the Jilin folktale data collation and dissemination and implement plans for the emergency repair.

© Springer Nature Singapore Pte Ltd. 2020
M. Atiquzzaman et al. (Eds.): BDCPS 2019, AISC 1117, pp. 1746–1751, 2020.
https://doi.org/10.1007/978-981-15-2568-1_246

The network information mining is the application of the data mining technology in the network information processing [3]. The network information mining is based on a large number of the training samples to get the internal characteristics of the data objects, and for the purpose of information extraction on this basis. The network information mining technology follows the excellent achievements of robots, full-text retrieval and other network information retrieval [4]. Meanwhile, based on the knowledge base technology, it comprehensively uses various technologies in the fields of the artificial intelligence, the pattern recognition and the neural network. The intelligent search engine system based on the network information mining technology can obtain the personalized information needs of users and search the information purposefully on the network or in the information base according to the target feature information [5].

The network information mining can be roughly divided into four steps: the resource discovery, i.e. searching the required network documents; the information selection and preprocessing, i.e. automatically selecting and preprocessing from the retrieved network resources to get the special information; generalization, i.e. discovering the common patterns from a single web site and among multiple sites, and analysis, i.e. confirming or explaining the mined patterns. According to different mining objects, the network information mining can be divided into the network content mining, the network structure mining and the network usage mining.

2 The Value of the Collation and Dissemination of Jilin Folktale Data Based on the Network Resource Mining

In essence, the collation and dissemination of Jilin folktale data based on the network resource mining is a kind of the "micro" dissemination activity based on the mature application of the information technology. The core feature of the micro communication is "micro". That is to say, the content of the communication is the "micro content" (short videos, and Vlog and so on), and the communication experience is the "micro action" (you can express your opinions through the simple finger activities), while the communication channel is the "micro medium" (mobile phones and so on). The direct purpose of the communication is to transfer the contents of the communication to the addressee, while the micro communication makes the best use of its fragmentation, individuation and rapidity to become the main mode of the communication. As a modern mainstream mode of the communication, relying on the new media platform and the "Internet +" information exchange network, create a new way to express the self character and promote the communication.

According to Raymond Williams, the culture includes the culture of "ideal", "literature" and "society". The traditional village culture is the epitome of the social culture. In recent years, the state has vigorously implemented the strategy of the "rural revitalization", carried forward the excellent traditional village cultures of the nation and the region, advocated the integration with the development of the modern civilization, and carried out the pioneering transformation and development. In the context of the micro communication, the traditional village culture conflicts with the foreign new culture, and the communication and the inheritance face new difficulties. The distribution of the traditional culture is relatively scattered. Based on this feature, we can effectively integrate the communication contents with the help of the characteristics of the micro communication, comprehensively summarize the similar traditional cultures to form their own communication advantages, and then maximize the impact of the traditional village cultural communication.

For regions outside Jilin, the folktale material is a good medium to publicize the traditional folk customs, cultures and histories of Jilin, which can form a high-low interaction with the old and the new media, with the effect of interpenetrating coverage. At the same time, because the reading group of the folk story materials is clear, a Jilin business card can be formed among the above groups. During the process of accompanying children to read, parents can intuitively understand Jilin's ancient and modern times, myths and legends, industrial process and other information, and understand Jilin's cultural dissemination and promotion of Jilin's traditional regional cultures. Let the readers of other cultural backgrounds understand the knowledge of the Jilin history, folk customs and cultures.

The driving force to promote the development and exchange of Jilin's cultural industry, with the full spread of the cultural confidence, the derivative industry chain with the folk story materials as the main body, will spring up, and the production and distribution of the folk story materials and books, the production and sale of the folk story materials and handicrafts, the development and sale of the tourist attractions and tourist attractions based on the contents of the folk story materials are bound to promote the development of Jilin tourism economy.

Throughout Jilin folk art, it has a long history of various forms. For example, Jilin folk cloth and stick painting is also called the cloth and pile painting. Wang's cloth and stick painting is the first batch of Jilin City to enter the intangible cultural heritage list of Jilin Province. Jilin Dongfeng farmer painting is a form of the folk painting in as early as 2008, and Dongfeng County of Jilin Province is also named "the hometown of the Chinese folk culture and art" by the Ministry of Culture. Jilin paper-cut has the original color of the regional nationality, and Baishan paper-cut and Man paper-cut have their own characteristics.

Establish the role image of the Jilin characteristic folktale materials, and mainly draw on the modeling characteristics of Jilin Dongfeng farmers' paintings and Jilin paper-cut, that is, being generous, rough and simple. The four limbs of the figures in Dongfeng farmer's paintings are all in the arc shape, which is highly decorative. It shows the strong and free vitality of the people. Therefore, based on these two folk arts, the role image should exaggerate the proportion of the body. The arms and legs are strong, and the legs are long and the shoulders are wide, while the face is square, and the five senses can be painted as a triangle according to the characteristics of Jilin paper-cut, and the eyes and mouth can be treated as an arc. People's impression of the inherent colors in Northeast China is mostly bright red with big green. The typical one is the northeast big flower cloth. In Dongfeng's peasant paintings, the colors break through the local flavor and the color concept, and emphasize the contrast of brightness and purity.

3 Jilin Folktale Data Arrangement and Communication Technology Based on the Network Resource Mining

3.1 Feature Extraction of the Target Samples

In the network information mining system, the vector space model is used, and the feature terms and their weights are used to represent the target information. In the information matching, these features are used to evaluate the correlation between the unknown text and the target samples. The selection of the feature terms and their weights is called the feature extraction of the target samples. The quality of the feature extraction algorithm will directly affect the operation effect of the system. The frequency distribution of entries in different content documents is different, so the feature extraction and the weight evaluation can be carried out according to the frequency characteristics of entries. In order to improve the efficiency of the operation, the system reduces the dimension of the feature vector, and only keeps the terms with the higher weight as the feature term of the document, thus forming the target feature vector with the lower dimension.

3.2 Chinese Word Segmentation

The information we are dealing with is mainly the text information. In order to extract the topic information of documents accurately and build the feature model better, we need to build the main thesaurus, the synonym thesaurus, the implicative thesaurus and other dictionaries, and take them as the extraction topics. A good professional dictionary will greatly improve the accuracy of the topic extraction. Chinese word segmentation is one of the key technologies in the network information mining. Because of the integrity and standardization of its subject system, Chinese Classified Thesaurus is undoubtedly very suitable for thesaurus. For the data mining with the higher professional requirements and the places that do not meet the requirements in the actual use, we can expand and modify the thesaurus on the basis of the thesaurus. Here we introduce the idea of the post control in Library Science, that is, to control the accuracy of the URL indexing through the specification of the thesaurus.

3.3 Get the Dynamic Information in the Network

Robot is an important part of the traditional search engine. It reads the Web pages according to the HTTP protocol and automatically roams on the WWW according to the hyperlink in the HTML document. Robot is also called spider, worm or crawler. However, robot can only get the static pages on the web, and the valuable information is often stored in the network database. People can't get these data through the search engine, only log in to the professional information website, submit the query request by using the query interface provided by the website, and get and browse the dynamic pages generated by the system. The network information mining system traverses the Jilin folktale information in the network database through the query interface provided by the website, and automatically analyzes and arranges the traversal results according to the professional knowledge base, and finally imports the local information base.

4 Jilin Folktale Data Arrangement and Communication Path Based on the Network Resource Mining

4.1 Highlight the Image Micro Propagation Effect

On the one hand, the image can record the history, and on the other hand, it can also beautify and refresh people's mind. In the era of the micro communication, the audience receives a large number of videos every day, and the construction of the world has gradually changed from the original naked eyes to the reconstruction through videos. In other words, the current audience sees the world through the mobile media. As a type of the film and television communication, the documentary really shows the historical status quo, and plays an educational role by combining the oral statements of the elderly in the traditional villages with the scene on the ground to truly reproduce the traditional cultural environment. Film is also the art of the time and space, which includes the expression of some subjective colors. Jump out of the traditional propaganda circle, open the innovative thinking combined with the background of the micro communication era, and achieve the greater communication effect.

4.2 Relying on the New Media to Build the Micro Communication Network

With the popularity of the mobile devices, the communication network based on the mobile phones, computers and other media has gradually expanded. It has become a new trend of the communication to build a micro communication network by relying on the new media, to make the social resources reasonably allocated, and to form a broader new form of the economic development with the new media as the infrastructure and the Internet big data network as the guidance. Under the background of the Internet, the information communication is very extensive, and the big data also provides more guidance and convenience for people's production and life. The new media is different from the traditional media in many aspects of the single propaganda, the multiple forms of propaganda and more reasonable allocation of propaganda resources, so that the propaganda can achieve the maximum effect. The relative fragmentation of the traditional culture is just in line with the characteristics of the small and precise communication of the new media, which can contribute to the development of the traditional culture. Through the micro blog small video communication, form a mobile phone to the network and then to the mobile phone communication network, to achieve the purpose of communication relying on the majority of the audience with the micro communication.

4.3 Government Support and Attention

Compared with the modern culture, the traditional culture is relatively weak and needs more protection and support from the external forces. The government should strengthen the publicity and education of the villagers to make them understand the importance of the traditional culture to the nation, form the awareness of protection and inheritance from themselves, cultivate the awareness of publicity of the villagers, establish the information exchange platforms such as websites, give full play to the villagers' sense of "master", actively participate in the process of the micro communication, and make joint efforts to contribute to the protection of the traditional culture.

4.4   Build an Online and Offline Integrated Mechanism

Nowadays, the ways of spreading the folk story materials are mainly divided into the online and the offline. One of them is the We-Chat Subscription of Jilin folk stories. After investigation, there is no We-Chat public account of Jilin folk tales. By regularly pushing Jilin folktales with the exquisite illustrations, we can expand the reading mass base because children's reading materials are all selected by adults, so as to deepen the impression of parents. Second, we can rely on the online bookstore to promote the distribution of orders for reading. The offline is mainly divided into two aspects. One is to actively extend to the school, expand students' learning of folktale materials, and the scope of the reading contents can be used as the classroom contents. The second is the Jilin tourist attraction gift shops and bookstore sales. In the upsurge of the cultural and creative tourism, people not only want to visit here, but also hope to gain something during the journey.

5   Conclusion

In the era of the intelligent information technology, the dissemination of the information is rich and convenient, which greatly improves the speed of the dissemination of the folktale materials. The folk story materials and derivatives are a good choice for tourists to acquire knowledge as well as characteristics. By using the technology of the network resource mining, we can effectively organize Jilin folktale materials, spread Jilin traditional cultures, and realize the efficient application of the traditional folktale materials.

Acknowledgement. Fund Project: Jilin Provincial Department of Education "13th Five-Year Plan" Social Science Project; Project Name: Research on Innovation and Cultural Communication of Jilin Folk Tales; Item Number: JJkh20190364SK.

References

1. Luqiang, Y.: Review of the research literature on the computer audit methods based on the data mining. Chin. Foreign Entrep. **10**, 54–55 (2012)
2. Biping, M., Tengjiao, W., Hongyan, L., et al.: Piecewise bitmap index: An auxiliary index mechanism for the cloud data management. Chin. J. Comput. **11**, 2306–2316 (2012)
3. Guangqiang, X.: Design of the telecommunication network resource management system. Comput. Knowl. Technol. (Acad. Exch.) **2**, 85–86 (2006)
4. Wen, Z.: A data mining method based on network resources. Digit. Commun. **5**, 84–87 (2013)
5. Yue, W.: A study of man beliefs and customs in Changbai mountain folk tales and local operas. Drama Lit. **11**, 123–128 (2016)

The Design for Electric Marketing Service Data Middle-Platform

Jinzhi Wang[1(✉)], Daoqiang Xu[2], Yuan Wang[3], Shijie Gao[4], Quangui Hu[5], and Xiao Ding[2]

[1] State Grid Corporation of China, Beijing 100031, China
shuxiao4617683@163.com
[2] State Grid Jiangsu Electric Power Co., Ltd., Nanjing 210024, China
[3] China Realtime Database Co., Ltd., Nanjing 210012, China
[4] Beijing China-Power Information Technology Co., Ltd., Beijing 100192, China
[5] Beijing SGITG-Accenture Information Technology Co., Ltd., Beijing 100031, China

Abstract. In order to comprehensively improve the electric power marketing business data integration level, enhance the data value mining and innovation driving ability, and promote the business-oriented level of data, it is an effective way to build the electric power marketing service data middle-platform. Referring to the middle-platform design concept and engineering practices of e-business companies, combined with the characteristics of the electric power marketing, this paper analyzed the design principles of the data middle-platform, designed the overall structure of the marketing data middle-platform consisting of the gathering data center, the public data center, the data extraction center, the service ability center and the supporting system, and studied the key technology framework of the data middle-platform. In the end, taking the electric service applying timing limit analysis as an example, the application scene of the data middle-platform is introduced.

Keywords: Electric marketing · Data middle-platform · Middle-platform structure · Key technology

1 Introduction

After more than ten years of development, the electric power marketing informatization has played a significant role in supporting the marketing quality service, the business development and the management improvement. However, with the continuous deepening of the electric power marketization reform, the rapid development of the comprehensive energy service market and emerging business, the connotation and development mode of the electric power marketing business are also undergoing profound changes. The problem that the marketing information system does not adapt to the business development gradually appears and they are more prominent in the data fusion and sharing and the data value mining. The main performance includes the following. First, the construction of the data model is not complete, unified and

M. Atiquzzaman et al. (Eds.): BDCPS 2019, AISC 1117, pp. 1752–1760, 2020.
https://doi.org/10.1007/978-981-15-2568-1_247

standard. The current data model is not fully built around the concept of the customer-center, resulting in a lot of the model deficiencies, with a lack of the organic unity between models. Second, the data cannot be effectively shared and the business integration is difficult. At present, a large amount of the data has been accumulated in the marketing business application, the power consumption information collection and other systems, but the data resources are widely distributed in various heterogeneous systems, and the data consistency and liquidity are poor, and their sharing is difficult. Third, the data utilization capacity is insufficient, and the data value is not fully reflected. The use of the data is mainly based on the statistical reports, with a lack of the data mining. Through the data analysis, the effect of achieving the lean management, the fine business, and the quality and efficiency improvement is not obvious, so that the data assets are not fully utilized, and the data value is not fully mined.

In order to improve the level of the data fusion, enhance the ability of the data value mining and the innovation driving, and promote the data re-business, building the data middle-platform is an effective way. The core value of the data in the middle-platform is the data empowerment. That is, the data integration and the data capability precipitation across the business domains are realized through the data modeling. The data encapsulation and the open sharing are realized through unified data service, which can meet the application requirements quickly and flexibly. The personalized data and the application requirements are met through the data development tools, which can realize the service of the data application and promote the data operation. The idea and system of middle-platform is introduced by some Internet e-commerce companies such as Alibaba in the process of the IT architecture transformation. Literature [1] discusses the business value, the evolution process, the service system construction and the related cases of the middle-platform in the Internet enterprises. Literature [2] gives the top-level design of the middle-platform in the cloud data, and describes the construction and industry of the middle-platform in the cloud data through cases and the formation process of business model. Based on the business and engineering practice of the Internet e-commerce enterprises, the above literature comprehensively expounds the strategy, thought, design and operation and maintenance system of the middle-platform, which has certain reference significance for the implementation of the middle-platform in the marketing service system. In the face of the requirements of the digital transformation, the relevant enterprises including the power industry have also conducted the relevant research on the application of the middle-platform in the marketing informatization. Literature [3] takes the data-drive as the starting point, and constructs the enterprise middle-platform through the big data mining, the product close association, the omni-channel coverage and the scenario marketing, so as to meet the personalized needs of customers and improve the customer experience. Literature [4] proposes the overall structure of the multi-channel customer service middle platform of the Internet of things, and describes the business contents of the core components of the user center, the working order center, the bill center and other shared service centers. Literature [5] designs the architecture of the omni-channel operation support platform based on the middle-platform technology, and studies the deployment mode and the business contents of the operation support platform for the power marketing.

To sum up, combined with the actual power marketing business, with reference to the existing research results, the research on the design of the marketing service data in the middle-platform has strong engineering practical value for the application of the middle-platform system in the power marketing service system.

2 Design Principles of the Middle-Platform in the Marketing Service Data

The core of the data middle-platform is the empowerment. Through the collection of the enterprise data capabilities and the support of the new IT technologies, it can effectively enhance the ability of the enterprise operation, extension and business creation. With the data flow as the connection, and the market-oriented response as the driving force, and the rapid data feedback in the whole domain, it can make the trial and error costs lower, and the response faster, and help enterprises build a business ecosystem. The data middle-platform is the driving force of the digital transformation for the market-oriented competitive enterprises, which constantly evolves new enterprise innovation services and products. Combined with the cooperation and the linkage of the business middle-platform, it effectively promotes the two-dimensional growth of the business capacity and the operation efficiency of enterprises, constructs a new supporting mode of "setting up a platform to sing a play", promotes the expansion of businesses, and efficiently helps the intelligent operation. Different from the business focus on the integration and efficient processing of the business capabilities, the data middle-plat focuses on the multi-perspective data aggregation and the data value mining, use different technical routes to process and store the data, and jointly support the rapid construction and the iterative innovation of the front-end business applications through the service-oriented way.

The following principles should be followed in the overall design of the power marketing service data center. (1) Principle of clarity. When describing the business meaning of tables and fields in the physical model of the data in the middle-platform, the business terms should be used to make the description clear, concise and accurate which is the core value of the data model. The professional vocabulary and the international standard vocabulary shall be adopted as far as possible, including the general industry terms such as IEC general electric power standard, China electric power industry standard, IASB international accounting standards of International Accounting Standards Association. (2) Integrity principle. The data middle-platform model design should follow the standardized design results of the enterprise information construction, including the main data table and the business table corresponding to the actual business, meet the needs of the management analysis and the auxiliary decision-making, and fully support the application of the enterprise big data analysis. (3) The principle of consistency. The construction of the data middle-platform comes from the analysis of businesses. It needs to guide the implementation of the data persistence and keep the consistency of the business requirements, the data model and the technology implementation. (4) Standardization principle. The design of the data center should follow the corresponding international, domestic and industrial standards, and follow the unified table naming specification, the field type, the extended field, the data homology requirements and other standards. (5) Applicability principle. To adapt to the current business situation and the development needs of the electric power enterprises, meet the reasonable business differences, face the market development, and reflect the application requirements of parameterization and configuration. (6) The principle of the continuous iteration. The construction of the data middle platform is not achieved overnight, and it needs continuous evolution and iteration. At the same time, the establishment of an organization and team in line with the strategy of the middle-platform also requires the business to continuously nourish the data middle-platform, precipitate the common model of the enterprise data, give full play to the value of the enterprise data and back feed the businesses.

3 Overall Framework of the Marketing Service Data Middle-Plat

The middle-platform design of the marketing service data should be consistent with the standards and technical routes of the enterprise strategy and the information construction, and guided by the business needs, realize the unified collection and introduction of the data, the unified construction and R & D of the data model, the unified management of the data assets, the unified theme service, the standardization of the data, the visualization and standardization of the data model, and the flexibility of the application construction. Further reduce the management cost of the data and improve the business value of the data.

Referring to the design concept of the middle-platform system and the construction practice of the Internet e-commerce enterprises, combined with the characteristics of the marketing business and the foundation and construction experience of the marketing informatization, this paper designs the overall framework of the marketing service data middle-platform, which is composed of the data center, the public data center, the extraction data center, the data enabling center, the unified data service, the data asset management and the data operation. The enterprise data operation system for the marketing business is established.

3.1 Converging Data Center

The function of the data center is to collect the heterogeneous data such as the business systems and the foreign data into a unified computing storage space for the subsequent processing and analysis. The data aggregation is a dynamic process. In response to different data update frequency and the different historical data traceability requirements, it needs to be compatible with the batch and streaming data input methods, and support the full and incremental data reading methods. The marketing service data center needs to collect the business data, the scene application data, the user behavior log data, the Internet of things and the measurement data, weather, economy, credit and other external data and so on. While we can collect and summarize the data, all kinds of the models are completely expressed through the metadata, so as to meet the observation, understanding and mastering of the business data.

The data aggregation needs to choose the appropriate aggregation method according to the needs of the actual application scenarios. Generally speaking, the data aggregation can be divided into six modes: the batch aggregation, the flow aggregation, the local incremental aggregation, the flow to batch aggregation, the flow to the local incremental aggregation, and the flow to zipper the table aggregation. The comparison and use scenarios of various methods are shown in Table 1.

Table 1. Comparison of the data aggregation methods

Convergence mode	Read frequency	Read type	Write frequency	Data redundancy	Difficulty in development and operation	Change history	Applicable scenario
Batch integration	Fixed time	Total quantity	Fixed time	Great	Low	Yes	Data with small volume or low timeliness requirements
Local increment	Fixed time	Increment	Fixed time	None	Middle	No	There is a large table with clear business cycle, well maintained timestamp and no requirements for data traceability
Streaming integration	Real time	Increment	Real time	Middle	Middle	Yes	There are data with high requirements for timeliness, data that are not easy to modify or delete, and self describing data without status
Circulation batch	Real time	Increment	Fixed time	Great	High	Yes	Medium volume, frequently updated data with no clear business cycle
Local increment of circulation	Real time	Increment	Fixed time	None	High	No	Data with large volume, frequent update and no clear business cycle
Circulation zipper table	Real time	Increment	Fixed time	Middle	Middle	Yes	Data with large volume and no clear business cycle

3.2   Public Data Center

The main function of the public data center is to reduce the redundancy, increase the reuse, have a large number of the data storage capacity, and support the rapid loading, the fast computing, the fast response and other user needs. Through the methodology of the data unified system, on the basis of the standard design of the middle-platform, the data of the different business subject areas (such as customers and equipment and so on) are organized to form the unique and the unified standard public data layer of the enterprises. The same data asset, the same customer and the same business management object in different business phases, different positions and different application systems are managed in a unified and standardized way and the same dimension is used as much as possible to complete different business management operations or associate asset information in different business phases. Guided by the business requirements and complying with the international IEC 61970/61968/62325 standard, the data model design of the public data center is carried out by adopting the combination mode of "business requirements driven from top to bottom" and "status driven from bottom to top". Aiming at the traditional stable business, the mode of "bottom-up driven by current situation" is extended according to the model design method. For the traditional change business and the emerging business, adopt the "business demand driven top-down" mode, modify or develop the business processes, the business activities and the business links according to the business demand changes or the new business needs, and then abstract the data objects.

3.3   Data Extraction Center

The main function of extracting the data center is to extract the data from the public data center according to the business requirements and build a data center for the business oriented data light summary. The goal of the data extraction is to support the business personnel to build the data analysis model, to quickly find the relationship between the data and the data, and to build the data analysis application based on the association data model.

There are three ways to extract the data middle-plat from the marketing service data. First, extract the data driven by the marketing business process with reference to the mature methodology of the Internet enterprises, that is, to sort out the relationship between entities in the public data center, complete the logic table of the entity relationship, and provide the clear data context for the subsequent extraction analysis. Second, with the data extraction driven by the operation or management indicators, sort out the traceability information of the data extraction indicators by analyzing the business requirements. Then, around the traceability information of the indicators, the key processes such as the data extraction task development and the data extraction task execution are carried out in turn to complete the data extraction, and the results of the indicators are precipitated to the extraction center in the data center to provide the data basis for the upper level services. The third is the business analysis demand oriented extraction. That is, with the data query, the data analysis, the data mining and other requirements as the core, comb from the functional level, clarify the application direction, the business value, the application scenario and the application scenario description of the demand analysis, effectively extract the source data, combto form the analysis model, and then precipitate to the extraction center in the form of labels, models and indicators and so on, to provide support for the upper service.

3.4   Data Middle Platform Support System

The data middle platform support system includes the data operation, the data asset management and the data development and operation.

The operation of the data center mainly includes the index management, the label operation, the strategy management, the self-service analysis, the model operation, and the knowledge operation and so on. Its function is to support the continuous service output of the data center, improve the service quality, and promote the sustainable development of the data center.

The data asset management includes the data directory management, the metadata management, the data security management, and the data quality management and so on, which can meet the requirements of forming a data integration management platform, support all functions of the entire process of the data integration, processing, management, monitoring and output services, realize the visual workflow design, support the multi-person cooperative operation mechanism, and carry out the task opening in different functions, such as dispatching, online dispatching, operation and maintenance, and the data authority management and so on, so that the data and tasks can complete the complex operation process without landing. Only care about the data calculation logic, and the functional way to define the statistical indicators, and we can complete the modeling research and development, so that the system automatically generates the code execution operation and other development modes.

The data development and the operation and maintenance include the data synchronization, the data development, the task scheduling, and the monitoring alarm and so on, which can support the graphical index calculation and analysis, the icon analysis, and the big data model analysis and so on, to achieve the rapid precipitation of the analysis results, including labels, portraits, analysis views, and analysis models and so on. It can develop all links of the data flow, to achieve the data access, the processing flow, and result storage, and the scheduling and management of the data tasks. Support the convenient data operation and maintenance functions, monitor the data flow links, find and warn problems in time, and ensure the reliable implementation of the data tasks.

4   Key Technologies in the Marketing Service Data Middle-Platform

4.1   Selection Principle of the Data Middle-Platform Technology

In order to ensure the stable and efficient operation of the data middle-platform, it is necessary to clarify the relevant technical requirements of the data middle-platform. The technical selection of each component of the data middle-platform shall first meet the existing relevant technical standards and guidelines of the enterprise, and shall comply with the following general principles in terms of availability, openness, scalability and technical support. (1) In terms of the component availability, based on fully meeting the needs of the business management, the product related functions and performances are fully verified and determined. (2) In terms of the openness of components, components shall provide flexible interactive interfaces or secondary development capabilities to meet the personalized needs in the process of the business development. (3) In terms of the component scalability, with the continuous development of the marketing business, components should have enough scalability to meet the rapid growth demand of the business processing volume in the process of the business development. (4) In terms of the manufacturer's supporting strength, there are inevitably various operation and use problems in the use of components. The component manufacturer shall have fast and positive service response ability and rich on-site support experience.

4.2   Technical Capability Framework of the Data Middle-Platform

Based on the overall structure of the marketing service data in the middle-platform, and referring to the experience of the digital operation leading industry, a customer-side data in the middle-platform technical capability framework is constructed, which includes five layers of the data collection and preprocessing, the data storage and calculation, the data analysis and mining, the data capability output, and the data management and control.

In order to ensure the stable and efficient operation of the marketing service data center, it is necessary to further clarify the relevant technical requirements of the data center. The capability descriptions of each part of the technical capability framework are shown in Table 2.

Table 2. Technical capacity of the data middle-platform

Technological capability	Specification
Data acquisition and preprocessing	According to different data source interface types and application timeliness requirements, the data collection of the heterogeneous data sources is realized through the file transmission, the synchronous replication, the message bus, the log collection, the crawler collection and other ways. By providing the corresponding data transformation and cleaning tools for the structured and unstructured data, the rapid and efficient data preprocessing can be realized
Data storage and calculation	Provide a variety of the data storage methods to meet the storage requirements of the structured data, the semi-structured data, the unstructured data, and the graph data, and provide the batch processing, the real-time query, the stream computing and other data processing architectures to meet different data computing needs, to achieve the high-efficiency storage and calculation of the massive heterogeneous data
Data governance	Through the protection of the sensitive data, the data audit, the data encryption and other means to ensure the data security in the data center, at the same time, through the data quality monitoring, effectively find the data quality problems, and provide the unified coordination services and job flow management, to achieve the overall control of the data, to ensure the safe and orderly operation of the data center
Data analysis mining	Support the efficient data mining modeling through mining the modeling tools, the AI computing framework, the algorithm database and the feature database, manage and operate the data processing and the extraction results effectively through the model database, the index database, the tag database, the strategy database and the knowledge database, and provide the visualization, the data query, the data analysis, the information search and other data application capabilities, so as to realize the full mining and sharing of the value of the data
Data capability output	Provide the data interface, the data file, the data presentation and application and other data service methods to meet the needs of the diversified data services. Through the front-end interaction and the back-end service development technology, build the customized data products and provide the complex data applications for specific application scenarios

5 Conclusion

In this paper, aiming at the problems of the scattered data resources, incomplete underlying model, inadequate sharing, and weak data value mining ability of the power marketing system, referring to the design concept and the relevant engineering practice of the middle platform of enterprises, and combined with the characteristics of the power marketing business, and the marketing service data center manager, including the data center, the public data center and the data extraction center capability center, is designed. This paper analyzes the key technical framework needed to build the data middle-platform, and validates the data middle-platform architecture proposed in this paper based on the typical application scenarios of the data middle-platform.

References

1. Hongfu, Z.: Building the enterprise data middle-platform and promoting the enterprise intelligent operation. C-Enterp. Manag. (02), 32–33 (2018)
2. Gang, X.: Research on the omni-channel scenario operation based on the enterprise middle-platform. Inf. Commun. (08), 268–269 (2017)
3. Lin, H., Fang, X., Yuan, B., Ou, Y.: Strategic research and design of the multi-channel customer service middle-platform of power Internet of things. Distrib. Util. (06), 39–45 (2019)
4. Zhao, G., Zhang, C., Ou, Y., Hong, Y.: Research on the architecture design of the omni-channel operation support platform based on the business middle platform. Distrib. Util. (06), 67–71 (2019)
5. Li, B., Hu, Q., Chen, X., et al.: Research and design of the data center in power grid enterprises. Electric Power Inf. Commun. Technol. 17(7), 29–34 (2019)

The Implementation of the Chinese Language and Character Recognition System Based on the Deep Learning

Yanwen Wang[✉]

School of Literature, Shandong University, Ji'nan 250100, Shandong, China
haixiang33749546@163.com

Abstract. The Chinese language is an important symbol of the Chinese traditional cultures, and also an important manifestation of the integration of the information technology and the traditional cultures. How to use the intelligent technology to promote the Chinese language and writing has also become an important trend of the development of the current era. In the process of the Chinese character recognition, the feature learning method and the DLQDF classifier proposed in this paper can obtain the performance of the deep convolution neural network (deep CNN). Therefore, the computational cost of the recognition is lower than that of the deep convolution neural network.

Keywords: In-depth learning · Chinese language and characters · Recognition system · Research · Implementation

The deep learning is a complex machine learning algorithm, which has achieved much better results in the speech and image recognition than the previous related technologies [1]. In view of this, this paper improves the performances of the Chinese character recognition by accelerating the classifier training and the large-scale feature learning on the basis of the traditional recognition methods.

1 The Theory Connotation and Application of the Deep Learning

Deep learning (DL) is a new research direction in the field of the machine learning. It has been introduced into the machine learning to make it closer to the original goal of AI (Artificial Intelligence). Deep learning is the inherent law and representation level of the learning sample data [2]. The information obtained during the learning process is very helpful to the interpretation of the data such as texts, images and sounds. Its ultimate goal is to enable machines to have the same analytical learning ability as human beings, and to recognize the data such as texts, images and sounds.

The Chinese character recognition technology can be applied in many industries of the national economy, such as the postal address recognition, which makes the automatic mail sorting possible, and can save a lot of manpower [3]. The identification of bank bills, tax receipts, books and manuscripts and so on can make the document

© Springer Nature Singapore Pte Ltd. 2020
M. Atiquzzaman et al. (Eds.): BDCPS 2019, AISC 1117, pp. 1761–1765, 2020.
https://doi.org/10.1007/978-981-15-2568-1_248

image electronic and facilitate the management, search and transmission in the future [4]. As a basic component of the handwritten Chinese document recognition, the Chinese character recognition has attracted many researchers' attention for a long time [5]. The main difficulty of the Chinese character recognition is that the set of the character categories is very large, and there are many similar characters. The writing style differences lead to the great distortion of similar characters written by different writers. These difficulties make the performance of the freehand Chinese character recognition unsatisfactory.

The parallel acceleration of the training process is performed using the graphics processor (GPU). In order to improve the generalization performance of the classifier, increasing the size of the training set is a common method. However, the large-scale training sets pose challenges to the training of some classifiers, and especially some training methods based on the discriminant learning. GPU has a large number of the floating-point computing units, which is suitable for the large-scale parallel computing. The training process of the discriminant feature extraction (DFE) and the discriminant learning quadratic discriminant function (DLQDF) classifier is accelerated by GPU. The training speed of these classifiers is increased by 30 times and 10 times respectively, which makes it possible to train the large-scale data sets. A large-scale feature learning method is proposed to improve the recognition performance. In order to improve the discriminant ability of the features, this paper uses the correlation information between the features to carry out the second dimension lifting on the basis of the original gradient direction histogram features, and obtains tens of thousands of secondary features. Then, in the high-dimensional feature space composed of the quadratic and gradient features, the discriminant learning is used to obtain the low-dimensional feature subspace.

2   Research on the Chinese Character Recognition System Based on the Deep Learning

The traditional Chinese character recognition generally includes the pretreatment (such as normalization), the feature extraction, the feature dimensionality reduction, and the classifier design and so on. But the introduction of the convolutional neural network (CNN) makes it possible to design an end-to-end handwritten character recognition system without complicated pretreatment, the feature extraction and the feature reduction. In addition, some researchers have found that although the end-to-end CNN handwriting recognition method can achieve better performances than the traditional method, the performance of the conventional CNN handwriting recognition system can still be further improved by combining the traditional domain knowledge. In addition, in view of the characteristics of the Chinese character recognition, many CNN training improvement methods have been proposed in recent years.

CNN has been widely used in the handwritten English recognition since it was proposed in 1990s. For example, in the field of the handwritten English recognition, the recognition rate of the LeNet5CNN model proposed by LeCun and others in 1998 on MINIST has reached 99.05%. With the training of the deformed samples, the recognition rate can be further increased to 99.2%. In 2003, Microsoft Research Institute's

Simard and others introduced the Elastic distortion and the Affine distortion, two kinds of the data argumentation technologies. Using a network structure similar to CNN, they achieved 99.6% excellent performance on MINIST, which was significantly ahead of machine learning methods including SVM, Boosting, and multi-layer sensors and so on.

The IDSIA Laboratory proposed a multi-column CNN model. The basic idea is to train the multiple CNN networks using GPU (each CNN network contains four convolution layers, four Pooling layers, and one full connection layer). Then, the output of all CNNs is integrated on average, and the on-line and off-line handwritten samples are converted into the image input into the CNN network, without any feature extraction or feature selection. The output of CNN directly acts as the final recognition result, which is a typical end-to-end solution. MCDNN achieves the most advanced recognition results both offline and online. On CASIA-OLHWDB 1.1 online handwritten Chinese data set, MCDNN reduces the error rate greatly by using only binary picture information of the online handwritten trajectory (ignoring timing information). Since then, the in-depth learning based on CNN has gradually attracted the attention of scholars in the field of the handwritten Chinese recognition.

3 Implementation of the Chinese Character Recognition System Based on the Deep Learning

Data generation technology: In order to avoid over fitting and improve the recognition performance of the CNN model in the training process, in addition to using some classical methods such as Dropout, obtaining sufficient training samples is the premise to guarantee the high performance of the CNN model. But in many cases, the training samples we get are very limited. For example, the excellent data sets such as CASIA-HWDB/CASIA-OLHWDB, whose writers are still limited and the collection area is limited, are not enough to cover the statistical distribution of the Chinese character writers nationwide. Therefore, the data argumentation technology is a very important technology to enhance the robustness and the generalization ability of the CNN system. LeCun and others proposed the methods of translation, scaling, rotation, horizontal and vertical stretching deformation (Squeezing, Shearing) to deform the data, which effectively improved the recognition performance. On the basis of the affine transformation, two data generation technologies, the elastic distortion and so on, were further proposed in the literature. CNN was used in MINIST. 99.60% of the performance is obtained. By using the non-linear function of the trigonometric function, 24 kinds of the orthogonal transformation methods for the Chinese characters are proposed, which can also be regarded as an effective method for generating the pseudo-sample data. For the NIST-19 handwritten data set, more than 819 million samples are generated, which effectively solved the problem of the insufficient training sample data. In addition to the handwritten character recognition, in the field of the text detection and recognition in the natural scenes, the data generation technology is often proved to be a very important and effective technologist to improve the performances of the deep learning model.

Using a deeper CNN network (15 layers, 9 convolution layers) and combined with the three-dimensional random deformation technology, the large-scale data generation can generate various deformation modes, simulate the changes of light, shadow and strokes, and learn from the introduction of the multiple output layers in the middle layer of the network by Google LeNet for the implementation of the multi-level supervised learning. The online and offline handwritten text line recognition is still an unsolved problem. Some research directions that deserve our attention include: the fusion of the segmentation-based + CNN-based word recognition + path optimization algorithm, among which many key technologies need to be solved, such as how to improve the reliability of CNN, whether the CNN method can be used for the character segmentation, and the sliding window recognition method based on RNN/LSTM/BLSTM without segmentation. As for the small Latin text, it has been recognized as the best method, but whether the large Chinese character recognition problem is optimal is still worth discussing and exploring, and an end-to-end pure in-depth learning solution based on CNN+LSTM.

Unconstrained handwritten character recognition: One of the issues worth attention is the rotation-independent handwritten recognition. The handwritten input software and equipment must be able to recognize the handwritten samples. However, most of the mainstream input methods in the market cannot meet this requirement. Although some researchers have noticed this problem and many preliminary explorations have been carried out, in the overall sense, the problem is still far from being effectively solved. It is believed that the emergence of the new technology of the in-depth learning will provide new ideas and technical means to solve this problem. The current research work is mostly limited to solving the simple problems, such as the single character recognition or the simple text line recognition. There are still few reports on the online handwritten overlapping text line recognition, the hybrid handwritten word/text line/overlapping and the Chinese character recognition from the arbitrary and unconstrained writing on the whole screen, which is a subject worthy of study.

The ultra-large categories of the Chinese character recognition: At present, the main types of the Chinese characters recognized by the research reports on the Chinese character recognition are 3755 types of the Chinese characters in the first-class national standard font library. There are few reports on the practical handwriting recognition research that can recognize more than 10,000 categories in the practical application scenarios, and there is a lack of the open super-large categories (such as 27533 training and test data sets supporting the GB8010-2000 standard). In such a large category, how to develop a practical solution based on the in-depth learning with the fast processing speed and the small model parameters will become very challenging. According to the evaluation results of the ICDAR 2013 Handwritten Chinese Contest, the solution based on the in-depth learning is still available in terms of the storage and the recognition speed. Therefore, the construction of the large-scale data sets, the fast algorithms for various deep learning models, and the parameter compression techniques for the deep learning models still need to be explored and solved by researchers.

Research on the application of the new deep learning model in the Chinese character recognition: At present, the deep learning model which is better than the traditional method in the field of the Chinese character recognition is mainly based on CNN and its various improved methods. Other deep learning models such as DBN, RNN, LSTM/BLSTM/MDLSTM and DRN are applied in the field of the Chinese character recognition. The research work of the large-scale Chinese character recognition has not been carried out much, and the research on the interrelation and fusion application of various deep learning models is not in-depth. We look forward to other in-depth learning models as well as the future updates and the better in-depth models for the character recognition, and make breakthroughs in the field of the Chinese character recognition, so as to promote the research and development in this field.

The text detection and recognition in natural scenes: In recent years, with the explosive growth of a large number of the Internet pictures, the text detection and recognition in natural scenes has become a very important and widely concerned hot research topic in the field of the text recognition and the computer vision. The emergence and development of the deep learning theory and technology provide a good solution to this challenging problem. In recent years, there have been a lot of research results. However, compared with the traditional MSER framework, the deep learning method has a slow processing speed and a large storage of the model parameters.

4 Conclusion

Generally speaking, the in-depth learning provides the new concepts and technologies for solving the Chinese character recognition. In recent years, a lot of research results have been achieved in this field, but there are still many research problems worthy of our further study. This paper reviews and discusses the research progress in the related fields, hoping to bring researchers in this field the new information and research ideas to promote the further development and prosperity of the handwritten Chinese character recognition and the related document analysis and recognition.

References

1. Xin, H.: Research on the text recognition method based on the AI machine learning. Telecommunications **13**, 234 (2016)
2. Li, M., Pan, C., Huang Qijun, C.: Design of the embedded text recognition system based on SoPC. Appl. Electron. Tech. **37**(9), 15–17 (2011)
3. Jie, L., Muyun, F.: Research on the character and similarity measurement in the character recognition. J. Yancheng Inst. Technol. (Nat. Sci. Ed.) **29**(4), 42–46 (2016)
4. Xiuguo, Z., Shixiu, Z.: Design of the character recognition experimental system based on DM6437. Res. Lab. Work Coll. Univ. **02**, 48–50 (2012)
5. Weiwei, J., Liden, B.: Offline handwritten character recognition based on the improved BP neural network. Video Eng. **13**, 200–202 (2014)

Online Handwritten Chinese Character Recognition Simulation System Based on Web

Yanwen Wang[(⊠)]

School of Literature, Shandong University, Ji'nan 250100, Shandong, China
haixiang33749546@163.com

Abstract. The online handwritten Chinese character recognition simulation system is based on the web development and the Tensor Flow experimental platform. The CASIA-OLHWDB data set is used to obtain the classified data, and the Python technology is used to build the Web server, and the real-time and automatic data refresh is realized on the Web side. According to the actual results of the simulation system, the system can accurately identify the online handwritten Chinese characters.

Keywords: Web · Online · Handwritten Chinese characters · Recognition simulation system · Management research

At present, the computer vision has become a hot research direction in the academic circles, and the pattern recognition is an important part of the research. It aims to segment and recognize regions of interest in pictures or videos using the image processing technology on the computer processing platform. Among many kinds of the pattern recognitions, the character recognition technology is one of the most important. It uses the computers to automatically segment and recognize characters [1]. It is applied to the text input, the automatic driving and other aspects, and has broad social and economic values.

1 Background Analysis of the Handwritten Chinese Character Recognition Simulation System

The Chinese characters have a history of thousands of years, and are also the most widely used words in the world. They have made indelible contributions to the formation and development of the splendid cultures of the Chinese nation, and will continue to play an important and irreplaceable role in other forms of writings [2]. However, the Chinese characters are non-alphabetic. In today's highly informationized society, how to input the Chinese characters into computers quickly and efficiently has become an important bottleneck affecting the efficiency of the human-computer communication information, and also related to whether computers can really be popularized and applied in our country. Around this problem, people put forward various solutions [3]. At present, the Chinese character input is mainly divided into the manual keyboard input and automatic machine recognition input.

© Springer Nature Singapore Pte Ltd. 2020
M. Atiquzzaman et al. (Eds.): BDCPS 2019, AISC 1117, pp. 1766–1771, 2020.
https://doi.org/10.1007/978-981-15-2568-1_249

The automatic recognition input is divided into the speech recognition and the character recognition [4]. The Chinese character recognition is an important branch of the pattern recognition and the most difficult problem in the field of the character recognition. It involves the pattern recognition, the image processing, the digital signal processing, the natural language understanding, the artificial intelligence, the fuzzy mathematics, the information theory, the computers, the Chinese information processing and other disciplines [5]. It is a comprehensive technology, which has important practical values and theoretical significances in the Chinese character information processing, the office automation, the machine translation, the artificial intelligence and other high-tech fields.

The Chinese character recognition technology can be divided into two categories: the printed Chinese character recognition and the handwritten Chinese character recognition. The latter can be divided into the online handwritten Chinese character recognition and the offline handwritten Chinese character recognition. From the perspective of the recognition, the handwriting recognition is more difficult than the print recognition, while the offline handwriting recognition is more difficult than the online handwriting recognition. Fortunately, through the efforts of the scientific researchers, our country has already sold the printed Chinese character recognition and the online handwritten Chinese character recognition products. At present, there are hundreds of schools of thought contending and flowers blooming, but the offline handwritten Chinese character recognition is still in the stage of the laboratory research. In the field of the off-line handwritten Chinese character recognition, the off-line handwritten Chinese character recognition for non-specific people is difficult for the specific handwritten Chinese character recognition.

2   Problems in the Handwritten Chinese Character Recognition

The handwritten Chinese character recognition is an important method that can be used to input the handwritten manuscripts into the computer at any time. It is also the most difficult task in the machine character recognition. These difficulties and problems are manifested in the following. There are many types of the Chinese characters. Kangxi Dictionary alone contains more than 49000 Chinese characters, and there are more than 4000 commonly used Chinese characters. Therefore, the problem of the Chinese character recognition belongs to a large category (or the super multi-category) pattern recognition, which is of great significance in the research of the pattern recognition theory and methods. The structure of the Chinese characters is complex. There are many similar characters in the set of the Chinese characters. Some Chinese characters differ only a little or a stroke. Due to the existence of the handwriting distortion, it is much more difficult to distinguish the similar characters in handwriting than in print.

Due to the differences of the writing styles of different people, the handwritten Chinese characters have great changes, which are embodied in the following aspects. The basic stroke will change. The horizontal line is uneven and the vertical line is not straight. The straight pen turns, and the corner of folding pen turns into arc and so on. The strokes are blurred and irregular. They should not be connected, but they should be connected. The positions between strokes and between parts and components will change. The inclination angle of the stroke, the length of the stroke and the size of the components will change. For the offline handwritten Chinese characters, the use of different pens by different people may cause changes in the size of the strokes. Among them, the change of the handwritten Chinese characters is the most difficult problem to solve.

3 Research Background of the Web-Based Online Handwritten Chinese Character Recognition Simulation System Management

The online handwritten Chinese character recognition is easier than the offline handwritten Chinese character recognition. The on-line handwritten Chinese character recognition is a real-time method to input the Chinese characters into a computer manually. It uses a writing board to convert the strokes into the one-dimensional electrical signals. The strokes trajectory represented by a sequence of the coordinate points is input into the computer. Therefore, the one-dimensional line (strokes) strings are processed, which contain the number of strokes. These lines contain the number of strokes, stroke directions, stroke orders and writing speeds. The offline handwritten Chinese character recognition only deals with the two-dimensional Chinese character lattice images, which is the last very difficult problem in the field of the Chinese character recognition. At present, it is still in the stage of the laboratory research.

Thus, for the off-line handwritten Chinese character recognition of the non-specific people, if there is no restriction on the handwritten Chinese character writing (that is, the free handwritten Chinese characters), it will be very difficult to recognize. In the scientific research, people always follow the rule from easy to difficult, and start with simple problems to find a breakthrough. Because the free handwritten Chinese character recognition is too difficult, people put forward the handwritten Chinese character recognition, which has become the main research object of the offline handwritten Chinese character recognition. The so-called handwritten print refers to the regular script handwriting, which requires the writers to write neatly, with as few continuous strokes as possible.

After synthesizing the advantages and disadvantages of various algorithms, an image handwritten character segmentation and recognition algorithm based on the adjustment factor is proposed. Firstly, the character image to be recognized is segmented into the region, the character region and the background region. Then the feature quantity of the binary character image is extracted and matched with each standard character feature quantity saved in the library beforehand. The recognition character with the smallest matching difference is selected. The experimental results show that the algorithm has the low computational complexity, and can segment and detect the contents of characters in a relatively short time. The recognition rate is high. For some characters written abnormally, it can also achieve the better segmentation and recognition results.

4   Research on the Management of the Online Handwritten Chinese Character Recognition Simulation System Based on Web

When designing and developing an on-line handwritten Chinese character recognition system based on the web, we can collect the information of the input devices such as the touch screen, and use the timing information or the dot matrix image information to recognize. But in the early stage of the system design, the cost of using the hardware equipment is slightly higher, so we can use the mouse as the input device for the development of the simulation system.

4.1   Introduction of the Online Handwriting Recognition Process

The process of the on-line handwriting recognition is basically the same as that of the general pattern recognition. It consists of four steps: the data acquisition and preprocessing, the feature extraction, the classification and recognition, and the post-processing. In the data acquisition and pre-processing stage, the sensor is used to collect the original physical information, and the common ones are the acceleration, speed, displacement, pen starting and pen writing. Then, the original information is pre-processed by the sensor correction and denoising. The feature extraction is one of the most important steps in the handwriting recognition. It has an important impact on the design of the classifier and the classification results. Choosing the appropriate features can not only improve the recognition rate, but also save the computing memory, the computing time and the feature extraction costs. The acceleration, displacement and DCT transform are the common feature extraction methods in the online handwriting recognition.

The classification and recognition is the core stage of the handwritten recognition. Most classifiers need to use the training samples to train the classifiers before they can be used in the practical classifications, and constantly revise the feature extraction methods and schemes, the decision rules and the parameters of the classifiers. At present, the training stage of the classification and recognition needs the manual intervention to achieve the best recognition rate. Some recognition systems use the post-processing to further improve the recognition rate after the classification and recognition.

4.2   On-Line Handwriting Recognition Experiment

In this paper, the methods of Bahlmann and others and Bothe and others are used in the online handwriting recognition experiments. The sample database used is the free online handwritten database UJIpenchars2. It uses Toshiba M400 Tablet PC to collect a total of 11640 handwritten samples of 60 writers. These samples contain ASCII, Latin and Spanish characters, and each character contains 80 training samples and 140 test samples. Each sample consists of one or more strokes, and the database provides the coordinate sequence of each stroke.

The coordinate sequence consists of the horizontal coordinate xi and the vertical coordinate yi of the pen tip collected at the equal intervals. In the experiment, the coordinate sequence of the sample does not undergo any pre-processing such as denoising. For each coordinate point, the column vectors are calculated by using the center of the gravity of the character (μx, μy) and the variance of the vertical coordinates σy. The ang is the function of calculating the imaginary phase angle. The feature vector of each character sample is T = (t1,... tNT), where NT is the number of the coordinate points collected, that is, the dimension of the eigenvector, and the NT of each character sample can be different. The online handwriting recognition experiment in this paper assumes that each character can be recognized without analyzing its context. Therefore, after the feature extraction, the classification algorithm described in this paper is used for the classification and recognition, and the output result is regarded as the final recognition result without any post-processing.

4.3   Support the Vector Machine

Suppose that the linear classifier has an input eigenvector x = {x1, x2,... xn} (n is the number of samples), and output Y = {y1, y2,... yn}. Among them, xi, I RN, and N are the dimensions of eigenvectors: $yi \in \{-1,1\}$. yi = −1 indicates that the samples (xi, yi) belong to the first category, and yi = 1 indicates that the samples (xi, yi) belong to another category. In order to minimize the structural risk, SVM calculates the optimal classification hyperplane that maximizes the distance from the training sample set to the classification hyperplane. It is equivalent to solving the convex quadratic programming problems by the pairwise formulation. That is to find the optimal alignment path to minimize the average distance. The smaller the DTW distance, the more similar the samples represented by T and R. The dynamic programming algorithm can be used to calculate the optimal alignment path and the DTW distance.

4.4   Web Terminal Design

The simulation system uses Python as the development language. At present, the main Python Web frameworks are Django, Flask and Tornado, among which Django is more representative. Many APPs and websites are based on Django, with perfect Django documents and the highest market share. Flask is a lightweight Web application framework. Its WSGI toolbox uses Werkzeug and its template engine uses Jinja2. Tornado is a powerful and extensible Web server written in Python. It is robust enough to handle the severe network traffic and can be used in a large number of applications and tools. The simulation system server uses the Flask framework, which supports the relational database and the NoSQL database.

The basic process of the Web server design is to first create a Flask instance, that is, an object of Flask class, and then the application instance starts Flask's integrated development Web server with the object run method. In order to separate the business logic from the presentation logic, the system uses a template, which is stored in the Templates folder and named index.html. The front-end of the prototype system is a Web interface. Canvas implements the function of the handwriting board. The drawing board uses an open source mouse Sketchpad plug-in drawingboard.js. The system retrieves the mouse handwritten Chinese characters from the board and transfers them to the background Chinese recognition module for recognition. The recognition results generated pictures are asynchronously returned to the Web end through the Ajax technology for the result display.

5   Conclusion

Aiming at the online handwritten Chinese character recognition system, the network graph structure is designed by using the deep learning open source framework Tensorflow, and the HWDB data set is trained to generate the model parameters. In the aspect of the result visualization, the Flask Web framework is used to build the Web server, and the Ajax technology is used to refresh and display the recognition results in the real time. The system has basically completed the recognition of the handwritten Chinese characters. This paper does not consider the timing information of the input Chinese characters, and the recognition rate needs to be improved. Combining the convolutional neural network with the timing information will be the next research work.

References

1. Jingnan, C., Changzheng, X.: Online handwritten Chinese character recognition based on the free Bishun and connected bits. Comput. Syst. Appl. **18**(5), 29–33 (2009)
2. Zhitao, G., Jinli, Y., Xiujun, Z., Shurui, F.: Handwritten Chinese character recognition based on the improved PSO neural network. J. Hebei Univ. Technol. **36**(4), 65–69 (2007)
3. Hua, Yu., Liang, C., Qiyuan, L.: Improvement of the BP neural network and its application in the handwritten Chinese character recognition. J. Jiangxi Normal Univ. (Nat. Sci. Ed.) **33**(5), 598–603 (2009)
4. Yu, Y., Jiang, W., Xu, M.: Prediction of the chlorophyll a in water by the BP neural network based on the PSO algorithm. Res. Environ. Sci. **24**(5), 526–532 (2011)
5. Junxue, W.: BP neural network research based on the PSO-EO algorithms optimization. Sci. Technol. Eng. **10**(24), 6047–6049 (2010)

Study on an Extreme Classification of Cost - Sensitive Classification Algorithm

Yu Wang and Nan Wang[✉]

School of Mathematical Science, Heilongjiang University, Harbin 150080,
People's Republic of China
hongqi25642100@126.com

Abstract. Based on the principle of cost sensitivity, this paper takes the cost sensitivity algorithm of neural network as the classifier algorithm, by using the idea of iteration, we can find a misclassification cost which can make the misclassification number of minority class samples to be zero. And compared with the evaluation indexes of some unbalanced data classification methods when the number of classification errors of minority class is non-zero, this paper hopes to realize the assumption that the minority class in the unbalanced data set will not be misclassified.

1 Introduction

With the rapid development of information technology, machine learning and data mining technology has been applied to various fields. As one of the key techniques of machine learning, classification algorithm has a direct impact on the classification of data. However, the emergence of unbalanced data brings great challenges to classification algorithms. Until now, unbalanced data classification algorithms have been widely used in many fields, such as biomedicine [1], financial industry [2], information security [3], industry [4] and computer vision [5]. In addition, the classification of unbalanced data has attracted extensive attention from international scholars [6–8]. In general, the unbalanced data classification problem can be divided into three parallel parts from different aspects of data classification: Preprocessing of unbalanced data [9–12], Feature analysis and extraction of unbalanced data [13, 14], classifier algorithm of unbalanced data [15–18]. In recent years, there have been a lot of related research articles [19–22] based on these three aspects. When using classifier algorithm to study the problem of unbalanced data classification, we will encounter a common technical method. That is to add the cost sensitive factor in the appropriate position of the classifier in order to realize the cost sensitive classification algorithm. The traditional data classification is based on the premise of balanced data, that is, the number of all kinds of samples in the data set is equal, and the cost of all kinds of misclassification is equal. In the real world, however, this assumption is often not formed, namely in the field of medical suppose there are two kinds of people, one kind is healthy people, the other kind of people are suffering from a disease, if the "healthy" as "disease" and the "disease" are judged to be healthy, the cost of the two kinds of people is obviously not equal. The cost of the former is economic loss, while the latter is fatal. Thus, we can see that the "imbalance" in the classification of

© Springer Nature Singapore Pte Ltd. 2020
M. Atiquzzaman et al. (Eds.): BDCPS 2019, AISC 1117, pp. 1772–1782, 2020.
https://doi.org/10.1007/978-981-15-2568-1_250

unbalanced data has both the imbalance between the number of various samples and the imbalance between the misclassification costs of various samples. In fact, the cost sensitive classification algorithm has many mature practical applications and excellent research results. For example, the technology roadmap mentioned in the project of MLnetII [23], the cost sensitive factors are placed in different positions of the artificial neural network algorithm and the results obtained are compared [24], and the cost sensitive idea is combined with the sampling technology [25].

However, as the core of the cost sensitive classification method, misclassification cost is difficult to be estimated accurately in most cases. This can not well reflect the distribution of real classes of unbalanced data, and cannot guarantee the effect of cost sensitive learning and directly affect the classification of unbalanced data. At present, the evaluation of cost - sensitive learning is based on the best classification effect of majority and minority classes. Based on above, this paper starts from the classifier algorithm of unbalanced data and combines the basic theory of cost sensitivity and the idea of iteration to find a special "misclassification cost" to realize an extreme classification of cost sensitive classification algorithm, i.e., the number of misclassification of minority samples are reduced to zero. Four groups of samples with different degrees of imbalance in the UCI data set were used for experiments, and the experimental results were compared with the evaluation indexes of "average cost" [24] and ROC curves generated by the cost sensitive neural network algorithm. Finally, we used a real data set of a project to experiment with the above method. Although some of evaluation indexes are not necessarily inclined to the method in this paper, but there is a practical significance to reduce the number of misclassification of minority samples to zero. The organization of the paper is as follows. In Sect. 2, we will give some basic definitions to be used in this paper, the basic idea of cost sensitive learning and the process of back propagation, and make some cost sensitive modifications to the process of back propagation learning. Section 3 introduces experimental work to evaluate different approaches. In Sect. 4, we show an example of how to apply cost sensitive neural network learning to practical problems of ROC curve analysis. Section 5 gives some conclusions and points out the possibilities for further work.

2 Learning Method

2.1 Definitions

At the beginning of this section, we will introduce some basic definitions, which will run through the learning method of this article. These basic definitions are: misclassification cost, cost matrix, cost vector, average cost, error rate and accuracy, confusion matrix, ROC curve and AUC.

The cost of misclassification is essentially a function of the prediction class and the actual class of a certain sample in a data set. The function is in the following form: Cost (actual class, prediction class), which also represents the cost matrix. The cost matrix will be inserted into the learning process as an additional term and it is also used as an evaluation of the cost sensitive neural network's ability to reduce the number of misclassified samples. The cost matrix is defined as follows:

- Cost [i, j] = The cost of assigning an example of "class i" to "class j"
- Cost [i, j] = 0 (cost of correct classification)

From the definition of the cost matrix, we can know the definition of the cost matrix under the condition of consistency of them is classification cost among various classes in the balanced data set:

- $\forall$i,j: when i $\neq$ j then Cost[i, j] = 1 when i = j then Cost[i, j] = 0

The cost vector represents the expected cost of the sample being mistakenly assigned to class i, and its expression is as follows:

-

$$\text{Cost Vector[i]} = \frac{1}{1 - p(i)} \sum_{j \neq i} p(j) \cos t[i,j]$$

Where P(i) represents the prior probability that the example in the sample belongs to class i. In the balanced data set, when the cost of classification between classes is equal, its cost vector is:

- $\forall$i : Cost Vector[i] = 1

As the evaluation standard of unbalanced classification algorithm, the average cost replaces the error rate and accuracy rate in the traditional balanced data classification algorithm. Its expression is as follows:

- Average cost $= \frac{1}{N} \sum_{i=1}^{N} \cos t$ [actual class(i), predicted class(i)] here N is the number of test samples.
- Error rate = # of incorrectly classified examples/N
- Accuracy = 1 – Error rate

When dealing with binary classification problems, we often use confusion matrix to observe the classification results. We usually define minority class as positive and majority class as negative. The confusion matrix is shown in Table 1.

Table 1. Confusion matrix

Classes	Positive	Negatives
Positive	True positive(TP)	False Negatives(FN)
Negatives	False positive(FP)	True Negatives(TN)

Where TP means positive sample prediction is still positive, FN means positive sample prediction result is negative, FP means negative sample prediction is positive, TN means negative sample prediction is still negative.

Receiver Operating Characteristic curve (ROC) is a graph that can be used to evaluate the classification effect of classifiers. When drawing the ROC curve, the classification error rate of negative samples in the test set is taken as the abscissa and the classification accuracy rate of positive samples as the ordinate. Such abscissa and ordinate can reveal the relative change of the probability that the positive sample is predicted to be positive and the probability that the negative sample is predicted to be positive when the internal parameters of the classifier change. From this, we can see that the closer the ROC curve is to the upper left corner, the better the performance of the classifier is.

However, the ROC curve cannot quantitatively describe the performance of the classifier. In view of this problem, scholars have proposed another index that can quantitatively evaluate the classification performance of the classifier, that is, AUC is the area under the ROC curve. We can easily see that the higher the value of AUC is, the better the performance of the classifier is.

2.2 Back Propagation Learning Process

The basic idea of BP neural network algorithm is that the learning process is composed of the forward propagation of the input signal and the back propagation of the error signal. In the forward propagation, the training sample enters from the input layer and is processed by each hidden layer before being passed into the output layer. Then the incoming signal processed by the output layer will be compared as the actual output and the expected output. If it is inconsistent, the error will enter the process of back propagation. The back propagation of the error is to propagate the error to the input layer by layer through the hidden in some form, and distribute the error to all the units in each layer, so as to obtain the error signal of the units in each layer. Then the error signal will be used as the basis for correcting the weights of each unit.

Generally, the gradient descent algorithm is implemented in the weight space for the purpose of minimizing the squared error:

$$E = \sum_{Examples} \frac{1}{2} \sum_{i \in Output} (y_i - o_i)^2$$

Where o_i is the actual output of the ith output neuron and y_i is the expected output. In addition, the weight is updated as follows:

$$\Delta \omega_{ji}^{(n+1)} = \alpha \, \Delta \omega_{ji}^{(n)} + \eta \, \delta_j \, y_j$$

$$\omega_{ji}^{(n+1)} = \omega_{ji}^{(n)} + \Delta \omega_{ji}^{(n+1)}$$

Where, n is the moment when the weight is adjusted, η is the rate of learning, α is called the momentum coefficient, the momentum term can reflect the accumulation of adjustment experience before the moment of n + 1, It damps the adjustment at time n + 1. Here, the calculation of δ is expressed differently in different positions of the hidden layer and the output layer:

$$\delta_j = \begin{cases} (y_j - o_j) * o_j(1 - o_j) & for \quad output \quad neurons \\ o_j(1 - o_j) \sum_k \delta_k \omega_{kj} & for \quad hidden \quad neurons \end{cases}$$

2.3   Cost - Sensitive Classification of Back Propagation Learning Algorithm

Throughout the learning process of the back-propagation learning algorithm, it is not difficult to know that each neuron in the output layer of the neural network represents a possible class in the sample, and each class corresponding to an example in the sample has a maximum output value of the neuron in the output layer corresponding to it. In other words, the output of the neural network can also be viewed from the perspective of probability, Standardize o_i:

$$\forall_{i \in Class} : p(i) = \frac{o_i}{\sum\limits_{j \in Output} o_j}$$

P(i) here can be the estimated probability that an example in the sample belongs to class i.

In conceiving the modification, an important goal is that modified algorithm should be consistent with the original back propagation algorithm in the running process. Then, the most direct way to reduce the cost of misclassification is to modify only the probability estimation of the neural network when classifying the samples in the test sample while maintaining the integrity of the learning process. Replace the probability P(i) of the sample belonging to class i with the changed probability [24], this probability takes into account the expected cost of misclassification:

$$p'(i) = \frac{CostVector[i]p(i)}{\sum_j CostVector[j]p(j)}$$

It can be seen from this that when the cost of misclassification is increased, the new estimated probability is correspondingly increased, and the probability of misclassification is also reduced, which is beneficial to the class with high cost of expected misclassification.

3   Experiment

3.1   Experimental Data

Our experiment was performed on several well-known data sets in UCI database [26], and the characteristics of the data sets used in the experiment are shown in Table 2.

Table 2. Characteristics of data sets used in the experiment

Domain name	examp	attrib	classes	Degree of unbalance
Data_for_UCI_name	10000	13	2	1.8
Spambase	4601	57	2	1.53
Wilt	4339	5	2	57
HTRU_2	17898	8	2	10
App	5814	23	2	28

3.2   Experimental Process

The specific process of this experiment is as follows:

First, according to the imbalance degree of the overall samples in each data set, 20% of the data from positive samples and negative samples are randomly selected as the prediction samples. Second step, in accordance with the same method in the remaining ten percent samples randomly selected as training samples, the training sample = (total sample − prediction sample) * 10%, put into the neural network training, at this moment, the learning rate is equal to 0.5, the momentum of the parameter value is 0.3, cost matrix is set to the initial state, namely the misclassification cost in the cost matrix is equal. In the third step, 10% of the test samples were selected from the remaining data, and 50 random test samples were selected repeatedly. The mathematical model obtained in the second step was used to test the 50 test samples. The fourth step is to judge whether the number of misclassification of positive samples is zero. If it is not zero (generally not equal to zero), the value of the corresponding positive samples in the cost matrix is iteratively updated, that is, the second and third steps are repeated after adding "1" to the value of each generation until the number of misclassification of positive samples is zero. In the fifth step, we obtained 50 values that make the number of misclassification of positive samples equal to zero, and calculated their maximum value, median value and average value respectively. In the sixth step, the predicted samples were randomly selected for 10 times, and 40% of the predicted samples were selected each time and the maximum, median and average values obtained in step 5 were used for prediction.

3.3   Experimental Result

After the above experiments, the experimental results are as follows:

First, the cost of misclassification that can make the number of positive classes being misclassified as zero in the test sample after n iterations is obtained, and their maximum, median and average values are calculated respectively, as shown in Table 3.

Table 3. Correlation values of misclassification cost

	Data_for_UCI_name	Spambase	Wilt	HTRU_2
Max	18	**116**	3	62
Median	4.5	53.5	1	14
Mean	5.62	57.18	1.12	17.32

After obtaining the relevant data of misclassification cost, the expected cost of misclassification can be generated and applied to the actual algorithm to predict 10 groups of data randomly selected from the prediction samples, and 10 groups of confusion matrix can be obtained from each prediction sample, as shown in Tables 4, 5, 6 and 7.

Table 4. Data_for_UCI_name confusion matrix

Data_for_ UCI_name		Max		Median		Mean	
		class1	class2	class1	class2	class1	class2
(1)	class1	472	38	496	14	495	15
	class2	0	284	1	283	0	284
(2)	class1	475	35	495	15	489	21
	class2	0	284	0	284	0	284
(3)	class1	470	40	499	11	496	14
	class2	0	284	0	284	0	284
(4)	class1	470	40	496	14	487	23
	class2	0	284	0	284	0	284
(5)	class1	475	35	502	8	489	21
	class2	0	284	1	283	0	284
(6)	class1	470	40	495	15	493	17
	class2	0	284	1	283	0	284
(7)	class1	468	42	489	21	495	15
	class2	0	284	0	284	0	284
(8)	class1	466	44	497	13	490	20
	class2	0	284	0	284	0	284
(9)	class1	466	44	496	14	497	13
	class2	0	284	0	284	0	284
(10)	class1	471	39	496	14	490	20
	class2	0	284	1	283	0	284

Table 5. Spambase confusion matrix

Spambase		Max		Median		Mean	
		class1	class2	class1	class2	class1	class2
(1)	class1	24	199	21	202	25	198
	class2	0	145	0	145	0	145
(2)	class1	16	207	27	196	27	196
	class2	0	145	0	145	0	145
(3)	class1	24	199	22	201	28	195
	class2	0	145	0	145	0	145
(4)	class1	25	198	31	192	29	194
	class2	0	145	0	145	0	145
(5)	class1	22	201	25	198	33	190
	class2	0	145	0	145	0	145
(6)	class1	17	206	30	193	26	197
	class2	0	145	0	145	0	145
(7)	class1	18	205	34	189	30	193
	class2	0	145	0	145	0	145
(8)	class1	17	206	28	195	32	191
	class2	0	145	0	145	0	145
(9)	class1	22	201	34	189	29	194
	class2	0	145	0	145	0	145
(10)	class1	23	200	28	195	23	200
	class2	0	145	0	145	0	145

Table 6. Wilt confusion matrix

Wilt		Max		Median		Mean	
		class1	class2	class1	class2	class1	class2
(1)	class1	341	0	340	1	341	0
	class2	0	6	1	5	0	6
(2)	class1	341	0	340	1	341	0
	class2	0	6	0	6	0	6
(3)	class1	340	1	341	0	341	0
	class2	0	6	1	5	1	5
(4)	class1	341	0	341	0	340	1
	class2	0	6	1	5	0	6
(5)	class1	340	1	341	0	340	1
	class2	0	6	0	6	1	5
(6)	class1	340	1	341	0	340	1
	class2	0	6	0	6	0	6
(7)	class1	341	0	340	1	341	0
	class2	0	6	0	6	0	6
(8)	class1	341	0	341	0	341	0
	class2	0	6	0	6	1	5
(9)	class1	340	1	340	1	341	0
	class2	0	6	0	6	0	6
(10)	class1	341	0	341	0	341	0
	class2	0	6	0	6	0	6

Table 7. HTRU_2 confusion matrix

HTRU_2		Max		Median		Mean	
		class1	class2	class1	class2	class1	class2
(1)	class1	1251	50	1288	13	1283	18
	class2	0	131	0	131	0	131
(2)	class1	1249	52	1286	15	1280	21
	class2	0	131	0	131	0	131
(3)	class1	1240	61	1284	17	1284	17
	class2	0	131	0	131	0	131
(4)	class1	1241	60	1284	17	1281	20
	class2	0	131	0	131	0	131
(5)	class1	1242	59	1282	19	1275	26
	class2	0	131	0	131	0	131
(6)	class1	1241	60	1285	16	1284	17
	class2	0	131	0	131	0	131
(7)	class1	1236	65	1278	23	1281	20
	class2	0	131	0	131	0	131
(8)	class1	1245	56	1286	15	1280	21
	class2	0	131	0	131	0	131
(9)	class1	1239	62	1284	17	1279	22
	class2	0	131	0	131	0	131
(10)	class1	1240	61	1282	19	1282	19
	class2	0	131	0	131	0	131

As can be seen from the above experimental results, when median and mean are used as the cost of misclassification for minority samples, the number of misclassification of minority samples of Data_for_name and wilt data sets will not be reduced to zero. When using Max as the cost of misclassification, the above experimental data show that the number of misclassification of minority class for four groups of data, namely Data_for_name, spambase, wilt and HTRU_2, has been reduced to zero. In addition, although this method can search a value from the unknown cost, and this value is used as the cost in the cost sensitive algorithm, it can reduce the number of misclassification of positive class samples to zero. But there is a cost to this approach, because when the number of misclassification of positive samples is reduced to zero, the error rate of negative samples is improved, The degree of imbalance of the data is also an important factor affecting the cost of this method. From the experimental data, it can be seen intuitively that, the unbalance of spambase sample is smaller than that of other samples. Although minority class has not been misclassification in spambase samples, but can know from Table 3, when the algorithm used in spambase data set, the number of iterations is more than other samples, the cost found is too great, the number of negative samples misclassified is high, In other words, this method is too costly for the data approaching equilibrium, and may lose its practical significance.

In most cases, it is difficult to make an accurate estimate of the true cost. At present, the determination of cost is closely related to the number of all kinds of samples. Usually, the proportion of all kinds of samples is taken as the cost or a cost is generated directly and randomly, which cannot better reflect the real class distribution characteristics of data and guarantee the effect of cost sensitive learning. However, the idea in this paper is to search for a cost of misclassification that can make the number of misclassification of minority class samples to be zero in an iterative way, but it may not guarantee to minimize the average cost index. To deal with this "double-edged sword" problem it depends on our preferences in practical applications.

4 Summarize

The problem of unbalanced data classification has important research significance and application value in the field of machine learning. With the development of sparse representation and deep learning, a series of new research results have appeared on the problem of unbalanced data classification. The method described in this paper gives up the traditional idea of unbalanced data classification with the minimum average cost as the standard, and focuses on the basic purpose of reducing the number of misclassification of minority class sample to zero, and makes horizontal and vertical comparison of the results generated by different misclassification cost values. At the same time, it should be noted that this method also has a certain cost, which becomes more and more obvious as the data set tends to be balanced. However, the degree to which we accept this cost depends on a detailed analysis of the practical problems applicable to the method described in this paper, so as to obtain the most appropriate value for use.

References

1. Bhattacharya, S., Rajan, V., Shrivastava, H.: ICU mortality prediction: a classification algorithm for imbalanced datasets. In: AAAI, pp. 1288–1294 (2017)
2. Zakaryazad, A., Duman, E.: A profit-driven Artificial Neural Network (ANN) with applications to fraud detection and direct marketing. Neurocomputing **175**, 121–131 (2016)
3. Zhong, W., Raahemi, B., Liu, J.: Classifying peer-to-peer applications using imbalanced concept-adapting very fast decision tree on IP data stream. Peer-to-Peer Netw. Appl. **6**(3), 233–246 (2013)
4. Martin-Diaz, I., Morinigo-Sotelo, D., Duque-Perez, O., et al.: Early fault detection in induction motors using AdaBoost with imbalanced small data and optimized sampling. IEEE Trans. Ind. Appl. **53**(3), 3066–3075 (2017)
5. Pouyanfar, S., Chen, S.C.: Automatic video event detection for imbalance data using enhanced ensemble deep learning. Int. J. Semant. Comput. **11**(1), 85–109 (2017)
6. Daniels, Z.A., Metaxas, D.N.: Addressing imbalance in multi-label classification using structured hellinger forests. In: AAAI, pp. 1826–1832 (2017)
7. Buda, M., Maki, A., Mazurowski, M.A.: A systematic study of the class imbalance problem in convolutional neural networks. Neural Netw. **106**, 249–259 (2018)
8. Wagner, C., Saalmann, P., Hellingrath, B.: Machine condition monitoring and fault diagnostics with imbalanced data sets based on the KDD process. IFAC-PapersOnLine **49**(30), 296–301 (2016)
9. Oquab, M., Bottou, L., Laptev, I., et al.: Learning and transferring mid-level image representations using convolutional neural networks. In: IEEE Conference on Computer Vision and Pattern Recognition, pp. 1717–1724. IEEE Computer Society (2014)
10. Lin, W.C., Tsai, C.F., Hu, Y.H., et al.: Clustering-based undersampling in class-imbalanced data. Inf. Sci. **409**, 17–26 (2017)
11. Zhu, T., Lin, Y., Liu, Y.: Synthetic minority oversampling technique for multiclass imbalance problems. Pattern Recogn. **72**, 327–340 (2017)
12. Li, J., Fong, S., Wong, R.K., et al.: Adaptive multi-objective swarm fusion for imbalanced data classification. Inf. Fusion **39**, 1–24 (2018)
13. Hou, X., Zhang, T., Ji, L., et al.: Combating highly imbalanced steganalysis with small training samples using feature selection. J. Vis. Commun. Image Represent. **49**, 243–256 (2017)
14. Moayedikia, A., Ong, K.L., Boo, Y.L., et al.: Feature selection for high dimensional imbalanced class data using harmony search. Eng. Appl. Artif. Intell. **57**, 38–49 (2017)
15. Zhang, Z.L., Luo, X.G., Garca, S.: Cost-Sensitive back-propagation neural networks with binarization techniques in addressing multi-class problems and non-competent classifiers. Appl. Soft Comput. **56**(C), 357–367 (2017)
16. Zhou, Z.H., Liu, X.Y.: On multi-classcost-sensitive learning. In: Proceedings of the 21st National Conference on Artificial Intelligence. AAAI-06, pp. 567–572 (2006)
17. Chaki, S., Verma, A.K., Routray, A., et al.: A One class Classifier based Framework using SVDD: Application to an Imbalanced Geological Dataset (2016). arXiv preprint arXiv:1612.01349
18. Liu, X.Y., Wu, J., Zhou, Z.H.: Exploratory undersampling for class-imbalance learning. IEEE Trans. Syst. Man Cybern. Part B **39**(2), 539–550 (2009)
19. Krawczyk, B.: Learning from imbalanced data: open challenges and future directions. Prog. Artif. Intell. **5**(4), 221–232 (2016)
20. Fernández, A., del Río, S., Chawla, N.V., et al.: An insight into imbalanced big data classification: outcomes and challenge. Complex Intell. Syst. **3**(2), 105–120 (2017)

21. Haixiang, G., Yijing, L., Shang, J., et al.: Learning from class-imbalanced data: Review of methods and applications. Exp. Syst. Appl. **73**, 220–239 (2017)
22. He, H., Garcia, E.A.: Learning from imbalanced data. IEEE Trans. Knowl. Data Eng. **21**(9), 1263–1284 (2009)
23. Saitta, L. (ed.): Machine Learning - A Technological Roadmap. University of Amsterdam, The Netherland (2000)
24. Kukar, M., Kononenko, I.: Cost-sensitive learning with neural networks. In: Proceedings of the 13th European Conference on Artificial Intelligence, Brighton, UK, pp. 445–449 (1998)
25. Zhou, Z., Liu, X.: Training cost-sensitive neural networks with methods addressing the class imbalance problem. IEEE Trans. Knowl. Data Eng. **18**(1), 63–77 (2006)
26. Blake, C., Keogh, E., Merz, C.J.: UCI repository of machine learning databases, Department of Information and Computer Science, University of California, Irvine, CA (1998). [http://www.ics.uci.edu/~mlearn/MLRepository.html]

The Optimization Model of Comprehensive Maintenance Plan Based on Multi Demand Service Supply Chain

Jian-hua Yang[✉] and Long Guo

Donlinks School of Economics and Management,
University of Science and Technology, Beijing 100083, China
2482292135@qq.com

Abstract. Aiming at maintenance service chain with multiple service point requirement, this paper proposes an integrated maintenance plan optimization model in the condition of equipment maintenance by outsourcing companies. It analyzes resource allocation under multiple service point requirement from the perspective of service providers, explores the method of reasonable allocation of service personnel and devices used for maintenance under the circumstance of maintenance by outsourcing companies while meeting the constraint of different duration in multiple service points, and builds a resource allocation optimization model in multiple service points from the viewpoint of service demand enterprise. The model takes into consideration such constraints as service provider's trust degree, service duration, service cost and resource tightness, and proves with cases the effectiveness of resource allocation optimization model in multiple service points.

Keywords: Service resource allocation · Multiple service points · Maintenance

The group company generally has multiple or dozens of branch companies. As a result of diversified operation, business scopes of branches are rather complex, such as mining, clothing, and electronics. Large-scale production equipment is highly complex and professional, thereby increasing the complexity of maintenance. When problems occur, outsourcing companies are needed to provide maintenance services. In order to reduce overall maintenance cost and improve the reliability of maintenance service, the group company may perform unified management on equipment maintenance service. As branch companies are different in production order duration and pieces of equipment need to be maintained, there is the need of ensuring the lowest overall maintenance cost under the premise of not affecting the normal production of branch companies. Service resource allocation involves the dynamic scheduling of maintenance personnel, testing devices, spare parts and other resources among branch companies. As maintenance staff have different skills and are of different service levels, it is necessary to meet the production duration requirements of branch companies and ensure rational personnel allocation in order to reduce as much as possible the waste of human resource. According to learning curve, while maintaining multiple similar pieces of equipment, maintenance personnel with lower service level can gradually increase maintenance efficiency and reduce maintenance time. As there are

© Springer Nature Singapore Pte Ltd. 2020
M. Atiquzzaman et al. (Eds.): BDCPS 2019, AISC 1117, pp. 1783–1802, 2020.
https://doi.org/10.1007/978-981-15-2568-1_251

uncertainties with equipment maintenance, the estimation of maintenance time and cost is a very important problem in service resource allocation. In multiple service points, different equipment types and using time will affect the accuracy of estimated maintenance time and cost.

The resource allocation of outsourcing companies in multiple service points covers the reasonable scheduling of vehicles, devices used for maintenance and personnel provided by service provider. In this paper, devices used for maintenance and maintenance staff are taken as a work unit and scheduled as a whole. Different ways of personnel allocation can also affect maintenance time and cost. At present, equipment maintenance resource scheduling has attracted the attention of scholars at home and abroad. In similar situations, service demand comes from a number of branch companies, and service resource includes personnel, devices used for maintenance, and spare parts. Li [1] proposed an optimization model for stochastic resource allocation in manufacturing enterprises that can be widely used in logistics, production and operation. The model takes into account the characteristics of distribution problem, and studies model convergence and performance in simulation. With this model, enterprises can make effective decisions. Ekpenyong [2] proposed a model of generator maintenance program with reliable criteria. This model includes a time window for new constraint. Through empirical analysis, it is found that the model can get more reliable solutions. Bruni [3] analyzed resource-constrained project scheduling problem in uncertain activity period, where resource are reusable and activity period is indefinite. In order to solve the problem of flexible resource constrained product development project scheduling, Minmei [4] considered precedence constraint relations of tasks, necessary skills and flexible resource constraints in the process of project scheduling, and found solutions for the flexible resource allocation scheme of each task with maximum network flow theory. In order to solve the problem of shared resource allocation in multiple project management, Ning [5] determined equivalent efficiency conversion coefficient of enterprise resource in multiple projects with stochastic theory, and established a resource allocation efficiency model, realizing the rational allocation of resources. As for MRO service provider scheduling, Haibo [6] proposed an optimization maintenance scheduling method based on anticipated problems with equipment. In references [7–10], optimization allocation of maintenance resources in different fields was analyzed from the perspective of the availability of maintenance resources. In references [11] and [12], specific methods of maintenance strategies in preventive maintenance were put forward. In references [13, 14], collaborative resource was optimally allocated to complete tasks at scheduled time when the service provider is faced with an order.

It can be seen from previous research results that most researches did not take into consideration labor, devices used for maintenance and integrated vehicle scheduling during the service process. But actually, personnel, devices used for maintenance and transport vehicles in equipment maintenance by outsourcing companies are regarded as a whole, the lack of any of which may lead to the result that the whole process is optimal. In terms of resource allocation, existing researches also did not consider the degree of resource tightness. The tenser maintenance resource of outsourcing company is, resource allocation in service chain is more unstable, which makes it easy to have service recovery, thus affecting service effectiveness. At the same time, due to the uncertainty of equipment maintenance time and cost, a reasonable time and cost estimation is also helpful for making maintenance plan in resource allocation. Therefore, this paper takes into account above factors. By evaluating service time and cost in the view of service providers with case-based reasoning, this paper takes personnel and devices used for maintenance as an integrated service unit, and maintenance process as time advancement, so service resource can be allocated according to different time slices. In the view of service demand enterprise, the paper considers service provider's maintenance resource allocation plan for maintenance by company and by outsourcing companies, meets the trust, duration, cost and tightness of maintenance resource, and establishes a multi-facet requirement oriented integrated maintenance plan model in service chain.

1 Problem Description

In order to control maintenance cost, the group company initiates maintenance service request to branch companies at a regular interval t_0. Branch companies report the number of equipment to be maintained according to the operation status of equipment. The time advancement of equipment maintenance is shown in Fig. 1, where the horizontal axis means time. The time range before maintenance service starts is reporting period. When the reporting period ends, the group company selects appropriate maintenance service providers for its branches according to the number and type of equipment to be maintained. The group company needs to not only meet the operation requirements of branches, but also reduce overall service cost. In Fig. 1, A_i reports equipment maintenance at moment t_i, where $i = 1, 2, 3$. And t_0 to t_4 is resource allocation period, t_4 to t_5 equipment maintenance period, and time after t_5 production period after equipment maintenance.

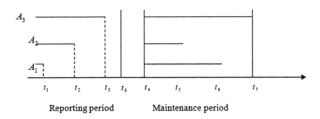

Fig. 1. Time for resource allocation in multiple service points

For resource network allocation and optimization of multiple service providers in multiple service point requirement, set requirement node set $A = \{a_{11}, a_{12}, \cdots, a_{mn}\}$, in which $i = 1, 2, \ldots m, j = 1, 2, \ldots n$, and a_{ij} means that component number j of branch company number i in maintenance service network needs to be maintained. Component is the smallest unit of maintenance task. Service provider node set

$S_l = \{s_{11}, s_{12}, \dots s_{lh}\}$, where s_{ij} means service provider number i provides resource number j. In multiple service point requirement, the total number of service providers is divided according to major types of equipment to be maintained. In the same service point, equipment to be maintained may belong to different categories. In different service points, even if pieces of equipment to be maintained are the same, components of equipment need for maintenance are not necessarily the same. So service tasks can be divided into several kinds. (1) In three or more service points, maintenance tasks of the same equipment are exactly the same and dependent on one another. (2) In three or more service points, maintenance tasks of the same equipment are exactly the same and independent from one another. (3) In three or more service points, both equipment to be maintained and maintenance tasks are different. The entire service resource allocation network is shown as Fig. 2. The figure is divided into three layers, with the top layer being the group company, the middle layer being branch companies, and the bottom layer being service providers. There are a total of three service providers and three demand subsidiaries. When service tasks are delivered, maintenance methods are determined according to the complexity of tasks. As there may be the same service tasks in different service points, when service resource is limited, personnel of independent equipment maintenance need to move among different service demand points, which may improve the efficiency of service resource.

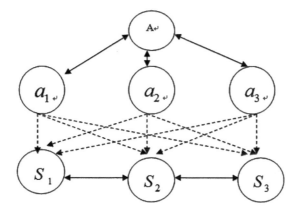

Fig. 2. Resource allocation network in multiple service points

When components need to be maintained or replaced are not particularly complex, service providers may send personnel to service demand points with testing devices, maintenance tools, and replaceable components. But when the process of components need to be maintained or replaced are rather complex, needing more testing devices and maintenance tools, it is more suitable for base maintenance, namely taking components to service supply point and after maintenance, sending them back to braches companies for installation and testing. The entire service process is shown in Fig. 3.

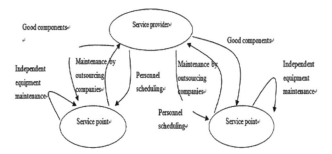

Fig. 3. Process of independent equipment maintenance and maintenance by outsourcing companies

2 Resource Scheduling of Maintenance by Outsourcing Companies in Multiple Service Points from the Perspective of Service Providers

As in many cases, the maintenance process of large-scale equipment is rather complex, professional testing tools and devices used for equipment are needed, and complex problems cannot be solved by sending maintenance personnel. Specific service process is shown in Fig. 4. In Fig. 4, there are pieces of equipment for maintenance by outsourcing companies in three service points. Dark squares show large-scale equipment to be maintained and light-colored squares small pieces of equipment to be maintained. Circles indicate non-critical equipment for maintenance by outsourcing companies, and rectangles indicate critical equipment for maintenance by outsourcing companies.

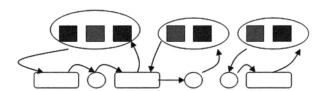

Fig. 4. Process of resource scheduling in maintenance by outsourcing companies

2.1 Premises

In formatting your page, set top margin to 25 mm and bottom margin to 31 mm. Left and right margins should be 20 mm. Page header should be 13 mm and page footer should be 20 mm. In document gridding menu, choose no gridding. Use a two-column format where each column is 21.95 characters wide and spacing 2 characters.

(1) Critical service resource of service providers is only available for one piece of equipment in one kind maintenance task. Non-critical service resources are unrestricted and meet the needs of all tasks in maintenance by outsourcing companies.

(2) In order to simplify the calculation, the load of medium-sized transport vehicle is set as one piece of large-scale equipment and one piece of small equipment. The loading capacity of a large transport vehicle is three pieces of large-scale equipment.
(3) Time needed for vehicle loading and unloading is not accounted.
(4) The durations of all equipment for maintenance by outsourcing companies in the same service point are the same.
(5) The use of resources cannot be interrupted in the process of maintenance. Other maintenance activities can be carried out only after the current maintenance task is completed.
(6) Time required for the transfer of equipment to be maintained between service resources is not taken into account.
(7) Due to transport vehicle constraint, available vehicles cannot transport all equipment to be maintained in one shot.
8) Lateral transport of vehicles among service points is not taken into consideration.

2.2   Basic Concepts

Critical service resource and non-critical service resource offered by service providers in maintenance by outsourcing companies

Definition 1: Task set of maintenance by outsourcing companies in service point A_i

$$A_i^0 = \{a_{i1}^0, a_{i2}^0, \ldots, a_{in}^0\} \tag{1}$$

Definition 2: Minimum completion time of maintenance by outsourcing companies

$$T_{ij}^{\min} = T_{ij}^0 + 3T_{ij}^1 \tag{2}$$

Definition 3: Equipment D_{ij} must be transported for the first time in service point A_i. The following conditions must be met:

$$T_{ij}^0 + 3T_{ij}^1 \leq T_i \tag{3}$$

$$T_{ij}^0 + 5T_{ij}^1 \geq T_i \tag{4}$$

In above formulas, T_{ij}^0 is the maintenance time of equipment number j for maintenance by outsourcing companies in service point A_i, T_{ij}^1 the first transportation time of equipment number j for maintenance by outsourcing companies in service point A_i, T_i completion time of maintenance by outsourcing companies, and $T_{ij}^0 + 3T_{ij}^1$ minimum completion time of equipment number j for maintenance by outsourcing companies in service point A_i.

2.3   Steps for Solution

Step 1 Service providers determine the number of equipment to be maintained and available vehicles at each service point based on customer's request for maintenance. Step 2 Estimate minimum completion time of all equipment for maintenance by outsourcing companies based on the number of equipment and maintenance tasks. The number of equipment to be transported for the first time must exceed the loading capacity of available vehicles offered by service provider, or else the maintenance task cannot be undertaken.
Step 3 The setting of time slice is also determined by the specific situation of the maintenance task, while considering maintenance time and the distance between service provider and multiple service points.
Step 4 For the first time slice, calculate key task sets in all service points, and assign key task sets without predecessor tasks to the task queue to be assigned, which will be allocated transport vehicle preferentially. Depending on the number of transport vehicles, resource scheduling can be divided into two cases. One is that only one transport vehicle is needed for the first time of transport. In this case, pieces of equipment for maintenance that have reached service point may occupy the same service resource, when the completion rate of critical task sets can be compared. If completion rates are the same, then compare the time of occupying critical service resources, the lesser of which will be allocated resources for equipment maintenance first. The other is that two transport vehicles are needed for the first time of transport, with different arrival time of equipment to be maintained. In this case, maintenance task that will be allocated maintenance resource and maintenance task on the way may occupy the same service resource. Completion rate can also be compared, with the transport time of completion rate calculation of equipment to be maintained not reaching service point included. And then non-key task sets with no predecessor tasks are put into the task queue to be allocated. For the selection of equipment not necessarily transported for the first time, the priority of respective task sets and occupancy of the same service resource of must be transported equipment should be taken into account. First select equipment that do not have to be transported for the first time and that occupies different resources with must be transported equipment, and then equipment with high task priority that do not have to be transported for the first time.
Step 5 From the second time slice on, service resource scheduling can be done in accordance with the method proposed in 2.2.
Step 6 When all maintenance tasks are completed, resource scheduling scheme in multiple service points is formed.

3   Service Cost Estimation from the Perspective of Service Providers

In the case of multiple service points, service provider undertakes different types of equipment maintenance services. In order to achieve customer satisfaction and ensure service benefits, service provider needs to make reasonable cost estimation for maintenance service. If the offer is too high, customers may choose other service providers. And if the offer is too low, the service provider cannot guarantee his own interests. As types and models of equipment used in the production process are different, it is a difficult problem to reasonably estimate problems with equipment and give customer a reasonable quotation. Traditional cost estimation of equipment maintenance mainly relies on expert experience, which is not accurate and reliable. But case-based reasoning is an important research direction in the field of artificial intelligence. Its main working principle is using similar past solutions to analyze current problems, and making certain adjustments according to differences between previous and current situation, thus getting the specific solution to the current problem. Case-based reasoning makes information acquisition more simple and feasible, and can improve the accuracy and reliability of problem-solving. Currently, case-based reasoning has been widely used in logistics outsourcing, management decision-making, software design and other fields, and is attracting more and more attention from experts in related fields. Core problems of this method are setting case attribute index, setting similarity measurement method, and adjusting estimated result. In this paper, case-based reasoning is applied to the cost evaluation of equipment maintenance. It is found that this method can effectively estimate the cost of different kinds of equipment and complex problems.

3.1 Steps for Equipment Maintenance Cost Estimation with Case-Based Reasoning

CBR technology takes equipment maintenance projects in case database as references, retrieves the most similar cases with current problem according to similarity method, and makes adjustments to estimated result based on specific spare parts and personnel allocation to get the results of estimated cost of current project.

Step 1 Take current equipment maintenance cost to be estimated target case, and attribute value to target case by experts.
Step 2 Decompose target case to equipment level, and then retrieve case database to find a case similar to current equipment to be maintained.
Step 3 Retrieve similar cases according to adjustment principle, modify quotation based on similar cases, and obtain the quotation of current project.
Step 4 Save new case to case database.

3.2 Case Description

Ways of case description affect the accuracy of retrieval. Commonly used case description methods are text description, list description and structured description. This paper uses mathematical structure to describe cases. The top layer is the industry where service providers lie, such as electrical equipment, transportation equipment, machinery manufacturing equipment, and garment processing equipment. The second layer is the scope of services. Take service processing field as an example, its service scope includes CAD software for garment design, NC cutting for cutting pieces, and intelligent clothing hanging system for production. The third layer is equipment brands under different service scopes, such as Sweden ETON brand in smart hanging system in garment production. The fourth layer is problems encountered with different equipment for current service, such as main track problems, receiving orbital problems, intelligent terminal problems, and transmission chain problems in suspension system. The fifth layer is main reasons for the existence of problems. There may be many reasons. The sixth layer is using different methods for different reasons. The seventh layer is offering quotation based on different methods. Specific methods include personnel allocation and the use of spare parts in current case. The whole case can be divided into three levels, namely case level, device level, and component level. Quotations stored here are also for different components. Many types of equipment are involved in service process, so problems will occur in different parts of equipment. Therefore, the number of cases meeting search requirements is very limited by retrieving in full accordance with case similarity.

3.3 Measurement of Case Similarity

The core of case retrieval technology is the similarity between current project and previous cases. At present, similarity measurement method is mainly K-nearest neighbor algorithm. This paper uses this method to analyze the similarity between projects. This method is an independent method based on distance. It first sets weight to the attribute of target case and then uses weighted method to estimate the similarity between target project and previous cases. In most literature studies, the similarity between two projects was used to select case. In this study, similarity analysis is conducted at equipment level to ensure the accuracy of maintenance quotation.

Commonly used distance-based metric functions include Euclidean distance, Manhattan distance and Minkowski distance. Specific functions are as follows:
Euclidean distance:

$$Dis\tan ce(P_a, P_b) = \sqrt{\sum_{i=1}^{m} (f_{xi} - f_{yi})^2} \tag{5}$$

Manhattan distance:

$$Dis\tan ce(P_a, P_b) = \sum_{i=1}^{m} |f_{xi} - f_{yi}|^2 \tag{6}$$

Minkowski distance

$$Dis\tan ce(P_a, P_b) = \sqrt[e]{\sum_{i=1}^{m} |f_{xi} - f_{yi}|^e} \tag{7}$$

In existing researches, a great part of case-based reasoning used Euclidean distance to set the weight of case attribute index. The European Distance of Case P_a and Case P_b is

$$Dis\tan ce(P_a, P_b) = \sqrt{\sum_{s=1}^{S} (W_s \times |f_{xs} - f_{ys}|)^2} \tag{8}$$

Similarity measurement between cases is as follow:

$$SIM_{pi}(P_a, P_b) = 1 - d_{pi}(P_a, P_b) \tag{9}$$

Based on above researches, this paper presents the similarity method of equipment maintenance service cost estimation, and divides maintenance service into three levels, i.e., case level, equipment level and component level. In equipment maintenance, component level is usually employed for problem analysis, personnel allocation, and component replacement. In cost estimation, the whole equipment is usually taken as calculation unit, so the whole equipment is also used as the unit of case similarity calculation. In case database, there are few cases that have exactly the same equipment and problems as those in the current case, and equipment similarity is calculated by integrating the similarity of each component. Therefore, when calculating the similarity of two pieces of equipment to be maintained, it is necessary to determine the similarity of each attribute index of components. The similarity of components is first calculated with Euclidean distance, and then equipment similarity is comprehensively calculated in accordance with component-level problems corresponding to equipment brand.

$$SIM_{pi}(P_a, P_b) = 1 - d_{pi}(P_a, P_b) \tag{10}$$

$$SIM_{dm}(P_a, P_b) = \sum_{i=1}^{n} \frac{SIM_{pi}(P_a, P_b)}{b_{pi}} \tag{11}$$

In Eq. (10), $SIM_{pi}(P_a, P_b)$ is the similarity between component number i to be maintained in case P_a and P_b at component level, and $d_{pi}(P_a, P_b)$ the distance between P_a and P_b in component i at component level. In Eq. (11), $SIM_{dm}(P_a, P_b)$ means the similarity between equipment number m between P_a and P_b at equipment level, and b_{pi} brand value of the equipment that component number i belongs to.

3.4   Case Adjustment

Case adjustment is a very important step in service provider cost evaluation. It is mainly targeted at case cost adjustment. Cost evaluation is usually calculated at equipment level. Current method for case adjustment mainly employs knowledge updating and strategy combination in the field. Knowledge updating is to filter unrelated attributes. Strategy combination is to seek solution with several similar equipment-level cost combinations in case database.

In this study, the combination of strategies is used to estimate the cost of a project because it is very unlikely that equipment in current project and project in case database are exactly the same. First of all, the total number of equipment to be maintained in current project and all problems with each piece of equipment are obtained. For each piece of equipment to be maintained in the current project, equipment retrieval is carried out in case database to obtain related equipment set with the highest similarity to current equipment. Then the cost of current equipment maintenance is estimated and the cost of all equipment in the current project is estimated, thus getting estimated cost of current project. Changes in maintenance cost are also highly relevant to current number of available staff and brands of spare parts. And cost of different brands of spare parts may change, too.

Estimated cost of equipment maintenance is:

$$C_d = \sum_{k=1}^{K} \frac{sim(d, d_k)}{\sum\limits_{k=1}^{K} sim(d, d_k)} (C_{d_k} + C_r - C_{kr} + \sum_{j=1}^{e} (C_{sj} - C_{ksj})) \tag{12}$$

In Eq. (12), C_d is cost of current equipment to be estimated, d current equipment to be estimated, d_k case equipment number k that is similar with current equipment to be estimated, C_{d_k} maintenance cost of case number k, and $sim(d, d_k)$ the similarity between current equipment and case equipment. Besides, C_r refers to personnel cost estimated based on available number of personnel for current equipment maintenance, C_{kr} personnel cost of available number of personnel for case equipment maintenance, e the number of problems with current equipment, C_{sj} cost of available spare parts for problem number j in current equipment maintenance, and C_{ksj} cost of available spare parts for problem number j in case equipment maintenance.

Estimated cost of maintenance project is:

$$C_p = \sum_{i=1}^{m} C_{di} \tag{13}$$

4   Resource Tightness

In the process of service chain resource allocation, in order to prevent the delay of service resulting from resource interruption, resource tightness needs to be considered. The tighter service resources are, the more likely service interruption occurs. Resource tightness is widely used in the process of project management. ZHU Chunchao employed peak ratio of resources used to measure the tension of resources:

$$\alpha_i = \max \sum_{k=1}^{m} r_{kt}/R_t, \ t \in [ST_i, \ ST_i + D_i] \tag{14}$$

In project buffer setting, Qian considered resource tightness, which was described with a fuzzy set. In the process of service chain resource allocation, this paper considers the tightness of transportation vehicles, service personnel, service resource and spare parts in maintenance by outsourcing companies.

Assume i types of vehicles are used in maintenance by outsourcing companies in multiple service points, vehicle resource tightness is:

$$\alpha_c = \max\left(\frac{c_i}{ac_i} - n_i\right) \tag{15}$$

c_i is the number of vehicles of type i used in multiple service demand points, ac_i the number of available vehicles of type i used by service provider, and n_i the substitutability of type i vehicles, where $0 \leq n_i \leq 1$.

Assume j kinds of service personnel are called in independent equipment maintenance in multiple service demand points, the tightness of service personnel is:

$$\alpha_r = \max\left(\frac{r_j}{ar_j} - l_j\right) \tag{16}$$

r_j is the number of type j personnel distributed in multiple service demand points, ar_j the number of available type j vehicles used by service provider, and l_j the substitutability of type j vehicles, where $0 \leq n_i \leq 1$.

Resource tightness of maintenance by outsourcing companies is:

$$\alpha_m = \max\left(\frac{m_s}{am_s}\right) \tag{17}$$

Spare parts tightness is also closely related to the number of spare parts, but estimated number of spare parts provided by service demand enterprise to service provider is not very accurate, which can be evaluated by experts according to the scale of service provider:

$$\alpha_{sp} = \alpha_{sp}^b \tag{18}$$

b is scale of specific enterprise and is segmented according to specific industries. Overall resource tightness of service provider is:

$$\alpha = \omega_1\alpha_c + \omega_2\alpha_r + \omega_3\alpha_m + \omega_4\alpha_{sp} \tag{19}$$

ω_i means the weight of the tightness of resource type number i.

5   Trust Based Resource Allocation Model in Multiple Service Points

Based on the train of thought and process of service provider resource allocation in the case of multiple service demand points in above-mentioned sections, this section establishes a corresponding model according to the requirements of different trust degrees, different service tasks, different completion times and different costs, which constitute the framework of service chain resource allocation model under the whole multiple service point requirement. The assumptions of the model are as follows:

(1) The distance between each service demand point and service supply point is known.
(2) Service chain resource allocation includes independent maintenance and maintenance by outsourcing companies.
(3) Service demand enterprise can effectively obtain the resources tightness of service provider's vehicles, personnel, service resource and spare parts.
(4) Service demand enterprise can effectively decompose maintenance task to equipment to be maintained in accordance with maintenance standards.
(5) The service durations of independent maintenance and maintenance by out sourcing companies in service demand points are the same.
(6) All service tasks of the same type of equipment are performed by one service provider, regardless of the transport and maintenance tasks of the same type of service equipment between different service providers.
(7) The subsidiaries of the group company are located in the same large service area, that is, the distance between subsidiaries is close, which makes resource allocation in multiple service points have practical significance.

According to the service demand of each subsidiary company, the group company regularly carries out service tender and obtains relevant information of each service provider from the bidding, thus establishing service chain resource allocation model in multiple service points. In the process of modeling, not only independent maintenance cost, cost of maintenance by outsourcing companies, service time and trust degree of potential service providers, but also the sufficiency of personnel, devices used for maintenance, spare parts, transport vehicles, that is, resource tightness should be taken into consideration. The meanings of the following symbols are:

MTP_i is service type number i in all service demand points, where $i = 1, 2, \ldots, n$. Equipment to be maintained of the same service type of is performed by one service provider;

$P_i = \{p_{i1}, p_{i2}, \ldots, p_{ik_i}\}$ means the set of bidders that assume all maintenance tasks of type MTP_i;

k_i is the number of bidders that assume all maintenance tasks of type MTP_i;

$p_{ir} \in \{0, 1\}$:1 are maintenance tasks of type MTP_i that bidder p_{ir} is chosen to assume. Bidder p_{ir} does not assume any maintenance task;

c_{ijr} is the cost of completing service task number j of type MTP_i by independent maintenance personnel sent by bidder p_{ir}, where $j = 1, 2, \ldots, m$;

dm_{ir} refers to personnel transfer cost after all independent maintenance tasks of type MTP_i are completed by p_{ir};

dc_{ir} is vehicle transport cost after all tasks of type MTP_i in maintenance by outsourcing companies are completed by p_{ir};

mc_{ibr} means cost of task number b in maintenance by outsourcing companies when all tasks of type MTP_i in maintenance by outsourcing companies are completed by p_{ir};

TP_{ir} is the trust degree of p_{ir} when all service tasks of MTP_i are completed by p_{ir};

MT_{sir} means the final completion time of all independent maintenance tasks of type MTP_i in service point A_s assumed by p_{ir};

LT_{sir} means the final completion time of all tasks of type MTP_i in maintenance by outsourcing companies in service point A_s assumed by p_{ir};

α_{ir} is the resource tightness of p_{ir} when all service tasks of MTP_i are completed by p_{ir};

ADT_{si} is service completion time of type MTP_i in service point A_s;

MST is the minimum overall trust degree of service chain;

E is the maximum overall resource tightness of service chain in multiple service points

Based on descriptions and assumptions of above model, a resource chain resource allocation model in multiple service points with the lowest transportation cost and maintenance cost can be established.

Objective functions are:

$$minC = \sum_{i=1}^{n}\sum_{j=1}^{m}\sum_{r=1}^{k_i} c_{ijr}p_{ir} + \sum_{i=1}^{n}(dm_{ir}p_{ir} + dc_{ir}p_{ir}) + \sum_{i=1}^{n}\sum_{b=1}^{l}\sum_{r=1}^{k_i} mc_{ibr}p_{ir} \quad (20)$$

$$\sum_{r=1}^{k_i} p_{ir} = 1, p_{ir} \in \{0,1\}, p_{ir} \in P_i \quad (21)$$

$$\prod_{i=1}^{n}\sum_{r=1}^{k_i} TP_{ir}p_{ir} \geq MST \quad (22)$$

$$\max\left(\sum_{r=1}^{k_i} MT_{sir}p_{ir}, \sum_{r=1}^{k_i} LT_{sir}p_{ir}\right) \leq ADT_{si} \quad (23)$$

$$\sum_{i=1}^{n}\sum_{r=1}^{k_i} \alpha_{ir}p_{ir} \leq E \quad (24)$$

Objective function (20) is the maintenance cost and transportation cost of independent maintenance and maintenance by outsourcing companies in the service chain of multiple service points. Constraint (21) means that only one candidate resource is guaranteed for each maintenance task and is added to the service chain. Constraint (22) means that the overall trust degree of service chain should be greater than set minimum value. Constraint (23) ensures that tasks in each service point are completed on time. And constraint (24) shows that the overall resource chain tension should be less than the set maximum value.

6   Case Analysis

The group company has 3 subsidiaries in the same region. During the same tender period, subsidiary companies report maintenance service requirements to the group company, and provides detailed list of equipment to be maintained, including key problems with equipment, required performance and so on. According to the specific situations of equipment to be maintained and service type, the group company classifies service demands of subsidiaries into five categories. Each type of service task is assumed by one service provider. First of all, internal resource allocations of service providers are analyzed from the perspective of service providers. According to customers' service needs, a total of two transport vehicles are provided, including one large truck and one medium truck. The group company has two maintenance demand points A_1 and A_2. A total of three pieces of large equipment and two pieces of small equipment need maintenance by outsourcing companies, whose duration is 23 time slices. The distance between A_1 and service point is 2 time slices. Two pieces of large equipment and one small equipment need maintenance by outsourcing companies in A_2, whose duration is 25 time slices. The distance between A_2 and service point is 2 time slices. M_{11}, M_{12}, M_{13}, M_{21} and M_{22} are large equipment. M_{14}, M_{15} and M_{23} are small equipment. Maintenance task can be expressed as a_{ij}^{nl}, sign of service point as i, sign of equipment to be maintained j, task sign as n, and sign of resource need to occupied by corresponding task as l. There are five kinds of resources in service point, of which the first and the second kind are key service resources, and the third, fourth and fifth kind are sufficient service resources. All maintenance tasks can be represented as follows:

$$M_{11} = \{(a_{11}^{11}, 5), (a_{11}^{23}, 4), (a_{11}^{35}, 6)\}, \quad M_{12} = \{(a_{12}^{12}, 2), (a_{12}^{23}, 8), (a_{12}^{34}, 5)\},$$
$$M_{13} = \{(a_{13}^{12}, 6)\}$$
$$M_{14} = \{(a_{14}^{11}, 2), (a_{14}^{23}, 3)\}, \quad M_{15} = \{(a_{15}^{13}, 3), (a_{15}^{25}, 4)\}$$
$$M_{21} = \{(a_{21}^{12}, 2), (a_{21}^{24}, 4)\}, \quad M_{22} = \{(a_{22}^{13}, 2), (a_{22}^{25}, 5)\},$$
$$M_{23} = \{(a_{23}^{11}, 2), (a_{23}^{23}, 3)\}$$

With above formulae, total maintenance time in all service points is obtained.
$M_1 = \{(a_{11}, 15), (a_{12}, 15), (a_{13}, 6), (a_{14}, 5), (a_{15}, 7)\}$
$M_2 = \{(a_{21}, 6), (a_{22}, 7), (a_{23}, 5)\}$. According to Definition 3, a_{11} and a_{12} are equipment must be transported for the first time in A_1. There are no equipment that must be transported for the first time in A_2. The selection of equipment transported for the first time is based on the work priority of task sets and the occupancy of service equipment in service demand points. Both a_{11} and a_{12} are equipment must be transported for the first time and have a higher priority of resource occupancy. As a_{11} and

a_{12} are large equipment, the first transport of A_1 needs large transport vehicles to load equipment a_{11}, a_{12} and. The first task of a_{11} needs to occupy the first critical service resource, and the first task of a_{12} the second critical service resource. Therefore, medium-sized transport vehicle is needed for the first transportation of A_2. Taking into account the loading of vehicles, loaded pieces of equipment tobe maintained are a_{22} and a_{23}.

The first to the fourth time slice is vehicle transport time, namely time needed for transport vehicles to drive from service demand point, load equipment, and drive back to service point.

(1) The 5th time slice

Maintenance tasks a_{11}, a_{12}, a_{15}, a_{22} and a_{23} enter task queue to be allocated, after which equipment to be maintained will be allocated with service resources. As both a_{11} and a_{23} occupy the first key service resource, through calculating key task set, the priority of service resource allocation is given to a_{11}. Based on task queue to be allocated in the 5th time slice, allocation results for maintenance tasks are shown in Table 1. After being unloaded, vehicles need to transport again equipment for maintenance in service demand point. Since two vehicles can meet remaining transportation requirements in A_1 and A_2, there will be no vehicle priority selection.

Table 1. Resource allocation for the 5th time slice

Task	a_{11}^{11}	a_{12}^{12}	a_{15}^{13}	a_{22}^{13}	a_{23}^{11}
Service resource	s_1^0	s_2^0	s_3^1	s_3^1	NULL
Proportion of completion	$\frac{1}{5}$	$\frac{1}{2}$	$\frac{1}{3}$	$\frac{1}{2}$	0

(2) The 8th time slice

When the 8th time slice ends, transport vehicles will transport remaining equipment in demand point to service point. At this time, the completion amount of equipment for which resources have been allocated is shown in Table 2.

Table 2. Resource allocation for the 8th time slice

Task	a_{11}^{11}	a_{12}^{12}	a_{15}^{13}	a_{22}^{13}	a_{23}^{11}
Service resource	s_1^0	s_2^0	s_3^1	s_3^1	NULL
Proportion of completion	$\frac{4}{5}$	1	1	1	0

(3) The 9th time slice

Completed maintenance tasks release resources it occupied. Subsequent tasks and remaining equipment that will soon arrive at service point are put into task queue to be allocated, waiting for corresponding resource allocation. Both a_{13} and a_{21} need to occupy the second key service resource. With priority comparison through calculating task sets, a_{13} and a_{21} are equal, so the next step is to compare the priority of subsequent tasks after a_{13} and a_{21}. In the first case, there are subsequent tasks for both a_{13} and a_{21}, so the task with subsequent task needing to occupy key resources will be given priority to resource allocation. In the second case, one of them has an immediate subsequent task and will be preferentially allocated with resources. In the third case, there is no subsequent task for both a_{13} and a_{21}, so the priority is given to large equipment. Task a_{14} will wait for the first kind of critical service resource (Tables 3 and 4).

Table 3. Resource allocation for the 9th time slice

Task	a_{11}^{11}	a_{12}^{23}	a_{13}^{12}	a_{14}^{11}	a_{15}^{25}	a_{21}^{12}	a_{22}^{25}	a_{23}^{11}
Service resource	s_1^0	s_3^1	NULL	NULL	s_5^1	s_2^0	s_5^1	NULL
Proportion of completion	1	$\frac{1}{8}$	0	0	$\frac{1}{4}$	$\frac{1}{2}$	$\frac{1}{5}$	0

(4) The 10th time slice

When task a_{11}^{11} is completed, the first kind of critical service resource is released. But both a_{14}^{11} and a_{23}^{11} need to occupy the first kind of critical service resource. After comparing critical task sets of equipment for maintenance, a_{14}^{11} is chosen to enter resource allocation queue.

Table 4. Resource allocation for the 10th time slice

Task	a_{11}^{23}	a_{12}^{23}	a_{13}^{12}	a_{14}^{11}	a_{15}^{25}	a_{21}^{12}	a_{22}^{25}	a_{23}^{11}
Service resource	s_3^1	s_3^1	NULL	s_1^0	s_5^1	s_2^0	s_5^1	NULL
Proportion of completion	$\frac{1}{4}$	$\frac{2}{8}$	0	$\frac{1}{2}$	$\frac{2}{4}$	1	$\frac{2}{5}$	0

Table 5 shows the completion time of all equipment for maintenance with above mentioned method. After the 16th time slice ends, large vehicle will transport equipment M_{21}, M_{22} and M_{23} to demand point A_2. The final completion time of demand point A_2 is the 18th time slice. When the 21st time slice ends, large and medium vehicles will deliver M_{11}, M_{12}, M_{13}, M_{14} and M_{15} to demand point A_1. The final completion time of demand point A_1 is the 23rd time slice.

Table 5. Maintenance completion time for all equipment

Completion time slice	12th	13th	14th	16th	19th	21st
Equipment maintained	M_{15}	M_{22}	M_{14} M_{21}	M_{13} M_{23}	M_{11}	M_{12}

Service provider allocates personnel, vehicles, and devices used for maintenance according to resource allocation method presented in this paper, and submits service cost and time calculated with this distribution strategy to the group company. In 5 types of tasks, there are 3 independent maintenance tasks and 2 maintenance tasks by outsourcing companies in the first type, 2 independent maintenance tasks and 4 maintenance tasks by outsourcing companies in the second type, and 3 independent maintenance tasks and 3 maintenance tasks by outsourcing companies in the third type. Service providers meeting service requirements can be expressed as [3, 4], with 3 standby service resources in the first task type, 3 standby service resources in the second task type, and 4 standby service resources in the third task type. Due to length limit, this article only lists service provider data in service demand point A_1, whose details are shown in Tables 5, 6 and 7. Genetic algorithm put forward by Aiello [7] is used to find solution. The optimal allocation of resources in service chain is [1, 2] (Table 8).

Table 6. Service costs of service providers in service demand point A_1

Independent maintenance	
T_{11}^1	$(p_{11}, 1.8)$ $(p_{12}, 1.5)$ $(p_{13}, 2.3)$
T_{12}^1	$(p_{11}, 0.3)$ $(p_{12}, 0.55)$ $(p_{13}, 0.32)$
T_{13}^1	$(p_{11}, 0.78)$ $(p_{12}, 1.2)$ $(p_{13}, 1.5)$
T_{21}^1	$(p_{21}, 1.8)$ $(p_{22}, 2.3)$ $(p_{23}, 1.2)$
T_{22}^1	$(p_{21}, 3.5)$ $(p_{22}, 3.1)$ $(p_{23}, 3.5)$
T_{31}^1	$(p_{31}, 6.9)(p_{32}, 7.2)(p_{33}, 6.9)(p_{34}, 5.3)$
T_{32}^1	$(p_{31}, 3.2)(p_{32}, 3.9)(p_{33}, 6.9)(p_{34}, 5.3)$
T_{33}^1	$(p_{31}, 10.3)(p_{32}, 11.6)(p_{33}, 8.9)(p_{34}, 9.6)$
Maintenance by outsourcing companies	
T_{14}^1	$(p_{11}, 12.6)$ $(p_{12}, 15)$ $(p_{13}, 11.6)$
T_{15}^1	$(p_{11}, 16.7)$ $(p_{12}, 18)$ $(p_{13}, 15.7)$
T_{23}^1	$(p_{21}, 2.7)$ $(p_{22}, 3.1)$ $(p_{23}, 2.1)$
T_{24}^1	$(p_{21}, 17.1)$ $(p_{22}, 15.8)$ $(p_{23}, 17.2)$
T_{25}^1	$(p_{21}, 5.2)$ $(p_{22}, 7.3)$ $(p_{23}, 5.1)$
T_{26}^1	$(p_{21}, 20)$ $(p_{22}, 15)$ $(p_{23}, 22)$
T_{31}^1	$(p_{31}, 3.8)(p_{32}, 2.8)(p_{33}, 4.3)(p_{34}, 3.7)$
T_{32}^1	$(p_{31}, 17.1)(p_{32}, 19)(p_{33}, 31)(p_{34}, 23)$

Table 7. Service time and duration of service providers in service demand point A_1

Independent maintenance	
$(T_1^1, 15)$	$(p_{11}, 10)$ $(p_{12}, 12)$ $(p_{13}, 9)$
$(T_2^1, 26)$	$(p_{21}, 17)$ $(p_{22}, 21)$ $(p_{23}, 25)$
$(T_3^1, 10)$	$(p_{31}, 5)$ $(p_{32}, 3)$ $(p_{33}, 3)$ $(p_{34}, 8)$
Maintenance by outsourcing companies	
$(T_1^1, 15)$	$(p_{11}, 11)$ $(p_{12}, 15)$ $(p_{13}, 8)$
$(T_2^1, 26)$	$(p_{21}, 15)$ $(p_{22}, 23)$ $(p_{23}, 22)$
$(T_3^1, 10)$	$(p_{31}, 7)$ $(p_{32}, 5)$ $(p_{33}, 9)$

Table 8. Service chain resource related data

Personnel transfer cost	Vehicle transportation cost
$(p_{11}, 0.2)$ $(p_{12}, 0.5)$ $(p_{13}, 0.3)$	$(p_{11}, 2)(p_{12}, 3.2)(p_{13}, 1.6)$
$(p_{21}, 3.1)$ $(p_{22}, 1.9)$ $(p_{23}, 3.9)$	$(p_{21}, 5.6)(p_{22}, 3.5)(p_{23}, 3.7)$
$(p_{31}, 0.7)$ $(p_{32}, 0.5)$ $(p_{33}, 1.2)$	$(p_{31}, 1.9)$ $(p_{32}, 3.3)$ $(p_{33}, 1.6)$
Trust degree	Resource tightness
$(p_{11}, 0.52)(p_{12}, 0.31)(p_{13}, 0.72)$	$(p_{11}, 0.03)(p_{12}, 0.38)(p_{13}, 0.15)$
$(p_{21}, 0.76)(p_{22}, 0.83)(p_{23}, 0.61)$	$(p_{21}, 0.59)(p_{22}, 0.17)(p_{23}, 0.62)$
$(p_{31}, 0.9)(p_{32}, 0.38)(p_{33}, 0.79)$	$(p_{31}, 0.32)(p_{32}, 0.53)(p_{33}, 0.16)$

7 Conclusions

In order to meet group companies' need of regular equipment maintenance in multiple service points and reduce their overall maintenance cost, this paper discusses service chain resource allocation in terms of multi-facet requirements instead of a single requirement, analyzes the rational allocation of vehicles, personnel and devices used for maintenance in independent maintenance and maintenance by outsourcing companies from the perspective of service providers, takes into account the differences between non-critical maintenance resources and key maintenance resources, and proposes a rational resource allocation method. The paper estimates service cost and duration with case-based reasoning method, and establishes a multi-facet requirement oriented integrated resource allocation optimization model in multiple service points in the view of service demand enterprise. The model takes into consideration such constraints as service provider's resource tightness, trust degree, service duration, and service cost, and proves the effectiveness of optimization model of integrated equipment by outsourcing companies in multiple service points, which has certain practical value. In order to simplify the model and process of finding solution, this paper puts limits on service providers' key resources and transportation vehicles. Future researches may consider more general resource-constrained maintenance plan optimization.

References

1. Li, F., Lida, X., Jin, C., Wang, H.: Random assignment method based on genetic algorithms and its application in resource allocation. Expert Syst. Appl. **39**, 12213–12219 (2012)
2. Ekpenyong, U.E., Zhang, J., Xia, X.: An improved robust model for generator maintenance scheduling. Electr. Power Syst. Res. **92**, 29–36 (2012)
3. Bruni, M.E., Beraldi, P., Guerriero, F., Pinto, E.: A heuristic approach for resource constrained project scheduling with uncertain activity durations. Comput. Oper. Res. **38**, 1305–1318 (2011)
4. Huang, M., Luo, R.: Study on optimal scheduling of product development projects under flexible resource constraints. J. Ind. Eng. Eng. Manag. **24**(4), 143–154 (2010)
5. Chen, N., Zhang, X., Wu, Z., Chen, S.: Multi-project resource allocation model based on stochastic theory and its application. Chin. J. Manag. **14**(4), 75–80 (2006)

6. Tang, H., Ye, C.: Pre-maintenance scheduling of equipment based on MRO service providers. J. Syst. Manag. **21**(3), 336–351 (2012)
7. Jha, M.K., Shariat, S., Abdullah, J., Devkota, B.: Maximinzing resource effectiveness of highway infrastructure maintenance inspection and scheduling for efficient city logistics operations. Procedia-Soc. Behav. Sci. **39**, 831–844 (2012)
8. Martorell, S., Villamizar, M., Carlos, S., Sanchez, A.: Maintenance modeling and optimization integrating human and material resources. Reliab. Eng. Syst. Saf. **95**, 1293–1299 (2010)
9. Ashayeri, J.: Development of computer-aided maintenance resources planning: a case of multiple CNC machining centers. Robot. Comput.-Integr. Manuf. **23**, 614–623 (2007)
10. de Castro, H.F., Cavalca, K.L.: Maintenance resources optimization applied to a manufacturing system. Reliab. Eng. Syst. Saf. **91**, 413–420 (2006)
11. Mollahassani-pour, M., Abdollahi, A., Rashidinejad, M.: Investigation of market-based demand response impacts on security-constrained preventive maintenance scheduling. IEEE Syst. J. **9**(4), 1496–1506 (2015)
12. Regattieri, A., Giazzi, A., Gamberi, M., et al.: An innovative method to optimize the maintenance policies in an aircraft: general framework and case study. J. Air Transp. Manag. **44**, 8–20 (2015)
13. Huang, J.-J., Chen, C.-Y., Liu, H.-H., Tzeng, G.-H.: A multiobjective programming model for partner selection-perspectives of objective synergies and resource allocations. Expert Syst. Appl. **37**, 3530–3536 (2010)
14. Macchion, L., Moretto, A., Caniato, F., et al.: Production and supply network strategies within the fashion industry. Int. J. Prod. Econ. **163**, 173–188 (2015)

The Improvement of News Communication Mode by Data Mining and Visualization Technology

Jiao Yang[(✉)]

Dianchi College of Yunnan University, Kunming 650228, Yunnan, China
guanjuchen@126.com

Abstract. With the development of technologies, the emergence of the data mining and the visualization technology is quietly affecting people's daily life. How to make better use of this technology to make rational use of the information has become a problem faced by all walks of life. At the same time, the mode of news communication has also changed, from the contents of the report and the ways of presentation to the production process. Based on this, this paper studies the improvement of the news communication mode by the data mining and the visualization technology from the aspects of accelerating the process of the media integration, using the social TV products to attract audiences, the digital communication and the network communication, strengthening the visual news communication, and increasing the interactive participation and so on. The use of the data mining and the visualization technology for the traditional news communication has brought a new revolution, and also has brought new vitality for the impact of the new media in the declining news industry.

Keywords: Data mining · Visualization technology · News communication · Mode

CLC Number: G206, Document Identification Code: A

1 Introduction

Nowadays, a new mode of the news broadcasting comes into being from the transformation of the news industry – data news visualization. It not only enriches the original form of the news broadcasting, but also enables the audience to feel the beauty of the rationality and depth from the news reports. It also promotes the development of the news communication and provides greater practice space for the new data journalism [1]. At the same time, by using the data mining and the visualization technology, we can analyze and process the information with the huge amount of data and disorder more accurately, so as to reveal the deeper social relationship of the news characters and dig out the stories behind the news, so as to get more able to reflect the essence of the events and predict the development trend of the news events [2]. The data mining and the visualization technology can show the complex relationship between different variables in the complex data set and the development relationship of the whole society, using the visual language for the audience. Through such a more

© Springer Nature Singapore Pte Ltd. 2020
M. Atiquzzaman et al. (Eds.): BDCPS 2019, AISC 1117, pp. 1803–1808, 2020.
https://doi.org/10.1007/978-981-15-2568-1_252

intuitive way of reporting, the audience can be inspired to actively discuss and participate in the public affairs. However, the addition of the data mining and the visualization technology does not mean that the traditional mode of the news communication will be replaced, so there will be more constraints and challenges on the way of the combination of the news communication, the data mining and the visualization technology, which is worthy of further study by researchers. At the same time, the data news puts forward higher requirements for journalists. In the new era, the news media people should establish the awareness of data, and combine the traditional role of the media monitoring, service and communication, so as to create a new team with the data mining and the visualization technology. As a new mode of the news communication, what profound influence will it have on the future news industry? How should we use the visual technology to present the traditional narrative ways of news with words? All these are the problems that should be considered in the current information age.

2 Accelerate the Process of the Media Integration

In the era of the big data, people pay more attention to the new data mining and the visualization technology, and the traditional news media are facing unprecedented challenges, as well as new opportunities. With the continuous acceleration of the media integration and reissue, the communication mode of news should also be improved.

First of all, we should speed up the process of the media integration.

The media convergence can be divided into two types. One is to use various media institutions to merge or reorganize them, so as to establish a larger media group. Through integration, the original advantages can be expanded, and the overall competitive strength can be gradually improved, forming a more group advantage. The other is to integrate the advanced technology and the traditional media technology in the modern information industry. By using the traditional and the modern media technologies for reference or replacement, a new mode of the news communication is formed.

For the news media, the influence of the broad integration and the technology integration in the media integration depends on which form of the integration can bring greater changes [3]. The media group generated through the integration is a greater challenge for every person who is originally in the media organization, and may even affect the change of their own attributes. Therefore, the media integration is more able to test the mentality of the practitioners and the coping abilities when the system changes. When the data mining and the visualization technology join in, it really reflects the qualitative change. Through media integration, the gap between the media will be broken, providing a common sharing platform for them, so that the resources obtained from different places will be shared, and the collection and editing process in the news has become a public and shared work. In such a platform, firstly, data and news contents should be collected, and the collected resources should be integrated and screened. According to different audiences, different forms of the news types are formulated, and the later soft packaging that is consistent with its own value is applied, and then it is spread through different platforms [4]. Through this mode of the news communication, combined with the advantages of different media, we can get the maximum advantage, accelerate the speed and effect of the news communication, improve the working efficiency of the news industry, and save a lot of the cost investment.

Through the data mining and the visualization technology, the media convergence can be more full and in-depth. In the era of the big data, the digital equipment is constantly updated, and the technology has become an important factor affecting the media convergence. Therefore, more new forms of the media convergence will appear in the future, and the communication industry will gradually become digital, which can make the development of the news media stronger through the continuous integration of newspapers, radio, television, mobile phone and other technologies.

3   Increase the Process of the Information Dissemination

The social TV is a brand-new term emerging in recent years, and it is also the product of the times and the integrated product of various media platforms under the background of the big data. There are TV stations, sponsors, digital TV operators, and technology companies, and so on, participating in the production. And through the survey, it is found that more than 30 companies have obtained huge venture capitals, and the income is expected to get greater growth in the next few years.

Social TV products transform the social activities in the real life into the virtual network environment, so as to meet the needs of the human social activities. But this new way is still through the TV media, and still belongs to the TV news program, accounting for a certain share of the audience. Because it makes the audience participate in it and increases the participation of the audience, this way of the news communication is more attractive to the audience. In the era of the social media, the audience is not only the audience of communication, but also the communicator. Social TV attracts the audience with "participation", which also promotes the secondary communication of news.

The audience who use the social TV products and services will be investigated, and the ratings of the TV news programs can be calculated by using the data mining technology. Through a survey of a TV guide service company, it is found that 21% of the viewers start to watch news contents of a channel through the recommendation of friends on the social TV, and more than 40% of them will continue to watch [5]. Through the platform established by the social TV, the recommendation and watching of friends has become a new and personalized electronic program guide. Customers can quickly find out the programs they want to watch from hundreds of programs on this platform, so they can quickly find the value of the program contents. Especially for some minority news channels, social TV plays an important role in attracting the audiences, and to a certain extent, it accelerates the process of the news dissemination.

With the continuous development of the mobile terminals, people have been able to replace most of the computer functions through the mobile phones, including the online video and the image services. Therefore, with the development of the data mining and the visualization technology, the TV news media industry should seize the opportunity to develop the new news program delivery technology on the mobile terminals, so that the audience of the news programs can watch the news anytime and anywhere through the mobile terminals. According to the survey, more than 20% of the audience will use the mobile phones to search the relevant information of the news events while watching the news on TV. This accompanying and multi-task viewing state can satisfy the audience's curiosity psychology, which can increase the viscosity of the news communication.

4   Optimize the Mode of the Digital News Communication

In the era of the rapid development of the information technology, different media in the traditional media also have different degrees of changes, among which the newspaper industry is the most impacted one, and its influence scope is constantly reduced, while the TV media has gradually formed a social and network development mode, and the news communication can get better development through continuous improvement [6]. According to incomplete statistics, at present, 25% of the people watch news on TV nationwide, compared with 80% in earlier years. At the same time, according to the Internet Information Center, as many as 35 million people have stopped watching news programs on TV in the past year. The ratings of the TV news should be reduced by more than 10% every year, and the ratings of watching news through the network video are increasing at the rate of about 25% every year.

At present, most of the TV news programs have synchronized it to the live network, and added many functions of interaction and comments. Some news channels have also launched a variety of types of the network news programs. Relying on the data mining and the visualization technology, we can effectively optimize the current mode of the data and network news communication, expand the channels of the news communication, help the audience to mine the news they are concerned about, and enhance the real-time nature of news. At present, a new kind of the micro radio communication mode is also favored by people. By using the form of the interaction platform and radio, it attracts a large number of audiences. The mode of the news communication can also be used for reference in the form of the radio communication. Through the way of combining radio with the network, relying on the accumulation of the audience in the Internet, it will gradually expand the dissemination of news.

5   Strengthen the Visual News Communication

From the form of news, the data news is a visual form of presentation through the information images, which constitutes the visual communication of the news contents. Through the way of the visual news report, we can classify and analyze the original complex text information visually, change the way of reading word by word, and save the readers to get the relevant content information faster. In short, the visual mode of the news communication is to transform concepts and data into the form of images for communication. In the traditional news communication, the chart information is a kind of the visual communication mode, but the image at this time only exists by the affiliation of the texts, so it can only be regarded as the primary stage of the visual news communication. However, with the development of the data news, data can be transformed into the information images through the data mining and the visualization technology, so that images become the main body of the news reports, thus forming a real sense of the visual communication.

In the process of the news communication, news enters the circulation link through the code of the disseminator, and is transmitted to the audience group through the channel. When the audience receives the corresponding code, they will make different understandings according to their own situations. However, due to the different cognition and background differences between the disseminators and the receivers, the information understood by the audience will be misunderstood [7]. However, through the data mining and visualization technology, the news data visualization can be presented to the audience, omitting the coding and decoding links in the communication process, so that the audience can directly receive the source of the information, reducing the possibility of misunderstanding of the news contents. The visual mode of the news communication can make full use of the advantages of the visual communication, aggregate and mine the data, transform the original abstract data information into an intuitive image form, and speed up the audience's ability to identify and process the information, so as to reduce the audience's reading pressure. In the same way, concreting the abstract information can also help the audience to understand and recognize the news information, which is more conducive to the audience's memory of the news contents.

6   Increase the Interactive Participation

Based on the analysis of the news audience's perception, the data mining and visualization technology can change the traditional one-way communication form of the news reports. In the dynamic news information image, the audience can choose the news contents they want to watch independently by clicking on different dimensions and indicators. Because people's attention and interest are different, they will watch the news according to their own preferences. Therefore, the same visual news product may bring different information interpretations to different audiences. But after joining the interactive participation mode, the audience can actively become the data source of the news by uploading the data, and truly participate in the production and dissemination of the news by forwarding on the social media, bringing more experience to the audience, and getting rid of the original role of the passive acceptance of information.

7   Conclusion

Nowadays, with the rapid development of the big data information, the news media is experiencing the survival and transformation of digitalization. At the same time, it also brings new opportunities and challenges for the development of the digital news. Data based news has become a way of the news reporting for all kinds of the media to seek breakthrough and transformation. A large amount of the news information can really show its unique charm only after collection and interpretation, and the data mining and visualization technology is the most effective way to solve the tedious information. This technology can be used to analyze and deal with the useful information from a large number of the complex information, which is of great significance to reveal the laws of the news development and predict the development of the news events.

References

1. Mou, Z., Yu, X., Wu, F.: Visual analysis of the current situation of the international education data mining research: hot spots and trends. E-educ. Res. **65**(04), 110–116 (2017)
2. Dang, Y.: Research on the music curriculum mode under the cloud computing and the data mining technology. Autom. Instrum. **75**(09), 058–064 (2018)
3. Liu, L.: Research on methods and technologies of the visual data mining. Telecommunications **75**(06), 296–297 (2017)
4. Wang, N., Dan, K., Chen, E., et al.: Research and application of the backtracking method based on the data mining and visualization technology for the causes of the power grid accidents. Autom. Appl. **54**(02), 082–086 (2019)
5. Lei, G., Wang, M., Chen, W.: Generation logic of the real-time data news: knowledge mining and visual design. Decoration **24**(03), 054–056 (2019)
6. Zhao, R., Yu, B.: International data mining research hotspot and frontier visualization analysis. J. Mod. Inf. **324**(06), 130–139 (2018)
7. Liu, J., Xu, T.: Index and weight of the news communication effect evaluation in the new media environment. J. Commun. Rev. **71**(04), 042–058 (2018)

Implementation of Chinese Character Extraction from Pictures Based on the Neural Network

Pan Yang[✉]

Zhengzhou University of Science and Technology,
Zhengzhou 450064, Henan, China
mengzhan714317@126.com

Abstract. The research of the Chinese character recognition technology is a key factor in the development of the social sciences and technologies. In the era of the high informationization, how to make computers recognize so many Chinese characters efficiently, and especially the printed Chinese characters, is an important problem in the field of the Chinese character recognition. The research on the method of extracting Chinese characters from pictures based on the neural network conforms to the practical value and application significance of the current information technology application.

Keywords: Neural network · Picture extraction · Chinese character method · Implementation mechanism

Through the calculation of the aggregation of the new features of the same Chinese characters and the discreteness between the new features of different Chinese characters in the feature library of the handwritten and printed Chinese characters, the validity of the new features of the Chinese characters is verified experimentally, and the robustness of the new features to the sizes, fonts and illumination of Chinese characters is analyzed [1]. The new feature has the good application value in extracting Chinese characters from pictures.

1 Overview of the Neural Network Technology

The neural network image recognition technology is a new image recognition technology with the development of the modern computer technologies, the image processing, the artificial intelligence, and the pattern recognition theory and so on. It is an image recognition method based on the traditional image recognition method and fusion of the neural network algorithm [2]. This paper analyses the commonly used image recognition methods based on the neural network. According to the characteristics of the image recognition, two kinds of the neural network image recognition models, the genetic algorithm and the BP network fusion, and the support vector machine and so on, are proposed. The learning algorithm and the specific application technology of the two models are given respectively [3].

© Springer Nature Singapore Pte Ltd. 2020
M. Atiquzzaman et al. (Eds.): BDCPS 2019, AISC 1117, pp. 1809–1814, 2020.
https://doi.org/10.1007/978-981-15-2568-1_253

The artificial neural network (ANN) is a mathematical model for the information processing using structures similar to the synaptic connections in the brain. It is usually referred to as ANN, and it is a signal processing model of the bionic nerves. In the early 1940s, people began to study the neural network [4]. After decades of development, the neural network also produced a series of breakthroughs. At present, the Hopfield model and the BP algorithm are most widely used.

The general model of the neural network generally includes ten aspects: environment, processing units, propagation rules, and states of the neural network, interconnection modes, stable states, operation modes, active rules, activation functions, and learning algorithms. Among them, the neuron, the interconnection mode and the learning algorithm are the three key factors in the neural network model [5]. An important content of the neural network is learning. Its learning methods can be divided into the supervised learning and the unsupervised learning. Its learning process generally follows the Hebb rule, the error correction learning algorithm and the winner-dominated learning rule. The Hebb rule is the most basic rule in the neural network learning.

The artificial neural network has the unique advantages. Firstly, it has the function of active learning. In the process of the Chinese character recognition, the template of the Chinese characters and the possible recognition results are input into the neural network. The neural network can recognize the Chinese characters through its own learning process. The self-learning function is very important for the prediction function of the neural network. Secondly, the neural network system has the associative memory functions, and its feedback functions can realize this association.

2   Research on the Methods of Extracting the Chinese Characters from Pictures Based on the Neural Network

The Chinese character recognition belongs to the large category pattern recognition. The artificial neural network can recognize the Chinese characters through three ways of the function approximation, the data classification, the data clustering and the special pattern of association. The Hopfield neural network is a kind of the feedback network. Its self-associative memory network enables the system to recognize the Chinese characters without much training. Therefore, the Hopfield neural network has unique advantages for the Chinese character recognition. The discrete Hopfield neural network can make its feedback process very stable through the serial a synchronism and the parallel synchronization, and the correct association can be realized only when the network is stable in a certain attractor state through the continuous evolution. The neural network is a network structure that simulates the human brain neurons. It is an adaptive non-linear dynamic system composed of a large number of the simple neurons connected to each other. Compared with the traditional multi-layer perceptrons, the rectified linear units are used as the activation function of the network, and the Softmax function is used as the output of the network, and the cross-entropy is used to measure the training loss. The classical stochastic gradient descent (SGD) method is used to solve the optimal solution of the objective function, while the back-propagation BP algorithm is to continuously adjust the weights of the network. To prevent the overfitting, the Dropout technology is used to improve the generalization ability of the network.

The online handwriting recognition can be divided into the training stage and the recognition stage. The training process is as follows: image preprocessing of the standard writing characters, feature extracting and constructing the feature library, establishing the Hopfield network model, training network and saving the weights. The process of the recognition stage is as follows. The coordinate sequence is transformed into the BMP image, and the test sample is preprocessed, and the feature is extracted and put into the network operation, and run the network to the equilibrium state, and analyze the result values. According to the workflow of the on-line handwriting recognition and the theory of the Hopfield network model, the on-line handwriting recognition system based on the Hopfield neural network is simulated in the environment of MATLAB, and the effect is very ideal.

Based on the numbers in the images, after a series of the processing of the images, the numbers contained in the images are clearly displayed. As an image recognition system, it can be used not only independently, but also as a core software application in other recognition systems. It has the high flexibility and versatility. The specific operation process of the system is as follows. After the system is started, the recognized image will be input into the system, and then the image is processed by the image preprocessing system, followed by the character segmentation, the feature extraction, and finally the character recognition. After that, the image recognition is completed.

3   Implementation Mechanism of the Chinese Character Extraction from Pictures Based on the Neural Network

The image preprocessing and the neural network character recognition are the flow of the image recognition system. Therefore, the image preprocessing, the feature extraction and the neural network recognition are three important modules of the image recognition system. The functions of each module are as follows. In the image preprocessing module, the image uploaded to the system is mainly processed. In the feature extraction module, there are many extraction methods, and the main function is to extract the grid features, the horizontal and vertical features, and the percentage of pixels of the image characters. In the neural network recognition module, the method of recognizing characters is BP after the optimization. The main functions of the algorithm include the network training, the data reading, the character determination and the result output.

3.1   Image Preprocessing Module

Read the image data. The image data is the basis of the image analysis and processing, so the first step of the preprocessing is to read the image data, including the width, the height and the color values of each pixel. When storing the pictures, there are many formats, such as JPG and GIF and so on. In order to reduce the amount of the data, the system chooses 256-color BMP format when storing the pictures. Dibapi.h and dibapi. cpp of the Microsoft image function library are used to read the image data.

Preprocess the images. Firstly, the contents of the 2256-color bitmap palette are very complex. Most of the algorithms cannot be applied in the image processing. Therefore, the gray processing is needed. In the color image, the R, G and B values of the pixels are different. By assigning each of these three points a weighting coefficient, the component of each point can be obtained. The values are the same, so that the gray level can be realized. Secondly, the gray image is binarized. The gray value of the gray image pixels is between 0 and 255. The bigger the gray value is, the brighter the image will be. In order to improve the convenience of the image processing, the image is binarized. According to the adaptive threshold method, the gray image pixels are divided into the black or white colors.

Segment the images. First, adjust the overall tilt of the image, scan the image according to the order from bottom to top, and record the first black pixels encountered. Then, change the direction of the scanning to top-down, and record the first black pixels encountered. The distance between the two points is roughly the height of the image. Secondly, image analysis is carried out, and the first black pixel is scanned from left to right. After the first black pixel is found, the point is used as the starting point of the image segmentation, and then the image is scanned. When there is no black pixel, the image segmentation stops. At the same time, the image segmentation ends.

Carry out the normalization of images. After the scanning, the characters in the image can be displayed, but the size of each character is different, which affects the standard and the accuracy of the recognition. Therefore, the size of the characters should be adjusted to make them the same size. Comparing the height required by the system and the character height, the coefficients to be transformed are calculated, and the corresponding height and width are transformed according to the coefficients. According to the interpolation method, the points in the new image are mapped to the original image to achieve the uniform size. Fifth, implement the image compression rearrangement. After the same size, the position of characters in the image is uncertain, which affects the accuracy of the feature extraction. The new image characters are formed by the compact rearrangement.

3.2   Feature Extraction Module

After the image preprocessing module, the characters in the image become the same size and orderly arrangement. Then, the feature vectors which can represent the characteristics of the characters are extracted. The image characters are brought into the BP network and the network training is carried out. The feature vectors to be recognized in the samples are extracted, and then the character recognition is carried out. The BP neural network is used as the training model. The excitation function of each layer is logsig, and the target error is set to 0.05, and the learning rate is 0.2. The model has 12 input nodes, 5 hidden layer nodes and one output node. Among them, 12 input parameters are: the length of the text per line, the total length of characters per line, the text density per line, the correlation between the contents of each line and the title, the four values of the previous line and the four values of the next line. The specific steps are as follows:

Get the training set and mark it. Preprocess the source files of the web page and generate the corresponding DOM tree. Read a line of the text from the DOM tree, count the corresponding values, and get the input vector and the expected output. The input vector passes through the transfer function of the hidden layer node and the output layer node to get the actual output. The errors of the actual and the expected output vectors are calculated, and the output errors and the hidden layer node errors are calculated. Read the next line of the text from the DOM tree and proceed. If the error is not within the allowable error range, the adjustment amount of each weight is calculated according to the calculated error term, and the weight is adjusted. Continue iterating until the errors of the actual output vector and the expected output vector meet the requirements. Read back the next line and continue learning. The label tree is traversed and the training is finished. The correlation values of the DOM tree pairs of elements are used as the inputs of the neural network, and the results of the sample marking are used as the outputs. The extraction rules are automatically generated by the learning algorithm, and the new page application extraction rules are tested.

3.3   Image Recognition Module of the BP Neural Network

The pattern recognition is an important use of the BP neural network. In the design of this module, the emphasis is to train a feasible and efficient BP network so as to accurately identify the 10 numbers of 0–9. After the image preprocessing and the feature extraction, it is necessary to use this module for the training and recognition. In the design of the BP network, the key is to determine the number of neurons in the three layers of the BP network. In this module, the first step is to train the BP network by using the known training samples. After the image preprocessing, the dimension of the feature is the number of nodes in the input layer. In the hidden layer, the number of nodes is not rigidly defined. Usually, the more neurons in this layer, the higher the spiritual level of the BP network is. The training time should be prolonged, but the number of neurons should not be too large to avoid the negative effects. The standard output setting determines the number of nodes in the input layer, that is to say, the number of nodes in the input layer is determined by the target desired encoding method. After the number of neurons is determined, character recognition is needed. After BP network training, the form of weight preservation is defined as file form. After training, character recognition is started. The number of hidden layer nodes, the minimum mean square error, the correlation coefficient and the training step are input into the work. After the training, the work of the character recognition begins. The number of the hidden layer nodes, the least mean square error, the correlation coefficient and the training step parameters are input. After the training, the character recognition results are obtained.

4   Conclusion

Using the related equipment, take multiple pictures and store them in the BMP format. Then, the images are input into the image recognition system designed in this paper. After the image preprocessing, the gray image is formed, and the digits in the image are standardized. Then, the feature vectors are extracted by using the feature extraction module. Then, the image recognition of the BP neural network is used to recognize the numbers in the pictures. The recognition results show that the image recognition system can accurately identify the numbers in the pictures.

References

1. Ma, W., Wang, X.: Robot vision image processing based on the smooth denoising block K-means algorithm. J. SSSRI **12**, 51–53 (2016)
2. Zheng, F.: Fault diagnosis of the transformer oil circuit circulation system based on the BP neural network. Heilongjiang Sci. Technol. Inf. **08**, 103–104 (2014)
3. Linu, Y., Xia, C.: Application of the deep learning neural network in the speech recognition. Netw. Secur. Technol. Appl. **12**, 104–105 (2014)
4. Liu, Y., Zhu, W., Pan, Y., Yin, J.: Multi-source fusion link prediction algorithm based on the low-rank sum sparse matrix decomposition. Comput. Res. Devel. **02**, 88–89 (2015)
5. Wang, L., Yang, J., Chen, L., Lin, W.: Parallel optimization algorithm of the Chinese language model based on the cyclic neural network. J. Appl. Sci. **05**, 148–149 (2015)

Computer Security Strategy and Management in the Tax Information Management System Under the Background of the Belt and Road

Yunxia Yao[1] and Yuru Hu[2(✉)]

[1] Anhui Vocational College of Finance and Trade, Hefei 230061, China
[2] Academy of Accounting, Anhui Wenda University of Information Engineering, Hefei 230032, China
peili490749@126.com

Abstract. Under the background of "the Belt and Road", China's foreign development and exchanges have entered a new stage. In the process of the tax information management, we should base on the data sharing and exchange, do a good job of the computer security management, and effectively ensure the effective implementation of the tax information management activities. As far as the tax information security management is concerned, its management effect directly affects the functions of the state, the modernization of the tax collection and the management and protection of the taxpayers' rights. Therefore, it is very important to explore the security strategy and the management mechanism of the system.

Keywords: The Belt and Road · Tax information management system · Computer · Security strategy · Management strategy

The application of the Internet technology in the tax system is feasible and can improve the quality of the work to the greatest extent. However, the application of the Internet technology in the tax system is also a double-edged sword, which results in some problems, and especially security problems, which are not conducive to the implementation of the tax work. To solve this problem, it is necessary to study the measures of the tax information management and strengthen the safety management so as to lay a good foundation for the development of the tax information management.

1 Background Analysis of the Belt and Road

The Belt and Road (abbreviated as B & R) is the abbreviation of the Silk Road Economic Belt and the Marine Silk Road of the 21st century. In September and October 2013, Chinese President Xi Jinping proposed the cooperative initiatives of building the New Silk Road Economic Belt and the Marine Silk Road of the 21st century respectively. It will fully rely on the existing bilateral and multilateral mechanisms between China and other countries concerned, and with the help of the existing and effective regional cooperation platforms, with the aim of borrowing the historical symbols of the ancient Silk Road, holding high the banner of the peaceful development and actively developing the economic partnerships with countries along the Silk Road.

M. Atiquzzaman et al. (Eds.): BDCPS 2019, AISC 1117, pp. 1815–1821, 2020.
https://doi.org/10.1007/978-981-15-2568-1_254

We will work together to build "the Belt and Road" and make efforts to inter-connect the Asian and European non-continental and adjacent oceans. We should establish and strengthen the partnership among all the countries along the border, and build an all-directional, multi-level and composite interconnection network so as to realize the pluralistic, autonomous, balanced and sustainable development of all the countries along the route. The interconnection project of "the Belt and Road" will promote the docking and coupling of the development strategies of various countries along the border, explore the potential of regional markets, promote the investment and consumption, create the demand and employment, and enhance the cultural exchanges and mutual learning between peoples of various countries along the route.

The countries along "the Belt and Road" should further strengthen the government statistical exchanges and cooperation, and strive to provide the accurate and reliable statistical data for the sustainable development of all countries. The information interconnection is the foundation of the economic interconnection and win-win situation. "The Belt and Road" action will promote the intergovernmental statistical cooperation and the information exchange, and provide the decision-making basis and support for the pragmatic cooperation and the mutual benefit. Statistical departments will actively carry out the statistics and monitoring of the indicators related to the sustainable development and will vigorously promote the construction of the modern statistical system. They will strive to provide the authoritative statistical data of China's economic and social development with a more active and open attitude, further improve the international comparability of China's statistical data, and share the reform and development practice of Chinese statistics with other countries. We will work together with the government statistical institutions along the "the Belt and Road" to strengthen the statistical exchanges and cooperation, and study the establishment of a mechanism for sharing the statistical data.

2 Risks of the Tax Information Management Under the Background of "the Belt and Road"

The tax information security management is an important part of the tax information security. However, under the new situation, the tax authorities (tax personnel) face the multiple risks such as the technology, administration and disputes due to their weak awareness of the information security management, lagging mode, a lack of the system and the defective risk management mechanism. In order to deal with these risks and promote and realize the tax informatization, the tax authorities should strengthen the information security management education and training, update the concept of the tax information security management, integrate the internal and the external control means to achieve the diversification of the tax information security management methods, improve the tax information security management framework and improve the tax information security management system, implement the tax information classification management, and improve the tax information security risk management mechanism and other measures.

There are security risks of the Internet access. At present, all units have opened the Internet for their work needs, but due to the lack of the security awareness and other reasons, they have not strictly implemented the physical isolation rules between the internal and the external networks. Some adopt the method of installing the double network cards on a computer or switching the network lines back and forth, both on the internal network and on the external network. Some use notebooks to access the Internet, but in addition, laptops are sometimes connected to the intranet. These unsafe and non-standard Internet access methods are easy to introduce viruses and Trojans on the Internet into the internal network of the system, and then spread on the internal network, which poses a serious threat to the safe operation of the internal network of the system.

There are the security risks in LAN. The LAN security risks are mainly manifested in the following aspects. First, there is a lack of the effective LAN access management and the technical monitoring means. Outsiders are able to connect the devices to the LAN at will, without the effective monitoring of the mobile devices such as notebooks, mobile hard disks, and U disks and so on. Second, the management of the computer IP address and the computer name is not standardized. Each unit lacks the unified planning of the IP address, and the IP address file is incomplete, and the name of the computer equipment is meaningless, and it cannot locate the specific equipment and the related responsible persons after the security incidents.

3 The Technology of the Tax Information Management System Under the Background of "the Belt and Road"

3.1 Firewall Technology

In order to ensure the information security and prevent the data from being destroyed in the tax system, firewalls are often used to prevent the illegal invasion of the tax bureau data center. The so-called firewall is the general term of a kind of the preventive measures. It means to set up a barrier between the protected intranet of an enterprise and the open network of the public, to analyze all the information to enter the Intranet or to authenticate the access users, to prevent the harmful information and the illegal intrusion from outside into the protected network, and in order to protect the security of the internal system, it prevents the illegal operation on one node of the Intranet itself and the spread of the harmful data to the outside. The essence of the firewall is the software precautionary measures to implement the filtering technology. Firewalls can be divided into different types. The most common are the IP layer firewalls based on the routers and the application layer firewalls based on the hosts. Two kinds of the firewalls have their own merits. The IP layer firewalls have good transparency to users, and the application layer firewalls have greater flexibility and security. In the practice, as long as there is the fund permit, the two firewalls are often used together to complement each other and ensure the security of the network. In addition, there are virus firewalls specially designed to filter the viruses, which can detect and kill the viruses for users at any time and protect the system.

3.2   Information Encryption Technology

The information encryption includes the cryptographic design, the cryptanalysis, the key management, and the verification and so on. By using the encryption technology, some important information or data can be converted from the plaintext to the ciphertext, and then the ciphertext can be restored to plaintext after the transmission through the line to the end user. The encryption of the data is an effective means to prevent the information leakage. Properly increasing the length of the key and more advanced key algorithm can greatly increase the difficulty of decoding. There are two kinds of the encryption methods: the private key encryption system that uses the same password when encrypting and decrypting. The private key cryptosystem includes the block cipher and the sequence cipher. The block ciphers group the plaintext symbols according to their fixed size, and then encrypt them group by group. The sequence cipher converts the plaintext symbols into the ciphertext symbols immediately, which is faster and safer. In the public key encryption system, the encryption key and the decryption key are divided into two different keys, one for encrypting information and the other for decrypting the encrypted information. These two keys are a pair of the interdependent keys.

3.3   Information Authentication Technology

The digital signature can confirm the identity of the sender and the authenticity of the information. It has four characteristics: unforgeability, authenticity, immutability and non-repeatability. The digital signature is realized by encrypting and decrypting the data by the cryptographic algorithm. The main way is that the sender first generates a message digest by running the hash function, and then encrypts the message digest with the private key to form the sender's digital signature. The digital signatures will be sent to the recipient of the message together with the message as an attachment to the message. After receiving the message, the receiver first runs the same hash function as the sender to generate the message digest, then decrypts the digital signature attached to the message with the sender's public key, and generates the message digest of the original message. By comparing the two digests, the sender can be confirmed, as well as the correctness of the messages.

3.4   Antivirus Technology

The virus prevention is the most common and easily overlooked part of the computer security. We propose to adopt this anti-virus measure, which combines the single-machine anti-virus and the network anti-virus, to maximize the end-to-end anti-virus architecture of the network. In addition, the anti-virus system and measures constitute a complete anti-virus system.

4   Computer Security Strategy and Management in the Tax Information Management System Under the Background of "the Belt and Road"

With the completion of the task of the tax information management system construction, the tax authorities have established a relatively complete network and the information security protection system, every virus infection incident, the security attack incident happened in the unit and so on. Therefore, we can make full use of the internal websites, the official documents, meetings, face-to-face exchanges and other forms to publicize the work of the network and information security, enhance the awareness of the security and the prevention of the whole staff, enhance the sense of responsibility, urgency and pressure of doing a good job of the network and the information security, firmly establish the idea of "security first" and tighten up a string of "Security" at all times, resolutely overcome the paralysis and the fluky mentality, and actively take various effective measures to resolutely prevent the major network and the information security incidents.

Improve the system, and carry out the strict assessment. We should strengthen the construction of the internal network security management system of the tax system, formulate and improve a series of the security management systems, such as the computer room management, the network management, the application system management, the data management, the emergency response and the assessment methods, and establish a relatively complete network and information security management system, so as to achieve all kinds of the security management workers. There are rules to follow. We should strictly manage and strengthen the assessment so as to promote the implementation of the assessment, effectively standardize the daily operation behaviors of the computer operators, find problems and deal with them in time, eradicate all kinds of the potential safety hazards in the bud, and ensure the implementation of the safety management rules and regulations.

Strict the Internet access, and ensure the security of the Internet access. In terms of the management, we should strictly implement the physical isolation rules of the internal and the external networks, adopt a single machine to access the Internet, strictly dedicate the special computers, completely realize the physical isolation of the internal and the external networks, and resolutely prevent the introduction of the viruses and Trojan Horse programs that cause the information leakage into the internal network of the system. Technically, we should use the function of the LAN desktop security protection system to detect the illegal outreach, strengthen the monitoring of the illegal access to the Internet, automatically lock computers that access the Internet in violation of regulations, record the violations, and strictly assess and resolutely prevent the occurrence of the illegal access to the Internet.

Strengthen the LAN management to ensure the LAN security. In terms of the management, the first is to strengthen the access management of LAN. Strict computer intranet access system should be established. All computers connected to Intranet must register at all levels of the information centers, uniformly assign the IP address, gateway and other parameters, install the uniform network version anti-virus software, and prohibit anyone from privately accessing the computer equipment to the Intranet operation. The second is to strengthen the management of the IP address usage. Formulate the IP address management scheme, establish all computer IP address use files, prohibit the users from changing the IP address privately, and ensure that each computer device has a unique IP identity. Third, we should standardize the naming of the computer equipment names. Establish the naming rules for the computer equipment names. All computer devices are named in accordance with the regulations to ensure that after the security incidents occur, they can quickly locate the specific personnel according to the name of the computer equipment. The safety management of the terminal equipment should be strengthened.

Enable the binding function of the IP address and the MAC address to effectively prevent the security risks caused by the users changing the IP address privately. Enable the hardware resource management function, strengthen the computer name checking, and strictly enforce the naming rules of "department name + user". For those who do not name the computer equipment according to the naming rules, strict examination shall be made to resolutely prevent the problem of being unable to locate the responsible persons after the security incidents occur due to the irregular computer names. The virus surveillance and management should be strengthened. All computer equipments are equipped with the network version anti-virus system, which is managed by the server and upgraded in time to ensure the effectiveness of the anti-virus system. The virus administrator monitors the virus infection of the computer equipment in his unit in the real time, and when the virus cannot be detected and killed, the virus will be disconnected in time to prevent the virus from spreading further in the network.

5   Conclusion

With the continuous development of the information technology, the computer and the network technologies are widely used in the tax system, which opens up a broad space for the tax management and services to achieve electronization, networking, verification and efficiency. At the same time, in the context of the belt and road, with the maturity of the international economic and the trade exchanges and the increasing frequency of the information exchange, the security of the computer network in the electronic tax system has become more prominent and important.

Acknowledgement. 2017 Anhui University of Finance and Trade Vocational College Connotation Promotion All Staff Action Plan Major Scientific Research Projects: Study on Tax Policy and Tax Planning of Export-oriented Enterprises in Anhui Province in the Context of "Belt and Road" (2017nhrwa01).

References

1. Wei, Q.: On the legal path of taxpayers' tax information protection. Tax. Econ. **4**, 81–88 (2012)
2. Yang, G.: Scheme design of the tax decision support system based on DW and OLAP. Mod. Comput. (Prof. Ed.) **5**, 152–154 (2008)

3. Ma, G.: Research on the usage principle of the dimension in star mode of data warehouse based on the relational database. Comput. Eng. Des. **26**(1), 216–217, 246 (2005)
4. She, M., Wang, T., Liu, H., Wang, W.: Design of the tax data warehouse system based on WEB. Fujian Comput. **4**, 122–123 (2007)
5. Fu, S.: Design of the tax data warehouse system based on WEB. Comput. Knowl. Technol. **2**(11), 211–213 (2008)

The Manufacturing Transfer in Shanghai Based on Big Data

Dongyu Yu[✉]

Shanghai University of Engineering Science, Shanghai 200437, China
ningbao708186@126.com

Abstract. Big data is the basis of intelligent manufacturing. Its application in mass customization of manufacturing includes data collection, data management, order management, intelligent manufacturing, customized platform, etc. The core is customized platform. Big data applications can be implemented with custom data reaching an order of magnitude. Through the mining of big data, we can achieve more applications such as popular forecasting, accurate matching, fashion management, social applications, marketing push and so on. With the economic transformation of Shanghai and the gradual maturity of the economic environment in other parts of the country, the trend of manufacturing transfer from Shanghai has increased. This transfer will generate economic development in Shanghai and production services that have a certain dependence on manufacturing. What impact? The transfer of manufacturing does not mean that the Shanghai region does not attach importance to manufacturing, but rather promotes the transformation and upgrading of the manufacturing industry and the development of high value-added industries. So what is the overall industrial structure and industrial structure in Shanghai? What is the degree of transfer and impact of manufacturing? Based on the data of Shanghai Statistical Yearbook and Statistical Bulletin, this paper uses location entropy, regional specialization and agglomeration index to measure the direction and degree of manufacturing industry transfer.

Keywords: Big data · Manufacturing · Industrial transfer measurement

1 Introduction

The information society has entered the era of big data. The emergence of big data has changed the way people live and work, and it has also changed the mode of operation of manufacturing companies. In recent years, with the rapid development of information technology and communication technology such as the Internet, Internet of Things, cloud computing, etc., the skyrocketing data volume has become a serious challenge and valuable opportunity faced by many industries [1–3]. With the advancement of manufacturing technology and the popularization of modern management concepts, the operation of manufacturing enterprises is increasingly dependent on information technology. Today, the entire life cycle of manufacturing, the entire life cycle of manufacturing products, involves a lot of data. At the same time, the data of manufacturing companies also showed an explosive growth trend. Big data is the

© Springer Nature Singapore Pte Ltd. 2020
M. Atiquzzaman et al. (Eds.): BDCPS 2019, AISC 1117, pp. 1822–1828, 2020.
https://doi.org/10.1007/978-981-15-2568-1_255

lifeblood of the industrial Internet. The great value that big data can bring is being recognized by traditional industries [4]. It is presented to business managers and participants through technological innovation and development, as well as the comprehensive perception, collection, analysis and sharing of data. Look at the new perspective of the manufacturing value chain.

2 Internet Big Data Overview

"Big data" refers to a large data set collected from many sources in multiple forms, often in real time. In the case of business-to-business sales, this data may come from social networks, e-commerce sites, customer visits, and many other sources. These data are not the normal data sets of the company's customer relationship management database [5, 6]. From a technical point of view, the relationship between big data and cloud computing is as inseparable as the front and back of a coin. Big data must not be processed by a single computer, and a distributed computing architecture must be used. It features mining of massive amounts of data, but it must rely on cloud computing for distributed processing, distributed databases, cloud storage, and/or virtualization technologies [7]. In the Big Data Era written by Victor Meyer-Schonberg and Kenneth Cooke, big data refers to shortcuts that do not use stochastic analysis (sample surveys), but all data methods) big data Features: Volume, Velocity, Variety, Value. The value of big data is reflected in the following aspects: (1) Enterprises that provide products or services to a large number of consumers can use big data for precise marketing. (2) Small and medium-sized enterprises can use big data for service transformation. (3) Traditional enterprises that must be transformed under the pressure of the Internet need to keep up with the times and make full use of the value of big data.

3 Analysis of Industrial Structure in Shanghai

As the economic center of China and the core city of the Yangtze River Delta region, Shanghai's own economic development, the transformation of economic growth mode and the optimization of industrial structure are of great significance to the national economy. In 2018, Shanghai's annual gross domestic product (GDP) reached 326.987 billion yuan, an increase of 6.6% over the previous year. Among them, the added value of the primary industry was 10.437 billion yuan, down 6.9%; the added value of the secondary industry was 973.254 billion yuan, an increase. 1.8%; the added value of the tertiary industry was 2,284.496 billion yuan, an increase of 8.7%. The added value of the tertiary industry accounted for 69.9% of Shanghai's GDP. Since 2012, the proportion of the tertiary industry's added value to Shanghai's GDP has reached 60% for the first time, and the proportion of the service industry has gradually increased. Although the proportion of Shanghai's tertiary industry is still not up to the level of developed countries, it is already at the leading level in China and the industrial structure is dynamically adjusting. Industrial transfer is the process of industry re-selection of spatial location. Industrial transfer not only promotes the adjustment and upgrading of industrial structure in developed regions, but also is a strategic choice for backward regions to catch up with developed regions. Chen [7] believes that the level of economic development in different regions is different, and the level of regional economic development determines the price of production factors in the region. That is to say, the difference in the price of production factors between regions is the driving force for the migration of industries in space. He [6] believes that the reason for the phenomenon of industrial transfer is mainly due to the difference in industrial competitive advantages between different countries and regions, and the competitive advantage comes from industrial clusters. Therefore, the difference in industrial clusters will affect industrial transfer. Zhao, Shi (2013) believes that industrial transfer is mainly affected by differences in factors such as labor, capital and technology, market size, and government policies. Hu (2014) used the data of three-digit manufacturing industry output value in 2003 and 2009 and found that China's manufacturing industry has experienced a large-scale transfer from the eastern region to the central and western regions. The impact of industrial transfer on the removal of land will take into account both positive and negative aspects, that is, it may promote the upgrading of the industrial structure of the place of removal, but it may also reduce the employment opportunities in the place of removal or lead to the phenomenon of "industry hollowing out".

According to the Shanghai Statistical Yearbook and the Statistical Bulletin, the proportion of the primary industry and the secondary industry in total output has continued to decline since 2000, and the proportion of the tertiary industry is gradually increasing. Shanghai's 13th Five-Year Development Plan proposes that the manufacturing industry should make full use of the Internet+, implement intelligent transformation of facilities and equipment, and accelerate the transformation of production methods into digital, networked, intelligent and flexible. Implement strategic emerging industry projects and promote the deep integration of information technology and manufacturing technology. From 2002 to 2018, the total output value of high-tech enterprises in Shanghai generally showed an upward trend, but there was a certain fluctuation in the process of rising, while the proportion of high-tech industries showed a downward trend. See Fig. 1.

In Shanghai, it is necessary to continuously improve the new industrial system, which is dominated by modern service industry, guided by strategic emerging industries and supported by advanced manufacturing. By 2020, the added value of the service industry will account for about 70% of the city's GDP, and the manufacturing industry will increase. The value of the city's GDP has remained at around 25%, and the development of strategic emerging industries is very important. Since 2003, the absolute value-added of strategic emerging industries and the proportion of Shanghai have shown an upward trend. In strategic emerging industries, the added value and proportion of services have exceeded that of industry. Does this mean that industries are moving outward? What is the extent of the transfer?

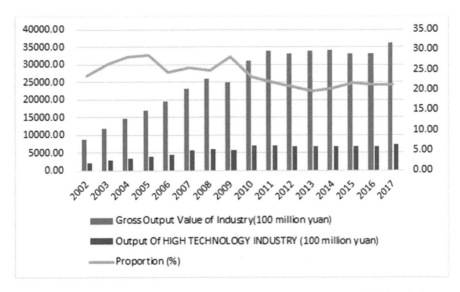

Fig. 1. Total output value and proportion of high-tech enterprises in Shanghai

4   Analysis of Industrial Transfer Measurement in Shanghai

4.1   Research Design

In the study of measurement industry transfer, the methods of deviation-share, gravity model, input-output method, and agglomeration index are mainly used. This paper draws on the methods of Jing (2010), [3] and Fan [1], using location entropy, regional specialization and agglomeration index to measure the direction and extent of manufacturing transfer. As a core city in the Yangtze River Delta, due to regional integration and geographical factors, this paper selects the Yangtze River Delta region to analyze the transfer of manufacturing. The regional relative specialization index is expressed as:

$$K_i = \sum_k \left| s_i^k - s_i^{\bar{\bar{k}}} \right|, \text{ among them, } s_i^{\bar{\bar{k}}} = \sum E_i^k / \sum_k \sum E_i^k. \tag{1}$$

Where i j k are area i, area j and industry k, respectively. E_i^k the total value of the industry k representing the region I. This formula represents the sum of the specialization coefficient of each industry in a certain region and the absolute value of the specialization coefficient of the corresponding industry in the rest of the country. The measure is the degree of manufacturing structure difference between the i-th region and the rest of the region, that is, the level of specialization in the region.

Then calculate the location entropy and agglomeration index of the manufacturing industry. The calculation formula for location entropy is:

$$Q_{ij} = (P_{ij}/E_i)/(P_j/E) \qquad (2)$$

Among them, Q_{ij} Indicates the location entropy of the region j industry i; P_{ij} For the production value of area j industry i; (P_j For regional j manufacturing or service industry total output value; E_i is the output value of high-level regional industry i; E is the high-level regional manufacturing or service industry gross output value.

Calculate the dynamic agglomeration index based on location entropy,

$$\alpha_{ij(0-t)} = v_{ij(0-t)} / \sum v_{ij(0-t)} \qquad (3)$$

Among them, $\alpha_{ij(0-t)}$ Represents the dynamic agglomeration index of industry i in region j during (0−t) time, $v_{ij(0-t)}$ indicates the growth rate of the output value of the industry i in the region j during the period (0–t), $v_{ij(0-t)}$ indicates the average of the industry i in the (0–t) time in Shanghai, Jiangsu and Zhejiang. Growth. Output value growth rate $v_{ij} = \frac{1}{n} \ln \frac{Y_{it}}{Y_{i0}}$, Y_{it} and Y_{i0} The output value of the industry i at the end of the period and the beginning of the period, n is the number of years. The data is compiled using the original data of the 2007–2017 Shanghai Statistical Yearbook, Jiangsu Statistical Yearbook, Zhejiang Statistical Yearbook and China Industrial Economics Statistical Yearbook, and the price is reduced. The manufacturing industry is consolidated into 16 categories based on industry characteristics.

4.2   Empirical Test Analysis

Using the regional specialization formula (1), the specialization indexes of Shanghai, Jiangsu, and Zhejiang provinces in 2007, 2012, and 2017 are measured. The relative specialization indexes of various provinces and cities are generally on the rise, indicating that the three regions form their respective industries. Division of labor, mutual cooperation. The integration of the Yangtze River Delta can enhance the division of labor between regions, thus achieving the scale advantage of various industries in space, of which Shanghai has the highest level of specialization, as shown in Table 1.

Table 1. Specialization index of the three provinces of the Yangtze River Delta in 2007–2018

	2007	2012	2017
Shanghai	0.52	0.61	0.67
Zhejiang	0.43	0.48	0.53
Jiangsu	0.35	0.39	0.44

In 2012, the proportion of the tertiary industry in Shanghai exceeded 60%. At this time, the pattern of manufacturing industry dominated the industrial structure dominated by the service industry, while the labor-intensive manufacturing industry moved to surrounding areas and inland provinces and cities. The transfer will mainly develop capital-intensive, high-tech, and strategically emerging industries and services.

Through the calculation and analysis of the location entropy index formula (2), among the 16 manufacturing industry categories, the location entropy of three industries in Shanghai has a tendency to become larger; the entropy of 8 industries is becoming smaller, indicating that Shanghai is a manufacturing agglomeration area. The advantages of the industry are weakening; the location entropy of five manufacturing industries in Shanghai is below 0.5, while the corresponding entropy of the corresponding industries in Jiangsu and Zhejiang provinces is above or close to 1, indicating that Shanghai has lost comparisons in such manufacturing industries. Advantages, the outward transfer of manufacturing industry is obviously conducive to enhancing the degree of specialization of production in such industries, in line with the requirements of industrial structure optimization. The Shanghai area is positioned to build "four centers", and due to higher housing prices, rising labor costs, and smaller geographic areas, resource-intensive and labor-intensive manufacturing industries are not suitable for agglomeration and development, and there is a need for outward transfer.

Using the industrial dynamic agglomeration index formula (3), it is concluded that only three industry dynamic agglomeration indices in Shanghai are greater than 1 during this period, most of which are between 0–1, and the dynamic agglomeration index of chemical fiber manufacturing is less than 0; Jiangsu has nearly 11 The dynamic agglomeration index of each industry is greater than 1, only 5 industries are greater than 0, less than 1, and there is no industry less than 0; Zhejiang has 7 industries with dynamic agglomeration index greater than 1, 9 industries greater than 0 and less than 1, both greater than 0. It can be seen that most of the manufacturing industries in Shanghai are slower than other regions in the Yangtze River Delta and are shifting to the outside, especially in the chemical fiber manufacturing industry. The situation in Zhejiang is similar, and most industries are showing The situation of external relative transfer; Jiangsu presents the characteristics of most industries accumulating internally. It is expected that with the economic development of Jiangsu and Zhejiang, more manufacturing industries will be transferred to inland provinces and cities in the future.

5   Conclusion

By analyzing the industrial structure of Shanghai and comparing the industrial characteristics of Jiangsu and Zhejiang, using regional specialization and location entropy indicators, dynamic agglomeration indicators and other calculations, the current Shanghai region has a tendency to shift manufacturing, and labor-intensive manufacturing and The general capital-intensive manufacturing industry has a relatively large outward transfer. This kind of industrial transfer is in line with the industrial development gradient theory and the regional comparative advantage theory. At present, the service industry in Shanghai is nearly 70%, and the manufacturing transfer is in line with the characteristics of economic development. However, the Shanghai region should adjust the manufacturing industry structure, increase support for high-tech industries and strategic emerging industries, and use the advantages of the Shanghai Free Trade Zone to deepen reforms, promote the healthy development of industry, and avoid the phenomenon of "hollowing" after industrial transfer.

Acknowledgment. Fund Project: Special Fund for Shanghai University Young Teachers Funding Program. Number: ZZGCD15128.

References

1. Fan, J.: Integration of the Yangtze river Delta, regional specialization and manufacturing space transfer. Manag. World (11), 77–84 (2004)
2. Jing, X.: Analysis of manufacturing transfer and agglomeration in the Yangtze river Delta region. Nanjing Soc. Sci. (3), 9–13 (2004)
3. Jing, X.: The impact of regional central cities on the spatial distribution of FDI in the hinterland——based on empirical analysis of Shanghai and the Yangtze river Delta region. J. Zhengzhou Univ. (Philos. Soc. Sci. Ed.) **2**, 53–56 (2013)
4. Yu, C.: Will manufacturing transfer inhibit the development of local producer service industry?—Taking Shanghai as an example. Econ. Jingwei (6), 71–76 (2014)
5. Zhao, R., Shi, M.: Analysis and countermeasure of industrial agglomeration transfer and its influencing factors in China. Econ. Manag. **12**, 73–76 (2013)
6. He, L.: Characteristics, problems and countermeasures of inter-regional industrial transfer in China. Econ. Aspects **09**, 55–58 (2009)
7. Chen, H.: The connotation, mechanism and effect of inter-regional industrial transfer. Inner Mongolia Soc. Sci. (Chin. Ed.) **01**, 16–18 (2002)

The Relationship Between Shanghai-Centered Producer Services and Manufacturing Industry Based on Simultaneous Equation Model in Cloud Computing Environment

Dongyu Yu[✉]

Shanghai University of Engineering Science, Shanghai 200437, China
ningbao708186@126.com

Abstract. In the cloud computing environment, based on the simultaneous equation model combined with the calculation index of influence coefficient and sensitivity coefficient, the simultaneous equation model is constructed to empirically analyze the degree of interaction between the two regions. There is a two-way causal relationship between producer services and manufacturing, but the degree of interaction between producer services and manufacturing in several regions is different. Taking Shanghai as the central economic circle, we must pay full attention to and play an interactive mechanism between the producer service industry and the manufacturing industry, and promote the benign and coordinated development of the producer service industry and the manufacturing industry.

Keywords: Cloud computing · Shanghai · Producer service industry · Manufacturing · Simultaneous equation model

1 Introduction

At present, one of the most economically viable regions in China, with Shanghai as the center, relying on the accumulation of manufacturing bases along the Yangtze River, has become the most mature economy in the domestic production service industry and manufacturing industry and the most potential for interaction between the two area. On the one hand, under the acceleration of the process of economic globalization, the phenomenon of non-industrialization has continued to occur, leading to the prosperity of cities with production services as the core. On the other hand, the incentive competition between local and foreign-funded enterprises in the development of an export-oriented economy centered on Shanghai, the foundation of the development of productive services in the region, and the relatively mature manufacturing outsourcing experience have all contributed to the promotion of the producer service industry [1]. The manufacturing upgrade process plays a greater role. Therefore, through the interaction and coordinated development of the producer service industry and the manufacturing industry, the improvement of the input structure of manufacturing factors in Shanghai as the central region and the realization of the global value chain of the manufacturing industry are the realization of the modern service industry and the

M. Atiquzzaman et al. (Eds.): BDCPS 2019, AISC 1117, pp. 1829–1837, 2020.
https://doi.org/10.1007/978-981-15-2568-1_256

advanced manufacturing industry [2]. It is an important measure to drive and implement the reform and development plan centered on Shanghai. Based on such an era background, the study of the interaction between producer services and manufacturing in Shanghai as the center has more important theoretical and practical significance and should be given full attention.

2   Cloud Computing Concepts, Features and Functions

Since the beginning of the 21st century, network technology has been continuously developed. As the latest development of distributed computing, parallel processing (Parallel Computing) and grid computing (Grid Computing), cloud computing has been paid more and more attention.

2.1   About the Expression of Cloud Computing

Academia and different enterprises have different representations of cloud computing, but most of them are based on the full use of networked computing and storage resources to achieve high-efficiency and low-cost computing goals, hoping to better integrate information on the Internet and different devices. And applications, all computing and storage resources are linked together to achieve maximum scope of collaboration and resource sharing [3]. For the majority of users, cloud computing is more about creating a convenient and efficient cloud computing environment with unlimited energy for the future.

2.2   Characteristics of Cloud Computing

Safe and reliable data storage. Cloud computing adopts measures such as data multi-copy fault tolerance and computational node isomorphism interchangeability to ensure high reliability of services, so that users no longer have to worry about data loss and virus intrusion.

On-demand service, timely and convenient, and low cost. "Cloud" is a huge resource pool. It has the lowest requirements on the user's equipment. The cloud can be purchased and billed on demand like tap water, electricity and gas. It is very convenient to use.

Strong versatility, high scalability and very powerful data sharing. Cloud computing is not targeted at specific applications. Under the support of "cloud", it can construct ever-changing applications, which can easily realize data and application sharing between different devices. The size of the "cloud" can be dynamically scaled, and the same "cloud" can simultaneously support different application operations to meet the needs of different applications and different user scale growth.

Very large scale, unlimited possibilities. "Cloud" is quite large, Google Cloud Computing has more than 1 million servers, and Amazon, IBM, Microsoft, Yahoo and other "clouds" have hundreds of thousands of servers. Enterprise private clouds typically have hundreds of thousands of servers. "Cloud" can give users unprecedented computing power, providing us with almost unlimited possibilities for using the network.

2.3   The Function of Cloud Computing

The elusive "cloud computing" is to connect all computing applications and information resources with the Internet to form a virtual environment for cloud computing, for personal and business users to access and share at any time, management, and use—related applications and resources are available through any server and data center worldwide. In such a cloud computing environment, the main functions that can be implemented are:

Provide resources, including computing, storage, and network resources. Infinitely stitched with existing hardware and software technologies, cloud service providers set up a large-scale global database and storage center for users, enabling massive cloud storage, excellent security, high privacy and reliability, and It is efficient, low cost and energy efficient.

Provide dynamic data services. Data includes raw data, semi-structured data, and processed structured data. In the era of data as the king, a good cloud computing architecture must have the intelligence to provide large-scale data storage, sharing, management, mining, search, analysis and service. Can solve the bottleneck problem of concurrent access to the greatest extent.

Provide a computing platform, including software development APIs, environments, and tools. Only in this way can cloud computing truly form an "ecosystem" that is vital, attractive and sustainable.

3   Estimation of the Interaction Between Production Service Industry and Manufacturing Industry with Shanghai as the Central Region

The influence coefficient and the sensitivity coefficient are the two most analytical tools that fully reflect the characteristics of industrial interaction. This section uses these two indicators to analyze the degree of interaction between the producer service industry and the manufacturing industry centered on Shanghai.

3.1   Measurement of the Influence Coefficient Between the Producer Service Industry and the Manufacturing Industry

The influence coefficient is the degree of demand that affects the various sectors of the national economy when the final use of a unit in a certain sector of the national economy is reflected. Since the influence coefficient is traced from the consumption department to the impact of the final demand change on each department, reflecting its degree of association with each subsequent production department, it is also called the backward correlation coefficient. Its calculation formula is:

$$T_j = \frac{1}{n}\sum_{i=1}^{n}A_{ij} \Big/ \frac{1}{n^2}\sum_{i=1}^{n}\sum_{j=1}^{n}A_{ij}, \quad (i, j = 1, 2, 3, \cdots n)$$

Among them, T_j is the coefficient of influence of the j industry department on other industrial sectors, and A_{ij} is the coefficient of the i-th row and the j-th column in the Lyon Wolfov inverse matrix $(I - A) - 1$. The influence coefficient of an industry is greater than 1 or less than 1, respectively, indicating that the influence of the industry is above or below the average level in all industries. The industry with the larger influence coefficient indicates that the industry has a greater pulling effect on other sectors. The manufacturing industry has a positive role in promoting the development of the producer service industry. More specifically, the influence coefficient of Jiangsu and Zhejiang manufacturing on the producer service industry is higher than that of Shanghai, and the manufacturing effect of the two provinces on the producer service industry is greater. From the average level of the influence coefficient of the manufacturing service industry to the manufacturing industry, the pulling effect of the producer service industry on the manufacturing industry is constantly increasing. However, the average value of the influence coefficient of the manufacturing service industry in the three provinces and cities is less than 1, indicating that the production service industry has a relatively small pulling effect on the manufacturing industry, and its radiation and pulling effect on the manufacturing industry needs to be further improve (Table 1).

Table 1. .

A particular year		Sensitivity coefficient of manufacturing industry to service industry	Sensitivity coefficient of service industry to manufacturing industry
2008	Shanghai	1.154	0.907
	Jiangsu	1.241	0.856
	Zhejiang	1.235	0.795
2013	Shanghai	1.221	0.965
	Jiangsu	1.352	0.897
	Zhejiang	1.152	0.743
2018	Shanghai	1.065	0.926
	Jiangsu	1.156	0.816
	Zhejiang	1.228	0.765

As shown in Table 2, from the overall point of view, in 2008–2018, the influence coefficient of the manufacturing industry in the three provinces and municipalities on the productive service industry is much higher than the influence coefficient of the production service industry on the manufacturing industry. The ripple effect of the producer service industry is higher. Specifically, from the average level of the influence coefficient of the manufacturing industry on the producer service industry, the average coefficient of influence of the manufacturing industry on the productive service industry in Shanghai, Jiangsu and Zhejiang provinces in 2008–2018 is greater than 1, indicating The manufacturing industry has a positive role in promoting the development of the producer service industry. More specifically, the influence coefficient of

Jiangsu and Zhejiang manufacturing on the producer service industry is higher than that of Shanghai, and the manufacturing effect of the two provinces on the producer service industry is even greater. From the average level of the impact coefficient of the manufacturing service industry on the manufacturing industry, the influence coefficient of the manufacturing service industry in Shanghai, Jiangsu and Zhejiang on the manufacturing industry in 2008–2018 is on the rise, indicating that the production service industry is the pulling effect of the manufacturing industry is constantly increasing. However, the average value of the influence coefficient of the manufacturing service industry in the three provinces and cities is less than 1, indicating that the production service industry has a relatively small pulling effect on the manufacturing industry, and its radiation and pulling effect on the manufacturing industry needs to be further improve.

3.2   Measurement of the Sensitivity Coefficient Between the Producer Service Industry and the Manufacturing Industry

The sensitivity coefficient refers to the demand-sensing group that a department receives for each additional unit of each unit of the national economy, that is, the output that the department needs to produce for other departments. Since the sensitivity coefficient is a situation in which the intermediate product department is affected by changes in various departments from the production department, it is also called the forward correlation coefficient. Its calculation formula is:

$$E_i = \frac{1}{n}\sum_{j=1}^{n}A_{ij} \bigg/ \frac{1}{n^2}\sum_{j=1}^{n}\sum_{j=1}^{n}A_j, \quad (i, j = 1, 2, 3, \cdots n)$$

Among them, E_i is the coefficient of sensitivity of the j industry department to other industrial sectors, and A_{ij} is the coefficient of the i-th row and the j-th column of the Lyon Wolf's inverse matrix $(I - A) - 1$. The sensitivity coefficient of an industry is greater than 1 or less than 1, respectively, indicating that the sensitivity of the industry is above or below the average level in all industries. The industry with higher sensitivity coefficient indicates that the national economy has a greater pulling effect on the industry. At present, there is a certain interaction between the producer service industry and the manufacturing industry in the Yangtze River Delta region. Specifically, the influence coefficient and the sensitivity coefficient of the manufacturing industry to the producer service industry are higher than the influence coefficient and the sensitivity coefficient of the producer service industry to the manufacturing industry, which indicates the ripple effect of the manufacturing industry on the producer service industry. And the demand gravitational effect is greater than the sweeping effect and demand gravitational effect of the producer service industry on the manufacturing industry. However, the role of the producer service industry in the manufacturing industry is increasing, and based on the results of the author's sub-industry, the industries such as the financial and insurance industry and the wholesale and retail industry are more effective in manufacturing, indicating that the development of these industries is direct. It affects the development level of the manufacturing industry and should be highly valued by the government and relevant departments.

Table 2. .

A particular year		Sensitivity coefficient of manufacturing industry to service industry	Sensitivity coefficient of service industry to manufacturing industry
2008	Shanghai	0.997	0.807
	Jiangsu	1.241	0.956
	Zhejiang	1.235	0.795
2013	Shanghai	1.221	0.965
	Jiangsu	1.352	0.697
	Zhejiang	1.152	0.742
2018	Shanghai	0.965	0.921
	Jiangsu	1.156	0.816
	Zhejiang	1.228	0.868

As shown in Table 3, from the overall point of view, in 2008–2018, the sensitivity coefficient of manufacturing in the three provinces and municipalities to the productive service industry is higher than the inductive coefficient of the manufacturing service industry to the manufacturing industry, indicating that the manufacturing industry is productive. The service industry is more dependent. Specifically, from the average level of the manufacturing coefficient of the manufacturing service industry, the average coefficient of influence of the manufacturing industries in Shanghai, Jiangsu and Zhejiang provinces on the production service industry in 2008–2018 is almost greater than 1, which is It shows that the manufacturing industry is more dependent on the production service industry than the average level. Apart from the slight fluctuation of Shanghai, it basically shows a gradual increase. More specifically, similar to the measurement of the influence coefficient, the manufacturing coefficient of the manufacturing industry in Jiangsu and Zhejiang provinces is higher than that in Shanghai, indicating that the manufacturing industries in these two provinces are more dependent on the producer service industry. From the average level of the manufacturing service industry's sensitivity coefficient to the manufacturing industry, the average value of the manufacturing service industry in the three provinces and cities in 2008–2018 is less than 1, indicating that the production service industry is the gravitational pull of demand in manufacturing needs to be strengthened. Combined with the calculation results of Tables 2 and 3, it can be seen that there is a certain interaction between the producer service industry and the manufacturing industry in the Yangtze River Delta region. Specifically, the influence coefficient and the sensitivity coefficient of the manufacturing industry to the producer service industry are higher than the influence coefficient and the sensitivity coefficient of the producer service industry to the manufacturing industry, which indicates the ripple effect of the manufacturing industry on the producer service industry. And the demand gravitational effect is greater than the sweeping effect and demand gravitational effect of the producer service industry on the manufacturing industry. However, the role of the producer service industry in the manufacturing industry is increasing, and based on the results of the author's sub-industry, the industries such as the financial and insurance industry and the wholesale and retail industry are more effective in manufacturing, indicating that the development of these industries is direct. It affects the development level of the manufacturing industry and should be highly valued by the government and relevant departments.

4 The Interaction Between Producer Service Industry and Manufacturing Industry

4.1 Estimation and Analysis of the Model

1. Unit Root Test of Data In order to ensure the validity of the results of the measurement model, it is necessary to analyze the stability of the variables before performing the analysis. In this paper, the ADF and PP unit root test method is used to test the stability of each variable. 2. The endogeneity test and identification of the model. Because of the two-way causal relationship between the producer service industry and the manufacturing industry, only the single equation is used to analyze the two. It will produce endogeneity problems of variables, which will lead to biased and non-consistent empirical results. In order to determine whether the producer service industry and the manufacturing industry interact, it is necessary to carry out a variable endogeneity test, that is, to test the parallelism of the equation. The commonly used test method is the Hausman test. The specific process is as follows: Firstly, lnpro performs auxiliary regression on all exogenous variables of the whole equation group, and obtains the estimated value of lnpro and the residual value resid (3).

$$\ln pro = \lambda_0 + \lambda_1 \ln com_t + \lambda_2 \ln emplind_t + \lambda_3 \ln cap_t + \\ \lambda_4 \ln emplser_t + \lambda_5 \ln city, + \lambda_6 \ln gdp_t + resid_t$$

4.2 Selection and Estimation Results of Regression Equation Estimation Methods

In this paper, the Generalized Method of Moment (GMM) is chosen as the estimation method of the simultaneous equation model. The GMM method defines the criterion function as a correlation function between the instrument variable and the disturbance term, so that it minimizes the estimated value of the parameter. It allows heteroscedasticity and sequence correlation for random error terms. At the same time, the GMM method does not need to know the exact distribution of the perturbation terms, so the GMM estimator is more robust than other methods. As an advanced production factor input, the producer service industry has actively promoted the improvement of the international competitiveness of the manufacturing industry. Specifically, for every 1% point increase in the efficiency of productive services in Shanghai, Jiangsu and Zhejiang provinces, manufacturing efficiency can increase by

0.562, 0.612 and 0.0.501% points respectively, with Jiangsu being the highest and Zhejiang the lowest. An important reason for this may be because the Yangtze River Delta region has formed a service industry development center centered on Shanghai, while the surrounding Jiangsu and Zhejiang have formed a pattern of manufacturing periphery. As the center of the Yangtze River Delta, Shanghai's development of its service industry not only satisfies the region, but also spreads to the surrounding areas, so it may weaken its role in manufacturing in the region. This from another angle also shows that the production service industry in Jiangsu Province may be overestimated in the manufacturing industry.

Table 3.

	Shanghai		Jiangsu		Zhejiang	
	Lnpro	Lnser	Lnpro	Lnser	Lnpro	Lnser
	7.854	4.655	1.889	3.655	3.551	8.052
	0.003	0.007	0.021	0.008	0.052	0.001
Lnser	0.562		0.612		0.501	
	0.031		0.031		0.065	
Lngdp	0.358	0.203	0.063	0.092	0.123	0.327
	0.045	0.003	0.003	0.052	0.031	0.022
Lncom	0.456		0.162		0.411	
	0.058		0.004		0.036	
Lnemplind	0.865		0.001		0.233	
	0.002		0.005		0.085	
Lncap	0.185		0.285		0.203	
	0.006		0.006		0.095	
Lnpro		0.621		0.883		0.778
		0.001		0.045		0.051
Lnemplser		0.132		0.421		0.652
		0.002		0.011		0.001
Lncity		0.882		0.533		0.065
		0.002		0.021		0.077
R^2	0.932	0.885	0.989	0.983	0.965	0.822
Adj R^2	0.925	0.862	0.822	0.882	0.921	0.944

Estimation results of the simultaneous equation model of Shanghai, Jiangsu and Zhejiang.

5 Conclusion

Based on the analysis of the development status of the producer service industry and the manufacturing industry, this paper tests the interaction between the two by constructing a simultaneous equation model of the producer service industry and the manufacturing industry. The research shows that there is interaction, interaction and common development between the productive service industry and the manufacturing industry in the Shanghai-centered economic circle. However, this interaction is still relatively small and needs to be further enhanced. Therefore, in the future, it is necessary to adjust the industrial development ideas of the region. At the same time as the introduction of processing trade-type foreign capital to develop manufacturing industries, measures should be taken to develop the productive service industry in the long-term region, paying full attention to and utilizing the productive service industry and the manufacturing industry. The interactive mechanism promotes the sound and coordinated development of the producer service industry and the manufacturing industry.

Acknowledgment. Fund Project: Special Fund for Shanghai University Young Teachers Funding Program. Number: ZZGCD15128.

References

1. Zhou, Q., Luo, J.: The study on evaluation method of urban network security in the big data era. Intell. Autom. Soft Comput. **24**(1), 133–138 (2018). https://doi.org/10.1080/10798587.2016.1267444
2. Zhou, Q., Xu, Z., Yen, N.Y.: User sentiment analysis based on social network information and its application in consumer reconstruction intention. Comput. Hum. Behav. **100**, 177–183 (2019). https://doi.org/10.1016/j.chb.2018.07.006
3. Zhou, Q., Lou, J., Jiang, Y.: Optimization of energy consumption of green data center in e-commerce. Sustain. Comput.: Inform. Syst. **23**, 103–110 (2019)

The Smart Grid System Design Based on the Big Data Technology

Hui Zhang[1](✉), Wentao Yan[2], Wei Wang[2], Qi Su[2], Wei Zhou[2], and Bin Zhang[2]

[1] Beijing China Power Information Technology Co., Ltd.,
Beijing 100085, China
buwen325052@126.com
[2] State Grid Shandong Electric Power Company Information
Communication Co. Ltd., Ji'nan 250000, Shandong, China

Abstract. The design and application of the smart grid system based on the big data technology has achieved some results, but no system has been formed. Therefore, the design, research and application of the smart grid system are still in the exploratory stage. With the continuous maturity of the big data technology application, the application of the big data technology in the smart grid system design will provide a more intelligent and perfect management mechanism for the grid operation, so as to improve management efficiency and achieve the expected goal of the power grid safe operation.

Keywords: Big data technology · Smart grid · Management system · Design research

The smart grid has produced a lot of data. The application of the big data technology can effectively improve the operation and management level of the power grid and the service level for the society and the users. At the same time, it also brings great challenges [1]. It is necessary for all parties to reach a consensus on the basis of fully understanding the significant benefits that the big data technology can bring and jointly promote the research and the technical development of the smart grid big data.

1 Design Background of the Smart Grid System Based on the Big Data Technology

Both the smart grid and the big data technology are the emerging products in recent years [2]. They are the high-tech developed with the popularization of the computer technology, the communication technology, the network technology and the digital technology.

1.1 Smart Grid

The smart grid is the intelligence of the power grids. It refers to the use of a large number of the intelligent devices connected to the power system to realize the

M. Atiquzzaman et al. (Eds.): BDCPS 2019, AISC 1117, pp. 1838–1843, 2020.
https://doi.org/10.1007/978-981-15-2568-1_257

intelligent and automatic power grid operation. It is a new product in the 21st century. In the smart grid, the sensor technology, the decision support system technology, the intelligent automation control technology and the measurement technology are integrated, which are realized based on the multiple communication networks [3]. Its core system consists of the intelligent substation, the intelligent distribution network, the intelligent dispatching system, and the intelligent energy meter and so on [4]. The smart grid can reduce the operation cost of the power grid, improve the strength of the power grid, reduce the loss of the power grid, and realize the economic, safe and reliable operation of the power grid [5]. Moreover, the smart grid allows the multiple forms of the power generation access, which provides convenience for the new energy utilization, has certain self-healing function and attack resistance ability, and effectively reduces the grid failure rate.

1.2   Big Data Technology

In the 21st century, the human society has entered the information age, and the information technology has been integrated into various fields, and the amount of the data generated in the social activities is increasing. The big data technology has been paid attention to and applied. Under the big data technology, there are many types of the data and information, with the large amount of data, the fast transmission speed, the high efficiency, and the wide sources of the information, and the unit of the measurement is at least P, and the form of the information transmission is mainly two-way or multi-directional communication. The data types can be audio, coding, or images and the heterogeneous data transmission can be realized. The data audience is very large, but the information value density will decline. The data purification is required.

1.3   Application of the Big Data Technology in the Smart Grid System

With the development of the power informatization and the construction and application of the intelligent substation, the intelligent electricity meter, the real-time monitoring system, the on-site mobile maintenance system, the integrated measurement and control system and a large number of the information management systems serving various disciplines, the scale and types of the data are growing rapidly, which together constitute the big data of the smart grid. These data are not completely independent, and they are interrelated and influenced by each other, and there is a more complex relationship. The GIS data of the electric power enterprises must take the municipal planning data as reference. In addition, these data structures are complex and various. In addition to the traditional structured data, they also contain a large number of the semi-structured and unstructured data, such as the voice data of the service system, the waveform data in the detection data, and the image data taken in the helicopter patrol inspection. And the sampling frequency and the life cycle of these data are also different, from the microsecond level to the minute level, and even to the annual level.

2   The Demand of the Smart Grid for the Data Transmission and Communication

Obviously, the smart grid is different from the traditional power grid, which requires the high data communication technology and has certain particularity. If the communication quality cannot be guaranteed, it will inevitably cause negative impact on the power grid system. The safe and efficient operation of the smart grid is inseparable from the support of the communication network. In the process of the power grid operation, it is necessary to deal with, control and transmit the power production, transmission, dispatching, consumption and other related information technologies in an efficient and timely manner, so the communication network is required to meet the requirements of stability, efficiency, real-time, and bidirectional nature, involving the data business, the graphics business and the special business.

According to the different transmission rates and the information types, it can be divided into the backbone communication network and the terminal access network. The network structure may be divided into the tree type and the line type. The smart grid has the high requirements for the communication quality. The communication coding error may cause the system disoperation, wrong operation and other problems, resulting in the grid fluctuation, and even the unplanned power outage, affecting the power supply stability and reliability. The data flow of the electric power communication is large, and the flow direction changes a lot. The communication nodes are complex, and the traffic volume is very large, which is very suitable for the application of the big data technology. The integration of the big data technology enhances the business connection, reduces the difficulty of the system control, improves the system stability and the data transmission rate, improves the network vulnerability, and avoids the communication failure caused by the failure of a node.

The smart grid has the higher requirements for the communication and the data transmission. The traditional data technology obviously cannot meet the communication needs, and its limitations are more and more obvious. The integration of the big data technology enhances the security, stability and reliability of the smart grid operation, and solves the communication problem of the smart grid. Under the traditional data communication technology, the smart grid communication is based on the static routing algorithm, which has a certain lag. When the network is expanded, it needs to be updated, so the communication speed is slow. The routing table data in the multi-node communication is very large, which will directly affect the addressing speed, resulting in the decrease of the communication speed. The big data technology integrates the dynamic multi-path routing algorithm and optimizes the addressing mode. In the big data and multi-node communication, it can still address quickly and adopts the hybrid driving addressing mode to reduce the addressing delay.

3 Smart Grid System Design Based on the Big Data Technology

3.1 Data Integration Technology

The smart grid big data has the characteristics of decentralization, diversity and complexity, which bring great challenges to the big data processing. In order to deal with the big data of the smart grid, it is necessary to integrate the data of many data sources, and establish the correct, complete, consistent, complete and effective big data of smart grids through the data extraction, conversion, elimination, correction and other processing. At present, the commonly used data integration models include the data federation, the middleware based model and the data warehouse. The data extraction is to extract the data needed by the target data source system from the targeting data source system. The data conversion is to convert the data obtained from the targeting data source into the form required by the targeting data source according to the business requirements, and to clean and process the wrong and inconsistent data. The data loading is to load the converted data into the targeting data source.

3.2 Data Storage Technology

In the smart grid big data, the vast majority of the data are the structured data, but there are also unstructured or semi-structured data such as texts, images, audios, and videos and so on. For the unstructured data, the distributed file system can be used for storage, and for the semi-structured data with the loose structure and no mode, the distributed database can be used, and for the massive structured data, the traditional relational database system or the distributed parallel database can be used. Hadoop is a solution to the big data, which can realize the storage, analysis and management of the big data. The HDFS is a distributed file system, which is a family member of the open source project Hadoop. HDFS divides the large-scale data into 64 megabytes data blocks, which are stored in a distributed cluster composed of the multiple data nodes. When the data scale increases, only more data nodes need to be added in the cluster, which has the strong scalability. At the same time, each data block will store multiple copies in different nodes, which has high fault tolerance. Because the data is stored in a distributed manner, it has the high throughput data access ability. New technologies are needed for the data storage, management, query and analysis in the big data environment. There are bottlenecks in the traditional database in the data storage scale, the throughput, the data type and the supporting application.

3.3 Data Processing Technology

There are many application types of the big data in smart grids, so different data processing technologies need to be adopted according to different business needs. According to the data characteristics and the calculation requirements of the big data, the big data processing technologies include shunting, batch processing, memory computing, and graph computing and so on.

The processing mode of the flow processing regards the data as a flow, and the continuous data constitutes the data flow. When the new data comes, it immediately processes and returns the required results. The data flow itself has the characteristics of continuous reaching, fast speed and huge scale, so it is not always permanent storage of all the data, and the data environment is in constant changes, so it is difficult for the system to accurately grasp the full picture of the data. At present, the widely used stream processing systems are Twitter Storm and Yahoo S4. Storm is a distributed real-time computing system, which is mainly used for the stream data processing. It can process a large number of the data streams simply, efficiently and reliably. It can process the continuous flow of the information, and then write the results to certain storage. The advantage of Storm is the full memory computing, because the memory addressing speed is more than one million times that of the hard disk, so Storm's speed is faster. Storm makes up for the real-time requirements that Hadoop batch processing cannot meet. It is often used in the real-time analysis, the online machine learning, the continuous computing, the distributed remote call and the ETL and other fields. Compared with the Hadoop Map-Reduce batch processing, the memory computing can provide the high-performance big data analysis and processing capabilities. The memory computing is an architectural solution, which can be combined with various computing modes, including the batch processing, the stream processing, and the graph computing and so on. For example, Spark is a typical parallel computing framework for the distributed memory computing. The distributed computing implemented by Spark based on Map-Reduce algorithm has the advantages of Hadoop Map-Reduce. But unlike Map-Reduce, the intermediate output of Job can be saved in memory, so there is no need to read and write HDFS. Therefore, Spark has better performances and is suitable for the Map-Reduce algorithm that needs the iteration such as the data mining and the machine learning.

3.4 Data Analysis Technology

The data analysis is the core of the big data processing in smart grids. Because of the characteristics of the big data, such as the massive, complex and fast changing, many traditional small data analysis algorithms in the big data environment are no longer applicable, and need to adopt the new data analysis methods or improve the existing data analysis methods. The data mining methods mainly include classification, association analysis, clustering, anomaly detection, and regression analysis and so on, each of which includes many algorithms. Classification includes the support vector machine, the decision tree, Bayesian, the neural network and other technologies. The association analysis includes Apriori, FP-growth and other algorithms. The clustering analysis is divided into the division method, the hierarchy method, the density method, the graph theory method, the model method and other specific algorithms, such as the k-means algorithm, the K-MEDOIDS algorithm, the Clara algorithm, the Clarans algorithm, the SOM neural network, and the FCM clustering algorithm and so on. The anomaly detection includes statistics, distance, deviation, density and other methods. In the application of smart grids, the existing algorithms need to be optimized and parallelized to achieve the distributed processing.

The machine learning is a task-oriented computer program based on the experience extraction model to achieve the optimal solution design. Through the rule of the experience learning, it is generally applied in the field of the lack of the theoretical model guidance but experience observation. The machine learning is divided into the induction learning, the analysis learning, the analogy learning, the genetic algorithm, the connection learning, and the reinforcement learning and so on. The deep learning is a new field in the machine learning research. It aims to build a neural network to simulate the human brain for the analysis and learning. At present, the deep learning has been applied in the speech recognition, the image recognition, the machine translation and other fields, and achieved good results. The smart grid big data mining is mainly the structured data, but there are also texts, images, audios, videos and other data. In the smart grid big data application, we need to adopt the appropriate data analysis methods for the specific businesses.

4   Conclusion

Each functional module in the platform is independent of each other and exchanges information and data through the bus. As a key link in the smart grid system, the scheduling link needs to process the massive data information, including the storage, analysis and processing, and engine search and so on. The big data technology can achieve the efficient processing of the large-scale data, while the resource sharing function and the powerful computing ability of the cloud computing provide the strong technical support for the calculation, analysis and operation of the data information of the intelligent grid dispatching system. While ensuring the reliability and accuracy of the system, in order to improve the real-time responsiveness of the cloud computing, the improved genetic algorithm is introduced into the cloud computing, and the platform resources are distributed scientifically and efficiently through the real-time monitoring and the reasonable coordination.

References

1. Zhang, H., Liu, L.: Design of the smart micro grid energy management system. J. Shenyang Inst. Eng. (Nat. Sci. Ed.) **08**(3), 201–203 (2012)
2. Wen, K.: Research and analysis of the micro grid and its energy management system. New Bus. Wkly. (5), 133 (2016)
3. Wang, S., Zu, Q., Niu, Y.: Research on the micro grid energy management. Electr. Eng. Autom. **38**(4), 80–82115 (2016)
4. Chen, M., Zhang, K.: Research on the optimal allocation of the energy storage capacity of wind and solar energy storage combined power supply system based on the improved cuckoo algorithm. Power Syst. Clean Energy (8), 141–146 (2016)
5. Xiao, F., Xiao, H.: Analysis of the joint power supply mode of the household solar power generation and the municipal power grid. West Leather (24), 18 (2016)

Design of the Interactive Intelligent Machine-Aided Translation Platform Based on Multi-strategies

Shuo Zhang[✉]

Liaoning University of International Business and Economics,
Dalian 116052, Liaoning, China
dingding8993930@126.com

Abstract. The interactive intelligent machine-aided translation platform based on multi-strategies can meet the needs of users for the fast translation, and use various intelligent tools to assist the manual translation, so as to effectively improve the efficiency and the quality of the translation. By adopting the leading technology of the human-computer interactive machine translation in the industry, integrating many advanced technologies such as the neural network machine translation, the statistical machine translation, the input method, the semantic understanding, and the data mining and so on, and cooperating with the bilingual parallel data of the bilingual level, we can provide users with the real-time intelligent translation assistance to help them complete the task of the translation better and faster. The purpose of the product is to pay homage to the artificial translation, assist the artificial translation to complete tasks faster and better, and then explore a new way of the AI enabling translation industry.

Keywords: Multi-strategy · Interactive · Intelligent machine · Assistant translation · Intelligent platform design

At present, due to the limitation of the machine translation technology, the accuracy of the machine translation is not high, so the real-time translation suggestions cannot be too long. Otherwise, the efficiency of the manual translation will be reduced. The difficulty of the real-time translation suggestions is how to provide phrases or clauses with the moderate length and the high accuracy [1]. To this end, the multi-strategy interactive intelligent machine-aided translation intelligent platform implements the intelligent translation recommendation algorithm, integrates the statistical machine translation, the neural machine translation and other technologies, and tries to provide users with the most appropriate translation suggestions [2].

© Springer Nature Singapore Pte Ltd. 2020
M. Atiquzzaman et al. (Eds.): BDCPS 2019, AISC 1117, pp. 1844–1849, 2020.
https://doi.org/10.1007/978-981-15-2568-1_258

1 Design Background of the Interactive Machine-Aided Translation Intelligent Platform Based on the Multi-strategies

In order to improve the translation processing effect of the practical machine translation system, an overall design scheme of the interactive intelligent assistant translation platform based on the multi-strategies is proposed [3]. The system is based on the multi-knowledge integrated representation, the multi-translation processing strategies, the multi-translation knowledge acquisition methods, and the multi-strategy translation selection and so on [4]. By using the object-oriented multi-type knowledge database management, the interactive intelligent processing of the multi-strategy and multi-knowledge is effectively realized. The platform realizes the integrated processing of the multi-translation modes based on the rule analysis, the analogical reasoning and the statistical knowledge, provides the human-computer interaction interface, and realizes the manual intervention of the translation results, and object-oriented engineering task management and the user management.

The framework used in the machine translation is the same, mainly for the detail processing or the improvement of some problems themselves. The current structure is the structure of the encoder and the decoder. This framework should be familiar with the in-depth learning. The most difficult one is to code the original sentence through RNN or other means through every word, and finally form a sentence vector [5]. At the decoding end, the blank is generated from the sentence according to the sentence vector. For example, the basic principle is very simple. At the end of the decoder, from the beginning of the sentence to the end of the sentence, it stops the translation process. If it does not encounter the beginning and the end of the sentence, it will continue the translation process.

In recent years, with the rapid development of the machine translation technology based on the neural networks, the quality of the machine translation has been continuously improved, and the argument that the machine translation replaces the manual translation has become very popular. In practice, the machine translation is often full of errors and omissions at this stage. In the absence of the human intervention, the accuracy of the machine translation cannot meet the translation requirements of the specific application areas, such as the commercial contracts, the legal provisions, the professional books, the academic literature, and the tourism texts and so on. A qualified professional translator needs not only the long-term professional and the hard training, but also a lot of practical experience, in order to be able to compete in the specific professional fields of the translation work. In the actual application scenario, the accuracy of the machine translation is still not comparable to that of the professional translation.

However, due to the enhancement of the computing power, the innovation of the machine translation model, and the large scale of the corpus, the quality of the machine translation has been improved significantly, and the translation speed is much faster than that of the manual translation. At present, the machine translation can provide the translation reference, help the manual translation to complete some relatively mechanical tasks, the free manual translation from the complicated and inefficient manual work of typing and word search, and devote the energy to a higher level of the translation creation. In the process of the artificial intelligence technology such as the machine translation assisting human translation, the human translation will obtain the personalized machine translation results. Through the interaction between the humans and the machines, the efficiency and the quality of the translation can be further improved. The machine translation cannot replace the manual translation, but it will reconstruct the process of the manual translation. The interactive intelligent machine-aided translation platform based on the multi-strategies is a technological innovation product developed on this idea, aiming at improving the efficiency and quality of the manual translation and meeting the growing demand for the translation.

2 Key Points of the Interactive Intelligent Machine-Aided Translation Platform Design Based on the Multi-strategies

The interactive machine translation (IMT) is a constrained translation and decoding technology based on the autonomous implementation. While improving the accuracy of the translation, the decoding speed is also optimized to meet the requirements of the real-time interaction in the Internet environment. The real-time decoding speed is a key factor affecting the landing of the interactive machine translation products. Combined with the existing accumulation, an interactive intelligent machine-aided translation platform based on the multi-strategy is developed and a neural machine translation system dedicated to human-computer interaction is implemented. Unlike the ordinary machine translation, the challenge of the interactive machine translation lies in the inability to predict the user actions, which makes it difficult to speed up the response by establishing caches in the original text.

Compared with the traditional assistant translation software, the source of the intelligent translation suggestions is no longer limited to the terminology library imported by users, but the comprehensive integration of the massive Internet data. Tens of millions of the professional terminology translations and billions of the bilingual examples are excavated from the Internet texts to provide the users with the translation reference information. Then, the reference information such as the terminology library and the example sentence library is displayed with the sentence as the dimension correlation to meet users' multi-domain and multi-style translation needs. In a multi-strategy machine translation system, the rule knowledge, the statistical knowledge, the dictionary knowledge, the pattern knowledge, and the case knowledge based on the surface strings are involved. These kinds of the knowledge and other knowledge based on the knowledge are complex and relevant in various processing links, such as the prior knowledge storage, the translation processing, the editing modification and the post-translation knowledge storage. The quality and management efficiency of the knowledge base are very important to the processing performance of the system. Therefore, it is necessary to organize and process the knowledge by using the object-oriented knowledge representation and organization structures.

The machine translation has a unique advantage in solving the language barriers on the Internet. However, due to the inherent difficulties of the machine translation, the current fully automatic translation system is still difficult to meet the user's requirements for the quality of the translation. The translation is often full of errors, unsatisfactory. Another way to solve the language barrier on the Internet is through some translation agencies, which download some foreign language pages regularly and upload the translated texts after being translated by the professional translators. This method can guarantee the quality of the translation, but the translation is completely done by hand, and the efficiency is difficult to guarantee. In order to solve the above two problems, we put forward the idea of the interactive multi-strategy machine translation (IHSMT). The system integrates the advantages of the above two methods and provides a set of the perfect human-computer interaction translation mechanism, so that it can get more accurate translation results quickly and efficiently.

The knowledge in the interactive translation platform exists in the bilingual form, and its content is changeable and complex. It needs reasonable knowledge representation and unified storage to improve the applicability of the knowledge description and knowledge representation. The system uses the object-oriented technology to effectively store and manage the knowledge and improve the operability of the knowledge and the consistency of the storage. Considering the difficulty of the semantics division and the inexhaustibility of the theory, the system takes the bilingual entry pairs in the dictionary as the object representation based on the corresponding semantic dictionary. The system adopts the object-oriented technology and adopts examples with the inheritance relation for various forms of patterns. The schema class representation improves the operability of the schema and the consistency of the storage.

3 Design of the Interactive Intelligent Machine-Aided Translation Platform Based on the Multi-strategies

The translation engineering refers to the collection of documents with a common theme or category, which are related to each other. The object of the translation engineering includes the attributes and methods of the translation engineering. The attributes of the object of the translation include the source of the project, the translation time, the translation type, the language, the specialty, and the path and so on. The object method includes the storage translation documents, the translation information, the engineering complexity analysis and other operations. Through the management of the project, putting the similar professional tasks into a project can not only realize the sharing of the knowledge, but also effectively reduce the scope of the ambiguity, maximize the use of the existing translation results, and effectively improve the reuse of the knowledge. In addition to the translation process, the system also realizes the management of the tasks and the personnel, and sets different privileges for different types of the engineering users to improve the safety of the engineering management.

The system can translate various kinds of projects, and there are many connections among them. Most of them have the inheritance relations. Using the object-oriented method to describe the translation engineering objects not only makes the relationship between projects more reasonable, but also manages the translated projects effectively, and makes full use of the translated projects, to make effective use of the translation knowledge, contents and other information. Through the management of the project, putting the similar professional tasks into a project, the knowledge sharing can be achieved, and the existing translation results can be used as much as possible, and a lot of the duplication of work can be avoided, thus greatly improving the working efficiency.

There are a large number of examples in the case-based schema library, which need to be tagged. It not only provides the data for the higher-level language processing, but also provides the accurate and reliable materials for the linguistic research, and provides the lexical and statistical support for the further language processing. By explaining the words or the structural attributes in the database, it is based on the schema features. The corpus is annotated, which can be divided into part-of-speech, grammar and semantic annotation according to the form of the annotation, the automatic annotation and interactive annotation according to the degree of the automation of annotation, the single-user and the multi-user annotation according to the scale of users, and the complete annotation and partial annotation according to the degree of the annotation.

All kinds of the meta-knowledge bases are relatively independent in physics, and different knowledge bases are relatively centralized, but the representation of the knowledge (related knowledge) among them is managed by semantics. Because of the many levels of the knowledge in the system, the system takes the special dictionary (semantics) as the key value of the object, and contains all the relevant rule knowledge and system in the object attributes, the accounting knowledge, the schema knowledge and the case knowledge. Since the semantic knowledge cannot be exhausted, in the specific processing, if more comprehensive knowledge retrieval is needed, it can be retrieved in the corresponding independent meta-knowledge base.

Most of the existing practical machine translation systems are based on rule analysis and transformation. However, due to the complexity and changeability of the natural language and the great randomness of its use, this kind of the translation system not only needs to establish a huge rule system to describe the complex and diverse language phenomena, but also needs to constantly add the individual rules to enhance the adaptability of the system. When the number of the rules increases to a certain extent, it will inevitably lead to redundancy, conflict and other phenomena, which makes the construction and maintenance of the high-quality machine translation knowledge base difficult and costly. The contradiction between the infinite language phenomena and the enumeration nature of the rule-based machine translation has certain limitations.

The intelligent machine-aided translation algorithm runs fast, is easy to implement, and can automatically process the noisy data. Its common uses include:

(1) Prediction: This is the most important application of the intelligent machine-aided translation algorithm. The existing data sets are used to train the classification or regression models to predict the future situations.

(2) Feature selection: When the intelligent machine-aided translation algorithm is used for the classification and regression problems, its basis is to select the best features step by step. In this process, the interrelated features will be filtered, and the features of little value to the problem will also be filtered. Therefore, the extension of the intelligent machine-aided translation method is applied. It is to extract the most useful features for solving problems.

(3) Evaluate the relative importance of the features: After identifying a set of the problem-related features, the relative importance of each feature can be assessed through the intelligent machine-aided translation model. The feasibility of this hairstyle is based on the fact that the intelligent machine-aided translation gives priority to the use of relatively more important features for prediction.

(4) Deal with the missing values: In the process of the feature acquisition, not all samples can acquire all features. Whether the data records with the missing feature or the missing value feature are discarded, bias will be introduced into the samples, resulting in the distortion of the results. The intelligent machine-aided translation algorithm can classify the missing value features into the separate classifications which can be distinguished from other classifications. It can also set the missing value features as the predictive targets of the decision tree model to predict, and replace these missing values with the predictive values of the model.

4 Conclusion

The visualization of the intelligent machine-aided translation will give us an intuitive understanding of the problem-solving process, which is very helpful for the data analysis. In the case of restricting the depth of the intelligent machine-aided translation, the visualized tree model will show the features, feature boundaries and data set partitioning of the model at the beginning, and provide the intuitive basis and information for the analysis of problems.

Acknowledgement. This paper is a phased achievement of the undergraduate quality project of Liaoning Institute of Foreign Trade and Economics. The title of the project is "Research and Practice of the English Translation Course Teaching Mode Reform Based on CDIO 'Multidimensional' Innovative Education Practice Platform", Project Number 2018XJGYB24.

References

1. Sun, H., Li, J., Gong, Y.: Design of the knowledge collaborative management system based on the SharePoint platform. Comput. CD Softw. Appl. (15), 204 (2011)
2. Huang, Y.: Research on the distribution collaborative design platform based on the knowledge management. Mark. Forum (4), 40–41 (2012)
3. Kong, B.: Library knowledge management and knowledge navigation based on the modern information technology. Speed Read. (8), 262–263 (2014)
4. Feng, Q, Cui, Q.: Post-translational editorial research: focus dialysis and development trends. Shanghai Transl. (6), 27–33 (2016)
5. Xia, Z.: Exploration of the Google online library aided English translation. J. Hotan Teach. Coll. (3), 44–47 (2015)

Design and Implementation of the Interactive Chinese Painting Display System

Jingsi Zhu[✉]

Zhengzhou University of Science and Technology,
Zhengzhou 450000, Henan, China
j363wg@yeah.net

Abstract. The purpose of this study is to realize the interactive display system of Chinese paintings combining with the traditional Chinese painting art. The appropriate elements are selected as the input devices and the input instructions, and the projector is used as the output device to display the layout art of Chinese paintings. By using the tablet, visitors can sign or draw on the new layout paintings, and encourage the recreation, thus realizing the application value of the interactive Chinese painting display system.

Keywords: Interactive · Chinese painting · Display system · Design · Implementation

The core technology of the interactive Chinese painting display system is the "visualization 3D restore technology", which integrates the 3D data collection, the virtual reality, the color restore, the real-time rendering and other scientific and technological means [1]. This technology has laid a solid data foundation for the protection and restoration of the cultural heritage, research and display, and communication and education.

1 Definition and Classification of the Interactive Demonstration System

The interactive whiteboard demonstration system is a product or solution that takes the human-computer interaction as its tenet and integrates a variety of the high-tech products such as cutting-edge electronics, induction, projection, and application software and so on, through the principle of the touch sensing, so as to create the interactive and collaborative space for users [2]. The interactive whiteboard demonstration system consists of the projection system, the interactive whiteboard, the audio system, the whiteboard application program in the offline state, the whiteboard application software, the connecting cable, and the control panel and remote control device. At the same time, the system can be connected with computers, cameras, video recorders, DVD players, USB storage and other devices.

The interactive whiteboard is connected with the computer and the audio system [3]. The projector projects the computer images onto the whiteboard. The whiteboard

M. Atiquzzaman et al. (Eds.): BDCPS 2019, AISC 1117, pp. 1850–1855, 2020.
https://doi.org/10.1007/978-981-15-2568-1_259

becomes a large screen interactive operation platform. The audio system transmits the sound files. The control panel and the remote control device can operate any equipment and external equipment in the system. Using the hand or the pointer instead of the mouse to click or browse the computer contents on the whiteboard, we can write, draw, edit, annotate or operate any application program in the real time, and can save files in various ways such as pictures, PowerPoint, web pages and PDF [4]. The offline whiteboard application supports users to open the projection system to write and save on the whiteboard without connecting to the computer.

Integrated interactive demonstration system: The integrated interactive demonstration system refers to an interactive demonstration system integrating the demonstration system, the projection system, the audio system and the operation control program [5]. On the one hand, this kind of system has powerful functions of all subsystems and the strong compatibility, and on the other hand, it avoids laying and connecting equipment cables, which saves space and is easy to install. It can also write and save without connecting to the computer, which is more convenient. The simple installation and the rich powerful functions make the integrated interactive demonstration system become the first choice of many users.

Combined interactive demonstration system: The combined interactive demonstration system refers to the assembly and combination of different types of the demonstration system, the projection system, the audio playback system and the operation control program according to different needs of users. This kind of solution needs to be fully considered and designed in the compatibility between various devices. The installation process is complex and tedious, and it does not have the functions of operation and storage when leaving the computer.

2 Design Requirements of the Interactive Chinese Painting Display System

"Brush and ink should follow the times", which seems to have become the motto of the Chinese painting innovation. The sensitive painters of the new generation, when they reflect on their own national painting history and current situation for thousands of years, find that no matter whether it is the so-called conservative or the innovative school of painting in the history of painting, in the end, they are just repeating their predecessors' brush and ink techniques and experience. That is to say, on the whole, the form and essence of the traditional Chinese painting is stable and has not changed with the times. Especially after the Song and the Yuan Dynasties, the so-called innovation and creation of paintings is actually the arrangement and combination of lines and ink colors. The "four kings" of the Qing Dynasty advocated the restoration of the ancients. In order to achieve the painting styles consistent with the original meanings of the ancient paintings, they comprehensively sorted out the excellent painting techniques and forms since the Song and the Yuan Dynasties to a certain extent, making them more mature and playing a relatively stable role. Shi Tao, Zhu Da and the eight monsters of Yangzhou took the road of innovation. Their style of paintings is livelier than the "four kings". We can regard them as the opposition between the innovation and the conservatism, but in fact, they are complementary to each other, because Shi Tao, Zhu Da and the eight monsters of Yangzhou are still interlinked with the "four kings" in the spirits of writing and ink, and strive to maintain and carry forward the basic characteristics of the excellent painting tradition of China in the form.

As a visual art, painting has formed two schools with different styles in the world. Chinese painting takes the ink lines as its essential feature, while the western painting takes the block colors as its main feature. With the progress of the human civilization, science and technology in the West has developed much faster than in the East. The western paintings have the lifelike charm, as well as the systematic theories such as the optical principles and the space geometry theorem, which seem to be "more scientific" and "more advanced" than the Chinese paintings. Therefore, some people are amazed and ashamed. In the recent decades, the communication between Chinese and the western painting schools has become more and more frequent, with the trend of the overwhelming lines with blocks and replacing ink with colors. This is not that "the ink should follow the times", but that "the ink should disappear with the times". It seems that as long as the traditional Chinese aesthetic psychology is maintained, it is enough. Although Kant affirmed the treasure of the Oriental painting, the Chinese painting, hundreds of years ago (though not consciously), he thinks that the color appeals to the pleasure of feelings, and that the line appeals to rational contemplation. The line is a more meaningful form than the color, but there is still a trend of thought about the crisis of the Chinese painting brush and ink caused by many factors. Therefore, it is necessary to analyze and study the formal system of the Chinese paintings from a scientific and holistic perspective to distinguish right from wrong.

As a visual art, painting deals with the visual information of the object images, accompanied by a series of the high-level nervous system activities such as people's thoughts and feelings. People's visual information is mainly processed in parallel, and the combination of the non-linear way and the human thoughts, feelings, cultivation and other comprehensive information constitutes a network type feedback way. Therefore, the information capacity of the formal structure of the Chinese and the Western painting schools far exceeds that of the visual system. This is the reason why the painting will flourish without any scruples after the production of the camera imitating the human eyes. In this sense, the main function of the painting art is not feeling but perception.

3   Design Framework of the Interactive Chinese Painting Display System

3.1   Design Goal

The main design goal of the system is to establish a three-dimensional platform for the full simulation of the real world, which is used for the display of Chinese paintings, in a virtual environment, so that the system can view and follow the movements of Chinese paintings and the sports models, and at the same time, with the help of the interactive click, click on the objects of Chinese paintings, and we can view the specific properties of Chinese paintings and the related instructions. It can also view all kinds of Chinese paintings in the Chinese painting library through the click operation, and click different icons to realize the virtual interactive dressing process, which reflects that nowadays the computer technology can completely provide assistance to the Chinese painting design, and has more powerful functions. This system is a subversion of the previous Chinese painting design, which has been continuously studied in recent years, and is concerned by the relevant fields, which can be said to play a great role in promoting the progress of the Chinese painting design.

3.2   Functional Structure Design

First of all, we need to build the corresponding model of the Chinese painting system with the help of the modeling tools, and the 3D Chinese painting also needs to build the corresponding model. The human body action needs to be set, and after setting the model and actions, we can import the processed actions and models to the Virtools engine of the virtual platform for the data import, so that the Virtools engine can load the relevant model and attribute the data information, and establish the corresponding visual database. With the help of the interaction plate, the interaction settings can be realized, and then the scene optimization function. The so-called scene optimization, in fact, is to render the effect and optimize the scene model, so that users can easily browse the effect of Chinese paintings through the system.

3.3   Virtual Display Technologies

The virtual reality display design based on the 3D modeling is based on the 3D scene and the 3D object model. For example, the scene and products in the exhibition are completed with the help of the professional modeling software (such as SolidWorks and 3DMAX and so on), and the built model is 3D, so that the real world can be expressed conveniently and realistically in the built scene. For example, it is easy to open the 3D display. You just need to add a stereoscopic display yourself or call this function from the OSG (OpenScene Graph) library. This system uses 3DMAX to build the 3D model, and reads the location information of the model through the XML files.

3.4   Characteristics of OSG

OSG is a cross platform application program interface based on the C++ language, which enables programmers to create the high-performance, cross-platform and interactive graphics programs quickly and conveniently. The technology consists of two parts. One is to organize, manage and traverse the scene, and the other is to render the scene and realize the continuous level details of the scene model. For the organization and management of the scenes, OSG adopts the data structure of the scene graph to organize scenes and their attributes into graphs. OSG uses the hierarchy to represent the scene, and the nodes in the scene are the basic elements of the scene graph.

4 The Realization of the Interactive Chinese Painting Display System

The application of this technology in the field of the cultural heritage can make all kinds of the cultural heritage, such as cultural relics, ancient buildings, and ancient sites and so on, accurately restored on the computer. As the carrier of the technology, this touch interaction system realizes the digitization and visualization of the cultural heritage. Users can not only appreciate the artistic beauty of China's precious cultural heritage, but also understand the development history of the Chinese civilization through the relevant operation and viewing experience on the touch screen. At the same time, they can fully feel the organic integration of the modern technologies and the historical cultures.

Compared with the exhibition of the traditional cultural relics, the main innovative features of the "cultural relics show you China" exhibition project are: interactive interest, close contact and mobile cultural relics for the audience to learn more stories contained in the cultural relics. Content renewability: the digital contents in the equipment can be expanded and updated according to the actual needs, design different themes, and constantly innovate the exhibition. Display mode: the sustainability of the impact, without the limitation of the exhibition period of the traditional exhibitions, can rely on various platforms for the long-term display, forming a continuous impact on the audience.

In terms of the product function, the Chinese cultural relics displayed by the visualization 3D digital reduction technology are rich in images, vivid and lifelike, clear at a glance, full of digital charm. It has a strong impact on the audience, which makes them fully feel the integration of China's advanced technologies and historical cultures. This way of communication makes the perceptual impact of the cultural relics and the rational summary of the historical knowledge reach a high degree of unity. It can not only enjoy the aesthetic taste brought by precious cultural relics, but also seek the historical wisdom. It is worth mentioning that this technology has completely independent intellectual property rights and is also at the leading level in the world. With the rapid development of the information technology, the multimedia technology and the sensor technology, the contents of the dispatching and monitoring center become more and more diversified and integrated. In particular, more requirements are put forward for the display contents and the display modes, including the ultra-high resolution, the 3D data visualization, the human and display contents, interaction, equipment linkage, and display contents and so on.

Based on the real-time image display contents, various forms of the human-computer interaction can be realized by adding the detection equipment. Presentations can be controlled in the real time via the touch screen, iPad, iPhone or PC. You can also use the airscan technology, which uses the gesture recognition technology to operate the required presentation directly on the large screen. The new 3D real-time data visualization interactive studio system based on the real-time images, database links and interactive devices makes full use of the large screen system. The customized design and perfect interactive experience make the customized presentation of the key contents in the system impressive to visitors. The network IT updates, the new firmware, and even the hardware failures can cause the system to need help.

5 Conclusion

In the traditional text interactive Chinese painting display system, "four Wang paintings" works are selected. In the heritage display, the effect of using the new interactive equipment is usually unable to be reprocessed with the digital paintings required by the system by all the experts, and the various environments of the separated paintings are kept in coordination, resulting in the visitors' feeling of abruptness. In this interactive Chinese painting display system, visitors can choose to appreciate different paintings or change different layout forms of the same painting by playing the strings, and can sign or redraw their own artistic creation on the painting with the writing board. The new layout or innovative projector is output to a white glass curtain wall, and the effect is like a hanging Chinese painting, so as to give visitors a sense of reality.

References

1. Ling, Y.: Analysis of the application of the new technology in the display design – taking the interactive virtual display design as an example. Art Sci. Technol. (10), 129–137 (2016)
2. Zhang, L., Liu, C.: Design and implementation of the filling production line simulation system based on EON. Comput. Knowl. Technol. **08**(19), 4766–4769 (2012)
3. Chen, N., Wu, X., Zhang, F.: Design and implementation of iPhone public transport information query system based on WebGIS. Sci. Surv. Mapp. **34**(6), 276–278 (2009)
4. Liu, Z., Shen, J., Liu, Y., Liu, M., Zhang, H., Shang, J.: Acoustic vibration characteristics of spruce wood for musical instrument sound board. Sci. Silvae Sin. **43**(8), 100–105 (2007)
5. Leng, J., Zhao, T., Fang, H., Li, X., Li, B.: Musical instrument audio segmentation algorithm based on the variance stability measurement. Comput. Eng. Des. **37**(3), 768–772 (2016)

The Data System of the Public Platform of Industrial Economic Informatization Under the Strategy of Rural Revitalization

Kunyan Zhu[✉]

Zhengzhou University of Science and Technology,
Zhengzhou 450064, Henan, China
f41T7Z@yeah.net

Abstract. In the process of promoting the strategy of the rural revitalization in an all-round way, we should take breaking the dual structure of the urban and rural areas, accelerating the flow of the urban and rural resources and integrating the urban and rural development as the main support points, give full play to the advantages of the big data technology, and promote the rural revitalization. Through the construction of the data system of the public platform of the industrial economic informatization, including the cooperation of all parties, the big business data and the big farmer data, an ecological layout with the farmer families as the core, covering new ecologies, new businesses and new technologies, a super platform in line with the rural development is formed, so that the farmers can truly share the achievements brought by the digital economy.

Keywords: Rural revitalization strategy · Industrial economy · Information public platform · Data system · Construction strategy

To speed up the construction of the "digital village" is an urgent need to realize the rural revitalization. Building a modern new rural area with "prosperous industry, livable ecology, civilized countryside, effective governance and rich life" is faced with many constraints, such as the resource environment and the personnel technology and so on, which cannot be solved by the traditional means [1]. Under the background of the information revolution, we should make full use of the Internet thinking and the information technology to seek unconventional and unconventional ways to promote the development of the agricultural and rural areas in China to achieve the quality change, the efficiency change and the power change.

1 Objectives and Development Requirements of the Strategy of the Rural Revitalization

The rural area is a regional complex with the natural, social and economic characteristics, which has multiple functions such as the production, life, ecology and culture [2]. It promotes and coexists with cities and towns to form the main space of human activities. The prosperity of the countryside means the prosperity of the nation, while

© Springer Nature Singapore Pte Ltd. 2020
M. Atiquzzaman et al. (Eds.): BDCPS 2019, AISC 1117, pp. 1856–1862, 2020.
https://doi.org/10.1007/978-981-15-2568-1_260

the decline of the countryside means the decline of the nation [3]. The contradiction between the growing needs of our people for a better life and the unbalanced and inadequate development is the most prominent in the countryside [4]. Our country is still in and will be in the primary stage of socialism for a long time. To build a moderately prosperous society in an all-round way and a strong socialist modernization country in an all-round way, the most arduous and toughest task lies in the countryside, and the most extensive and profound foundation lies in the countryside, and the greatest potential lies in the countryside [5]. The implementation of the rural revitalization strategy is an inevitable requirement to solve the major social contradictions in the new era, realize the two centenary goals and realize the Chinese dream of the great rejuvenation of the Chinese nation. It has great practical and far-reaching historical significance.

We should give priority to the development of agriculture and the rural areas, establish and improve the system, mechanism and policy system for the integrated development of the urban and rural areas in accordance with the general requirements of the industrial prosperity, the ecological livability, the rural style civilization, the effective governance and the rich life, promote the rural economic construction, the political construction, the cultural construction, the social construction, the ecological civilization construction and the Party construction as a whole, and accelerate the modernization of the implementation of the rural governance system and capacity, accelerate the modernization of the agricultural and rural areas, take the road of rejuvenating the socialist countryside with Chinese characteristics, make agriculture a leading industry, make farmers an attractive career, and make the countryside a beautiful home for living and working in peace.

2 Problems in the Construction of the Data System of the Public Platform of Industrial Economic Informatization Under the Strategy of the Rural Revitalization

In the development and utilization of the information resources, the popularization and application of the information technology and the cultivation of the information talents, the "digital gap" between the urban and the rural areas is also very obvious, and still shows an expanding trend. The main problems in the rural informatization construction are as follows:

2.1 Weak Rural Information Infrastructure

China's rural areas are vast, but it is difficult to form a city level system of the information infrastructure. The local government despises the influence of the informatization development and the construction on the rural agriculture, and the investment and attention in this aspect are not enough. In addition, the use of the information awareness of farmers is low, and the degree of their participation is not high, and the information construction is more ignored by the local governments. In addition, the decentralization · of farmers' living areas and agricultural operations has affected the spread of the agricultural information, and the inconvenient transportation in the rural areas has hindered the establishment and maintenance of the modern information system to a certain extent. To a large extent, this has affected the promotion of the rural informatization.

2.2   Lack of the Effective Overall Planning and Coordination of the Rural Grass-Roots Information Work

The lack of the effective overall planning and coordination management in the rural informatization construction will inevitably lead to the situation of "divide and rule, and do their own thing", and fail to enable a large number of farmers to enjoy the informatization convenience brought by the digitalization and networking. Although they provide comprehensive infrastructure for each village's information-based education project, and the computer room and the network are all laid in place, these equipment are either idle, or only play a limited function, which fail to play an effective docking with the local agricultural information resources and the relevant service institutions, and fail to form an effective role of complementation and promotion, and all kinds of the rural grass-roots social functional departments are in administration. The village's work did not make full use of the advanced and convenient information management methods, resulting in a large number of the human, material and financial waste.

2.3   The Information Level of the Rural Residents Is not High

Most of the farmers do not have a high education level, and there has never been a training course for farmers in the local areas. Planting depends on experience. Basically no one will get information from the network or plant new methods and practice them in the actual planting process. According to the research data, nearly half of the rural residents are lower than the junior high school education level, and the low overall quality of the rural residents' labor force is the major reason for the weak awareness of farmers' informatization. The limited education received by the rural residents and the low informatization ability of farmers have seriously affected the popularization of the agricultural informatization.

2.4   Imperfect Rural Information System and Relevant Policies

Although in recent years the government investment has increased year by year, it has not really managed the construction funds well. The government investment is still focused on the construction of the agricultural informatization hardware facilities, and the personnel who can operate the facilities are insufficient. Most city, county and township governments have not yet formulated the informatization policies suitable for the economic development of the region, and only apply the model construction, resulting in the unclear definition of the informatization, and the problems in the process of the rural informatization construction cannot be solved in time, reducing the efficiency of the informatization construction.

3 The Construction Requirements of the Data System of the Public Platform of the Industrial Economic Informatization Under the Strategy of the Rural Revitalization

The industrial economy information public platform data system is working with the major counties. Taking the "Internet + targeted poverty alleviation + agricultural products ascending" as the breakthrough point, through the docking of the agricultural products upward, we will promote the development of the new agricultural management entities such as the large grower, the family farms, and cooperatives and so on, and effectively promote the increase of farmers' income and get rid of poverty and become rich, which is a successful practice of the rural revitalization strategy of "cultivating the new agricultural business subjects" and "increasing farmers' income through multiple channels".

In the promotion of the agricultural products, we should pay special attention to highlighting the regional brands, extending the industrial chain, expanding various socialized services such as agricultural materials, processing, and logistics and so on, actively explore the building of the comprehensive trading center for the agricultural industrialization consortia and agricultural products, and vigorously promote the construction of the modern agricultural industry system, production system, and business system. This is a successful practice of the rural revitalization strategy of "promoting the innovative development of agriculture" and "promoting the integrated development of the primary, secondary and tertiary industries in the rural areas".

In cooperation with the major counties, taking the Internet + precision poverty alleviation + agricultural products upward as the breakthrough point, relying on the leading industry of the agricultural products, the ecommerce providers and the agricultural big data in the community network, provide the trading platform for farmers to sell goods, provide the decision support for the county's poverty alleviation, provide farmers with various types of the rural e-commerce training, and help the qualified new farmers to develop their electricity suppliers, to innovate the way of the agricultural management, which is the practice of "Internet plus agriculture" for the rural revitalization strategy.

By setting up a large-scale agricultural products uplink team in the fields, linking up the service links after the accurate matching of the online supply and demand information, dispatching the one-stop services such as inspection, goods collection, receiving, grading, loading and unloading, and the car distribution throughout the process, promote the employment of the rural population and provide the social services for farmers, and help farmers directly sell goods, effectively promoting the poverty alleviation work.

4 The Construction Strategy of the Public Platform Data System of the Industrial Economic Informatization Under the Strategy of the Rural Revitalization

4.1 Construction of the Platform Contents

The public platform of the economic informatization is mainly to exhibit the research work of "3 + 3" key industry standard information collection. It is committed to building the key pillar industries and the key emerging industries covering the rural areas, integrating the functions of the domestic and the foreign relevant standard collection, standard free reading, standard free downloading, WTO/TBT technical trade barrier information, product inspection and certification, and experts and enterprise interaction. It is a professional platform with source, practicability and research, so as to be close to the enterprises, embody characteristics, authority, effectiveness, pertinence and dynamic continuity.

4.2 Application System Construction

The application system construction of the economic information public platform is mainly to build the information standard data management system, which provides the data package management, warehousing, classification, retrieval and other functions of the standard literature record data (China National Standard GB, and the industry standard) on the platform. Through the search engine, the data retrieval files are automatically generated when the title data is released, which provides a fast industry standard information retrieval service for Nantong "3 + 3" key industry standard information disclosure service website. At the same time, in the application system, the documents can only be viewed online, and it is not allowed to download and install the reading plug-in. Through the plug-in, the purchased standard documents and the bibliographic materials can be read. The displayed documents include the watermark, the screen capture tracking code, the printing prohibition, the time lock (the time limit for reading and using the purchased documents), and reading record traces, which ensures the reliability of the platform information.

4.3 Construction of the Data Sharing and Exchange System

A data sharing and exchange system is built on the public platform of the economic information to provide the data reception and the synchronization services for the data of the standard literature bibliography and the application system. At the same time, it provides the perfect industry standard data services for the later smart city information platform. The design of the data sharing and exchange system can support all kinds of the file data, update and modify the file data, and query and trace. It has the function of the automatic recovery for faults, such as the network faults. Once recovered, it should be able to automatically handle the remaining data exchange tasks. For faults that cannot be automatically recovered, it should have the alarm prompt and the processing building. For example, the length of the field is not enough. In case of the network terminal, the database crash and other abnormal conditions, the error can be sent to the system administrator in time.

4.4   Construction of the Safety System

The security system is built on the public platform of the economic information, and the physical separation is used to ensure the ability of the system and the data to resist the catastrophic events. The disaster recovery refers to the establishment of a set of the data system, which is the real-time replication of the key application data of the local system. In case of a disaster, the disaster recovery system can quickly replace the local system to ensure the continuity of the business, so as to ensure the security of the economic information public platform. Integrate the high-quality resources on the upstream supply side, take the intelligent retail stores as the contact point, build the rural whole-industry chain transaction service platform, help the township circulation entities to upgrade to the new retail subject serving the rural market locally, and the big data generates the social value for the rural areas, provide the comprehensive digital transformation around the three dimensions of "people", "goods" and "market", and provide the new impetus for the rural economy.

4.5   Solidly Promote the Construction of the Digital Countryside

The digital village refers to a new economic form in which the rural areas rely on the development of the digital economy, take the modern information network as an important carrier, take the digital technology innovation as the core driving force of the rural revitalization, realize the digitalization of the rural production, governance and life, continuously improve the digitalization and intelligence level of the traditional industries, and accelerate the reconstruction of the economic development and the rural governance mode. The industry digitization refers to the implementation of the rural industry digitization construction project. Through the implementation of the rural industry digital construction project, we will promote the reform of the rural economic efficiency, and drive the agriculture from the production orientation to the demand orientation. By vigorously developing the digital agriculture, building and improving the "sky earth" integrated data acquisition system such as satellite, aviation, and ground wireless sensor and so on, accelerate the construction of the agricultural and rural digital resource system, and build the quality and safety traceability system of agricultural products. Use the modern information technology to strengthen the network monitoring and the risk warning of the environment of the agricultural products and the use of inputs, the safety guarantee of the agricultural products and the risk warning. Diagnose, realize the entire process monitoring of the agricultural products from "field" to "tip of the tongue", and solve the problems of the homogeneous competition of the agricultural products and "increasing the production without increasing the income". Give full play to the role of the Internet in connecting people, businesses and industries, develop various functions of agriculture, extend the industrial chain, enhance the value chain, improve the interest chain, accelerate the integration of the primary, secondary and tertiary industries in the rural areas, and let farmers reasonably share the added income of the whole industrial chain.

5   Conclusion

In the digitization of life, we should implement the rural life digitization construction project. Promote the "Internet + education", "Internet + medical care", and "Internet + culture" and so on, let farmers run less and data more, and continuously upgrade the level of the equalization, popularization and convenience of public services. Persist in the problem orientation, seize the outstanding contradictions and problems in the people's livelihood, strengthen people's livelihood services, make up for the short board of people's livelihood, and promote the popularization and application of the big data in education, employment, social security, medicine and health, housing, transportation and other fields, and deeply develop all kinds of the convenient applications. Strengthen the application of the big data in the field of the targeted poverty alleviation and the ecological environment, and help to win the battle against poverty and accelerate the improvement of the ecological environment. Upgrade the digital literacy to a strategic level, implement the "digital literacy project", and improve the ability of villagers to use the digital resources and the digital tools, and expand the demand for the digital use.

Acknowledgements. The 2018 Henan provincial government decision making research, Study on the Effective Utilization Model and Realization Path of Farmers' Unused Homestead in Henan Province under the Separation of Three Powers [2018B219]; Henan University Key Research Projects. Research on the Mechanism of Operation of the Trading Market of Rural Land Management Right under the Division of Three Powers [18A790033]; Henan University Key Research Projects. Research on the Modernizing Mode and Realization Path of Smallholder with Characteristics in Henan Province [19A630035]; The 2020 key Scientific Research Projects of Higher Education Institutions in Henan Province, Research on High-quality Development Model and Realization Path of Unused Homestead in Henan Rural Areas under the Trend of Separation of Three Rights [20B790018].

References

1. Lu, B., Li, Y., Zhang, W.: Research and design of the intelligent transportation data analysis platform system based on the big data technology. J. Hubei Univ. Sci. Technol. **36**(5), 33–38 (2016)
2. Zhao, M., Zhou, R., Zhang, A.: Analysis of the urgent problems in the intensive construction of the transportation information system based on the cloud computing public service platform. Logist. Sci-Tech **38**(12), 11–17 (2015)
3. Zhang, F., Li, J., Liu, B.: Discussion on the design of logistics public information platform under the cloud computing architecture. Commer. Times (22), 31–33 (2011)
4. Zhao, W., Lei, M., Wang, L.: Research and design of the sports logistics data service platform based on the SOA architecture. Logist. Technol. **33**(6), 22–29 (2014)
5. Zhang, J.: Design and implementation of the data mining platform based on Map Reduce. Netw. Secur. Technol. Appl. (11), 49–52 (2014)

Design of Community Rehabilitation Service System and APP for Cerebral Stroke Elderly

Ruibo Song[✉] and Shuaijun Feng

The Design Art School, Shandong Youth University of Political Science,
Jinan 250103, Shandong, China
xE011J@yeah.net

Abstract. To solve the problem that cerebral stroke elderly cannot recover effectively, we should pay attention to the rehabilitation experience of the old people with cerebral apoplexy, so that the old people with cerebral apoplexy can keep a positive attitude for rehabilitation training. By using service design tools such as user portrait and user journey map, the rehabilitation of elderly stroke patients was studied, and the contact points of products and people were analyzed and studied in one day of rehabilitation training activities, and problems, pain points and demands in the rehabilitation process were found. The new experience mode of community rehabilitation for the elderly with cerebral apoplexy can strengthen the closeness between the elderly and the community, eliminate the inferiority complex in the heart of the elderly, strengthen self-identity, and promote the rehabilitation of neurological function and limb motor function.

Keywords: Cerebral stroke elderly · Community rehabilitation · Service design · APP

With the continuous improvement of medical technology, human beings have greater confidence in facing diseases. However, there are still many people suffering from diseases. Stroke, with the title of "top health killer" for the elderly, will have a significant impact on the human nervous system once it comes on, resulting in a lack of physical coordination and inability to take care of themselves [1]. Even if effective treatment is obtained, it will still have a long-term and profound impact on patients. Patients' persistent training and keeping a good state of mind are of great significance for rehabilitation.

1 Research Background

1.1 Current Status of Stroke in China

Cerebral stroke elderly, also known as apoplexy. It mainly includes ischemic stroke (cerebral infarction) and hemorrhagic stroke [2]. The incidence of stroke in China ranks the first in the world, and has become the leading cause of death and disability in China. Cerebral stroke elderly is characterized by high incidence, mortality and disability. The average age of stroke patients aged 40 years and above was 60.9–63.4

© Springer Nature Singapore Pte Ltd. 2020
M. Atiquzzaman et al. (Eds.): BDCPS 2019, AISC 1117, pp. 1863–1872, 2020.
https://doi.org/10.1007/978-981-15-2568-1_261

years old for the first time, and the disease of patients could also be affected by different disease locations, different onset time and other factors, resulting in varying degrees of motor loss.

According to the stroke association of hunan province in 2018 study, patients with new onset in about 2.5 million people a year in our country, the incidence is given priority to with the elderly in the population, the number is increasing in successive years, and is on the rise, with the increase of aging population, China's population will reach new heights, cerebral stroke elderly as the growth of the age at risk are also growing.

1.2   Existing Rehabilitation Measures for Cerebral Stroke Elderly in China

The rehabilitation of cerebral apoplexy in China is mostly concentrated in the stage of primary rehabilitation and secondary rehabilitation [3–5]. After the patient's condition is under control, the corresponding rehabilitation training and drug treatment are carried out after the evaluation of the condition by doctors. After return to lower limb has some support in patients entered the stage of tertiary recovery, tertiary recovery in domestic also has the very big space, tertiary recovery in the main stressed the patients can return to the family or community rehabilitation, community and family under different environments, such as the correct rehabilitation activities, can promote the patient's ability to adapt to the environment [6]. Most of the people surveyed are concentrated in their homes. Among the patients with stroke 1 year later, only 60% had received early rehabilitation training, while 42.4% had never taken any rehabilitation measures. Most people don't have an effective or timely and correct recovery process when they go home, thus worsening their condition [7]. In recent years, some community medical services for people with stroke have appeared in China, but they are all focused on how to better care for people with severe stroke and lack of systematic service measures.

1.3   The Importance of Community Rehabilitation Training

Rehabilitation training is a painful process, higher body muscle tension and pain seriously affected joint of voluntary movement, if patients with abnormal limb muscle tension and pain refused to rehabilitation training or not timely healing will lead to limb muscle atrophy, joint contracture, deformation, resulting in limited joint activities, eventually lead to the body of waste in. Drug therapy and physical rehabilitation training for stroke patients are inseparable. Three-level rehabilitation in the rehabilitation system can greatly improve the living ability of patients and increase the possibility of patients returning to society. Rehabilitation training in the community can reduce patients' sense of fear and inferiority and improve their self-confidence. Therefore, the elderly in the community need more care. Patients with limb motor nerve injury are the main objects in stroke population. NIHSS (see Table 1) rating table divides the degree of disease of patients into 6 parts.

Table 1. National institutes of health stroke scale

Score	Severity
0–1	Normal or nearly normal
2–4	Mild stroke/minor stroke
5–15	Moderate stroke
16–20	Moderate to severe stroke
20–42	Severe stroke

Note: the more serious the injury, the higher the score; The blue part is the target population of this study, among which the patients with moderate to severe stroke have certain support in lower limbs.

2 Field Investigation and Analysis of Cerebral Stroke Elderly in Community

Respectively for the community old man and the old man out of research in the hospital as a result, the old man of the most members in the hospital without support, the restoration process in strict accordance with the specified by doctors to plan, the community elderly relative freedom, however, most patients with self-discipline after discharge variation, gradually will gradually give up exercise, or exercise time reduced and etc. The existing problems of elderly people with community stroke are shown in Table 2.

Table 2. User community rehabilitation problem analysis table

Perspective	Problem statement
Their own factors	Physical inconvenience, unable to go out independently The environment in the home is relatively closed, leading to depression No one to accompany, feel lonely Lack of motivation to train Did not see the effect of recovery
Environmental factors	Children to work busy affairs Lack of professional rehabilitation equipment in the community The lack of contact between patients in the same community

3 The Process of Constructing Community Rehabilitation Service System for Cerebral Stroke Elderly

Combined with the previous research, the general characteristics of this kind of elderly patients were summarized, and the specific image of this kind of people was drawn through the user portrait tool (Fig. 4). "Uncle li used to be a doctor in a clinic.

However, I had a sudden stroke two months ago. Due to timely medical treatment, my condition was under control. However, my right limb basically lost its mobility, and I repeated rehabilitation training at home every day. Uncle li is a patient with moderate stroke, which is the most representative group in the process of rehabilitation. Patients by the hospital treatment, has met the tertiary recovery standards, still unceasingly in the home rehabilitation training after discharge, but many problems appeared in the process of its training, the old man at the beginning of this ages during self-training, generally present a more positive attitude, but through the training for more than a week later, the patient reached the cyclical fatigue, irritability period basically has disappeared, after a long time of training feedback still don't get obvious improvement, patients prone to resistance. After interviewing several patients, it was found that this problem was common. During the boring rehabilitation training process, patients were unable to perceive their progress, which would seriously affect patients' mood and adversely affect their recovery.

Through communication with patients, the following conclusions are drawn: "patients can't experience happiness in the process of rehabilitation training, and they are resistant to it, which will bring more trouble to the later rehabilitation training." Further interviews and observations were conducted to investigate the question "why are patients unhappy? What is the cause of unhappiness?" And so on. Summarized four problems of patients: first, taking medicine is a very painful thing; But always bored at home; Three, don't want to have boring rehabilitation training; Fourth, low efficiency, easy to lazy.

According to the user portrait, the potential contacts of such people are sorted out, and the stakeholder map is drawn, through which the strong and weak interest relations between patients and various social roles are clearly distinguished. For patients in the community, family members, doctors and residents in the community (neighbors) have a greater correlation with the intensity of benefits.

After the human model is built, the user's journey diagram is drawn based on the patient's behavior, and all the behaviors and activities of the patient in a day are listed in chronological order. Through the behavior point, the contact point, the pain point, the opportunity point has carried on the discussion and the induction. With the deepening of the design research, it is found that the user's journey diagram drawn involves a wide range of aspects, from which it can be found that many problem points are in parallel, and each problem point covers aspects with a strong ductility.

The user's journey map is further elaborated in three major directions: training, entertainment and dining. From the selected training aspect, the daily training users' journey chart was redrawn for patients' activities (as shown in Fig. 1), and all activities conducive to self-training were connected in series to get a complete time line: getting up – eating – going to the toilet – massage – rehabilitation training – going out for a walk – sleeping.

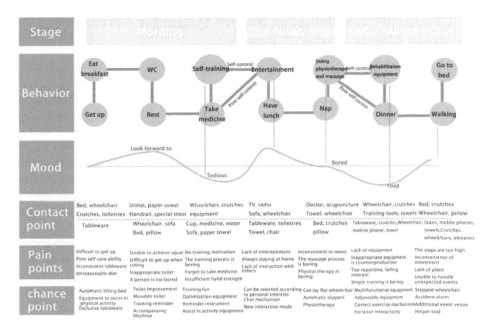

Fig. 1. User journey diagram

Contact point in the journey through the user research, and discovered the user of a series of problems, and carries on the sorting, finally through the problem point and curve, the precise positioning, finding out the most pain in patients with three points, including: how to make the patients' rehabilitation training on voluntary, how to make patients when no one is under the supervision of independent completion of training and in the process of training without company.

Opportunity point, excavation, in the form of "one to many" (i.e., offers multiple points) for a problem, as far as possible put forward many solutions, and then made a number of important points, including: app and teaching service, mutual entertainment venues, the score type training mechanism, the network community youth volunteers, rehabilitation training, community organizations with rehabilitation.

Through continuous simulation of the above solutions, the model of product plus APP was finalized. Elderly products should follow the appearance of generous, simple operation and other characteristics, abandon the consistent screen design, use interactive voice design. The use of voice mode for binding information will greatly reduce its security, so considering the APP, the main provider is children.

4   Community Rehabilitation Service System for Cerebral Stroke Elderly Produced Results

4.1   Design Concept Presentation

Firstly, we need to consider the old people's own reasons. The old people's ability to adapt to the surrounding environment has gradually become worse. Secondly, we should try our best to adapt the products to the old people rather than the old people to the products, so as to reduce the discomfort of the old people. Making products for the elderly should fully understand the daily life of the sick elderly. Products for the sick elderly should have no screen, easy to operate, portable and other characteristics. Through the construction of the overall social service rehabilitation system, the central product is a product similar to a badge. The back of the product adopts a powerful magnet, which can be attached to the clothes. However, the input of some privacy information such as password and id number in the form of voice broadcast will greatly reduce the security of the product. In order to avoid this kind of situation and enhance the safety performance of products, we adopt the form of APP to form the mode of APP plus products. The service target of the product is the elderly, and the service target of the APP is children. Rehabilitation equipment is the branch, and the sick old people in the whole society are carried out by product node, and the service system covers the whole community.

4.2   Design Content

Through the conception of the concept, further structure, formed the community cerebral stroke elderly service system diagram (Fig. 2). The badge is used as a point through the whole system diagram, spreading to the surrounding system. The badge mainly involves three major aspects: hospital, community and home. Children through the APP to the binding of patient information, after the success of the binding badge will follow each APP, kids can through the APP store now recovering recent cases information sent to the hospital, the hospital will reduce the cases of information sent to the good at the doctor, the doctor after formulate corresponding plans back to the APP, because the APP is connected with badges, so training program will be the first time will be synchronized to the badge, patients can according to the plan for training. The children can also communicate with the APP online. The doctor will arrange the online follow-up visit time or the offline face-to-face visit time or the physical therapy and massage time. The badge will remind the patient to conduct the offline consultation after the specified time.

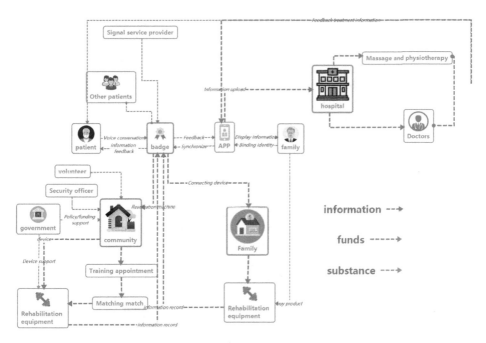

Fig. 2. Service system diagram

Patients can reserve time for community exercise through badges, which will send signals to the community, and the community will automatically match the idle machines during this time, effectively avoiding the waste of resources. If patients want to play with patients, can undertake to make an appointment through the badge, the community will automatically match rivals (or two met independently), the community will also provide spare machine at the same time, due to differences in condition, the system will be given according to the personal performance is a certain score, score the most wins. Upon arrival at the competition venue, the badge will automatically connect to the reserved machine. After the rehabilitation training, the rehabilitation machine will record the rehabilitation status into the badge. Of course, if you can afford to buy a machine at home, you can also do rehabilitation training at home, and the rehabilitation data will also be recorded in the badge.

Through the system diagram, we can see the goal we want to achieve. The badge is like a big network, which can establish the contact between patients in the community, solve the problem that the elderly patients are reluctant to go out, promote the inter-action among patients, make them exercise consciously, eliminate the prejudice of patients on their own defects, and eliminate the fear of patients. Surveys show that many people think the disease is common and normal and do not discriminate against it. In the survey, it is found that patients will actively carry out rehabilitation training when they first fall ill. However, since rehabilitation training is a long-term process, it is difficult to see big changes in a short time, which will lead to the occurrence of patients' negative emotions. Once the negative situation happens, it will have a great impact on the rehabilitation training. Through badge to quantify the value of the movement of the processing, can be intuitive show in the form of percentage recovery (such as: badges will be credited today and yesterday's training data, calculated by the background on the right side of the fail today than it was yesterday leg strength increased by 0.6%, if appear negative growth, the system will be the cause of the points of negative growth, adjust remind patients), so that we can more clearly make patients feel their progress, let patients have continued power and information.

In order to ensure the safety of patients in the region and increase the channels for patients to communicate with others, volunteer activities are set up. Volunteers can accompany the elderly to recover in the community. In order to deal with the emergence of emergencies, there are professional safety officers in the community to escort patients.

To the future design product, has carried on the use flow the verification. In the validation process encountered in the relevant product and service process deficiencies, and make up. It is hoped that through the application of this series of service processes and products, users can experience the happiness in the process of rehabilitation training, so as to better enable users to recover physically.

The training equipment that can be connected with the badge is designed for the community. Based on the original equipment, it is modified to adapt the equipment to the elderly. Two kinds of rehabilitation equipment are produced, one for upper limb training equipment and the other for lower limb training equipment. The device can be connected to the badge, and the badge will synchronize the exercise data in real time to provide information tips for users. All the rehabilitation instruments are equipped with anti-spasticity system. Once the limb spasm of the patient is felt, the instrument will reverse to relieve the spasm. If the limb of the patient is still not improved, the instrument will automatically stop and send a prompt sound to remind the surrounding companions. In the upper limb exercise equipment, patients can choose to stand if they have some leg support, which can stabilize the strength of the lower limbs while exercising the upper limbs. Of course, patients can also sit for upper limb rehabilitation. The lower end of the device is equipped with a "shaking leg" device, the main purpose is to prevent patients when sitting exercise, the possibility of lean muscle atrophy of the lower extremities, the principle and function of massage. Monitors are installed on the top of the equipment, which can show the progress of exercise in real time for the minder to check. At the same time, an emergency stop is installed, so that the equipment can be immediately stopped in case of special situations.

The role of the emblem is particularly important in the design, but the operation is very simple, need only children to bring the old man in the chest, middle part of the product is only one button switch products, in the side of the product also equipped with charging mouth, phone card socket and a small button to reset the product, to prevent patients by accident, button in the form of the concave.

Due to the limitation of patients' mobility, all products for patients adopt adjustable and simple operation to create rehabilitation conditions for patients to the greatest extent. The children's APP (as shown in Fig. 3 service APP design) adopts the form of WeChat small program to avoid the adverse impact of complicated apps on users. In the APP, children can regularly upload data synchronized from badges, and communicate and consult with doctors online. Children can check the old people's sports market at any time through their mobile phones. Since there is a mobile phone card in product, children can contact the old people by phone at any time.

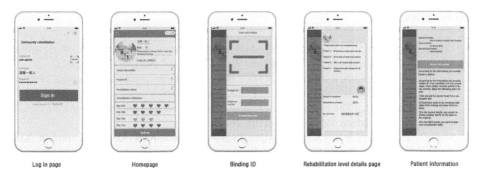

| Log in page | Homepage | Binding ID | Rehabilitation level details page | Patient information |

Fig. 3. Service APP design

Through the service blueprint, we can see the activity status of the elderly in a day and the connection status of dacon with various systems. Patients can set their own time for taking medicine, and patients can also conduct voice interaction with dakan to set the time for them to know their goals for the day. As can be seen from the table, the third part of user behavior is "chatting with others". Product has the function of friends circle built in. Patients can chat with others while exercising at home, as well as when doing acupuncture or physical therapy, mainly to solve the boredom of the elderly when they are alone.

5   Conclusion

Take the form of present products add APP, build up the community service system, as a "network" is closely connected, each patient are all nodes of the "net", products like various nodes connection path, relate these patients within the community, allow patients to exercise together, make progress together, in order to increase the chances of patients to an activity, to eliminate the old man on his own, inferiority complex, to strengthen the identity of self. Improve the mental state of stroke patients, improve self-confidence, promote the rehabilitation of neurological function and limb motor function, and improve the quality of life of patients. Rehabilitation training as a kind of "entertainment" for patients, there is no psychological burden, so that patients urge each other, consciously exercise. The emergence of product can effectively solve the problem that patients are unwilling to communicate with outsiders, and enhance the contact between patients through interactive forms such as the Internet of things. We will vigorously promote the establishment of community rehabilitation centers, provide reasonable training places for patients, and establish a more scientific and professional community rehabilitation system.

Acknowledgment. This achievement is funded by the 2018 shandong province art education special project entitled "design art innovation research based on problem thinking".

References

1. Zhang, H., Zhang, J.: A comparative study of the reform and innovation of classroom teaching mode in colleges and universities. World Educ. Inf. **30**(12), 62–65 (2017)
2. Yu, S.: New proposition of design: service design - taking community clothing cleaning service design as an example. Decoration (10), 80–81 (2008)
3. Song, R., Zheng, L.: The essence of customized home is based on the whole process of user experience. Sales Mark. (Manag. Ed.) (10), 76–78 (2018)
4. Wei, C.: Design and research of taxi waiting pavilion based on service design concept. Xian Engineering University (2017)
5. Shen, D.: Design of service system for classification and treatment of urban domestic waste in Guangzhou. Decoration **06**, 142–143 (2017)
6. Yu, C.-Y.: Research on the service design of internet + experience agriculture. Age Agric. **46**(02), 84–85+106 (2019)
7. Jing, P., He, L.: Enlightenment of complex adaptive system theory on sustainable design process. Design (09), 64–65 (2016)

Design of Community Service System and APP for the Elderly with Diabetes

Ruibo Song[(⊠)] and Xurong Huang

The Design Art School, Shandong Youth University of Political Science,
Jinan 250103, Shandong, China
xE011J@yeah.net

Abstract. The aggravation of the aging problem and the imbalance of the national diet structure lead to the increasing proportion of the elderly people suffering from diabetes, and improper self-management is likely to cause a variety of complications. Use literature research and field research to understand user needs, combine with service design to dig out user pain points, comprehensive and accurate user characteristics of the system, improve the experience of contact points and design prototypes, and integrate stakeholders to form a brand new service system. The design concept of "with sugar" to "without sugar" is put forward. The whole service system, from the perspective of blood sugar, diet and exercise, helps the elderly with pre-type II diabetes develop a healthy lifestyle, stabilize the disease, and form a new systematic service process.

Keywords: Elderly people with diabetes · Community · Service system design · APP

Diabetes as a long-term incurable diseases have a sophisticated management methods, China's existing medical more applicable to handle sudden disease, that need in the life, in the face of diabetes to achieve good control effect of diseases, medical resources and diabetes treatment services especially not matching problem [1, 2]. In this paper, the method of service design is used to systematically study the elderly with diabetes.

1 Community Medical System Service Design

1.1 Research Background

Current Situation of Community Medical Services in China
Since the 18th national congress of the communist party of China (CPC), the CPC central committee with comrade xi jinping as the core knows that only by maintaining the health of the whole people can the long-term development goal be achieved, and medical and health services are directly related to people's health. By 2020, we will strive to expand the contract service of family doctors to the whole population, form a long-term and stable contractual service relationship with residents, and basically realize the full coverage of the contract service system of family doctors [3–5]. Family doctor contract service is a kind of comprehensive medical and health service with

© Springer Nature Singapore Pte Ltd. 2020
M. Atiquzzaman et al. (Eds.): BDCPS 2019, AISC 1117, pp. 1873–1881, 2020.
https://doi.org/10.1007/978-981-15-2568-1_262

general practitioners as the carrier, community as the scope, family as the unit and overall health management as the goal. Through the form of contract, it provides continuous, safe, effective and appropriate comprehensive medical and health care for families and their members.

At present, China has become an aging country. Due to the degeneration of physical function and other reasons, the middle-aged and elderly people have the characteristics of high prevalence rate and high disability rate, and there are many difficulties in health management [6]. The contracted service mode of family doctors can guide the healthy life of the elderly and reduce the incidence of diseases, which is an effective health management mode.

Current Status of Diabetes Treatment in the Elderly in China

Diabetes, as a chronic disease that cannot be cured for a long time, has a complicated self-management method. In the face of diabetes, which needs to achieve good control effect in life, the problem of mismatch between medical resources and diabetes treatment services is particularly prominent. At present, few researchers apply the methods and tools of service design to study the systematic service mode of chronic diseases as a whole [7].

1.2   Purpose and Significance of Community Medical System Service Design

With the promotion of national policies, family doctors have appeared in people's sight. The service design of community medical system provides users with more effective methods to analyze the overall situation and situation, and provides targeted solutions to help users reduce the pain associated with symptoms, and forms a virtuous circle between community medical care and users by combining various stakeholders, so as to stimulate users' sense of participation in community medical care.

2   Design Method and Process

2.1   Design Research

The number of adults with diabetes worldwide is estimated at 450 million and is expected to reach 629 million by 2045. The total number of people with diabetes worldwide is shown in Fig. 1.

Although diabetes has the title of "chronic disease", its harm to human body is only second to cancer, which is disheard-of by ordinary people, and it has a higher direct mortality rate. In Africa and southeast Asia, the direct death rate from diabetes is more than one percent, followed closely by the relatively developed economies of Europe and the western Pacific.

With the improvement of people's living standards in China, people's living pressure continues to increase, and long-term mental tension will lead to fluctuations in the range of blood sugar in the body. At the same time, due to some bad living and eating habits, blood sugar is constantly rising. In addition, the aging of the population is becoming more and more serious, the proportion of middle-aged and elderly people is

The total number of people with diabetes worldwide

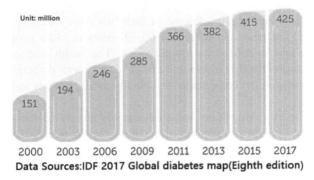

Data Sources:IDF 2017 Global diabetes map(Eighth edition)

Fig. 1. Total number of people with diabetes worldwide

increasing, and the probability of diabetes among the elderly is greater, which is one of the common diseases of the elderly. For the elderly, there is no cure for diabetes and they are prone to complications.

Based on the above situation, this paper collected data by means of field research, and interviewed a total of 30 respondents in one senior apartment and one community in jinan, including medical staff in senior apartment, senior citizens in senior apartment, doctors in huasen senior apartment community clinic, and doctors in jianda garden community hospital. The main problems include the proportion of elderly people with diabetes in the apartment, age, fixed examination in community hospitals, drug dosage of elderly people with diabetes, differences between elderly people with diabetes and normal elderly people in life, guidance methods for doctors to control the condition of elderly people with diabetes, and control of sugar content in diet of diabetic people.

The apartment is surrounded by a medical clinic, facilities and staff. According to the communication with the medical staff, it is found that there are more than 200 elderly people in the apartment, and most of the elderly in the apartment suffer from diseases, but with different degrees of disease. Among them, the proportion of diabetes is about 35%, and the medical staff will control the diet and exercise of the elderly with diabetes. In this community, jianda garden is divided into 8 districts, with about 50 buildings and a large number of people. In its community health service hospital, doctors and nurses found that the drug cost of early diabetes patients was not expensive, mainly relying on diet and exercise to control, to form a good lifestyle, can effectively stabilize the disease.

2.2 Target User Definition

According to the classification standard of the elderly in China, 45–59 years old is the prophase of old age, namely the middle-aged and elderly; 60–89 years old is old age, that is, old people; Longevity was observed between 90 and 99 years old. Longevity is defined as those who are 100 years old or older. Referring to the classification standard, this paper will target the retired elderly aged from 60 to 89 years old. The social status of the age group in this period will change, including their living habits, social status, economic income and other aspects.

The main manifestations are the narrowing of social circle, the shift of life focus and the abundance of time. Meanwhile, the main causes of type 2 diabetes are obesity, hyperlipidemia and low activity. People in their 50s have difficulty controlling their blood sugar, and the peak age is around 60. Diabetes in older people is difficult to control and can lead to dangerous complications. The health care staff in the elderly apartment will control the diet and exercise of the elderly with diabetes, and the patients with type 2 diabetes in the early stage are able to move, they can stabilize the disease through exercise.

Therefore, the user orientation was a 62-year-old elderly person with early type 2 diabetes and mobility. The primary topic direction is "sugar-free life for the elderly with sugar", and the service goal to be achieved in the future is to form good living habits from diet and exercise, and stabilize the condition of the 62 year old with type 2 diabetes.

Through the above positioning, the method of user portrait is adopted to clarify user characteristics and requirements. User portrait takes basic information, pain points and wishes as the main elements to ensure full understanding of user characteristics and psychology, so as to take users as the center and dig deep needs of users. The following is a user portrait of a typical elderly person with type 2 diabetes. User portrait is shown in Fig. 2.

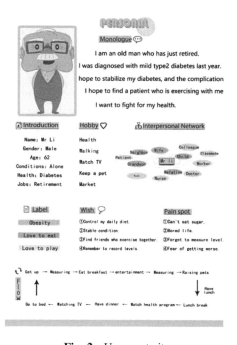

Fig. 2. User portrait

Such users are retirees with early type 2 diabetes who live with their spouses. Such users have high requirements for social contact, interest, activities and other aspects of leisure life, and they hope to control the diabetes condition. However, the methods to control diabetes are difficult for such users to measure and cannot effectively control such diseases. Meanwhile, the most direct stakeholder, the spouse, cannot effectively care for such users, and stakeholders affect users all the time, especially those who are closely related to users. Meanwhile, the participation of patients and doctors is crucial for such users.

2.3   User Journey Analysis

User journey diagram requires considering the systematic relationship between "people – things – environment", and analyzing the pain points and opportunities generated by the contact points in the user journey, so as to help users find solutions, improve the "useful – usable – easy to use" experience, solve problems more targeted, and service process more systematic.

The behavior observation was summarized, and the vertical axis of user journey diagram was divided into six aspects: behavior stage, user behavior, contact point, mood, pain point and opportunity point for analysis and discussion. The user behavior stage was divided into getting up, measuring blood glucose, taking medicine, eating, measuring blood glucose and entertainment on the horizontal axis. Users have different pain points under the action of different behavior stages. For example, in the wake up stage, there are three contact points – toilet, toiletries and cups. At this stage of blood glucose measurement, there are four contact points: blood glucose meter, test paper, blood sampling needle, cotton swab and book. Pain points include pain of blood sampling, complicated use method, inaccurate measurement data and forgetting to record, etc. See Fig. 3 for user experience map.

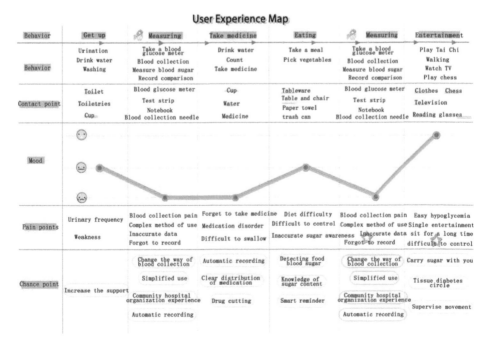

Fig. 3. User experience map

According to the analysis and discussion, grandpa li forgot to record the blood sugar value, inaccurate cognition of sugar amount and difficulty in controlling exercise were selected as the three pain points. Through digging out the opportunity points of pain points, users can realize the sustainability of their daily life. During the blood glucose measurement stage, the opportunity points include changing the blood collection method, simplifying the use method, organizing physical examination in community hospitals, and automatic recording, etc. During the meal stage, there are opportunities to test food glucose, know the amount of popular sugar, and remind, test and supervise, etc. In the recreational stage, there are opportunities to carry sugar with you, form a diabetes circle, and urge you to exercise.

2.4   Definition of Contact Points

According to the preliminary investigation and analysis, the service design based on the healthy habit formation of the elderly with type II diabetes in the community can be concluded as follows:

Mobile APP

Connect users with community hospitals, main functions include blood glucose analysis, production of diet plan and exercise plan. See Fig. 4 Service APP design.

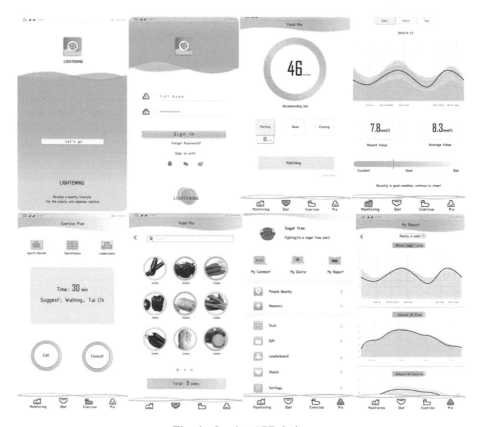

Fig. 4. Service APP design

Lood Glucose Meter
With the function of blood glucose detection and transmission, the APP can provide the blood glucose data of the elderly with diabetes, and provide the basis for diet and exercise plans.

Earable Devices
It provides the function of reminding exercise, real-time monitoring of patients' exercise, setting a button to call for help, transmitting data to the APP, and adding a layer of guarantee for the health of the elderly.

3 Design of Service System for the Healthy Habit Formation of the Elderly with Diabetes in the Community

The whole community elderly diabetes health habits to develop service system, community hospital as initiator, for the service system to provide the carrier and the reward mechanism, the user's doctor, children and doctor as stakeholders, using APP play the role of supervision, at the same time the doctor to provide advisory services, platform through the doctor suggested and large background data to provide personalized service. Users have the qualification to redeem gifts after completing the goal, which promotes the healthy habit formation of users. From the perspective of blood glucose, diet and exercise, the service system helps the elderly with pre-type II diabetes develop a healthy lifestyle, stabilize the disease, and form a brand new service process, as shown in Fig. 5.

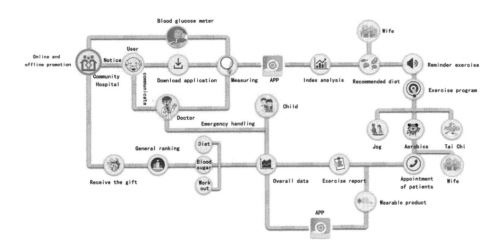

Fig. 5. Service system for health habit formation of elderly people with diabetes in community

As a medium for communication between users and doctors, APP is an important carrier throughout the service system. Blood glucose monitoring, diet matching and exercise testing are the main functions. In the blood glucose monitoring, blood glucose data of the elderly with diabetes are obtained from the blood glucose meter, and the blood glucose value is analyzed. According to the blood sugar level, medical advice and background big data of elderly diabetics, food collocation, recommendation and exercise plan are generated. The wearable device will upload the monitoring of the elderly's exercise status to the APP, which will form the overall analysis report (blood glucose level, exercise, exercise) at the end of the day, and upload it to the electronic patient file of the community hospital, so that doctors can provide more accurate treatment for patients based on the patient file. In this process, the contact point runs through the service system to detect users' lives in real time, effectively establish the contact between stakeholders' doctors and users, and improve the interaction between users and community hospitals, so as to effectively manage users' healthy lifestyle.

An APP for the elderly with pre-diabetes to develop a healthy lifestyle – light sugar, the Logo mainly USES graphics to reflect the concept of management, mainly green and blue, showing a healthy and positive attitude. The design of ICONS fully considers the characteristics of users, who need concise and clear ICONS. Therefore, the design of ICONS adopts the combination of lines and surfaces. The linear ICONS have a strong visual sense, highlighting the functional entrance, and the lines and surfaces reflect a light and lively feeling.

APP interface design adopts a concise and clear style, mainly pictures and tables, clear interface design makes it easy for users to use. In the interface design of the APP, the standard graphics are curved and the overall colors are green and blue, highlighting the concept of "light sugar", improving users' experience and fostering a healthy lifestyle for the elderly with type II diabetes. The blood glucose meter is the contact point in the service system process. During the blood glucose measurement stage, the user and the sugar friend go to the community hospital to test the blood glucose, and the blood glucose value is automatically transmitted to the APP and analyzed, which is conducive to the follow-up plan of diet and exercise. Meanwhile, users can communicate with doctors offline while taking blood glucose readings, and doctors can play a supervisory role to achieve the combination of online and offline activities and enhance users' sense of participation and confidence.

The intelligent medical glucose meter, based on which has the function of general automatic printing and with mobile terminals can also be interconnected, base of the slope is more convenient to watch display, and between the subject and the base of the product is a set design, easier to hold, its color is given priority to with white attune, blue and yellow as the ornament is concise and bright, more in line with the concept of health.

Wearable intelligent device using the user in the movement phase, users need to formulate a reasonable exercise plan to control blood sugar, blood sugar value analysis was used in the APP users appropriate exercise plan pushed to users, at the same time users can be a key to call friends, friends with sugar exercise together, play the role of mutual supervision, raise the enthusiasm of the user to exercise at the same time.

The wearable device is used on the user's waist, in the same position as insulin injection, to alert users. At the same time, this product has the function of reminding exercise, real-time monitoring of patients' exercise, setting the function of one-click call for help, and then transmitting the data to the APP. This product is mainly made of clean white with yellow as an ornament to create a relaxed and pleasant mood and guarantee the health of the elderly.

4 Conclusion

By integrating the service system into the design, the pain points of users can be explored comprehensively and systematically, so as to help the elderly with type II diabetes develop a healthy lifestyle, stabilize the disease and form a brand new system service process. The service system for the healthy living habits of the elderly with diabetes in the whole community, with the users' doctors, children and doctors as the main stakeholders, helps the users to control the disease more effectively. As the initiator, the community hospital provides the carrier and incentive mechanism for the service system and promotes the users' enthusiasm. The system service design can analyze the solution from the overall situation and situation, balance the system rela-tionship of "people – things – environment", improve the user experience, be user-centered, realize the new service process, and provide more effective solutions for users. Service design can guide various stakeholders to quickly and effectively integrate existing resources into a complex but effective solution, which can play an increasingly important role in solving the prominent contradiction in the current medical service construction.

Acknowledgment. Note: This achievement is funded by the 2018 shandong province art education special project entitled "design art innovation research based on problem thinking".

References

1. Gao, B., Duan, Z.: Study on service design for community treatment of diabetes in the elderly. Packag. Eng. **38**(10), 42–47 (2017)
2. Guo, F., Yu, F.: Innovative design of old-age community service based on sustainable concept. Packag. Works **40**(4), 203–208 (2019)
3. Song, R., Zheng, L.: The essence of customized home is based on the whole process of user experience. Sales Mark. (Manag. Ed.) (10), 76–78 (2018)
4. Wang, G.: Service Design and Innovation, pp. 57–70. China Construction Industry Press, Beijing (2015)
5. Zhang, H., Zhang, J.: A comparative study of the reform and innovation of classroom teaching mode in colleges and universities. World Educ. Inf. **30**(12), 62–65 (2017)
6. Yu, S.: New proposition of design: service design - taking community clothing cleaning service design as an example. Decoration (10), 80–81 (2008)
7. Zhang, P., Ding, X.: Research on interactive design of aged intelligent products under compensation mechanism. J. Graph. **39**(04), 700–705 (2018)

NVH Analysis and Optimization of Driving System in Battery Electric Vehicle

Xiang-huan Liu[1(⊠)], De-fu Liu[1], Chao Li[2], and Yin-cheng Sun[1]

[1] College of Mechanical and Electrical Engineering, Central South University,
Changsha 410083, China
zhilin87445667@126.com
[2] Zhuzhou Gear Co., Ltd., Zhuzhou 412000, China

Abstract. Vehicles have currently entered electrical age with high-speed transmission, and the output speed of battery electric vehicle (BEV) exceeds 10,000 rpm. With the increasing integration of BEV driving system, high-speed integrated electric driving system is becoming a trend in the development of BEV. However, higher integration makes it more difficult to determine the cause of noise, vibration and harshness (NVH) problems of the driving system. This paper used SMT/MASTA software to simulate and analyze the integrated electric driving system. With the comparison of test results and CAE results of the driving system, the corresponding parts with problems could be found. Then the finite element method was used to optimize the system response index, and the correctness of NVH optimization scheme was verified by comparison tests. Our work provides an effective NVH optimization method for BEV driving system.

Keywords: BEV driving system · System response · Resonance · NVH optimization

1 Introduction

With the continuous improvement of the social economy, the number of domestic car ownership continues to rise. The widespread use of electric vehicles can reduce vehicle exhaust emissions and is of great significance for energy security, energy conservation and emission reduction, air pollution prevention, and automobile technology development [1]. The electric vehicle industry started late in China. Compared with developed countries, China's automobile industry has a weak foundation, especially in the aspects of automobile design and key parts manufacturing. In addition, the key technology of NVH (noise, vibration and harshness) has not been overcome, which will become a major obstacle to the industrialization of new energy vehicles.

The development of the BEV industry has driven the integration of the electric driving system. The "two-in-one" and "three-in-one" integrated drive systems of domestic component manufacturers are on the market. The problems generated by integrated electric drive systems are becoming more and more complex, including NVH problems in powertrains, efficiency issues, and comprehensive durability issues. For transmission noise and vibration, Nakamura et al. studied the nonlinear dynamics of gear system clearance in 1967 [2]. In 1977, based on the "impact pair" model of

M. Atiquzzaman et al. (Eds.): BDCPS 2019, AISC 1117, pp. 1882–1886, 2020.
https://doi.org/10.1007/978-981-15-2568-1_263

gears, Azar et al. performed numerical calculations on the nonlinear problem of the spur gear system, considering the effects of gear inertia, time-varying stiffness, and tooth surface friction [3]. They found that for light-duty gear systems, the system produces large vibrations due to the effect of the backlash when the meshing frequency is 1/2 of the natural frequency of the output shaft. Lin et al. studied the influence of gear meshing stiffness on the vibration instability of the transmission system by establishing a mathematical model of the secondary gear transmission system [4]. Wang et al. researched the nonlinear dynamics and NVH performance of the engine-transmission coupling system, which provided a method for BEV NVH performance analysis [5]. James et al. believed that the presence of the reducer changed the vibration and noise characteristics of the motor [6]. Fang et al. analyzed the modal characteristic changes of the integrated reducer motor through the mutual coupling of the motor and the reducer, and studied the influence of the existence of the reducer on the vibration and noise characteristics of the motor [7].

2   Gear Whine Analysis of Electric Driving System

The electric driving system of BEV usually consists of a motor and a reducer, mostly using a combination of a permanent magnet synchronous motor and a two-stage reducer. The cause of gear whine in electric driving system is complicated, mainly including motor electromagnetic excitation, reducer system resonance and system coupled modal resonance. In the vehicle test study, we found that the electric driving system had structural resonance problems. Based on the MASTA software, this paper mainly simulated and analyzed the powertrain to find out the causes and correction methods of structural resonance in the driving system.

In the vehicle NVH test, we collected near-field noise data in the car through LMS data acquisition equipment. The collected data was analyzed by noise analysis of near-field noise by the data analysis software of LMS Test Lab to find the corresponding order of gear whine. Finally, the excitation source of the gear whine noise was determined through the order analysis of gear whine noise.

Under the WOT condition of the whole vehicle, when the input end speed was 1600–2000 rpm (586.6–733.3 Hz), the electric driving system had the 22nd order gear whine caused by resonance. According to the structure of the electric driving system, we basically determined that it was the noise generated by the high speed gear of the reducer in the driving system. Therefore, the driving system has a system structural resonance near 696 Hz, and it is necessary to adjust the system structure to improve the situation.

3   Reducer Order Noise Analysis

3.1   Establishment of MASTA Reducer Analysis Model

The MASTA analysis model based on the electric driving system. The gear parameters of the electric driving system are shown in Table 1.

3.2   MASTA Software Analysis System Model

First, the macro parameters and micro-shaping parameters of the gear pair were entered in the software. It can be seen that the peak-to-peak calculation result of the high-speed gear pair transmission error satisfies the design requirements, and the frequency domain amplitude of the transmission error is not large under the WOT condition of the whole vehicle. The results show that the order excitation of the 22nd order of electric driving system has little effect, and the gear whine should be caused by system resonance.

Table 1. Macro parameters of the gear pair in reducer

Basic parameters	High-level gear	High-level large gear	Low-level small gear	Low-level large gear
Number of gear teeth	22	68	23	68
Normal module (mm)	1.777	1.777	2.738	2.738
Width of tooth (mm)	36	32	44	38
Normal pressure angle (°)				
Center distance (mm)	95		140	
Rotation	Right	Left	Left	Right

In the range of 1000–2000 rpm, the 22nd order excitation has multiple potential resonance points with the system natural frequency. The 13th order (651.1 Hz) of the system coupling mode is just within the scope of this gear whine and can be determined as the problem frequency.

Through the simulation results of MASTA software, we analyzed the 13th order (651.1 Hz) of the system coupling mode. It can be seen that the main problem is that the dynamic response energy of the intermediate shaft high-speed large gear in the transmission process is too large, accounting for more than 40% of the dynamic energy of the system response. After analyzing the problem parts, we found that the gear web of the high-speed large gear has insufficient rigidity and a large degree of deformation, resulting in an eccentric load of the gear mesh. The dynamic response at the front bearing of the output of the housing is too high, even as high as 1.744 μm at 651.1 Hz. In view of the above situation, we have considered increasing the thickness of the high-speed large gear web. The comparison results are shown in Table 2.

Table 2. Dynamic response at the front bearing housing of the output (651.1 Hz)

Housing	Before rectification	Increase 5 mm	Increase 10 mm
Response amplitude (μm)	1.7440	0.6708	0.2460

According to the data in Table 2, after increasing the thickness of the high-speed large gear web, the stiffness of the system is significantly improved. At 651.1 Hz, the dynamic response at the front bearing of the housing output is significantly improved. In addition, after the thickness of the web is increased by 5 mm, the dynamic response of the casing at the front bearing of the output is reduced from 1.744 to 0.6708 μm, which has reached the standard in the vehicle test standard. Considering the NVH performance requirements of the electric driving system and the manufacturing cost of the assembly parts, we determined to increase the thickness of the high-speed large gear web by 5 mm.

4   Reducer Loading Test

In order to verify the above product optimization results and the correctness of the software analysis results, we carried out the loading test of the optimized integrated products and compared the test results with the results before optimization. Under the WOT condition of the whole vehicle, the subjective test effect of the optimized gearbox loading noise is obviously better than that before the optimization. The noise analysis of the near-field noise is performed by the LMS Test Lab data analysis software to verify the loading results. The red curve is the 22nd order interior noise order slice of the reducer assembly before optimization, and the green is the optimized 22nd order noise order slice.

In the range of 2000–4000 rpm, the vibration is obviously reduced, which is consistent with the trend of interior noise. Above 4000 rpm, the rate of increase of the speed of the test vehicle is inconsistent, so no comparison is made. Within 2000 rpm, the improvement is not obvious. In addition, the comparison data of two times does not show that the 22nd order noise curve has structural resonance around 650 Hz. However, according to the subjective evaluation result of interior noise, the gear whine is in the range of 1600–2000 rpm. There is indeed improvement.

5   Conclusion

(1) For the NVH gear whine problem of electric driving system in the project, this paper uses MASTA software to carry out finite element calculation analysis of electric driving system and find the corresponding solution. Although this solution solves the problem of exceeding the standard order noise, it does not solve the problem of low frequency gear whine. Test methods and analytical methods still require further optimization. Through simulation calculation and experimental verification, the following conclusions are obtained:

   (1) The electric driving system has a high degree of integration, which causes the coupling mode of the system to change, which in turn causes the NVH problem to become more complicated;

   (2) The adjustment of a component in the system affects the coupling mode of the entire system;

   (3) The dynamic response of the system can be reduced by increasing the stiffness of certain components, and the NVH gear whine problem caused by the transmission path can be solved;

   (4) Adjusting the stiffness of the system only reduces the NVH gear whine problem from the noise transmission path, and the transmission error is the

key to solving the NVH problem [8]. Reducing the transmission error during the work of the gear pair can reduce the vibration excitation source and fundamentally solve the NVH gear whine problem.

Acknowledgment. Fund Project: National Key Research and Development Project for New Energy Vehicles "High-performance Precision Integrated Drive Motor System Development Project" (2018YFB0104901).

References

1. Blue Book of New Energy Vehicles. Social Sciences Academic Press, pp. 1–2 (2018)
2. Azar, R.C., Crossley, F.: Digital simulation of impact phenomenon in spur gear systems. J. Eng. Ind. **99**(3), 792–798 (1977)
3. Iida, H., Tamura, A., Yamada, Y.: Vibrational characteristics of friction between gear teeth. Bull. JSME **28**(241), 1512–1519 (1985)
4. Lin, J., Parker, R.G.: Mesh stiffness variation instabilities in two-stage gear systems. J. Vib. Acoust. **124**(1), 68–76 (2002)
5. Wang, L.: Research on nonlinear dynamics and NVH performance of engine-transmission coupling system, pp. 137–139. Zhejiang University (2014)
6. James, B., Hofmann, A., Doncker, R.W.D.: Reducing noise in an electric vehicle powertrain by means of numerical simulation. In: Automotive NVH Technology, pp. 29–46 (2016)
7. Yuan, F., Tong, Z., Peng, Y., Rong, G.: Research on vibration and noise characteristics of permanent magnet synchronous motor with integrated drive structure. J. Tongji Univ. Nat. Sci. **43**(7), 1070 (2015)
8. Pan, X., Liu, X., Li, C.: Research on NVH performance optimization of BEV high-speedgear transmission. J. Chongqing Univ. Technol. Nat. Sci. 25–31 (2017)

Research and Design of College Teachers' Performance and Salary Collaborative Management System

Qiaoyue Zhao[✉]

Sichuan Vocational and Technical College, Suining, Sichuan, China
ZhaoQiaoyue1989@haoxueshu.com

Abstract. Collaborative management of performance and salary of college teachers is one of the core ways of university management informationization. The design of college teachers' performance payroll management system benefits from the organic combination of management science and information technology. This paper automatically generates the teacher's performance allowance through the division of labor and process of several main functional modules, combined with the parameter setting and calculation of the system. Thereby it achieves the optimal allocation of limited resources in colleges and universities and the rational distribution of performance pay. While motivating faculty and staff to be dedicated and self-fulfilling, we will achieve a win-win situation in which teachers and colleges complement each other and achieve harmonious progress.

Keywords: College teacher · Performance salary · Collaborative management

1 Introduction

Performance-based payroll management in higher education institutions is the planning, organization, coordination, execution and supervision of the payment of performance pay of the majority of faculty and staff, and the query, statistics and analysis of relevant data, according to national and local relevant policies, laws and regulations [1]. The rapid development of modern management science and network information technology has greatly improved the level of personnel and salary management in all walks of life [2]. Full application of modern computer technology and contemporary advanced management concepts in the human resources management of colleges and universities will play an effective role in promoting personnel management in colleges and universities, improve management and better serve the development of schools [3]. Especially with the continuous expansion of the scale of higher education, the scale of teachers in various universities is also rapidly expanding, the categories of faculty and staff are becoming more and more complex, and the management of faculty and staff performance has become more complicated.

At present, the market mainstream performance salary management system mainly includes the general manager family salary management system, Yishen personnel salary management system, Xinchuang general personnel salary management system,

© Springer Nature Singapore Pte Ltd. 2020
M. Atiquzzaman et al. (Eds.): BDCPS 2019, AISC 1117, pp. 1887–1894, 2020.
https://doi.org/10.1007/978-981-15-2568-1_264

Kingdee financial management software [4]. The general manager family salary management system includes two parts: the personnel management system and the salary management system, which can realize the functions of personnel information management, salary management, and introduction and export of personnel compensation data [5]. Yishen personnel salary management system also includes personnel file management and salary management, which can realize the management of various items in personnel file information, management of job titles, assessment attendance, designing and calculating salary data items and calculating calculation methods for individual taxes [6]. Formulas, import and export staff salary statistics, reports, payrolls and other functions. The above software is mainly aimed at the salary management system developed by the needs of enterprise personnel salary management [7]. The salary management mainly focuses on hourly wages and piece-rate wages, and is not suitable for college performance payroll management [8].

In view of the above analysis of the performance-based salary management system in terms of salary incentives, this paper introduces the system into the performance salary management of college teachers, in order to stimulate the enthusiasm of college teachers for teaching and research, give full play to the creativity of college teachers, improve the efficiency of research and the quality of teaching, and ultimately It is particularly urgent to realize the benign interaction and win-win development between teachers and schools. Through the effective combination of management science and information technology, the design and implementation of college teachers' performance salary management system based on fairness and efficiency are discussed.

2   Overall System Design

The college teacher performance payroll management system is an integrated management platform, which is a scientific combination of information technology and management technology [9]. The implementation of the performance-based payroll management process using management information systems not only improves the scientific and efficient performance management, reduces manual intervention in the management process, but also achieves fairness, fairness, openness, and data sharing and management of performance management [10].

The overall design of the performance-based payroll management system of colleges and universities adopts a structured and modular design scheme, which is the realization of the design goal of the system at the macro level. In the system design process, the main basis for completing the design work is the results of system requirements analysis and foreseeable system functions [11]. According to the design goal of the performance-based salary management system of ordinary colleges and universities, the system functions according to the concept of module design [12]. The system function modules can be divided into: personnel management, salary management, attendance management, subsidy management, income inquiry, user management, Eight modules, such as interactive communication and system management, establish data exchange rules between functional modules to realize the functions of the performance salary management system.

In order to construct a performance salary management model that adapts to the characteristics of Chengde Petroleum College, to achieve the system design goals, the overall framework of the performance salary management system of ordinary colleges and universities is constructed, as shown in Fig. 1.

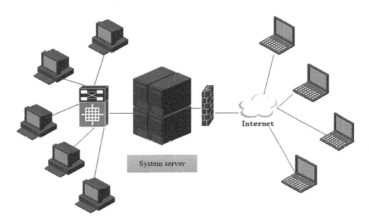

Fig. 1. System framework diagram

The overall framework design of the performance payroll management system fully utilizes the modern advanced computer technology to meet the needs of the faculty and staff and various system administrators to use the system, which fully meets the requirements for the design and implementation of the performance payroll management system of ordinary colleges and universities.

3   System Structure and Function Design

The system is based on the theory and technology of human capital theory, fairness theory, demand hierarchy theory, efficiency compensation theory, etc., and has undergone multi-layer demand analysis and design. The system function module mainly includes eight modules of authority management, employee management, organization management, post management, performance salary management, information management, statistical management, and security management, as shown in Fig. 2.

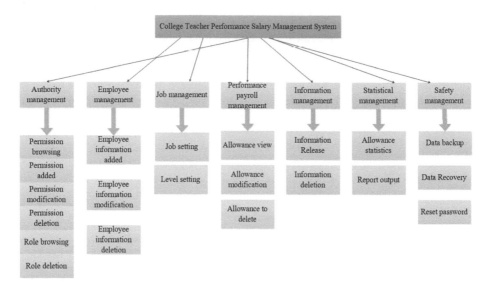

Fig. 2. System module

3.1 System Use Case Model Design

In this system, there are eight modules including authority management, employee management, position management, performance salary management, statistical management, security management, organization management and information management. The super administrator has the authority to manage these eight modules. The Science and Technology Department administrator, personnel office administrator, finance department administrator, academic office administrator and each college administrator belong to the secondary system administrator, who have different user rights. The personnel of the Personnel Department mainly manages the post allowances in employees, positions, and performance pay. Specifically, it can have the functions of post setting, setting post performance pay, feedback employee appeal, modifying employee information, and issuing personnel information. The Manager of the Technology and Technology Department manages the research and development awards, research projects and research-related information. The administrator of the Academic Affairs Office manages the teaching workload, teaching performance and information related to teaching and teaching. The Finance Department administrator manages, compares, taxes, distributes, and finances information related to teacher performance pay. The college administrators mainly conduct preliminary assessments and statistics on teachers' teaching performance and work performance. The system use case model is shown in Fig. 3.

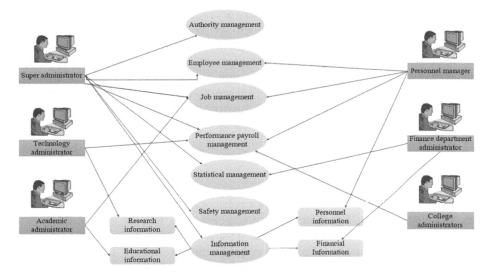

Fig. 3. System use case model

3.2 Sequence Model Design

After the employee enters the user name and password on the web page, the login information is sent to the system, and then the data information is queried through the

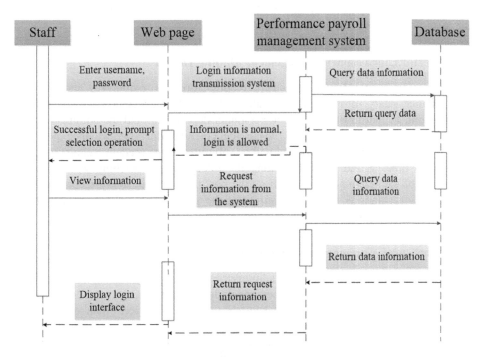

Fig. 4. System sequence model

database, and the database returns the query data to the system, that is, the employee information is verified correctly, and the login success information is returned, and then the employee can Follow the system prompts to make the appropriate selection. The employee requests information from the system while browsing the information. The database returns the corresponding information according to the employee's request and displays it on the web page. When the employee exits the system after the operation is completed, an exit request is sent, the database automatically clears the session information, and returns an exit confirmation message. The system sequence model is shown in Fig. 4.

3.3   Collaborative Model Design

This system is a typical multi-administrator collaborative office system, so each module works collaboratively. When an ordinary employee logs into the system, the system first sends a request to the system, and the system returns the corresponding operation prompt, and the employee can perform corresponding operations. The system administrator and department administrator give feedback through the corresponding module function operation request system. The system returns the corresponding operational data according to the request. For example, when requesting a post modification, first request the post modification information from the system, and perform the corresponding post modification operation according to the system feedback. When requesting performance salary setting, the system further determines which department administrator's authority is based on which performance salary the user requests. Therefore, the system collaboration model reflects a collaborative relationship between the system and various users and administrators.

3.4   Performance Salary Management Module Design

In the performance salary management module, ordinary employees can view the performance salary information. The administrators of the Science and Technology Department, the administrators of the personnel department, and the administrators at all levels can view, modify, and delete the corresponding performance pay. The workload, and the teaching workload will be submitted to the academic affairs office for review on time, as the basis for setting the salary for the teaching workload. The personnel of the Personnel Office will set the post performance salary according to the post setting regulations. The administrator of the science and technology department will provide the document according to the scientific research award.

3.5   Statistical Management Module Design

The statistical module belongs to the administrator's operation authority. Since the post performance salary is set by the personnel department, the personnel department administrator is responsible for the summary of the post performance salary; the Academic Affairs Office evaluates the teacher's workload and then follows the corresponding assessment criteria. Workload performance is evaluated accordingly to aggregate teacher workload performance pay; the Science and Technology Department is responsible for the aggregation of research reward performance. At this point, the three performance payrolls are each summarized, and then submitted to the Finance Department for review, and preliminary statistics are carried out, and finally summarized, and the final summary results are released for publication.

4   Conclusion

This paper combines the professional and personnel management characteristics of college teachers to design a performance salary management system for college staff to improve the fairness and efficiency of personnel management level and performance salary distribution in colleges and universities. Under the premise of promoting the goal management of colleges and universities, realizing the rational allocation of resources and the scientific and efficient distribution of performance, combined with the performance subsidy allocation system that is compatible with the development of colleges and universities, the goal completion and performance subsidy are directly linked to achieve fairness and openness. Remuneration, fully mobilize the enthusiasm and scientific research creativity of the faculty and staff, with a view to improving the quality of teaching and work efficiency, and promoting the realization and harmonious development of the university's established goals.

Acknowledgement. This project is supported by Foundation for Humanities and Social Sciences Projects of Sichuan Provincial Department of Education (Project No.: 18SB0703).

References

1. Saqr, M., Fors, U., Tedre, M.: How the study of online collaborative learning can guide teachers and predict students' performance in a medical course. BMC Med. Educ. **18**(1), 24 (2018)
2. Arkhipenka, V., Dawson, S., Fitriyah, S., et al.: Practice and performance: changing perspectives of teachers through collaborative enquiry. Educ. Res. **60**(1), 97–112 (2018)
3. Wan, S., Li, D., Gao, J., et al.: A collaborative machine tool maintenance planning system based on content management technologies. Int. J. Adv. Manuf. Technol. **34**(2), 10–12 (2016)
4. Wang, S.L., Hong, H.T.: The roles of collective task value and collaborative behaviors in collaborative performance through collaborative creation in CSCL. Educ. Technol. Res. Dev. **59**(2), 89–92 (2018)
5. Zharova, A., Tellinger-Rice, J., Härdle, W.K.: How to measure the performance of a Collaborative Research Center. Scientometrics **87**(21), 70–73 (2018)
6. Zhang, L., Warren, Z., Swanson, A., et al.: Understanding performance and verbal-communication of children with ASD in a collaborative virtual environment. J. Autism Dev. Disord. **9**(5), 1–11 (2018)
7. Browning, C.A., Harris, C.B., Bergen, P.V., et al.: Collaboration and prospective memory: comparing nominal and collaborative group performance in strangers and couples. Memory **26**(9), 1–14 (2018)
8. Wu, L., Sun, P., Hong, R., et al.: Collaborative neural social recommendation. IEEE Trans. Syst. Man Cybern.: Syst. **6**(99), 1–13 (2018)

9. Zhao, L.L., Jiang, X.L., Li, L.M., Zeng, G.Q., Liu, H.J.: Optimization of a robust collaborative-relay beamforming design for simultaneous wireless information and power transfer. Frontiers Inf. Technol. Electron. Eng. **19**(11), 134–145 (2018)
10. Zulfiqar, S., Zhou, R., Asmi, F., et al.: Using simulation system for collaborative learning to enhance learner's performance. Cognet Educ. **8**(7), 112–117 (2018)
11. Zhang, N., Henderson, C.N.: Requiring students to justify answer changes during collaborative testing may be necessary for improved academic performance. J. Chiropractic Educ. **31**(2), 45–47 (2017)
12. King, K.G., Lange, T.K., Lange, T.K.: Measuring teamwork and team performance in collaborative work environments. In: Evidence-Based HRM: A Global Forum for Empirical Scholarship, vol. 12, no. 9, pp. 34–38 (2017)

Relationship Model of Green Energy and Environmental Change Based on Computer Simulation

Yongzhi Chen[✉] and Yingxing Lin

The School of Economics and Management, Fuzhou University, Fuzhou, Fujian, China
Yongzhi_Chen1982@haoxueshu.com

Abstract. The massive use of energy has led to a serious deterioration in environmental quality. This paper takes green energy as the object and applies computer simulation technology to study the relationship between green energy and environmental changes, and then establish a model. Firstly, according to the principle of support vector machine, the green technology innovation strategy selection model of support vector machine is established. Then, the grid search method is used to optimize the penalty parameters and kernel function parameters to achieve better generalization ability. The simulation results show that by using computer simulation technology to study the relationship between green energy and environmental change, we can better choose green strategy, improve energy resource productivity, reduce environmental pollution, help formulate green energy innovation strategy, and improve strategy. The accuracy of the selection is of great significance to the environmental protection and the sustainable development of green energy.

Keywords: The relationship model · Green energy · Environmental change · Computer simulation

1 Introduction

In recent years, environmental pollution and energy depletion have become the top problems in the world. Nowadays, with the lack of natural resources and increasing attention to environmental protection issues, the development model that pursues only economic growth is no longer in line with current development needs. What humans need is a scientific, rational, environmentally friendly and sustainable economic development model. The use of green energy instead of non-renewable resources has become the current economic development strategy [1]. Management innovation and technological innovation aiming at protecting the environment are generally referred to as green technology innovation. At present, the commonly used green technology innovation method is to use computer simulation technology to study the relationship between green energy and environmental changes, so that the economic, social and ecological benefits of the enterprise can be coordinated and unified, and finally achieve sustainable economic development [2, 3].

M. Atiquzzaman et al. (Eds.): BDCPS 2019, AISC 1117, pp. 1895–1902, 2020.
https://doi.org/10.1007/978-981-15-2568-1_265

The relationship between green energy and environmental change is a complex nonlinear problem. There are many indicators involved, and there are many description methods such as quantitative and qualitative indicators. It is difficult to establish a satisfactory evaluation model. In recent years, people have noticed the superiority of computer simulation technology. Many studies have applied the theory and method of computational intelligence from different angles and different aspects. Zhou et al. [4]. constructed a creative evaluation model based on the support function vector machine based on radial basis function, and analyzed the relationship between green energy and environmental change, and realized the optimization of kernel function parameters. Gu et al. [5] designed a model to optimize the allocation of green resources and environmental resources through big data and decision-making methods. Cai [6] and Wang [7] also applied neural networks to the study of the relationship between green energy and environmental change. Because of the ability of the neural network in mathematical mapping, the neural network can be identified through learning to achieve the purpose of assisting decision-making. However, the neural network-based method is slow to train, and it is difficult to reconcile the contradiction between overfitting and generalization, and it is easy to converge to the local best. The support vector machine [8] created by Vapnik is based on the principle of structural risk minimization, has good generalization performance and simple mathematical form, and is a new machine learning method.

Based on the above problems, compared with the neural network, the theoretical basis of the support vector machine is more perfect, and the parameters that need to be set are relatively few [9]. Furthermore, the local optimal solution of its algorithm must be a global optimal solution, which is not available in neural networks. Therefore, this paper uses support vector machine to study the relationship between green energy and environmental changes. Case studies show the effectiveness of the method.

2 Methodology

2.1 Support Vector Machine

SVM can achieve the best generalization and profile by looking for balance, which has the advantage of overcoming the problems of small sample size, nonlinearity and high dimensional data. Classification is the identification of the type of sample by identifying the model. Gravitational wave noise event recognition belongs to two types of problems. As an example, there are two categories that introduce the basic principles of SVM [10].

A set of training samples (x_i, y_i) is assumed, where $x_i \in V^d$ stand for some point of view a d-dimensional feature vector, and $y_i \in [-1, +1]$ stand for some point of view a sample category of x_i. The SVM solves the smallest classification error by maximizing the hyper-plane and the last two sampling intervals. The central idea is to establish the largest interval hyper-plane in the high dimensional space.

The SVM handles the binary classification problem, as shown in Fig. 1, where the maximum interval of the hyper-plane is indicated by a dashed line. In most cases, the sample cannot be separated directly. Therefore, the SVM needs to separate the training samples of the original indivisible space by mapping to high dimensions to separate the space. The SVM then establishes a maximum interval hyper-plane $\omega * x + b = 0$. Where ω represents the normal vector and b represents the offset. On the left side of Fig. 1, the case of linear separability is shown. The following can correctly distinguish between two types of samples: $\omega * x + b \geq +1$ when $y_i = +1$, $\omega * x + b \leq -1$ when $y_i = -1$. The unified expression is shown in the following formula 1:

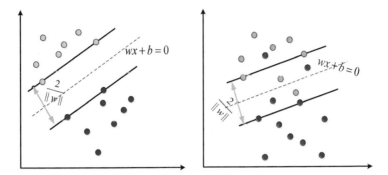

Fig. 1. SVM processing classification problem

$$y_i((\omega, x) + b) - 1 \geq 0 \tag{1}$$

The SVM solves the quadratic programming problem by establishing a maximum interval hyper-plane, as follows:

$$\min_{\omega, b, \xi} \frac{1}{2} \omega^T \omega + C \sum_{i=1}^{n} \xi_i \quad s.t \quad y_i(\omega^T \Phi(x_i) + b) \geq 1 - \xi_i \quad \xi_i \geq 0, i = 1, 2, \ldots, n \tag{2}$$

2.2 SVM-Based Green Energy and Environmental Change Model

The relationship between green energy and environmental change can be abstracted as a mapping from environmental space to energy space. Based on the reference data [11, 12], five influencing factors were selected: economic environment (X1), green technology level (X2), environmental equipment investment ratio (X3), energy consumption reduction rate (X4), Three wastes treatment rate (X5).

The selection principles of the SVM-based environment and the green energy change model are as follows: First, the input vector of the support vector machine uses information describing the characteristics of the green energy and environmental changes, and outputs a class label indicating the correspondence; the sample training support vector classifier, So that different input vectors can get the corresponding output category. Finally, the trained support vector classifier can be used as an effective tool to evaluate objects outside the sample mode. In this way, the established SVM-based green resource and environmental change model can not only simulate the relationship between the two offline, but also avoid human error in the decision process.

In the model study, there are both qualitative and quantitative indicators, and each indicator must be standardized so that the indicators are comparable across the system. The specific processing method is as follows:

(1) When the target value is larger, the evaluation is better.

$$F_j = (x_j - x_{j\min})/(x_{j\max} - x_{j\min}) \tag{3}$$

(2) Evaluate with a smaller target value as a better standard.

$$F_j = 1 - (x_j - x_{j\min})/(x_{j\max} - x_{j\min}) \tag{4}$$

Where F_j is the normalized value of the target value x_j, $x_{j\min}$ is the minimum value of the predetermined j-th index, $x_{j\max}$ is the maximum value of the predetermined j-th index, and j is the number of evaluation indicators.

(3) Qualitative indicators are determined by expert scoring methods. In order to maintain comparability with quantitative indicators, it must be standardized and processed in the same way as quantitative indicators.

2.3   Simulation Experiment

Taking natural gas green energy as an example, computer simulation is used for analysis. The training samples consisted of 10 groups, as shown in Table 1.

Table 1. Training samples

Number	X1	X2	X3	X4	X5
1	0.51	0.55	0.43	0.61	0.12
2	0.51	0.54	0.44	0.64	0.34
3	0.53	0.53	0.43	0.67	0.61
4	0.49	0.54	0.44	0.58	0.33
5	0.50	0.55	0.45	0.60	0.2
6	0.50	0.55	0.43	0.61	0.39
7	0.51	0.56	0.43	0.35	0.25
8	0.49	0.53	0.45	0.67	0.14
9	0.52	0.54	0.44	0.81	0.78
10	0.51	0.55	0.45	0.34	0.66

Because of the existence of multiple kernel functions, one of the main elements of designing an SVM is to choose a kernel function and appropriate parameters. There are three main types of commonly used kernel functions: polynomial kernel functions, radial basis kernel functions, and multilayer perceptron kernel functions. Different kernel functions have almost no effect on SVM performance, which is the discovery of Vapnik. And the choice of penalty coefficients and kernel function parameters is critical to the performance of SVM. This paper uses the most widely used radial basis kernel function:

$$K(x, \ x_i) = e^{\left(-\gamma|x-x_i|^2\right)} \tag{5}$$

C and γ are cross-validated using a grid search method. Each basic pair is tried, and then the highest accuracy is selected in the cross-validation. Figure 2 is the parameter optimization result graph.

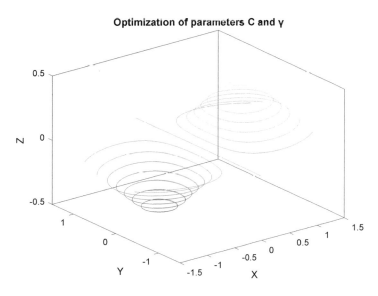

Fig. 2. Parameter optimization result graph

By using computer simulation technology and support vector machine to construct a green energy and environmental change model, the following results can be obtained, as shown in Figs. 3 and 4. It can be seen from the analysis that the decline in the use of non-renewable resources will lead to a reduction in the amount of various pollutants. With the increase in the use rate of non-renewable resources, non-renewable resources will increase the amount of treatment of five kinds of pollutants. Therefore, the growth rate of the five types of pollution is very fast, showing an increasing trend. With the use of green energy, the use of non-renewable resources has declined, and the rate of pollutants in 5 has gradually become slow until it is flat. This trend indicates that the use of green energy has had a significant improvement in environmental change.

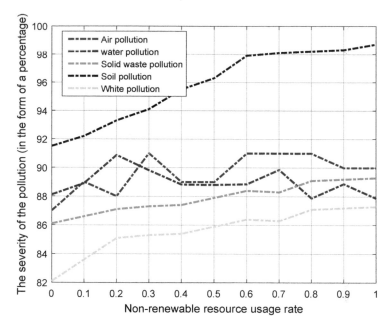

Fig. 3. A graph of the impact of the use of non-renewable resources on environmental changes

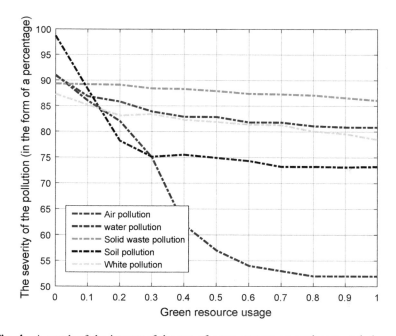

Fig. 4. A graph of the impact of the use of green energy on environmental changes

3 Summary

This paper uses computer simulation technology to build a model of the relationship between green energy and environmental change. Based on statistical learning theory, SVM has a strict mathematical theory foundation and in view of the principle of minimize structural risk, thus ensuring that learning machines have good generalization ability. In this paper, grid search and cross-validation are used to optimize the penalty parameters and kernel function parameters to avoid the problem of over-fitting or under-fitting in training. Finally, the simulation analysis is carried out in combination with actual cases. Simulation studies show that the established SVM model can accurately estimate the relationship between green energy and environmental changes. Through the analysis of the model, combined with the actual situation of non-renewable energy and green energy, the relationship between environmental pollution and energy use and the existence of energy consumption is obtained. Increasing the production of green energy will reduce environmental pollution and the use of non-renewable resources. With the strengthening of environmental protection awareness in China, environmental pollution has stimulated the development of green energy. Therefore, it is very meaningful to use computer simulation technology to build the relationship between green energy and environmental change.

References

1. Wang, D., Ghannouchi, F.M., Yi, D., et al.: 70% energy saving in wireless positioning systems: non-data-bearing OFDM transmission replaces non-pulse-shaping PN transmission. IEEE Syst. J. **9**(3), 664–674 (2017)
2. Stephenson, A.L., Dupree, P., Scott, S.A., et al.: The environmental and economic sustainability of potential bioethanol from willow in the UK. Bioresour. Technol. **101**(24), 9612–9623 (2010)
3. Campisi, D., Morea, D., Farinelli, E.: Economic sustainability of ground mounted photovoltaic systems: an Italian case study. Int. J. Energy Sector Manag. **9**(2), 156–175 (2015)
4. Zhou, F.Z.: On evaluation model of green technology innovation capability of pulp and paper enterprise based on support vector machines. Adv. Mater. Res. **886**, 285–288 (2014)
5. Gu, W., Saaty, T.L., Wei, L., et al.: Evaluating and optimizing technological innovation efficiency of industrial enterprises based on both data and judgments. Int. J. Inf. Technol. Decis. Making (IJITDM) **17**(1), 9–43 (2018)
6. Cai, L., Liu, Q., Zhu, X., et al.: Market orientation and technological innovation: the moderating role of entrepreneurial support policies. Int. Entrepreneurship Manag. J. **11**(3), 645–671 (2015)
7. Wang, B., Wang, X., Wang, J., et al.: Construction and empirical analysis of agricultural science and technology enterprises investment risk evaluation index system. Ieri Procedia **2**(4), 485–491 (2012)
8. Zhang, Y.D., Yang, Z.J., Lu, H.M., et al.: Facial emotion recognition based on biorthogonal wavelet entropy, fuzzy support vector machine, and stratified cross validation. IEEE Access **4**(99), 8375–8385 (2017)

9. Chen, W., Duan, Q., Wei, G., et al.: An evaluation of adaptive surrogate modeling based optimization with two benchmark problems. Environ. Modell. Softw. **60**(76), 167–179 (2014)
10. Chandra, M.A., Bedi, S.S.: Survey on SVM and their application in image classification. Int. J. Inf. Technol. **2**, 1–11 (2018)
11. Tanveer, M., Shubham, K.: A regularization on Lagrangian twin support vector regression. Int. J. Mach. Learn. Cybern. **8**(3), 807–821 (2017)
12. Sharp, B.E., Miller, S.A.: The potential for integrating diffusion of innovation principles into life cycle assessment of emerging technologies. Environ. Sci. Technol. **50**(6), 2771–2781 (2016)

Risks of the Blockchain Technology

Jinlan Guo[✉]

Artificial Intelligence Law School, Shanghai University of Political Science
and Law, Shanghai, China
guojinlan1976@haoxueshu.com

Abstract. This paper comprehensively analyses various risks of the blockchain
technology in circulation, including wallet security risks, code vulnerability,
risks on the validity of the consensus mechanism, Hash algorithm risk, 51%
attack risk and exchange risk. At present, with the vigorous development of the
blockchain technology and the absence of government and legal supervision,
users can only rely on themselves to fully identify risks to effectively avoid loss.

Keywords: Blockchain · Technical risks · Risk research

1 Introduction

The concept of blockchain was first mentioned by Stuart Haber and W. Scott Stornetta
in a paper written in 1991. This paper describes a digital system called blockchain,
which uses digital timestamps in business transactions. This theory became a legend in
the hands of Stoshi Nakamoto 17 years later. Blockchain, as the core underlying
technology of Bitcoin, has attracted attention from the finance, insurance, trade and
other fields outside the IT industry. This technology pursues trustfree, de-centralization
and tamper-proof, and can be executed without the third-party of trusted authority [1].
It is considered to be the leader of science and technology in the future, and can bring
great changes to the production relations in today's society. Its subversiveness will far
exceed the technologies of cloud computing, big data, Internet of Things and artificial
intelligence.

Bitcoin is considered to be the first generation of blockchain application. It has
become one of the most popular payment methods, because the system is super safe
[2]. It completely offsets the possibility of fraudulent transactions and double-spending.
The second generation of blockchain application, represented by Ethereum, carries part
of its commercial development and business expansion through smart contracts. The
third generation of blockchain application, represented by EOS, NEO, QTUM, AE and
Finchain financial chains, have unlimited capacity for commercial applications, and can
access to thousands of commercial entities [3]. They also expand their advantages, such
as lower transaction costs, faster transaction verification, as well as greater mobility
compatibility and higher developer friendliness [4].

Although the blockchain technology has been developing and evolving, its
essential characteristics are always decentralization, non-tampering and traceability.
These characteristics are determined by the data structure, the data storage method and
the date flow mode. From the point of view of data structure, blockchain data are linked

© Springer Nature Singapore Pte Ltd. 2020
M. Atiquzzaman et al. (Eds.): BDCPS 2019, AISC 1117, pp. 1903–1909, 2020.
https://doi.org/10.1007/978-981-15-2568-1_266

in blocks through the (Hash) chain structure, and there is a sequence that links data blocks. The sequence cannot be tampered. From the point of view of data storage, each node in the blockchain can save the whole blockchain, which is an implementation of the "de-centralized database" technology [5]. From the perspective of openness degrees or node access rules, blockchains can be divided into three categories: public blockchains, alliance chains and private blockchains. In addition, the blockchain has a programmable feature, which provides a technical way to build smart contracts and establish virtual autonomous organizations [6].

In the future, the blockchain technology will be widely used in following platforms and systems: smart contract application; e-money systems such as the earliest bitcoins and Libra released by Facebook; decentralized application development and deployment platforms, such as e-commerce platforms based on the blockchain technology [7]. Today, e-commerce platforms such as Taobao and Jingdong are centralized platforms; the data of buyers and sellers are stored in the center; the accounts of buyers and sellers are controlled by the center. For users, there is a danger of being banned or blocked at any time. Sellers need to pay a lot of money to the platform and are constrained in many aspects. The smart contract application is a code protocol designed to disseminate, verify and implement the contract through information. It can reduce intermediate links and save costs. The blockchain technology can also be applied in the digital asset management network and business applications developed for other scenarios.

Blockchain is a destructive and innovative technology [8], which may completely destroy the old industrial ecology or rebuild a new one. Therefore, security is the lifeblood of the technology. Before introducing the blockchain into a variety of commercial applications, we must make a comprehensive study on the inherent risks of this technology [9]. The blockchain technology is complex. It is relatively difficult for ordinary participants to understand. This information asymmetry also leads to risks. Once there are technological loopholes, the consequences are incalculable. The public discussion of various technical issues among enthusiasts can play a role of risk warning. Some algorithm geeks are also working hard constantly to verify the technical limitations of digital currency, such as the resolution of Bitcoin bifurcations. The powerful error correction mechanism composed of community opinions and the computer voting shows the self-healing ability of digital currencies. However, the risk of blockchain technology has not been completely ruled out in the whole circulation process [10].

2 Wallet Security Risks

From Bitcoins of the first generation, to ETH developed by Ethereum of the second generation and EOS of the third generation, for the blockchain, wallets always play a vital role in circulation. The official Bitcoin wallet, bitcoin-Qt is taken as an example. The file that clients store the private key of Bitcoin is wallet. dat. Generally, the storage path in Windows 8 system is c: c:\users\AppData\Roaming\bitcoin (AppDate is a hidden folder). Wallet. dat is essentially a private key that holds all the addresses of the wallet. With this file, the user can prove that these bitcoins to him. So the risk of wallet is the risk of files named wallet. dat. For example, some people delete their wallets by mistake, losing bitcoins worth tens of thousands of dollars. Some people do not backup their wallet files correctly, resulting in the total loss of money traded over a period of time. Some people's computers are infected with Trojans. Wallet files are stolen and all bitcoins are gone.

The corresponding solution is to pay attention to the security of the computer. Users need to prevent hackers and Trojan Horses from intruding, properly save the wallet. dat files and backup them regularly. Paper wallets, brain wallets and money wallets can also be considered, but these wallets also take their own risks and need to be used with caution.

For ordinary users, it is still recommended to put the information in a large exchange. Even if the password is forgotten or stolen, there are still ways to retrieve or roll back. Cold wallet is a better choice for advanced users who hold a lot of bitcoins. The advantages of cold wallet are obvious. It is basically not connected to the Internet, and the private key will not leave the cold wallet device, so the security level is higher. But it shifts the risk from exchange security to user's preservation of cold wallet equipment and mnemonics.

Another risk is that the wallet's private key can be stolen. For digital currencies like bitcoin, Ethernet and derivative EOS, the core of trust is the user's private key (a sequence of random numbers generated based on the elliptic curve asymmetric algorithm). The money will lose if the private key is lost. Hackers of course know this point very well. With the rise of the value of digital money, attacking and thieving private keys become their favorite work.

Recently, nccgroup, an American information security organization, has found vulnerabilities in some open source algorithm libraries. Hackers can acquire vulnerabilities of the ECDSA or DSA private keys through side channel attacks. Side Channel Attack (SCA) is a method of attacking encryption devices by utilizing the leakage of side channel information such as the time consumption, the power consumption and the electromagnetic radiation when the encryption application is in operation. This new type of attack is much more effective than the mathematical method of cryptanalysis, so it poses a serious threat to cryptographic applications. For instance, the changes of architecture running state, power consumption or electromagnetic radiation in processors all provide attack sources for SCA. Flush + Reload is a kind of Cache side-channel attack, which is an attacking method using the time difference of cache introducing. When accessing memory, the data which have been accessed by the cache will be very fast, but data which have not been accessed will be slow. Cache side-channel attack uses this time difference to steal data. It is worth mentioning that this situation generally occurs when the attacker and the victim share the server, or they use cloud-based virtual machines which are scheduled on the same physical hardware.

3 Code Vulnerability and Risks on the Validity of the Consensus Mechanism

The blockchain technology is complex in itself [4]. Without perfect codes, code vulnerabilities can cause great losses even if there's no problem on the technical framework.

Firstly, devices, systems, software or programs may still have backdoors, and the risks caused by them remain. For alliance chains and private chains, data are more sensitive; the impacts are greater. Secondly, without perfect codes, there are obvious potential risks in the actual implementation of the trust project. In 2016, The DAO, a smart contract application of the Ethereum, was attacked by hackers (The DAO smart contract code has a loophole). The event provided a clear answer, and directly led to the subsequent hard bifurcation of Ethereum.

The consensus mechanism is very important for the shared maintenance and control in a decentralized environment. Consensus is embodied both on the blockchain (the consensus algorithm) and off the blockchain (consensus rules). Even if the consensus algorithm is safe and risk-free, consensus rules may need to be changed over time. However, it is clearly unlikely that consensus rules can be reached all the time. Therefore, we should fully recognize the potential risks of consensus and the risks of hard and soft bifurcation.

For alliance chains, the failure of consensus rules may lead to the exit of important nodes. Therefore, the support of alliance chains for a safe and risk-free exit mechanism is particularly important [5].

4 Hash Algorithm Risk

The security of most digital currencies is guaranteed based on specific Hash algorithms. If the algorithm is cracked, an attacker can deduce the same message digest from two different sets of information. That is to say, he can find random numbers that meet the requirements in a very short period time, and then he can quickly grab the right of keeping accounts and generate new blocks. In this case, the security mechanism based on the proof of work will be useless.

At present, blockchain technology involves many cryptographic algorithms, such as the random number generation algorithm, the Hash algorithm and the digital signature algorithm. Therefore, blockchain is highly dependent on cryptographic technology. Loopholes in cryptographic algorithms are fatal to the blockchain system. The potential risks are huge. Data show that NSA has created backdoor programs to add encryption standards (which are used to generate random numbers) approved by NIST. This behavior caused controversy. Therefore, the direct adoption of cryptographic algorithm standards issued by other countries or other organizations of standardization (the algorithmic design may have built-in backdoors) poses a huge security risk, which highlights the importance of designing and selecting standardized cryptographic algorithms with independent intellectual property rights. In addition, the specific implementation of algorithms also faces serious troubles of backdoors and loopholes. Code review, test evaluation and other procedures should be carried out for key applications to eliminate implementation risks like backdoors and loopholes.

With the progress of cryptographic analysis technology and the gradual improvement of human computing power, many cryptographic algorithms will expose their weaknesses. Especially for the Hash algorithm, collisions are certain and inevitable in theory, but it is difficult to find the collision. For example, in 2004, Xiao-yun Wang, a cryptographer in China, discovered the weakness of MD5 Hash algorithm, and revealed that the collision cracking of Hash functions will happen sooner or later. Secondly, most digital signature algorithms are provably secure based on mathematical difficulties, such as factorization and discrete logarithm. With the advancement of quantum computing research, these difficult problems will become easy to solve under the quantum computing model, so it is very important to carry out the research or construction of quantum cryptography algorithm.

With the introduction of cryptographic protocols such as the zero-knowledge proof, the commitment protocol and the secure multi-party computation, protocol risks will become an important issue in blockchain security like algorithm risks.

5   51% Attack Risk

51% attacks have been feared since the first day of Bitcoin's birth. It actually happened before. 51% attack means, if an attacker controls more than 50% of the network's computing power, it will be able for him to modify his own transaction records. He can realize double-spending, prevent blocks from sealing certain or all transactions, and prevent some or all miners from finding valid blocks.

The way to prevent 51% attacks is to increase the computing power of the whole network quickly, which can make it difficult for individual attackers to have more than 50% of the computing power. Adopting the dynamic checkpoint technology is also a feasible solution. For Bitcoin, it is almost impossible to launch 51% attacks since the whole network computing power exceeds 1200 PHhash/s. The dynamic checkpoint technology enhances the centralization requirement. For other digital currencies with weak computing powers, the risk of 51% attacks always exists.

It is worth pointing out that if an exchange is attacked by hackers who want to remove attacking transactions which have occurred by rolling back the transaction, that is to say, to return to the old block, this rollback transaction will be equivalent to a 51% power attack. Similar situations occurred in 2016, in the community response after the DAO attack in Ethereum, which directly led to the hard bifurcation of the Ethereum network (the emergence of the Ethereum classic). Compared with BTC lost by the exchange, such treatment may cause more damage to the Bitcoin network and to people's confidence in virtual currency. So in 2019, after Binance was attacked by hackers and lost 70,000 BTC, it did not use rollback trading.

6   Exchange Security Risk

Since 2019, at least four central exchanges have been attacked: Cryptopia, DragonEx, Bithumb and Binance. The attackers are all profit-seeking. Centralized exchanges are always the target of hacker attacks, since the number and value of currencies held by centralized exchanges are far greater than those held by decentralized exchanges. Meanwhile, the assets of de-centralized exchanges are managed by contracts, and most of them are open source. Fully open source smart contracts enable programmers around the world to help users check whether there are loopholes in the project. The transparency of the entire trading process is high, which decreases the probability of problems. Therefore, de-centralized exchanges are relatively safer. However, it should be noted that decentralized exchanges are not as good as centralized exchanges in terms of user experience, the transaction depth and the user threshold.

Generally, trading centers set up mechanisms to protect users' digital currency assets, so as to protect users from losses. But it is difficult to recover all the stolen assets. After the attack, the stolen digital assets can be tracked; the address of the hacker's wallet can be monitored in real time; possible methods of money laundering can be analyzed. However, in order to recover the stolen assets, cooperation and support from various parties in the ecosystem are needed. Exchanges need to freeze hacker's recharge timely; collaboration among super nodes or corresponding governance organizations (e.g. ECAF) is also needed. Moreover, due to the absence of government and legal supervision, the attacker can evade civil or criminal liability, which is also one of the reasons why hackers are rampant [11].

7   Conclusion

This paper analyzes risks of the blockchain technology in circulation, including the wallet risk, the code vulnerability, the risk of validity of consensus mechanisms, the risk of algorithm cracking, as well as the 51% attack risk and the exchange security risk. The vast number of blockchain users should fully understand these potential risks. More importantly, they need to call on the government to participate in supervision and regulation. Regulators in China must fully recognize the importance of technical supervision and the formulation of technical rules, and make adequate preparations for the possible large-scale application of the blockchain technology in the capital market through optimizing the design of risk management, reducing the cost of risk management, and improving the regulation efficiency and quality. They need to set up appropriate regulatory nodes on the blockchain, so as to timely grasp the transaction situation on the whole chain, and establish a more perfect defense mechanism to deal with possible large-scale risk transmission. At the same time, regulators should make strict records of relevant applications and expand the scope of filing from token issuance projects to all blockchain projects. Through these methods, they can know the status of technical operation as well as the update and iterations of algorithms of various projects in a timely manner, so as to maintain the stability of the market, and prevent enterprises from bringing negative impacts to the blockchain market due to technological attacks.

Our country should cultivate and attract relevant talents of the blockchain technology, actively apply the innovative training mode of "industry, research and education", improve the regulatory capacity of relevant departments, and constantly explore the practical application of the blockchain technology in supervision and regulation, so as to actively respond to challenges brought by emerging technologies.

References

1. Xu, D.Q.: The nature of "destructive innovation" in financial science and technology and new ideas of regulation science and technology. Oriental Law (2) 2018
2. Liu, Y.H., Zhou, S.Q.: Exploration of the application, challenges and regulatory measures of security BlockChain. Financial Regulation Research (4) 2017
3. Wang, X.F.: Application of distributed book-keeping technology based on blockchain in financial field and suggestions for supervision. Bus. Econ. (4) (2017)
4. Zhuang, L., Zhao, C.G.: Evolution of digital money under technological innovation of block chain: theory and framework. Economist (5) (2017)
5. Friedrich, H.: The impact of blockchain technology on business models in the payments industry. In: International Conference on Wirtschaftsinformatik (2017)
6. McWaters, J.R.: The future of financial infrastructure: an ambitious look at how blockchain can reshape financial services. In: World Economic Forum (2016)
7. Gausdal, A.H., Czachorowski, K.V., Solesvik, M.Z.: Applying blockchain technology: evidence from norwegian companies. Sustainability (6) (2018)
8. Hossein, K., Nicolette, K.D.S., Bart, C.: The blockchain revolution: an analysis of regulation and technology related to distributed ledger technologies SSRN. https://papers.ssrn.com/sol3/papers.cfm?abstract_id=2849251. Accessed 05 Jan 2017
9. Kevin, B., Matthieu, L., et al.: Beyond the hype: blockchains in capital markets. Investment Financing and Trade (2) 2016
10. Zhao, L., Chen, X.J., Dai, M.Y.: Technological risk of blockchain: an empirical study of regulation. Journal of Shanghai University of Foreign Trade and Economics (5) 2019
11. Stratiev, O.: Cryptocurrency and blockchain: how to regulate something we do not understand. Bank. Financ. Law Rev. 33(2), 173–212 (2018)

Effects of the Computer-Assisted Gallery Activity Program on Student Motivation and Engagement in College Classes

Ming Li[✉]

School of Foreign Languages, Shanghai University of Engineering Science,
Shanghai, China
LiMing_SH@haoxueshu.com

Abstract. Gallery activity has been regarded as an ideal strategy when American college professors use it to redesign their instruction. When computers and Internet are becoming more and more popular, the computer-assisted gallery activity program demonstrates more advantages in China's colleges. The purpose of this paper is to compare the key components of the computer-assisted gallery activity program with the MUSIC Model of academic motivation, which is a framework of instructional strategies based on the current motivation research and theory. Also, this paper illustrates how computer and Internet help gallery activity program run more smoothly through the online network than the traditional one. The results of the analysis indicates that the computer-assisted gallery activity program is consistent with the five principles of the MUSIC model of motivation. This paper implies that college professors could use the gallery activity in their teaching to motivate their students to engage in their learning.

Keywords: The computer-assisted gallery activity program · College students · Motivation · Engagement · The MUSIC model of motivation

1 Introduction

In the year 2016, China's Ministry of Education called for the reform of engineering education in higher education and this reform was called the reform of Emerging Engineering Education (simplified as 3E). The focus of this reform is to improve teaching quality and let the students in engineering become the focus in class. In January 2018, China' Ministry of Education issued the national standard of college education quality focusing on the three principles, that is, student-centered, output-oriented, and long-term mechanism. In educational psychology, gallery activity program is proved to be effective when it is used for college professors' instructional design. And in this program, the focus of class is students and students' comprehensive capabilities such as collaboration, leadership, and problem-solving are enhanced. Therefore, Chinese college professors tried to use the gallery activity program in the college classes in a university of engineering science in east China. This computer-assisted gallery activity program demonstrates the strength of the strategies in students' learning process. The major strength of this program includes: the students in the class

© Springer Nature Singapore Pte Ltd. 2020
M. Atiquzzaman et al. (Eds.): BDCPS 2019, AISC 1117, pp. 1910–1917, 2020.
https://doi.org/10.1007/978-981-15-2568-1_267

are divided into several groups, five or six, and the students have the rights to join the group he or she likes best. Then the teacher asks the group students to preview the text in the textbook. The teacher provides the students with a rubric, explaining the rules for the students' poster and presentation of the text. As for the turn of the presentation, the groups have the right to make decision when to give the presentation for example, on week two or week three or week six in the whole semester. The group students finish a poster whose content is the main idea of the given text in the textbook. After that, each group gives presentation of the poster in front of the classroom; then each group post the poster on the wall of the classrooms; all the students in the class will walk, watch, and write down their feedback on the posters by using stars, the number of the stars for each poster symbolizing the quality of the poster. Before the gallery activity, the instructor creates a QQ community inviting all his or her students to the online community and uploading the rubric for the poster presentation and the relevant reference materials for students' further understanding of the text. The instructor is available to the students all the time online and offline. In this paper, the author chooses students enrolled in the college English course as the subject in the research.

From the qualitative research data with the students participated in the computer-assisted gallery activity program, the author found that the students welcome the gallery activity program very much. Therefore they are more likely to be motivated to engage in their coursework. The purpose of this research is to use the MUSIC model of motivation to examine the strength of the computer-assisted gallery activity program from the motivation perspective. The author hopes that college professors in engineering programs can make use of the MUSIC model principles to redesign their instruction in order to motivate their students to engage in their learning.

There are a lot of research and theories concerning motivation in the field of educational psychology. For example, the theory of self-determination confirms that teachers should give students opportunity in order to make decisions about some aspects of their coursework [1]. The self-efficacy theory claims that teachers should let the students have the sense of success by designing some coursework that his not beyond students' capabilities [2], the expectancy-value theory indicates that teachers should let the students feel that what they are learning and doing are useful for their short-term or long-term aims [3], the dominant interest theory says that teachers can cultivate students' situational interest by designing very interesting course activities [4]. The attachment and caring theories hold that teachers should make sure that the students feel that their teachers and peers care about both their learning and well-being [7, 8]. However, college professors outside the realm of educational psychology are not familiar with these theories. Specifically, most college professors in engineering are not familiar to these theories and principles in motivation research and theory. It is very hard for college professors in engineering to design the student-centered class effectively. Thus, it is a real challenge for them to meet the needs of the 3E reform and the national standard of college education quality. Based on such problems, this paper plans to use the MUSIC model of motivation to analyze the success of the gallery activity program in the college English classes in an engineering university [5].

MUSIC is an acronym of the five words, eMpowerment, Usefulness, Success, Interest and Caring, which are the key principles of motivational strategies. Jones holds that: (1) students should have some rights to make decisions over their coursework, (2) students should feel that the content is useful for their short-term or long-term goals, (3) students believe that they will be successful if they put the required effort, (4) students feel that the class is interesting either the content or the class activities, and (5) students feel that the teachers and peers care about their learning and well-being [7, 8]. This paper conducts the comparison between the computer-assisted gallery activity program and the five elements of the MUSIC model of motivation. After that this paper analyzes why college professors in engineering can make use of the MUSIC model in class to make their students have higher motivation and engagement.

2   Comparing the Computer-Assisted Gallery Activity Program to the MUSIC Model of Motivation

2.1   Teachers Become Assistants and Students Become Main Speakers in Class in the Gallery Activity Program Who Can Make Decisions About the Form and Content of the Poster and the Presentation

The first element of the MUSIC model is empowerment. This point emphasizes that instructors should give students the power to have choice and make decision about some aspects of their coursework [1]. This point is derived from Deci and Ryan's self-determination theory (SDT) [1]. They hold that if one is given choice about a task, he or she tends to engage in the task that looks interesting to them [1]. Therefore, students will feel empowered when they can decide some aspect of the coursework so that they will be motivated to their leaning.

In the computer-assisted gallery activity program, college students are always main speakers and actors for the coursework of the college English classes. First, after learning the text, teachers will ask the students to form several small group at their own will. Then teachers introduce the term concept map to the student, and then the students will use the computer or cellphone to find the relevant materials to get further understanding of the term concept map. Next, students will decide what kind of structure they will use for their concept map of the text. Then, the students will decide when to give the presentation in class, and one of the group members will go to the front of the classroom and write down their own number on the blackboard, which symbolizing their turn in class for presentation. For example, if the group writes down number one on the blackboard, it means that this group chooses to be the first group to present their concept map in class. The group will be the last for presentation if they write down the number nine on the blackboard because there are a total of nine groups in class. The above mentioned activities will make the students feel empowered because they have the rights to make decision when and how to present their concept map in the gallery activity program. Figure 1 demonstrates how the students are working together for their concept map task and then how they decide one as the frame of their gallery activity. When we interview the students in this gallery activity, one group says that, "We love this activity very much because we have so much power in class. And there is a strong contrast between this class and the traditional class. We are passive listeners and note-takers in the tradition class while we are active leaders in this class."

2.2   The Usefulness of the Gallery Activity Program for Students' Comprehensive Capabilities

The second point of the MUSIC model is usefulness. It means that students feel that they understand why the content, both the text materials and the class activities, are useful for their present or future aims. The future time perspective theory and the expectancy-value theory were the foundation of this point. The former one holds that students are more likely to engage in learning if they can understand that the course-work or activity are of use for their future. The latter one holds that if students can understand the value of task they usually contribute more effort and persistence to it. Dr. Brett D. Jones claimed that instructors should be patient to tell the students why the course and the relevant gallery activity are useful to students' daily lives or near future [6].

In the computer-assisted gallery activity program, the teacher designs three activities, showing the usefulness of the gallery activity. The first activity is that teachers demonstrate part of several video programs which are from TED talk program, cases of how to deal with job interviews, or clips of gallery activities from top universities such as Harvard and Yale before teachers give the college students the assignment of gallery activities. The teacher also share part of the gallery activity from Harvard with the groups in class and ask the students to discuss and then summarize the usefulness of the gallery activity. After these warming up exercises for gallery activity, teachers will upload the file of gallery activities to the online network, which was created for the gallery activity. In this way, students can understand the usefulness of the gallery activity both in class and out of class, and they even can share them with their family members or peers. Next, teachers ask the students to make a short video in which he or she talks with his or her friends or other teachers demonstrating the usefulness of the gallery activity. Finally, the teacher shares the letters written by the upper students once joined the gallery activity to share his or her perceptions of the gallery activity. The teachers upload the upper class students' gallery activity perception report to the online network, so that the students in this experiment will read it in order to prepare for their own gallery activity in the most effective way Throughout the above mentioned activities the college students will thoroughly understand the usefulness of the galley activity. When asked the students' perceptions of the usefulness point, some of them say, "The upper class students impressed us very much for their wonderful performance in the presentation, and we know the capabilities, such as cooperation, leadership, and problem-solving, shown in the presentation are very useful for our future. Therefore we try to work hard to give the same good job in our activity."

Fig. 1. The first one is that the college students are discussing the design of the concept map. The second is that these students are presenting their posters. The last one is that these students are conducting the gallery activity

2.3   The Sense of Success that Students Can Obtain When They Are Busy with Their Posters in the Gallery Activity

The third element of the MUSIC model is success, which means the students will put forth more efforts if they believe that they can be successful in their assignments, class activities or the test [2]. This component is based on two theories, one is the self-efficacy theory [4] and the other is the expectancy-value theory [6]. According to the former theory, one tends to work hard if he or she believes that he or she is able to finish the task successfully. The latter one focuses that the individual tends to work hard if he or she has the high expectancy toward the task [2]. Jones also argued that if the teacher tells the students they could be successful if they follow the teacher's guidance to put forth enough efforts in the task, the students usually would work more diligently than before [5]. The point of success in the MUSIC model implies that when teachers design the instruction, they should grasp that the distance between students' present capability and the contents that are to be taught in class. Another implication is that teachers should give students feedback honestly and regularly so that the students will be clear about their success and improvement in their learning [6].

The college teacher in the computer-assisted gallery activity program improves students' sense of success in that the students can choose the approach they like best when they decide the form and design of their concept map of the text. For example, one group chooses the form of family tree to summarize the outline of the text. The other group uses the cartoon pictures and arrow lines to summarize the text without using any English words. And in the final step of the gallery activity program, this one without any English words is selected as the best poster by all the classmates. So that students in the group with cartoon pictures feel a strong sense of success in this gallery activity. The teacher also identifies the advantages of the poster in each group in order to increase all the students' sense of success. Next, the teacher asks the groups to give their presentation in front of the classroom and the teacher makes video for each group whey they are presenting their poster. Finally, the teacher uploads the video of the presentation to the online network. In the first few weeks of the first semester, the teacher just asks the group students to prepare for the galley activity based on the new words appeared in the text and the teacher will be available to the students when they have any problems about the activity. Step by step the teacher put forward the higher level gallery activity assignments to the students such as the complicated long sentence analysis and the summary of the paragraphs. In this way students is able to feel that they will be successful in the different level of gallery activity gradually. In addition, the teacher asks the student to submit the draft of the gallery poster to the online network or send it to teacher via Wechat system. Thus, the teacher will identify the mistakes or problems before their presentation. According to the teacher's suggestions, the group students will work together to correct the mistake in their poster. Therefore, when the group students are giving a gallery activity presentation in class, the teacher is standing in class and listening to the presentation attentively and will applaud for their perfect performance. After the poster presentation, all the groups will stick their posters on the wall of the classroom, so that all the peers will walk and comment the posters with their group members and then draw stars on the poster(more stars mean higher scores and the top one will get five stars). At that moment, the peers are judge for the competition among the posters. Also, the teacher takes photos and video during students' presentation. The students will feel successful for their group work and will work more actively for next assignment.

When we interview students about the sense of success, one of the team member says, "We definitely have the sense of success because we discuss the key points of the poster not only with our instructor and the peers but also with our high school grammar teacher in order to make a perfect performance in class. W have never had such sense of success before."

The gallery activity is interesting to the students.

The fourth point in the MUSIC model is interest. This point is mainly related to Hidi and Renninger's interest theory. Hidi and Renninger claimed that students' situational interest can be cultivated by the instructors if they create the proper learning environment [4]. Therefore, it is necessary for instructors to design relatively interesting climate for their instruction [4].

The computer-assisted gallery activity program has several advantages in the aspect of interest. First, the form and content of the poster was totally decided and designed by the students. In contrast to the traditional teacher-centered class, the posters designed by the group students themselves are more interesting than the boring list of new vocabulary and the common sentences printed simply in the textbook. Second, the teacher provides students with relevant information to their poster such as pictures and reference reading materials and the students will add these pictures or other things to their gallery activity text before the gallery activity time. In contrast to the formal paper test, the gallery activity, the class presentation, tend to interest students more for the freshness in forms and contents.

As for the interesting aspect of the gallery activity, some of the students say, "This activity is very interesting because we have never done the gallery activity before in our various courses. The big sheet of paper, the colorful makers, the peers evaluation on the sheet, the photos and videos, contribute to the interesting aspect of the gallery activity."

2.4   Students Feel that They Are Cared for by Teachers and Peers During the Gallery Activity Program

The caring aspect in the MUSIC model shows that when students feel that their teachers and peers care about whether they are successful in their academic tasks and about their sense of happiness, they are more likely to be motivated to engage in their learning [7]. Bergin and Bergin and Wentzel confirms that when students have close relationships with their peers and teachers they will be active in class and like putting more efforts in their assignments [8]. Noddings, the main representative of the caring theory, holds that teachers should not only to pay attention to and improve students' academic achievement but also to care about the students' sense of happiness [7]. What we can learn from Nodding is that whenever students need teacher's help, the teacher should be available. Another suggestion is that teachers should make sure that teachers and students respect each other in class [7].

The computer-assisted gallery activity program shows four points of advantages in this perspective. First, the network created both by the instructor and the administration of teaching affairs, where the teacher uploads the relevant materials such as the syllabus, the rubric for presentation, the reference books and reading materials. And the other is the online network which is mainly used to store all the class activities materials such as the pictures of the posters, the video of the presentation, the records of the presentation scores for the gallery activity program. The teacher can assign the task to the groups that were randomly form according to students' own choice. When the college students meet any problems in their preparation for the gallery activity program, they can seek for help from their teacher through email, Wechat, and QQ community. That's to say, the teacher is available all the time both online and offline. When the groups upload their materials relevant to the gallery activity to the online QQ community, their teacher will examine the materials very carefully and then gives them feedback respectively. The teacher takes photos when the group is presenting their poster. Then the teacher will upload them to the on. Third, the teacher will accept any student's invitation to be a friend in the wechat system or QQ system. In this way, the students can communicate with their teacher personally. Usually the teacher gives a quick and honest feedback to the student's question. Sometimes, the teacher will give the relevant link of websites to the student in order to provide rich information for further understanding of the question or problems.

When the students are interviewed for the caring point, many of them express their strong feelings about it. One of them says that she can not believe the instructor always gives her reply in a few minutes when she asks question in the QQ chatroom. The other student says that she is surprised and appreciated that the instructor agrees that she can ask for leave without losing a point for her regular attendance when she has to leave for her elder brother's wedding ceremony.

3   Conclusion

The MUSIC model of motivation is an ideal framework when teachers want to examine the quality of their instructional design [9, 10]. This paper uses it to examine the quality of the computer-assisted gallery activity program. From the above mentioned analysis, we can conclude that this gallery activity empowered students by providing students a lot of choices when they prepare and present their presentation. By providing upper class gallery videos, the students in this experiment realize that this activity is very useful for their current learning and future job hunting. It is no doubt that the class presentation makes the students feel that they are successful in this coursework. The diversity of the form of the concept map, the online and offline group discussions, the availability of the teacher, and the video and photos about the group poster presentations, and the walking and assessment of the posters on the wall in the classroom, make the class very interesting. Finally, the instructor and the students in the gallery activity care about each other both in the learning process and their sense of well-being.

Acknowledgment. This paper was part of the achievement of the project entitled First-class Discipline Construction of Foreign Language and literature and it was supported by Shanghai University of Engineering Science with the grant No. 2019XKZX011.

References

1. Deci, R.M., Ryan, R.M.: A motivational approach to self: integration in personality. In: University of Nebraska (Lincoln campus). Department of Psychology. Nebraska Symposium on Motivation, vol. 38, pp. 237–288 (1991)
2. Bandura, A.: Social Foundations of Thought and Action: A Social Cognitive Theory. Prentice-Hall, Englewood Cliffs (1986)
3. Wigfield, A., Eccles, J.S.: Expectancy–value theory of achievement motivation. Contemp. Educ. Psychol. **25**(1), 68–81 (2000)
4. Hidi, S., Renninger, K.A.: The four-phase model of interest development. Educ. Psychol. **41**(2), 111–127 (2006). https://doi.org/10.1207/s15326985ep4102_4
5. Jones, B.D.: Motivating students to engage in learning: the MUSIC model of academic motivation. Int. J. Teach. Learn. High. Educ. **21**(2), 272–285 (2009)
6. Jones, B.D.: Motivating Students by Design: Practical Strategies for Professors. CreateSpace, Charleston (2015)
7. Noddings, N.: The Challenge to Care in Schools: An Alternative Approach to Education. Teachers College Press, New York (1992)
8. Bergin, C., Bergin, D.: Attachment in the classroom. Educ. Psychol. Rev. **21**(2), 141–170 (2009)
9. Gardner, A.F., Jones, B.D.: Examining the Reggio Emilia approach: keys to understanding why it motivates students. Electr. J. Res. Educ. Psychol. **14**(3), 602–625 (2016)
10. Jones., B.D., Epler, C.M., Mokri, P., Bryant, L.H., Paretti, M.C.: The effects of a collaborative problem-based learning experience on students' motivation in engineering capstone courses. Interdiscip. J. Probl.-Based Learn. **7**(2), 2 (2013)

Construction of Creative Writing in College English Teaching in the Age of "Internet+"

Ying Liu[✉]

Shandong Vocational College of Light Industry, Zibo, Shandong, China
LiuYing1983@haoxueshu.com

Abstract. In the era of the Internet, many things will change its original appearance, and the creative writing of college students will usher in a new atmosphere under the "Internet+" era. Faced with the interconnection of information, the expansion of the public, and the new situation of cross-border integration, the creative writing training among college students is more oriented towards the innovation of methods and the broadening of the horizon, adapting to the era of interactive integration due to the Internet. As a new idea of English writing teaching reform, creative writing has gradually attracted attention. Creative writing has the effect of stimulating students' writing motivation, improving students' writing ability and improving students' comprehensive ability in English study. Based on the "Internet+" era, the construction of creative writing in the process of college English teaching provides ideas for the reform of college English teaching, it also can stimulate students' interest in English learning and improve their comprehensive ability of English application.

Keywords: Internet+ · Creative writing · Teaching reform

1 Introduction

In campus, English writing course is a comprehensive course, which can improve students' English language ability in an all-round way. Not only does it require a large number of English language input, but it also challenges students' native language writing over the years whether they can get rid of the influence of native language writing, or integrate them on the basis of native language writing to write authentic English works. However, writing teaching should not and should not be divorced from students' life, nor should it conform to students' cognitive rules [1]. How to turn the dull English writing class into a creative class that students like is a question that the author has been thinking about all the time. Writing should be the training of thinking, the externalization of self and the outlet of releasing self. Writing in acquisition language allows them to say things that cannot be expressed in their mother tongue [2]. Students' writing in English can better express what they don't want to express when they write in Chinese, and there is more room for them to write [3].

The concept and application of "Internet +" has promoted the development and leap of emerging education models such as MOOC, micro-class, and flip classroom. Among them, many advantages highlighted by online education, such as fragmented learning, data sharing, entertainment and social networking are particularly compatible with the

© Springer Nature Singapore Pte Ltd. 2020
M. Atiquzzaman et al. (Eds.): BDCPS 2019, AISC 1117, pp. 1918–1925, 2020.
https://doi.org/10.1007/978-981-15-2568-1_268

core concept of creative writing [4]. The teaching mode embodies the organic combination of task-based teaching method and cooperative teaching method, and maximizes the advantage through the "Internet +" kinetic energy engine [5].

The creative writing teaching mode enhances students' creative thinking, cultivates students' love of literature and English language writing, and promotes students' ability to improve their sense of innovation. In the environment that students take English as a foreign language or a second language, creative writing usually means that it can stimulate Imaginative task writing, such as writing of poetry, stories, dramas, etc [6]. Compared with other texts, creative writing laid particular emphasis on the free expression of intuition, imagination and personal emotions [7]. To this end, this paper explores the construction of creative writing in college English teaching based on the "Internet +" era, analyzes the influence of creative writing on students' writing motivation, writing ability and comprehensive ability, and can cultivate students with problems of discovery. The ability to analyze problems and solve problems has a global concept and an international vision [8].

2 The Construction of Creative Writing in the Process of College English Teaching

2.1 The Training of Students' Thinking in Creative Writing

Creative writing can rebuild students' positive cognitive and response patterns, and it is a kind of benign stimulation and improvement for the current college students' learning habits and thinking patterns. Creative writing can help students clear up the blockage of personal consciousness, unconsciousness and collective unconsciousness, open up the channels of memory, Association and imagination, train reverse thinking, divergent thinking, and expand the depth and breadth of thinking. Commonly used creative thinking training methods include Brainstorming, Mind Mapping, Mandala Thinking, Reverse Thinking, Synectics Method, Attribute Listing Technology, Hope Point Listing, Advantage and Disadvantage Listing, Checklist Me. Thod, 5W2H, compulsory association, Creative Problem Solving, etc [9].

Among the much creative thinking training, the author is most commonly used in the polymerization method. "Aggregation is a creative production device that goes into the right brain by disintegrating linear thinking. You draw circles, not lines of logic, causality, and context [10]. Then, ideas flow into your brain, with a word or Write it down in one sentence, circle them, and radiate them by connecting keywords." Traditional writing training allows students to conceive ideas from topical sentences, supporting details, and linear thinking of summing up sentences. The training allows students to jump out of the imprisonment of thinking, start with words or sentences that can inspire them [7], and divergent thinking, which is of great benefit to their thinking development training (Fig. 1).

Fig. 1. The teaching process of creative writing in college English

2.2   Training Model of Creative Writing

Different from the traditional teacher's "one-word" writing teaching mode, creative writing usually adopts the mode of workshop. Generally, teachers are the keynote speakers, and 10–20 students are guided by the speaker, through reading, analysis, short comment, etc. Way to start the discussion. In the creative writing workshop, students and teachers form a cooperative group, each student reads his work in class, and other students propose amendments. The workshops are flexible and free from space and time [4]. Teachers and students can conduct round-table discussions in the classroom, or walk out of the classroom and take the form of "fields and winds". At the same time, teachers and students can establish various discussion channels and establish online discussions. Groups, WeChat groups or e-newsletters, share, communicate and communicate at any time. As the New Yorker columnist Luis Mannand said: "A group of students who have never published poetry can teach another group of students who have never published poetry to write a poem that can be published."

The author tried the "writing class round table" mode many times in classroom, applied student-oriented teaching mode, letting them think positively and speak freely. Throughout the traditional writing training: the students hand over the written works to the teachers, the teachers judge, and the students are passively revised according to the teacher's revised opinions. Such a model undoubtedly obliterates students' thinking ability and regards students as passive acceptance terminals, which leads students' writing ability to stay in the initial stage of imitation. At the same time, the author invited the students to the home, and everyone completed the discussion on the writing theme of Food Safari in the process of cooking each other. From buying ingredients to cooking and then sharing food, students' minds are activated and inspired by the story of food. Everyone in the "writing table meeting" to carry out imagination, repeated discussion, modification, reset, the story and the story can collide with each other [6].

At the same time, the author deeply understands each student's personality and writing characteristics in the process of student discussion, and is more conducive to modifying their articles. Modifying the article is by no means picking up the grammatical mistakes of the students, but giving more advice on the students' ideas and writings. "The Secret of the Turtle" was born in the "Writing Table Meeting". The grandfather in the story came from Weinan and gave his grandmother a turtle soup for a lifetime. When Grandpa died, Grandma arranged the turtle shells collected and wrote what they wanted to say. At Grandpa's funeral, the friends who arrived at the seat received the corresponding turtle shell according to the seat number. The turtle shell that the grandson received said: "Do you know? I never eat turtles." The story is inspired only by the table. A chat by a student: "My grandfather is a Minnan. He used to have a turtle when I was a child. Now the turtle shell is still at home." The author guides students to focus on the "Grandpa - Grandma - Turtle" thread. Talking about the taste of life and the feelings between grandparents and grandmothers into the pot to cook, made the "sweet turtle" this exciting "dish".

2.3   Mixed Teaching of Creative Writing and Basic Writing

Writing training, like building a house, requires a solid foundation to build a tall building. Only when the cornerstone of basic writing is well established can creative writing soar into the clouds. Basic English writing is a common course in Colleges and universities. It focuses on paragraph writing and the writing methods of different types of articles. It opens the door for students to write regularly. In basic English writing, English teachers should help students establish a top-down model, that is, starting with the production of complete texts, to solve common writing problems, such as misuse of words, disorder of organization and inadequate expansion of content. Basic writing begins with a paragraph, which is just like a mini-article. It should be consistent, coherent and properly developed [3]. After the training of paragraph writing is mature, teachers guide students to write complete compositions, including time and space sequential writing, expository writing, process analysis, comparative writing, causal analysis, definitional writing, classification and summary writing. However, if the teaching of basic writing is limited to the classroom, the students' writing ability will not make great progress. Good articles must be interesting and fresh. Creative writing teaching can make up for the shortcomings of basic writing teaching, liberate students' fixed thinking mode, and add creativity and interest to writing and their life. The teaching of creative writing needs a deep understanding between teachers and students, and they need to achieve a certain tacit understanding before teaching can be carried out. Therefore, in the teaching of basic writing, teachers must shine their eyes, observe students' preferences and create a certain platform for teachers and students, so as to pave the way for creative thinking and creative writing training.

In the English writing class, the author arranges the first half of the semester as the basic writing training and the second half as the creative writing training. In the training of basic writing, the author teaches from the form of manuscript, wording, sentence-making, paragraph to the complete composition, so as to lay a solid foundation for students. In the creative writing training in the second half of the semester, the author encourages students to go out of the classroom, interview interested people around them, and gain more rich life experience and writing inspiration. After a semester of intensive teaching and training, students' critical thinking ability and innovative ability have been improved, and they have more confidence in English writing. The effective integration of creative writing and basic writing effectively expands students' thinking, stimulates students' innovative ability, and closely combines English learning with students' daily life. While stimulating students to learn English, it encourages them to step out of the small circle of their lives and move towards a larger stage (Fig. 2).

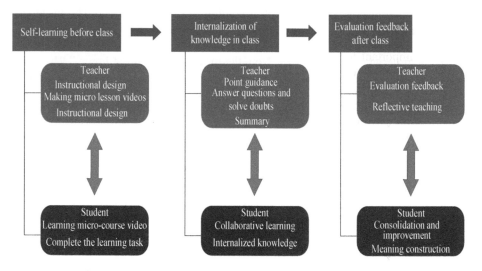

Fig. 2. College English creative writing mode based on internet+

3   The Impact of Creative Writing on Students' Comprehensive Writing Ability

Although creative writing started earlier and developed more maturely in English-speaking countries, the introduction of creative writing into foreign language teaching classes is not the focus of attention of scholars and teachers in English-speaking countries. Therefore, the author mainly cites the research results of creative writing applied in foreign language teaching in China and non-English-speaking countries [8]. It is concluded that creative writing can help foreign language learners, especially

English as a foreign language learner. Compared with traditional writing, the advantages of creative writing are highlighted as follows:

3.1   Creative Writing Helps to Enhance Students' Vocabulary and Syntax

Language output activities can cause learners to pay attention to the grammatical structure and lexical appropriateness of their use, and promote the automation of their language use. As one of the forms of understandable output, creative writing has an obvious help in improving language learners' language skills. The primary purpose of teachers to introduce creative writing into foreign language classrooms is to assist language teaching. Students' training in creation is not their original intention. Creative writing is a practical tool for learning English in class, which can be used to help students increase their vocabulary. Quantitative, auxiliary syntactic teaching and improve students' grammatical accuracy. It can be seen that creative writing is not only about writing ability, but the language ability that non-English majors generally pay attention to can also be effectively improved through creative writing.

3.2   Creative Writing Can Increase the Writing Motivation of Students

It is very important to stimulate students' motivation in teaching activities, as motivation is the direct cause of behavior. There are many ways to improve students' motivation, including stimulating curiosity, fostering confidence, and building a sense of accomplishment. Some scholars have similar views on the effects of creative writing in stimulating writing motivation. some research points out that to explore learners' potential motivation is the principal cause why creative writing is increasingly favored in English as a foreign or second language environment, because creative writing respects the free expression of individual thoughts and advocates student-centeredness. It is good at guiding students to discover their inner world. This way is more conducive to the active acquisition of language and personal growth. Learning to write poetry or storytelling by the language you have learned can bring great sense of achievement to students, make them proud of their works and hope to share them with others. Creative writing is just like language games, which creates more fun of language learning for students. Moreover, help students to find more readers who appreciate their work and publish their work. This is not only a recognition of the student's work, but also allows them to enjoy the writing process (Fig. 3).

3.3   Creative Writing Can Promote the Cultivation of Students' Writing Ability

Creative writing promotes the acquisition of all levels of the student's language, such as vocabulary, grammar, and text, which is the groundwork of reading. Creative writing, such as poetry, stories, short play, etc., requires students to use the language appropriately and accurately, express their thoughts and intentions clearly.

In addition, the combination of creative writing with daily English teaching and learning can help students get more exercise, and improved their writing ability. At the same time, the positive writing motivation brought by creative writing makes students

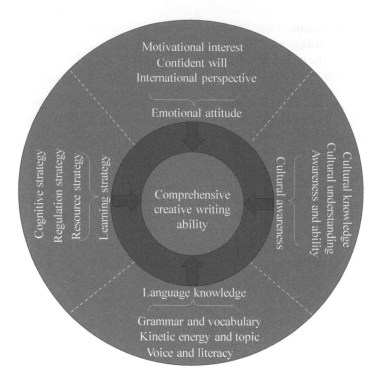

Fig. 3. The Influence of creative writing on students' comprehensive writing ability

happy to write and take the initiative to write, which is also a necessary condition for improving writing ability.

3.4   Creative Writing Can Improve Students' Comprehensive Ability

To improve the quality of education in Colleges and universities, the key is to cultivate students' comprehensive ability. In addition to providing them with modern expertise, they are more important to develop and improve their ability to innovate, to discover and analyze problems, to solve problems, to learn, and to work with others. Maley pointed out that creative writing can promote creative reading and make good use of the development of right brain, because the humans' right brain contains such information as imagination, creativity, memory, understanding, intuition ability. Creative writing training, such as cooperative writing or mutual modification, also trains students' ability to cooperate with others.

4   Conclusion

Creative writing teaching, as a new mode of English writing teaching reform, is effective way in college English teaching. Well-designed tasks, effective guidance of teachers, and the characteristics of creative writing itself, such as fun and innovation, have cultivated students' enthusiasm for writing to a large extent, enhanced their motivation for learning, and exercised their ability to think and imagine positively. Teachers can use the complementary functions of creative writing and basic writing to stimulate students' enthusiasm for English learning, and guide them to turn English writing into a part of life, express themselves in writing, explore the joy of life, and gradually become rich in inner life. Interesting English learners. Creative writing which change passive writing into active writing, as a new method that is conducive to cultivating students' comprehensive ability and worth promoting.

References

1. Jordan-Baker, C.: The philosophy of creative writing. New Writ. **12**(2), 1–11 (2015)
2. Dai, F.: Teaching creative writing in English in the Chinese context. World Engl. **34**(2), 247–259 (2015)
3. Babine, K.: Imagination in the classroom: teaching and learning creative writing in Ireland ed. Anne Fogarty, Éilís Ní Dhuibhne, and Eibhear Walshe (review). New Hibernia Rev. **20** (1), 148–151 (2016)
4. Charon, R., Hermann, N., Devlin, M.J., Michael, J.: Close reading and creative writing in clinical education. Acad. Med. **91**(3), 345–350 (2016)
5. Jaafar, Z.B., Yusof, N.M., Ibrahim, N.: Negotiating memory and creativity: choices of image-text representations in the creative writing classroom. Procedia - Soc. Behav. Sci. **118**, 190–197 (2014)
6. Kornilov, S.A., Kornilova, T.V., Grigorenko, E.L.: The cross-cultural invariance of creative cognition: a case study of creative writing in U.S. and Russian college students. New Dir. Child Adolesc. Dev. **2016**(151), 47–59 (2016)
7. Mazur, A.E.: Developing discoursal selves: academic writing in a linguistically diverse Puente English class. Diss. Theses-Gradworks **5**, 805–810 (2014)
8. Dockrell, J.E., Marshall, C.R., Wyse, D.: Teachers' reported practices for teaching writing in England. Read. Writ. **29**(3), 409–434 (2016)
9. Gilbert, F.: Aesthetic learning, creative writing and English teaching. Chang. Engl. Stud. Cult. Educ. **23**(3), 257–268 (2016)
10. Frawley, E.: No time for the "Airy Fairy": teacher perspectives on creative writing in high stakes environments. Engl. Aust. **49**(1), 17–26 (2014)

Application of Computer Simulation Technology in Food Biotechnology Teaching

Lisha Feng[1,2(✉)]

[1] Yunnan Open University, Kunming, Yunnan, China
Lisha_Feng81@haoxueshu.com
[2] Yunnan Vocational and Technical College of National Defence Industry,
Kunming, Yunnan, China

Abstract. Computer simulation use the computer modeling and simulation technology to realistically display the internal structure, working principle and related process flow of engineering technology, bioreactor, process equipment and other related equipment in a dynamic form. The professional course of food biotechnology is a very practical course. Computer simulation technology will undoubtedly have a positive impact on traditional teaching methods. However, there is currently no report on the application of computer simulation technology to pharmaceutical engineering related courses. This paper takes the teaching of food biotechnology course as the research object, explores the mode of applying simulation technology to teach; promotes the teaching reform of food biotechnology course and improves the teaching level; lays a foundation for broadening the application of computer simulation technology in the course of food biotechnology.

Keywords: Computer simulation technology · Food biology · Simulation

1 Introduction

Since the first industrial revolution, factory-based mechanical production has gradually replaced manpower as the main force of industrial production with its advantages of high efficiency and high output, and then spread to all walks of life [1]. Entering the 21st Century, with the development of computer technology, food biotechnology has gradually entered a new era of intelligent industry. Through the design and validation of the relevant processing parameters in food biotechnology by computer simulation technology, we can clearly understand the key control points and optimal parameters in the process of food biotechnology, and lay a foundation for the realization of the intelligent production of food biotechnology industry in terms of precise division of labor, optimal production, collaboration and efficient output [2].

Food biotechnology is an important basic course of food science and engineering, food quality and safety. The purpose of this course is to enable students to understand the structure and properties of the main ingredients in food materials, their interactions, their changes in food processing and preservation, and their effects on food color, aroma, taste, texture, and nutrition and preservation stability [3]. At the same time, it provides a broad theoretical basis for students to engage in food processing,

M. Atiquzzaman et al. (Eds.): BDCPS 2019, AISC 1117, pp. 1926–1933, 2020.
https://doi.org/10.1007/978-981-15-2568-1_269

preservation and development of new products, as well as an important basis for students to understand new theories, new technologies and new research methods in food processing and preservation [4].

In teaching, how to make students quickly understand abstract theoretical concepts, various reaction phenomena, and smoothly realize the transition from theory to practice is often the focus and difficulty of teaching. As a modern teaching method, computer simulation technology can integrate sound, image, video and text because of its diversity, visualization, richness and flexibility. It can display abstract theoretical concepts and experiments in an image and intuitive form, and help students from perceptual knowledge to perceptual knowledge [5]. The transition of rational understanding can greatly enrich classroom teaching, promote students' understanding and memory of knowledge, cultivate students' thinking ability, improve students' quality and greatly improve the teaching effect [6].

2   The Principle and Advantages of Computer Simulation Technology

Computer simulation technology is a way to find out the rules according to the characteristics of the object and use computer language to render it on the computer. Most of the research is based on mathematical models established by mathematical theory, using mathematical expressions to parse a process and simulate it on a computer. The model is divided into various forms, such as the mechanism model, the experimental model and the mixed model according to the mathematical simulation process; the steady state model and the dynamic model are divided according to the nature of the tense; and the centralized parameter model and the distributed parameter model are divided according to the parameter distribution; According to determinism, it is divided into determining model and stochastic model [7]. For example, zein is enzymatically hydrolyzed after ultrasonic pretreatment, and the protease is automatically screened by a simulation software for protein digestion process on a computer. This model belongs to the mechanism model and reflects the nature of the actual process [8].

Process systems that do not apply to mathematical model representations are generally represented using artificial neural networks. Based on bionics, ANN simulates the neural network structure and functional characteristics of the human brain. Through engineering techniques, many primitive and uncertain units are connected into a nonlinear and adaptive information processing system. The advantage of ANN is that it can effectively perform nonlinear adaptive information processing on uncertain data, and has good automatic adjustment capability and adaptability [9]. The use of computer simulation technology in the experiment can simplify and shorten the progress of expanding the experimental results to industrial production, and effectively avoid the problems that some variables are difficult to control, difficult to calculate and have large errors during the actual simulation, making the experiment more economical. Faster and more detailed, it is beneficial to study the dynamic nature and control mode of the experimental object, and it is more convenient to screen out the optimal solution [10].

3   Application of Computer Simulation Technology in Food Biotechnology Teaching

3.1   Thermal Food Technology

The main changes of food during heating are heat transfer, phase change, shrinkage and mass transfer. The traditional heating method is heat transfer, which makes food reach similar temperature through high temperature. At present, resistance heating, vacuum low-temperature frying, radio frequency heating and other heating methods are gradually replacing the traditional heating methods in food applications, and computer simulation technology plays an important role in the development and application of these technologies. Through computer simulation technology, relevant heating models are established, as is shown in Fig. 1. In the field of food drying, hot air drying, vacuum drying, infrared drying, microwave drying and other drying technologies are constantly developed. According to different drying methods, the drying numerical models such as microwave drying model, infrared drying model, vacuum drying model and hot air drying model are established by using computer numerical simulation technology.

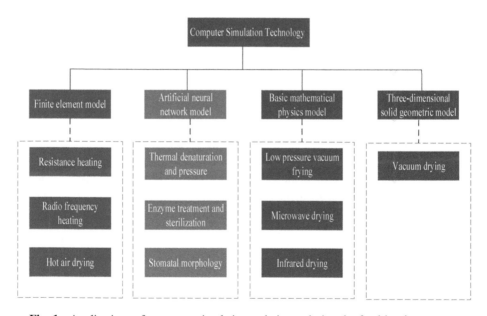

Fig. 1. Applications of computer simulation techniques during the food heating process

3.2   Separation of Active Ingredients in Food

In order to obtain specific proteins, amino acids, active peptides and other functional active substances, the utilization of enzymes and enzymatic hydrolysis process are very important in food processing. In recent years, this technology has made great progress. With the help of computer molecular docking technology, the ACE inhibitory active polypeptides in the enzymatic hydrolysates of clams were screened, the interaction sites between the polypeptides and ACE protein were determined, and the ACE inhibitory active polypeptides were accurately prepared. Based on SalS1 protein source of salmon protein, according to the simulated digestion of Sals1 protein sequence by protease recommended in Peptide Cutter software, an egg hydrolysis model was established to help screen suitable combinations. The results showed that the combination of neutral protease and alkaline protease could remove the allergen list to the greatest extent. Phase, or even toxic peptide segment. In addition, ANN plays an important role in the separation of active ingredients in food [11]. With the help of computer simulation technology, the separation process of effective ingredients in food can be predicted quickly and accurately, and the work efficiency can be effectively improved.

3.3   Food 3D Printing

3D printing technology is a kind of rapid prototyping technology. It divides the 3D digital model of computer-aided design into multi-layer planar slices, and then the 3D printer divides the adhesive materials into layers by layering and superposition molding. The technology of adding materials to generate three-dimensional entities is a cutting-edge technology that combines digital modeling technology, information technology, electromechanical control technology, materials science and chemistry. The nozzle height, nozzle speed and extrusion rate play a key role in the quality of 3D printed chocolate molding. 3D printing technology is also applied to foods [12]. For example, food adhesives and seasonings are used as inks. The shape of 3D modeling is used to print artificial meat. The taste is similar to the taste of ordinary meat, and the shape can be customized. The processing methods and varieties of meat products also meet the needs of different consumers; because the raw materials of 3D printed foods are easy to carry, dry powdered carbohydrates, proteins and trace elements can be stored for 30 years, so it is especially suitable for aerospace food (Fig. 2).

Fig. 2. Food printed by computer simulation technology

3.4 Food Testing

Food sensory evaluation is the most intuitive indicator for describing and judging food quality. It is a discipline for developing, measuring, analyzing and explaining human perception through visual, olfactory, gustatory, tactile and auditory senses. Through computer simulation technology, equipments such as texture analyzers, electronic tongues, electronic noses, and electronic eyes that can be used for sensory evaluation of foods have been developed to evaluate foods more accurately, quantitatively, and qualitatively [8].

The olfactory visualization system based on color-sensitive sensor array success-fully identified vinegar vinegar age. Molecular models before and after the reaction of metalloporphyrins with volatile gases were established. A computer-based olfactory visualization sensor for detecting the freshness of brambles was developed. By using electronic nose and electronic tongue system, the samples of chicken, pork, mutton, mutton with different freeze-thaw times and adulterated mutton mince mixed with chicken and pork were detected and identified. The electronic nose, electronic tongue and their combined signals were successfully distinguished and identified for different kinds of samples, and the effective prediction of mutton mince was established. The model of freeze-thaw times of pork, chicken and mutton mixed in it. In computer vision technology, the color information of extruded food collected by computer vision system and the texture characteristic value obtained by texture analyzer are used to analyze the correlation of texture characteristics of extruded food by linear fitting model. Through the ANN model, the color information collected by visual system is directly used. The relationship between pizza quality and image quality was clarified by processing pizza image information [6]. The automatic detection technology of pizza quality and safety was established to realize non-destructive and fast detection of pizza quality. Computer vision technology was applied to fish freshness detection. The image characteristics of iris, gill, color and texture of the fish eye were collected and analyzed by image acquisition system under 4 C storage condition. The storage days of fish were predicted by ANN model. The accuracy rate reached 85%. Based on the analysis of the color change of Atlantic salmon, the grading criteria for judging the quality of fish meat were established and automatic grading was realized. The effects of drying temperature and drying medium rate on the color change of shrimp during the drying process were studied by computer vision technology, which indicated that the quality attributes of dried products could be realized through computer vision system. Time Prediction, Control and Optimization [10].

4 Analysis and Comparison of Teaching Effect

Comparing the teaching effects of computer simulation technology teaching and traditional teaching mode, the test results of the two classes are shown in the following tables. Which applied traditional teaching mode, its result is shown in Table 1.

Examination results of students using computer simulation technology in teaching food biotechnology course are shown in Table 2.

Table 1. The examination achievements of using traditional teaching model

Score interval	Number	Proportion
90–100	1	2.7%
80–89	15	40.5%
70–79	17	46%
60–69	4	10.8%

Table 2. The examination achievements of using computer simulation technology

Score interval	Number	Proportion
90–100	10	25%
80–89	15	37.5%
70–79	13	32.5%
60–69	2	5%

It can be seen that the teaching effect of the computer simulation technology teaching mode has been significantly improved, and the number of outstanding people has increased by 9 people, an increase of 22.3%. The proportion of students with more than 80 points has also increased significantly, with an increase of 19.3%. And the overall average score increased by 9.7 points (Table 3).

Table 3. A comparative analysis of the examination results

Teaching model	The proportion of students with more than 80 points	The average score	The highest score	The lowest mark	Passing rate
Traditional teaching model	43.2%	77.4	90	60	100%
Applied computer simulation technology teaching model	62.5%	87.1	94	67	100%
Difference value	22.3	9.7	4	7	0

The result of the inquiry examination in the final examination is further analyzed. The total score of the examination is 20 points. The average score of the students in the class teaching by computer simulation technology is 14.9, the score rate is 74.5%, and the standard deviation is 3.37. Compared with the previous students, the overall effect of the students' answers has been greatly improved, and the overall score rate has reached 14.9. 72%. In addition, the usual assessment results of computer simulation technology teaching and learning as the assessment content are analyzed. The lowest score is 80 points, the highest score is 92 points, with an average score of 85.2 points. It can be seen that computer simulation technology teaching has achieved good learning results.

5 The Effect of Computer Simulation Technology in the Teaching of Food Biotechnology

5.1 Arouse Learning Interest, Stimulate Learning Motivation

Many students feel that multimedia food biotechnology is interesting and novel. It is much easier to read the contents of books than to listen to the lectures. What is the reason? Originally, multimedia food biotechnology is based on interactivity, through pictures, texts, Sound, like stimulating the senses of students, causes students to be interested in theoretical and experimental lessons. Its interesting experimental quiz, small games and entertainment into the teaching, greatly stimulate students' motivation.

5.2 Make Food Biotechnology Teaching More Intuitive and Macroscopic

All substances are inseparable from atoms and molecules. In terms of molecular structure, the molecular structure of organic substances is much more complex than that of inorganic substances. Using two-dimensional and three-dimensional image technology of food biotechnology, all the molecular structures of organic substances can be vividly displayed on the display screen, and an abstract content can be visualized. For example, when we talk about the molecular structure of carbohydrates such as monosaccharides, disaccharides and polysaccharides, it is very difficult for students to understand and accept the monotonous elemental letters and the molecular structure composed of one molecule bond in class. In food biotechnology courseware, we can color these uninteresting molecular structures. In this way, these molecular formulas become more intuitive and three-dimensional, arouse students' interest in learning, and deepen students' understanding and memory of molecular structure.

5.3 Shorten Teaching Time and Improve Teaching Quality

Computer simulation technology is suitable for many functions, such as shape, light, sound and color. It plays an active role in assisting students' learning and tutor's teaching. Under the assisted learning of food biotechnology courseware, students' interest in learning is greatly increased, and their pressure is reduced. Their enthusiasm is improved, and students really grasp food biology. The content and key points of chemistry have enhanced the ability of experiment. For the tutor, the burden has been reduced, and the quality of teaching has been improved with the help of food biotechnology courseware.

6 Conclusion

The ultimate goal of teaching reform is to maximize students' learning efficiency and cultivate high-quality talents. Food biotechnology teaching course is a practical professional course. We should enrich, perfect and enrich food biotechnology teaching with computer simulation technology, so as to improve the vividness and Enlightenment of teaching. With students' participation, cultivate students' thinking methods, guide students to make comprehensive use of book knowledge and network resources for self-learning, which is conducive to grasping practical knowledge. In the actual teaching process, the advantages of computer simulation technology are better integrated with food biology teaching, so as to make classroom teaching better. Only by becoming more vivid, making teaching more orderly and better, and maximizing the development of students, can we improve the quality of teaching. To maximize their learning efficiency and train high-level food biology professionals for the society.

References

1. Schaick, P.G., Zwietering, M.H.: The use of computer simulation for biotechnology education. Biotechnol. Educ. **4**, 23–27 (2016)
2. Pan, R.X., Gao, Y., Chen, W.L., et al.: Dissolution testing combined with computer simulation technology to evaluate the bioequivalence of domestic amoxicillin capsule. Acta Pharm. Sin. **49**(8), 1155–1157 (2014)
3. Keller, R., Dunn, I.J.: Computer simulation of the biomass production rate of cyclic fed batch continuous culture. J. Chem. Technol. Biotechnol. **28**(11), 784–790 (2016)
4. Ng, B.I.L., Yap, K.C., Yin, K.H.: Students' perception of interdisciplinary, problem-based learning in a food biotechnology course. J. Food Sci. Educ. **10**(1), 4–8 (2014)
5. Vanderschuren, H., Heinzmann, D., Faso, C., et al.: A cross-sectional study of biotechnology awareness and teaching in European high schools. New Biotechnol. **27**(6), 822–828 (2016)
6. Llave, Y., Liu, S., Fukuoka, M., et al.: Computer simulation of radiofrequency defrosting of frozen foods. J. Food Eng. **152**, 32–42 (2015)
7. Hao, L., Mellor, S., Seaman, O., et al.: Material characterization and process development for chocolate additive layer manufacturing. Virtual Phys. Prototyp. **5**(2), 57–64 (2015)
8. Melchiade, D., Foroni, I., Corrado, G., et al.: Authentication of the 'Annurca' apple in agro-food chain by amplification of microsatellite loci. Food Biotechnol. **21**(1), 33–43 (2017)
9. Raso, J., Frey, W., Ferrari, G., et al.: Recommendations guidelines on the key information to be reported in studies of application of PEF technology in food and biotechnological processes. Innov. Food Sci. Emerg. Technol. **37**, 312–321 (2016)
10. Reges de Sena, A., de Barros dos Santos, A.C., Gouveia, M.J., et al.: Production, characterization and application of a thermostable tannase from Pestalotiopsis guepinii URM 7114. Food Technol. Biotechnol. **52**(4), 459 (2014)
11. Le Feunteun, S., Barbé, F., Rémond, D., et al.: Impact of the dairy matrix structure on milk protein digestion kinetics: mechanistic modelling based on mini-pig in vivo data. Food Bioprocess Technol. **7**(4), 1099–1113 (2014)
12. Anubhav, S., Hosahalli, R.: Effect of can orientation on heat transfer coefficients associated with liquid particulate mixtures during reciprocation agitation thermal processing. Food Bioprocess Technol. **8**(7), 1405–1418 (2015)

Design and Optimization of the Architecture for High Performance Seismic Exploration Computers

Shuren Liu[✉], Chaomin Feng, Changning Cai, and Li Fan

Institute of Computer Technology, Northwest Branch of PetroChina Research
Institute of Petroleum Exploration and Development, Lanzhou, Gansu, China
LiuShuren88@haoxueshu.com

Abstract. The deep application of "two widths and one height" acquisition technology in the petroleum exploration puts forward higher requirements for the ability of seismic data parallel processing. This paper analyses the calculation flow of the special application module of seismic processing in the petroleum exploration, finds out its demands for server resources, and configures high-performance server clusters rationally. Three special application modes of memory sensitive, network sensitive and hard disk sensitive are analyzed. Finally, according to Linpack benchmark test results and the comparison of actual application module operation, the rationality of the architecture designed in this paper is determined. From the data, it can be seen that this computing framework improves the operation efficiency significantly, and is suitable for seismic data processing in exploration data center. The design has certain universality.

Keywords: Cluster NAS · Linpack · IOZONE · HPC · Petroleum exploration · Seismic data processing

1 Introduction

At present, seismic processing is an important part in the whole petroleum exploration project. Most companies in the petroleum industry use High Performance Computer Cluster [1, 2], high-speed and large-capacity storage system and high-speed network system as seismic processing platform. Now many exploration projects adopt the high-density and wide-azimuth seismic data acquisition methods [3], which lead to the increasing data acquisition amount. The original data amount has been upgraded from the level of GB to the level of TB; the date of some projects can reach tens of Tera Bytes or even hundreds of Tera Bytes. The increasing amount of data directly leads to the increase of calculated quantity. The different demands of various applications for computer resources make the configuration and optimization of high-performance seismic exploration architecture particularly important.

© Springer Nature Singapore Pte Ltd. 2020
M. Atiquzzaman et al. (Eds.): BDCPS 2019, AISC 1117, pp. 1934–1943, 2020.
https://doi.org/10.1007/978-981-15-2568-1_270

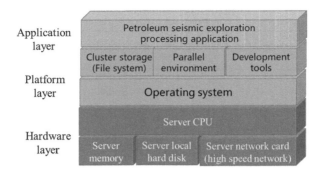

Fig. 1. Layered architecture for high performance computing

Figure 1 shows a layered architecture for high performance computing. In the hardware layer, the server CPU performance can be maximized by reasonable hardware configuration strategy. In the platform layer, the cluster storage can run efficiently, while the parallel environment and the development environment are normal through optimizing the operating system. In the application layer, clusters can be configured reasonably by analyzing the operation characteristics of the application.

In this paper, according to characteristics of the hardware layer and the application layer, the accuracy of cluster hardware configuration is determined through benchmarking and actual task operation according to the requirements of relevant modules in seismic data processing application and the characteristics of mainstream hardware at present (CPU, memory, network and hard disk).

2 Analysis on the Characteristics of High Performance Seismic Applications in Petroleum Exploration and Hardware Configuration

According to the computing characteristics of high performance applications for seismic processing in the petroleum exploration, the applications are divided into following types: CPU sensitive, memory sensitive, local hard disk sensitive and network sensitive. It is self-evident that all kinds of applications are sensitive to CPU; CPU performance plays an irreplaceable role in improving computing efficiency, which has been discussed in previous studies. This paper will analyze three other points that may have been neglected in the past.

2.1 Effects of the Local Hard Disk on Operational Efficiency

Researchers use the term "local hard disk" in this paper is because that in the petroleum exploration data center, in addition to the hard disk of the server, the cluster storage (NAS) [4, 5] is also used as the storage medium for the final raw data and results data. In seismic processing applications, there are many applications sensitive to the local hard disk. High IOPS Application module is one of the representative applications, and the most representative one is Sort on Trace Head Literal.

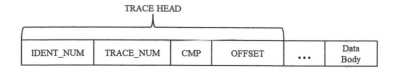

TRACE HEAD

IDENT_NUM	TRACE_NUM	CMP	OFFSET	...	Data Body

Fig. 2. Trace data format

Figure 2 shows the format of trace data. The trace data is composed of the trace head and the data body. There are many parameters for the trace head, which are the attribute items of the trace data. The process of Sort on Trace Head Literal means to sort the trace data according to these attribute items.

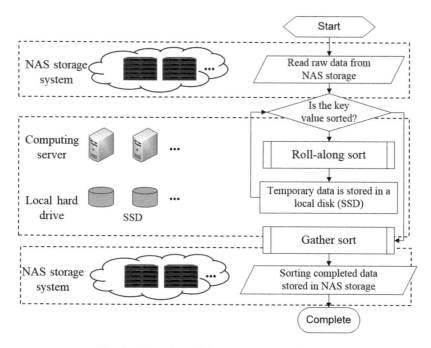

Fig. 3. Flow chart of Sort on Trace Head Literal

Figure 3 shows the process of Sorting on Trace Head Literal. The specific process is determined based on the parameters of the trace head selected by processors. The Roll-along Sort process runs according to the parameters of each trace head. After the sorting process, relevant data are stored as intermediate data. After all the parameters are sorted, the Gather Sort process runs to complete the final sorting. Data flows from NAS storage to the SSD hard disk of the server [6, 7], and then back to the NAS storage. Distribution and merging of data are accomplished on each server; the storage protocol is NFS V4.

In the process, in addition to the CPU and memory resources, the writing and reading of intermediate data after each sorting also affect the job efficiency. This is because there are multiple sorting processes in the operation and the sorting is serial.

In summary, the most important feature of this kind of processing application is high IOPS. That is to say, storage needs to accept a large number of access requests issued by the server in a short period of time, but the amount of data accessed each time is small, which requires a high access rate to ensure that the application can run quickly. Therefore, in the process of server configuration, the local hard disk adopts the SSD hard disk, and stores the intermediate data in the process on the SSD disk, so as to reduce the time of reading and writing the intermediate data.

2.2   Effects of the Memory on Operational Efficiency

In seismic exploration processing applications, there are many memory-sensitive applications. Figure 4 shows the flow chart of a computing server in the cluster parallel operation of the grid tomography algorithm module. As can be seen from the figure, the data is put into memory after being read from the NAS storage; the operations of establishing matrix, solving matrix and modifying speed files are completed in the memory. The size of the matrix (data grid) determines the demand for the server memory. In the practical application, 800*800 grid (the smaller the grid, the larger the computing resources) requires more than 200 GB of memory.

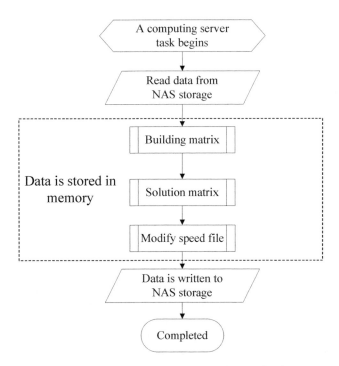

Fig. 4. Flow chart for memory sensitive applications

The most important feature of this kind of application is the high memory occupancy rate. When configuring high-performance computing cluster, the memory configuration of each server becomes a crucial link. In practice, we configure large memory servers to run related applications in order to improve the operational efficiency.

2.3   The Effect of Network on Operational Efficiency

This paper chooses the time migration arithmetic module which is used in the seismic exploration processing of many network-sensitive applications. Figure 5 shows the operation flow of a server in cluster parallel computation. The module is mainly based on calculation, but the calculation is divided into several parts: calculating travel time, adding travel time paths and imaging overlay. In the process of computing, the original data still needs to be read from NAS storage, which increases the IO requirement of the server. At the same time, because of the need to communicate with other computing servers, the module is ultimately more sensitive to the network performance of the server.

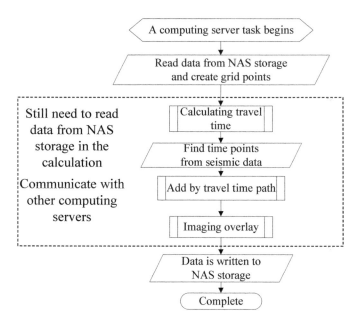

Fig. 5. Flow chart for network sensitive applications

The most important feature of this application is more data exchange, which has higher requirements for network configuration between cluster servers. When configuring high performance computing clusters, each server network configuration becomes an important factor that impacts the computing efficiency. In reality, we use high performance servers to configure the high speed network (IB or 25G network), in order to better improve the efficiency of computing.

3   Experiments and Results Analysis

In this paper, the benchmark test of the operational efficiency is carried out based on Linpack [8–10]. The test kit is parallel_studio_xe_2016 and over. The AVX is opened in the server. The basic models of the test include SMP protocol used between different processes in the cluster server and the MPI protocol used between servers.

The IOZONE k [11, 12] tool is adopted for the benchmark of hard disks. The application modules in this paper are related modules in Omega and Paradigm software.

3.1   The Impact of Local Hard Disk Configuration on the Operational Efficiency of Relevant Applications

3.1.1   IOZONE Benchmark Comparison

The benchmark test in this paper is carried out based on HP DL360 Gen9, with two Intel E5 2680 CPUs (14 cores) and 256 GB (32*8 GB) memory. The local hard disk is configured on the server as the RAID5 mode.

A. The local hard disk is the MU-3 12 Gbps SSD.
B. The local hard disk is the 15 K 12 Gbps SAS hard disk.
C. The hard disk is the 10K 12 Gbps SAS hard disk.

IOZONE is a benchmark tool to test the performance of systems on reading and writing files. It can test the hard disk performance through different modes. Most seismic data processing applications have similar characteristics in access to storage.

Figures 6 and 7 show the results of IOZONE tests for three types of hard disks. In terms of read and write performance, SSD has obvious advantages, with the write performance of 1.4 GB. It has greatly improved the efficiency of traditional hard disks with SATA interfaces and SAS interfaces. Therefore, the usage of SSD in the local hard disk can improve the performance of cluster configurations in the oil exploration.

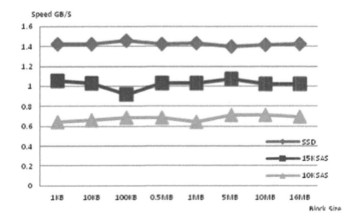

Fig. 6. IOZONE write performance of three hard disks

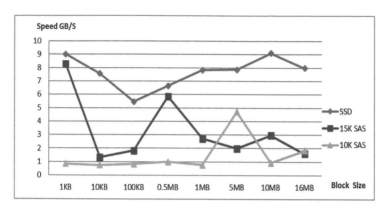

Fig. 7. IOZONE read performance of three hard disks

3.1.2   Operation Effects of Actual Sort on Trace Head Literal

Table 1 shows the actual operation effects of the Sort on Trace Head Literal process. In order to test the rationality of the design, the SSD hard disk (RAID 0), the cluster NAS storage (Huawei Ocean Store) and the SAS hard disk (RAID 0) are used to store intermediate data on the HPDL360 Gen9 server with the Intel E5 2680 CPU (28 cores) and the memory of 256 GB. The process is sorting files with the size of 410 GB and 20 lines (Trace). The operation efficiencies are shown in Table 1. The table shows that the SSD hard disk is the best one. Under the condition of high IOPS, the cluster NAS stores has the lowest operation efficiency. Therefore, the SSD scheme for temporary data storage in sorting operation has a good application effect.

Table 1. Operating effects of Sort on Trace Head Literal under various conditions

Time consuming (min)	SSD	Clustered NAS storage	SAS 10K
Data 1	35.53	264.06	72
Data 2	27.26	240.8	75
Data 3	33.2	229.1	65
Data 4	30.6	220.1	80
Data 5	31.2	210.5	82
Average value	31.558	232.9	74.8

3.2   The Impact of Memory Configuration on the Operational Efficiency of Relevant Applications

3.2.1   Linpack Benchmark Comparison

The Linpack comparative tests are carried out for two clusters of configuration; both of them are equipped with the Intel Gold 6132 CPU, and the whole system adopts the rack-mounted server. The first cluster has 64 servers with 384 GB of memory, and the second cluster has 64 servers with 192 GB of memory. the RedHat 7.4 is used in both cluster operating systems. As can be seen from Fig. 8, with the increase of memory, the cluster's operation efficiency gradually increases, but when it reaches a certain level (such as 192 GB in this paper), the impact begin to decrease.

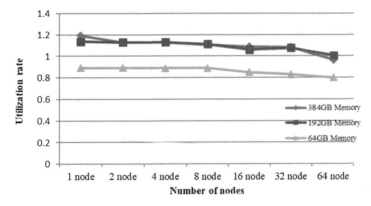

Fig. 8. Comparison of cluster linpack tests with two different memory configurations

3.2.2   Comparison of Efficiency of Actual Grid Tomography

For the three-dimensional project in Dachuan, which has 2000 km^2 of coverage area and 7.79 TB data before operation, researchers run the grid tomography module on two different clusters. Because of the small memory, when cluster 1 has a higher parameter (three layers of data layers and the 800*500 data grid), the application will exit because of the insufficient memory. With the reduction of operation parameters (4 layers of data layers and 800*800 of data grid), the operation speed of cluster 1 with low memory is still lower than that of cluster 2 with high memory (Table 2).

Table 2. Effect of memory on grid tomography efficiency

		Cluster 1	Cluster 2
Server configuration	CPU	Intel Gold 6132 (2*14)	Intel Gold 6132 (2*14)
	memory size	128 GB	192 GB
Operational parameter	Data hierarchy	4	3
	Grid size	800*800	800*500
Number of servers used		16	16
50 line calculations take time		9 h	7 h

Therefore, according to benchmark test results and actual application performances, in actual practice, researchers can configure high memories for memory-sensitive applications appropriately. But for other applications, memories beyond a certain range cannot improve the operation performance significantly.

3.3 The Impact of Network Configuration on the Operational Efficiency of Related Applications

3.3.1 Linpack Benchmark Comparison

Comparing the two clusters, the CPU model is Intel Gold 613. Other parameters of the two clusters are as following. The 25G network card cluster has 192 GB memory; the local hard disk is SAS. The 10G network card cluster has 256 GB memory; the local hard disk is SSD. In the case of poor memory and local hard disk, the Linpack efficiency of the 25G network card cluster has been greatly improved. When the number of cluster servers is more than 8, the performance of cluster decreases slightly (Fig. 9).

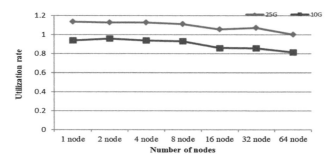

Fig. 9. Comparison of cluster linpack tests of two different network configurations

3.3.2 Comparison of Actual Time Migration Efficiency

For a joint project with a total area of 780 km^2 and raw data of 6 TB, the time migration module runs on two different clusters. Because of the different network card speed, the speed of cluster 1 in calculating a line can reach three times that of cluster 2 (Table 3).

Table 3. Actual time migration efficiency

		Cluster 1	Cluster 2
Operational parameter	CPU	Intel Gold 6132 (2*14)	Intel Gold 6132 (2*14)
	memory size	192 GB	192 GB
	NIC speed	10G	25G
Number of servers used		16	16
1 line operation takes time		18 min	6 min

In summary, high-speed network card can improve cluster computing efficiency. In network-sensitive applications, high-speed network cards can better improve the actual operation effects, which is very helpful to the actual project.

4 Conclusion

In view of the increasing demand for high-performance cluster computing resources caused by the increasing seismic data processing amount in petroleum exploration projects, this paper analyses the calculation flow of the special application module of seismic processing in the petroleum exploration, finds out its demands for server resources, and configures high-performance server clusters rationally. In this paper, the three special application modes of memory sensitive, network sensitive and hard disk sensitive are analyzed. The rationality of the server architecture designed in this paper is determined through Linpack benchmark tests and comparative tests of the actual application modules. From data obtained, it can be seen that the design and optimization of the architecture for high-performance seismic exploration computing have a certain promotion significance.

References

1. Fox, A., Griffith, R., Joseph, A., Katz, R., Konwinski, A., Lee, G., Stoica, I.: Above the clouds: a berkeley view of cloud computing. Dept. Electrical Eng. and Comput. Sciences, University of California, Berkeley, Rep. UCB/EECS **28**(13) (2009)
2. Qiu, Q., Wu, Q., Bishop, M., et al.: A parallel neuromorphic text recognition system and its implementation on a heterogeneous high-performance computing cluster. IEEE Trans. Comput. **62**(5), 886–899 (2013)
3. Du, J.H., Yang, T., Li, X.: Oil and gas exploration and discovery of PetroChina Company Limited during the 12th Five-Year Plan and the prospect during the 13th Five-Year Plan. China Pet. Explor. **21**(2), 1–15 (2016)
4. Leong, D.: A new revolution in enterprise storage architecture. IEEE Potentials **28**(6), 32–33 (2009)
5. Miller, S.D., Hawkins, J.D., Kent, J., et al.: NexSat: previewing NPOESS/VIIRS imagery capabilities. Bull. Am. Meteorol. Soc. **87**(4), 433–446 (2006)
6. Jung, M., Choi, W., Shalf, J., Kandemir, M. T.: Triple-A: a Non-SSD based autonomic all-flash array for high performance storage systems. In: ACM SIGARCH Computer Architecture News, vol. 42, no. 1, pp. 441-454. ACM (2014)
7. Tavakkol, A., Arjomand, M., Sarbazi-Azad, H.: Network-on-SSD: a scalable and high-performance communication design paradigm for SSDs. IEEE Comput. Arch. Lett. **12**(1), 5–8 (2013)
8. Havlak, P., Kennedy, K.: An implementation of interprocedural bounded regular section analysis. IEEE Trans. Parallel Distrib. Syst. (3), 350–360 (1991)
9. Gan, X., Hu, Y., Liu, J., et al.: Customizing the HPL for China accelerator. Sci. China Inf. Sci. **4**, 042102 (2018)
10. Bönisch, T., Resch, M., Schwitalla, T., et al.: Hazel Hen leading HPC technology and its impact on science in Germany and Europe. Parallel Comput. **64**, 3–11 (2017)
11. Seo, B., Kang, S., Choi, J., et al.: IO workload characterization revisited: a data-mining approach. IEEE Trans. Comput. **63**(12), 3026–3038 (2014)
12. Biardzki, C., Ludwig, T.: Analyzing metadata performance in distributed file systems. In: Lecture Notes in Computer Science, vol. 5698, pp. 8–18 (2009)

Current Situation and Countermeasure Analysis of College Etiquette Education in the Era of "Internet+"

Bingjie Han[✉]

Shandong Vocational College of Light Industry, Zibo, Shandong, China
Bingjie_Han81@haoxueshu.com

Abstract. Etiquette is an important yardstick to measure people's temperament and personality, and a concentrated reflection of individual quality. However, with the deepening of the market economy and the influence of social ideology, uncivilized behavior in university campuses is often seen, and the status quo of civilized etiquette cultivation of some college students is not optimistic. As the coming of the "internet+" era and the wide application of new media technology, the etiquette education in Colleges is faced with the disadvantages of weakening teachers' influence, combining closely with ideological & political education, and lacking relative content of Internet etiquette education. This paper puts forward some countermeasures, such as using the new media platform to carry out etiquette education activities, strengthening the content of traditional moral core education in etiquette curriculum, promoting the close integration of etiquette education and ideological & political education, and giving scope to the practical approach and ideological guidance function of etiquette education.

Keywords: Etiquette education · Internet+ · Countermeasure analysis

1 Introduction

Etiquette is a kind of behavioral norm for dealing with people. It is also an art of communication. With the progress of human civilization, etiquette has gradually become a bridge for the establishment of harmonious relations between human beings. Confucius said: Do not learn to ritual, no standing [1]. In his writings, there are as many as 72 records about "rituals", which shows that etiquette education plays an important role in society. China has always been a state of etiquette. Etiquette is not only a part of spiritual civilization construction, but also a manifestation of a person's spiritual outlook [2]. In more depth, a person's etiquette culture literacy can map the degree of civilization of a country.

As the most dynamic and lively college students in the whole society, they are the future of the country. They also bear the responsibility of inheriting the civilization of the great powers. However, in today's rapid development of China's economic strength, the civilized courtesy of contemporary college students and the comprehensive quality of the whole people are not directly proportional to our investment in education [3]. Uncivilized behavior and inconsistencies with the city's highly modernized development abound.

M. Atiquzzaman et al. (Eds.): BDCPS 2019, AISC 1117, pp. 1944–1951, 2020.
https://doi.org/10.1007/978-981-15-2568-1_271

With the socioeconomic development and the improvement of spiritual civilization, in modern society, etiquette has penetrated into all aspects of people's work and life. This requires that college students should receive formal etiquette education and strengthen their etiquette training in order to invest in the increasingly professional and standardized work and life. In the era of "Internet+", new media technology has shown an explosive development [4]. The Internet as a new information dissemination carrier has changed the way of education activities to a certain extent. "Internet+ education" will become a technological change in the field of education. Hotspots, in the "Internet +" era, how to better carry out the etiquette education for college students has become a challenge to be solved in college etiquette education [5].

2   Problems in Etiquette Education in Colleges

2.1   The Weakening of Etiquette Course Teachers' Influences on Students

Under the background of "Internet+", the three-centered theory of traditional teaching proposed by Herbart has been questioned. With the development of new technology of media, students' access to etiquette knowledge is becoming more convenient and diversified [6]. Compared with the systematic etiquette curriculum, students are more inclined to use the Internet to acquire the required etiquette knowledge, such as public relations etiquette, job-hunting etiquette, which undoubtedly weakens the influence of etiquette curriculum teachers, and thus triggers a series of consequences.

The authority of ceremonial curriculum teachers is weakened, and college etiquette education is difficult to systematically carry out. The existence of teacher authority is the premise of the effective development of teaching activities. The weakening of the authority status of teachers in college etiquette courses will undoubtedly affect the development of etiquette education in colleges [7].

The etiquette knowledge acquired by students is mixed and fragmented. Etiquette education for college students should be multi-faceted. This requires the cooperation of daily etiquette education and systematic etiquette courses. With the wide application of Internet technology, students are more inclined to use the Internet to obtain the etiquette knowledge they need [8]. A process often has utilitarianism, and the etiquette knowledge acquired is usually related to specific social activities such as job hunting and interviews. The etiquette knowledge acquired by students through the Internet is increasingly fragmented and cannot guarantee the systematic and accurate etiquette knowledge [9].

2.2   Etiquette Education Is Not Closely Integrated with Ideological & Political Education

The three-level goal of etiquette education in colleges and universities determines that the etiquette education in colleges is not only an education about etiquette techniques such as etiquette norms and courtesy etiquette, but also an education of moral cultivation and moral improvement. Therefore, etiquette education is an important part of moral education. Ideological & political education in colleges and universities is an education practice carried out by college students as the main body of education [10]. Its purpose is the ideological and moral requirements for the cultivation of talents in modernization. Etiquette education and ideological & political education in college have the same goal. The ideological and moral requirements of college students' ideological & political education should also be the requirements of college etiquette education. College ideological & political education can be regarded as college etiquette education. Soul, the content of etiquette education must serve ideological & political education to achieve its social value [11].

2.3   Relatively Lack of Network Etiquette Education

In the era of "Internet+", Internet etiquette has become a "new frontier" of etiquette education in Colleges and universities with the increasingly close contact between college students and the Internet. In the United States, where the Internet has developed earlier, there has been a mature paradigm of online etiquette education, such as CyberSmart, which provides perfect information ethics course content and learning activities, and Netiquette, which provides online etiquette testing and online preview of online etiquette books [5]. Some universities in Japan and South Korea take the course of "Internet Etiquette" as one of the general courses for students. Compared with western developed countries, the education of network etiquette in our country is still in the exploratory stage. The relative lack of network etiquette education for college students and the long-term lack of network legislation lead to the lack of legal consciousness of network subjects in network behavior.

3   Investigation and Analysis of Etiquette Accomplishment

This paper divides the structure of etiquette cultivation into three dimensions: etiquette cognition, etiquette emotional attitude and etiquette behavior habits. The results and analysis are as follows.

Table 1. Analysis of differences in etiquette cultivation in different grades

	Grade	N	Average	Standard deviation	F	P
Etiquette cognition	Freshman	29	3.1862	1.04599	0.301	0.825
	Sophomore	111	3.2721	0.90686		
	Junior	145	3.2524	0.93601		
	Senior	115	3.1670	0.91342		
Etiquette attitude	Freshman	29	3.3736	0.87928	0.374	0.772
	Sophomore	111	3.1682	0.94775		
	Junior	145	3.2057	0.96326		
	Senior	115	3.1870	0.94672		
Etiquette behavior	Freshman	29	3.4109	0.79897	0.744	0.526
	Sophomore	111	3.1704	0.97619		
	Junior	145	3.1874	0.89325		
	Senior	115	3.1297	0.89489		

It can be seen from Table 1 that according to the standard of $X = 0.05$, there is no statistically significant difference in etiquette, etiquette attitude and etiquette behavior scores of different grades ($P > 0.05$). The results of different levels of etiquette level analysis data do not have statistics. The meaning of learning, so we can say that there is no obvious grade difference in the level of etiquette for the freshman to senior. We can see that from the admission of college students as freshmen to the graduation of seniors, the four years of higher education and the culture of university campus culture have not significantly improved the improvement of students' etiquette.

It can be seen from Table 2 that according to the standard of a = 0.05, the differences in the scores of cognition and etiquette of college students of different majors are not significant ($p > 0.05$), and the scores of etiquette behaviors of college students of different majors are significant ($p < 0.05$), the comparison results of the two pairs can be seen that the ritual behavior of literature and history is higher than that of science and engineering, and the ritual behavior score of art is higher than that of science and engineering.

Table 2. Analysis of differences in etiquette cultivation between different majors

	Grade	N	Average	Standard deviation	F	P
Etiquette cognition	Science and engineering	206	3.1466	0.90875	1.647	0.179
	Literature and history	151	3.3033	0.96162		
	Art and sports	28	3.4857	0.79723		
	Medical	15	3.1200	0.99875		
Etiquette attitude	Science and engineering	206	3.1013	0.91187	2.146	0.095
	Literature and history	151	3.2835	0.96874		
	Art and sports	28	3.5067	0.93024		
	Medical	15	3.1789	1.10213		
Etiquette behavior	Science and engineering	206	3.0601	0.90321	3.821	0.011
	Literature and history	151	3.2865	0.92046		
	Art and sports	28	3.5779	0.65487		
	Medical	15	3.0670	1.05326		

Table 3. Analysis of correlation result

		Etiquette cognition	Etiquette attitude	Etiquette behavior
Etiquette cognition	Pearson correlation	1	0.840	0.800
	Significance		0.000	0.000
	N	400	400	400
Etiquette attitude	Pearson correlation	0.840	1	0.799
	Significance	0.000		0.000
	N	400	400	400
Etiquette behavior	Pearson correlation	0.800	0.799	1
	Significance	0.000	0.000	
	N	400	400	400

It can be seen from Table 3 that the etiquette behavior and the etiquette cognition are significantly positively correlated at the 1% level, and the correlation coefficient is 0.800; the etiquette behavior and the etiquette emotional attitude are significantly positively correlated at the 1% level, and the correlation coefficient is 0.799; Etiquette emotional attitude and ceremonial cognition were significantly positively correlated at the 1% level, with a correlation coefficient of 0.840. There is a significant positive correlation between etiquette cognition, etiquette and attitude, and etiquette behavior. The moral structure related to etiquette needs to conform to or agree with the etiquette of etiquette and the emotional attitude of etiquette culture. The level of etiquette is here. The acquisition and integration of the three are achieved. The construction of this process needs to constantly strengthen the etiquette of etiquette, expand the experience of etiquette and emotion, and internalize the compliance behavior of etiquette norms into the individual's own needs.

4   Countermeasures of Etiquette Education in Colleges in the Age of "Internet+"

4.1   Using New Media Platform to Develop Rich Etiquette Education Activities

As an important part of moral education, etiquette education can be divided into explicit education and implicit education. The explicit etiquette education is mainly realized through systematic etiquette curriculum, which has a certain compulsory color. In the past, etiquette education practice in Colleges and universities mostly equated etiquette education with explicit etiquette education [7]. Education, that is, etiquette course education, the etiquette education courses offered by colleges and universities have professional compulsory courses, professional elective courses, public compulsory courses and other forms. The advantages of this kind of etiquette education form lie in its systematization and convenient consideration of students' achievements. However, in the course of development, students' attendance rate and elective rate are often low [8]. The course of etiquette education is not attractive enough because of its dull content and other problems.

As an emerging information media, Internet new media technology has reduced the influence of college etiquette education teachers and provided a new model of "Internet+" etiquette education for the development of etiquette education in colleges and universities. The "Internet+" etiquette education based on new media technology is rich in forms of hidden etiquette education. Educators can use the new media platform such as WeChat public account to guide students to establish ceremonial interest associations, give play to the educational guidance of peer groups, and make use of new The media platform forms a ceremonial education atmosphere covering colleges and universities, which helps the college students to internalize the etiquette education and promote the etiquette norms to their own moral cultivation [4]. Through the etiquette education activities of interest to such student groups, the etiquette knowledge acquired in the classroom is applied to the category of moral practice. College etiquette educators can also use the form of MOOC and webcast to guide the correct ideology on the Internet, and push the case of morality and etiquette education with hot events as the carrier in the student life, so that the college students feel in their daily life. The atmosphere of etiquette education can play the role of etiquette education more sustainably and effectively [10] (Fig. 1).

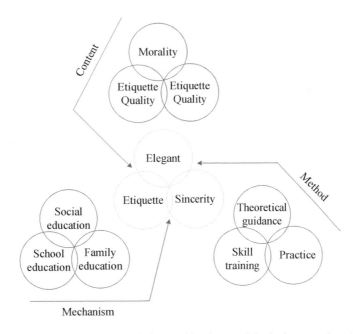

Fig. 1. The basic structure of the combination model of etiquette education

4.2   Promoting the Combination of Etiquette Education and Ideological & Political Education

Etiquette education and ideological guidance function of etiquette education have goal consistency in a certain sense. In the process of carrying out educational practice, this goal consistency is reflected in etiquette education is the practical way of Ideological & political education in Colleges and universities. Students' requirements and norms can be embodied in their specific words and deeds through etiquette education, which promotes the organic combination of etiquette education and ideological & political education in practice.

In the field of etiquette curriculum education, some universities have incorporated etiquette education as a public basic course into the content of Ideological & political education of College students. College ideological and political teachers act as etiquette as a public basic course. Teachers of etiquette education curriculum infiltrate the concept of Ideological & political education in the process of etiquette education, and the course results are incorporated into the evaluation system of Ideological & political education curriculum in Colleges and universities [6].

The network civility etiquette construction is taken as an integral part of college students' network ideology safety education, which is included in the ideological & political education course of college students and the etiquette education course of colleges and universities. In the era of "Internet+", with the rise of new media, colleges and universities have become the frontier battlefield in the ideological field. Under this background, college etiquette education should play its role as the practical way and ideology guiding of ideological & political education in colleges and universities. Function, adding the content of network civility and etiquette construction in explicit etiquette education courses and daily hidden etiquette education, not only makes up for the blank of network etiquette education in colleges and universities in the era of "Internet+", but also guides students to pay attention to network ideology security. To improve students' ability to identify false values and decadent concepts on the Internet, and to effectively play the ideology guiding function of etiquette education.

5   Conclusion

To build a college etiquette course with the traditional etiquette culture as the core, to a certain extent, it is to adhere to the tradition, but more to meet the needs of modern development. In the etiquette education of colleges, we should pay full attention to the penetration of Chinese traditional culture. In the era of "Internet+", with the rise of new media, use the new media platform to carry out etiquette education activities, strengthen the traditional moral core education content in etiquette courses and promote etiquette education. It is closely integrated with ideological & political education, and plays a coping strategy such as the practical approach of etiquette education and the guiding function of ideology. It can make up for the blank of etiquette education in campus in the age of "Internet+". At the same time, it can also guide students to show solicitude for the security of network ideology, improve students' ability to identify false values and decadent concepts on the Internet, and effectively play the ideology guiding function of etiquette education. Thereby enriching the connotation of etiquette education, inheriting Chinese traditional culture, helping students to establish a correct world outlook, life and values.

References

1. Lewin-Jones, J., Mason, V.: Understanding style, language and etiquette in email communication in higher education: a survey. Res. Post-Compuls. Educ. **19**, 75–90 (2014)
2. Peeters, M.J., Harpe, S.E.: Numbers etiquette in reports of pharmacy education scholarship. Curr. Pharm. Teach. Learn. **8**, 896–904 (2016)
3. Filippone, M., Survinski, M.: The importance of etiquette in school email. Am. Second. Educ. **5**, 456–462 (2016)
4. Kim, D.H., Yoon, H.B., Yoo, D.M., Lee, S.M., Jung, H.Y., Kim, S.J., Yim, J.J.: Etiquette for medical students' email communication with faculty members: a single-institution study. BMC Med. Educ. **16**(1), 129 (2016)
5. Santovec, M.L.: Business etiquette: not just for black dress events. Women High. Educ. **22**, 22 (2014)
6. Choi, J.S., Kim, K.M.: Predictors of respiratory hygiene/cough etiquette in a large community in Korea: a descriptive study. Am. J. Infect. Control **44**, 271–273 (2016)
7. Nutt, J., Mehdian, R., Kellett, C.: Theatre etiquette course: students' experiences. Clin. Teach. **11**, 131–135 (2014)
8. Zhu, X., Wei, M., Chen, R., et al.: A study to analyze the effectiveness of video-feedback for teaching nursing etiquette. In: Lecture Notes in Electrical Engineering, vol. 269, pp. 1315–1320 (2014)
9. Clark, A.: Disability awareness and etiquette: transforming perceptions through a series of experiential exercises. J. Creat. Ment. Health **10**, 456–470 (2015)
10. Baranyuk, V.: Experience of forming professional and communicative competency of future social workers in education systems of Western European Countries. Comp. Prof. Pedagogy **5**, 109–114 (2015)
11. Dearn, L.K., Pitts, S.E.: Popular music and young audiences: exploring the classical chamber music concert from the perspective of young adult listeners. J. Popular Music Educ. **1**, 43–62 (2017)

Innovative Teaching of University Management Course Under the Background of "Internet+"

Weiguo Dong$^{(\boxtimes)}$ and Tingting Liu

Yinchuan University of Energy, Yinchuan, Ningxia, China
Weiguo_Dong76@haoxueshu.com

Abstract. The rapid development of Internet technology leads and promotes the reform of college teaching mode and talent training mode. The management course has high requirements for theoretical knowledge and practical operation skills. In the traditional teaching mode, students' learning enthusiasm, initiative, Class participation is poor and can't be used. In order to integrate theoretical knowledge with practical teaching in the rapid development of Internet information technology, it is necessary to carry out innovation and practice in management teaching, and to promote the management teaching with advanced Internet technology as a powerful support for classroom teaching. This paper will study the mode of management practice teaching from multiple levels. The cultivation of students should be comprehensive and multi-form, especially in practice teaching. It is hoped that the diversified practice teaching will achieve the connection between the training goal and the social career, and meet the society's demand for student employment.

Keywords: "Internet+" · Modern teaching technology · Management · Innovative teaching

1 Introduction

In the context of the "Internet+" era, China's Internet technology has developed very rapidly, driving the rapid development of management theory [1]. At present, many people in the society who are engaged in management work face more challenges. Because society has higher requirements for managers, this kind of social status makes teachers who are engaged in management teaching in universities face many challenges. In the era of big data, university education in the context of Internet information technology requires innovation in educational methods and educational ideas. Universities need to make full use of Internet information in the era of big data to provide more effective teaching methods for university courses [2], thereby improving the quality and effectiveness of university teaching.

Management is the most basic discipline in the management discipline group. It mainly teaches the basic theories and methods of management. It is theoretical, practical and comprehensive. Practical teaching is a kind of teaching method. It advocates that in the process of theoretical teaching and classroom teaching, teaching should be

M. Atiquzzaman et al. (Eds.): BDCPS 2019, AISC 1117, pp. 1952–1958, 2020.
https://doi.org/10.1007/978-981-15-2568-1_272

organized by practical methods, so as to change the subject and object status of teachers and students and create practical opportunities for students [3]. "Internet+" has become an important guiding ideology of education and teaching reform, and Internet technology has become an essential tool for students to learn and live. Therefore, under the guidance of the idea of "practical teaching" and the background of the Internet, it is of great significance to study the practical teaching methods of management course and how to cultivate students' professional skills and transferable abilities through this course, so as to improve the teaching level of the course and the quality of personnel training [4].

The development of Internet technology has an impact on the teaching mode of management, which greatly promoted the development of management theory teaching. In today's society, the requirements for managers are increasing, so the pressure on management professional employment is also growing. This kind of situation also makes the teachers who are engaged in management teaching in the university face many challenges [5]. They have to adapt to the rapid development of information technology, constantly improve and innovate the teaching mode, and innovate more efficient teaching methods for the school, thus improving the teaching quality. This paper starts from the practice of modern teaching methods based on "Internet+" in the teaching of university management courses in recent years, and carries out innovative research on the teaching of this course [6].

2 The Influence of Modern Teaching Technology on University Teaching Under the Background of "Internet+"

Internet technology is a hot topic in this era, and it is also widely used. Among them, universities have also introduced Internet technology into the campus. This new teaching model is completely different from the traditional teaching model, which not only collects rich teaching resources, but also Efficient teaching is possible. Due to the use of Internet, a large amount of information is uploaded to the cloud, and teachers can directly access relevant data through various search engines during the preparation of lessons [7]. Compared with the traditional teaching mode, the traditional teaching materials are all derived from the textbooks. Not only the materials are less and the reading speed is slow, but also the students can not understand them intuitively for some abstract knowledge points, which greatly reduces the efficiency of teaching.

With the Internet, the way of college classes is no longer limited to classroom teaching, from a single classroom teaching to a diversified online teaching. For example, there are "micro-courses" and "Mu classes". This kind of teaching method enables students to study at any time and at any place, making full use of the students' free time, so that students can solve the problems encountered in daily learning [8]. Moreover, students can communicate with teachers through some chat software on the Internet. Because they can't see each other on the Internet, students can boldly ask the teacher what they want to ask, instead of asking questions and fearing teachers in classroom teaching. The situation has greatly stimulated the initiative of students to learn.

The Internet age is an era of information explosion. Most of the resources are shared. In addition to the materials collected by the teachers, students can collect some information themselves, but there is also a problem that the information on the network is one side [9], if the students blindly collect, they will mislead the students, so the teacher needs to teach the students how to distinguish the reliability of the data, and tell the students to collect the information to be collected on some regular websites (Fig. 1).

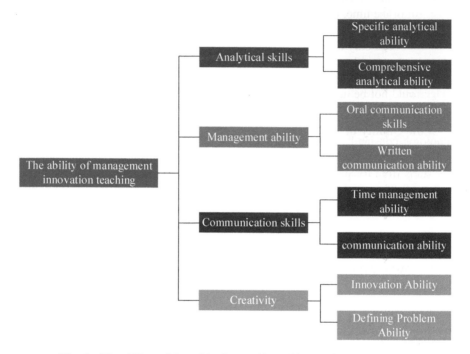

Fig. 1. The ability cultivated in the teaching of innovative management

3   Problems in the Teaching of University Management Courses in the Era of "Internet+"

Under the background of "Internet+" era, the management concepts, communication methods and management methods of enterprises are developing in a diversified direction. This requires high management knowledge of university graduates, and puts forward the effect and quality of management teaching. Influenced by traditional teaching, the research on management practice teaching is still lagging behind, and there are still many shortcomings in teaching.

3.1   Lack of Practice in Curriculum

Many professional courses in Chinese universities pay too much attention to theoretical knowledge and lack practical teaching links. Some courses are not innovative and the teaching content is stale. Because there are many theoretical tests in China, the teachers of the university pay too much attention to the teaching of curriculum theory, while ignoring the cultivation of students' practical ability. Although some universities set up practical courses, there is still no specific implementation process, so students can't effectively digest theoretical knowledge in the process of practice, which is a great hindrance to improve students' ability to innovate. There is almost no opportunity for students to practice. There is not much time for training in the teaching. In the practice class, most of the time, it is also taught by teachers [10]. The students passively absorb it. There is no facility for students to experiment and practice training. This cramming teaching mode is only a passive acceptance of students unilaterally. It cannot mobilize students' interest in learning professional knowledge, and students' hands-on practical ability cannot be strengthened. The goal of training students' innovative hands-on ability should not be achieved [11].

3.2 Low Quality of Teachers

In the process of rapid development of the Internet, the quality requirements for teachers are also constantly improving. University teachers are usually taught directly after graduating from university, so there is less practical experience in teaching. Because the teacher's overall quality is generally low, in order to complete the teaching tasks, the teacher will choose the textbooks that he is good at, which is too backward. Therefore, in order to carry out classroom theory and practical teaching, teachers need to improve their professional quality, because the comprehensive quality of teachers has an important role in promoting the development of the school and has a great influence on the development of students [12].

3.3 Choosing the Wrong Teaching Case

Some universities still use teacher-infused teaching methods, so many teachers use teaching methods and teaching models are still very simple, classroom interaction and practice links are less, which is very unsuitable in the Internet age, especially some teachers are When choosing teaching cases, because they are not familiar with the current situation of contemporary management, the cases they choose lack timeliness, and even run counter to China's national conditions and the development of the times [6].

4 Innovation of Management Course Teaching Under the Background of "Internet+"

4.1 Transformation of Teaching Ideas

"Internet+" is not only the application of Internet technology and advanced equipment in classroom teaching, but more importantly, teachers should change their teaching thinking, design teaching ideas according to students' learning situation and hobbies, and realize Internet thinking and teaching process. For example, teachers can publish student's management assignments in the form of self-media, and forward it through the media platform through WeChat, Weibo, etc., collect evaluations and praises, and give certain rewards for the highest evaluation and most praised assignments [8]. As a result, the students' fears of management work will be greatly improved, and the enthusiasm for doing homework will be greatly improved, thereby achieving the purpose of improving work quality and work efficiency. In the era of "Internet+", the traditional teacher-instilled and passively accepted teaching mode has fallen behind, students have gradually become the main body of learning, teachers have become instructors from indoctrins, and students have learned independently under the guidance of teachers. Class time is more used for answering questions and discussion [3]. This makes the students' enthusiasm for learning effectively mobilized, strengthens the interaction between teachers and students, and improves the learning efficiency of students while realizing the teaching and learning (Fig. 2).

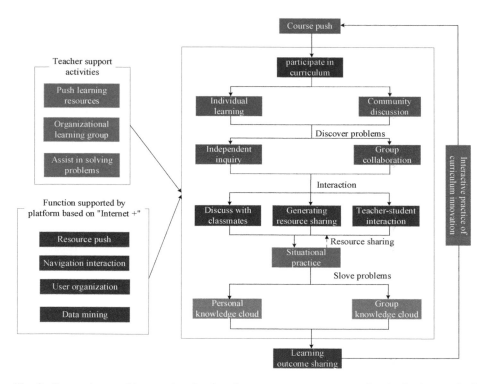

Fig. 2. Innovative teaching mode of university management course under the background of "internet+"

4.2   Innovation in the Process of Teaching

For the discipline of management, students need to be familiar with theoretical knowledge before applying it to practice. So how do you use classroom time efficiently to familiarize students with theoretical knowledge? Teaching can be done from the following methods: First, transform textbook teaching into multimedia teaching. The teaching of traditional textbook teaching is slow, and it is just listening to the teacher. The students don't feel the visual effect. When it comes to abstract theory, there is no actual model for reference [5]. If the language of the teacher's lecture is not very vivid, it will affect. The attention of students in class reduces the efficiency and quality of teaching to a large extent. The use of multimedia for teaching can solve the problems in these traditional teachings. Multimedia teaching is mostly done with PPT, a large amount of data can be collected from the network, such as teaching cases and teaching videos. At the same time, when the teacher is following the PPT, the students will have an intuitive visual perception. They can follow the teacher's pace to learn. When encountering more important and difficult theoretical knowledge, they can teach by playing video and let students observe the theory. How to combine with practice, through specific examples to deepen students' understanding of knowledge points and improve the quality of teaching.

4.3   Innovation in Practice

For the discipline of management, theoretical knowledge is very important, because only with rich theoretical knowledge can we solve problems encountered according to theoretical knowledge. The practice link is equally important, but practice is a link that many universities do not pay enough attention to, resulting in a solid theoretical knowledge of students who graduate from college, but their practical ability is weak. The university should pay attention to the practice of management course. It is necessary to carry out the construction of the practice base for the management major. The construction practice base can cooperate with the enterprise, and the students can go to the enterprise, according to their own profession, to different management. The level goes to the trainee, so that students continue to enrich their practical experience (Table 1).

Table 1. Comparison of teaching effects between traditional teaching methods and innovative teaching methods

Evaluation dimension	Traditional teaching method	Innovative teaching method
Improve management philosophy	68.7%	77.5%
Master management concepts	93.2%	26.3%
Improving the way of thinking	31.4%	87.9%
Gain management theory knowledge	90.1%	7.8%
Learning to analyze the problem	17.8%	92.7%
Improve problem-solving ability	6.2%	91.4%
Exercise communication skills	5.6%	74.8%

5   Conclusion

In the context of "Internet+", the university's management programs have become more vivid and diverse. Through the Internet, university teachers can make use of the associated online classrooms to make management classroom teaching more novel, students can come along and receive a large amount of management knowledge through self-learning, which is very effective in improving the effectiveness of the university management course. This innovative teaching method and teaching model can not only effectively promote the development of university teaching reform, but also be very meaningful to the individualized training of students.

References

1. Lin, W.: Research on the teaching management of track and field web course based on software programming method. Cluster Comput. **3**, 1–12 (2018)
2. Jianguang, C., Hua, T., Sigui, W., et al.: Research-oriented teaching pattern of the exercise physiology course based on the internet. J. Appl. Sci. **14**, 2386–2390 (2014)
3. Joly, M., Rocha, R., Sousa, L.C.F., et al.: The strategic importance of teaching operations research for achieving high performance in the petroleum refining business. Educ. Chem. Eng. **10**, 1–19 (2015)
4. Rong, R.: Research on website construction of quality course based on network teaching platform. Int. J. Technol. Manag. **1**, 89–92 (2014)
5. Yu, J., Ye, B.: Research on connotative development strategies of experimental teaching based on the cultivation of innovative talents. Res. Explor. Lab. **5**, 78–82 (2014)
6. Chen, R.Q.: Teaching innovation and exploration of principles and applications of transducer course based on CDIO idea. In: Advanced Materials Research, vol. 5, pp. 2193–2196 (2014)
7. Miranda, S., Marzano, A., Lytras, M.D.: A research initiative on the construction of innovative environments for teaching and learning: montessori and munari based psycho-pedagogical insights in computers and human behavior for the "new school". Comput. Hum. Behav. **66**, 282–290 (2017)
8. Lincoln, M., Hines, M., Fairweather, C., et al.: Multiple stakeholder perspectives on teletherapy delivery of speech pathology services in rural schools: a preliminary. Qual. Invest. Int. J. Telerehabilitation **6**, 65–74 (2014)
9. Mathison, K.: Effects of the performance management context on Australian academics' engagement with the scholarship of teaching and learning: a pilot study. Aust. Educ. Res. **42**, 97–116 (2015)
10. Kendhammer, L.K., Murphy, K.L.: Innovative uses of assessments for teaching and research. In: ACS Symposium, vol. 1182, pp. 1–4 (2014)
11. Wu, H., Patel, C.: Adoption of Anglo-American Models of Corporate Governance and Financial Reporting in China. Studies in Managerial & Financial Accounting, vol. 29, pp. 291–255 (2015)
12. Washington, C.H., Tyler, F.J., Davis, J., et al.: Trauma training course: innovative teaching models and methods for training health workers in active conflict zones of Eastern Myanmar. Int. J. Emerg. Med. **7**, 46 (2014)

Chinese-Foreign Cooperative Education Management System Based on Cloud Platform

Haitao Chi[✉]

Lambton College, Jilin University, Changchun, Jilin, China
HaitaoChi1981@haoxueshu.com

Abstract. In recent years, cloud computing technology has been widely used in the field of education in the country, especially in the field of higher vocational education. The education cloud platform based on the World University City has taken initial shape. And some teaching reforms based on cloud space are also in full swing. On the basis of introducing the concept of cloud computing and education cloud platform, and we give the application status of the current cloud platform. Some measures for applying the cloud education platform for teaching and management reform are proposed. Taking learning management as an example, the in-depth study of the Chinese-foreign cooperative education management system can achieve the purpose of improving teaching effectiveness and improving the quality of personnel training.

Keywords: Cloud computing · Education cloud platform · Management system · Education reform

1 Introduction

At present, there are three main forms of Chinese-foreign cooperative education in China's higher education [1]. First, there are 7 institutions with independent legal person qualifications. Second, there are 65 institutions that do not have independent legal personality. The third is the Chinese-foreign cooperative education project, a total of 1679 [2–4]. Through the collection of information on the above-mentioned institutions and projects, a systematic analysis of the regional distribution, school-running level and situation of Chinese-foreign cooperative education were conducted.

1.1 Regional Distribution

From the perspective of the scale of running a school, the annual enrollment scale of Chinese-foreign cooperative education projects alone reached 131,000. More than half of the Chinese-foreign cooperative education projects are concentrated in the eastern region, accounting for 970, accounting for 57.8% of the total number of projects. The enrollment scale is about 72,500, accounting for 55.2% of the total enrollment. In contrast, in the western region, there are 165 school-run projects, only one-sixth of the eastern region; the enrollment scale is 14,000, only one-fifth of the eastern region; Tibet, Qinghai and Ningxia have no Chinese-foreign cooperative education projects [5–7].

© Springer Nature Singapore Pte Ltd. 2020
M. Atiquzzaman et al. (Eds.): BDCPS 2019, AISC 1117, pp. 1959–1966, 2020.
https://doi.org/10.1007/978-981-15-2568-1_273

Table 1. Regional distribution of Chinese-foreign cooperative education projects and institutions (sorted by the enrollment scale of school-running projects)

Sort	Area	Number of school projects	Project enrollment scale	Undergraduate enrollment ratio	Number of institutions	Institutional scale
1	Jiangsu	285	17227	34.68%	8	20260
2	Henan	89	10710	88.70%	3	–
3	Shanghai	150	10512	63.47%	10	14220
4	Heilongjiang	175	10040	99.00%	0	0
5	Zhejiang	106	9170	43.40%	5	11825
9	Shandong	93	8380	75.78%	9	5250
7	Beijing	109	7304	86.99%	8	6775
8	Hubei	89	7078	64.26%	2	2100
9	Guangdong	60	5600	29.38%	3	5310
10	Jilin	43	5010	79.24%	2	1900
11	Hebei	57	4635	24.38%	2	800
12	Hunan	44	4415	37.94%	1	800
13	Tianjin	41	4189	87.35%	1	–
14	Jiangxi	45	3640	43.41%	0	0
15	Sichuan	44	3580	25.42%	1	300
16	Anhui	38	2600	16.73%	1	300
17	Liaoning	35	2212	97.20%	8	5380
18	Chongqing	27	1945	64.52%	4	2200
19	Shaanxi	19	1770	45.76%	0	0
20	Guangxi	16	1735	70.89%	0	0
21	Fujian	16	1680	64.29%	1	2000
22	Hainan	18	1590	7.55%	0	0
23	Inner Mongolia	20	1540	53.90%	0	0
24	Guizhou	15	1460	17.81%	0	0
25	Yunnan	15	1390	71.22%	0	0
26	Shanxi	21	1325	5.66%	3	2900
27	Xinjiang	8	470	17.02%	0	0
28	Gansu	1	60	100.00%	0	0

From the statistical data in Table 1, it can be analyzed that the distribution of Chinese and foreign schools in the central and western regions of China is shown in Table 2.

Table 2. Scale of schooling in the central and western regions

Region	Number of school projects	Project enrollment scale	Undergraduate enrollment ratio	Number of institutions	Institutional scale
Eastern	970	72499	53.95%	55	71820
Central	544	44818	70.78%	12	8000
West	165	13950	46.06%	5	2500

1.2   School Level

From the perspective of different levels of school-running projects and their enrollment scale, the number of undergraduate programs is the largest, and the enrollment scale is the largest. A total of 757 school-running projects enroll 65,300 students each year, accounting for 45.09% and 49.72% of the total number of school-run projects and the total enrollment of school-run projects. There are 728 high-level vocational schools, accounting for 43.36%, and the enrollment quota is about 54,000, accounting for 41.14% of the total size, which is slightly inferior to the undergraduate. The enrollment of the master's program is about 11,800, and the enrollment of the doctoral program is less than 200, which together account for about 9% of the total. It is worth noting that more than 98% of the enrollment of higher vocational colleges and undergraduate programs are included in the national enrollment plan: the enrollment of high vocational colleges only accounts for 1.46% of the enrollment, and the undergraduate program accounts for 1.91%. The proportion of independent enrollment in master's and doctoral programs has reached more than 80%, of which 94.11% are self-enrolled for master's programs and 80.49% for doctors. In summary, with the improvement of the academic level, the enrollment scale of Chinese-foreign cooperative education is decreasing; most of the enrollment of higher vocational colleges and undergraduates are included in the national enrollment plan, while the enrollment of master's and doctoral enrollment accounts for more than 80% (see Table 3 for details) [8].

Table 3. Number of enrollment institutions, number of projects, enrollment scale and enrollment methods in different sections

Highest education	Enrollment method	Number of institutions	Number of school projects	School project Enrollment scale
Vocational college, college	Self-enrollment	0	10	790
	National admissions plan	23	718	53214
Bachelor	Self-enrollment	3	14	1245
	National admissions plan	27	743	64015
Master's degree	Self-enrollment	1	166	11103
	National admissions plan	10	19	695
Doctor	Self-enrollment	4	7	165
	National admissions plan	4	2	40

2   Chinese-Foreign Cooperative Education Management System

2.1   Teaching Management Reform Based on Education Cloud Platform

First of all, teachers are actively reforming existing teaching methods. First, teachers should change from the concept, find or adapt to the new space-based teaching methods; second, teachers should actively change roles, from the original classroom protagonist to the classroom learning process managers and organizers, learning task placement. The instructor of the implementation of the learning task and the inspector of the completion of the learning task are needed [9].

Second, students actively change existing learning methods. First, students should change the original passive learning style, actively seek learning resources, and actively communicate with teachers. Second, students should adapt to the team learning style, form a suitable learning team, cooperate and communicate with each other in the process of exploring problems, and play the team. The third advantage is to record the learning process through personal space, accumulate learning experience, and submit learning outcomes.

Third, it is to reform the existing assessment methods. First, reform the evaluation method of learning effects through a single test method; secondly, pay attention to self-evaluation and peer-to-peer evaluation of students, promote self-improvement and mutual improvement; third, introduce social evaluation, and evaluate the assessment by means of social certification; the fourth is the reform of the deep chemical system and the elective system, mutual recognition of credits across schools and unified evaluation criteria.

2.2 Teaching Management Reform Based on Education Cloud Platform

First of all, it is the change of the workflow of the teaching management department. First, establish and improve the teaching management work space, standardize the teaching management system and process, and uniformly publish relevant teaching management documents, notices, and notifications; second, establish and improve the space-based teaching management process, the teacher's personal space is carried out, and each teaching link of teaching is required to be reflected in the teaching space as much as possible.

Secondly, it is to give full play to the role of teaching managers, teachers, counselors and student cadres in teaching management [10].

3 The Construction of the Chinese-Foreign Cooperative Education Model of the Platform Is Implemented in the Implementation

3.1 Cloud Platform

The so-called cloud computing is a computing model (wiki definition) that provides dynamically scalable virtualized resources through the Internet. Resources in this mode are shared by all cloud computing users, and can be easily accessed through the network. Users do not need to master the technology of cloud computing, and only need to rent according to the needs of individuals or groups. The migration of cloud computing in the field of education is called "education cloud". It is the infrastructure of future education informationization, including all the hardware computing resources necessary for education informatization. The platform formed by virtualization after these resources is that: Education Cloud Platform, which provides a variety of teaching services and educational resources for the majority of schools, teachers, students and parents. At present, there are many educational cloud platforms at home and abroad, such as the large open online course in the United States (http://www.mooc.net/) and the cloud platform of the Asian Education Network (http://www.aedu.cn/), National Education Resources Public Service Platform (http://www.eduyun.cn/), World University City (http://www.worlduc.com), etc. This article is a follow-up discussion of the World University City Cloud Platform.

3.2 Cloud Platform Construction

There may be redundant data or redundant connections in the preliminary model. Redundant data refers to data that can be derived from basic data, and redundant connections refer to links that can be derived from other contacts. The existence of redundancy is likely to damage the integrity of the database, adding difficulties to the maintenance of the database and should be eliminated. A preliminary model that eliminates unnecessary redundancy is called a basic model.

The database includes teaching plans, teaching tasks, course information, course information, classroom information, etc., the overall database map, as shown in Fig. 1.

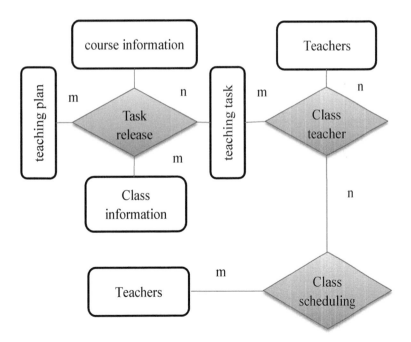

Fig. 1. Graph data overall map

Learning activities are the sum of teacher-student behaviors to achieve specific learning goals. Different learning activity sequences constitute different learning processes or learning modes, reflecting different teaching strategies, and different teaching strategies point to different teaching objectives. According to activity theory, instructional design based on the smart education cloud platform can be seen as a sequence of activities with specific purposes (not necessarily linear). A complete teaching and learning process necessarily includes multiple learning activities, and an activity may contain multiple activity tasks or only one activity task. In the goal-oriented learning activity design, the task of the activity is first determined according to the target; secondly, the corresponding activity organization strategy is selected according to the characteristics of the activity task, and accordingly, the activity sequence is arranged accordingly, and finally each design is gradually refined. The sequence of classroom activities based on the Smart Education Cloud Platform (hereinafter referred to as the "Platform") is shown in Fig. 2.

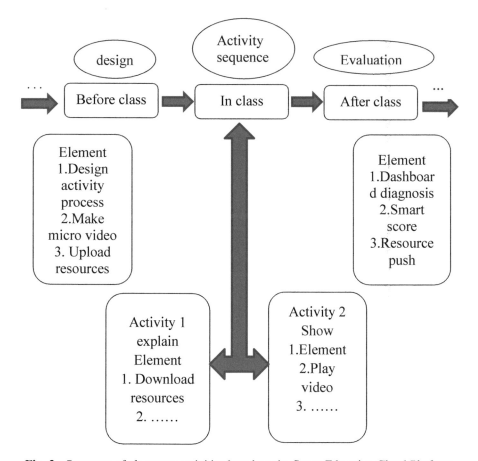

Fig. 2. Sequence of classroom activities based on the Smart Education Cloud Platform

3.3 Student Learning Activities Participation Statistics

The statistics of job completion statistics based on the cloud platform are intuitive and objective.

Through the data shown in Fig. 3, and through relevant knowledge, it is known that the assignments 1, 2, and 4 are necessary experimental questions, and the assignment 3 is a thinking question (optional). It can be seen that students have higher completion of the required questions and less enthusiasm for thinking questions. In response to this situation, we can improve the enthusiasm of everyone by taking measures such as rewarding regular time and class recognition.

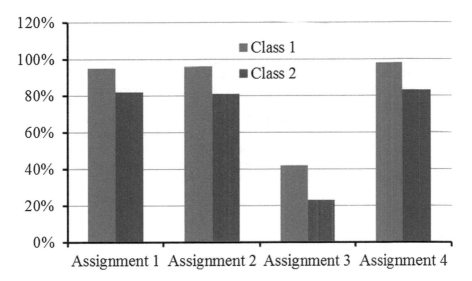

Fig. 3. Job completion statistics

4 Conclusion

In short, with the revolutionary influence of informatization on education, all departments and colleges must change their concepts, recognize the development situation, actively adopt effective measures. And teachers can fully apply information technology, especially the application of cloud platforms. Continuously deepen the reform of teaching management, teacher must actively explore new teaching methods, improve the level of running schools and improve the quality of personnel training.

Acknowledgement. Project Fund: 1. "Blended Learning Research In The Context Of Internet+" (Project Number: EIJYB2017-073), 2017 Annual Project Of Education Management Information Center, Ministry Of Education; 2. "Research On Independent And Cooperative Learning Strategies Of College Students Under The Model Of Chinese-Foreign Cooperative School-Running" (Project No. 2018060109), University Foreign Language Teaching And Research Project, Foreign Language Teaching And Research Press.

References

1. Jun, Z.: Developing the cloud platform supporting the operations of the Internet of Things. Telecommun. Sci. **6**(1), 57–68 (2010)
2. Jin, H., Wang, X., Wu, S., et al.: Towards optimized fine-grained pricing of IaaS cloud platform. IEEE Trans. Cloud Comput. **3**(4), 436–448 (2015)
3. Deng, Z.: Research and application of intelligent grinding cloud platform based on cloud computing. China Mech. Eng. **23**(1), 54–65 (2012)
4. Adams, J., Song, H.: Key developments and future challenges in Chinese-foreign cooperation in higher education. J. Knowl.-Based Innov. China **1**(3), 185–205 (2009)
5. Camba, A.: Inter-state relations and state capacity: the rise and fall of Chinese foreign direct investment in the Philippines. Palgrave Commun. **3**(1), 41 (2017). Article number: 41
6. Zhou, J., Jiang, Y., Yao, Y.: The investigation on critical thinking ability in EFL reading class. Engl. Lang. Teach. **8**(1), 83–94 (2014)
7. Li, T., Lei, T., Xie, Z., Zhang, T.: Determinants of basic public health services provision by village doctors in China: using non-communicable diseases management as an example. BMC Health Serv. Res. **16**(1), 42 (2015). Article number: 42
8. Onsman, A.: International students at Chinese joint venture universities: factors influencing decisions to enrol. Aust. Univ. Rev. **55**, 15–23 (2013)
9. Qiubo, Y., Shibin, W., Zha, Q.: Canada's industry-university co-op education accreditation system and its inspiration for the evaluation of China's industry-university-institute cooperative education. Chin. Educ. Soc. **49**(3), 182–197 (2016)
10. Armony, A.C., Velásquez, N.: Anti-Chinese sentiment in Latin America: an analysis of online discourse. J. Chin. Polit. Sci. **20**(3), 1–28 (2015)

SAR and Optical Remote Sensing Image Registration Based on an Improved Point Feature

Yanfeng Shang[1(✉)], Jie Qin[2], and Guo Cao[2]

[1] The Third Research Institute of Ministry of Public Security,
Shanghai 200120, China
aysyf@126.com
[2] School of Computer Science and Engineering, Nanjing University of Science
and Technology, Nanjing 210094, China

Abstract. This paper proposes a simple and stable point feature-based registration method for synthetic aperture radar (SAR) and optical remote sensing images. First, we extend Harris detector's response function from linear item to quadratic form and build a new weight function by combining spatial and intensity information of pixels, which enable the location of corners more precisely. Next, we create a structural feature descriptor using both the amplitude and orientation of corners to provide more distinctive local image features. Finally, we set up the correspondence based on the generated point features, and map all pixels in the sensed image to the reference. Experimental results demonstrate that the improved detector can achieve better detection performance compared with conventional Harris corner detector. In addition, registration with SAR and optical images demonstrate the efficiency and accuracy of the proposed approach.

Index Terms: Image registration · SAR image · Optical remote sensing image · Harris corner detector

1 Introduction

As a fundamental step in remote sensing image processing, image registration is the process of spatially aligning two images of a scene so that corresponding pixels assume the same coordinates [1]. For multimodal images, the relationship between intensity values of corresponding pixels is complex and unknown [2]. Therefore, multimodal image registration is a challenging task especially for optical and SAR images. During the processes of image registration, one image will be referred to as the reference image and the other images will be referred to as the sensed images. The reference image is kept unchanged and the sensed images are transformed to take the geometry and spatial coordinates of the reference image. The general steps involved in image registration are listed as follows: (1) feature extraction; (2) feature pattern matching; (3) transformation model estimation; (4) resampling. Multi-modal image registration methods can be classified into two categories: intensity-based and point-based.

© Springer Nature Singapore Pte Ltd. 2020
M. Atiquzzaman et al. (Eds.): BDCPS 2019, AISC 1117, pp. 1967–1975, 2020.
https://doi.org/10.1007/978-981-15-2568-1_274

Intensity-based methods usually need to optimize a global objective function. For images from different modalities, the mutual information (MI) and cross-cumulative residual entropy (CCRE) are widely used to be optimized. Both MI [3] and CCRE [4] need to compute joint probability function and marginal probability function in each iteration, which may lead to high computational complexity with the increase of image size or template window size.

Point-based methods first extract features from the reference image and sensed images, then match them by some similarity standard, and last reach the goal of registration. Conventional image features include point features [5], line features [6], structure and shape features [7], and region features [8]. Scale-invariant feature transform (SIFT) [9] has been proven to be the most effective feature descriptor. However, the SIFT algorithm and its variants [10] do not perform well on optical-SAR registration.

In this paper, a novel structure feature named histogram of improved orientated Harris corner point (HIOHC) is proposed for SAR and optical image registration. The HIOHC descriptor can be extracted for each image separately and then use angle cosine as similarity metric for matching between detected corner points. Finally, images can be aligned after finding enough matched points.

2   HIOHC

In this section, Harris corner detector is improved and descriptor of the detected corners is presented.

2.1   Expansion of Harris Corner Response Function

Let I be image intensities and $E(u, v)$ be the intensity changes that caused by a shift (u, v), then $E(u, v)$ can be defined as follows:

$$E(u,v) = \sum_{x,y} w(x,y)[I(x+u,y+v) - I(x,y)]^2 \tag{1}$$

where w is weight matrix.

Conventional approaches use the Taylor expansion up to linear term to get $E(u, v)$, then thresholding response values to get corner points. These corners are not robust against noise in SAR images and difficult to create their structural feature descriptors for matching accordingly. To address the problems, a quadratic Taylor expansion is considered in corner response function, and $E(u, v)$ can be rewritten as follows:

$$E(u,v) = \sum_{x,y} w(x,y)\left[I_x u + I_y v + 1/2I_{xx}u^2 + I_{xy}uv + 1/2I_{yy}v^2\right]^2$$

$$= \begin{bmatrix} u^2 & u & uv & v & v^2 \end{bmatrix} \sum_{x,y} w(x,y) \begin{bmatrix} A & B & C & D & E \\ B & F & G & H & U \\ C & G & J & K & L \\ D & H & K & M & N \\ E & U & L & N & O \end{bmatrix} \begin{bmatrix} u^2 \\ u \\ uv \\ v \\ v^2 \end{bmatrix} \qquad (2)$$

$$= \begin{bmatrix} u^2 & u & uv & v & v^2 \end{bmatrix} P \begin{bmatrix} u^2 & u & uv & v & v^2 \end{bmatrix}^T = X^T P X$$

where P is a 5×5 symmetric matrix, $w(x,y)$ is a weight,

$$I_x = \partial I(x,y)/\partial x, I_y = \partial I(x,y)/\partial y, I_{xx} = \partial^2 I(x,y)/\partial x^2,$$

$$I_{yy} = \partial^2 I(x,y)/\partial y^2, I_{xy} = \partial^2 I(x,y)/\partial x \partial y, A = 1/4I_{xx}^2,$$

$$B = 1/2I_{xx}I_x, C = 1/2I_{xx}I_y, D = 1/2I_{xx}I_y, E = 1/4I_{xx}I_{yy},$$

$$F = I_x^2, G = I_x I_y, H = I_x I_y, U = 1/2I_x I_{yy}, J = I_{xy}^2,$$

$$K = I_y I_{xy}, L = 1/2I_{yy}I_{xy}, M = I_y^2, N = 1/2I_{yy}I_y, O = 1/4I_{yy}^2$$

Suppose $\alpha_1, \alpha_2, \alpha_3, \alpha_4, \alpha_5$ are the coefficients of quadratic form $E(u,v)$, which are the eigenvalues of P with descending order. That means α_1 is the maximum eigenvalue and α_2 is second and so on. Then, $E(u,v)$ can also be rewritten as follows:

$$E(u,v) = \sum_{i=1}^{5} \alpha_i y_i^2 \qquad (3)$$

where $Y = \begin{bmatrix} y_1 & y_2 & y_3 & y_4 & y_5 \end{bmatrix}^T$, $X = VY$, V is the unit orthogonal eigenvectors corresponding to the eigenvalues of P. All eigenvalues would be very small if the image patch is smooth. α_1, α_2 are relatively large, while $\alpha_3, \alpha_4, \alpha_5$ would be very small if the image patch is located on image edges. $\alpha_1, \alpha_2, \alpha_3, \alpha_4$ are relatively large, α_5 would be very small if image patch is located at corners.

Based on the above analysis, we can define a new response function as follows (Fig. 1):

$$R = \sum_{i=1}^{4} \alpha_i - \alpha_5 \qquad (4)$$

For SAR images, Conventional Harris corner response function is susceptible to noise and may lead to misdetection of corners. In contrast, the proposed corner point detection as shown in Fig. 2 is robust against noise and beneficial for construction of structural feature descriptors in the following sections.

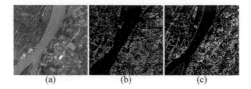

(a)   (b)   (c)

Fig. 1. Comparison of corner response images. (a) Original image. (b) Harris detector. (c) Results using the improved method.

Original image					Weight			Response value		
204	204	204	204	204						
204	204	204	204	204	1	1	1	0.108	0.102	0.071
204	204	255	255	255	1	1	1	0.102	0.234	0.277
204	204	255	204	204	1	1	1	0.071	0.277	0.234
204	204	255	204	204						
204	204	204	204	204						
204	204	204	204	204	62	95	95	0.020	0.134	0.240
204	204	255	255	255	95	142	138	0.215	0.576	0.635
204	204	255	204	204	95	138	110	0.543	0.840	0.860
204	204	255	204	204						

Fig. 2. Response values with different weight functions, where the dotted line indicates the image patch to be detected.

2.2 The Weight of Harris Corner Response Function

The weight of conventional Harris detector's response function is an indicator function, which is defined as:

$$w(x, y) = \begin{cases} 1, & (x,y)\ \text{in the window} \\ 0, & \text{otherwise} \end{cases} \tag{5}$$

The contributions of all pixels within the window are equal, which may not help locate corner point correctly. Accordingly, it would be a better choice to construct an asymmetric weight function by incorporating both spatial and intensity information of pixels. An improved weight function can be defined as follows:

$$w(x, y) = \begin{cases} \sqrt{I_x * I_x + I_y * I_y} \otimes G, & (x,y)\ \text{in the window} \\ 0, & \text{otherwise} \end{cases} \tag{6}$$

where I_x, I_y denotes x, y directional derivatives, respectively. G is a Gaussian kernel with 3×3 window size and standard deviation of 1.5. $*$ denotes hadamard product. $\otimes$ denotes convolution. Figures 2 and 3 show response values with different weight functions in real data.

From the figures, we find that the improved weight function can detect the exact location of corner point.

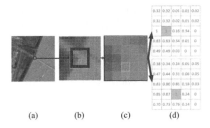

(a)   (b)   (c)   (d)

Fig. 3. Response values with different weight functions on real data, where the red box indicates the interest region, the yellow box indicates the corner point (a) Original image. (b) Zoom in the local image patch. (c) Input image patch. (d) Original weight function (top), improved weight function (down).

2.3 The Orientation of Harris Corner Point

The Harris detector only considers corner amplitudes and ignores corner orientations. Consequently, it cannot effectively describe the structure information of image patches. Therefore, we supplement orientation representation to the improved Harris detector to construct a robust feature descriptor.

For a given image window with size 3×3, there are 8 directions and we define direction code from 0 to 7. We can calculate $E(u, v)$ of eight directions, and then select two smallest values to compute the orientation of corner points. The orientation of corner point can be defined as follows:

$$\theta = \frac{\vartheta(\min_{j}\{U - \min_{i}\{U\}\} + \vartheta(\min\{U\})}{2} \quad (7)$$

$U = \{E_0, E_1, \ldots E_7\}$, where i and j denote the direction code, $\vartheta(.)$ denotes the angle of direction code, U denotes the sets of $E(u, v)$ with different directions. Figure 4 illustrates the orientation of corner points in the range $(0°, 180°]$.

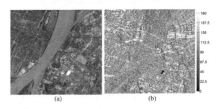

(a)   (b)

Fig. 4. The orientation of corner points. (a) Original remote sensing image. (b) Orientation image.

2.4 Structural Feature Descriptor

A structural feature descriptor named HIOHC is presented, which uses both amplitude and orientation of corner point to describe local image features. Orientation of corner points in the descriptor reflect structural information as gradients in SIFT method. The main steps involved in the proposed HIOHC descriptor and the relation between these steps are shown in Fig. 5. The detailed steps of HIOHC are as follows:

(1) Select a window with a certain size in an image and compute the amplitude and orientation of corner points for each pixel within the window.
(2) Divide the window into overlapping blocks (overlapping rate is β, where each block consists of $m \times m$ cells, and each cell containing $n \times n$ pixels.
(3) Filter each pixel's direction with a Gaussian function, then accumulate them into orientation histograms (γ bins), and weight the histograms using corner amplitude with trilinear interpolation method.
(4) Collect the HIOHC descriptors of all the blocks in the window to form a vector, which can be used in matching.

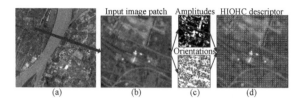

(a)   (b)   (c)   (d)

Fig. 5. A HIOHC descriptor is created first by computing the corner point amplitude (c) and orientation (c) at input image patch (b) in the original image (a). The orientation is filtered by a Gaussian function, and then interpolated using amplitude. Orientation histograms of cells are normalized using L2 norm to form the final HIOHC descriptor for the block (d).

3   Experimental Results

In this section, we use a group of data sets to evaluate our proposed method, and registration method based on $HOPC_{ncc}$ [7], manual registration method and registration method based on SIFT [9] are adopted as the comparison algorithms.

3.1   Description of Data Set

The proposed technique was applied on ten pairs of remote sensing images (GF-1, GF-2 and GF-3), including ten full-polar SAR images and ten optical images, acquired over Shanghai city and Changshu city in the east of China from different time, sensors, look angles and orbits. The tested image pairs cover different terrains including urban, suburban, river, and mountain areas.

3.2   Experimental Results

In manual registration, 10–30 control points were selected evenly over the reference image and the sensed image. For the SIFT-based method, SIFT algorithm was first used to extract the points in the reference and sensed image, and then Random Sample

Consensus was used to estimate the affine transformation mode to achieve image registration. We used *RMSE* to assess the image registration accuracy.

Figure 6 shows the registration results of a few test data sets as the length limit. From this figure, we can see that the registration results of the proposed method are satisfactory. Figure 7 shows the registration accuracy of data sets with different methods. For most of testing image pairs, the proposed registration method achieve the best performance compared with the other methods. $HOPC_{ncc}$ reaches the second smallest RMSE value, followed by manual, and the method based on SIFT is failed (the RMSE value is infinity (Inf)). We can observe that the optical image and SAR image have large radiometric differences as shown in Fig. 6, the SIFT algorithm is not able to effectively extract highly repeatable features in the optical image and SAR image at the same time. Furthermore, $HOPC_{ncc}$ can also capture the structural information of local point region so that the RMSE values are also relative low.

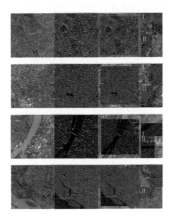

Fig. 6. Registration results for four test image pairs. Each image pair accounts for four columns, the first column are optical images, the second column are SAR images, the third column are registration results, and the fourth column are zoomed in regions highlighted by white box.

The standard deviation of the proposed method is 0.2585, the method based on $HOPC_{ncc}$ is 0.4576, the manual method is 0.4444, and the method based on SIFT fails in the experiments. Therefore, the proposed method is very stable.

There are four parameters need to determined in HIOHC. They are overlap rate β, block size $m \times m$, cell size $n \times n$ and orientation bins γ. According to [7], we use three levels orthogonal array L_{27} (3^{13}) to determine the parameter values. Finally, the best parameter settings are $n = 3, m = 3, \beta = 0.75, \gamma = 8$, which have been used in the above experiments.

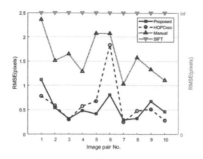

Fig. 7. RMSE values for all the data sets in different methods.

4 Conclusion

This paper proposed a novel structural feature descriptor named HIOHC for automatic SAR and optical image registration. Harris corner detector is improved by extending the response function from linear form to quadratic item, which makes the detector robust against noise. In addition, a weight function is added to the detector model considering the spatial and intensity information of neighbor pixels, which enable the detector to find the exact location of corner points. Compared with conventional Harris corner detector, our experimental results show that the proposed detector provides superior performance to classical Harris corner detector. Inspired by the works of SIFT, a structural feature descriptor is generated. Registration results validate the efficiency and accuracy of our proposed method.

Acknowledgments. This work has been partially supported by the National key Research and Development Program of China (2016YFC0801304, 2017YFC0803705).

References

1. Goshtasby, A.A.: 2-D and 3-D Image Registration: for Medical, Remote Sensing, and Industrial Applications. Wiley, Hoboken (2005)
2. Mani, V.R.S., Arivazhagan, S.: A new phase based approach to multimodal image registration. Elixir Int. J. Adv. Eng. Inf. 13146–13150 (2013)
3. Suri, S., Reinartz, P.: Mutual-information-based registration of TerraSAR-X and Ikonos imagery in urban areas. IEEE Trans. Geosci. Remote Sens. **48**(2), 939–949 (2010)
4. Hasan, M., Pickering, M.R., Jia, X.: Robust automatic registration of multimodal satellite images using CCRE with partial volume interpolation. IEEE Trans. Geosci. Remote Sens. **50** (10), 4050–4061 (2012)
5. Yu, L., Zhang, D., Holden, E.J.: A fast and fully automatic registration approach based on point features for multi-source remote-sensing images. Comput. Geosci. **34**(7), 838–848 (2008)
6. Sui, H., et al.: Automatic optical-to-SAR image registration by iterative line extraction and Voronoi integrated spectral point matching. IEEE Trans. Geosci. Remote Sens. **53**(11), 6058–6072 (2015)

7. Ye, Y., et al.: Robust registration of multimodal remote sensing images based on structural similarity. IEEE Trans. Geosci. Remote Sens. **55**(5), 2941–2958 (2017)
8. Goncalves, H., Goncalves, J.A., Corte-Real, L.: HAIRIS: a method for automatic image registration through histogram-based image segmentation. IEEE Trans. Image Process. **20**(3), 776–789 (2011)
9. Lowe, D.G.: Distinctive image features from scale-invariant keypoints. Int. J. Comput. Vis. **60**(2), 91–110 (2004)
10. Rublee, E., et al.: ORB: an efficient alternative to SIFT or SURF. In: International Conference on Computer Vision, pp. 2564–2571. IEEE (2012)
11. Zhang, G., Zhu, X.: A study of the RPC model of TerraSAR-X and COSMO-SkyMed SAR imagery. Int. Arch. Photogram. Remote Sens. Spat. Inf. Sci. 321–324 (2008)

Dispatching and Control Cloud Based Power Grid Operation Data Association Analysis Method

Liyuan Zhang[1], Jie Zhang[2], Dapeng Li[3], Zhenyu Chen[3(✉)],
Ying Zhang[1], Mingyu Wang[1], Dandan Xiao[1], and Fangchun Di[3]

[1] State Grid Tianjin Chengxi Electric Power Company, Tianjin 300190, China
[2] State Grid Tianjin Electric Power Company, Tianjin 300010, China
[3] China Electric Power Research Institute, Beijing 100192, China
chenzhenyu@epri.com.cn

Abstract. In order to adapt to the integrated operation characteristics of the power grid, some advanced IT technologies (cloud computing, big data and artificial intelligence, etc.) are applied to build the dispatching and control cloud. Based on the aggregated various data resource on the cloud, this paper proposes the power grid operation data association analysis method, which mainly describes the relationship between the dispatching and control operation data information and the health status of the equipment. According to the historical data, the obtained comprehensive score for a certain equipment exceeds a certain threshold, indicating that power grid operation data reflects that the device repeatedly has similar defects, and the device can be diagnosed.

Keywords: Dispatching and control cloud · Power grid · Operation data · Association analysis

1 Introduction

With the continuous deepening of the construction of UHV interconnected power grids in China, the scale of power grid has developed rapidly, the structure of power grid has become increasingly complex. Accordingly, higher requirements have been placed on the integrated operation level of large-scale power grid, and the support capabilities of dispatching and control system needs to be further strengthened, and to continuously improve the level of real-time sharing of dispatching and control information, enhance data processing, application computing and service capabilities, so as to achieve on-demand access to better meet the needs of large-scale power grid's security, [1–3] researches [4–11] and a variety of application services [12–18].

In order to adapt to the integrated operation characteristics of the power grid, its operation and dispatching management businesses are demand-oriented, which relies on IT technologies such as cloud computing, big data and mobile Internet to build the dispatching and control cloud, and gradually form the technology support system of "resource virtualization, data standardization, application servitization". It aims to improve the integrated coordination control ability, information support ability and

© Springer Nature Singapore Pte Ltd. 2020
M. Atiquzzaman et al. (Eds.): BDCPS 2019, AISC 1117, pp. 1976–1982, 2020.
https://doi.org/10.1007/978-981-15-2568-1_275

global resource sharing ability of the online analysis and calculation system, realize offline and online applications, deepen the value of power grid dispatching and control data, and realize lean security analysis of power grid, also aims to promote intelligent decision-making in power grid operation, promote the transformation of management mode for dispatching and control management businesses, and comprehensively ensure the safe and high-quality operation of power grid and the lean and efficient operation of dispatching management.

The power grid dispatching monitoring system generates massive operational data every day, but these data structures are not uniform with their coding, the data volume is huge, and there is no unified operational data processing center, which results in a large number of interfaces between various systems, even forms lots of information islands. Many valuable power grid operation and management data have not been effectively integrated and fully utilized, and issues such as data sharing and integration applications have become increasingly prominent. At the same time, the dispatching center information system can only support simple operations such as saving massive historical power grid operation data, measurement data, fault alarms, scheduling plans, dispatching management, and online analysis result data, and there is no professional system for conducting effective classification management and correlation analysis these large amounts of historical information, which related to effectively utilize these valuable data resources and mine their values. In the fields of power grid simulation calculation and online analysis, the joint analysis of cross-professional calculation results is mainly carried out manually. Although analysts can obtain some important information by means of knowledge and experience, it is inevitable that information will be lost due to data simplification. The automatic fusion and reanalysis of various dispatching operation data and online analysis results through data correlation analysis technology, can not only promote the in-depth study of complex power grid problems, but also the ultimate goal is to conduct comprehensive analysis or conclusion verification based on all possible situations.

2   Association Rules and Classification

2.1   Association Rules

Let $I = \{i_1, i_2, i_3, \ldots, i_m\}$ be a collection of items, D is a transaction set, T is each transaction, is also a collection of items, $T \subseteq I$, and each transaction has a unique identifier TID. X is called the itemset and is a collection of items in T. If $X \subseteq T$, then transaction T contains X. The association rule is an implication $X \Longrightarrow Y$, where $X \subseteq I$, $Y \subseteq I$ and $X \cap Y = \varnothing$.

The association rule expresses an association relationship, which is meaningful under certain probability constraints. For $X \Longrightarrow Y$, this probability refers to the probability that the combination of X and Y events which appears in the total transaction record. This probability is called support degree. In practical applications, a domain expert often sets a threshold, and the event combination will be studied when exceeds the threshold. This threshold is called the minimum support degree min_S. The support degree is represented by $S(X \Longrightarrow Y) = |\{T| X \cup Y \subseteq T, T \in D\}|/|D|$.

Another probability constraint for the association $X \Longrightarrow Y$, is the number of occurrences of the X and Y event combination divided by the number of occurrences of the X event, which is called the confidence level. It is also set a threshold by the domain expert. If below the threshold, the X event occurs insufficiently to cause the occurrence of the Y event, which is called the minimum confidence level min_C. The confidence level is expressed by the formula $C(X \Longrightarrow Y) = |\{T|X \cup Y \subseteq T, T \in D\}|/|\{T|X \subseteq T, T \in D\}|$.

The set of items is called an item set, the item set that satisfies the minimum support degree is called a frequent item set. The frequent item set containing K items is called a frequent K-item set. Association rule mining is to generate association rules whose support degree and confidence level are greater than minimum support degree and minimum confidence level, is also known as the strong rule.

2.2   Classification of Association Rules

According to different situations, the association rules can be classified as follows:

(1) Based on the categories of variables processed in the rules, association rules can be classified into categorical types and numerical types. The values processed by the type association rules are discrete and typed; and the numerical association rules can be combined with multi-dimensional associations or multi-layer association rules to process numeric fields, while numerical association rules can also include types variable.
(2) Based on the abstraction level of data in the rules, it can be divided into single-layer association rules and multi-layer association rules. In the single-layer association rule, all variables don't take into account that the actual data has multiple different levels; in the multi-level association rules, the multi-layered nature of the data has been fully considered.
(3) Based on the dimension of the data involved in the rule, the association rules can be divided into single-dimensional and multi-dimensional. In a one-dimensional association rule, we only involve one dimension of the data; in a multidimensional association rule, the data to be processed will involve multiple dimensions.

According to the above-mentioned association rule classification, in the data analysis process of multi-level power grid dispatching and control, the method of combining rules can be selected according to the data type to perform data pre-processing and mining.

3   Association Rules Mining

3.1   Association Rules Mining

The association rule mining problem is the process of finding the appropriate association rules in a given transaction database by specifying the minimum support degree and minimum confidence level. In general, the association rule mining problem can be divided into two sub-problems:

(1) Discover frequent itemsets

Through the min_S given by the user, all frequent itemsets are found, that is, the project set with the support degree not less than min_S is satisfied. In fact, these frequent itemsets may have containment relationships. Generally speaking, only the collection of so-called frequent large itemsets, which are not included in other frequent itemsets, are the basis for forming association rules.

(2) Generate association rules

Through the min_C given by the user, an association rule with a confidence level not less than min_C is found in each of the largest frequent itemsets.

For above sub-question (1), it is the research focus of the association rule mining algorithm. Its core principle is that the subset of frequent itemsets is the frequent item set, and the superset of infrequent itemsets is a non-frequent item set.

For above sub-question (2), it is relatively simple to match the rules one by one in each frequent large item set and perform $Confidence(I_1 \implies I_2) > min_C$.

3.2 Association Analysis Mining Model

According to the equipment model, measurement data, alarm information, protection information, faults and events related to power grid operation, the grid-level and equipment-level characteristics of equipment alarms and other dispatching and control related data are extracted, to realize the fusion of grid-level and equipment-level model characteristics. The frequent item set mining algorithm is used to establish the correlation model between the data and the running state of equipment, and the same equipment with similar defects repeatedly is mined. According to the mining result, the deep correlation analysis method is used to judge the running state of the equipment, and give the equipment alarm according to results of the correlation and impact analysis of information and operational equipment failures.

As shown in Fig. 1, the association analysis mining model mainly describes the relationship between the dispatching and control operation data information and the health status of equipment. According to the historical data, the comprehensive score obtained by regression calculation for a certain equipment exceeds a certain threshold, it indicates that power grid operation data reflects that the device repeatedly has similar defects, and this device can be diagnosed as a case of defects, which is necessary to pay attention; further, if the device issues alarm data and is highly correlated with the health status of device, then the device is pre-judged to determine whether the device will be stopped, and the result of prediction is notified to the dispatcher.

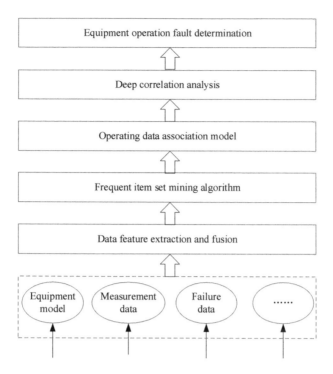

Fig. 1. Association analysis mining model

4   Comprehensive Analysis

The FP-Growth algorithm is used to mine frequent itemsets. Based on the power grid operation event data and historical data obeyed by the structured design of the general data object of dispatching and control cloud, the association analysis is carried out to establish the equipment defect and fault probability evaluation model. The massive operation data based on the power grid operation data platform is provided. The quantitative analysis of the operating state trend of substation equipment, combining with alarm information, online monitoring data and equipment family information, establishes a behavior recognition model of typical faulty equipment, and realizes the operational status and fault correlation mining and prediction analysis of power grid dispatching and control equipment.

Through comprehensive analysis of different factors and fault types, it is found which factors have a high commonality relationship with equipment faults. In order to find more rules, you need to lower the threshold. The main analysis factors include the relationship between weather, cause type, and equipment category for fault types. The configuration interface of the algorithm verification is shown in Fig. 2. Consequently, we get some typical equipment fault association rule, for example, the probability of line accidents occurring in the working day of the third quarter is the highest.

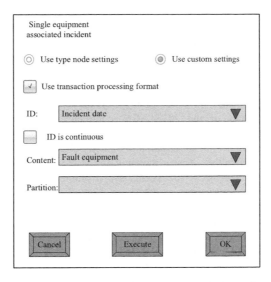

Fig. 2. Multi-factor associated failure analysis settings

5   Conclusions

This paper is based on the dispatching and control cloud system, utilizes association rule classification and mining algorithm to analysis the data association for massive data of power grid operation. According to multi-source historical data obtained from the dispatching monitoring system, including device models, measurement data, alarm information, protection information, faults and events, to establish the association model between data and device status by means of integration, classification and aggregation. This method provides technical support for power gird operation risk prediction and evaluation analysis based on dispatching and control cloud.

Acknowledgment. This work was supported by Science and Technology Program of State Grid Tianjin Electric Power Company under Grant No. KJ19-1-12 (Research on Support Technology of Power Grid Operation Risk Analysis and Assessment Base on Dispatching and Control Cloud).

References

1. Ding, F., Zhu, H., Zhang, M.: The construction of SCADA disaster backup system for TCSC region power grid dispatching. Power Distrib. Util. **22**(2), 41–43 (2005)
2. Zhang, J.: The solution of data synchronization and acquisition for power grid reserved dispatching system. Power Syst. Commun. **30**(8), 47–50 (2009)
3. Mai, S., Liang, S.: The solution of data synchronization for reserved dispatching EMS system. Power Syst. Commun. **31**(7), 46–49 (2010)
4. Li, D., Chen, Z., Deng, Z., et al.: A wide area service oriented architecture design for plug and play of power grid equipment. Proc. Comput. Sci. **129**, 353–357 (2018)

5. Chen, Z., Li, D., Deng, Z., et al.: The application of power grid equipment plug and play based on wide area SOA. In: Proceedings of 2nd IEEE International Conference on Energy Internet, pp. 19–23. IEEE, Beijing (2018)

6. Chen, Z., Chen, Y., Gao, X., et al.: Unobtrusive sensing incremental social contexts using fuzzy class incremental learning. In: Proceedings of International Conference on Data Mining, pp. 71–80. IEEE, USA (2015)

7. Chen, Z., Chen, Y., Wang, S., et al.: Inferring social contextual behavior from bluetooth traces. In: Proceedings of the 2013 ACM conference on Pervasive and ubiquitous computing, pp. 267–270. ACM, USA (2013)

8. Gao, X., Chen, Z., Tang, S., et al.: Adaptive weighted imbalance learning with application to abnormal activity recognition. Neurocomputing **173**, 1927–1935 (2016)

9. Gao, X., Hoi, S.C., Zhang, Y., et al.: SOML: sparse online metric learning with application to image retrieval. In: Proceedings of AAAI, pp. 1206–1212, USA (2014)

10. Gao, X., Hoi, S.C., Zhang, Y., et al.: Sparse online learning of image similarity. ACM Trans. Intell. Syst. Technol. (TIST) **8**(5), 64 (2017)

11. Xiang, Z., Chen, Z., Gao, X., et al.: Solving large-scale tsp using a fast wedging insertion partitioning approach. Math. Prob. Eng. **2015**, 1–9 (2015)

12. Zhang, H., Yuan, J., Gao, X., et al.: Boosting cross-media retrieval via visual-auditory feature analysis and relevance feedback. In: Proceedings of the 22nd ACM International Conference on Multimedia, pp. 953–956. ACM (2014)

13. Chen, Z., Chen, Y., Hu, L., et al.: ContextSense: unobtrusive discovery of incremental social context using dynamic bluetooth data. In: Proceedings of the 2014 ACM Conference on Pervasive and Ubiquitous Computing, pp. 23–26. ACM, USA (2014)

14. Wang, R., Chen, F., Chen, Z., et al.: StudentLife: assessing mental health, academic performance and behavioral trends of college students using smartphones. In: Proceedings of the 2014 ACM Conference on Pervasive and Ubiquitous Computing, pp. 3–14. ACM, USA (2014)

15. Chen, Z., Wang, S., Shen, Z., et al.: Online sequential ELM based transfer learning for transportation mode recognition. In: Proceedings of the 6th IEEE International Conference on Cybernetics and Intelligent Systems, pp. 78–83. ACM, USA (2014)

16. Chen, Z., Lin, M., Chen, F., et al.: Unobtrusive sleep monitoring using smartphones. In: Proceedings of the 7th International ICST Conference on Pervasive Computing Technologies for Healthcare, pp. 145–152. ICST, Venice (2013)

17. Chen, Z., Wang, S., Chen, Y., et al.: InferLoc: calibration free based location inference for temporal and spatial fine-granularity magnitude. In: Proceedings of the 10th IEEE International Conference on Embedded and Ubiquitous Computing, pp. 453–460. IEEE, Paphos (2012)

18. Chen, Y., Chen, Z., Liu, J., et al.: Surrounding context and episode awareness using dynamic bluetooth data. In: Proceedings of the 2012 ACM Conference on Pervasive and Ubiquitous Computing, pp. 629–630, ACM, USA (2012)

An Equipment Association Failure Analysis Method of Power Grid Based on Dispatching and Control Cloud

Lingxu Guo[1], Xinyu Tong[2], Zhenyu Chen[3(✉)], Dapeng Li[3],
Jian Chen[1], Zi Wang[1], Zhipeng Li[2], and Yunhao Huang[3]

[1] State Grid Tianjin Electric Power Company, Tianjin 300010, China
[2] State Grid Tianjin Chengxi Electric Power Company, Tianjin 300190, China
[3] China Electric Power Research Institute, Beijing 100192, China
chenzhenyu@epri.com.cn

Abstract. The power grid dispatching and control system undertakes the monitoring and control tasks of the power grid equipment. In order to monitor and analyze power grid equipment failures, we need to diagnose the equipment faulty conditions in time, and execute corresponding maintenance, which aims to seriously decrease failure occurrence probability, so as to ensure large-scale power grid's safe and stable operation. This paper proposes an equipment association failure analysis method of power grid based on dispatching and control cloud, according to the mining and analysis of the equipment fault information based on the power grid model and operational data. Consequently, the method is able to establish the relationship between the faults of different equipment, and analyze the equipment influence range and degree, which is further propitious to build the operation and maintenance prevention decision-making mechanism.

Keywords: Dispatching and control cloud · Power grid · Operation data · Association analysis

1 Introduction

As the power grid system continues to develop in large-scale operation and large-capacity directions, its safe and stable operation has an increasingly greater impact on the national economy and people's livelihood. As an important part of the power grid system, if there is a sudden power outage, it will cause huge economic losses and adverse social impacts. The power grid dispatching and control system undertakes the monitoring and control tasks of the power grid equipment, monitors and analyzes the faults of the power grid equipment, diagnoses the faulty conditions in time, and performs corresponding check, repair and operation maintenance according to the scientific inspection strategy. It can greatly reduce the probability of sudden failures, which is of great significance for the safe and stable operation of the overall power grid system, so as to better meet the needs of large-scale power grid studies and security [1–5], even a large number of application services [6–18].

© Springer Nature Singapore Pte Ltd. 2020
M. Atiquzzaman et al. (Eds.): BDCPS 2019, AISC 1117, pp. 1983–1988, 2020.
https://doi.org/10.1007/978-981-15-2568-1_276

The monitoring analysis and evaluation of the power grid control system equipment is to measure the state of the power equipment during operation, analyze and infer the cause of the bad state, and determine the serious state of the fault. With the long-term operation of power equipment, the safe operation state of the equipment is declining, and the probability of failure of the power equipment is further increased. In order to ensure that the power control system operates at a safe and healthy level, it is necessary to conduct state monitoring and fault diagnosis of the power equipment, to diagnose various faults of the equipment at an early stage, and to timely repair or update the equipment to avoid more serious accidents [19–22].

2 Methodology of Device Association Failure Analysis

2.1 Device Association Failure Analysis

Based on the influencing factors such as the time period, area, and type of accident of the equipment accident, the equipment failure related factors are analyzed, and the equipment association accident analysis is performed based on the above analysis results, and analyzed the association relationship of the same accident attribution, also the associated relationship between different attributions, including the area related accident analysis, equipment type associated accident analysis, single equipment failure accident analysis, mixed correlation analysis and so on.

2.2 Equipment Accident Correlation Feature

Depending on different data types, select the associated features from the following aspects:

(1) Equipment characteristics
    It is mainly from power grid equipment account and equipment failure data, which is used to analyze the distribution rule of heavy overload stations under different types of equipment, different fault types and different user composition ratios;
(2) Timing characteristics
    It is used to analyze the cause of the equipment failure time, important weather, holiday changes with time, association trend, and so on.
(3) Associated characteristics
    It is used to analyze the associated factors related to equipment failure, including the relationship with equipment category, weather, cause type, etc., and the degree of association with each dimension.

The purpose of equipment accident correlation analysis is to obtain frequent itemsets and equipment accidents associated with equipment incidents, which are based on the faulty equipment object, the time of the accident, the end of the accident, the type of accident, the type of accident, the tripping of the line, and the coincidence of the line.

2.3 Equipment Accident Correlation Evaluation Analysis

Based on the above-mentioned equipment accident occurrence frequent itemsets and equipment accident correlation examples, the following indicators are used to evaluate the degree of association:

(1) Relationship support degree

$$Support(X \rightarrow Y) = P(X, Y)/P(I) = P(X \cup Y)/P(I) = num(XUY)/num(I)$$

Where I represents the total transaction set; $num()$ represents the number of occurrences of a particular item set in the transaction set, $num(I)$ represents the total number of transaction sets, and $num(XUY)$ represents the number of transaction sets containing $\{X, Y\}$.

(2) Relationship confidence degree

$$Confidence(X \rightarrow Y) = P(Y|X) = P(X, Y)/P(X) - P(XUY)/P(X)$$

(3) Relationship promotion degree

$$Lift(X \rightarrow Y) = P(Y|X)/P(Y)$$

(4) Derivative comprehensive score

The Z-standardization (or 0-1 normalization) is able to separately perform according to the evaluation indexes such as the relationship support degree, the relationship confidence degree, and the relationship promotion degree of the above-mentioned association relationship, and the standardized value is substituted for the value before the calculation. After standardization, the relationship support degree, confidence degree, and promotion degree are summed up to obtain a new derivative comprehensive score.

3   Equipment Association Failure Analysis

The results of the equipment association failure analysis are evaluated according to the equipment accident correlation relationship, as follows:

(1) The relationship between weather, cause type and equipment category for accident type.
   The correlation results of weather, cause type and equipment category for accident type are shown in Fig. 1. The figure uses different colors to distinguish four related dimensions: category name, accident type, scene weather, and cause type. According to the analysis results in the figure, it can be seen that among various types of causes, The strong correlation causes of line accidents are fire burned mountain, lightning stroke, typhoon, external force damage and foreign matters and so on.

(2) Equipment type related accident analysis

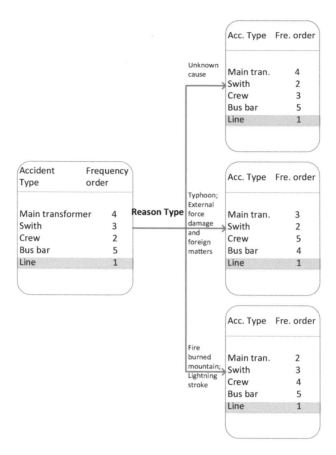

Fig. 1. Schematic diagram of the association of weather, cause type and equipment category with accident type

As for the equipment type associated accident, after an accident occurs, various types of equipment (such as lines, bus bars, circuit breakers, transformers, various generator sets) will have cascade failure response.

4 Conclusions

This paper is based on the power grid model and operational data provided by the dispatching and control cloud system. Through the mining and analysis of the equipment fault information of the power grid dispatching and control system, the relationship between the faults of different equipment is found, and the influence range and influence degree of the equipment are analyzed to find other equipment, which may occur equipment failure, so as to establish an operation and maintenance prevention decision-making mechanism.

Acknowledgment. This work was supported by Science and Technology Program of State Grid Tianjin Electric Power Company under Grant No. KJ19-1-12 (Research on Support Technology of Power Grid Operation Risk Analysis and Assessment Base on Dispatching and Control Cloud).

References

1. Song, Y., Zhou, G., Zhu, Y.: Present status and challenges of big data processing in smart grid. Power Syst. Technol. **37**(4), 927–935 (2013)
2. Cleveland, F.: IntelliGrid architecture: power system functions and strategic vision. Utility Consulting International (2005)
3. Li, Z., Pang, B., Li, G., et al.: Development of unified European electricity market and Its implications for China. Autom. Electric Power Syst. **41**(24), 2–9 (2017)
4. Li, D., Chen, Z., Deng, Z., et al.: A wide area service oriented architecture design for plug and play of power grid equipment. Proc. Comput. Sci. **129**, 353–357 (2018)
5. Chen, Z., Li, D., Deng, Z., et al.: The application of power grid equipment plug and play based on wide area SOA. In: Proceedings of 2nd IEEE International Conference on Energy Internet, pp. 19–23. IEEE, Beijing (2018)
6. Chen, Z., Chen, Y., Gao, X., et al.: Unobtrusive sensing incremental social contexts using fuzzy class incremental learning. In: Proceedings of International Conference on Data Mining, pp. 71–80. IEEE, USA (2015)
7. Zhang, M., Xu, H., Wang, X., et al.: Google tensorflow machine learning framework and applications. Microcomput. Appl. **36**(10), 58–60 (2017)
8. Gao, X., Chen, Z., Tang, S., et al.: Adaptive weighted imbalance learning with application to abnormal activity recognition. Neurocomputing **173**, 1927–1935 (2016)
9. Gao, X., Hoi, S.C., Zhang, Y., et al.: SOML: Sparse online metric learning with application to image retrieval. In: Proceedings of AAAI, pp. 1206–1212, USA (2014)
10. Gao, X., Hoi, S.C., Zhang, Y., et al.: Sparse online learning of image similarity. ACM Trans. Intell. Syst. Technol. (TIST) **8**(5), 64 (2017)
11. Xiang, Z., Chen, Z., Gao, X., et al.: Solving large-scale tsp using a fast wedging insertion partitioning approach. Math. Prob. Eng. **2015**, 1–9 (2015)
12. Zhang, H., Yuan, J., Gao, X., et al.: Boosting cross-media retrieval via visual-auditory feature analysis and relevance feedback. In: Proceedings of the 22nd ACM International Conference on Multimedia, pp. 953–956. ACM (2014)
13. Chen, Z., Chen, Y., Hu, L., et al.: ContextSense: unobtrusive discovery of incremental social context using dynamic bluetooth data. In: Proceedings of the 2014 ACM Conference on Pervasive and Ubiquitous Computing, pp. 23–26. ACM, USA (2014)
14. Wang, R., Chen, F., Chen, Z., et al.: StudentLife: assessing mental health, academic performance and behavioral trends of college students using smartphones. In: Proceedings of the 2014 ACM Conference on Pervasive and Ubiquitous Computing, pp. 3–14. ACM, USA (2014)
15. Chen, Z., Wang, S., Shen, Z., et al.: Online sequential ELM based transfer learning for transportation mode recognition. In: Proceedings of the 6th IEEE International Conference on Cybernetics and Intelligent Systems, pp. 78–83. ACM, USA (2014)
16. Chen, Z., Lin, M., Chen, F., et al.: Unobtrusive sleep monitoring using smartphones. In: Proceedings of the 7th International ICST Conference on Pervasive Computing Technologies for Healthcare, pp. 145–152. ICST, Venice (2013)

17. Chen, Z., Wang, S., Chen, Y., et al.: InferLoc: calibration free based location inference for temporal and spatial fine-granularity magnitude. In: Proceedings of the 10th IEEE International Conference on Embedded and Ubiquitous Computing, pp. 453–460. IEEE, Paphos (2012)
18. Chen, Y., Chen, Z., Liu, J., et al.: Surrounding context and episode awareness using dynamic bluetooth data. In: Proceedings of the 2012 ACM Conference on PERVASIVE and Ubiquitous Computing, pp. 629–630. ACM, USA (2012)
19. Wang, J., Sheng, W., Wang, J., et al.: Design and implementation of a unified data acquisition and monitoring system for medium and low voltage distribution networks. Autom. Electric Power Syst. **36**(18), 72–76 (2012)
20. Wang, L., Tao, J., Rajiv, R., et al.: G-hadoop: map reduce across distributed data centers for dataintensive computing. Futur. Gener. Comput. Syst. **29**(3), 739–750 (2013)
21. Li, W., Lang, B.: A tetrahedron data model of unstructured database. Sci. China **40**(8), 1039–1053 (2010)
22. Cai, Y., Fu, T., Ni, S., et al.: Study on key technology of unstructured data modeling features. Power Syst. Clean Energy **33**(1), 13–17 (2017)

Design of AC Motor Efficiency Optimization Control System

Yuqin Zhu[1]([✉]), Fang Han[1], Kui Zhao[1], Jinghua Zhao[3],
and Zheng Xu[2]

[1] School of Mechanical and Electrical Engineering, Huainan Normal University,
Huainan, Anhui, China
zhuyuqin3148@sina.com
[2] Guangxi Key Laboratory of Cryptography and Information Security,
Guilin, China
[3] University of Shanghai for Science and Technology, Shanghai 200082, China

Abstract. The uncertainty and strong coupling of the AC motor control model can easily cause high core loss and low efficiency during light-load operation. To solve this problem, the AC motor auto-disturbance rejection control system is designed for optimization to improve efficiency. This system consists of a tracking differentiator, extended state observer, and nonlinear error feedback controller. The error feedback method is used to estimate the velocity in real time and make compensation. Simulation results indicate that the system has strong adaptability and robustness.

Keywords: Motor · Auto-disturbance rejection · Core loss · Efficiency

1 Introduction

Most AC motor vector control systems adopt the traditional PID controller, which can be implemented easily with a stable dynamic response [1]. However, overshoot and lag problems can lead to the poor stability and anti-interference capability of the system. The auto-disturbance regulator is added to the vector control system, and the active disturbance rejection method is used to search for and adjust the parameters of the PID speed regulator. The advantages of auto-disturbance rejection and traditional PID adjustments are combined to enhance the adaptability and robustness of the system [2].

2 Establishment of the System Control Plan

The vector control model includes speed regulation, current hysteresis regulation, current linkage calculation, coordinate system conversion, excitation current, and torque current calculation. The stator currents on the A, B, and C axes are is A, is B, and is C, respectively, and they can be converted into alternating currents is α and is β on the DQ axis through three-phase to two-phase conversion. After the two-phase stationary rotation, they can be converted into direct currents is d and is q on the rotating axis

© Springer Nature Singapore Pte Ltd. 2020
M. Atiquzzaman et al. (Eds.): BDCPS 2019, AISC 1117, pp. 1989–1995, 2020.
https://doi.org/10.1007/978-981-15-2568-1_277

system. Figure 1 shows the system control structure which lays a good foundation for the optimal control of motor efficiency [3].

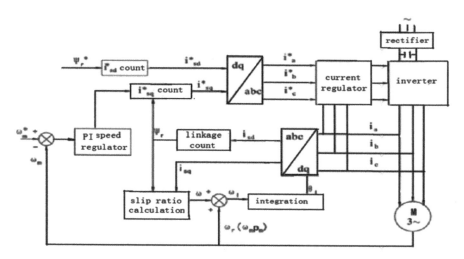

Fig. 1. Structural diagram of the vector control system of AC motor speed

3 Optimization Design of System Structure

The auto-disturbance controller module is added to the vector control system. The module includes a tracking differentiator, extended state observer, and nonlinear error feedback controller, and the core component is the extended state observer, which is mainly for linearizing the nonlinear automatic control system and making an adjustment by using the nonlinear error feedback strategy. This structure not only increases the robustness of the system and controls the overshoot, but also improves the dynamic response speed [4, 5].

3.1 Tracking Differentiator

The tracking differentiator in the auto-disturbance controller is mainly used to process signal transition and separate the nonlinear differential signal [6]. The nonlinear differential tracking process is as follows.

Given that the input of the automatic control system is $v(t)$, the tracker changes to $v_1(t)$, $v_2(t)$, $v_1(t)$ changes with the signal $v(t)$, and $v_2(t)$ is the differential form of $v_1(t)$.

Two trackers, TD_1 and TD_2, are used in the nonlinear PID. TD_1 is used to plan properly for $v_1(t)$ and $v_2(t)$ on the basis of $v(t)$. TD_2 is used for measuring the values of $y_1(t)$ and differential signal $y_2(t)$ after $y(t)$ is filtered, and error signal $e_1(t)$ and differential form $e_2(t)$ are obtained through $v_1(t) - y_1(t)$ and $v_2(t) - y_2(t)$. The second-order system is set to be.

$$\begin{cases} \dot{x}_1 = x_2 \\ \dot{x}_2 = u, |u| \le r \end{cases}. \tag{1}$$

The fast optimal adjustment model is as follows.

$$\begin{cases} x = x_2 \\ x_2 = -rsign(\dot{x}_1 - v(t) + \frac{x_2|x_2|}{2r}) \end{cases}. \tag{2}$$

Further optimization of the above formula yields discrete forms as.

$$\begin{cases} x_1(k+1) = x_1(k) + h.x_2(k) \\ x_2(k+1) = x_2(k) + h.u(k), |u(k)| \le r \end{cases} \tag{3}$$

The corresponding feedback speed control system is obtained as.

$$\begin{cases} fh + fhan(v_1(k) - v(k), v_2(k), r_0, h_1) \\ v_1(k+1) = v_1(k) + h.v_2(k) \\ v_2(k+1) = v_2(k) + h.fh \end{cases} \tag{4}$$

Among them, h is the step length, r_0 is the speed variable, h_1 filter variable, h_1 is larger, the better the effect of the filter TD. After a series of optimization processes, the optimal mathematical function of the tracking differentiator is finally obtained.

$$\begin{cases} d = rh^2, a_0 = hx_2 \\ y = x_1 + a \\ a_1 = \sqrt{d(d + 8r|y|)} \\ a_2 = a_0 + sign(y)(a_1 - d)/2 \\ s_y = (sign(a+d) - sign(a-d))/2 \\ a = (a_0 + y - a_2)s_y + a_2 \\ s_a = (sign(a+d) - sign(a-d))/2 \\ fhan = -r(\frac{a}{d} - sign(a))s_a - rsign(a) \end{cases}. \tag{5}$$

The bandpass phase shift of the tracking differentiator model is small and without resonance; thus, it can effectively solve the lag and overshoot of the linear automatic control system. The model can also be equivalent to a filter because it has a good filtering effect.

3.2   Observer Under Expansion State

If some types of external interference have a certain disturbance on the system output, then such disturbance can be observed through the outlet end using the "feedforward compensation" method [7]. Compensation is made by examining the disturbance to eliminate its effect on the entire control system, and the simple modeling process is as follows:

Let the function of a nonlinear controlled object be

$$x^{(n)}(t) = f(x, \dot{x}, \ddot{x}, \ldots x^{(n-1)}, t) + bu, y = x. \tag{6}$$

Then, the corresponding extended state space expression is as follows:

$$\begin{cases} \dot{x}_1 = x_2 \\ \cdots \\ \dot{x}_{n-1} = x_n \\ \dot{x}_n = x_{n+1} + bu \\ \dot{x}_{n+1} = g(x_1, x_2, \ldots, x_n, t) \\ y = x_1 \end{cases}, \tag{7}$$

where y is the output amount of the system and x_1 is the input. Through n differentials, $x_2, x_3, \ldots, x_n$ can be obtained. When $f(x_1, x_2, \ldots, x_n, t)$ is uncertain, observer estimates are formed by the nonlinear feedback. The optimal model of the nonlinear feedback observer is as follows:

$$\begin{cases} \varepsilon_1 = z_1 - y \\ \dot{z}_1 = z_2 - \beta_{01}\varepsilon_1 \\ \dot{z}_2 = z_3 - \beta_{02}f(\varepsilon_1, a_1, \delta) \\ \cdots \cdots \\ \dot{z}_n = z_{n+1} - \beta_{0n}f(\varepsilon_1, a_{n-1}, \delta) + bu \\ \dot{z}_{n+1} = -\beta_{0n+1}f(\varepsilon_1, a_n, \delta) \end{cases}. \tag{8}$$

In the above state observer, $Z_1 \rightarrow x_1 \ldots z_n \rightarrow x_n, x_{n+1} \rightarrow f(x, x_2, \ldots, x_n, t)$, under the expanded state, acts on the uncertain perturbation $f(x_1, x_2, \ldots, x_n, t)$, which can obtain a desirable estimation. Such a disturbance controller in a control system is referred to as an expansion state observer.

3.3 Error Feedback Control Law Under Nonlinear State

On the basis of the above models, the nonlinear feedback regulation law of the system can be used to classify the internal and external disturbance as the total disturbance and as the binding force of the system. By comparing the actual motion behavior of the motor with the given trajectory, the PID factor is searched, estimated, and supplemented and not limited by the parameters of the motor itself, that is, the "nonlinear state error feedback control [8] (ESO)." The error feedback controller is constructed using the estimation formula $z_{n+1} = a(t) = f(x, \dot{x}, \ddot{x}, \ldots, x^{(n-1)}, t)$ in ESO. Let $u = u_0 - \frac{z_{n+1}}{b}$, z_{n+1} track a(t), then

$$a(t) = x \begin{cases} \dot{x}_1 = x_2 \\ \dot{x}_2 = x_3 \\ \cdots \\ \dot{x}_n = bu_0 \\ y = x_1 \end{cases}. \tag{9}$$

Let $u_0 = \sum_i^n \beta_i f(v_i - z_i, \alpha_i, \delta)$, then the compensation disturbance control quantity u (k) is as follows:

$$u(k) = \frac{u_0(t) - z_3(k)}{b_0}. \qquad (10)$$

4   System Simulation Analysis

This System Uses the DSP Chip
TMS320F2812 produced by TI Company as the control core to build the simulation experiment platform. The motor parameters are as follows: n_N = 445r/min, U = 380 V, f = 50 Hz, P = 4.0 kW, P = 3, R_s = 0.68 Ω, L_m = 0.1486 H, L_{sl} = 0042 H, L_{rl} = 0.0042 H, R_r = 0.45 Ω, J = 0.05 kg*m², and R_m = 45.0 Ω. Subsequently, the traditional PID and auto-disturbance rejection controllers are used to simulate and compare the systems.

(1)   Under the traditional PID controller, the given speed is 1045 r/min, and the load torque is increased from 10 N*m to 50 N*m at 5 s. Figure 2 shows the simulation waveform.

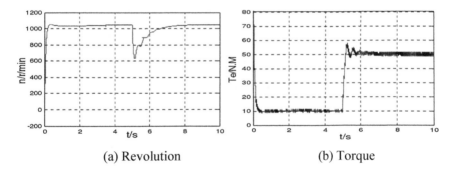

(a) Revolution   (b) Torque

Fig. 2.   Simulation waveform under the traditional PID controller mode

(2)   Under the auto-disturbance rejection controller, the given speed is 450 r/min, the load torque is 10 N*m, and the parameters are adjusted by self-tuning. Figure 3 shows the simulation waveform.

Figure 2 shows that when changing the load torque, the electromagnetic torque waveform changes smoothly until it is equal to the given value. The motor speed can return to the given value after a slight fluctuation, and the torque is substantially kept stable with a slight increase. Figure 3 presents that under the regulation of the auto-disturbance rejection controller, the system response time is short, the adjustment effect is enhanced, the overshoot amount is significantly reduced, and the reaction rapidity is

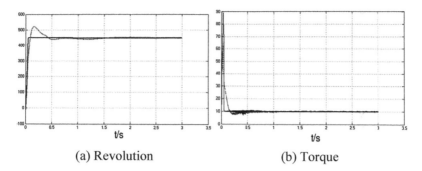

(a) Revolution   (b) Torque

Fig. 3. Simulation effect of the system under the auto-disturbance rejection controller mode

greatly improved, thereby overcoming the overshoot of the conventional PID adjustment device and the shortcoming of long adjustment time. These findings provide ideas for energy-saving control devices.

5   Conclusion

The PID controller is widely applied in motor speed control systems, and the basis of the auto-disturbance rejection control technology is traditional PID adjustment theory. This technology integrates the advanced methods of digital and modern adjustment and applies nonlinear technology to adjust the speed parameters in real time. It does not rely on motor variables and has high adaptability and robustness. Its use of an auto-disturbance rejection controller instead of the traditional PID controller can improve the overshoot and lag of the traditional PID controller.

Acknowledgement. This work was support by Key Projects of Natural Science in Huainan Normal University (2018xj16zd) and key Projects of Natural Science in Anhui Higher Education Institutions (KJ2018A0466), and supported by Guangxi Key Laboratory of Cryptography and Information Security (No. GCIS201719).

References

1. Gao, M.: Research on the efficiency optimization control of induction motor vector control variable frequency speed regulation system. Doctor of Changsha Central South University (2012)
2. Yang, J.: Study on servo control strategy of permanent magnet synchronous motor based on self-disturbance resistance and fractional order PID control. Doctor of Beijing Institute of Technology (2015)
3. He, G., Yujian, J., Jingping: Use Fibonacci method to achieve optimal efficiency control of AC motors. J. Control Decis. Making **14**(4), 349–353 (2010)
4. Lei, J., Dou, M.: Fuzzy adaptive optimization control of high altitude motor efficiency based on loss analysis. J. Central South Univ. (Nat. Sci. Ed.) **45**(3), 742–747 (2014)

5. Liu, C., Fan, K.: Self-disturbance resistance controller is used to solve the parameter robustness problem of asynchronous motor speed regulation system. J. Qingdao Univ. Sci. Technol. (Nat. Sci. Ed.) **33**(2), 193–196 (2012)
6. Xie, F.: Research on transient disturbance control method of asynchronous motor. Doctor of Anhui University (2015)
7. Fang, K.: Room corning-research on optimization control of asynchronous motor efficiency based on vector control. Doctor of China University of Mining and Technology (2014)
8. Huang, Q., Huang, S., Wu, Q.: Permanent magnet synchronous motor position servo system based on fuzzy self-disturbance controller. J. Electr. Technol. **28**(9), 294–301 (2013)

Control Algorithm and Analysis of Logistic Automatic Sorting Line

Fang Han[1(✉)], Yuqin Zhu[1], Wenyan Nie[1], Jinghua Zhao[3],
and Zheng Xu[2]

[1] School of Mechanical and Electrical Engineering, Huainan Normal University,
Huainan, Anhui, China
hanfang0554@126.com
[2] Guangxi Key Laboratory of Cryptography and Information Security,
Guilin, China
[3] University of Shanghai for Science and Technology, Shanghai 200082, China

Abstract. The control system of automatic logistic sorting line is designed to improve the accuracy of logistic sorting with the Siemens S7-200 series CPU224 PLC as the control core. The internal model control strategy is used to control the cylinder model, and the tracking and anti-interference performance of the pneumatic system is analyzed under the internal model control. MATLAB simulation results show that the internal model control can adjust the parameters and closed-loop response of the system and is more capable of target tracking than the conventional PID controller.

Keywords: Pneumatical system · PLC · Internal model control · MATLAB

1 Introduction

Capital and manpower consumed in various logistic links of the logistic distribution and sorting process account for more than 50% of the total consumption. Efficient and automatic sorting mechanism has become the main technical bottleneck in the logistic industry [1]. The trial production and exploration application of intelligent automatic sorting equipment is significant. The control system of logistic automatic sorting line adopts advanced technologies, such as machine, electricity, and light, and constitutes an efficient, reliable, and scalable sorting system combined with an intelligent control algorithm of pneumatic system to reduce costs and increase efficiency [2].

Compressed air serves as the power source in the pneumatic device to drive the mechanical device to move. In actual work, the sensor needs a certain time to detect the movement of push cylinder. Different weights of the pieces to be sorted, the gravity of the pneumatic device, the friction of the piston, and the external environmental force may lead to the incapability of the pneumatic device to push the logistic piece into the designated area [3]. High-speed response of pneumatic devices and high-precision force and displacement control are crucial.

M. Atiquzzaman et al. (Eds.): BDCPS 2019, AISC 1117, pp. 1996–2003, 2020.
https://doi.org/10.1007/978-981-15-2568-1_278

2   System Structure Design

The sorting device adopts a desktop structure with the conveyor belt as an actuator, which is driven by an asynchronous machine. Three slide receiving vats are on each side of the transport tape. Each receiving vat corresponds to a color sensor and pneumatic device. The system control core is Siemens S7-200 series CPU224 PLC. The automatic sorting system adopts the centralized control mode of the controlled object and the designated I/O port. PLC DC24V DC power is supplied from the touch screen and communicates with the PLC through the RS-232 serial port. The analog button of the touch screen controls the PLC to start the inverter and manage the speed of the three-phase asynchronous machine. In accordance with the set charging conditions, the PLC drives the solenoid valve to control the operation of the pneumatic device and pushes the logistic component into the corresponding position. Figure 1 presents the schematic of the system structure control.

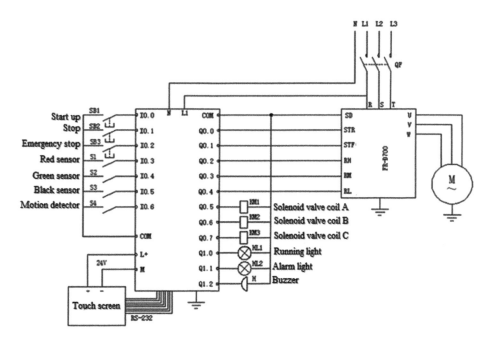

Fig. 1. Schematic of the system control structure

3   System Building

3.1   Pneumatic Model

The output force and displacement of the pneumatic system are subjected to two stages as switching actions of switching valve and filling and discharging of cylinder a and b chambers. The flow status of the gas in the pipeline is complicated. The simplified model can be expressed by differential equation in the case of gas leakage [4]:

$$\frac{dp}{dt} = \frac{k}{V}\left(q_m RT - p\frac{dV}{dt}\right),\tag{1}$$

where p is the pressure in the cavity, V is the volume of the cavity, and q_m is the change of the flow rate in the cavity. The specific heat capacity of the gas is $k = Cp/Cv$, which is the ratio of the specific heat of constant pressure to the specific heat of constant volume. R and T are the gas constant and gas temperature, respectively.

Considering the gas leakage, a and b cavity pressures Pa and Pb are expressed as

$$\frac{dp_a}{dt} = \frac{k}{V_{a0} + A_a x}\left(q_{hsv} RT - RTq_x - A_a p_a \frac{dx}{dt}\right),\tag{2}$$

$$\frac{dp_b}{dt} = \frac{k}{V_{b0} - A_b x}\left(-q_{hsv} RT + RTq_x + A_b p_b \frac{dx}{dt}\right),\tag{3}$$

where A is the effective area of cavity, $q_x = k_c(p_a - p_b)$ is the flow leaking from a and b cavities, and k_c is the gas leakage coefficient. The mean output flow of the switching valve is $q_{hsv} = k_{eq} + k_\tau \tau + k_{pa} p_a + k_{pb} p_b$ if a linear relationship exists in response to PWM signal occupancy rio τ [5]. x is the piston displacement. The cylinder piston moves slightly near its initial intermediate position:

$$\begin{cases} V_a = V_{a0} + A_a \bar{x} \\ V_b = V_{b0} - A_b \bar{x} \\ p_a = p_{a0} + \bar{p}_a \\ p_b = p_{b0} + \bar{p}_b \end{cases}.\tag{4}$$

The transfer function of the force and displacement in the pneumatic system is as follows:

$$\begin{aligned} F(s) &= \frac{(A_a b_1 + A_b a_1)k_\tau (Ms^2 + k_v s + k_l)}{h_3 s^3 + h_2 s^2 + h_1 s + h_0}\tau(s) + \frac{(A_a a_2 b_1 + A_b a_1 a_2)s + \frac{F_0}{RT}(A_a k_2 - A_b k_1)}{h_3 s^3 + h_2 s^2 + h_1 s + h_0}G(s) \\ &\quad + \frac{(A_a b_1 + A_b a_1)k_{eq}(Ms^2 + k_v s + k_l)}{h_3 s^3 + h_2 s^2 + h_1 s + h_0} \\ X(s) &= \frac{(A_a b_1 + A_b a_1)k_\tau}{h_3 s^3 + h_2 s^2 + h_1 s + h_0}\tau(s) - \frac{a_1 b_1 s + a_1 k_2 + b_1 k_1}{h_3 s^3 + h_2 s^2 + h_1 s + h_0}G(s) + \frac{(A_a b_1 + A_b a_1)k_{eq}}{h_3 s^3 + h_2 s^2 + h_1 s + h_0}, \end{aligned}\tag{5}$$

where

$$\begin{cases} h_3 = Ma_1 b_1 \\ h_2 = Ma_1 k_2 + Mb_1 k_1 + k_v a_1 b_1 \\ h_1 = k_v a_1 k_2 + k_v b_1 k_1 + k_l a_1 b_1 + A_a a_2 b_1 + A_b a_1 b_2 \\ h_0 = k_l(a_1 k_2 + b_1 k_1) + (A_a k_2 - A_b k_1)F_0/RT \end{cases}\tag{6}$$

where M is the cylinder piston quality, k_l is the environment stiffness, k_v is the viscosity friction coefficient, $a_1 = \frac{V_{a0}+A_a\bar{x}}{RTk}$, $b_1 = \frac{V_{b0}-A_b\bar{x}}{RTk}$, $k_1 = k_c - k_{pa}$, $k_2 = k_c + k_b$, $F_0 = A_a p_{a0} - A_b p_{b0}$, $a_2 = \frac{A_a p_{a0}}{RT}$, and $b_2 = \frac{A_b p_{b0}}{RT}$.

The relationship between the open-loop output force and displacement of the pneumatic system and the input duty ratio τ is a third-order system [6].

3.2   Internal Model Control

The internal model control of the pneumatic system is mainly composed of the model, controller, and feedback link; it is a control strategy based on the process object model design controller. This strategy is called the internal model control, because the control system contains the internal model. The design method is to connect the object model with the actual object in which the controller approximates the dynamic inverse of the model. For the univariate system, the internal model controller is used as the inverse of the minimum phase part of the model [7]. To improve the robust stability of the closed-loop system, a low-pass filter F(s) with a static gain of 1 is added to the internal model controller to obtain an extended internal model controller [8], as shown in Fig. 2.

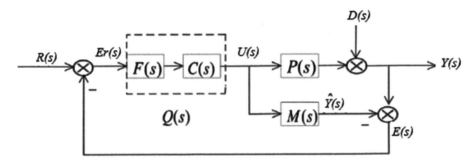

Fig. 2. Block diagram of internal model control with filter

P(s) is the control object, M(s) is the internal model of the object model, and C(s) is the internal model controller. R(s), U(s), Y(s), and $\hat{Y}(s)$ are the reference input, control quantity, object output, and model output, respectively. D(s) is the external interference, and E(s) is the model error. F(s) is a low-pass filter, and F(0) = 1. T(s) = R(s) − Y(s) represents the tracking error; then,

$$Y(s) = \frac{C(s)P(s)}{1+C(s)[P(s)-M(s)]}R(s) + \frac{1-C(s)M(s)}{1+C(s)[P(s)-M(s)]}D(s)$$

$$T(s) = \frac{1-C(s)M(s)}{1+C(s)[P(s)-M(s)]}[R(s)-D(s)]$$

(7)

4   System Experiment

4.1   Test Plan Setting

Figure 3 presents a flowchart of the sorting system control, where A, B, and C are three sensors with different colors. The flowchart shows that the operation of the sorting system is controlled by the PLC ladder program. The PLC sequentially controls the operation unit through the ladder diagram, which mainly consists of self-test, operation, and emergency stop programs.

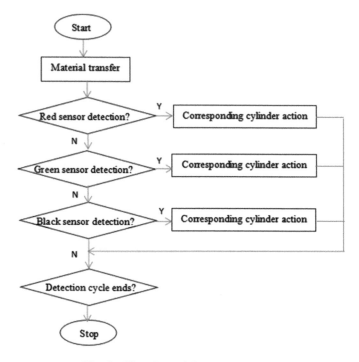

Fig. 3. Flowchart of the system control

4.2   System Simulation Tests

The system parameters are selected from the product manual and substituted into Eq. (5) [9] to obtain the force F and linear transfer function of the displacement x with respect to the input duty cycle τ. The step response of $F(s)/\tau(s)$ is drawn in MATLAB, as shown in Fig. 4(a). Considering the impact interference, $G(s) = 50/s$ and the duty ratio $\tau(s) = 0.6/s$ are taken. Figure 4(b) presents the full response.

The steady-state value of F is approximately 200 N. However, dynamic characteristics are poor, and step response possesses (24%) overshoot. Instant peak value can reach -2×10^4 N due to constant and impact interference. Although the duration is less than 0.03 s, the impact on the actuator is greatly impaired.

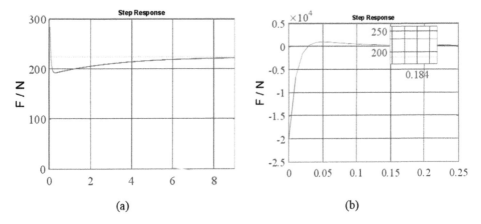

Fig. 4. Output response diagram of the pneumatic system

Time constant is set as $\lambda = 0.1$ upon intimal control simulation. $F1 = 5\ t^2$ is input for acceleration at time 0–20 s with reference to input force F; $F2 = 2000$ is input for step at 20–40 s; $F3 = -50\ t$ is input for slope at 40–60 s; $F4 = 1000$ is input for step at 60–80 s. Figure 5 presents the tracking error. Step, slope, and acc indicate that the external interference $D(s)$ parameters are step, ramp, and acceleration signals, respectively. The constant value interference poses no influence on the steady-state error of the pneumatic system. When the interference parameters are step and ramp signals, the tracking effect is improved. However, when the interference is an acceleration signal, the steady-state error of the system is great.

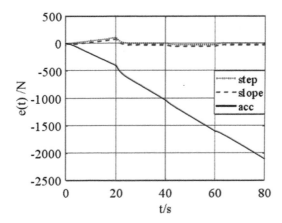

Fig. 5. Simulation diagram on the anti-interference performance of the pneumatic system

Figure 6 shows the step response curve of the displacement x. The IMC controller can enable the system to produce a smaller overshoot with a shorter adjustment time than the conventional PID controller.

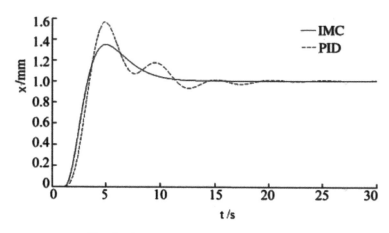

Fig. 6. Step response curve of displacement

5   Conclusion

The automatic sorting line control of logistics is designed by combining PLC control and pneumatic technologies. The system layout is flexible, the program development is simple, and the work is stable and reliable. In comparison with the traditional PID feedback control algorithm, the intimate control algorithm with convenient parameter adjustment and robustness maintains the stability of the cylinder operation and the accuracy of the piston position. This algorithm also effectively inhibits the interference when the modeling model is mismatched with the actual object by adjusting the controller parameters.

Acknowledgement. This work was support by Key Projects of Natural Science in Anhui Higher Education Institutions (KJ2018A0466), Anhui Natural Science Foundation (1808085MF166) and Key Projects of Natural Science in Huainan Normal University (2018xj16zd), and supported by Guangxi Key Laboratory of Cryptography and Information Security (No. GCIS201719).

References

1. Gao, P.: Scheme design of logistics distribution center system. Nanjing University of Science and Technology, Nanjing (2013)
2. Chen, C., Wang, T., et al.: Electrical Control and PLC Application. Electronic Industry Press, Beijing (2018)
3. Rui, D.: Pneumatic Control Technology. China Labor and Social Security Press, Beijing (2016)
4. Rivera, D.E., Morari, M., Skogestad, S.: Internal model control: PID controller design. Ind. Eng. Chem. Process Des. Dev. **25**(1), 2163–2163 (1986)
5. Shang, Y.: Research on cylinder control strategy based on high-speed on-off valve. Lanzhou University of Technology, Lanzhou (2016)

6. Gu, X.: Research on compliance force control for robot end pneumatic actuators. Harbin Institute of Technology, Harbin (2018)
7. Zhou, Y., Hu, Q., Hu, W.: New Development of internal model control research. Control Theory Appl. **55**(6), 475–476 (2004)
8. Carlos, C.E., Morari, M.: Internal model control 3 multivariable control law computation and tuning guideline. Ind. Eng. Chem. Processes Des. Dev. **24**(2), 484–494 (1985)
9. Yao, J., Wang, L., Ke, H., et al.: Improvement of air density and gas universal constant measurement experiment. Phys. Exp. **31**(12), 24–26 (2011)

Efficiency Loss Model for Asynchronous Motor Vector Control

Fang Han[1(✉)], Yuqin Zhu[1], Wenyan Nie[1], and Zheng Xu[2]

[1] School of Mechanical and Electrical Engineering, Huainan Normal University,
Huainan, Anhui, China
hanfang0554@126.com
[2] Guangxi Key Laboratory of Cryptography and Information Security,
Guilin, China

Abstract. In view of problems related to large loss, low efficiency, and poor performance of asynchronous motor in light load, the influence of rotor iron loss on motor operating efficiency is analyzed. A mathematical model of asynchronous motor vector control is established by considering iron loss, and a maximum efficiency control strategy based on optimal flux linkage is studied. Parameters, such as rotor flux and input power under different speeds and load torques, are tested under the condition of iron loss using Matlab/Simulink to build a system simulation platform. Results show that the system optimization scheme is feasible and effective.

Keywords: Asynchronous motor · Vector control · Loss · Efficiency optimization

1 Introduction

Asynchronous motors are widely used in industrial, agricultural, and national defense fields, and their total electricity consumption accounts for more than 60% of China's industrial electricity consumption [1]. Using variable frequency speed control remarkably improves the performance and efficiency of the speed control system of the asynchronous motor, which is efficient when running near rated loads. For instance, a motor with a rated power of 1–75 kW has a rated operating point efficiency between 76% and 93.6% [2]. However, for asynchronous motors with long-term light-load operation or wide-load change range, the operating efficiency and power factor will be considerably reduced. Thus, optimizing the operating efficiency of asynchronous motors is important for energy conservation, system cooling, and controlling environmental pollution.

Establishing a reliable and accurate loss model is key to maximum efficiency control. Reference [3] established a simple motor loss model, where the relationship between the optimal flux linkage and load and speed at the highest efficiency was deduced. However, a large error occurred because the leakage pressure drop at high speed and light-load conditions was ignored. Reference [4] proposed a leakage motor loss model for a fixed rotor, but rotor iron loss was neglected. Reference [5] considered the close correlation between the accuracy of motor control based on a loss model and

© Springer Nature Singapore Pte Ltd. 2020
M. Atiquzzaman et al. (Eds.): BDCPS 2019, AISC 1117, pp. 2004–2010, 2020.
https://doi.org/10.1007/978-981-15-2568-1_279

motor parameters. This study investigated the real-time self-adjustment of the parameters under the control of maximum efficiency of the motor. The design established the flux linkage closed-loop observer of the asynchronous motor rotor based on the loss model, and the optimal rotor flux was calculated in real time according to load and speed size to achieve optimal motor efficiency.

2   Construction of System Vector Control Model

The mathematical model of an asynchronous motor constructed under a three-phase stationary A–B–C coordinate system is nonlinear. Particularly, the flux linkage equation is highly complex, and solving and analyzing the nonlinear equation are difficult. Coordinate transformation can be used, and Clark transformation can be introduced. The mathematical model under the three-phase stationary A–B–C coordinate system can be transformed into a mathematical model under a two-phase stationary a–p coordinate system. Here, the physical quantity of the motor remains AC. Park transformation is further introduced, and the mathematical model under the two-phase stationary a–p coordinate system is transformed into a mathematical model under a two-phase rotation d–q coordinate system. A structural model of the asynchronous motor efficiency optimization control system rotating at the synchronous angle of the stator magnetic field is obtained [6], as shown in Fig. 1.

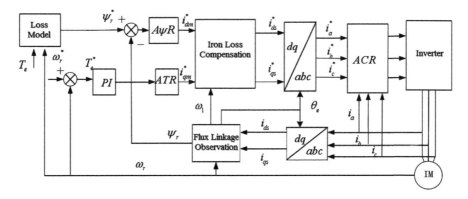

Fig. 1. Schematics of efficiency optimization vector control system for asynchronous motors

3   Modeling of System Efficiency Optimization

Given the condition of iron loss, the optimal model of system efficiency is constructed. The traditional flux linkage observation method is no longer applicable due to the change in the AC motor model. Therefore, the flux linkage observer should be redesigned. The design uses the observation method of the closed-loop flux linkage closed loop to improve the observation accuracy of the rotor flux linkage and increase its anti-interference capacity [7].

On the basis of the induction motor model column when considering iron loss, the state equation with stator current, iron loss current, and rotor flux linkage as state variables can be expressed as follows:

$$\frac{d}{dt}x = Ax + Bu. \tag{1}$$

The output equation is

$$y = Cx, \tag{2}$$

where

$$A = \begin{bmatrix} -\frac{R_s}{L_{s\sigma}} & 0 & -\frac{R_m}{L_{s\sigma}} & 0 & 0 & 0 \\ 0 & -\frac{R_s}{L_{s\sigma}} & 0 & -\frac{R_m}{L_{s\sigma}} & 0 & 0 \\ \frac{R_s}{L_{s\sigma}} - \frac{M_m}{L_{r\sigma}T_r} & 0 & \frac{\sigma L_s L_r R_m}{M_m L_{s\sigma}L_{r\sigma}} + \frac{M_m}{L_{r\sigma}T_r} & 0 & \frac{1}{L_{r\sigma}T_r} & \frac{\omega_r}{L_{r\sigma}} \\ 0 & \frac{R_s}{L_{s\sigma}} - \frac{M_m}{L_{r\sigma}T_r} & 0 & \frac{\sigma L_s L_r R_m}{M_m L_{s\sigma}L_{r\sigma}} + \frac{M_m}{L_{r\sigma}T_r} & -\frac{\omega_r}{L_{r\sigma}} & \frac{1}{L_{r\sigma}T_r} \\ \frac{M_m}{T_r} & 0 & -\frac{M_m}{T_r} & 0 & -\frac{1}{T_r} & -\omega_r \\ 0 & \frac{M_m}{T_r} & 0 & -\frac{M_m}{T_r} & \omega_r & -\frac{1}{T_r} \end{bmatrix}$$

$$B = \begin{bmatrix} \frac{1}{L_{s\sigma}} & 0 \\ 0 & \frac{1}{L_{s\sigma}} \\ -\frac{1}{L_{s\sigma}} & 0 \\ 0 & -\frac{1}{L_{s\sigma}} \\ 0 & 0 \\ 0 & 0 \end{bmatrix};$$

$$x = \begin{bmatrix} i_{s\alpha} & i_{s\beta} & i_{Rm\alpha} & i_{Rm\beta} & \psi_{r\alpha} & \psi_{r\beta} \end{bmatrix}^T.$$

$$y = \begin{bmatrix} i_{s\alpha} & i_{s\beta} \end{bmatrix}^T$$

$$u = \begin{bmatrix} u_{s\alpha} & u_{s\beta} \end{bmatrix}^T$$

The full-order state observer with rotor flux linkage can be expressed as follows:

$$\dot{\hat{x}} = (A - GC)\hat{x} + Bu + Gy. \tag{3}$$

Subtracting Eq. (3) from Eq. (1) obtain yields

$$\frac{de}{dt} = \frac{d}{dt}(x - \hat{x}) = (A - GC)(x - \hat{x}) = (A - GC)e. \tag{4}$$

The convergence speed of error e is determined by the pole position of the matrix $(A–GC)$. The asymptomatic stability of error e can be established and converged to the origin at a sufficient speed to meet the requirements of the dynamic convergence speed of the system. The feedback gain matrix is determined by via pole configuration method, as shown as follows:

$$G = \begin{bmatrix} g_1I + g_2J \\ g_3I + g_4J \\ g_5I + g_6J \end{bmatrix}.$$ (5)

If set $a_{11} = -\frac{R_s}{L_{s\sigma}}$, $a_{13} = -\frac{R_m}{L_{s\sigma}}$, $a_{31} = \frac{R_s}{L_{s\sigma}} - \frac{M_m}{L_{r\sigma}T_r}$, $a_{33} = \frac{\sigma L_s L_r R_m}{M_m L_{s\sigma} L_{r\sigma}} + \frac{M_m}{L_{r\sigma}T_r}$, $a_{35} = \frac{1}{L_{r\sigma}T_r}$, $a_{36} = -\frac{\omega_r}{L_{r\sigma}}$, $a_{51} = \frac{M_m}{T_r}$, $a_{53} = -\frac{M_m}{T_r}$, $a_{55} = -\frac{1}{T_r}$, $a_{56} = \omega_r$, $I = \begin{bmatrix} 1 & 0 \\ 0 & 1 \end{bmatrix}$, and $J = \begin{bmatrix} 0 & -1 \\ 1 & 0 \end{bmatrix}$,

then

$$A - GC = \begin{bmatrix} (a_{11} - g_1)I - g_2J & a_{13}I & 0 \\ (a_{31} - g_3)I - g_4J & a_{33}I & a_{35}I + a_{36}J \\ (a_{51} - g_5)I - g_6J & a_{53}I & a_{55}I + a_{56}J \end{bmatrix}$$ (6)

In designing the observer, the system pole position should be located on the left side of the s plane, provided that the motor model is stable. To ensure that the observer state is stable and converges at a sufficient speed, the pole of the observer is generally proportional to the pole of the motor, and the proportional constant is set to $k \geq 1$. If the pole of the observer is located on the left side of the pole of the motor model, then a faster convergence speed than the motor model can be achieved. However, when k is extremely large, the observer pole will be located far from the origin, which amplifies noise during the measurement process and influences the stability of the system [8].

The characteristic equations of the motor and observer models, that is, $fm(s) = 0$ and $fo(s) = 0$, respectively, are given pole assignments to solve $g1$, $g2$, $g3$, $g4$, and $g6$. The assignments are substituted into corresponding motor parameters in Eq. (5). The full-order observer feedback gain matrix G of the rotor flux linkage is designed for completion. Thus, a closed-loop observer that considers iron loss in an asynchronous motor rotor flux linkage is established in this study [7].

4   System Simulation Analysis

The system uses Matlab/Simulink to construct a motor model that considers iron loss and simulates it at different speeds and load torques. The specific procedures are as follows.

(1) The motor has a given load torque of 2.92 Nm, speed of 1080 rpm, optimized efficiency control at 0.6 s, and speed step of 1.2 s at 1440 rpm. Figure 2 shows the simulation results. After adding efficiency optimization control at 0.6 s, the rotor flux and input power are reduced, whereas efficiency is increased. At this

time, nearly no fluctuations are observed in terms of speed and electromagnetic
torque. After mutating speed at 1.2 s, the speed can be used to quickly track a
given value, and electromagnetic torque returns to normal after an initial jitter.
The rotor flux curve indicates that for a given speed, the optimal rotor flux
decreases with the increase in speed.

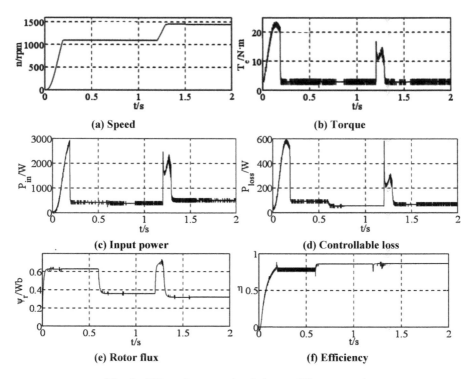

Fig. 2. Effect of system simulation at different speeds

(2) The given speed of the motor 1440 rpm, and the group of load torques are set to
     1.46, 2.92, 4.38, and 5.84 Nm. Figure 3 shows the start-up efficiency optimization
     control at 0.6 s and the change in input power.

As shown in Fig. 3, the efficiency optimization control based on the loss model can
reduce motor loss and improve motor operating efficiency to varying degrees. That is,
when the speed is fixed, the lighter the load is, the more effective the efficiency gains
will be.

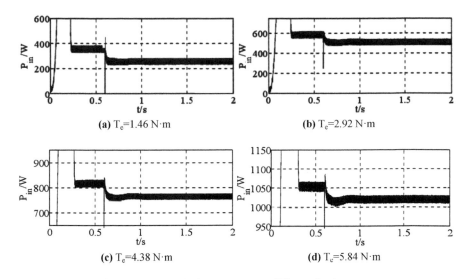

Fig. 3. Changes in input power at different load torques

5   Conclusion

This study presents the efficiency optimization vector control system for asynchronous motors. Efficiency optimization control theory is introduced into the vector control system because of the complexity and nonlinearity of the efficiency loss model. In view of motor iron loss, the efficiency optimization control algorithm based on the loss model is derived. The influence of change in motor parameters on efficiency optimization control is analyzed. Matlab/Simulink is used to construct the simulation platform. The simulation results show that the system can effectively reduce motor loss and improve motor efficiency.

Acknowledgement. This work was support by Key Projects of Natural Science in Anhui Higher Education Institutions (KJ2018A0466), Anhui Natural Science Foundation (1808085MF166) and Key Projects of Natural Science in Huainan Normal University (2018xj16zd), and supported by Guangxi Key Laboratory of Cryptography and Information Security (No. GCIS201719).

References

1. Zhao, Y., Qin, H.: Formulation of China's motor energy efficiency standards. Small Medium Sized Motors **29**(2), 58–62 (2002)
2. Liu, Y., Zhang, Z.: Research on efficiency optimization control of induction motor drive system. Micro Motor **44**(2), 67–70 (2016)
3. Haddioun, A., Benbouzid, M.E.H., Diallo, D., et al.: A loss-minimization DTC scheme for EV induction motors. IEEE Trans. Veh. Technol. **56**(1), 81–88 (2007)
4. Yu, J., Yu, H., Lin, C.: Fuzzy adaptive command filtering backstepping control for asynchronous motor considering iron loss. Control Decis. **31**(12), 2189–2194 (2016)

5. Li, Y., Tong, S., Liu, Y., et al.: Adaptive fuzzy robust output feedback control of nonlinear systems with unknown dead zones based on small-gain approach. IEEE Trans. Fuzzy Syst. **22**(1), 164–176 (2014)
6. Zhang, Z., Jiao, Y., Zhu, Y.: Research on Asynchronous Motor Model Considering Iron Loss and Magnetic Saturation. Electr. Power Sci. Eng. **2**(26–2), 28–31 (2010)
7. Yuan, S.: Research on flux linkage observation and speed identification method of induction motor. Beijing Institute of Technology (2013)
8. Zuo, H.: Research on efficiency optimization of asynchronous motor direct torque control system. Nanjing Tech University (2012)

Author Index

© Springer Nature Singapore Pte Ltd. 2020
M. Atiquzzaman et al. (Eds.): BDCPS 2019, AISC 1117, pp. 2011–2016, 2020.
https://doi.org/10.1007/978-981-15-2568-1

9789811525674VOL02